A Member of the International Code

INTERNATIONAL RESIDENTIAL CODE®

FOR ONE- AND TWO-FAMILY DWELLINGS

COMMENTARY - Vol. I

2003

First Printing

Publication Date: September, 2004

ISBN # 1-58001-133-0

COPYRIGHT © 2004

by

INTERNATIONAL CODE COUNCIL®, INC.

ALL RIGHTS RESERVED. This 2003 International Residential Code *Commentary—Volume I is a copyrighted work owned by the International Code Council, Inc. Without advance written permission from the copyright owner, no part of this book may be reproduced, distributed, or transmitted in any form or by any means, including, without limitation, electronic, optical or mechanical means (by way of example and not limitation, photocopying, or recording by or in an information storage retrieval system). For information on permission to copy material exceeding fair use, please contact: Publications, 4051 West Flossmoor Road, Country Club Hills, IL 60478-5795 (Phone 800-214-4321).*

Trademarks: "International Code Council," the International Code Council logo and the "International Residential Code" are trademarks of the International Code Council, Inc.

PRINTED IN THE U.S.A.

PREFACE

The principal purpose of the *Commentary* is to provide a basic volume of knowledge and facts relating to building construction as it pertains to the regulations set forth in the 2003 *International Residential Code®*. The person who is serious about effectively designing, constructing, and regulating buildings and structures will find the *Commentary* to be a reliable data source and reference for almost all components of the built environment.

As a follow-up to the *International Residential Code*, we offer a companion document, the International Residential Code *Commentary—Volume I*. This volume covers Chapters 1 through 11 of the 2003 *International Residential Code*. The basic appeal of the *Commentary* is that it provides in a small package and at reasonable cost thorough coverage of many issues likely to be dealt with when using the *International Residential Code*. It then supplements that coverage with historical and technical background. Reference lists, information sources, and bibliographies are also included.

Throughout the creation of this volume, strenuous effort has been made to keep the vast quantity of material accessible and its method of presentation useful. With a comprehensive yet concise summary of each section, the *Commentary* provides a convenient reference for regulations applicable to the construction of buildings and structures. In the chapters that follow, discussions focus on the full meaning and implications of the code text. Guidelines suggest the most effective method of application and the consequences of not adhering to the code text. Illustrations are provided to aid understanding, but they do not necessarily illustrate the only methods of achieving code compliance.

The format of the *Commentary* includes the full text of each IRC section, table, and figure in the code, followed immediately by the commentary for that text. At the time of printing, the *Commentary* reflects the most up-to-date text of the 2003 *International Residential Code*. Each chapter's narrative includes a section on "General Comments" and "Purpose," and each section's narrative usually includes a discussion about why the requirement commands the conditions set forth. All code text is reproduced as it appears in the *International Residential Code*, and all commentary is indented below the code text, beginning with the symbol ❖. All code figures and tables are reproduced as they appear in the IRC. Commentary figures and tables are identified in the text by the word "Commentary" (as in "see Commentary Figure 704.3"), and each has a full border.

The *Commentary* is to be used in conjunction with the *International Residential Code* and not as a substitute for the code. **The *Commentary* is advisory only;** the building official alone possesses the authority and responsibility for interpreting the code.

Comments and recommendations are encouraged, for through your input we can improve future editions. Please direct your comments to the Codes and Standards Development Department at the Chicago District Office.

TABLE OF CONTENTS

Part I—Administrative

CHAPTER 1   ADMINISTRATION .... 1-1 – 1-24

Part II—Definitions

CHAPTER 2   DEFINITIONS .... 2-1 – 2-44

Part III—Building Planning and Construction

CHAPTER 3   BUILDING PLANNING .... 3-1 – 3-110

CHAPTER 4   FOUNDATIONS .... 4-1 – 4-48

CHAPTER 5   FLOORS .... 5-1 – 5-54

CHAPTER 6   WALL CONSTRUCTION .... 6-1 – 6-136

CHAPTER 7   WALL COVERING .... 7-1 – 7-30

CHAPTER 8   ROOF-CEILING CONSTRUCTION .... 8-1 – 8-48

CHAPTER 9   ROOF ASSEMBLIES .... 9-1 – 9-38

CHAPTER 10   CHIMNEYS AND FIREPLACES .... 10-1 – 10-22

Part IV—Energy Conservation

CHAPTER 11   ENERGY EFFICIENCY .... 11-1 – 11-30

Part I — Administrative

Chapter 1: Administration

General Comments

Chapter 1 of the code is largely concerned with maintaining due process of law in enforcing the performance criteria contained in the body of the code. Only through careful observation of the administrative provisions can the building official reasonably hope to demonstrate that equal protection under the law has been provided. While it is generally assumed that the administrative and enforcement section of a code is addressed to the building official, this is not entirely true. The provisions also establish the rights and privileges of the design professional, the contractor, and the building owner. The position of the building official is to review the proposed and completed work and to determine if the residential structure conforms to the code requirements. The design professional, if one is used, is responsible for the design of the structure. The contractor is responsible for constructing the building in strict accordance with the code and any approved construction documents.

During the course of the construction of a building, the building official reviews the activity to make certain that the intent and letter of the law are being met and that the structure will provide adequate protection for the health, safety, and welfare of the users. As a public servant, the building official enforces the code in an unbiased, professional and honest manner. Every individual is guaranteed equal enforcement of the code. Furthermore, design professionals, contractors, and building owners have the right of due process for any requirement in the code.

Section R101 establishes the title, scope, and purpose of the document. Section R102 establishes the applicability of the code. Section R103 establishes the Department of Building Safety. Section R104 establishes the duties and powers of the building official. Section R105 addresses the requirements for permits. Section R106 establishes the requirements for construction documents. Section R107 addresses the topic of temporary structures and uses. Section R108 establishes permit fees, payment of fees, building permit valuations and related fees and refunds. Section R109 establishes the requirements for inspections. Section R110 establishes the requirements for occupancy, the issuance of an occupancy certificate and the revocation of an occupancy certificate. Section R111 regulates the connection and disconnection of utilities. Section R112 establishes the board of appeals and its authority. Section R113 addresses the topic of violations of the code. Section R114 establishes the authority for the building official to stop work.

Purpose

A construction code is intended to be adopted as a legally enforceable document that will safeguard health, safety, property and public welfare. A code cannot be effective without adequate provisions for its administration and enforcement. The official charged with the administration and enforcement of construction regulations has a great responsibility, and with this responsibility goes authority. No matter how detailed the code may be, the building official must exercise judgement in determining code compliance. He or she is responsible for assuring that the homes in which the citizens of the community reside are designed and constructed to be reasonably free from hazards associated with the building's use. The *International Residential Code*® (IRC®) establishes a minimum acceptable level of safety.

SECTION R101
TITLE, SCOPE AND PURPOSE

R101.1 Title. These provisions shall be known as the *Residential Code for One- and Two-Family Dwellings* of [NAME OF JURISDICTION], and shall be cited as such and will be referred to herein as "this code."

❖ This code is formally known as the *International Residential Code*® *for One- and Two-Family Dwellings*, generally referred to as the *International Residential Code*. Upon adoption by the jurisdiction, it is known as the *Residential Code for One- and Two-Family Dwellings* of the adopting jurisdiction, and in the document is often referred to as "this code." It is offered for adoption as a model document of prescriptive provisions to jurisdictions as a stand-alone residential code that establishes minimum regulations for one- and two-family dwellings and townhouses. The forum under which this code is developed encourages consistency of application of its provisions, and it is offered ready for adoption by all communities, large and small, internationally.

R101.2 Scope. The provisions of the *International Residential Code for One- and Two-Family Dwellings* shall apply to the construction, alteration, movement, enlargement, replacement, repair, equipment, use and occupancy, location, removal and demolition of detached one- and two-family dwellings and multiple single-family dwellings (townhouses) not more than three stories in height with a separate means of egress and their accessory structures.

Exception: Existing buildings undergoing repair, alteration or additions, and change of occupancy shall be permitted to comply with the *International Existing Building Code*.

❖ The provisions of the IRC apply to all aspects of construction for detached one-and two-family dwellings, multiple single-family dwellings defined as townhouses and all structures accessory to the dwellings and townhouses. This section sets a limitation in its scope of application to include only those townhouses and dwellings that are up to and including three stories. Additionally, the provisions require each two-family dwelling or townhouse to have separate egress systems for each of the dwelling units. Where a dwelling or townhouse exceeds the allowed height in stories, does not provide individual egress for each dwelling unit or does not conform to the prescriptive provisions of the code, the structures are then beyond the scope of the IRC and the provisions of the code cannot be applied. The building must then meet the provisions of the *International Building Code*® (IBC®)or other legally adopted building code of the jurisdiction. The actual height of the building, measured in feet and as applied to the height of each story, is limited by the governing provisions for each specific material as found in Chapter 6 of the code. The user of the code will discover that, depending upon which material is selected for the wall construction, the result may be buildings of different permitted heights. For instance, where the wall system is of insulating concrete form construction as prescribed in Section R611, the building is limited to two stories above grade and each story is limited to 10 feet (3048 mm) in height. If wood stud wall framing is used pursuant to the requirements of Section R602, the allowable story height and the overall building height will greatly exceed those permitted for the insulating concrete form wall construction method. The code does not limit the area of the building.

The provisions address all aspects of constructing, altering, repairing, maintaining, using, occupying, enlarging, locating, removing or demolishing any one-family dwelling, two-family dwelling, townhouse or accessory structure. The code regulates any and all activities that modify the buildings as well as any structures that are of incidental use to the main buildings and that are also located on the same lot. The code regulates all construction, plumbing, mechanical, electrical, equipment, fixture and gas piping installations that are done to the building and its operating systems as well as to other structures incidental to the main building and on the same lot. Even work that is specifically exempted from permits must comply with the requirements of the code.

R101.3 Purpose. The purpose of this code is to provide minimum requirements to safeguard the public safety, health and general welfare, through affordability, structural strength, means of egress facilities, stability, sanitation, light and ventilation, energy conservation and safety to life and property from fire and other hazards attributed to the built environment.

❖ With the adoption and establishment of a set of minimum construction standards, a community can impose reasonable standards for construction that will maintain the livability of the community while reducing factors that contribute to substandard and hazardous conditions that risk the public health, safety or welfare. Adoption of a modern construction code such as this one raises the level of safety and quality in the built environment, and it is a necessary instrument used to reduce substandard conditions or construction by establishing minimum levels of acceptable construction practice. A reduction in blighted and slum conditions benefits the general public welfare and contributes toward maintenance of a consistent base for the property tax assessments that local governments typically use to fund their general budgets. By applying minimum structural, health, sanitation, fire-safety and life-safety criteria that must be met through the prescriptive or performance provisions of the code, a standard is set that assures the public and individual building occupants they will not be exposed to construction that has gone unchecked or unregulated. The regulation and inspection of plumbing, electrical and mechanical installations also enhances safety for the public health and welfare. The imposition of construction requirements that are in excess of the minimum standards would, in most cases, be considered unreasonable and would most likely suffer from a lack of support, which in turn could undermine the purpose of construction regulations.

SECTION R102
APPLICABILITY

R102.1 General. Where, in any specific case, different sections of this code specify different materials, methods of construction or other requirements, the most restrictive shall govern. Where there is a conflict between a general requirement and a specific requirement, the specific requirement shall be applicable.

❖ The provisions of this section provide guidance to both building officials and other code users on the application of the IRC when different sections of the code specify different materials, methods of construction or other requirements. The importance of this section should not be understated. It resolves the question of how to handle conflicts between the general and specific provisions found in the code or those instances where different sections specify different requirements. This section provides a necessary hierarchy for application of code provisions and clarifies code applications that would otherwise leave persistent questions and lead to debate. The code requires that where different sections of the code apply but contain different requirements, the most restrictive provisions govern. The code also resolves conflicts between the general requirements of any particular issue with any specific requirements of the same issue by indicating that the specific requirements take precedence over the general requirements.

The following example illustrates the principle. Section R311.5 applies to all stairway types within the purview of the IRC. Section R311.5.3.1 limits the maximum height of risers to $7^3/_4$ inches (196 mm), thus providing a general requirement for stairway riser height. Section R311.5.8.1 requires risers within a spiral stairway to be a maximum height of $9^1/_2$ inches (241 mm). These provisions are specific to spiral stairways. At first it may appear that these two sections have requirements that are in conflict with one another. However, Sections R311.5.3.1 and R311.5.8.1 are subordinate requirements of Section R311.5. In this case, the specific requirements of Section R311.5.8.1 take precedence over the general requirements of Section R311.5.3.1 in those applications specific to spiral stairways.

Another example would be in relating the requirements for foam plastics to the requirements for wall and ceiling finishes. The code might be interpreted to state that foam plastic boards meeting the requirements of Section R314.1.1 with a maximum flame-spread rating of 75 could be used as the final surface finish for walls and ceilings, since Section R315 allows a flame spread classification for wall and ceiling finishes with a rating of up to 200. This, however, would be a mistake. The provisions of Section R314.1.2 require the foam plastic to be covered by a finish material equivalent to a thermal barrier that limits the average temperature rise of the unexposed surface to no more than 250°F (121°C) after 15 minutes of fire exposure to the ASTM E 119 standard time-temperature curve, or to be covered with minimum $^1/_2$-inch (12.7 mm) gypsum wallboard. In this case, the uncovered foam plastic must be covered to meet the requirements of Section R314.1.2 and have the thermal barrier installed. Additionally, the final surface finish material that is chosen must comply with Section R314.1.2, and it must also meet the required flame spread classification of 200 or less, as specified in Section R315.

To summarize, where several code sections apply to the use of a material or a method of construction, the most restrictive requirements apply.

R102.2 Other laws. The provisions of this code shall not be deemed to nullify any provisions of local, state or federal law.

❖ Compliance with the requirements of the IRC does not entail authorization, approval or permission to violate the regulations of other local, state or federal laws. Other laws, ordinances and regulations not regulated or enforced by the building official could be in existence and enforced by another authority having jurisdiction over those provisions. Although the requirements may have similar provisions to those of this code, the work must be in conformance with the other regulations.

R102.3 Application of references. References to chapter or section numbers, or to provisions not specifically identified by number, shall be construed to refer to such chapter, section or provision of this code.

❖ There are many instances in the IRC where a reference is merely a chapter number, section number, or in some cases, a provision not specified by number. In all such situations, these references are to the IRC and not some other code or publication.

R102.4 Referenced codes and standards. The codes and standards referenced in this code shall be considered part of the requirements of this code to the prescribed extent of each such reference. Where differences occur between provisions of this code and referenced codes and standards, the provisions of this code shall apply.

> **Exception:** Where enforcement of a code provision would violate the conditions of the listing of the equipment or appliance, the conditions of the listing and manufacturer's instructions shall apply.

❖ The IRC references numerous standards. Chapter 43 is a full list of these standards, which are extensions of the code and become law to the extent to which they are referenced. A standard is a published technical document that represents an industry consensus on how a material or assembly is to be designed, manufactured, tested or installed in order for a specific level of performance to be obtained. Although the code establishes the minimum quality and performance criteria for a material, installation or method of design, the code relies on the referenced standards to provide the criteria to determine if a material or method is in compliance with the code provisions.

Referenced standards in this code are considered part of the code, which allows them to be as equally enforceable as the code itself. The referenced standards, however, are enforceable only to the extent that the code prescribes. This means that only the specific portions of a standard that are specified apply, and the remaining portions of the standard should be considered only as advisory, voluntary and in some cases not applicable.

For example, the requirements in Section R703.2 for weather-resistant sheathing paper state that asphalt-saturated felt that is free from holes and breaks, weighs at least 14 pounds (16 kg) per 100 square feet (9 m^2) and complies with ASTM D 226 or any other approved weather-resistant material must be applied over studs or the sheathing of all exterior walls as required by Table R703.4. The referenced standard specification covers asphalt-saturated felts, with or without perforations. It is clear that the code does not allow perforations of any kind, simply because the stated code language is "free from holes and breaks." Therefore, only non-perforated sheathing paper complying with ASTM D 226 fulfills the code requirements.

The provisions of the code take precedence over the provisions of a referenced standard where differences exist. However, this principle does not apply when enforcement of the provisions of the code would cause a piece of equipment to be installed in violation of the conditions of the listing for the equipment. The equipment and its installation must meet the conditions of the listing and the manufacturer's instructions.

R102.5 Appendices. Provisions in the appendices shall not apply unless specifically referenced in the adopting ordinance.

❖ Provisions of the appendix do not apply unless the jurisdiction has adopted the appendix by statute or ordinance.

R102.6 Partial invalidity. In the event any part or provision of this code is held to be illegal or void, this shall not have the effect of making void or illegal any of the other parts or provisions.

❖ There may be a situation where one or more specific provisions of the code are found to be void or illegal. This may be because a local, state or federal ordinance, statute or law has precedence over the adopted construction provisions. Under such conditions, only those specific provisions found to be void or illegal are affected; the rest of the code remains in force.

[EB] R102.7 Existing structures. The legal occupancy of any structure existing on the date of adoption of this code shall be permitted to continue without change, except as is specifically covered in this code, the *International Property Maintenance Code* or the *International Fire Code*, or as is deemed necessary by the building official for the general safety and welfare of the occupants and the public.

❖ Buildings that exist legally at the time the code is adopted are allowed to have their existing use and occupancy continued if it was also legally in existence. This means that as long as a structure or building remains in a safe and sanitary condition it need not be upgraded to meet the more current standards. However, any new construction, addition or remodeling will require such work to conform to the requirements of the new code. A change of occupancy of the building also will force the building to conform to the new standards.

The existence of a building prior to the adoption of a new edition of the code does not grant it the status of a legal existence. A building is thought of as being "grandfathered" under prior rules and not needing to be brought up to current requirements when there are records to show that it was constructed to meet the regulations of the jurisdiction in force at the time it was built. The most common way to demonstrate legal compliance with the construction codes of a community is through the public records. Copies of past building permits can be researched at the jurisdictional archives. Upon discovery that a building does not have a legal existence, corrective actions will be needed in order to bring the structure into compliance with the regulations of the jurisdiction at the time the building was built.

[EB] R102.7.1 Additions, alterations or repairs. Additions, alterations or repairs to any structure shall conform to that required for a new structure without requiring the existing structure to comply with all of the requirements of this code, unless otherwise stated. Additions, alterations or repairs shall not cause an existing structure to become unsafe or adversely affect the performance of the building.

❖ Buildings and structures that have additions, alterations or repairs made to them are required to meet the requirements of the currently adopted code. The existing building is not required to comply with all of the provisions for new buildings; however, the new construction is not allowed to create an unsafe condition in the existing structure. In this case, the existing building elements that would be negatively affected must be made to comply with the new requirements.

SECTION R103
DEPARTMENT OF BUILDING SAFETY

R103.1 Creation of enforcement agency. The department of building safety is hereby created and the official in charge thereof shall be known as the building official.

❖ This section establishes a building department that provides plan review and inspections for buildings regulated by this code. It also establishes the position of building official, who will be the administrator for the enforcement of the jurisdictional codes. The employees of the department may be given varying degrees of authority by the building official. The Department of Building Safety is charged with the responsibility for enforcing the provisions of the code.

R103.2 Appointment. The building official shall be appointed by the chief appointing authority of the jurisdiction.

❖ The building official is an appointed officer of the jurisdiction, charged with the administrative responsibilities of the Department of Building Safety. The building official is appointed to the position by the chief appointing authority of the jurisdiction. Typically the appointment is made by the mayor, council or commission and carried out through the city manager or other administrative authority.

R103.3 Deputies. In accordance with the prescribed procedures of this jurisdiction and with the concurrence of the appointing authority, the building official shall have the authority to appoint a deputy building official, the related technical officers, inspectors, plan examiners and other employees. Such employees shall have powers as delegated by the building official.

❖ The building official has the authority, acting in conjunction and in agreement with the appointing authority of the jurisdiction, to appoint officers and employees of the department. The building official can delegate certain powers of his or her authority to a deputy building official as well as to all technical officers and employees of the Department of Building Safety. This group of deputies typically includes inspectors and plans examiners.

SECTION R104
DUTIES AND POWERS OF THE
BUILDING OFFICIAL

R104.1 General. The building official is hereby authorized and directed to enforce the provisions of this code. The building official shall have the authority to render interpretations of this code and to adopt policies and procedures in order to clarify the application of its provisions. Such interpretations, policies and procedures shall be in conformance with the intent and purpose of this code. Such policies and procedures shall not have the effect of waiving requirements specifically provided for in this code.

❖ The building official is appointed by the legislative body of the jurisdiction to serve as the employee with the authority and responsibility for the proper administration of the code enforcement agency. The building official establishes policies and procedures that will clarify the applications of the code. The development of those policies and procedures should not be simply for the convenience of the jurisdiction's employees but should be viewed as a way to effectively communicate to all interested parties involved in the construction process how the department will process applications, review construction documents, make inspections, approve projects and determine and clarify the application of the code provisions. Properly developed, these policies and procedures can make the code enforcement department more predictable for those who are regulated and will also establish improved code compliance and public relations.

When interpretation of the code is needed, the building official is the one individual of the jurisdiction with the legal authority to interpret the code and determine how the provisions should be applied, in both general and specific cases. Some departments formalize the interpretation process and require that persons with a question should submit it in writing. Departments are encouraged to develop policies for both formal (written) and informal (verbal) requests for code interpretations. Any such interpretations are required to be in conformance with the intent and letter of the code and may not waive any requirements. It may be necessary in some cases for the building official to write these code interpretations into the permit with regard to the specific code provision in question.

R104.2 Applications and permits. The building official shall receive applications, review construction documents and issue permits for the erection and alteration of buildings and structures, inspect the premises for which such permits have been issued and enforce compliance with the provisions of this code.

❖ This section states that the building official must receive applications, review construction documents, issue permits, conduct inspections and enforce the provisions of the code. She or he is to provide the services required to carry the project from application for the permit to final approval. The building official is to accept all properly completed applications and not refuse the receipt of an application that meets the policy requirements. This same principle holds for the review of the construction documents, issuance of permits, inspections and for the enforcement of the code's provisions. The requirements of the code must be met, and approval will be granted only when compliance is verified.

R104.3 Notices and orders. The building official shall issue all necessary notices or orders to ensure compliance with this code.

❖ Building officials are to communicate in writing the disposition of their findings regarding code compliance. If an inspection is conducted and shows that the work fails to comply with the code provisions, the building official or technical officer who conducted the inspection must issue a written report noting the corrections that are needed. A copy of the report is to be provided to the permit holders or their agent.

R104.4 Inspections. The building official is authorized to make all of the required inspections, or the building official shall have the authority to accept reports of inspection by approved agencies or individuals. Reports of such inspections shall be in writing and be certified by a responsible officer of such approved agency or by the responsible individual. The building official is authorized to engage such expert opinion as deemed necessary to report upon unusual technical issues that arise, subject to the approval of the appointing authority.

❖ This code gives the building official the authority to conduct all required inspections. The building official also has the authority to accept reports from other inspection agencies or private inspectors who have been granted prior approval by the building official to conduct inspections and provide reports. The reports submitted by the approved inspection agencies or individuals must be in writing and must be certified by only those individuals who have been approved by the building official as being qualified to submit the reports. When unusual technical issues arise during the course of construction, the building official has the authority to hire the services of an expert to report on the conditions and technical issues germane to the subject at hand. Prior to hiring the expert for consulting services, the building official must seek approval from the appointing authority.

R104.5 Identification. The building official shall carry proper identification when inspecting structures or premises in the performance of duties under this code.

❖ When the building official and other employees of the jurisdiction are performing their duties and inspecting structures or premises of construction, they are required to carry and display identification that will identify them as employees of the jurisdiction. Commentary Figure R104.5 is an example of the proper identification.

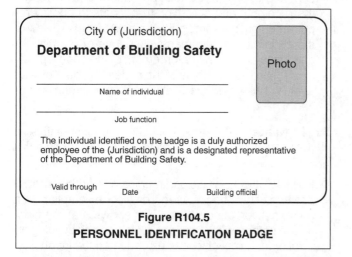

**Figure R104.5
PERSONNEL IDENTIFICATION BADGE**

R104.6 Right of entry. Where it is necessary to make an inspection to enforce the provisions of this code, or where the building official has reasonable cause to believe that there exists in a structure or upon a premises a condition which is contrary to or in violation of this code which makes the structure or premises unsafe, dangerous or hazardous, the building official or designee is authorized to enter the structure or premises at reasonable times to inspect or to perform the duties imposed by this code, provided that if such structure or premises be occupied that credentials be presented to the occupant and entry requested. If such structure or premises be unoccupied, the building official shall first make a reasonable effort to locate the owner or other person having charge or control of the structure or premises and request entry. If entry is refused, the building official shall have recourse to the remedies provided by law to secure entry.

❖ The building official and employees of the jurisdiction are authorized to conduct inspections in order to enforce the provisions of the jurisdiction's construction codes. Section R109 identifies specific progress points for inspections that must be conducted for both the intermediate phases of the work being done and the final inspection and approval. The officials of the jurisdiction also have the authority to conduct an inspection when the building official has reasonable cause to believe that conditions exist that constitute a violation of the code. The officials of the jurisdiction must take the necessary precautions so the constitutional rights of the owner or tenants are not violated.

The code authorizes the building official to enter and conduct an inspection at reasonable times.

Unless there is an immediate life-threatening condition or eminent immediate safety concern, the building official must secure an invitation or be requested to conduct a scheduled inspection. Additionally, the building official can conduct an inspection when permission has been requested and granted by the person with immediate control of the premises.

When permission to conduct an inspection is denied, the building official has the authority to seek an inspection warrant.

R104.7 Department records. The building official shall keep official records of applications received, permits and certificates issued, fees collected, reports of inspections, and notices and orders issued. Such records shall be retained in the official records for the period required for the retention of public records.

❖ An important function of the Department of Building Safety is the record keeping associated with the department's functions. It is the responsibility of the building official to maintain adequate and accurate records that can be referenced for a variety of reasons. For example, staffing levels and other budgetary concerns can be better addressed through a historical review of department activity. Also, the department can provide a wealth of information to assist in community development and planning procedures, and a sound record-keeping process best supports legal actions.

It is necessary to maintain the appropriate and complete department records for an adequate period of time, based on the value of the information being retained. In all cases, it is mandatory that those public records required to be retained by state or local law must be kept for the minimum time period required by such law. The legal representative for the jurisdiction should be consulted in regard to any questions or concerns on the minimum time period for the retention of official records.

R104.8 Liability. The building official, member of the board of appeals or employee charged with the enforcement of this code, while acting for the jurisdiction in good faith and without malice in the discharge of the duties required by this code or other pertinent law or ordinance, shall not thereby be rendered liable personally and is hereby relieved from personal liability for any damage accruing to persons or property as a result of any act or by reason of an act or omission in the discharge of official duties. Any suit instituted against an officer or employee because of an act performed by that officer or employee in the lawful discharge of duties and under the provisions of this code shall be defended by legal representative of the jurisdiction until the final termination of the proceedings. The building official or any subordinate shall not be liable for cost in any action, suit or proceeding that is instituted in pursuance of the provisions of this code.

❖ The building official, members of the appeals board and other employees charged with the duty of enforcing the provisions of the IRC are relieved from personal liability by the jurisdiction when they are acting in good faith during the performance of their official duties. This provision does not provide absolute immunity from all tort liability for employees of the jurisdiction in all cases; an employee acting maliciously is not relieved from his or her personal liability and will most likely not be defended by the jurisdiction. Building officials should not fear lawsuits even if their state does not guarantee them absolute immunity for their actions. Rather, the building official should understand those elements that a plaintiff must show in a lawsuit in order to prevail.

Public officials should familiarize themselves with the laws of their respective states regarding their exposure to tort liability. Only a few states grant absolute immunity from liability for any actions taken by public officials when those actions are within the scope of their employment.

R104.9 Approved materials and equipment. Materials, equipment and devices approved by the building official shall be constructed and installed in accordance with such approval.

❖ The code is a compilation of criteria with which materials, equipment, devices and systems must comply to be suitable for a particular application. Where the building official grants approval for the use of specific materials, equipment or devices as a part of the construction process, it is important that the approved items be constructed or installed in a manner consistent with the approval. For example, the manufacturer's instructions and recommendations are to be followed if the approval of the material was based even in part on those instructions and recommendations.

The approval authority given the building official is a significant responsibility and is a key to code compliance. The approval process is first technical and then administrative. For example, if data to determine code compliance are required, the data should be in the form of test reports or engineering analysis and not simply taken from a sales brochure.

R104.9.1 Used materials and equipment. Used materials, equipment and devices shall not be reused unless approved by the building official.

❖ In keeping with the authority of the building official to evaluate construction materials based on their equivalency to those specified in the code, the use of used materials and equipment is limited to those approved by the building official. Testing and materials technology has permitted the development of new criteria that old materials may not satisfy. As a result, used materials are often evaluated by the building official in the same way as new materials. It is a common practice to require that used materials and equipment be equivalent to those required by the code if they are to be utilized in a new installation.

R104.10 Modifications. Wherever there are practical difficulties involved in carrying out the provisions of this code, the building official shall have the authority to grant modifications for individual cases, provided the building official shall first find that special individual reason makes the strict letter of this code impractical and the modification is in compliance with the intent and purpose of this code and that such modification does not lessen health, life and fire safety requirements or structural. The details of action granting modifications shall be recorded and entered in the files of the department of building safety.

❖ The building official has the authority to accept modifications of the code provisions in specific cases. For the building official to allow a modification, he or she must first determine that the strict application of the code is impractical for a specific reason. When the building official grants a modification, it is not a waiver from the requirements. It should be thought of as fulfilling the requirements to the greatest extent possible but deviating from the requirements slightly in order for the intent of the provisions to be met. The modification must not lessen the health, fire-safety, life-safety or structural requirements of the code. All modification actions must be recorded in the files of the building department.

R104.10.1 Areas prone to flooding. The building official shall not grant modifications to any provision related to areas prone to flooding as established by Table R301.2(1) without the granting of a variance to such provisions by the board of appeals.

❖ Section R323 contains provisions for determining flood hazard areas. A modification cannot be granted by the building official for structures located in areas that are prone to flooding without the board of appeals first granting a variance to the provisions. The regulations of the National Flood Insurance Program (NFIP) 44 C.F.R. 60.3 require that proposals meet or exceed the minimum provisions of the program. Requests for modifications to any provision related to flood hazard areas are to be handled as formal variances. The criteria for issuance of variances are given in R112.2.2.

R104.11 Alternative materials, design and methods of construction and equipment. The provisions of this code are not intended to prevent the installation of any material or to prohibit any design or method of construction not specifically prescribed by this code, provided that any such alternative has been approved. An alternative material, design or method of construction shall be approved where the building official finds that the proposed design is satisfactory and complies with the intent of the provisions of this code, and that the material, method or work offered is, for the purpose intended, at least the equivalent of that prescribed in this code. Compliance with the specific performance-based provisions of the *International Codes* in lieu of specific requirements of this code shall also be permitted as an alternate.

❖ Although the code reflects current technologies, it is impossible to foresee all potential applications of new materials, construction techniques, or design methods. The code encourages the use of new materials and technologies by allowing them to be presented to the building official for approval. The building official must approve a proposed alternative when it is found to be satisfactory and in compliance with the intent of the provisions of the code and is equivalent to that prescribed by the code. Approval may also be granted for the use of any alternative that is in compliance with the performance-based provisions from the other members of the International Code family.

R104.11.1 Tests. Whenever there is insufficient evidence of compliance with the provisions of this code, or evidence that a material or method does not conform to the requirements of this code, or in order to substantiate claims for alternative materials or methods, the building official shall have the authority to require tests as evidence of compliance to be made at no expense to the jurisdiction. Test methods shall be as specified in this code or by other recognized test standards. In the absence of recognized and accepted test methods, the building official shall approve the testing procedures. Tests shall be performed by an approved agency. Reports of such tests shall be retained by the building official for the period required for retention of public records.

❖ The building official has the authority to require tests in order to substantiate the claim that an alternative is equivalent and meets the intent of the code. Any tests shall be in compliance with those specified in the code or other recognized test standards approved by the building official. The cost of any tests shall be borne by the proponent seeking the approval of the alternative.

SECTION R105
PERMITS

R105.1 Required. Any owner or authorized agent who intends to construct, enlarge, alter, repair, move, demolish, or change the occupancy of a building or structure, or to erect, install, enlarge, alter, repair, remove, convert or replace any electrical, gas, mechanical or plumbing system, the installation of which is regulated by this code, or to cause any such work to be done, shall first make application to the building official and obtain the required permit.

❖ This section lists the types of work or installations of equipment or utilities that will require an owner or authorized agent to obtain permits, which are to be obtained before work begins. In general, a permit is required for all activities that are regulated by the code, and these activities cannot begin until the permit is issued.

R105.2 Work exempt from permit. Permits shall not be required for the following. Exemption from the permit requirements of this code shall not be deemed to grant authorization for any work to be done in any manner in violation of the provisions of this code or any other laws or ordinances of this jurisdiction.

Building:

1. One-story detached accessory structures, provided the floor area does not exceed 200 square feet (18.58 m^2).
2. Fences not over 6 feet (1829 mm) high.
3. Retaining walls that are not over 4 feet (1219 mm) in height measured from the bottom of the footing to the top of the wall, unless supporting a surcharge.
4. Water tanks supported directly upon grade if the capacity does not exceed 5,000 gallons (18927 L) and the ratio of height to diameter or width does not exceed 2 to 1.
5. Sidewalks and driveways not more than 30 inches (762 mm) above adjacent grade and not over any basement or story below.
6. Painting, papering, tiling, carpeting, cabinets, counter tops and similar finish work.
7. Prefabricated swimming pools that are less than 24 inches (610 mm) deep.
8. Swings and other playground equipment accessory to a one or two-family dwelling.
9. Window awnings supported by an exterior wall which do not project more than 54 inches (1372 mm) from the exterior wall and do not require additional support.

Electrical:

Repairs and maintenance: A permit shall not be required for minor repair work, including the replacement of lamps or the connection of approved portable electrical equipment to approved permanently installed receptacles.

Gas:

1. Portable heating, cooking or clothes drying appliances.
2. Replacement of any minor part that does not alter approval of equipment or make such equipment unsafe.
3. Portable fuel cell appliances that are not connected to a fixed piping system and are not interconnected to a power grid.

Mechanical:

1. Portable heating appliance.
2. Portable ventilation appliances.
3. Portable cooling unit.
4. Steam, hot or chilled water piping within any heating or cooling equipment regulated by this code.
5. Replacement of any minor part that does not alter approval of equipment or make such equipment unsafe.
6. Portable evaporative cooler.
7. Self-contained refrigeration systems containing 10 pounds (4.54 kg) or less of refrigerant or that are actuated by motors of 1 horsepower (746 W) or less.

8. Portable fuel cell appliances that are not connected to a fixed piping system and are not interconnected to a power grid.

The stopping of leaks in drains, water, soil, waste or vent pipe; provided, however, that if any concealed trap, drainpipe, water, soil, waste or vent pipe becomes defective and it becomes necessary to remove and replace the same with new material, such work shall be considered as new work and a permit shall be obtained and inspection made as provided in this code.

The clearing of stoppages or the repairing of leaks in pipes, valves or fixtures, and the removal and reinstallation of water closets, provided such repairs do not involve or require the replacement or rearrangement of valves, pipes or fixtures.

❖ This section of the code lists the types of work in five categories—building, electrical, gas, mechanical and plumbing—that do not require permits. However, all work, even work that does not require a permit, must be done in a manner that will comply with the code requirements.

R105.2.1 Emergency repairs. Where equipment replacements and repairs must be performed in an emergency situation, the permit application shall be submitted within the next working business day to the building official.

❖ Occasionally repairs or replacement work must be done under emergency conditions. The code does not intend that such emergency work be held up until the necessary permits are secured. It is important, however, that the permit application be obtained in a timely manner once the emergency has been controlled. Any required permit must be applied for within the next business day following the emergency repair or replacement.

R105.2.2 Repairs. Application or notice to the building official is not required for ordinary repairs to structures, replacement of lamps or the connection of approved portable electrical equipment to approved permanently installed receptacles. Such repairs shall not include the cutting away of any wall, partition or portion thereof, the removal or cutting of any structural beam or load bearing support, or the removal or change of any required means of egress, or rearrangement of parts of a structure affecting the egress requirements; nor shall ordinary repairs include addition to, alteration of, replacement or relocation of any, water supply, sewer, drainage, drain leader, gas, soil, waste, vent or similar piping, electric wiring or mechanical or other work affecting public health or general safety.

❖ There is a variety of ordinary repair, replacement or connection work that is exempt from the permit application process. This section identifies a number of general situations in which a permit is not required. The provisions then state the types of repairs for which a permit is required. Repair work done without a permit must still comply with the applicable provisions of the code.

R105.2.3 Public service agencies. A permit shall not be required for the installation, alteration or repair of generation, transmission, distribution, metering or other related equipment that is under the ownership and control of public service agencies by established right.

❖ When the ownership and control of equipment is held by a public service agency, such as a county water district, permits are not required for any work that might be done on that equipment. The scope of this provision includes not only repair activities but also any installation or alteration work. It is clear from this section that public service agencies are self-regulating when it comes to work involving equipment used for generation, transmission, distribution and metering.

R105.3 Application for permit. To obtain a permit, the applicant shall first file an application therefor in writing on a form furnished by the department of building safety for that purpose. Such application shall:

1. Identify and describe the work to be covered by the permit for which application is made.
2. Describe the land on which the proposed work is to be done by legal description, street address or similar description that will readily identify and definitely locate the proposed building or work.
3. Indicate the use and occupancy for which the proposed work is intended.
4. Be accompanied by construction documents and other information as required in Section R106.1.
5. State the valuation of the proposed work.
6. Be signed by the applicant, or the applicant's authorized agent.
7. Give such other data and information as required by the building official.

❖ The code lists the minimum information required in an application for a permit. The owner or agent is to fully describe the location of the site, the type and nature of the work to be done and all other pertinent information regarding the job. This provides the jurisdiction with a clear understanding of what will actually be done under the permit. It is a common belief by some owners that all that is required of them is to complete an application form and the permit will be issued over the counter. Although issuance of over-the-counter permits may be practical when a water heater is to be installed or for other minor work, this is not the case for more complex work. The general public is often surprised when they discover a complete application includes the requirements to provide construction drawings as required by Section R106, detailed information about the property, engineering data and other information as required by the building department. The department should develop informational fliers and packets that will assist the applicant with understanding the requirements for obtaining a permit. Recognizing that incomplete applications do not serve the applicant's best interest (and create unnecessary delays with processing applications), many building departments have personnel who specialize in providing assistance to customers to ease them through the process.

R105.3.1 Action on application. The building official shall examine or cause to be examined applications for permits and amendments thereto within a reasonable time after filing. If the application or the construction documents do not conform to the requirements of pertinent laws, the building official shall reject such application in writing, stating the reasons therefor. If the building official is satisfied that the proposed work conforms to the requirements of this code and laws and ordinances applicable thereto, the building official shall issue a permit therefor as soon as practicable.

❖ When an application is determined to be incomplete, or when the construction documents do not show compliance with the code provisions or other requirements of the jurisdiction, the building official is to reject them. When the construction documents are rejected, the building official is required to provide to the applicant a list of reasons for the rejection so the applicant will know what action to take for subsequent approval.

When the application and construction documents are determined to be in compliance with the requirements of the code and other regulations of the jurisdiction, the building official must issue a permit. This section mandates the issuance of a permit when the plans and other data show compliance with all regulations of the jurisdiction, and it mandates that the permit be issued without delay.

R105.3.1.1 Substantially improved or substantially damaged existing buildings in areas prone to flooding. For applications for reconstruction, rehabilitation, addition, or other improvement of existing buildings or structures located in an area prone to flooding as established by Table R301.2(1), the building official shall examine or cause to be examined the construction documents and shall prepare a finding with regard to the value of the proposed work. For buildings that have sustained damage of any origin, the value of the proposed work shall include the cost to repair the building or structure to its predamage condition. If the building official finds that the value of proposed work equals or exceeds 50 percent of the market value of the building or structure before the damage has occurred or the improvement is started, the finding shall be provided to the board of appeals for a determination of substantial improvement or substantial damage. Applications determined by the board of appeals to constitute substantial improvement or substantial damage shall meet the requirements of Section R323.

❖ The definitions of "substantial damage" and "substantial improvement" are included in federal regulations (44 CFR 59.1 Definitions). The following is from the National Flood Insurance Program regulations:

Substantial damage means damage of any origin sustained by a structure whereby the cost of restoring the structure to its before-damaged condition would equal or exceed 50 percent of the market value of the structure before the damage occurred.

Substantial improvement means any reconstruction, rehabilitation, addition or other improvement of a structure, the cost of which equals or exceeds 50 percent of the market value of the structure before the start of construction of the improvement. This term includes structures [that] have incurred "substantial damage," regardless of the actual repair work performed. The term does not, however, include either:

1. Any project for improvement of a structure to correct existing violations of state or local health, sanitary or safety code specifications [that] have been identified by the local code enforcement official and [that] are the minimum necessary to assure safe living conditions, or
2. Any alteration of a historic structure, provided that the alteration will not preclude the structure's continued designation as an historic structure.

These definitions are consistent with the definitions in Section 1612.2 of the *International Building Code*. "Substantial improvement" is also defined in Section R112.2.1.

Long-term reduction in exposure to flood hazards, including exposure of older buildings, is one of the purposes for regulating floodplain development. Existing buildings or structures located in flood hazard areas are to be brought into compliance with the flood-resistance provisions of Section R323, when the value of improvements or the repair of damage exceeds a certain value.

Section R105.3 requires the applicant to state the valuation of the proposed work as part of the information submitted to obtain a permit. If the proposed work will be performed on existing buildings or structures in flood hazard areas, including restoration of damage from any cause, the building official is to prepare a finding to determine the value of the proposed work, including the value of the property owner's labor and the value of donated labor and materials. For damaged buildings, the value of the proposed work is the value of work necessary to restore the building to its predamage condition, even if the applicant is proposing less work.

To make a determination about whether a proposed repair, reconstruction, rehabilitation, addition or improvement of a building or structure will constitute a substantial improvement or correction of substantial damage, the cost of the proposed work is to be compared to the market value of the building or structure before the work is started. In order to determine market value, the building official may require the applicant to provide such information as allowed under R105.3. For additional guidance, refer to FEMA 213, *Answers to Questions about Substantially Damaged Buildings* and FEMA 311, *Guidance on Estimating*

Substantial Damage Using the NFIP Residential Substantial Damage Estimator.

The building official's finding is to be provided to the board of appeals if the value of the proposed work equals or exceeds 50 percent of the market value of the building, less land value. The board of appeals is to determine whether the proposed work constitutes a substantial improvement of the existing building or whether a damaged building sustained substantial damage. If the board finds that the work is a substantial improvement, then the existing building is to be brought into compliance with the flood-resistance provisions of Section R323.

R105.3.2 Time limitation of application. An application for a permit for any proposed work shall be deemed to have been abandoned 180 days after the date of filing, unless such application has been pursued in good faith or a permit has been issued; except that the building official is authorized to grant one or more extensions of time for additional periods not exceeding 180 days each. The extension shall be requested in writing and justifiable cause demonstrated.

❖ Applications for permits are considered valid for 180 days. The permit application and review process must be done in a timely manner within that period. The applicant must be responsive to requests for additional information made by the building department. The 180-day limitation is not intended to penalize an applicant for the lack of action on the part of the jurisdiction. It is merely a measure that is used to void an application when it is no longer reasonable to keep it active because the applicant is delaying the process and is not responding to legitimate requests for information. The building official can extend the time limit of an application in increments of 180 days, provided the applicant can show a valid reason for an extension. The applicant must make this request in writing.

R105.4 Validity of permit. The issuance or granting of a permit shall not be construed to be a permit for, or an approval of, any violation of any of the provisions of this code or of any other ordinance of the jurisdiction. Permits presuming to give authority to violate or cancel the provisions of this code or other ordinances of the jurisdiction shall not be valid. The issuance of a permit based on construction documents and other data shall not prevent the building official from requiring the correction of errors in the construction documents and other data. The building official is also authorized to prevent occupancy or use of a structure where in violation of this code or of any other ordinances of this jurisdiction.

❖ A permit authorizes the permit holder to proceed with work that complies with the code requirements. A permit is not valid if it is issued for work that does not comply with the code requirements. If, after the permit has been issued, errors in the plans or construction are discovered, the building official has the authority to require the correction of the plans or the work to comply with the requirements of the code.

R105.5 Expiration. Every permit issued shall become invalid unless the work authorized by such permit is commenced within 180 days after its issuance, or if the work authorized by such permit is suspended or abandoned for a period of 180 days after the time the work is commenced. The building official is authorized to grant, in writing, one or more extensions of time, for periods not more than 180 days each. The extension shall be requested in writing and justifiable cause demonstrated.

❖ Once a permit has been issued, the permit holder has 180 days to begin the construction work; otherwise, the permit will become invalid. If at any time the work stops for a period of 180 days or more, the permit is invalid. Extensions can be granted by the building official in writing in increments of 180 days when the applicant makes a written request.

R105.6 Suspension or revocation. The building official is authorized to suspend or revoke a permit issued under the provisions of this code wherever the permit is issued in error or on the basis of incorrect, inaccurate or incomplete information, or in violation of any ordinance or regulation or any of the provisions of this code.

❖ Any permit that has been issued based on false, misleading or incorrect information can be revoked or suspended by the authority of the building official. He or she also has the authority to suspend or revoke any permit issued when a violation exists with regard to the building code or any other ordinance, code, law or regulation that is legally in effect in the jurisdiction.

R105.7 Placement of permit. The building permit or copy thereof shall be kept on the site of the work until the completion of the project.

❖ The permit must be displayed at the work site until the certificate of occupancy has been issued. Because paperwork at a job site is sometimes lost, the code allows a copy of the permit to be kept at the site and the original to be retained in a more secure place. Keeping a record of permits at the project location satisfies the legal requirements set forth in the code, and any interested party can verify that a valid permit has been obtained.

R105.8 Responsibility. It shall be the duty of every person who performs work for the installation or repair of building, structure, electrical, gas, mechanical or plumbing systems, for which this code is applicable, to comply with this code.

❖ This is one of several code provisions that emphasize the required compliance with the code for every aspect of the project. While the permit holder is designated as having the primary responsibility for overall code compliance, it is the responsibility of each and every person working on the job to adhere to the requirements of the code.

SECTION R106
CONSTRUCTION DOCUMENTS

R106.1 Submittal documents. Construction documents, special inspection and structural observation programs, and other data shall be submitted in one or more sets with each application for a permit. The construction documents shall be prepared by a registered design professional where required by the statutes of the jurisdiction in which the project is to be constructed. Where special conditions exist, the building official is authorized to require additional construction documents to be prepared by a registered design professional.

> **Exception:** The building official is authorized to waive the submission of construction documents and other data not required to be prepared by a registered design professional if it is found that the nature of the work applied for is such that reviewing of construction documents is not necessary to obtain compliance with this code.

❖ This section provides the minimum requirements for construction documents that an applicant must provide along with the permit application form for the application package to be considered complete. Construction documents are not just a set of drawings. Construction documents are the entire set of all submitted forms and information necessary to accurately communicate the scope of the construction. The submittals may include written special inspection and structural observation programs, construction drawings and details, reports, calculations, specifications, shop drawings, manufacturer's installation instructions, site plans and other graphic and written forms that will describe the proposed work in detail. The building official can waive the submission of construction documents for types of work where the review of documents is not necessary to show compliance with the requirements of the code.

This code is prescriptive and makes possible the design of a dwelling or townhouse without the requirement for a licensed design professional. However, the construction documents must be prepared by a licensed design professional when required by the statutes of the state or jurisdiction. Additionally, the building official has the authority to require that plans be prepared by licensed design professionals when not otherwise required by the statutes of the jurisdiction if in the opinion of the building official special or unique conditions exist or if the design of a building does not meet the prescriptive provisions of the code.

R106.1.1 Information on construction documents. Construction documents shall be drawn upon suitable material. Electronic media documents are permitted to be submitted when approved by the building official. Construction documents shall be of sufficient clarity to indicate the location, nature and extent of the work proposed and show in detail that it will conform to the provisions of this code and relevant laws, ordinances, rules and regulations, as determined by the building official.

❖ The emphasis of this section is on the clarity, completeness and accuracy of the construction documents. A wide variety of individuals will be using the construction documents to perform their specific tasks; therefore, it is critical that there be no confusion about the intent of the designer based on the information in the plans and other documents.

Electronic submittal of construction documents is rapidly gaining popularity. Where the department of public safety has the means to review plans electronically, it may request such submittals. Many departments continue to use hard copy documents for the plan review process, with a request for electronic media documents copies for archival purposes.

R106.1.2 Manufacturer's installation instructions. Manufacturer's installation instructions, as required by this code, shall be available on the job site at the time of inspection.

❖ Throughout the IRC, the code directs that materials or equipment be installed in accordance with the manufacturer's installation instructions. An example is the installation of modified bitumen roofing as set forth in Section R905.11.3. The code recognizes that the manufacturer can best relate the specific installation requirements applicable to its specific product. Where the code mandates that the manufacturer's installation instructions be followed, those instructions must be available at the job site. In this way, both the installers and the inspector are able to see that the directives of the manufacturer are being followed.

R106.1.3 Information for construction in areas prone to flooding. For buildings and structures in flood hazard areas as established by Table R301.2(1), construction documents shall include:

1. Delineation of flood hazard areas, floodway boundaries, and flood zones, and the design flood elevation, as appropriate;
2. The elevation of the proposed lowest floor, including basement; in areas of shallow flooding (AO zones), the height of the proposed lowest floor, including basement, above the highest adjacent grade; and
3. The elevation of the bottom of the lowest horizontal structural member in coastal high hazard areas (V Zone); and
4. If design flood elevations are not included on the community's Flood Insurance Rate Map (FIRM), the building official and the applicant shall obtain and reasonably utilize any design flood elevation and floodway data available from other sources.

❖ This section details the information to be included in an application for a permit to build within a flood hazard area. The site plan is to show sufficient detail and information about the designated flood hazard area, including floodway and flood zones, to allow for a complete review of the proposed activities. Flood Insurance Rate Maps (FIRM) are flood hazard maps prepared by the Federal Emergency Management Agency (FEMA). FIRMs may show specific Base Flood Elevations (BFE). If the community adopts a flood hazard map other than the FIRM, then the Design Flood Elevations (DFE) must be at least as high as the BFEs [see Table R301.2(1)].

Construction documents are to include the proposed elevation of the lowest floor and the elevation of the bottom of the lowest horizontal structural member, which are to be at or above the minimum as given in Section R323. In flood hazard areas except coastal high hazard areas, applicants may propose placing fill with the intent of later constructing buildings with excavated basements. When excavated into fill, basements may be subject to damage, especially where waters remain high for more than a few hours. Fill materials can become saturated and provide inadequate support, or water pressure can collapse below-grade walls. Basements below residential buildings are not to be constructed below the DFE, even if excavated into fill that is placed above the DFE.

If elevated on individual fill pads in flood hazard areas, buildings will be surrounded by water during general conditions of flooding. Local emergency personnel responsible for evacuations should be consulted during the subdivision approval process. Many states and communities have provisions that require uninterrupted access to all buildings during a flood. These requirements may be administered by agencies responsible for permitting the construction of public rights-of-way or may be imposed as part of subdivision approval.

Many FIRMs show flood hazard areas without specifying the BFEs, indicating that FEMA has not prepared engineering analyses for those areas. These flood hazard areas are often referred to as unnumbered A zones. An important step in regulating development of these areas is the determination of the DFE. The building official and the applicant are to search for and use data from other sources, which may include the U.S. Army Corps of Engineers, the Natural Resources Conservation Service, the state, a local flood control agency or district or historical records. If flood elevation information is not available, the building official may require the applicant to develop the DFE in accordance with accepted engineering practices. Local officials unfamiliar with establishing DFEs are encouraged to contact the NFIP state coordinator or the appropriate FEMA regional office. For additional guidance, refer to FEMA 265, *Managing Floodplain Development in Approximate Zone A Areas: A Guide for Obtaining and Developing Base (100-Year) Flood Elevations*.

R106.2 Site plan. The construction documents submitted with the application for permit shall be accompanied by a site plan showing the size and location of new construction and existing structures on the site and distances from lot lines. In the case of demolition, the site plan shall show construction to be demolished and the location and size of existing structures and construction that are to remain on the site or plot.

❖ One valuable part of the construction documents is the site plan. As a part of reviewing the building's location on the site for conformance with the code, it is necessary for the building official to know the size and location of all other structures on the lot as well as their physical relationship to the new structure. The proximity of the new structure to lot lines and any public ways must also be shown. The distance between a building and the lot lines may trigger a variety of code requirements, the most notable being exterior wall fire-resistance rating and opening prohibition (see Section R302).

The code also requires that a site plan be submitted when a building or structure is to be demolished. The plan must identify the location of the building to be demolished as well as any surrounding structures that will remain in place.

R106.3 Examination of documents. The building official shall examine or cause to be examined construction documents for code compliance.

❖ The building official must review the construction documents or have qualified employees or consultants examine the plans to determine if they comply with the requirements of the jurisdiction.

R106.3.1 Approval of construction documents. When the building official issues a permit, the construction documents shall be approved in writing or by stamp. One set of construction documents so reviewed shall be retained by the building official. The other set shall be returned to the applicant, shall be kept at the site of work and shall be open to inspection by the building official or his or her authorized representative.

❖ Approval of the construction documents is the first in a series of reviews and approvals throughout the design and construction process. The building official or authorized representative must indicate that the construction documents are approved for construction, typically by stamping the plans and including any additional information that is necessary for the project. A set of the documents must be retained by the building department as its record for the life of the project. It is not uncommon throughout the job for questions to arise that require referencing the approved set of plans in the office. An efficient filing system should be developed to make the retrieval of construction documents a simple process.

In addition to the set of construction documents retained by the Department of Building Safety, at least one set of approved plans is to be returned to the permit holder. These construction documents must be maintained at the job site for reference purposes throughout the project. To avoid confusion, they must duplicate the documents that were approved and stamped. They must be available for review by building department personnel during the numerous inspections that may take place. Additionally, the contractor cannot determine compliance with approved plans unless they are available. Another reason to have the documents available is that these plans will generally indicate any special items or issues identified by the plans examiner that may not be shown on any other construction documents. The approved plans are usually required to be available at the job site prior to an inspection.

R106.3.2 Previous approvals. This code shall not require changes in the construction documents, construction or designated occupancy of a structure for which a lawful permit has been heretofore issued or otherwise lawfully authorized, and the construction of which has been pursued in good faith within 180 days after the effective date of this code and has not been abandoned.

❖ The code does not require work to be brought up to the requirements of a code that was adopted since the edition of the code under which the existing valid permit was issued. If the work is abandoned for 180 days or more, and an extension of the permit is not granted, the original permit is no longer valid, and a new permit must be sought under the provisions of the latest edition of the code.

R106.3.3 Phased approval. The building official is authorized to issue a permit for the construction of foundations or any other part of a building or structure before the construction documents for the whole building or structure have been submitted, provided that adequate information and detailed statements have been filed complying with pertinent requirements of this code. The holder of such permit for the foundation or other parts of a building or structure shall proceed at the holder's own risk with the building operation and without assurance that a permit for the entire structure will be granted.

❖ Phased approval is needed for projects that use the "fast-track" construction method, which allows construction to begin before completion of all of the plans and specifications. While it is preferable to issue permits for projects in their entirety, the building official has the authority to issue, at his or her discretion, a permit for a portion of the construction. The building official must be satisfied that the information provided shows in satisfactory detail that the partial construction will conform to the requirements of the code. In such a case, the permit holder proceeds at his or her own risk with the construction and has no assurance of the entire permit ever being issued.

R106.4 Amended construction documents. Work shall be installed in accordance with the approved construction documents, and any changes made during construction that are not in compliance with the approved construction documents shall be resubmitted for approval as an amended set of construction documents.

❖ The code requires all work to be done in accordance with the approved plans and other construction documents. Where the construction will not conform to the approved construction documents, the documents must be revised and resubmitted to the building official for review and approval. The building official must retain one set of the amended and approved plans. The other set is to be kept at the construction site, ready for use by the jurisdiction's inspection staff.

R106.5 Retention of construction documents. One set of approved construction documents shall be retained by the building official for a period of not less than 180 days from date of completion of the permitted work, or as required by state or local laws.

❖ Construction documents must be retained in case a dispute arises after completion of the project. Unless modified because of state of local statutes, the retention period for the approved construction documents is a minimum of 180 days following the completion of the work, typically the date the certificate of occupancy is issued. Any further retention of plans by the jurisdiction as an archival record of construction activity in the community is not required by the code.

SECTION R107
TEMPORARY STRUCTURES AND USES

R107.1 General. The building official is authorized to issue a permit for temporary structures and temporary uses. Such permits shall be limited as to time of service, but shall not be permitted for more than 180 days. The building official is authorized to grant extensions for demonstrated cause.

❖ The building official can authorize temporary use permits when the applicant has met all regulations governing such use. Permits for temporary structures may also be granted subject to the provisions of this section. While the permit is to be granted for a time period consistent with the time of service, 180 days is the maximum period of time for which a temporary use or structure can be valid. This code section often applies to structures that are commonly used for a short duration of time at a specific location, then easily dismantled and removed, often to be reconstructed at a different site. It is common for this type of structure to be used at events like street fairs, carnivals, circuses, parades, sporting events, weddings, concerts or revivals, or they may also be used to house construction offices. These structures range in size from small tents or shade structures, which may house a vendor at a street fair, to large tents for circuses or religious revivals. Due to the large size, large occupant load, and complexity that can result from smaller structures being grouped together, the fire- and life-safety concerns as covered in Section R107.2 must be duly considered before a permit is issued.

R107.2 Conformance. Temporary structures and uses shall conform to the structural strength, fire safety, means of egress, light, ventilation and sanitary requirements of this code as necessary to ensure the public health, safety and general welfare.

❖ This section gives the building official discretion in determining the specific criteria for conformance. The issues of structural strength, fire safety, egress, light, ventilation and sanitation are mentioned in this section as the key areas of concern for temporary structures or uses. The levels of required conformance must be determined by the building official to achieve the jurisdiction's required level of safety, health and welfare. Full compliance with the code is not required for a temporary structure or temporary use, but it is clear that the level of performance set forth by the code must be considered in the development of the requirements.

R107.3 Temporary power. The building official is authorized to give permission to temporarily supply and use power in part of an electric installation before such installation has been fully completed and the final certificate of completion has been issued. The part covered by the temporary certificate shall comply with the requirements specified for temporary lighting, heat or power in the ICC *Electrical Code*.

❖ The ICC *Electrical Code*® is referenced regarding temporary lighting, heat and power. The code allows temporary electrical service to be provided prior to the completion of the entire electrical system if the building official approves such a connection.

R107.4 Termination of approval. The building official is authorized to terminate such permit for a temporary structure or use and to order the temporary structure or use to be discontinued.

❖ Where the use of a temporary structure is not consistent with that approved by the building official or where an unsafe condition exists, the building official is authorized to terminate the temporary permit.

SECTION R108
FEES

R108.1 Payment of fees. A permit shall not be valid until the fees prescribed by law have been paid. Nor shall an amendment to a permit be released until the additional fee, if any, has been paid.

❖ This section addresses the costs necessary to operate the Department of Building Safety by creating a mechanism for fee collection. Such fees are typically set to provide enough funds to adequately pay for the costs of operating the various department functions, including administration, plans examination and inspection. All fees, including those for changes to the permit, are to be collected before a valid permit is issued.

R108.2 Schedule of permit fees. On buildings, structures, electrical, gas, mechanical, and plumbing systems or alterations requiring a permit, a fee for each permit shall be paid as required, in accordance with the schedule as established by the applicable governing authority.

❖ The code states that fees for activities regulated by the jurisdiction and administered by the building official should be equitably assessed and adequate to fund the administration, inspection and plan review services required by the code. Exorbitant fees are unreasonable and can often be a basis for some citizens' distrust of government agencies, while unrealistically low fees serve little purpose. In either situation, the effectiveness of the department suffers from either distrust and poor relations with the public or undertrained and understaffed departments that are hopelessly unable to retain the best-qualified employees. Either fee schedule does not serve the best interests of the public, which depends on the jurisdiction to enforce the code provisions effectively to maintain a minimum level of safety in the built environment.

R108.3 Building permit valuations. Building permit valuation shall include total value of the work for which a permit is being issued, such as electrical, gas, mechanical, plumbing equipment and other permanent systems, including materials and labor.

❖ Most jurisdictions develop a fee schedule based on the projected construction cost of the work to be done. Two methods are used to determine this cost valuation: (1) a "per-square-foot" factor based on the use or occupancy of the building and the type of construction involved or (2) the "bid cost" factor based on the total accepted bid price for doing the work. The valuation is determined by including the value of the construction process, including both materials and labor. It is important that a realistic valuation be determined for every project so that permit fees are applied fairly and accurately.

R108.4 Related fees. The payment of the fee for the construction, alteration, removal or demolition for work done in connection with or concurrently with the work authorized by a building permit shall not relieve the applicant or holder of the permit from the payment of other fees that are prescribed by law.

❖ A building permit and the fees attached to that permit do not necessarily cover all aspects of the work to be performed. All fees of the jurisdiction, including those for additional permits, reinspections, investigations or other departmental functions, must be paid along with those for a building permit.

R108.5 Refunds. The building official is authorized to establish a refund policy.

❖ The building official has the authority to develop policies for refunding permit fees. The refund policy should retain fees for which the jurisdiction has provided services; only that portion of a fee for which no service has been rendered should be refunded.

SECTION R109
INSPECTIONS

R109.1 Types of inspections. For onsite construction, from time to time the building official, upon notification from the permit holder or his agent, shall make or cause to be made any necessary inspections and shall either approve that portion of the construction as completed or shall notify the permit holder or his or her agent wherein the same fails to comply with this code.

❖ Inspections are necessary to verify that the construction conforms to the code requirements, and this section outlines the minimum required inspections. Besides the minimum required inspections that are specifically listed, the building official has the authority to require additional inspections so that compliance with the code can be determined. It is the duty of the permit holder or an authorized agent of the permit holder to notify the building department that some or all of the work covered by the permit is ready and available for inspection. At that point, the appropriate jurisdictional inspector performs the necessary on-site inspection. The inspector then must inform the permit holder or agent that the work has been inspected. This may be through a telephone call, an electronic message or in many cases a written record of the inspection posted at the job site.

R109.1.1 Foundation inspection. Inspection of the foundation shall be made after poles or piers are set or trenches or basement areas are excavated and any required forms erected and any required reinforcing steel is in place and prior to the placing of concrete. The foundation inspection shall include excavations for thickened slabs intended for the support of bearing walls, partitions, structural supports, or equipment and special requirements for wood foundations.

❖ The foundation inspection is typically the first jobsite inspection by a representative of the Department of Building Safety. At that time, the inspector will verify that the footing and/or foundation is located as shown on the approved plans and in accordance with the jurisdictional requirements for building setbacks and easements. Where a footing system is to be used, the footing trenches must be excavated and any required reinforcing steel must be in place. For foundation walls, the wall forms must be erected completely with the appropriate steel reinforcement placed within the forms. Column pads, thickened slabs and other foundation work must also be ready for inspection prior to concrete placement.

R109.1.2 Plumbing, mechanical, gas and electrical systems inspection. Rough inspection of plumbing, mechanical, gas and electrical systems shall be made prior to covering or concealment, before fixtures or appliances are set or installed, and prior to framing inspection.

Exception: Ground-source heat pump loop systems tested in accordance with Section M2105.1 shall be permitted to be backfilled prior to inspection.

❖ The various trade rough-in inspections must be completed prior to the framing inspection. The electrical, gas, mechanical and plumbing systems that are to be concealed must be inspected and approved prior to their concealment. In addition, the effect of these systems on the structural integrity of the building must be reviewed after they are completed and ready to be covered. All of the trade inspections must be completed and approvals granted prior to the inspection of the framing system.

R109.1.3 Floodplain inspections. For construction in areas prone to flooding as established by Table R301.2(1), upon placement of the lowest floor, including basement, and prior to further vertical construction, the building official shall require submission of documentation, prepared and sealed by a registered design professional, of the elevation of the lowest floor, including basement, required in Section R323.

❖ For any building constructed in a flood-prone area, the elevation of the lowest floor level must be established immediately after placement of the floor. Documentation of the elevation of the lowest floor is evidence of compliance (see the commentary for Section R323.1.9). If the documentation is submitted when the lowest floor level is established and before further vertical construction, errors in elevation may be corrected at the minimum cost. The elevation of the lowest floor is used by insurance agents to compute flood insurance premium costs. An error in the minimum required elevation may significantly increase the cost of flood insurance, which may be required by mortgage lenders. If the elevation of the lowest floor goes uncorrected, all future owners will incur the additional increase in cost.

R109.1.4 Frame and masonry inspection. Inspection of framing and masonry construction shall be made after the roof, masonry, all framing, firestopping, draftstopping and bracing are in place and after the plumbing, mechanical and electrical rough inspections are approved.

❖ The framing inspection is usually the final opportunity for the inspector to view all of the items that will be concealed within the structure. The inspection includes the structural framework of the building as well as any fireblocking or draftstopping that will be contained within concealed spaces. All of the electrical, mechanical, gas and plumbing inspections must be completed and approved prior to the framing inspection. This allows any framing members to be repaired while they are accessible.

R109.1.5 Other inspections. In addition to the called inspections above, the building official may make or require any other inspections to ascertain compliance with this code and other laws enforced by the building official.

❖ A variety of other inspections, such as those for gypsum board or insulation, may be mandated by the building official in order to verify compliance with the code or other city ordinances. These inspections are generally established as a result of local concern regarding a specific portion of the construction process. The procedure for such inspections should be consistent with the other provisions of this section.

R109.1.5.1 Fire-resistance-rated construction inspection. Where fire-resistance-rated construction is required between dwelling units or due to location on property, the building official shall require an inspection of such construction after all lathing and/or wallboard is in place, but before any plaster is applied, or before wallboard joints and fasteners are taped and finished.

❖ There are a limited number of situations in which fire-resistance-rated construction is required by this code. For example, where two or more dwelling units are located within the same structure, they are required by Section R317 to be completely separated from each other by a specified level of fire resistance. This is the case for both two-family dwellings and townhouses. A second example would be the proximity of the building to an adjoining property line. Section R302 requires exterior walls located less than 3 feet (914 mm) from a property line (unless abutting a public way) to be of minimum 1-hour fire-resistance-rated construction. If either of these conditions should occur, an inspection of the fire-resistance-rated construction is required.

The inspection for compliance with the fire resistance requirements of the IRC should be made at a point of construction when the membrane materials are in place but the fasteners are still exposed. This allows the inspector to verify the appropriate fastener type and location based on the specific fire-resistance listing of the portion of the building under consideration.

R109.1.6 Final inspection. Final inspection shall be made after the permitted work is complete and prior to occupancy.

❖ The final inspection should occur after all of the work addressed by the code is complete but prior to occupancy of the building. The issues addressed in the final inspection cover all aspects of construction, including fire safety, life safety and structural safety, as well as electrical, plumbing, gas and mechanical items. The final inspection must be approved before the issuance of a certificate of occupancy can be issued.

R109.2 Inspection agencies. The building official is authorized to accept reports of approved agencies, provided such agencies satisfy the requirements as to qualifications and reliability.

❖ It is not uncommon for the building official to rely on other agencies for informational or inspection reports regarding various aspects covering methods of the construction process and/or materials. This reliance should be based on the building official's approval of the qualifications and reliability possessed by the third-party inspection or testing service.

R109.3 Inspection requests. It shall be the duty of the permit holder or their agent to notify the building official that such work is ready for inspection. It shall be the duty of the person requesting any inspections required by this code to provide access to and means for inspection of such work.

❖ The individual doing the authorized work has the responsibility for notifying the building department when the work is ready for inspection. Each building department establishes its own procedures on how and when requests should be made. Once an inspection has been scheduled, access to the area ready for inspection must be provided. The individuals performing the work should make the inspection process run as smoothly as possible.

R109.4 Approval required. Work shall not be done beyond the point indicated in each successive inspection without first obtaining the approval of the building official. The building official upon notification, shall make the requested inspections and shall either indicate the portion of the construction that is satisfactory as completed, or shall notify the permit holder or an agent of the permit holder wherein the same fails to comply with this code. Any portions that do not comply shall be corrected and such portion shall not be covered or concealed until authorized by the building official.

❖ Work must not continue past the point of a required inspection until it has been approved by the building department. It is possible that if the work progresses beyond this point and is not in total compliance with the code, some of the work may have to be removed. It is critical that each individual stage of the project be approved prior to continuance of construction.

As indicated in Section R109.1, inspections must be performed when requested, and the inspector must indicate if the construction is either satisfactory or is not compliant. If the work is not approved, it must be corrected, and a reinspection must be requested. No work may be concealed until the building department approves it.

SECTION R110
CERTIFICATE OF OCCUPANCY

R110.1 Use and occupancy. No building or structure shall be used or occupied, and no change in the existing occupancy classification of a building or structure or portion thereof shall be made until the building official has issued a certificate of occupancy therefor as provided herein. Issuance of a certificate of occupancy shall not be construed as an approval of a violation of the provisions of this code or of other ordinances of the jurisdiction. Certificates presuming to give authority to violate or cancel the provisions of this code or other ordinances of the jurisdiction shall not be valid.

Exception: Certificates of occupancy are not required for work exempt from permits under Section R105.2.

❖ The tool the building official employs to control the uses and occupancies of the various buildings in a jurisdiction is the certificate of occupancy. This section establishes the conditions of a certificate of occupancy, commonly called a C.O., and identifies the information the certificate must contain. The building official must be satisfied that the structure meets the requirements of the code before a certificate of occupancy can be given, and the structure cannot be legally occupied until a certificate has been issued. If the occupancy classification of an existing building has changed, such as changing an old Victorian home to a small office building, a new certificate of occupancy must be issued. The certificate of occupancy is the legal notification from the Department of Building Safety that the building may be occupied for its intended purpose.

The granting of a certificate of occupancy does not necessarily indicate that no violations of the code or other jurisdictional laws exist. While the building official should make every effort to determine compliance with all applicable code provisions and other ordinances of the jurisdiction, it is important that violations must be corrected. It is possible that the certificate of occupancy will be revoked if it is found that it was issued in error.

[EB] R110.2 Change in use. Changes in the character or use of an existing structure shall not be made except as specified in Sections 3406 and 3407 of the *International Building Code*.

❖ When an existing building's character or use is modified, the provisions of Sections 3406 and 3407 of the *International Building Code* dealing with change of occupancy and historic buildings must be met. Since the *International Residential Code* is limited in scope to specific residential occupancies, any change of occupancy to a use beyond the scope of Section R101.2 will be governed by the IBC.

R110.3 Certificate issued. After the building official inspects the building or structure and finds no violations of the provisions of this code or other laws that are enforced by the department of building safety, the building official shall issue a certificate of occupancy which shall contain the following:

1. The building permit number.
2. The address of the structure.
3. The name and address of the owner.
4. A description of that portion of the structure for which the certificate is issued.
5. A statement that the described portion of the structure has been inspected for compliance with the requirements of this code.
6. The name of the building official.
7. The edition of the code under which the permit was issued.
8. If an automatic sprinkler system is provided.
9. Any special stipulations and conditions of the building permit.

❖ Prior to use or occupancy of the building, the building official shall perform a final inspection as addressed in Section R109.1.6. If the official finds no violations of the code and other laws enforced by the Department of Building Safety, he or she is required to issue a certificate of occupancy. Commentary Figure R110.3 illustrates the information that must be provided on the certificate of occupancy.

R110.4 Temporary occupancy. The building official is authorized to issue a temporary certificate of occupancy before the completion of the entire work covered by the permit, provided that such portion or portions shall be occupied safely. The building official shall set a time period during which the temporary certificate of occupancy is valid.

❖ Where a portion of a building is intended to be occupied prior to occupancy of the entire structure, the building official may issue a temporary certificate of occupancy. Prior to issuance of a temporary certificate of occupancy, it is critical that the building official determine that the portions to be occupied provide the minimum levels of safety required by the code. In addition, the building official must establish a definitive length of time for the temporary certificate of occupancy to be valid.

R110.5 Revocation. The building official shall, in writing, suspend or revoke a certificate of occupancy issued under the provisions of this code wherever the certificate is issued in error, or on the basis of incorrect information supplied, or where it is determined that the building or structure or portion thereof is in violation of any ordinance or regulation or any of the provisions of this code.

❖ In essence, the certificate of occupancy certifies that the described building or portion of that building complies with the requirements of the code for the intended use. However, any certificate of occupancy may be suspended or revoked by the building official under one of three conditions: (1) when the certificate is issued in error, (2) when incorrect information is supplied to the building official or (3) when it is determined that the building or a portion of the building is shown to be in violation of the code or any other ordinance or regulation of the jurisdiction.

SECTION R111
SERVICE UTILITIES

R111.1 Connection of service utilities. No person shall make connections from a utility, source of energy, fuel or power to any building or system that is regulated by this code for which a permit is required, until approved by the building official.

❖ This section addresses the connection and disconnection, either permanent or temporary, of any utilities

**Figure R110.3
CERTIFICATE OF OCCUPANCY**

that service a building or structure regulated by this code. The building official is authorized to control the connection for any service utility when the connection is to a building that is regulated by the code and requires a permit. Prior to the connection of a utility, source of energy, fuel or power, all conditions for the connection must be met and verified by required inspections.

R111.2 Temporary connection. The building official shall have the authority to authorize and approve the temporary connection of the building or system to the utility source of energy, fuel or power.

❖ Temporary service utility connections, such as for temporary electrical service during the construction process, are permitted when approved by the building official after the necessary inspections have been performed and any additional conditions met.

R111.3 Authority to disconnect service utilities. The building official shall have the authority to authorize disconnection of utility service to the building, structure or system regulated by this code and the referenced codes and standards set forth in Section R102.4 in case of emergency where necessary to eliminate an immediate hazard to life or property or when such utility connection has been made without the approval required by Section R111.1 or R111.2. The building official shall notify the serving utility and whenever possible the owner and occupant of the building, structure or service system of the decision to disconnect prior to taking such action if not notified prior to disconnection. The owner or occupant of the building, structure or service system shall be notified in writing as soon as practical thereafter.

❖ When an immediate hazard to life or property exists, the building official has the authority to order disconnection of the utility services. This can also occur when the utility service has been connected without the necessary approvals required by the code. Whenever possible, the building owner and the building occupant or occupants should be notified prior to the disconnection of the services. Then at the first practical opportunity, the building official is to formally notify the building owner in writing of the disconnection activities. As with all administrative functions, all aspects of due process must be followed.

SECTION R112
BOARD OF APPEALS

R112.1 General. In order to hear and decide appeals of orders, decisions or determinations made by the building official relative to the application and interpretation of this code, there shall be and is hereby created a board of appeals. The building official shall be an ex officio member of said board but shall have no vote on any matter before the board. The board of appeals shall be appointed by the governing body and shall hold office at its pleasure. The board shall adopt rules of procedure for conducting its business, and shall render all decisions and findings in writing to the appellant with a duplicate copy to the building official.

❖ This section holds that any aggrieved party with a material interest in the decision of the building official may appeal such a decision before a board of appeals. This provides a forum other than the court of the jurisdiction in which the building official's action can be reviewed. A board of appeals is to be created by the jurisdiction. The primary function of the board of appeals is hearing and acting on the appeal of orders or decisions the building official has made on the application or interpretation of the code. The building official is as an ex-officio member of the board of appeals, but he or she has no vote on any issue that comes before the board. The appellant is to be provided with a written copy of the board's findings and decisions, with a duplicate copy sent to the building official. Appendix B of the *International Building Code* supplies example rules of procedure.

R112.2 Limitations on authority. An application for appeal shall be based on a claim that the true intent of this code or the rules legally adopted thereunder have been incorrectly interpreted, the provisions of this code do not fully apply, or an equally good or better form of construction is proposed. The board shall have no authority to waive requirements of this code.

❖ The code does not grant authority to the board to hear appeals regarding the administrative provisions of the code, nor does it grant the board the authority to waive any code requirements. An appeal must be based on the claim that the provisions of the code have been misinterpreted, that the provisions do not apply to the appellant's circumstances, or that where an alternate method or technique of construction is used it has been shown to be at least equal to the code requirements.

R112.2.1 Determination of substantial improvement in areas prone to flooding. When the building official provides a finding required in Section R105.3.1.1, the board of appeals shall determine whether the value of the proposed work constitutes a substantial improvement. A substantial improvement means any repair, reconstruction, rehabilitation, addition, or improvement of a building or structure, the cost of which equals or exceeds 50 percent of the market value of the building or structure before the improvement or repair is started. If the building or structure has sustained substantial damage, all repairs are considered substantial improvement regardless of the actual repair work performed. The term does not include:

1. Improvements of a building or structure required to correct existing health, sanitary or safety code violations identified by the building official and which are the minimum necessary to assure safe living conditions; or

2. Any alteration of an historic building or structure provided that the alteration will not preclude the continued designation as an historic building or structure. For the purpose of this exclusion, an historic building is:

    2.1. Listed or preliminarily determined to be eligible for listing in the National Register of Historic Places; or

    2.2. Determined by the Secretary of the U.S. Department of Interior as contributing to the historical significance of a registered historic district or a district preliminarily determined to qualify as an historic district; or

    2.3. Designated as historic under a state or local historic preservation program that is approved by the Department of Interior.

❖ "Substantial improvement," found in R105.3.1.1 and R112.2.1, is defined in this section and is consistent with the definition used by the National Flood Insurance Program (NFIP) and Section 1612.2 of the *International Building Code*. One of the long-range objectives of the NFIP is to reduce the exposure of older buildings that were built in flood hazard areas before local jurisdictions adopted flood hazard area maps and regulations. Section R105.3 of the code directs the applicant to state the valuation of the proposed work as part of the application for a permit. To determine whether a proposed alteration, repair, addition or improvement of a building or structure constitutes a substantial improvement, the cost of the proposed work is to be compared to the market value of the building or structure before the work is started. To determine market value, the building official may require the applicant to provide the information listed in R105.3. For additional guidance, refer to FEMA 213, *Answers to Questions about Substantially Damaged Buildings* and FEMA 311, *Guidance on Estimating Substantial Damage Using the NFIP Residential Substantial Damage Estimator*.

When the board of appeals makes its determination, certain items are not included in the valuation of proposed work. Specifically, if certain health, sanitation or safety-code violations have been cited previously, then the cost of the minimum repairs required to correct those violations is not included in the determination.

Alteration of an historic structure is not considered a substantial improvement or repair of substantial damage, provided the proposed work does not alter the building to the extent that it would no longer qualify as an historic structure. The building official may require applicants to consult an appropriate historic preservation authority to determine whether a proposed alteration will jeopardize a structure's historical designation. The exception for historic structures does not apply to structures located within designated historic districts unless those structures are individually listed as historic structures.

R112.2.2 Criteria for issuance of a variance for areas prone to flooding. A variance shall only be issued upon:

1. A showing of good and sufficient cause that the unique characteristics of the size, configuration or topography of the site render the elevation standards in Section R323 inappropriate.

2. A determination that failure to grant the variance would result in exceptional hardship by rendering the lot undevelopable.

3. A determination that the granting of a variance will not result in increased flood heights, additional threats to public safety, extraordinary public expense, nor create nui-

sances, cause fraud on or victimization of the public, or conflict with existing local laws or ordinances.

4. A determination that the variance is the minimum necessary to afford relief, considering the flood hazard.
5. Submission to the applicant of written notice specifying the difference between the design flood elevation and the elevation to which the building is to be built, stating that the cost of flood insurance will be commensurate with the increased risk resulting from the reduced floor elevation, and stating that construction below the design flood elevation increases risks to life and property.

❖ All the criteria set forth in this section must be met in order for the board of appeals to consider granting a variance for construction in flood-prone areas and provide relief from selected provisions for flood-resistant construction. Granting of the variance must not cause additional public safety concerns beyond those already present.

The board of appeals is empowered to hear requests for variances from the flood hazard provisions of this code. Variances from these provisions may place people and property at significant risk. Therefore, communities are cautioned to carefully evaluate the impacts of issuing a variance, particularly variances to the requirements for elevating buildings to the Base Flood Elevation. The elements that are to be evaluated include impacts on the site, the applicant and other parties who may be affected, such as adjacent property owners and the community as a whole. Flood plain development that is not undertaken in accordance with the flood-resistance provisions of the code will be exposed to increased flood damages. As a consequence, flood insurance premium rates will be significantly higher. Variance decisions made by the board of appeals are to be based solely on technical justifications outlined in this section and not on the personal circumstances of an owner or applicant.

Applicants sometimes request variances to the minimum elevation requirements for the lowest floor of buildings in flood hazard areas in order to improve access for the disabled and the elderly. Generally, variances of this nature are not to be granted because these are personal circumstances that will change as the property changes ownership. Not only would persons of limited mobility be at risk from flooding, but a building built below the BFE would continue to be exposed to flood damage long after the personal need for a variance ends. More appropriate alternatives are to be considered to serve the needs of disabled or elderly persons, such as varying setbacks to allow construction on less flood-prone portions of sites or the installation of personal elevators.

All variances are to be the minimum necessary to afford relief. The board of appeals will address each listed condition for a variance, especially the requirement that it determine whether the failure to grant the variance would result in exceptional hardship by rendering the lot undevelopable. By itself, this determination may be insufficient to result in an exceptional hardship if other conditions for issuing a variance cannot be met. The determination of hardship is to be based on the unique characteristics of the site and not on the personal circumstances of the applicant.

In guidance materials, FEMA cautions that economic hardship alone is not an exceptional hardship. Building officials and boards of appeals are cautioned that granting a variance does not affect how the building will be rated for the purposes of NFIP flood insurance. Even if circumstances justify granting a variance to build a lowest floor that is below the Design Flood Elevation, the rate used to calculate the cost of a flood insurance policy will be based on the risk to the building. Flood insurance, required by certain mortgage lenders, may be extremely expensive. Although the applicant may not be required to purchase flood insurance, the requirement may be imposed on subsequent owners. The building official is to provide the applicant a written notice to this effect, along with the other cautions listed in this section.

R112.3 Qualifications. The board of appeals shall consist of members who are qualified by experience and training to pass on matters pertaining to building construction and are not employees of the jurisdiction.

❖ The members of the board of appeals are to be selected by the appointing authority of the jurisdiction and must be qualified to pass judgment on appeals associated with building construction. The purpose of the board is to provide a review of the appeal independent from that of the building department, so no jurisdictional employee is permitted to be a member. Additional information on suggested qualifications of members of the board of appeals can be found in Appendix B of the *International Building Code*.

R112.4 Administration. The building official shall take immediate action in accordance with the decision of the board.

❖ Decisions made by the board of appeals must be enacted as quickly as possible. The building official is to take whatever action is necessary to see that the orders of the board are carried out.

SECTION R113
VIOLATIONS

R113.1 Unlawful acts. It shall be unlawful for any person, firm or corporation to erect, construct, alter, extend, repair, move, remove, demolish or occupy any building, structure or equipment regulated by this code, or cause same to be done, in conflict with or in violation of any of the provisions of this code.

❖ This section describes the citing, recording and subsequent actions to be taken when code violations are found.

R113.2 Notice of violation. The building official is authorized to serve a notice of violation or order on the person responsible for the erection, construction, alteration, extension, repair, moving, removal, demolition or occupancy of a building or structure in violation of the provisions of this code, or in violation of a detail statement or a plan approved thereunder, or in violation of a permit or certificate issued under the provisions of this code. Such order shall direct the discontinuance of the illegal action or condition and the abatement of the violation.

❖ The building official is required to notify the person responsible for the construction or use of a building found to be in violation of the code. The section of the code that is being violated must be cited so that the responsible party can respond to the notice.

R113.3 Prosecution of violation. If the notice of violation is not complied with in the time prescribed by such notice, the building official is authorized to request the legal counsel of the jurisdiction to institute the appropriate proceeding at law or in equity to restrain, correct or abate such violation, or to require the removal or termination of the unlawful occupancy of the building or structure in violation of the provisions of this code or of the order or direction made pursuant thereto.

❖ When the building owner, the owner's agent or the tenant does not correct the condition causing the violation as directed, the building official must pursue, through the use of legal counsel of the jurisdiction, legal means to correct the violation. This is not optional.

Any extensions that allow the violations to be corrected voluntarily must be for a reasonable, bona fide cause, or the building official may be subject to criticism for "arbitrary and capricious" actions. In general, it is better to have a standard time limitation for correction of violations. Departures from this standard must be for a clear and reasonable purpose, usually stated in writing by the violator.

R113.4 Violation penalties. Any person who violates a provision of this code or fails to comply with any of the requirements thereof or who erects, constructs, alters or repairs a building or structure in violation of the approved construction documents or directive of the building official, or of a permit or certificate issued under the provisions of this code, shall be subject to penalties as prescribed by law.

❖ The jurisdiction must establish penalties for a variety of violations that may occur. Violations specifically addressed by the code include failure to comply with the code, approved plans or directives of the building official.

SECTION R114
STOP WORK ORDER

R114.1 Notice to owner. Upon notice from the building official that work on any building or structure is being prosecuted contrary to the provisions of this code or in an unsafe and dangerous manner, such work shall be immediately stopped. The stop work order shall be in writing and shall be given to the owner of the property involved, or to the owner's agent, or to the person doing the work; and shall state the conditions under which work will be permitted to resume.

❖ The stop work order is a tool authorized by this code that enables the building official to demand that work on a building or structure be temporarily suspended. Typically used under rare circumstances, this order may be issued where the work being performed is dangerous, unsafe or significantly contrary to the provisions of the code.

The stop work order is to be a written document indicating the reason or reasons for the suspension of work, identifying those conditions where compliance is necessary before work is allowed to resume. All work addressed by the order must cease immediately. The stop work order must be presented to either the owner of the subject property, the agent of the owner or the individual doing the work. Commentary Figure R114.1 is an example of a stop work order.

R114.2 Unlawful continuance. Any person who shall continue any work in or about the structure after having been served with a stop work order, except such work as that person is directed to perform to remove a violation or unsafe condition, shall be subject to penalties as prescribed by law.

❖ The only activity permitted in a building or a portion of a building subject to a stop work order is that work necessary to eliminate the violation or unsafe condition. Otherwise, penalties prescribed by laws of the jurisdiction must be imposed for illegal construction activity in defiance of a stop work order.

STOP WORK

DEPARTMENT OF BUILDING SAFETY
NOTICE

This building has been inspected and the
- ☐ Footing/foundation
- ☐ Concrete slab/underfloor
- ☐ Framing
- ☐ Masonry
- ☐ Lath/gypsum board
- ☐ Energy efficiency
- ☐ Rough electrical
- ☐ Rough fuel gas
- ☐ Rough plumbing
- ☐ Rough mechanical
- ☐ Fire-rated construction
- ☐ Final inspection

IS NOT ACCEPTABLE

Please call _____
before any further work is done.

_____   _____
Date   Inspector

Do not remove this notice.

- -

DETACH and bring this portion
of card with you.

Location: _____

Date: _____

_____
Building Official

_____
Phone

Figure R114.1
STOP WORK ORDER

Bibliography

The following resource materials are referenced in this chapter or are relevant to the subject matter addressed in this chapter.

FEMA 213, *Answers to Questions About Substantially Damaged Buildings.* Washington, DC: Federal Emergency Management Agency, 1991. (For ordering information, see the Bibliography for Chapter 3).

FEMA 257, *Mitigation of Flood and Erosion Damage to Residential Buildings in Coastal Areas.* Washington, DC: Federal Emergency Management Agency, 1994.

FEMA 265, *Managing Floodplain Development in Approximate Zone A Areas: A Guide for Obtaining and Developing Base (100-Year) Flood Elevations.* Washington, DC: Federal Emergency Management Agency, 1994.

FEMA 311, *Guidance on Estimating Substantial Damage Using the NFIP Residential Substantial Damage Estimator.* Washington, DC: Federal Emergency Management Agency, 1995.

FEMA 348, *Protecting Building Utilities from Flood Damage.* Washington, DC: Federal Emergency Management Agency, 1999.

IBC-2003, *International Building Code.* Falls Church, VA: International Code Council, Inc., 2003.

IFC-2003, *International Fire Code.* Falls Church, VA: International Code Council, Inc., 2003.

IPMC-2003, *International Property Maintenance Code.* Falls Church, VA: International Code Council, Inc., 2003.

Legal Aspects of Code Administration. Country Club Hills, IL: International Code Council, Inc., 2001.

O'Bannon, Robert E. *Building Department Administration.* Whittier, CA: International Conference of Building Officials, 1973.

Readings in Code Administration. Volume 1: History/Philosophy/Law. Country Club Hills, IL: Building Officials and Code Administrators International, Inc., 1974.

Part II — Definitions

Chapter 2: Definitions

General Comments

The code user should be familiar with the terms in this chapter because the definitions are essential to the correct interpretation of the *International Residential Code®*, and the user might not be aware that a particular term encountered in the text has the special definition found herein.

Section R201.1 contains the scope of the chapter. Section R201.2 establishes the interchangeability of the terms in the code. Section R201.3 establishes the use of terms defined in other chapters. Section R201.4 establishes the use of undefined terms, and Section R202 lists terms and their definition according to this code.

Purpose

Codes are technical documents, so literally every word, term, and punctuation mark can add to or change the meaning of the provision. Furthermore, the IRC, with its broad scope of applicability, includes terms inherent in a variety of construction disciplines. These terms can often have multiple meanings depending on the context or discipline being used at the time. For these reasons, it is necessary to maintain a consensus on the specific meaning of terms contained in the IRC. Chapter 2 performs this function by stating clearly what specific terms mean for the purpose of this code.

SECTION R201
GENERAL

R201.1 Scope. Unless otherwise expressly stated, the following words and terms shall, for the purposes of this code, have the meanings indicated in this chapter.

❖ The purpose of this section is to clarify the terminology used in the *International Residential Code*. The terms defined in the IRC often have very specific meanings, which can be different from their typical meanings. This section gives guidance to the use of the defined words relative to tense and gender and also provides the means to resolve those terms not defined.

R201.2 Interchangeability. Words used in the present tense include the future; words in the masculine gender include the feminine and neuter; the singular number includes the plural and the plural, the singular.

❖ While the definitions contained in Chapter 2 are to be taken literally, gender and tense are considered to be interchangeable; thus, any gender and tense inconsistencies within the code text should not hinder the understanding or enforcement of the requirements.

R201.3 Terms defined in other codes. Where terms are not defined in this code such terms shall have meanings ascribed to them as in other code publications of the International Code Council.

❖ When a word or term appears in the code that is not defined in this chapter, other code publications of the International Code Council® may be used to find its definition. These code documents include the *International Building Code, International Mechanical Code®, International Plumbing Code®, International Fuel Gas Code®, International Fire Code*, and others. These codes contain additional definitions (some parallel and duplicative) which may be used in the enforcement of this code or in the enforcement of the other ICC codes by reference. When using a definition from another code, keep in mind the admonition from "Purpose": "These terms can often have multiple meanings depending on the context or discipline being used at the time."

R201.4 Terms not defined. Where terms are not defined through the methods authorized by this section, such terms shall have ordinarily accepted meanings such as the context implies.

❖ Another possible source for the definitions of words or terms not defined in this code or in other codes is their "ordinarily accepted meanings." Dictionary definitions may suffice, provided that the definitions are in context.

Sometimes construction terms used throughout the code may not be defined in Chapter 2, in another code or in a dictionary. In such cases, one would first turn to the definitions contained in the referenced standards (see Chapter 43) and then to textbooks on the subject in question.

SECTION R202
DEFINITIONS

ACCESSIBLE. Signifies access that requires the removal of an access panel or similar removable obstruction.

❖ In general, where immediate access is not required due to the low level of hazard involved, an accessible method is all that is mandated, such as the removal of an access panel. The electrical definitions found in Chapter 34 contain two additional definitions for "accessible" that apply specifically to electrical wiring methods and electrical equipment. See Section 1102 of the *International Building Code* for the meaning of "accessible" as it applies to the requirements for "accessible" dwelling units in Section R326.

ACCESSIBLE, READY. Signifies access without the necessity for removing a panel or similar obstruction.

❖ Where this term is designated as a requirement by other sections of the code, it is intended that access to the device, controls, shutoff valves or other elements be extremely easy. "Readily accessible" means that the device must be able to be reached directly without a panel, door or equipment needing to be moved in order to gain access. Chapter 24, Fuel Gas, has a similar definition noted as "Ready Access (To)."

ACCESSORY STRUCTURE. In one- and two-family dwellings not more than three stories high with separate means of egress, a building, the use of which is incidental to that of the main building and which is located on the same lot.

❖ As it applies to the scope in Section R101.2, this term describes structures that are designed for accessory use to one- or two-family dwellings and multiple single-family townhouses. These structures are commonly used as garages, carports, cabanas, storage sheds, tool sheds, playhouses and garden structures. The structures all house uses that are incidental to the primary use, which is the dwelling unit, and the activities that take place in accessory structures occur as a result of the primary building. Their use is secondary or minor in importance to the primary residence.

[B] ADDITION. An extension or increase in floor area or height of a building or structure.

❖ Where a structure is increased in size vertically or horizontally, the increase is considered an addition. The addition will typically contribute additional floor area to the building, although an increase in roof height only is also an addition.

AIR ADMITTANCE VALVE. A one-way valve designed to allow air into the plumbing drainage system when a negative pressure develops in the piping. This device shall close by gravity and seal the terminal under conditions of zero differential pressure (no flow conditions) and under positive internal pressure.

❖ An air admittance valve is an alternative to traditional vent piping configurations. Such valves allow the venting of fixtures where conventional venting may be impractical. The valve is designed to open and admit air into the drainage system when negative pressures occur. The valve closes by gravity and seals the vent terminal when the internal drain pressure is equal to or exceeds atmospheric pressure. The seal prevents sewer gas from entering the building; see Commentary Figures R202(1) and R202(2).

AIR BREAK (DRAINAGE SYSTEM). An arrangement in which a discharge pipe from a fixture, appliance or device drains indirectly into a receptor below the flood-level rim of the receptor, and above the trap seal.

❖ An air break is an indirect drainage method by which waste discharges to the drainage system through piping that terminates below the flood level rim of an approved receptor. An air break is commonly used to protect mechanical equipment from sewage backup in the event that stoppage occurs. It also protects the drainage system from adverse pressure conditions caused by pumped discharge; see Commentary Figure R202(3).

AIR CIRCULATION, FORCED. A means of providing space conditioning utilizing movement of air through ducts or plenums by mechanical means.

❖ Forced-air systems use a central heating or cooling unit equipped with a fan to return unconditioned air through return ducts and to deliver conditioned air through supply ducts or plenums to occupied spaces.

AIR-CONDITIONING SYSTEM. A system that consists of heat exchangers, blowers, filters, supply, exhaust and return-air systems, and shall include any apparatus installed in connection therewith.

❖ This definition is limited to the components commonly used in a mechanical air-conditioning system. Additional parts of the air-conditioning system include thermostats, humidistats, dampers and any other controls needed for the system to operate properly.

AIR GAP, DRAINAGE SYSTEM. The unobstructed vertical distance through free atmosphere between the outlet of a waste pipe and the flood-level rim of the fixture or receptor into which it is discharging.

❖ In a drainage system air gap, waste is discharged to the drainage system through piping that terminates at a specified distance above the flood level rim of an approved receptor.
The air gap serves as an impossible barrier for sewage to overcome in the event that stoppage occurs in the receptor drain since sewage backup would overflow in the receptor drain flood level before it comes in contact with the drain line above.

AIR GAP, WATER-DISTRIBUTION SYSTEM. The unobstructed vertical distance through free atmosphere between the lowest opening from a water supply discharge to the flood-level rim of a plumbing fixture.

❖ An air gap is the most reliable and effective means of backflow protection. It is simply the vertical air space between the potable water supply outlet and the possible source of contamination. This air gap prevents possible contamination of the potable water supply by preventing the supply outlet, such as a faucet, from backsiphoning waste water from a basin, for example. Many manufacturers have developed air gap fittings

Figure R202(1)
AIR ADMITTANCE VALVE OPENS TO ADMIT AIR TO RELIEVE NEGATIVE PRESSURE

Figure R202(2)
AIR ADMITTANCE VALVE CLOSES AND SEALS VENT UNDER ZERO OR POSITIVE PRESSURES

that provide a rigid connection to the drainage system while maintaining the minimum level of protection from contamination.

[B] ALTERATION. Any construction or renovation to an existing structure other than repair or addition that requires a permit. Also, a change in a mechanical system that involves an extension, addition or change to the arrangement, type or purpose of the original installation that requires a permit.

❖ The modification of an existing structure without adding any floor area or height to the structure is an alteration. Section R105 specifies that a permit for the alteration work is required before work can begin. The term "alteration" also applies to mechanical work where the original installation is altered in a manner requiring a permit. The repairs described in Section R105.2.2 are not alterations since a permit is not required.

[B] ANCHORS. See "Supports."

❖ Anchors are the same types of devices as supports. Any device, such as a hanger used to support or secure piping and fixtures, is an anchor.

ANTISIPHON. A term applied to valves or mechanical devices that eliminate siphonage.

❖ The term refers to the function of certain valves and devices used to break or prevent the siphon effect that can be created in plumbing systems.

APPLIANCE. A device or apparatus that is manufactured and designed to utilize energy and for which this code provides specific requirements.

❖ An appliance is a manufactured component or assembly of components that converts one form of energy into another to serve a specific purpose. The term "appliance" generally refers to residential and commercial equipment that is manufactured in standardized sizes or types. The term is generally not associated with industrial equipment. For the application of the code provisions, the terms "appliance" and "equipment" are mutually exclusive.

Examples of appliances include furnaces; boilers; water heaters; room heaters; refrigeration units; cooking equipment; clothes dryers; wood stoves; pool, spa and hot tub heaters; unit heaters; ovens and similar fuel-fired or electrically operated appliances. See the definition of "Equipment."

APPROVED. Approved refers to approval by the building official as the result of investigation and tests conducted by him or her, or by reason of accepted principles or tests by nationally recognized organizations.

❖ Throughout the code, the term "approved" is used to describe a specific material or method of construction, such as the approved drainage system mentioned in Section R408.5. Where "approved" is used, it means that the design, material or method of construction is acceptable to the building official. It is imperative that the building official bases her or his decision of approval on the result of investigations, tests or accepted principles or practices.

APPROVED AGENCY. An established and recognized agency regularly engaged in conducting tests or furnishing inspection services, when such agency has been approved by the building official.

❖ The building official will occasionally rely on the expertise of others to assist in plan review and inspection activities, or the owner will employ an individual or com-

Figure R202(3)
AIR GAP, AIR BREAK

DEFINITIONS

pany to provide testing services for portions of the construction work. An agency qualified to perform such activities is an "approved agency." The building official will review an agency's qualifications, expertise and reliability to determine if the agency should be granted approval for a specific activity.

ASPECT RATIO. The ratio of the height to width (h/w) of a shear wall. The shear wall height is the maximum clear height from top of foundation or diaphragm to bottom of diaphragm framing above and the shear wall width is the sheathed dimension in the direction of applied force on the shear wall.

❖ The shear wall or braced wall panel height divided by the width or length of the shear wall or braced wall panel is defined as the aspect ratio. Generally, the code places limits on shear wall aspect ratios in order to limit shear wall deformation.

ATTIC. The unfinished space between the ceiling joists of the top story and the roof rafters.

❖ Several provisions apply to the attic area of a building, such as those relating to ventilation of the attic space. The code identifies an attic as the unfinished space between the ceiling joists of the top story and the roof rafters. Such a space would be the top story rather than the attic if it is finished and occupiable.

BACKFLOW, DRAINAGE. A reversal of flow in the drainage system.

❖ Backflow is a condition found in plumbing systems where the contents of the piping flow in a direction opposite of the intended direction.

BACKFLOW PREVENTER. A device or means to prevent backflow.

❖ Backflow preventers are designed for different applications, depending on the pressures and the degree of hazard posed by the cross connection. There are six basic types of backflow preventers that can be used to correct cross connections: air gaps; barometric loops; vacuum breakers, both atmospheric and pressure type; double check valve with intermediate atmospheric vent; double check-valve assembly and reduced pressure principle devices.

BACKFLOW PREVENTER, REDUCED-PRESSURE-ZONE TYPE. A backflow-prevention device consisting of two independently acting check valves, internally force loaded to a normally closed position and separated by an intermediate chamber (or zone) in which there is an automatic relief means of venting to atmosphere internally loaded to a normally open position between two tightly closing shutoff valves and with means for testing for tightness of the checks and opening of relief means.

❖ This device is one of the most reliable mechanical devices (second only to an air gap) for the prevention of backflow. It employs two spring-loaded check valves with a relief valve in between that monitors the system pressure upstream and downstream of the device. The zone between the check valves is maintained at a pressure that is less than the water supply pressure.

BACKFLOW, WATER DISTRIBUTION. The flow of water or other liquids into the potable water-supply piping from any sources other than its intended source. Backsiphonage is one type of backflow.

❖ There are primarily three types of backflow conditions that could occur in a water distribution system: (1) backpressure; (2) backpressure, low-head and (3) backsiphonage.

BACKPRESSURE. Pressure created by any means in the water distribution system, which by being in excess of the pressure in the water supply mains causes a potential backflow condition.

❖ Liquid in a pipe flows from a high pressure condition to a low pressure condition, such as when the pressure within the water distribution system is greater than that in the water service. The water will reverse its intended direction of flow and return to the water supply, unless a device to prevent backflow has been installed in the line. Elevated pressures can be created through mechanical means, such as a pumping system, or can be the result of a loss of pressure within the main water supply. Backpressure will often occur in a system or a portion of a system that is closed, i.e., it is not open to the atmosphere.

BACKPRESSURE, LOW HEAD. A pressure less than or equal to 4.33 psi (29.88 kPa) or the pressure exerted by a 10-foot (3048 mm) column of water.

❖ Low head backpressure is created where the source of backpressure comes from a hose elevated to a level that produces 10 feet (3048 mm) or less of water column at the outlet. A hose connection, such as a wall hydrant or faucet, is where low head backpressure is typically experienced. Also see the commentary to "Backpressure."

BACKSIPHONAGE. The flowing back of used or contaminated water from piping into a potable water-supply pipe due to a negative pressure in such pipe.

❖ Backsiphonage occurs when the pressure within a potable water distribution system drops below atmospheric pressure or negative gauge pressure. It, like backpressure, allows the normal direction of flow to be reversed. A siphon can result in an unprotected cross connection, causing contamination or pollution of the potable water supply. The main difference between backsiphonage and backpressure is the pressure in the system. Water distribution systems having a pressure less than zero gauge have the potential to create a siphon, resulting in backsiphonage. Backpressure typically occurs in a closed system, and backsiphonage typically occurs in an open system (open to the atmosphere).

BACKWATER VALVE. A device installed in a drain or pipe to prevent backflow of sewage.

❖ A backwater valve is a type of check valve designed for installation in drainage piping. The valve has a lower invert downstream of the flapper than upstream to help prevent solids from interfering with valve closure.

[B] BALCONY, EXTERIOR. An exterior floor projecting from and supported by a structure without additional independent supports.

❖ This term describes an exterior floor that is supported by the structure without the use of independent supports, such as posts or columns. A balcony typically cantilevers from the supporting structure, whereas a deck is supported not only by the main structure but also by independent columns or posts. An exterior balcony lacks the structural redundancy of a deck.

BALL COCK. A valve that is used inside a gravity-type water closet flush tank to control the supply of water into the tank. It may also be called a flush-tank fill valve or water control.

❖ A ball cock is a float-actuated valve used in a traditional gravity-flush water closet flush tank. The valve serves to control the water supply for refilling the flush tank and restoring the trap seal in the water-closet bowl. Ball cock valves are available with or without an anti-siphon device, depending on the installation requirements.

[B] BASEMENT. That portion of a building that is partly or completely below grade (see "Story above grade").

❖ Defined as "that portion of a building which is partially or totally below grade," a basement is further identified in the provisions in Section R202 for a "story above grade." Specific provisions, including the requirement for emergency escape and rescue openings in Section R310.1, are applicable for those floor levels meeting the criteria for basements. The presence of occupiable space below grade level causes various concerns that are addressed by the code.

BASEMENT WALL. The opaque portion of a wall that encloses one side of a basement and has an average below grade wall area that is 50 percent or more of the total opaque and nonopaque area of that enclosing side.

❖ A basement wall is defined in respect to the energy efficiency provisions of Chapter 11. A wall that does not qualify as an exterior wall is a basement wall.

BASIC WIND SPEED. Three-second gust speed at 33 feet (10 058 mm) above the ground in Exposure C (see Section R301.2.1) as given in Figure R301.2(4).

❖ Basic wind speed, the basis for a number of structural decisions in the IRC, is the 3-second gust speed measured at a point 33 feet (10 058 mm) above the ground. The assumed exposure used in determining the basic wind speed is that for generally open terrain having scattered obstructions.

BATHROOM GROUP. A group of fixtures, including or excluding a bidet, consisting of a water closet, lavatory, and bathtub or shower. Such fixtures are located together on the same floor level.

❖ Special consideration is given to a bathroom group when sizing water distribution systems and drain, waste and vent piping because of fixture usage. "Bathroom group" has historically referred to a single water closet, a lavatory and a bathtub or shower, all located within a single room. Such an arrangement would normally allow only one occupant in the room; therefore, the likelihood of simultaneous fixture discharge is remote. In order for the fixture-usage theory of the bathroom group to be valid, the fixtures in the group must be located within the same bathroom. A bidet has been added to the group of fixtures constituting a bathroom group.

BEND. A drainage fitting, designed to provide a change in direction of a drain pipe of less than the angle specified by the amount necessary to establish the desired slope of the line (see "Elbow" and "Sweep").

❖ This term denotes a one-way fitting (often referred to as an elbow) that is typically used in a drainage system to make a change in direction within a single run of pipe. Although bends (or elbows) are used in other plumbing and mechanical piping systems, in this instance the term refers to a drainage fitting that will meet the required radius of turn.

BOILER. A self-contained appliance from which hot water is circulated for heating purposes and then returned to the boiler, and which operates at water pressures not exceeding 160 pounds per square inch gage (psig) (1102 kPa gage) and at water temperatures not exceeding 250°F (121°C).

❖ Boilers are usually manufactured of steel or cast iron and are used to transfer heat (from the combustion of a fuel or from an electric-resistance element) to water for supplying steam or hot pressurized water for heating or other process or power purposes.

Boilers are usually installed in closed systems where the heat transfer medium is recirculated and retained within the system. Hot water supply boilers are normally part of open systems where the heated water is supplied and used externally to the boiler. Large domestic (potable) water heating systems often employ hot water supply boilers.

Boilers are required to be labeled and installed in accordance with the manufacturers' installation instructions and the applicable sections of the code. Boilers are rated in accordance with standards published by the American Society of Mechanical Engineers (ASME), the Hydronics Institute, the Steel Boiler Insti-

DEFINITIONS

tute (SBI), the American Gas Association (AGA) and the American Boiler Manufacturers Association (ABMA). Boilers can be classified in accordance with the following: working temperature, working pressure, type of fuel used (or electric boilers), materials of construction and whether or not the heat transfer medium changes from a liquid to a vapor.

[B] BOND BEAM. A horizontal grouted element within masonry in which reinforcement is embedded.

❖ Composed of masonry, grout and reinforcement, a bond beam is a horizontal structural member constructed to transfer loads while tying the wall together.

[B] BRACED WALL LINE. A series of braced wall panels in a single story constructed in accordance with Section R602.10 for wood framing or Section R603.7 or R301.1.1 for cold-formed steel framing to resist racking from seismic and wind forces.

❖ Commonly referred to as a shear wall, a braced wall line is used to resist the lateral racking forces created by seismic or wind loading. A number of braced wall panels work together to form a braced wall line. The code has specific requirements for the method of constructing a braced wall line. Where a building does not comply with the bracing requirements of Chapter 6, those portions of the building not meeting the requirements must be designed in accordance with accepted engineering practice.

[B] BRACED WALL PANEL. A section of a braced wall line constructed in accordance with Section R602.10 for wood framing or Section R603.7 or R301.1.1 for cold-formed steel framing, which extend the full height of the wall

❖ A braced wall line is composed of a series of braced wall panels, constructed in accordance with the provisions of Chapter 6. A braced wall panel is required to extend the full height of the wall.

BRANCH. Any part of the piping system other than a riser, main or stack.

❖ A branch refers to piping that connects to a riser, main or stack in a plumbing system. Branch piping is generalized as horizontal piping, can include vertical offsets and is smaller in size or capacity than the main artery or trunk from which it extends. It is analogous to branches of a tree.

BRANCH, FIXTURE. See "Fixture branch, drainage."

BRANCH, HORIZONTAL. See "Horizontal branch, drainage."

BRANCH, INTERVAL. A distance along a soil or waste stack corresponding to a story height, but not less than 8 feet (2438 mm), within which the horizontal branches from one floor or story of a building are connected to the stack.

❖ A branch interval is the vertical distance between a drainage stack's branch connections. In typical construction practice, the branch drain will connect to a stack near the floor level; therefore, the branch interval will correspond to the height of the story in which it is located. Depending on the design on the building, however, a branch interval may not correspond to the height of the story because stories are not limited in height. Therefore, the code requires a branch interval to be at least 8 feet (2438 mm) in length.

Branch intervals are a design factor in drainage pipe sizing and venting design. Drain, waste and vent system design must consider the nature of waste and airflow in a stack and the effects that branch connections have on such flow.

BRANCH, MAIN. A water-distribution pipe that extends horizontally off a main or riser to convey water to branches or fixture groups.

❖ This applies to the primary water supply system piping that extends horizontally off a main or riser to supply water to branches of groups of fixtures.

BRANCH, VENT. A vent connecting two or more individual vents with a vent stack or stack vent.

❖ See Commentary Figure R202(4).

BTU/H. The listed maximum capacity of an appliance, absorption unit or burner expressed in British thermal units input per hour.

❖ This term stands for "British thermal units per hour." Fuel-fired appliances and equipment are rated based on their Btu/h (W) input or output.

[B] BUILDING. Building shall mean any one- and two-family dwelling or portion thereof, including townhouses, that is used, or designed or intended to be used for human habitation, for living, sleeping, cooking or eating purposes, or any combination thereof, and shall include accessory structures thereto.

❖ A building according to the IRC may be a single-family dwelling, a two-family dwelling, a townhouse or an accessory structure to such buildings. The use of a building, excluding an accessory structure, is human habitation, which specifically includes living, sleeping, cooking or eating.

BUILDING DRAIN. The lowest piping that collects the discharge from all other drainage piping inside the house and extends 30 inches (762 mm) in developed length of pipe, beyond the exterior walls and conveys the drainage to the building sewer.

❖ A building drain is usually the main drain of a system within a structure. Building drains are horizontal (including vertical offsets) and are the portion of the drainage system that is at the lowest elevation in the structure. All horizontal drains above the elevation of the building drain are horizontal branches. The building drain terminates at the point where it exits the building. See Commentary Figure R202(5).

Figure R202(4)
BRANCH VENT

BUILDING, EXISTING. Existing building is a building erected prior to the adoption of this code, or one for which a legal building permit has been issued.

❖ There are two definitions for an existing building. The most general use of the term is for a building that has been constructed prior to the adoption of the 2000 IRC. In other words, any building that is in existence at the time the current code is adopted by the jurisdiction is "existing." A second definition relates to those buildings that have been issued a building permit. Once a building is under construction, it is an "existing building" and is regulated by the code under which the permit was issued.

BUILDING LINE. The line established by law, beyond which a building shall not extend, except as specifically provided by law.

❖ The building line is not established by the code but rather by the jurisdiction as the point beyond which the construction of a building is not permitted. Often known as a "setback line" or a "building setback," the building line is established to maintain order within the community's building stock. The IRC does not regulate construction with regard to any jurisdictional building line.

[B] BUILDING OFFICIAL. The officer or other designated authority charged with the administration and enforcement of this code.

❖ Regardless of title, the individual designated by the jurisdiction as the person who administers and enforces the *International Residential Code* is the building official. In addition, the building official may appoint various other individuals to assist in the activities of the Department of Building Safety. In many jurisdictions, the authority of the building official is extended to plans examiners and inspectors to some degree. Section R104 sets forth the duties and responsibilities of the building official.

BUILDING SEWER. That part of the drainage system that extends from the end of the building drain and conveys its discharge to a public sewer, private sewer, individual sewage-disposal system or other point of disposal.

❖ The building sewer is the extension of the building drain and is located entirely outside the building exterior walls.

BUILDING THERMAL ENVELOPE. The basement walls, exterior walls, floor, roof and any other building element that enclose conditioned spaces.

❖ The building thermal envelope includes the roof/ceiling assembly, wall assemblies and floor assemblies that surround a conditioned area, which is the space

Figure R202(5)
DRAINAGE SYSTEM COMPONENTS

that is being intentionally heated and/or cooled. The building envelope is the assembly that separates conditioned space from unconditioned space or the outdoors. For example, a wall between a conditioned dwelling and an unconditioned garage is part of the building thermal envelope. Other elements of the building envelope include attic kneewalls, the perimeter joist between two conditioned floors and skylight wells.

BUILT-UP ROOF COVERING. Two or more layers of felt cemented together and surfaced with a cap sheet, mineral aggregate, smooth coating or similar surfacing material.

❖ A common roof covering for buildings with relatively flat roofs, built-up roof covering uses two or more felt layers with a cap-sheet surfacing. A number of materials standards regulate the wide variety of built-up roofing coverings that are available.

CEILING HEIGHT. The clear vertical distance from the finished floor to the finished ceiling.

❖ Ceiling height as regulated by Section R305 is measured from the finished floor surface to the finished ceiling. Depending on the space under consideration, the minimum ceiling height required by the IRC can vary greatly.

[B] CHIMNEY. A primary vertical structure containing one or more flues, for the purpose of carrying gaseous products of combustion and air from a fuel-burning appliance to the outside atmosphere.

❖ Chimneys differ from vents in the materials they are made of and the type of appliance they are designed to serve. Chimneys are capable of venting flue gases of much higher temperatures than vents.

CHIMNEY CONNECTOR. A pipe that connects a fuel-burning appliance to a chimney.

❖ Chimney connectors are sections of pipe used to convey combustion products from an appliance flue outlet to a chimney inlet. Factory-built chimneys can connect directly to some appliances without the need for a connector; however, masonry chimneys cannot connect directly to an appliance because of the chimney's weight.

[B] CHIMNEY TYPES

Residential-type appliance. An approved chimney for removing the products of combustion from fuel-burning, residential-type appliances producing combustion gases not in excess of 1,000°F (538°C) under normal operating conditions, but capable of producing combustion gases of 1,400°F (760°C) during intermittent forces firing for periods up to 1 hour. All temperatures shall be measured at the appliance flue outlet. Residential-type appliance chimneys include masonry and factory-built types.

❖ The chimneys addressed in the IRC are generally for use with residential-type appliances. Such chimneys are limited in the temperature of the combustion gases they can exhaust. Both masonry and factory-built chimneys can be residential type.

CIRCUIT VENT. A vent that connects to a horizontal drainage branch and vents two traps to a maximum of eight traps or trapped fixtures connected into a battery.

❖ A circuit vent vents multiple fixtures using only one or two vents. The horizontal drainage branch actually serves as a wet vent for the fixtures located downstream of the circuit vent connection; see Section P3110.

CLADDING. The exterior materials that cover the surface of the building envelope that is directly loaded by the wind.

❖ Sheathing, siding or other materials used on the exterior envelope of a building are "cladding" for the purpose of wind design.

[B] CLEANOUT. An accessible opening in the drainage system used for the removal of possible obstruction.

❖ Cleanouts are broadly defined as any access point into the drainage system that allows the removal of a clog or other obstruction in the drain line. A cleanout provides convenient access to the piping interior without significant disassembly of the plumbing installation. A "P" trap with a removable U-bend is typically acceptable as a cleanout for the fixture drain or branch to which the trap discharges.

CLOSET. A small room or chamber used for storage.

❖ A closet is simply a small storage room. Closets are generally found throughout dwelling units but are not considered habitable spaces. The typical hazard related to closets is the probable storage of combustible materials.

COMBINATION WASTE AND VENT SYSTEM. A specially designed system of waste piping embodying the horizontal wet venting of one or more sinks or floor drains by means of a common waste and vent pipe adequately sized to provide free movement of air above the flow line of the drain.

❖ This special system is employed when no other conventional system of separate waste and vent is practical. The combination waste and vent system is used only for venting floor drains, standpipes, sinks and lavatories. The drainage piping is larger than required for draining purposes only, and the drainage branch and stack should be provided with vent piping. The system is commonly used where floor drains are installed in large open areas that cannot accommodate vertical vent risers from the floor drains; see Section P3111.

COMBUSTIBLE MATERIAL. Any material not defined as noncombustible.

❖ Any material that does not qualify as a noncombustible material by ASTM E 136 is considered combustible. The presence of sizable quantities of combustible material increases the potential fire hazard. Buildings containing or constructed of combustible materials are typically more highly regulated than those that are primarily noncombustible.

COMBUSTION AIR. The air provided to fuel-burning equipment including air for fuel combustion, draft hood dilution and ventilation of the equipment enclosure.

❖ The process of combustion requires a specific amount of oxygen to initiate and sustain the combustion reaction. Combustion air includes primary air, secondary air, draft hood dilution air and excess air. Combustion air is the amount of atmospheric air required for complete combustion of a fuel and is related to the molecular composition of the fuel being burned, the design of the fuel-burning equipment and the percentage of oxygen in the combustion air. Too little combustion air will result in incomplete combustion of a fuel and the possible formation of carbon deposits (soot), carbon monoxide, toxic alcohols, ketones, aldehydes, nitrous oxides and other byproducts. The required amount of combustion air is usually stated in terms of cubic feet per minute (m^3/s) or pounds per hour (kg/h).

COMMON VENT. A single pipe venting two trap arms within the same branch interval, either back-to-back or one above the other.

❖ Where two or more fixtures are connected to the same drainage pipe, a common vent could be employed. This common vent is sized and classified as an individ-

DEFINITIONS

ual vent. Any two fixtures may be common vented to either a vertical or horizontal drainage pipe.

A typical form of common venting connects two fixtures at the same level. Two fixtures connecting at different levels but within the same story are also common vented. When one of the fixtures connected is at a different level, the vertical drain between the fixtures is oversized because it functions as a drain for the upper fixture and a vent for the lower fixture.

CONDENSATE. The liquid that separates from a gas due to a reduction in temperature, e.g., water that condenses from flue gases and water that condenses from air circulating through the cooling coil in air conditioning equipment.

❖ Condensate forms when the temperature of a vapor is lowered to its dew-point temperature. Air-conditioning systems produce condensate when an airstream contacts cooling coils. The moisture in the air condenses on the cold surface of the coils and the air is dehumidified. Condensate also forms within improperly designed chimneys and vents when the products of combustion (which contain water vapor) contact the colder walls of the flue. If the temperature of the products of combustion is lowered to the dew-point temperature of the water vapor, condensate will form on the inside walls of the flue. Condensed steam in hydronic systems is also referred to as "condensate."

CONDENSING APPLIANCE. An appliance that condenses water generated by the burning of fuels.

❖ Such units include compressors and condenser heat exchangers; they convert refrigerant vapor to liquid.

CONDITIONED AIR. Air treated to control its temperature, relative humidity or quality.

❖ Conditioned air is air that is heated, cooled, humidified, dehumidified or decontaminated.

CONDITIONED AREA. That area within a building provided with heating and/or cooling systems or appliances capable of maintaining, through design or heat loss/gain, 68°F (20°C) during the heating season and/or 80°F (27°C) during the cooling season, or has a fixed opening directly adjacent to a conditioned area.

❖ The conditioned area is the space within the building that is provided with conditioned air. The space can be directly conditioned or indirectly conditioned by an adjacent conditioned space.

CONDITIONED FLOOR AREA. The horizontal projection of the floors associated with the conditioned space.

❖ The conditioned floor area is the floor area in square feet of a conditioned space.

CONDITIONED SPACE. For energy purposes, space within a building that is provided with heating and/or cooling equipment or systems capable of maintaining, through design or heat loss/gain, 50°F (10°C) during the heating season and 85°F (29°C) during the cooling season, or communicates directly with a conditioned space. For mechanical purposes, an area, room or space being heated or cooled by any equipment or appliance.

❖ If a space is heated, cooled or humidified/dehumidified, it is conditioned space. The building must have a system that is capable of maintaining the space at 50°F (10°C) or above for heating and 85°F (29°C) or below for cooling during normal operation. Also, a basement is conditioned if the floor/ceiling assembly between the basement and the conditioned first floor is uninsulated and there is enough heat transfer across the floor to maintain the space at 50°F (10°C) or above. In addition, uninsulated duct systems located in an unconditioned basement may raise the temperature of the space to 50°F (10°C) and above during normal operation of the system. The basement would then be conditioned space. If the space is conditioned, the building envelope that surrounds the space must meet the thermal requirements in Table N1102.1.

CONFINED SPACE. A room or space having a volume less than 50 cubic feet per 1,000 Btu/h (4.83 L/W) of the aggregate input rating of all fuel-burning appliances installed in that space.

❖ This term is used in conjunction with Chapter 20 and other sections related to combustion air.

[B] CONSTRUCTION DOCUMENTS. Written, graphic and pictorial documents prepared or assembled for describing the design, location and physical characteristics of the elements of a project necessary for obtaining a building permit. Construction drawings shall be drawn to an appropriate scale.

❖ Construction documents are the necessary drawings, specifications, and support materials created for the design and construction of a building. Addressed in Section R106, construction documents are required at the time of permit application unless the work is of such a minor nature that the documents are not necessary to provide compliance with the code. The construction documents are required to be drawn to an appropriate scale so they may be easily interpreted.

CONTAMINATION. An impairment of the quality of the potable water that creates an actual hazard to the public health through poisoning or through the spread of disease by sewage, industrial fluids or waste.

❖ The Environmental Protection Agency (EPA) sets maximum levels for various chemicals and bacteria in drinking water. When an unacceptable level of one or more contaminants is present, the water is considered to be nonpotable. One of the primary purposes of the code is to protect potable water supplies from contamination. Contamination consists of either hazardous chemicals or raw sewage and is considered by the code to represent a high hazard when present in the water supply. Refer to Table P2902.2 for the required potable protection requirements related to high-hazard situations.

The purpose of the water supply provisions of the code and other related regulations, such as those from the EPA, is to reduce, if not totally eliminate, the risk to the general public of poisoning or health impairment through sewage, industrial fluids or waste entering the potable water supply. These contaminants would have a harmful effect on the occupants of a building should such water be consumed. Contamination is quite different from pollution, which is a situation where the water supply is still considered to be nonpotable, but would not kill or harm if ingested. Refer to the definition of "Pollution" for additional information on that term.

CONTINUOUS WASTE. A drain from two or more similar adjacent fixtures connected to a single trap.

❖ Continuous waste is the piping described and regulated in Section P3201.6. The most common occurrences are where multiple compartments of a kitchen sink or double lavatories are connected to a single trap.

CONTROL, LIMIT. An automatic control responsive to changes in liquid flow or level, pressure, or temperature for limiting the operation of an appliance.

❖ Limit controls are safety devices used to protect equipment, appliances, property and persons. Such controls act at their setpoint to limit a condition such as temperature or pressure and include high and low limits. This definition applies to controls for all types of fuel: gas, liquid and solid.

CONTROL, PRIMARY SAFETY. A safety control responsive directly to flame properties that senses the presence or absence of flame and, in event of ignition failure or unintentional flame extinguishment, automatically causes shutdown of mechanical equipment.

❖ This is a device designed to shut down mechanical equipment in response to changes in liquid or gas flow rates that could extinguish a flame.

CONVECTOR. A system-incorporating heating element in an enclosure in which air enters an opening below the heating element, is heated and leaves the enclosure through an opening located above the heating element.

❖ See Commentary Figure R202(6) for an illustration of an electric hydronic convector.

[B] CORROSION RESISTANCE. The ability of a material to withstand deterioration of its surface or its properties when exposed to its environment.

❖ Under those conditions where corrosion of materials is detrimental to the integrity of the building's construction and use, the code mandates corrosion-resistant materials. Corrosion-resistant materials control or withstand the deterioration of their surfaces and retain their physical properties in their intended environment.

Figure R202(6)
ELECTRIC HYDRONIC CONVECTOR

Corrosion resistance can be provided by the material or by a process applied to the material.

[B] COURT. A space, open and unobstructed to the sky, located at or above grade level on a lot and bounded on three or more sides by walls or a building.

❖ An exterior area is a court if it is enclosed on at least three sides by exterior walls of the building or other enclosing elements and is open and unobstructed to the sky above. By virtue of being substantially open to the exterior and to the sky, a court may be used to obtain natural light and ventilation for the building.

CRIPPLE WALL. A framed wall extending from the top of the foundation to the underside of the floor framing of the first story above grade plane.

❖ Cripple walls are built on the top of footings or foundation walls. They can typically be found along the top of stepped foundation walls where the grade adjoining the structure changes height. Cripple walls must be properly braced to resist lateral forces. They are often treated the same as a first-story wall. Provisions for the bracing of cripple walls are in Section R602.10.2.

CROSS CONNECTION. Any connection between two otherwise separate piping systems whereby there may be a flow from one system to the other.

❖ Cross connections are the links through which it is possible for contaminated materials to enter a potable water supply. The contaminant enters the potable water supply when the pressure of the polluted source exceeds the pressure of the potable source. The action may be called "backsiphonage" or "backpressure."

Many serious outbreaks of illnesses and disease have been traced to cross connections. The intent of the code is to eliminate cross connections or prevent backflow where cross connections cannot be eliminated. Refer to the definitions and commentary for the terms "Backflow" and "Backsiphonage."

DEFINITIONS
R202

DALLE GLASS. A decorative composite glazing material made of individual pieces of glass that are embedded in a cast matrix of concrete or epoxy.

❖ Various types of decorative glazing materials are exempt from the safety glazing provisions when installed in side-hinged doors, adjacent to doors, or in large glazed openings. Dalle glass is one such material. Made up of multiple individual pieces of glass, dalle glass is exempt because of its decorative function.

DAMPER, VOLUME. A device that will restrict, retard or direct the flow of air in any duct, or the products of combustion of heat-producing equipment, vent connector, vent or chimney.

❖ A damper is used in an air distribution system as a restrictor to regulate airflow through ductwork. When used in flues venting combustion gases, a damper is used to regulate draft.

DEAD END. A branch leading from a DWV system terminating at a developed length of 2 feet (610 mm) or more. Dead ends shall be prohibited except as an approved part of a rough-in for future connection.

❖ The term "dead end" refers to horizontal piping that does not conduct waste flow but is connected to piping that does conduct such flow. A dead end will collect solid waste as a result of the normal flow depth in the drain pipe to which it is connected. A dead end will also collect solids as a result of drainage system stoppages.

[B] DEAD LOADS. The weight of all materials of construction in-corporated into the building, including but not limited to walls, floors, roofs, ceilings, stairways, built-in partitions, finishes, cladding, and other similarly incorporated architectural and structural items, and fixed service equipment.

❖ Dead loads are considered in the structural design of a building and are used in the span tables for sizing floor joists, ceiling joists and rafters. The weight of the materials of construction joists, sheathing, studs and gypsum board is counted as the dead load, as are stairways, cladding and fixed service equipment. The dead loads make up the fixed load that is placed in the building and are independent from the live loads.

[B] DECK. An exterior floor system supported on at least two opposing sides by an adjoining structure and/or posts, piers, or other independent supports.

❖ Where an exterior floor system is supported on a minimum of two opposing sides, it is a deck. The support may be totally by posts or piers or, as often occurs, by the dwelling on one side with posts or piers on the opposite side.

[B] DECORATIVE GLASS. A carved, leaded or Dalle glass or glazing material whose purpose is decorative or artistic, not functional; whose coloring, texture or other design qualities or components cannot be removed without destroying the glazing material; and whose surface, or assembly into which it is incorporated, is divided into segments.

❖ Decorative glass is not designed for functional purposes such as a vision panel, but rather is used for artistic reasons. Some types of decorative glass include carved, leaded and dalle glass. Decorative glass is typically made up of multiple pieces of glass arranged in some design or pattern. Fixed-in-place decorative glass is exempt from some of the requirements for glazing in hazardous locations.

DESIGN PROFESSIONAL. See definition of "Registered design professional."

❖ "Design professional" is another term for "registered design professional." See the definition of "Registered Design Professional."

DEVELOPED LENGTH. The length of a pipeline measured along the center line of the pipe and fittings.

❖ This term identifies a concept necessary for computing the actual length of piping in a plumbing system. The developed length is measured along the actual flow path of piping and includes the piping lengths in all offsets and changes in direction. Several code requirements are dependent on the actual length of plumbing piping.

DIAMETER. Unless specifically stated, the term "diameter" is the nominal diameter as designated by the approved material standard.

❖ "Diameter" is the length of a line passing though the center of a pipe from one side to the other. Depending on the pipe standard, this distance can be measured to the inside portions (inside diameter, or ID) or the outside portions (outside diameter, or OD) of the pipe.

[B] DIAPHRAGM. A horizontal or nearly horizontal system acting to transmit lateral forces to the vertical resisting elements. When the term "diaphragm" is used, it includes horizontal bracing systems.

❖ In the load path for forces transmitted through a building or structure, the diaphragm is that horizontal portion of the path where lateral forces are transferred to the vertical resisting elements. Roof diaphragms and floor diaphragms carry lateral loads to the shear walls.

DILUTION AIR. Air that enters a draft hood or draft regulator and mixes with flue gases.

❖ Dilution air is associated with draft-hood-equipped Category I appliances. Fan-assisted Category I appliances do not use dilution air, as all of the air brought in by the fan passes directly through the combustion chamber of the appliance. Dilution air causes lowering of the dew point and cooling of flue gases, but to a point that is still above the point when condensation will occur inside the vent.

DIRECT-VENT APPLIANCE. A fuel-burning appliance with a sealed combustion system that draws all air for combustion from the outside atmosphere and discharges all flue gases to the outside atmosphere.

❖ Such appliances are provided with independent exhaust and intake pipes or with concentric pipes that vent through the inner pipe and convey combustion air in the annular space between pipe walls; see the commentary for Sections R304.1 and R804.1.

DRAFT. The pressure difference existing between the appliance or any component part and the atmosphere, that causes a continuous flow of air and products of combustion through the gas passages of the appliance to the atmosphere.

Induced draft. The pressure difference created by the action of a fan, blower or ejector, that is located between the appliance and the chimney or vent termination.

Natural draft. The pressure difference created by a vent or chimney because of its height, and the temperature difference between the flue gases and the atmosphere.

❖ Draft is the negative static pressure measured relative to atmospheric pressure that is developed in chimneys, vents and in the flue-ways of fuel-burning appliances. Draft can be produced by hot flue-gas buoyancy ("stack effect"), mechanically by fans and exhausters or by a combination of both natural and mechanical means.

DRAFT HOOD. A device built into an appliance, or a part of the vent connector from an appliance, which is designed to provide for the ready escape of the flue gases from the appliance in the event of no draft, backdraft or stoppage beyond the draft hood; prevent a backdraft from entering the appliance; and neutralize the effect of stack action of the chimney or gas vent on the operation of the appliance.

❖ Draft hoods are integral to or supplied with natural-draft, atmospheric-burner, gas-fired appliances. Appliances equipped with draft hoods are classified by the manufacturer as Category I appliances. Because of new minimum-efficiency standards and the popularity of mechanical and other special proprietary venting systems, draft-hood-equipped appliances are becoming rare in the marketplace.

DRAFT REGULATOR. A device that functions to maintain a desired draft in the appliance by automatically reducing the draft to the desired value.

❖ Excessive draft reduces the efficiency of an appliance due to energy loss in the form of heat loss out the vent. Draft regulators automatically adjust the draft by allowing cooler air to enter the vent at a preset flow rate, resulting in reducing the draft.

[B] DRAFT STOP. A material, device or construction installed to restrict the movement of air within open spaces of concealed areas of building components such as crawl spaces, floor-ceiling assemblies, roof-ceiling assemblies and attics.

❖ Draft stops divide a large concealed space into smaller compartments in wood-frame floor construction. They limit the movement of air within the cavity, reducing the potential for rapid fire spread. Draftstopping materials include gypsum board, wood structural panels, particleboard or other substantial materials that are adequately supported.

DRAIN. Any pipe that carries soil and water-borne wastes in a building drainage system.

❖ A drain is any pipe in a plumbing system that carries sanitary waste, clear-water waste or storm water. In the case of wet vents, common vents, waste stack vents, circuit vents and combination drain and vents, the drain may also be conducting airflow.

DRAINAGE FITTING. A pipe fitting designed to provide connections in the drainage system that have provisions for establishing the desired slope in the system. These fittings are made from a variety of both metals and plastics. The methods of coupling provide for required slope in the system (see "Durham fitting").

❖ These fittings will accommodate the necessary slope and sweep required for a drain line. See Section P3005.1.

DUCT SYSTEM. A continuous passageway for the transmission of air which, in addition to ducts, includes duct fittings, dampers, plenums, fans and accessory air-handling equipment and appliances.

❖ Duct systems are part of an air distribution system and include supply, return and relief/exhaust air systems.

DURHAM FITTING. A special type of drainage fitting for use in the durham systems installations in which the joints are made with recessed and tapered threaded fittings, as opposed to bell and spigot lead/oakum or solvent/cemented or soldered joints. The tapping is at an angle (not 90 degrees) to provide for proper slope in otherwise rigid connections.

❖ This is a specific type of fitting used in Durham piping systems, which employ specially designed cast-iron threaded fittings and threaded galvanized steel piping.

DURHAM SYSTEM. A term used to describe soil or waste systems where all piping is of threaded pipe, tube or other such rigid construction using recessed drainage fittings to correspond to the types of piping.

❖ See "Durham Fitting."

DWELLING. Any building that contains one or two dwelling units used, intended, or designed to be built, used, rented, leased, let or hired out to be occupied, or that are occupied for living purposes.

❖ A dwelling is a building that contains either one or two dwelling units. The purpose of a dwelling is occupation for living purposes, regardless of the manner of ownership. Single-family houses and duplexes fall under the definition of dwelling. See also "Dwelling Unit."

DEFINITIONS R202

[B] DWELLING UNIT. A single unit providing complete independent living facilities for one or more persons, including permanent provisions for living, sleeping, eating, cooking and sanitation.

❖ The specific purpose of a dwelling unit is to provide the essential amenities necessary for complete and independent facilities. Commonly, dwelling units are thought of as single-family houses or individual living units in duplexes or townhouses.

DWV. Abbreviated term for drain, waste and vent piping as used in common plumbing practice.

❖ This term is used extensively to identify systems and materials suitable for drain, waste, and vent installations. It will often be labeled on specific materials approved for such use.

EFFECTIVE OPENING. The minimum cross-sectional area at the point of water-supply discharge, measured or expressed in terms of diameter of a circle and if the opening is not circular, the diameter of a circle of equivalent cross-sectional area. (This is applicable to air gap.)

❖ An effective opening is used to determine the minimum air gap required between a potable water supply outlet or opening and the flood level rim of a receptacle, fixture or other potential source of contamination.

ELBOW. A pressure pipe fitting designed to provide an exact change in direction of a pipe run. An elbow provides a sharp turn in the flow path (see "Bend" and "Sweep").

❖ See the commentary for the definition of "Bend."

EMERGENCY ESCAPE AND RESCUE OPENING. An operable window, door or similar device that provides for a means of escape and access for rescue in the event of an emergency.

❖ In the case of an emergency, particularly a fire, immediate action must be taken. Evacuation of the dwelling unit in a timely manner is often required. If occupants are sleeping at the time of the incident, the time for evacuation is extended, often to the point where normal egress cannot be accomplished. In such situations, a door or window to the exterior can be used. It is also possible to use such an exterior opening for rescue purposes. See Section R310 for code requirements.

EQUIPMENT. All piping, ducts, vents, control devices and other components of systems other than appliances that are permanently installed and integrated to provide control of environmental conditions for buildings. This definition shall also include other systems specifically regulated in this code.

❖ Throughout the code, the terms "equipment" and "appliance" have been used as necessary to match the terms with the intent and context of the code text. Appliances are not referred to as equipment and vise versa. Traditionally, the term "equipment" meant large machinery and specialized hardware not thought of as an "appliance." See the definition of "Appliance."

EQUIVALENT LENGTH. For determining friction losses in a piping system, the effect of a particular fitting equal to the friction loss through a straight piping length of the same nominal diameter.

❖ A straight run of piping will contain a friction loss [typically figured in friction loss per every 100 feet (30 480 mm)]. Where fittings are included, the result is additional friction loss. While it is possible to find such friction loss through calculations, the prescriptive method of water supply sizing in Chapter 29 uses a multiplier of 1.2 to provide an average friction-loss calculation. Therefore, the equivalent length of a 100-foot (30 480 mm) run of piping with fittings will be 120 feet (36 570 mm) to accommodate the expected friction loss. The equivalent length will be used in applying the sizing tables in Chapter 29. See Section P2903.7, Item 2.

ESSENTIALLY NONTOXIC TRANSFER FLUIDS. Fluids having a Gosselin rating of 1, including propylene glycol; mineral oil; polydimenthyoil oxane; hydrochlorofluorocarbon, chlorofluorocarbon and hydrofluorocarbon refrigerants; and FDA-approved boiler water additives for steam boilers.

❖ Transfer fluids are liquids or gases that transfer heat to or remove heat from another fluid, such as water. The exchange of heat energy takes place through a heat exchange material, which separates the transfer fluid from the fluid being heated or cooled.

ESSENTIALLY TOXIC TRANSFER FLUIDS. Soil, water or gray water and fluids having a Gosselin rating of 2 or more including ethylene glycol, hydrocarbon oils, ammonia refrigerants and hydrazine.

❖ See the commentary for the definition of "Essentially Nontoxic Transfer Fluids."

EVAPORATIVE COOLER. A device used for reducing air temperature by the process of evaporating water into an airstream.

❖ Also known as "swamp coolers," such units are used in arid climates and use water as a refrigerant. Such units substantially increase the humidity of the air being conditioned.

EXCESS AIR. Air that passes through the combustion chamber and the appliance flue in excess of that which is theoretically required for complete combustion.

❖ The nature of natural-draft, fuel-fired appliances is such that the introduction and efficient intermixing of primary and secondary combustion air is not precise and is coupled with the induced air inflow caused by the internal draft. In order to achieve complete oxidation (combustion), there will always be more air (excess) introduced into the appliances than is theoretically necessary for the complete combustion of the fuel.

EXHAUST HOOD, FULL OPENING. An exhaust hood with an opening at least equal to the diameter of the connecting vent.

❖ See Commentary Figure R202(7) for an illustration of a full opening exhaust hood.

EXISTING INSTALLATIONS. Any plumbing system regulated by this code that was legally installed prior to the effective date of this code, or for which a permit to install has been issued.

❖ "Existing installations" is a term that applies to all plumbing work that has been legally installed prior to the effective date of the code, as well as to all plumbing work for which a permit to install has been issued prior to the effective date of the code. Plumbing that has been illegally installed prior to the effective date of the code is not "existing" and is subject to all of the code requirements for new installations.

EXTERIOR INSULATION FINISH SYSTEMS (EIFS). Synthetic stucco cladding systems typically consisting of five layers: adhesive, insulation board, base coat into which fiberglass reinforcing mesh is embedded, and a finish coat in the desired color.

❖ A variety of popular exterior finish systems have been developed, with the typical construction consisting of five layers of material. The code has relatively few provisions regulating EIFS; most of the requirements are those of the manufacturer.

EXTERIOR WALL. An above-grade wall enclosing conditioned space. Includes between floor spandrels, peripheral edges of floors, roof and basement knee walls, dormer walls, gable end walls, walls enclosing a mansard roof, and basement walls with an average below grade wall area that is less than 50 percent of the total opaque and non-opaque area of that enclosing side.

❖ An exterior wall is any wall located above grade level enclosing conditioned space. Regulated by this definition for the purpose of energy efficiency, a basement wall is exterior where less than 50 percent of its enclosing surface area is below grade.

FACTORY-BUILT CHIMNEY. A listed and labeled chimney composed of factory-made components assembled in the field in accordance with the manufacturer's instructions and the conditions of the listing.

❖ A factory-built chimney is a manufactured, listed and labeled chimney that has been tested by an approved agency to determine its performance characteristics. Factory-built chimneys are manufactured in two basic designs: a double-wall insulated design or a triple-wall air-cooled design. Both designs employ stainless steel inner liners to resist the corrosive effects of combustion products.

FENESTRATION. Skylights, roof windows, vertical windows (whether fixed or moveable); opaque doors; glazed doors; glass block; and combination opaque/glazed doors.

❖ Exterior windows and doors are fenestrations in the exterior wall. Skylights and other sloped glazing are

Figure R202(7)
FULL OPENING EXHAUST HOOD

DEFINITIONS

also included in the definition, as are glass doors, opaque doors and opaque/glazed doors. Chapter 11 regulates fenestrations for energy efficiency.

FIBER CEMENT SIDING. A manufactured, fiber-reinforcing product made with an inorganic hydraulic or calcium silicate binder formed by chemical reaction and reinforced with organic or inorganic non-asbestos fibers, or both. Additives which enhance manufacturing or product performance are permitted. Fiber cement siding products have either smooth or textured faces and are intended for exterior wall and related applications.

❖ Fiber cement siding is produced from flat nonasbestos fiber cement sheets. It is intended for exterior application and consists of fiber reinforced cement that is formed either with or without pressure and cured either under natural or accelerated conditions. The surface of the sheet to be exposed is either smooth, granular or textured. This code permits its use as either panel siding or horizontal lap siding.

FIREBLOCKING. Building materials installed to resist the free passage of flame to other areas of the building through concealed spaces.

❖ Fireblocking materials include lumber, wood structural panels, particleboard, gypsum board, cement-based millboard, batts or blankets of mineral wool or glass fiber or any other approved material that will resist the passage of flame from one concealed area to another. Regulated by Section R602.8 for wood-frame construction, fireblocking is often used to isolate vertical cavities such as stud spaces from horizontal concealed areas such as attics or floor-ceiling assemblies.

[B] FIREPLACE. An assembly consisting of a hearth and fire chamber of noncombustible material and provided with a chimney, for use with solid fuels.

❖ Fireplaces burn solid fuels such as wood and coal and are not referred to as appliances in the code. Both masonry and factory-built fireplaces are regulated by the IRC.

Factory-built fireplace. A listed and labeled fireplace and chimney system composed of factory-made components, and assembled in the field in accordance with manufacturer's instructions and the conditions of the listing.

❖ Factory-built fireplaces are solid-fuel burning units having a fire chamber that is either open to the room or, if equipped with doors, operated with the doors either open or closed. The term "fireplace" describes a complete assembly, which includes the hearth, fire chamber and chimney. A factory-built fireplace is composed of factory-built components representative of the prototypes tested, and it is to be installed in accordance with the manufacturer's installation instructions to form a completed fireplace.

Masonry chimney. A field-constructed chimney composed of solid masonry units, bricks, stones or concrete.

❖ Masonry chimneys can have one or more flues within them and are field constructed of brick, stone, concrete, and fire-clay materials. A masonry chimney can stand alone or be part of a masonry fireplace.

Masonry fireplace. A field-constructed fireplace composed of solid masonry units, bricks, stones or concrete.

❖ Masonry fireplaces must be constructed in accordance with the requirements found in Section R1003. These specific requirements are based on tradition and field experience and describe the conventional fireplace that has proven to be reliable where properly constructed, used and maintained.

FIREPLACE STOVE. A free-standing, chimney-connected solid-fuel-burning heater designed to be operated with the fire chamber doors in either the open or closed position.

❖ Fireplace stoves are generally of the free-standing type and heat a space by direct radiation. There are various types of fireplace stoves, and their installation must be in compliance with the listing of the stove.

FIREPLACE THROAT. The opening between the top of the firebox and the smoke chamber.

❖ The fireplace throat is the point where the smoke and heat from the fireplace pass into the chimney.

FIRE SEPARATION DISTANCE. The distance measured from the building face to the closest interior lot line, to the centerline of a street, alley or public way, or to an imaginary line between two buildings on the property. The distance shall be measured at right angles from the lot line.

❖ This is the distance between the exterior surface of a building and one of three locations: the nearest interior lot line; the centerline of a street, alley or other public way or an imaginary line placed between two buildings on the same lot. The measurement is perpendicular to the lot line [see Commentary Figure R202(8)]. The fire separation distance is important in the determining exterior wall protection and the potential prohibition of exterior openings based on the proximity to the lot lines.

FIXTURE. See "Plumbing fixture."

FIXTURE BRANCH, DRAINAGE. A drain serving two or more fixtures that discharges into another portion of the drainage system.

❖ A drain serving two or more fixtures that discharges into another portion of the drainage system.

FIXTURE BRANCH, WATER-SUPPLY. A water-supply pipe between the fixture supply and a main water-distribution pipe or fixture group main.

❖ See "Branch."

Figure R202(8)
FIRE-SEPARATION DISTANCE

FIXTURE DRAIN. The drain from the trap of a fixture to the junction of that drain with any other drain pipe.

❖ Commonly referred to as a trap arm, a fixture drain is the horizontal section of pipe connecting the outlet weir of a trap to a stack, fixture branch, or any other drain. See Commentary Figures R202(9) and (10).

FIXTURE FITTING

Supply fitting. A fitting that controls the volume and/or directional flow of water and is either attached to or accessible from a fixture or is used with an open or atmospheric discharge.

Waste fitting. A combination of components that conveys the sanitary waste from the outlet of a fixture to the connection of the sanitary drainage system.

❖ This includes the various portions of a faucet from its connecting supply lines to the termination point where water is delivered. When applied to waste components, it includes tubing, fittings and traps used for the connection to the drainage system.

FIXTURE GROUP, MAIN. The main water-distribution pipe (or secondary branch) serving a plumbing fixture grouping such as a bath, kitchen or laundry area to which two or more individual fixture branch pipes are connected.

❖ This is the main or branch water-distribution pipe that extends to a group of fixtures (i.e. bathroom, laundry or kitchen).

FIXTURE SUPPLY. The water-supply pipe connecting a fixture or fixture fitting to a fixture branch.

❖ A fixture supply is a water supply pipe serving a single fixture.

FIXTURE UNIT, DRAINAGE (d.f.u.). A measure of probable discharge into the drainage system by various types of plumbing fixtures, used to size DWV piping systems. The drainage fixture-unit value for a particular fixture depends on its volume rate of drainage discharge, on the time duration of a single drainage operation and on the average time between successive operations.

❖ The conventional method of designing a sanitary drainage system is based on drainage fixture unit (d.f.u.) load values. The fixture-unit approach takes into consideration the probability of load on a drainage system. The d.f.u. is an arbitrary loading factor assigned to each fixture relative to its impact on the drainage system. D.f.u. values are determined based on the average rate of water discharge by a fixture, the duration of a single operation and the frequency of use or interval between each operation.

Because d.f.u. values have a built-in probability factor, they cannot be directly translated into flow rates or discharge rates.

A d.f.u. is not the same as the water supply fixture unit (w.s.f.u.) described in Chapter 29 and below.

FIXTURE UNIT, WATER-SUPPLY (w.s.f.u.). A measure of the probable hydraulic demand on the water supply by various types of plumbing fixtures used to size water-piping systems. The water-supply fixture-unit value for a particular fixture de-

**Figure R202(9)
FIXTURE AND HORIZONTAL BRANCH DRAIN**

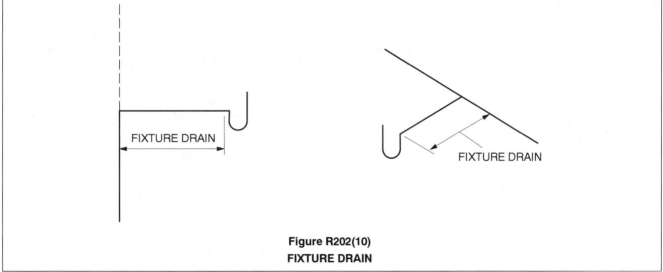

**Figure R202(10)
FIXTURE DRAIN**

pends on its volume rate of supply, on the time duration of a single supply operation and on the average time between successive operations.

❖ When estimating peak demand for water supply systems, sizing methods use water supply fixture units (w.s.f.u.). This is a numerical factor on an arbitrary scale assigned to intermittently used fixtures in order to calculate their load-producing effects on the water supply system.

The use of fixture units makes it possible to simplify the difficult task of calculating load-producing characteristics of any fixture to a common basis. The fixture units of different kinds of fixtures can be applied to a single basic probability curve found in various sizing methods.

[B] FLAME SPREAD. The propagation of flame over a surface.

❖ During a fire, flames will travel across a building's interior surfaces in different forms and at different speeds. The movement of the flames along the surface material is the flame spread.

[B] FLAME SPREAD INDEX. The numeric value assigned to a material tested in accordance with ASTM E 84.

❖ In order to regulate the spread of fire over an interior surface material, limitations are imposed on the characteristics of the material. The standardized test procedure described in ASTM E 84 measures the flame spread under specified conditions, and a numerical value is assigned to the material being tested. This value is the flame spread index.

FLOOD-LEVEL RIM. The edge of the receptor or fixture from which water overflows.

❖ The flood-level rim is the highest elevation that liquid can be contained in a receptacle without spilling over the side of the fixture or receptor.

FLOOR DRAIN. A plumbing fixture for recess in the floor having a floor-level strainer intended for the purpose of the collection and disposal of waste water used in cleaning the floor and for the collection and disposal of accidental spillage to the floor.

❖ Floor drains are typically installed as emergency fixtures, preventing the flooding of a room or space.

FLOOR FURNACE. A self-contained furnace suspended from the floor of the space being heated, taking air for combustion from outside such space, and with means for lighting the appliance from such space.

❖ Such units supply heat through a floor grille placed directly over the unit's heat exchanger. Typically, floor furnaces are classified as gravity-type furnaces. The circulation of air from a gravity-type floor furnace is accomplished by convection. Floor furnaces may be equipped with factory fans to circulate the air.

FLOW PRESSURE. The static pressure reading in the water-supply pipe near the faucet or water outlet while the faucet or water outlet is open and flowing at capacity.

❖ Each fixture requiring water has a minimum flow pressure necessary for the fixture to operate properly and to help protect the potable water supply from contamination. Flow pressure is a factor used in sizing the water supply system. The flow pressure of a plumbing system is less than the static pressure because energy is lost in order to put the fluid in motion.

FLUE. See "Vent."

FLUE, APPLIANCE. The passages within an appliance through which combustion products pass from the combustion chamber to the flue collar.

❖ This is the passage through an appliance used for the removal of combustion products.

FLUE COLLAR. The portion of a fuel-burning appliance designed for the attachment of a draft hood, vent connector or venting system.

❖ See the commentary for the definition of "Draft hood."

FLUE GASES. Products of combustion plus excess air in appliance flues or heat exchangers.

❖ The exact composition of flue gases depends on the materials being burned. The primary components of flue gases are nitrogen, carbon dioxide, water vapor, particulates, carbon monoxide and a myriad of compounds and trace elements that vary with the nature of the fuel.

FLUSH VALVE. A device located at the bottom of a flush tank that is operated to flush water closets.

❖ A flush valve consists of the flapper and the flush-valve seat in a flush-tank-type water closet.

FLUSHOMETER TANK. A device integrated within an air accumulator vessel that is designed to discharge a predetermined quantity of water to fixtures for flushing purposes.

❖ This device is a hybrid between a flushometer valve and a gravity flush tank. A flushometer tank employs a compression tank that holds water under a pressure equivalent to the water-supply pressure. The flushing action created by the flushometer tank is similar to that of the flushometer valve, and the amount of water used is limited as it is with a gravity flush tank.

FLUSHOMETER VALVE. A flushometer valve is a device that discharges a predetermined quantity of water to fixtures for flushing purposes and is actuated by direct water pressure.

❖ This flushing device is a type of metering valve and is used primarily in public occupancy applications because of its powerful flushing action.

[B] FOAM PLASTIC INSULATION. A plastic that is intentionally expanded by the use of a foaming agent to produce a reduced-density plastic consisting open or closed cells distributed throughout the plastic and that has a density less than 20 pounds per cubic foot (320 kg/m^3).

❖ An expanded plastic produced through use of a foaming agent, this reduced-density plastic contains voids consisting of open or closed cells distributed throughout the plastic, which provide thermal insulation and/or acoustic control. The density of the material is to be less than 20 pounds per cubic foot (320 kg/m^3).

FUEL-PIPING SYSTEM. All piping, tubing, valves and fittings used to connect fuel utilization equipment to the point of fuel delivery.

DEFINITIONS   R202

❖ As used in this code, this term includes the tubing and pipe used to convey fuel gas from the point of delivery to the appliance.

FULLWAY VALVE. A valve that in the full open position has an opening cross-sectional area equal to a minimum of 85 percent of the cross-sectional area of the connecting pipe.

❖ Gate valves and ball valves are typical fullway (full-open) valves. Unlike a globe valve, a fullway valve will provide the most open and straight run for the contents to flow through because the closing portion is perpendicular to the run of pipe.

FURNACE. A vented heating appliance designed or arranged to discharge heated air into a conditioned space or through a duct or ducts.

❖ The single most distinguishing characteristic of furnaces is that they use air as the heat transfer medium. Furnaces can be fueled by gas, oil, solid fuel or electricity and can employ fans, blowers and gravity (convection) to circulate the heated air to and from the unit. In the context of the code, the primary use of the term "furnace" refers to heating appliance units that combine a combustion chamber with related components, one or more heat exchangers and an air-handling system.

GLAZING AREA. The interior surface area of all glazed fenestration, including the area of sash, curbing or other framing elements, that enclose conditioned space. Includes the area of glazed fenestration assemblies in walls bounding conditioned basements.

❖ The glazing area includes not only the surface area of the exposed glazing but also the framing elements, including the sash and curbing. The amount of glazing area is regulated for natural light by Section R303; however, this definition applies to the energy efficiency provisions of Chapter 11.

GRADE. The finished ground level adjoining the building at all exterior walls.

❖ This is the point at which the finished exterior ground level intersects the exterior wall of the building. The grade around a building may remain relatively constant, such as on a flat site, or may change dramatically from one point to the next if the site is steeply sloping.

GRADE FLOOR OPENING. A window or other opening located such that the sill height of the opening is not more than 44 inches (1118 mm) above or below the finished ground level adjacent to the opening.

❖ In the requirements for emergency escape and rescue openings found in Section R310, the size of the openings may be reduced if they are grade floor openings. These are windows or other openings that are located within close proximity to the finished ground level. The sill of a grade floor opening may be located either above or below the adjacent ground level, provided it is located no more than 44 inches (1118 mm) vertically from the level of the ground.

GRADE, PIPING. See "Slope."

[B] GRADE PLANE. A reference plane representing the average of the finished ground level adjoining the building at all exterior walls. Where the finished ground level slopes away from the exterior walls, the reference plane shall be established by the lowest points within the area between the building and the lot line or, where the lot line is more than 6 ft (1829 mm) from the building between the structure and a point 6 ft (1829 mm) from the building.

❖ This definition can be important in determining the number of stories within a building as well as its height in feet. In some cases, the finished surface of the ground may be artificially raised with imported fill to create a higher grade plane around a building to decrease the number of stories or height. The definition requires that the lowest elevation within 6 feet (1829 mm) of the exterior wall be used to determine the grade plane.

GROSS AREA OF EXTERIOR WALLS. The normal projection of all exterior walls, including the area of all windows and doors installed therein.

❖ The calculation for determining the gross area of exterior walls for energy efficiency purposes is based on the total area of the entire exterior surface, including openings such as windows and doors.

GROUND-SOURCE HEAT PUMP LOOP SYSTEM. Piping buried in horizontal or vertical excavations or placed in a body of water for the purpose of transporting heat transfer liquid to and from a heat pump. Included in this definition are closed loop systems in which the liquid is recirculated and open loop systems in which the liquid is drawn from a well or other source.

❖ This definition assists the user with a ready means of distinguishing ground-source heat pump loop systems from other hydronic systems.

GUARD. A building component or a system of building components located near the open sides of elevated walking surfaces that minimizes the possibility of a fall from the walking surface to the lower level.

❖ A guard is a component or system of components whose function is the prevention of falls from an elevated area. Placed adjacent to an elevation change, a guard must be of adequate height, strength and configuration to help prevent people, especially small children, from falling over or through the guard to the area below.

[B] HABITABLE SPACE. A space in a building for living, sleeping, eating or cooking. Bathrooms, toilet rooms, closets, halls, storage or utility spaces and similar areas are not considered habitable spaces.

❖ An area within a building used for living, sleeping, dining or cooking purposes is a habitable space. Those areas not meeting this definition include bathrooms, closets, hallways and utility rooms. Habitable spaces are typically occupied, and as such they are more highly regulated than accessory use areas.

[B] HANDRAIL. A horizontal or sloping rail intended for grasping by the hand for guidance or support.

❖ Typically used in conjunction with a ramp or stairway, a handrail provides support for the user along the travel path. Through its height, size, shape, continuity and structural stability, a handrail assists users at elevation changes inside and outside a building. A handrail may also be used as a guide to direct the user in a specified direction.

HANGERS. See "Supports."

HAZARDOUS LOCATION. Any location considered to be a fire hazard for flammable vapors, dust, combustible fibers or other highly combustible substances.

❖ The environment in which mechanical equipment and appliances operate plays a significant role in the safe performance of the equipment installation. Locations that may contain ignitable or explosive atmospheres are classified as hazardous locations with respect to the installation of mechanical equipment and appliances. For example, repair garages can be classified as hazardous locations since they can contain gasoline vapors from vehicles stored within them as well as other volatile chemicals. Public and private garages can be considered as hazardous locations because of the presence of motor vehicles and because of the owners' propensity for storing paint, varnish, thinners, lawn and home maintenance products, and other chemicals within such space.

HEATING DEGREE DAYS (HDD). The sum, on an annual basis, of the difference between 65°F (18°C) and the mean temperature for each day as determined from "NOAA Annual Degree Days to Selected Bases Derived from the 1960-1990 Normals" or other weather data sources acceptable to the code official.

❖ Heating Degree Days are a measure of the heating requirements for buildings in a particular climate. The greater the number of HDD, the colder the climate. HDD is calculated by first determining the average temperature for each day of the year over a 24-hour period. The average temperature for each day is then subtracted from a base temperature of 65°F to determine the HDD value for the day. For days on which the average temperature is over 65°F, the HDD is zero. The HDD for each day are then totaled to determine the HDD value for the particular area. The average temperatures are calculated over a number of years. These values are published in the ASHRAE *Handbook of Fundamentals*. They can also be found in the NOAA *Annual Degree Days to Selected Bases Derived from the 1960-1990 Normals*.

Climate Zone numbers listed in Table N1101.2 can also be used to approximate the HDD value for a particular county. Each climate zone is equal to 500 HDD. For example, Climate Zone 1 is equal to 0-499 HDD, and Climate Zone 2 is equal to 500-999 HDD.

HEAT PUMP. An appliance having heating or heating/cooling capability and that uses refrigerants to extract heat from air, liquid or other sources.

❖ Heat pumps are referred to as reverse-cycle refrigeration systems. Special controls and valves allow the system to be used as a comfort heating system, comfort cooling system and potable water heating system. Auxiliary heat is often installed in heat pumps in the form of electric resistance heat or fuel-fired furnaces. In most cases, the design and installation of heat pumps is more critical than other types of comfort air-conditioning systems because of the heat-transfer process and the amount of air required to be circulated. Heat-pump types include water source (hydronic and wells), ground source (earth loop) and air source.

[B] HEIGHT, BUILDING. The vertical distance from grade plane to the average height of the highest roof surface.

❖ Once the elevation of the grade plane has been calculated, it is possible to determine a building's height, which is measured vertically from the grade plane to the average height of the highest roof surface. Commentary Figure R202(11) contains examples of this measurement.

If the building is stepped or terraced, it is logical that the height is the maximum height of any segment of the building. It may be appropriate under certain circumstances that the number of stories in a building be determined in the same manner. Each case should be judged individually based on the characteristics of the site and construction.

[B] HEIGHT, STORY. The vertical distance from top to top of two successive tiers of beams or finished floor surfaces; and, for the topmost story, from the top of the floor finish to the top of the ceiling joists or, where there is not a ceiling, to the top of the roof rafters.

❖ Within the building itself, the height of an individual story is determined by measuring the vertical distance between two successive finished floor surfaces. In a single-story building or for the uppermost floor in a multistory structure, the measurement is taken from the finished floor surface to the top of the ceiling joists or to the underside of the roof deck (top of the roof rafters) where a ceiling is not provided. See Commentary Figure R202 (12).

HIGH-TEMPERATURE (H.T.) CHIMNEY. A high temperature chimney complying with the requirements of UL 103. A Type H.T. chimney is identifiable by the markings "Type H.T." on each chimney pipe section.

❖ This is a factory-built chimney used to vent solid-fuel-burning appliances when a masonry chimney is not available. See Commentary Figure R202(13) for an illustration of a Type H.T. chimney.

HORIZONTAL BRANCH, DRAINAGE. A drain pipe extending laterally from a soil or waste stack or building drain, that receives the discharge from one or more fixture drains.

❖ See Commentary Figure R202 (9).

HORIZONTAL PIPE. Any pipe or fitting that makes an angle of less than 45 degrees (0.79 rad) with the horizontal.

❖ This definition is needed for the application of many code provisions that apply only to horizontal piping.

HOT WATER. Water at a temperature greater than or equal to 110°F (43°C).

❖ The code does not specify a maximum temperature for hot water; however, it should be limited to a temperature that will minimize 1) the risk of a user's burn and 2) thermal stress to plumbing system components. The designer, installer or owner has the choice of determining the temperature. Most water heaters are now shipped with a factory-set temperature of 120°F (49°C). The minimum temperature is established to determine what is considered hot water. Water temperature below the minimum temperature is either tempered or cold. Care must be used when installing a system that has temperature capability of exceeding 120°F (49°C) because of the possibility of a user being scalded.

HURRICANE-PRONE REGIONS. Areas vulnerable to hurricanes, defined as the U.S. Atlantic Ocean and Gulf of Mexico coasts where the basic wind speed is greater than 110 miles per hour (177 km/h), and Hawaii, Puerto Rico, Guam, Virgin Islands, and America Samoa.

❖ Those geographical areas subject to hurricanes are "hurricane-prone regions." The specific areas include

**Figure R202(11)
HEIGHT OF BUILDING**

**Figure R202(12)
STORY HEIGHT**

Figure R202(13)
HIGH-TEMPERATURE CHIMNEY

those having a minimum basic wind speed exceeding 110 miles per hour (177 km/h) along the United States coastline of the Atlantic Ocean, the coasts bordering the Gulf of Mexico and the indicated islands.

HYDROGEN GENERATING APPLIANCE. A self-contained package or factory-matched packages of integrated systems for generating gaseous hydrogen. Hydrogen generating appliances utilize electrolysis, reformation, chemical, or other processes to generate hydrogen.

DEFINITIONS

❖ Hydrogen is being generated to fuel a new line of developmental vehicles and to generate electric power in fuel cell power systems. There are several different processes being used to generate hydrogen. One such process removes hydrogen from tap water (H_2O) in an electrically powered appliance.

Installation requirements for these appliances are found in Chapter 7 of the *International Fuel Gas Code®,* and ventilation requirements for spaces containing the appliances are found in Section M1307.4 of this code.

IGNITION SOURCE. A flame, spark or hot surface capable of igniting flammable vapors or fumes. Such sources include appliance burners, burner ignitions and electrical switching devices.

❖ Any energized portion of an electrical system that can generate a spark or that can produce significant heat (often referred to as "glow," as in a heating element) is an ignition source should it come in contact with flammable gasses or liquids. Understandably, any open flame (such as a pilot light) is also an ignition source.

INDIRECT WASTE PIPE. A waste pipe that discharges into the drainage system through an air gap into a trap, fixture or receptor.

❖ Often referred to as "open site" or "safe waste," indirect waste pipe installations protect fixtures, equipment and systems from the backflow or backsiphonage of waste.

INDIVIDUAL SEWAGE DISPOSAL SYSTEM. A system for disposal of sewage by means of a septic tank or mechanical treatment, designed for use apart from a public sewer to serve a single establishment or building.

❖ When installed in accordance with the *International Private Sewage Disposal Code®* and approved by the administrative authority, these systems are acceptable as a point of disposal for sewage when a public sewer system is not available.

INDIVIDUAL VENT. A pipe installed to vent a single-fixture drain that connects with the vent system above or terminates independently outside the building.

❖ An individual vent is the simplest and most common method of venting a single plumbing fixture.

INDIVIDUAL WATER SUPPLY. A supply other than an approved public water supply that serves one or more families.

❖ This type of water supply includes wells, springs, streams and cisterns. Surface bodies of water and land cisterns can be used when properly treated and approved.

INSULATING CONCRETE FORM (ICF). A concrete forming system using stay-in-place forms of rigid foam plastic insulation, a hybrid of cement and foam insulation, a hybrid of cement and wood chips, or other insulating material for constructing cast-in-place concrete walls.

❖ This is a relatively new construction technique for foundation/basement walls and above-grade walls that use stay-in-place insulation forms and concrete. A wide variety of systems are available, with several types of insulating materials used for the formwork. Section R404.4 (foundation walls) and R611 (above-grade walls) address three different systems consisting of insulating concrete forms: flat, waffle grid and screen grid.

INSULATING SHEATHING. An insulating board having a minimum thermal resistance of R-2 of the core material.

❖ A sheathing board having an insulating core material is insulating sheathing if its thermal resistance is at least R-2.

JURISDICTION. The governmental unit that has adopted this code under due legislative authority.

❖ The governmental entity that adopts and enforces the IRC is the "jurisdiction." A jurisdiction can take on a variety of forms such as a state, city, town, county, fire district, improvement district or federal department. The jurisdiction appoints a building official to administer and enforce the provisions of the code.

KITCHEN. Kitchen shall mean an area used, or designated to be used, for the preparation of food.

❖ This is an area designed to be used for the preparation of food. A kitchen is a habitable space and is subject to a number of requirements.

[B] LABEL. An identification applied on a product by the manufacturer which contains the name of the manufacturer, the function and performance characteristics of the product or material, and the name and identification of an approved agency and that indicates that the representative sample of the product or material has been tested and evaluated by an approved agency. (See also "Manufacturer's designation" and "Mark.")

❖ A label identifies a product or material and provides other information that can be investigated if there is question as to the suitability of the product or material for a specific installation.

The applicable reference standard often states the minimum identifying information required on a label. The information on a label as mandated by the code includes the name of the manufacturer, the product's function or performance characteristics, the name and identification of the approved labeling agency and the approval of the testing agency. The product name, serial number, installation specifications or applicable tests and standards might be additional information that is provided.

LABELED. Devices, equipment or materials to which have been affixed a label, seal, symbol or other identifying mark of a testing laboratory, inspection agency or other organization concerned with product evaluation that maintains periodic inspection of the production of the above labeled items that attests to compliance with a specific standard.

❖ When a product is labeled, the label indicates that the material has been tested for conformance to an applicable standard and that the component is subject to a third-party inspection that verifies that the minimum level of quality required by the appropriate standard is maintained. Labeling provides a readily available source of information that is useful for field inspection of installed products.

The labeling agency performing the third-party inspection must be approved by the building official, and the basis for this approval may include, but is not necessarily limited to, the capacity and capability of the agency to perform the specific testing and inspection.

LIGHT-FRAMED CONSTRUCTION. A type of construction whose vertical and horizontal structural elements are primarily formed by a system of repetitive wood or light gage steel framing members.

❖ Light-framed construction is by far the predominant construction technique used for dwelling units. Consisting primarily of repetitive framing members such as studs, floor joists, ceiling joists and rafters, light-framed construction can use wood or light-gauge steel as framing members. The repetitive members work together as a system to provide the structural integrity of the building. The *International Residential Code* devotes much of its structural provisions to structures of light-framed construction.

LISTED AND LISTING. Terms referring to equipment that is shown in a list published by an approved testing agency qualified and equipped for experimental testing and maintaining an adequate periodic inspection of current productions and whose listing states that the equipment complies with nationally recognized standards when installed in accordance with the manufacturer's installation instructions.

❖ As stated in the definition, the listing states that either the equipment or material meets nationally recognized standards or it has been found suitable for use in a specific manner. The listing becomes part of the documentation the building official can use to approve or disapprove the equipment or appliance.

[B] LIVE LOADS. Those loads produced by the use and occupancy of the building or other structure and do not include construction or environmental loads such as wind load, snow load, rain load, earthquake load, flood load or dead load.

❖ A variety of loads are imposed on a structure that must be considered in the building design. One group is live loads, produced through the use of the building. Different types of uses or occupancies result in different live loads, due primarily to the number of people involved, the type of activity that takes place and the movable furnishings and equipment that are typically a part of the use.

Design loads that are not live loads include environmental loads such as wind loads, snow loads, rain loads, earthquake loads and flood loads, as well as construction loads. Fixed-in-place equipment is a dead load rather than a live load.

LIVING SPACE. Space within a dwelling unit utilized for living, sleeping, eating, cooking, bathing, washing and sanitation purposes.

❖ The living space within a dwelling extends beyond what is defined as habitable space. Living space includes not only those areas used for living, sleeping, eating and cooking but also includes bathing and washing areas such as bathrooms. Storage areas such as closets and garages are not a portion of the living space.

[B] LOT. A portion or parcel of land considered as a unit.

❖ A lot is a piece of property regulated as a single unit. The code regulates the construction of one or more buildings based on their location on a single lot. It is expected that owners can control what occurs on their own lots, with no control over any adjacent property. Therefore, several provisions are based upon the proximity of the building to the lot line.

[B] LOT LINE. A line dividing one lot from another, or from a street or any public place.

❖ The lot line defines the exterior perimeter of the lot. It is used in several provisions of the code as the basis for requirements dealing with the proximity of an exterior wall or opening to the lot line.

MACERATING TOILET SYSTEMS. A system comprised of a sump with macerating pump and with connections for a water closet and other plumbing fixtures, that is designed to accept, grind and pump wastes to an approved point of discharge.

❖ See Section P2723.

MAIN. The principal pipe artery to which branches may be connected.

❖ This generic term applies to the primary or principal pipe artery in either water-supply systems or drainage systems to which branches may be connected.

MAIN SEWER. See "Public sewer."

MANIFOLD WATER DISTRIBUTION SYSTEMS. A fabricated piping arrangement in which a large supply main is fitted with multiple branches in close proximity in which water is distributed separately to fixtures from each branch.

❖ A manifold is a multiple-opening header to which one or more branch lines connect. See Section P2903.8.

MANUFACTURED HOME. Manufactured home means a structure, transportable in one or more sections, which in the traveling mode is 8 body feet (2438 body mm) or more in width

DEFINITIONS

or 40 body feet (12 192 body mm) or more in length, or, when erected on site, is 320 square feet (30 m^2) or more, and which is built on a permanent chassis and designed to be used as a dwelling with or without a permanent foundation when connected to the required utilities, and includes the plumbing, heating, air-conditioning and electrical systems contained therein; except that such term shall include any structure that meets all the requirements of this paragraph except the size requirements and with respect to which the manufacturer voluntarily files a certification required by the secretary (HUD) and complies with the standards established under this title. For mobile homes built prior to June 15, 1976, a label certifying compliance to the Standard for Mobile Homes, NFPA 501, in effect at the time of manufacture is required. For the purpose of these provisions, a mobile home shall be considered a manufactured home.

❖ Appendix E of the *International Residential Code* contains provisions for the use of manufactured housing as dwellings. Appendix E is applicable only if so specified in the adoption ordinance of the jurisdiction. The provisions of this appendix are limited to three areas: (1) the construction, alteration or repair of a foundation system used to support a manufactured home unit; (2) the construction, installation, addition, alteration, repair or maintenance of building services equipment necessary for connecting the manufactured home to water, fuel, power supplies or sewage systems and (3) alterations, additions or repairs to existing manufactured homes.

The code describes a manufactured home as being transportable in one or more sections, sets forth minimum dimensions for the home and indicates that a manufactured home is constructed on a permanent chassis. Under the provisions of the IRC, a mobile home is a manufactured home.

[B] **MANUFACTURER'S DESIGNATION.** An identification applied on a product by the manufacturer indicating that a product or material complies with a specified standard or set of rules. (See also "Mark" and "Label.")

❖ A manufacturer's designation is identification applied on a product by the manufacturer indicating that a product or material complies with a specified standard or set of rules.

MANUFACTURER'S INSTALLATION INSTRUCTIONS. Printed instructions included with equipment as part of the conditions of listing and labeling.

❖ This term refers to a set of instructions provided by the manufacturer of a product that guides the user or installer through the necessary steps to ensure a complying installation. These instructions help ensure that an installer does not violate the listing of a product.

[B] **MARK.** An identification applied on a product by the manufacturer indicating the name of the manufacturer and the function of a product or material. (See also "Manufacturer's designation" and "Label.")

❖ A mark is identification applied on a product by the manufacturer indicating the name of the manufacturer and the function of a product or material.

MASONRY CHIMNEY. A field-constructed chimney composed of solid masonry units, bricks, stones or concrete.

❖ Masonry chimneys can have one or more flues within them and are field-constructed of brick, stone, concrete or fire-clay materials. Masonry chimneys can stand alone or be part of a masonry fireplace.

MASONRY HEATER. A masonry heater is a solid fuel burning heating appliance constructed predominantly of concrete or solid masonry having a mass of at least 1,100 lb. (500 kg), excluding the chimney and foundation. It is designed to absorb and store a substantial portion of heat from a fire built in the firebox by routing exhaust gases through internal heat exchange channels in which the flow path downstream of the firebox includes at least one 180-degree (3.14-rad) change in flow direction before entering the chimney and which deliver heat by radiation through the masonry surface of the heater.

❖ Masonry heaters are appliances designed to absorb and store heat from a relatively small fire and to radiate that heat into the building interior. They are thermally more efficient than traditional fireplaces because of their design. Interior passageways through the heater allow hot exhaust gases from the fire to transfer heat into the masonry, which then radiates into the building.

MASONRY, SOLID. Masonry consisting of solid masonry units laid contiguously with the joints between the units filled with mortar.

❖ In solid masonry construction, walls are made up of solid masonry units. The joints between the units are also solid, as they must be filled with mortar. Solid masonry walls may be either single-wythe or multiple-wythe construction. Where multiple wythes are used, the space between the wythes is filled with mortar or grout.

[B] **MASONRY UNIT.** Brick, tile, stone, glass block or concrete block conforming to the requirements specified in Section 2103 of the *International Building Code*.

❖ The *International Building Code* contains the material requirements for masonry units. Various materials are used to construct masonry units, including brick, tile, stone, glass and concrete.

Clay. A building unit larger in size than a brick, composed of burned clay, shale, fire clay or mixtures thereof.

❖ Clay masonry units are composed of burned clay, shale, fire clay or a combination of these materials. A clay masonry unit is somewhat larger than a brick.

Concrete. A building unit or block larger in size than 12 inches by 4 inches by 4 inches (305 mm by 102 mm by 102 mm) made of cement and suitable aggregates.

❖ Masonry units constructed of cement, aggregates, water and suitable admixtures are concrete masonry units. The minimum size for a concrete masonry unit is 12 inches by 4 inches by 4 inches (305 mm by 102 mm by 102 mm).

Glass. Nonload-bearing masonry composed of glass units bonded by mortar.

❖ Glass masonry units are unique in that they are only permitted for use in nonload-bearing conditions. The units, either solid or hollow, are connected with mortar in the manner described in Section R610.

Hollow. A masonry unit whose net cross-sectional area in any plane parallel to the loadbearing surface is less than 75 percent of its gross cross-sectional area measured in the same plane.

❖ The net cross-sectional area of a hollow masonry unit is less than 75 percent of its gross cross-sectional area measured in the same plane. The measurement is made parallel to the load-bearing surface. A unit having a greater ratio of net area to gross area than a hollow masonry unit is considered a solid masonry unit.

Solid. A masonry unit whose net cross-sectional area in every plane parallel to the loadbearing surface is 75 percent or more of its cross-sectional area measured in the same plane.

❖ The net cross-sectional area of a solid masonry unit is at least 75 percent of its gross cross-sectional area measured in the same plane. The measurement is made parallel to the load-bearing surface. A unit having a ratio of net area to gross area below 75 percent is a hollow masonry unit.

MASS WALL. Masonry or concrete walls having a mass greater than or equal to 30 pounds per square foot (146 kg/m^2), solid wood walls having a mass greater than or equal to 20 pounds per square foot (98 kg/m^2), and any other walls having a heat capacity greater than or equal to 6 Btu/ft^2 · °F [266 J/(m^2 · k)].

❖ A mass wall must have a heat capacity of 6 Btu/ft^2 · °F [266 J/(m^2 · k)]. In general terms, the heat capacity is a measure of how well a material stores heat. The higher the heat capacity the greater amount of heat stored. For example, a 6-inch (152 mm), heavyweight concrete wall has a heat capacity of 14 Btu/ft^2 · °F compared to a conventional 2-inch by 4-inch (51 mm by 102 mm) wood framed wall with a heat capacity of approximately 3 Btu/ft^2 · °F. The code defines a mass wall as a masonry or concrete wall having a mass greater than or equal to 30 lbs/ft^2 (14 kg/m^2), a solid wood wall (e.g., a log house) having a mass of greater than 20 lbs/ft^2 (198 kg/m^2), or any other walls having a heat capacity of greater than or equal to 6 Btu/ft^2 · °F [266 J/(m^2 · k)].

MEAN ROOF HEIGHT. The average of the roof eave height and the height to the highest point on the roof surface, except that eave height shall be used for roof angle of less than or equal to 10 degrees (0.18 rad).

❖ For a relatively flat roof (up to 10 degrees slope), the mean roof height is the height of the eave. For steeper roofs, the mean roof height is measured to the average point between the eave and the highest point on the roof surface, typically the highest ridge.

MECHANICAL DRAFT SYSTEM. A venting system designed to remove flue or vent gases by mechanical means, that consists of an induced draft portion under nonpositive static pressure or a forced draft portion under positive static pressure.

❖ Mechanical draft systems do not depend on draft; they use fans or blowers.

Forced-draft venting system. A portion of a venting system using a fan or other mechanical means to cause the removal of flue or vent gases under positive static pressure.

❖ Power exhausters and some power-burner systems are examples of forced-draft systems. Vents and chimneys must be listed for positive pressure applications where used with forced-draft systems.

Induced draft venting system. A portion of a venting system using a fan or other mechanical means to cause the removal of flue or vent gases under nonpositive static vent pressure.

❖ Induced-draft venting is commonly accomplished with field-installed inducer fans designed to supplement natural-draft chimneys or vents.

Power venting system. A portion of a venting system using a fan or other mechanical means to cause the removal of flue or vent gases under positive static vent pressure.

❖ See the commentary for "Forced-draft venting system" above.

MECHANICAL EXHAUST SYSTEM. A system for removing air from a room or space by mechanical means.

❖ A mechanical exhaust system uses a fan or other air-handling equipment to exhaust air to the outdoors. Mechanical exhaust systems include those used for hazardous exhaust and commercial kitchen exhaust. Mechanical exhaust systems may or may not employ ductwork as part of the system and may include air-cleaning or air-filtering equipment and fire-suppression equipment.

MECHANICAL SYSTEM. A system specifically addressed and regulated in this code and composed of components, devices, appliances and equipment.

❖ Mechanical systems include, among others, refrigeration systems, air-conditioning systems, exhaust systems, piping systems, duct systems, venting systems, hydronic systems and ventilation systems. A mechanical system may be any of the above systems or incorporate one or more of the systems. The system includes all of the equipment and appliances required to perform the function for which the system is designed.

[B] METAL ROOF PANEL. An interlocking metal sheet having a minimum installed weather exposure of at least 3 square feet (0.28 m^2) per sheet.

❖ A metal roof panel is specifically defined here in order to identify its difference from a metal roof shingle. Metal roof panels are each at least 3 square feet (.3 m^2) in surface area and interlock to provide a suitable roof covering. The provisions of Section R905.10 address roof coverings consisting of metal roof panels.

[B] METAL ROOF SHINGLE. An interlocking metal sheet having an installed weather exposure less than 3 square feet (0.28 m^2) per sheet.

❖ Section R905.4 regulates roof coverings consisting of metal roof shingles, which are smaller than 3 square feet (.3 m^2) per sheet.

[B] MEZZANINE, LOFT. An intermediate level or levels between the floor and ceiling of any story with an aggregate floor area of not more than one-third of the area of the room or space in which the level or levels are located.

❖ Floor levels consisting of relatively small floor areas opening into larger rooms are loft mezzanines. Located at an intermediate level within a high-ceiling space, a loft mezzanine is limited to one-third of the area of the room in which it is located. Because of its openness and relatively small size, a loft mezzanine is not an additional story.

[B] MODIFIED BITUMEN ROOF COVERING. One or more layers of polymer modified asphalt sheets. The sheet materials shall be fully adhered or mechanically attached to the substrate or held in place with an approved ballast layer.

❖ Polymer-modified asphalt sheets, applied in one or more layers, may be used as a roof covering when installed in accordance with Section R905.11. There are several approved methods for attaching the asphalt sheets, including (1) full adhesion to the substrate, (2) mechanical attachment to the substrate and (3) use of a layer of ballast.

MULTIPLE STATION SMOKE ALARM. Two or more single station alarm devices that are capable of interconnection such that actuation of one causes all integral or separate audible alarms to operate.

❖ A multiple station smoke alarm consists of two or more interconnected single station smoke alarms. When one of the smoke alarms is actuated, all of the audible alarms sound. Most dwelling units will require several smoke alarms throughout the unit; Section R317.1 requires that they be interconnected.

NATURAL DRAFT SYSTEM. A venting system designed to remove flue or vent gases under nonpositive static vent pressure entirely by natural draft.

❖ Natural-draft chimneys and vents do not rely on any mechanical means to convey combustion products to the outdoors. Draft is produced by the temperature difference between the combustion gases (flue gases) and the ambient atmosphere. Hot gases are less dense and more buoyant; therefore, they rise or produce a draft.

NONCOMBUSTIBLE MATERIAL. Materials that pass the test procedure for defining noncombustibility of elementary materials set forth in ASTM E 136.

❖ A material that successfully passes the ASTM E 136 test is considered to be noncombustible. The test determines whether a building material will act to aid combustion or add appreciable heat to a fire. A material may have a limited amount of combustible content but not contribute appreciably to a fire; thus, it may still qualify as noncombustible.

NONCONDITIONED SPACE. A space that is not a conditioned space by insulated walls, floors or ceilings.

❖ Defined for the purpose of regulating energy efficiency, this is an area of a building that does not qualify as conditioned. A nonconditioned space is not enclosed with the appropriate insulating materials of the kind that are considered in the definition of conditioned space.

[B] OCCUPIED SPACE. The total area of all buildings or structures on any lot or parcel of ground projected on a horizontal plane, excluding permitted projections as allowed by this code.

❖ The entire floor area of a building is generally the occupied space. It encompasses all floor levels, including basements. Any exterior space without exterior walls that is covered by a roof or floor above is occupied space. Only those areas beneath projections permitted by Section R302 do not contribute to the total occupied space.

OFFSET. A combination of fittings that makes two changes in direction bringing one section of the pipe out of line but into a line parallel with the other section.

❖ Offsets are necessary to route piping around or through structural elements, such as beams, joists, trusses and columns. An offset always involves at least two changes in direction; this is necessary to keep the piping heading in the same direction as the pipe before the offset. Any other arrangement would be considered a bend.

[B] OWNER. Any person, agent, firm or corporation having a legal or equitable interest in the property.

❖ The code places certain responsibilities on the owner in the construction and maintenance of a building. The owner is considered the individual, firm or corporation who, under applicable laws, has a legal interest in the property.

PELLET FUEL-BURNING APPLIANCE. A closed combustion, vented appliance equipped with a fuel feed mechanism for burning processed pellets of solid fuel of a specified size and composition.

❖ This is an appliance that uses solid fuel in the form of pellets in a closed combustion system. The pellets consist of a blend of wood waste products (sawdust and chips) mixed with resin (binders) and compressed into small pellets. The pellets are fed into the appliance in a regulated manner to provide consistent burning.

PELLET VENT. A vent listed and labeled for use with a listed pellet fuel-burning appliance.

❖ See the commentary for the definition of "Vent."

[B] PERMIT. An official document or certificate issued by the authority having jurisdiction that authorizes performance of a specified activity.

❖ The Department of Building Safety is authorized to require a permit according to Section R105. The permit is the official document issued by the jurisdiction authorizing construction-related activity to take place.

[B] PERSON. An individual, heirs, executors, administrators or assigns, and also includes a firm, partnership or corporation, its or their successors or assigns, or the agent of any of the aforesaid.

❖ As used in the IRC, a person is not only an individual but also any heirs, executors, administrators, firm, partnership or corporation, including its or their successors or assigns. An agent for any person as described is also a "person."

PITCH. See "Slope."

❖ The pitch of a pipe run is the fall of the pipe in relationship to a horizontal plane. It is also known as the slope and is measured as a ratio between change in horizontal units and vertical units along a length of pipe.

PLATFORM CONSTRUCTION. A method of construction by which floor framing bears on load bearing walls that are not continuous through the story levels or floor framing.

❖ In light-frame construction, platform construction is the technique of stacking floor systems on load-bearing stud walls. As opposed to walls in balloon construction, the load-bearing walls in platform construction do not continue beyond each individual story level. Platform construction is by far the most popular framing method in use today.

PLENUM. A chamber that forms part of an air-circulation system other than the occupied space being conditioned.

❖ A plenum can also be a room or space that supplies conditioned air to another room or space. Corridors cannot be used as plenums.

PLUMBING. For the purpose of this code, plumbing refers to those installations, repairs, maintenance and alterations regulated by Chapters 25 through 32.

❖ This term refers to the collective system of piping, fixtures, fittings, components, devices or appurtenances that transport potable water and liquid and solid wastes associated with cleaning, washing, bathing, food preparation, drinking and the elimination of bodily wastes. Plumbing also includes appliances, equipment and systems such as water conditioners, water heaters, storm drainage systems, water coolers, water filters and waste treatment systems.

The word "plumbing" comes from the Latin word for the element lead (plumbum), because lead was used extensively in the construction of piping systems. The practice of installing piping systems and the materials used in such systems became known as plumbing.

Piping systems such as fuel gas, fuel oil, heating, cooling, air conditioning, lawn irrigation and fire suppression systems are not plumbing, even though plumbers often perform the work necessary to install these systems. Chapters 25-32 regulate plumbing.

PLUMBING APPLIANCE. An energized household appliance with plumbing connections, such as a dishwasher, food-waste grinder, clothes washer or water heater.

❖ Examples of plumbing appliances include water heaters, hot water dispensers, garbage disposals, dishwashers, clothes washers, water purifiers and water softeners. Typically, an appliance is dependent on a source of energy for its operation.

PLUMBING APPURTENANCE. A device or assembly that is an adjunct to the basic plumbing system and demands no additional water supply nor adds any discharge load to the system. It is presumed that it performs some useful function in the operation, maintenance, servicing, economy or safety of the plumbing system. Examples include filters, relief valves and aerators.

❖ Examples of plumbing appurtenances include water closet seats, hand-held showers, manifolds, backflow preventers, water-hammer arrestors, strainers and filters. An appurtenance by default becomes part of the entire fixture.

PLUMBING FIXTURE. A receptor or device that requires both a water-supply connection and a discharge to the drainage system, such as water closets, lavatories, bathtubs and sinks. Plumbing appliances as a special class of fixture are further defined.

❖ Plumbing fixtures include water closets, bidets, lavatories, sinks, showers, bathtubs and floor drains.

A separate class of plumbing fixtures defined as plumbing appliances includes clothes washing machines, dishwashers, water heaters, water softeners, hot water dispensers, garbage disposals and water purifiers. See "Plumbing Appliance."

PLUMBING SYSTEM. Includes the water supply and distribution pipes, plumbing fixtures, supports and appurtenances; soil, waste and vent pipes; sanitary drains and building sewers to an approved point of disposal.

❖ The plumbing system includes all piping, fixtures and components that transport potable water and convey liquid and liquid-borne solid wastes. See the definition of "Plumbing." Devices that treat the water prior to its use are also included.

POLLUTION. An impairment of the quality of the potable water to a degree that does not create a hazard to the public health but that does adversely and unreasonably affect the aesthetic qualities of such potable water for domestic use.

❖ Books have been written in an attempt to define the term "pollution." For use in the code, pollution is anything that reduces the quality of the potable water supply so that it is undesirable for consumption or use. A pollutant in the water supply causes the water to look, smell or taste bad, but drinking the water would not be harmful. This does not change the fact that polluted water is still considered to be nonpotable. Pollutants are thus not considered potentially harmful, whereas a contaminant would be considered potentially harmful. Refer to the definition of the term "Contamination" to find additional information and the differences between "pollution" and "contamination."

The methods for determining potable and nonpotable water supplies are very closely regulated by the EPA. For the purposes of protecting a water supply, the code considers a polluted source to represent a low hazard when discussing the selection of a backflow preventer. Refer to Table P2902.2.1 for requirements related to low-hazard situations in a potable water supply.

Most forms of pollution are the direct result of some action by humans on their environment; thus, humans are the source for preventing this pollution.

PORTABLE FUEL CELL APPLIANCE. A fuel cell generator of electricity, which is not fixed in place. A portable fuel cell appliance utilizes a cord and plug connection to a grid-isolated load and has an integral fuel supply.

❖ These appliances are a smaller version of the stationary fuel cell power plants regulated by Section M1903 of this code. Fuel cells generate electricity from hydrocarbon fuels; when pure hydrogen is used, they release water and heat as the only by-products. Portable fuel cell appliances are generally fueled by gas tanks or some other local source and are only connected to the house current or are used to charge the batteries of electric cars. Connecting the appliance to the local electrical power grid or hard-piping fuel to the appliance would constitute a permanent installation and require permitting in accordance with Section 105.

[B] POSITIVE ROOF DRAINAGE. The drainage condition in which consideration has been made for all loading deflections of the roof deck, and additional slope has been provided to ensure drainage of the roof within 48 hours of precipitation.

❖ Unless a roof is designed to support a specified amount of water, adequate drainage must be provided to approved locations. In considering the appropriate drainage pattern, all of the possible roof deck deflections should be included. The code mandates that the water be removed from the roof within a period of 48 hours from the time of deposit.

POTABLE WATER. Water free from impurities present in amounts sufficient to cause disease or harmful physiological effects and conforming in bacteriological and chemical quality to the requirements of the public health authority having jurisdiction.

❖ In the United States, the EPA Clean Drinking Water Act defines the quality requirements for potable water. "Potable" means fit to drink.

PRESSURE-RELIEF VALVE. A pressure-actuated valve held closed by a spring or other means and designed to automatically relieve pressure at the pressure at which it is set.

❖ This valve is designed to be actuated by excessive pressures for the purpose of preventing the rupture or explosion of tanks, vessels or piping. The degree of valve opening is directly proportional to the pressure acting on the valve disk.

PUBLIC SEWER. A common sewer directly controlled by public authority.

❖ A public sewer is the piping system that receives the discharge from building sewers and conveys it to treatment facilities.

PUBLIC WATER MAIN. A water-supply pipe for public use controlled by public authority.

❖ A water main is owned and operated by municipalities, rural water districts, privately owned water purveyors and other such entities.

PUBLIC WAY. Any street, alley or other parcel of land open to the outside air leading to a public street, which has been deeded, dedicated or otherwise permanently appropriated to the public for public use and that has a clear width and height of not less than 10 feet (3048 mm).

❖ A public way is essentially a street, alley or any parcel of land that is permanently appropriated for public use. Therefore, the public's right to use such a parcel of land is guaranteed. A public way is open to the exterior and available for use as an egress path if necessary. To qualify as a public way, the space must be at least 10 feet (3048 mm) in width and must have a clear height of at least 10 feet (3048 mm).

PURGE. To clear of air, gas or other foreign substances.

❖ This term applies to a process used for clearing piping systems of impurities, making the system pure prior to the introduction of a desired substance. While the practice of flushing a system to remove debris is sometimes referred to as "purging the system," true purging involves a specific process of forcing a product into a system until impurities are completely removed. One such method would be the introduction of nitrous within a copper system serving a medical gas installation. Such a purge is required in order to clear the system of impurities that may have been left during installation.

QUICK-CLOSING VALVE. A valve or faucet that closes automatically when released manually or controlled by mechanical means for fast-action closing.

❖ A quick-closing valve is any type of solenoid-actuated valve; spring-loaded, self-closing faucets or any other device capable of instantaneously reducing water flow from full flow to no flow. This rapid change from a full-flow to a no-flow condition can cause water-hammer-related damage. Although most faucets can be manually closed fast enough to produce water hammer, the code does not designate manually closed valves or faucets as "quick closing."

Typical quick-closing valves include electrically actuated valves such as those found in dishwashing machines, clothes washing machines and boiler makeup water feeders. Some self-closing valves and faucets are not quick closing. For example, typical metering faucets and typical flushometer valves are slowly, automatically closed by a diaphragm/bleed orifice or pilot valve assembly. Also, a typical float-actuated valve is automatic yet slow in closing.

R-VALUE, THERMAL RESISTANCE. The inverse of the time rate of heat flow through a building thermal envelope element from one of its bounding surfaces to the other for a unit temperature difference between the two surfaces, under steady state conditions, per unit area (h · ft^2 · °F/Btu).

❖ The R-value is a unit of thermal resistance used for comparing insulating values of different materials. The greater the R-value of a material, the greater its insulating properties and the slower that heat flows through it. R-values are used to rate roof/ceiling, wall and floor insulation.

RAMP. A walking surface that has a running slope steeper than 1 unit vertical in 20 units horizontal (5-percent slope).

❖ A pedestrian travel surface that has a slope of 1:20 (5 percent) or less is a walking surface. A surface that is steeper than 1 unit vertical in 20 units horizontal is a ramp. Since steeply sloping walking surfaces present a considerable hazard to users, ramps are regulated for maximum slope, handrails and landings by Section R313.

RECEPTOR. A fixture or device that receives the discharge from indirect waste pipes.

❖ This term applies to a fixture used for collecting drainage from fixtures that are not directly connected to the drainage system. A floor sink is the most common type of receptor; however, a laundry sink that is used for collecting the drainage from a clothes washer could be defined as a receptor.

REFRIGERANT. A substance used to produce refrigeration by its expansion or evaporation.

❖ The refrigerant is the working fluid in refrigeration and air-conditioning systems. In vapor-refrigeration cycles, refrigerants absorb heat from the load side at the evaporator and reject heat at the condenser. The selection of a refrigerant must take into consideration suitable thermodynamic properties, chemical stability, flammability, toxicity and environmental compatibility. Refrigeration is a result of the physical laws of vaporization (evaporation) of liquids. Basically, evaporation of liquid refrigerant is an endothermic process, and condensing of vapors is an exothermic process.

REFRIGERANT COMPRESSOR. A specific machine, with or without accessories, for compressing a given refrigerant vapor.

❖ A compressor is the heart of mechanical refrigeration systems. It is used in a vapor-refrigeration cycle to raise the pressure and enthalpy of the refrigerant into the superheated vapor state, at which point the refrigerant vapor enters the condenser and transfers heat energy to a cooler medium.

REFRIGERATING SYSTEM. A combination of interconnected parts forming a closed circuit in which refrigerant is circulated for the purpose of extracting, then rejecting, heat. A direct refrigerating system is one in which the evaporator or condenser of the refrigerating system is in direct contact with the air or other substances to be cooled or heated. An indirect refrigerating system is one in which a secondary coolant cooled or heated by the refrigerating system is circulated to the air or other substance to be cooled or heated.

❖ Such systems include at minimum a pressure-imposing element or generator, an evaporator, a condenser and interconnecting piping. A single piece of equipment can contain multiple refrigeration systems (circuits).

[B] REGISTERED DESIGN PROFESSIONAL. An individual who is registered or licensed to practice their respective design profession as defined by the statutory requirements of the professional registration laws of the state or jurisdiction in which the project is to be constructed.

❖ All states have imposed some degree of registration or licensing requirements on design professionals, such as architects and engineers, who practice in their state. An individual who meets the statutory requirements in the laws of the state, and possibly even an individual jurisdiction, is a registered design professional.

RELIEF VALVE, VACUUM. A device to prevent excessive buildup of vacuum in a pressure vessel.

DEFINITIONS

❖ This device will automatically open to allow air into a vessel or system in the event of backsiphonage. Such a device is used where a water heater without an antisiphon dip tube is installed above the fixtures served. Where the water supply to the water heater is closed and the lower fixtures opened, the vacuum relief valve will open, thus breaking the siphoning effect.

[EB] REPAIR. The reconstruction or renewal of any part of an existing building for the purpose of its maintenance.

❖ When an existing building undergoes some form of remodel, reconstruction or renewal, the work is defined as a repair. A repair maintains a portion of the building in safe and sound working order, without reducing the intended level of protection or structural safety. Major repair work may be subject to a permit where required by Section R105.

[B] REROOFING. The process of recovering or replacing an existing roof covering. See "Roof recover."

❖ The description of reroofing includes two different processes: (1) recovering an existing roof covering and (2) replacing an existing roof covering. Recovering involves the installation of an additional roof covering over an existing roof covering without the removal of the existing roof covering. Replacement addresses those cases where the existing roof covering is completely removed prior to installation of the new roof covering.

RESIDENTIAL BUILDING TYPE. The type of residential building for determining building thermal envelope criteria. Detached one- and two-family dwellings are Type A-1. Townhouses are Type A-2.

❖ For the purposes of this code, Group R buildings include:

Type A-1—Detached one- and two-family dwellings:

Type A-1 residential buildings are detached one- and two-family dwellings, including duplexes (two-family dwellings). They are often referred to as "single-family dwellings," even though duplexes are included. A series of duplexes is a Type A-1 residential building.

Type A-2—All other residential buildings, three stories or fewer in height:

The classification "Type A-2 residential building" includes all other residential buildings three stories or fewer in height above grade that contain three or more attached dwelling units. Type A-2 buildings are often referred to as "low-rise multifamily buildings," and the classification includes apartments, condominiums, townhouses and rowhouses where three or more units are physically attached. The code defines a dwelling unit as a single housekeeping unit of one or more rooms providing complete, independent living facilities, including permanent provisions for living, sleeping, cooking and sanitation. Guestrooms in hotels, motels, barracks and nursing homes do not meet these criteria. Therefore, these buildings are considered commercial, even if they are three stories or fewer in height.

For Type A-2 residential buildings, compliance can be demonstrated for either the building as a whole or each dwelling unit.

Many code provisions address dwelling units specifically. All of the elements listed in the definition must be present for a dwelling unit to exist.

RETURN AIR. Air removed from an approved conditioned space or location and recirculated or exhausted.

❖ Return air is air that is being returned to the air handler. Only air in excess of the required ventilation air can be recirculated.

RISER. A water pipe that extends vertically one full story or more to convey water to branches or to a group of fixtures.

❖ See the commentary to the definition for "Water Pipe, Riser."

[B] ROOF ASSEMBLY. A system designed to provide weather protection and resistance to design loads. The system consists of a roof covering and roof deck or a single component serving as both the roof covering and the roof deck. A roof assembly includes the roof deck, vapor retarder, substrate or thermal barrier, insulation, vapor retarder, and roof covering.

❖ A roof assembly may be either a single component serving as both the roof covering and the deck or, as typically occurs, a combination of individual roof deck and roof covering components used together to form a complete assembly. A roof assembly may include the roof deck, vapor retarder, substrate, thermal barrier, insulation and roof covering.

[B] ROOF COVERING. The covering applied to the roof deck for weather resistance, fire classification or appearance.

❖ Roof covering provides a building with weather protection, fire retardancy or decoration.

[B] ROOF COVERING SYSTEM. See "Roof assembly."

❖ "Roof Covering System" is another term for "Roof assembly."

[B] ROOF DECK. The flat or sloped surface not including its supporting members or vertical supports.

❖ Often composed of wood structural panels, a roof deck is the flat or sloped surface on which roof covering is typically attached. Roof decks assist in both the weather protection and structural integrity of the building. The rafters and associated structural roof members are not a part of the roof deck, nor is the insulation or roof covering that is placed on the deck.

[B] ROOF RECOVER. The process of installing an additional roof covering over a prepared existing roof covering without removing the existing roof covering.

❖ Under the provisions of Section R907, it is acceptable under specific conditions to install new roof covering over an existing roof without first removing the old roof covering. In a roof recover, the existing roof covering must be prepared in an approved manner prior to installation of the new roof covering.

[B] ROOF REPAIR. Reconstruction or renewal of any part of an existing roof for the purposes of its maintenance.

❖ A roof repair occurs when only minor reconstruction or renewal work to an existing roof is necessary. The extent of a roof repair is limited by Section R907.1.

[B] ROOFTOP STRUCTURE. An enclosed structure on or above the roof of any part of a building.

❖ Often used to house mechanical equipment or installed for decorative purposes, a rooftop structure is any enclosed structure on or above the roof. A rooftop structure must be adequately supported and comply with the construction provisions applicable to the remainder of the building.

ROOM HEATER. A freestanding heating appliance installed in the space being heated and not connected to ducts.

❖ Room heaters are space-heating appliances that are not connected to ducts. They are also called "unvented room heaters."

ROUGH-IN. The installation of all parts of the plumbing system that must be completed prior to the installation of fixtures. This includes DWV, water supply and built-in fixture supports.

❖ Rough-in refers to that stage of construction prior to the installation of any building materials that would conceal plumbing installations in the building structure. This term generally refers to the installation of drain, waste, vent and water supply piping that will be concealed in floor, ceiling and wall cavities or concealed underground or under slab. At this stage of construction, the plumbing systems are tested for leaks and the entire plumbing rough-in is inspected for code compliance.

[B] RUNNING BOND. The placement of masonry units such that head joints in successive courses are horizontally offset at least one-quarter the unit length.

❖ A common method of placing masonry units, running bond describes the type of masonry construction where the units are horizontally offset on each successive course. The length of the offset must be at least one-fourth the length of the masonry unit, otherwise it is considered as stack bond. Commentary Figure R202(14) shows an example of running bond.

SANITARY SEWER. A sewer that carries sewage and excludes storm, surface and groundwater.

❖ A sanitary sewer extends from the sanitary building drain to the public sewer system or to a private sewage disposal system. Most jurisdictions arbitrarily set a distance such as 3 or 5 feet (914 or 1524 mm) outside of the building foundation as the point at which the building drain becomes the building sewer.

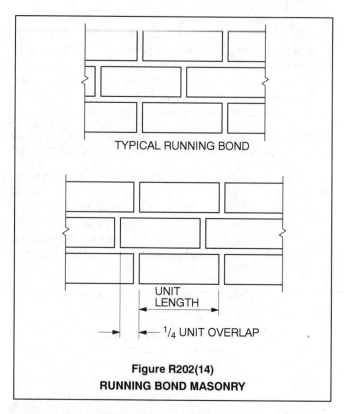

**Figure R202(14)
RUNNING BOND MASONRY**

[B] SCUPPER. An opening in a wall or parapet that allows water to drain from a roof.

❖ Various design methods are used to drain water from a roof. A common approach is to slope the roof toward one or more exterior walls and provide drainage over the edge of the roof, often into a gutter and downspout system. Another method is the use of roof drains that capture water from numerous low points of the roof and use interior drain lines to remove the water. A third procedure is the use of scuppers, which are openings in the exterior wall surface that allow water to flow off the roof.

Where scuppers are used, the code specifies conditions as to their size and height above the roof. Scuppers are often included as a portion of the overflow or emergency drain system, a system required by the code and designed to shed water when the primary roof drains are blocked.

SEISMIC DESIGN CATEGORY. A classification assigned to a structure based on its Seismic Group and the severity of the design earthquake ground motion at the site.

❖ The structural provisions of the IRC vary based on several factors, one of which is the probability of an earthquake causing damage to the building. The term used by the code to assign a rating or level of seismic risk is "Seismic Design Category." Classification into one of the categories is based on two factors: 1) the Seismic Group and 2) the severity of the ground motion at the

DEFINITIONS

site. The Seismic Group is the classification assigned based upon use of the structure. Residential construction is considered in Seismic Group I, with Groups II and III assigned to structures whose loss would create a substantial public hazard or those structures considered as essential facilities. The ground-motion criteria used in the IRC is based on Short Period Design Spectral Response Accelerations (0.2 second). Sections 1613 through 1623 of the *International Building Code* contain further explanation of the seismic considerations.

SEPTIC TANK. A water-tight receptor that receives the discharge of a building sanitary drainage system and is constructed so as to separate solids from the liquid, digest organic matter through a period of detention, and allow the liquids to discharge into the soil outside of the tank through a system of open joint or perforated piping or a seepage pit.

❖ This type of tank is typically used in private sewage disposal systems. See Appendix I.

SEWAGE. Any liquid waste containing animal matter, vegetable matter or other impurity in suspension or solution.

❖ This term refers to the discharge from all plumbing fixtures, which includes human bodily wastes and the wastes associated with cleaning, washing, bathing and food preparation.

SEWAGE PUMP. A permanently installed mechanical device for removing sewage or liquid waste from a sump.

❖ Sewage pumps are designed to transport liquid/solid mixtures without clogging and are typically equipped with large suction and discharging ports. See Section P3007.

SHALL. The term, when used in the code, is construed as mandatory.

❖ The term "shall" means that the requirement is a mandatory provision. Where a provision is allowed but not required, the term "is permitted" generally appears in the code.

[B] SHEAR WALL. A general term for walls that are designed and constructed to resist racking from seismic and wind by use of masonry, concrete, cold-formed steel or wood framing in accordance with Chapter 6 of this code and the associated limitations in Section R301.2 of this code.

❖ Shear walls are walls designed to carry lateral loads such as seismic or wind loads. Shear walls may be of a variety of materials; however, they must be designed in the specific manner required by the code. Shear walls are typically used in conjunction with diaphragms to provide a load path for transmitting lateral loads through the structure.

SIDE VENT. A vent connecting to the drain pipe through a fitting at an angle less than 45 degrees (0.79 rad) to the horizontal.

❖ A typical application of a side vent would be where the drain from a water closet flange extends downward vertically with the vent for the water closet, connecting to the vertical drain and extending horizontally, perpendicular to the drain.

SINGLE PLY MEMBRANE. A roofing membrane that is field applied using one layer of membrane material (either homogeneous or composite) rather than multiple layers.

❖ In contrast to a built-up roof, a single ply membrane roof system consists of one layer of membrane material. Chapter 9 addresses both thermoset and thermoplastic single-ply roofing systems.

SINGLE STATION SMOKE ALARM. An assembly incorporating the detector, control equipment and alarm sounding device in one unit that is operated from a power supply either in the unit or obtained at the point of installation.

❖ A single station smoke alarm contains all components for operation of the device, including the power supply in battery-operated units. For a hard-wired unit, the power is available directly at the alarm device. Where two or more single-station smoke alarms are interconnected, they are multiple-station alarms.

SKYLIGHT AND SLOPED GLAZING. See Section R308.6.1.

❖ Any glass or other transparent or translucent glazing installed in the building construction at a slope greater than 15 degrees (.3 rad) from the vertical is sloped glazing. Skylights are the most common type of sloped glazing, which may also include solariums, sun rooms, roofs and sloped walls.

SKYLIGHT, UNIT. See Section R308.6.1.

SLIP JOINT. A mechanical-type joint used primarily on fixture traps. The joint tightness is obtained by compressing a friction-type washer such as rubber, nylon, neoprene, lead or special packing material against the pipe by the tightening of a (slip) nut.

❖ A slip joint is commonly found in fixture tailpieces and in "P" traps. See the commentary for Section P2704.

SLOPE. The fall (pitch) of a line of pipe in reference to a horizonal plane. In drainage, the slope is expressed as the fall in units vertical per units horizontal (percent) for a length of pipe.

❖ Commonly referred to as pitch, fall or grade, slope is the gradual change in elevation of horizontal piping. Slope is necessary to cause waste flow by gravity.

SMOKE-DEVELOPED RATING. A numerical index indicating the relative density of smoke produced by burning assigned to a material tested in accordance with ASTM E 84.

❖ The ASTM E 84 test standard for flame spread also addresses the potential for smoke development of specific materials. The numerical value assigned to the material being tested, based on the density of the smoke produced during the test, is the smoke-developed rating, which is typically regulated in conjunction with the flame spread index.

SOIL STACK OR PIPE. A pipe that conveys sewage containing fecal material.

❖ In the plumbing trade, "soil pipe" is a common name for cast-iron drainage pipe. The term, however, is not specific to cast-iron pipe and refers to any drainpipe that conveys the discharge of water closets, urinals or any other fixture that receives human waste.

SOLAR HEAT GAIN COEFFICIENT (SHGC). The solar heat gain through a fenestration or glazing assembly relative to the incident solar radiation (Btu/h · ft^2 · °F).

❖ The solar heat gain coefficient (SHGC) measures the ratio of the solar radiation that passes through the window versus the total solar radiation that strikes the window. Sunlight, or solar radiation, striking a window contains visible light and heat. As the sunlight strikes the window, a portion of the visible light and heat pass through into the building. A portion of the heat is absorbed by the glass and window sash. Moreover, a portion of the visible light and heat is reflected to the outdoors. The lower the SHGC, the lower the solar radiation that is allowed to pass through the window. For example, a window with an SHGC value of 0.40 allows only 40 percent of the total solar radiation that strikes the window to pass through.

SOLID MASONRY. Load-bearing or nonload-bearing construction using masonry units where the net cross-sectional area of each unit in any plane parallel to the bearing surface is not less than 75 percent of its gross cross-sectional area. Solid masonry units shall conform to ASTM C 55, C 62, C 73, C 145 or C 216.

❖ Solid masonry can consist of either load-bearing or nonload-bearing construction made up of masonry units. While not required to be totally of solid material, solid masonry must have a net cross-sectional equal to at least 75 percent of its gross cross-sectional area. The percentage must be maintained in any plane parallel to the bearing surface.

STACK. Any main vertical DWV line, including offsets, that extends one or more stories as directly as possible to its vent terminal.

❖ Traditionally, the term "stack" has been used to describe vertical piping (90 degrees [91.57 rad] from the horizontal) that extends through one or more stories of a building and that constitutes a main to which branch piping connects. Vertical vent branch piping that extends to the open air or through two or more stories is a stack.

[B] STACK BOND. The placement of masonry units in a bond pattern is such that head joints in successive courses are vertically aligned. For the purpose of this code, requirements for stack bond shall apply to all masonry laid in other than running bond.

❖ Where the head joints in successive courses of masonry construction are not offset horizontally by more than one-fourth of the unit length, the method of placement is a stack bond. Vertical joints directly aligned on each course are typical in stack bond masonry.

STACK VENT. The extension of soil or waste stack above the highest horizontal drain connected.

❖ A stack vent is the dry extension of a soil or waste stack. Generally, a stack vent extends to the open air and can serve as a main vent to which branch vents connect.

STACK VENTING. A method of venting a fixture or fixtures through the soil or waste stack without individual fixture vents.

❖ This is a method of wet venting where a uniformly oversized vertical drain serves as the vent for one or more fixtures. The fixtures being vented are connected to the stack independently. The use of this venting method is limited to waste fixtures. It is also referred to as "Waste Stack Venting"; see Section P3109.

STANDARD TRUSS. Any construction that does not permit the roof/ceiling insulation to achieve the required R-value over the exterior walls.

❖ The term "standard truss" describes typical roof truss construction where the insulation cannot achieve the necessary *R*-value mandated for the roof/ceiling assembly.

STATIONARY FUEL CELL POWER PLANT. A self-contained package or factory-matched packages which constitute an automatically-operated assembly of integrated systems for generating useful electrical energy and recoverable thermal energy that is permanently connected and fixed in place.

❖ Fuel cells generate electricity from hydrocarbon fuels; when pure hydrogen is used, they release water and heat as the only by-products (see the commentary for Section M1903 for a description of the conversion process). A stationary fuel cell appliance cannot exceed 1000 kW of power output, must be listed and tested in accordance with ANSI Z21.83 and must be installed in accordance with NFPA 853 and the manufacturer's instructions. These appliances may be independent of or connected to the local electrical power grid and may be fueled by fuel tanks or permanent piping systems.

STORM SEWER, DRAIN. A pipe used for conveying rainwater, surface water, subsurface water and similar liquid waste.

❖ This is the general term used to describe the piping that conducts rainwater or surface-water waste from structures to the point of disposal.

DEFINITIONS

[B] STORY. That portion of a building included between the upper surface of a floor and the upper surface of the floor or roof next above.

❖ A story is that portion of a building from a floor surface to the floor surface or roof above. In the case of the topmost story, the height of the story is measured from the floor surface to the top of the ceiling joists of an attic. Where a ceiling does not create an attic, such as a cathedral ceiling, the story height is measured to the top of the roof rafters. See "Story Above Grade."

[B] STORY ABOVE GRADE. Any story having its finished floor surface entirely above grade, except that a basement shall be considered as a story above grade where the finished surface of the floor above the basement is:

1. More than 6 feet (1829 mm) above grade plane.
2. More than 6 feet (1829 mm) above the finished ground level for more than 50 percent of the total building perimeter.
3. More than 12 feet (3658 mm) above the finished ground level at any point.

❖ The code defines a "story above grade" as any story having its finished floor surface entirely above grade. However, the critical part of the definition involves whether or not a basement is a story above grade. A level that is a story above grade may be either an inhabited story or may be unused under-floor space. Three criteria are important to the determination of whether a given floor level is either a story above grade or a basement:
  1. If the finished floor above the level under consideration or above the under-floor space is more than 6 feet (1829 mm) above the grade plane as defined in Section R202, the level under consideration is a story above grade.
  2. If the finished floor level above the level under consideration or above the under-floor space is more than 6 feet (1829 mm) above the finished ground level for more than 50 percent of the total perimeter of the building, the level under consideration is a story above grade.
  3. It the finished floor level above the level under consideration or above the under-floor space is more than 12 feet (3658 mm) above the finished ground level at any point, the floor level under consideration or the under-floor space is a story above grade.

  Conversely, if the finished floor level above the level under consideration is 6 feet (1829 mm) or less above the grade plane, and is 6 feet (1829 mm) or less above the finished ground level for more than 50 percent of the perimeter and does not exceed 12 feet (3658 mm) at any point, the floor level under consideration is a basement. Or, described a bit differently, a basement is a floor level that does not qualify as a story above grade. Commentary Figure R202(15) illustrates the definition of "story above grade."

STRUCTURAL INSULATED PANELS (SIPS). Factory fabricated panels of solid core insulation with structural skins of oriented strand board (OSB) or plywood.

❖ Structural insulated panels are construction elements composed of solid-core insulation panels enclosed within structural wood-panel membranes. These panels are fabricated at the factory, then brought to the job site and installed.

STRUCTURE. That which is built or constructed.

❖ This definition is intentionally broad so as to include within its scope—and therefore the scope of the code (see Section 101.2)—everything that is built as an improvement to real property.

SUMP. A tank or pit that receives sewage or waste, located below the normal grade of the gravity system and that must be emptied by mechanical means.

❖ This is the receiving tank or vessel used to collect and store waste from drainage systems that are incapable of draining by gravity. A sump generally houses an ejector or pump used to evacuate the contents. A sump can refer to a receiver for either waste water or storm water.

SUMP PUMP. A pump installed to empty a sump. These pumps are used for removing storm water only. The pump is selected for the specific head and volume of the load and is usually operated by level controllers.

❖ These pumps are used for removing storm water.

SUNROOM ADDITION. A one-story structure added to an existing dwelling with a glazing area in excess of 40 percent of the gross area of the structure's exterior walls and roof.

❖ Sunroom additions are unique rooms that create a space that differs in character from that provided by conventional additions. Sunrooms and other highly glazed structures are sometimes called conservatories or solariums. This definition distinguishes a sunroom addition from other conventional additions because they are limited to one story in height and the total glazing area needs to be at least 40 percent of the exterior wall and roof area. This definition is important since Section R303.2 permits a sunroom addition to be considered for light and ventilation. See the commentary for "Thermal isolation."

SUPPLY AIR. Air delivered to a conditioned space through ducts or plenums from the heat exchanger of a heating, cooling or ventilating system.

❖ As opposed to return air, supply air is delivered to the conditioned space by an air handler and may or may not be returned to the air handler. Supply air can include ventilation air.

FIGURE R202(15) - R202

For SI: 1 inch = 25.4 mm, 1 foot = 304.8 mm.

Figure R202(15)
THREE-STORY BUILDING

SUPPORTS. Devices for supporting, hanging and securing pipes, fixtures and equipment.

❖ Types of supports include hangers, anchors, braces, brackets, strapping and any other material or device used to support plumbing piping and components.

SWEEP. A drainage fitting designed to provide a change in direction of a drain pipe of less than the angle specified by the amount necessary to establish the desired slope of the line. Sweeps provide a longer turning radius than bends and a less turbulent flow pattern (see "Bend" and "Elbow").

❖ See the definition of "Bend."

TEMPERATURE- AND PRESSURE-RELIEF (T AND P) VALVE. A combination relief valve designed to function as both a temperature-relief and pressure-relief valve.

❖ These devices combine the components of pressure-relief and temperature-relief valves and are designed

DEFINITIONS   R202

to discharge automatically at excessive pressures or temperatures or both.

TEMPERATURE-RELIEF VALVE. A temperature-actuated valve designed to discharge automatically at the temperature at which it is set.

❖ This valve is designed to discharge at excessive temperatures to prevent dangerously high water temperatures that can cause tank, vessel, or pipe explosions.

THERMAL ISOLATION. A separation of conditioned spaces, between a sunroom addition and a dwelling unit, consisting of existing or new wall(s), doors, and/or windows.

❖ Thermal isolation is conceptually similar to the separation provided by the building envelope, but instead of being between 1) conditioned space and the exterior or 2) conditioned space and unconditioned space, this separation occurs between two conditioned spaces. In this situation it is the separation between the existing dwelling and the new sunroom addition. Unless a new wall is constructed, the existing exterior wall of the dwelling will generally provide the thermal isolation between the dwelling and the sunroom.

THERMAL RESISTANCE, *R*-VALUE. The inverse of the time rate of heat flow through a body from one of its bounding surfaces to the other for a unit temperature difference between the two surfaces, under steady state conditions, per unit area (h · ft^2 · °F/Btu).

❖ *R*-value is a unit of thermal resistance used for comparing insulating values of different materials. The greater the *R*-value of a material, the greater its insulating properties and the slower heat flows through it. *R*-values are used to rate roof/ceiling, wall and floor insulation.

THERMAL TRANSMITTANCE, *U*-FACTOR. The coefficient of heat transmission (air to air) through a building envelope component or assembly, equal to the time rate of heat flow per unit area and unit temperature difference between the warm side and cold side air films (Btu/h · ft^2 · °F).

❖ *U*-factor is the rate of heat transfer in Btu per hour through a square foot of a surface when the difference between the air temperatures on either side is 1°F (-17°C). The *U*-factor is the reciprocal of the *R*-value. Vertical windows, skylights and doors are rated with *U*-factors. The lower the *U*-factor, the more thermally efficient the material or assembly.

TOWNHOUSE. A single-family dwelling unit constructed in a group of three or more attached units in which each unit extends from foundation to roof and with open space on at least two sides.

❖ A configuration of three or more single-family dwellings attached together in a single structure constitute a townhouse if both of the following conditions exist: 1) each unit extends vertically from the foundation to the roof (townhouses cannot be stacked) and 2) each unit is open to the exterior on at least two sides, providing some degree of independence from other units.

In order for townhouses to be designed and constructed under the *International Residential Code*, they must also comply with the limitations of Section R101.2. Townhouses are limited to three stories, and each unit must be provided with independent egress to the exterior.

TRAP. A fitting, either separate or built into a fixture, that provides a liquid seal to prevent the emission of sewer gases without materially affecting the flow of sewage or waste water through it.

❖ The sole purpose of a trap is to isolate the interior of occupiable spaces from the sanitary drainage and vent system. It traps liquid waste and retains it to form a seal or barrier through which gases and vapors in the drainage and vent system cannot pass under normal operating conditions. Traps are either separate devices or are integral with fixtures. Other than a grease trap or trap that is integral with a fixture, the only type of trap permitted by the code is the "P" type. See Commentary Figure R202(16) and "Fixture Drain."

TRAP ARM. That portion of a fixture drain between a trap weir and the vent fitting.

❖ This is the common name for a "Fixture Drain."

TRAP PRIMER. A device or system of piping to maintain a water seal in a trap, typically installed where infrequent use of the trap would result in evaporation of the trap seal, such as floor drains.

❖ This is a device or system of piping that maintains a water seal in a trap. See Section P3201.2.

TRAP SEAL. The trap seal is the maximum vertical depth of liquid that a trap will retain, measured between the crown weir and the top of the dip of the trap.

❖ A trap seal is the vertical depth of liquid held in the dip of a trap. Trap seals are required to have a minimum depth.

TRIM. Picture molds, chair rails, baseboards, handrails, door and window frames, and similar decorative or protective materials used in fixed applications.

❖ Where limited amounts of decorative or protective materials are installed as a part of the building construction, they are regulated as trim. Examples include chair rails, door and window frames, handrails, baseboards and cove molding.

TRUSS DESIGN DRAWING. The graphic depiction of an individual truss, which describes the design and physical characteristics of the truss.

❖ If a floor or roof truss system is to be used in a building's construction, truss design drawings must be prepared and submitted to the building official prior to

Figure R202(16)
DEPTH OF WATER SEAL

installation. The truss design drawings should graphically illustrate each specific type of truss to be installed and include a variety of other information as required in the code.

TYPE L VENT. A listed and labeled vent conforming to UL 641 for venting oil-burning appliances listed for use with Type L vents or with gas appliances listed for use with Type B vents.

❖ See the commentary for the definition of "Vent."

U-FACTOR, THERMAL TRANSMITTANCE. The coefficient of heat transmission (air to air) through a building envelope component or assembly, equal to the time rate of heat flow per unit area and unit temperature difference between the warm side and cold side air films (Btu/h · ft^2 · °F).

❖ See "Thermal Transmittance, U-factor."

UNCONFINED SPACE. A space having a volume not less than 50 cubic feet per 1,000 Btu/h (4.8 m^3/kW) of the aggregate input rating of all appliances installed in that space. Rooms communicating directly with the space in which the appliances are installed, through openings not furnished with doors, are considered a part of the unconfined space.

❖ For simplicity of calculation, the volume per Btu/h (W) input rating ratio can be expressed as 1 cubic foot per 20 Btu/h (4.8 m^3/kW).

[B] UNDERLAYMENT. One or more layers of felt, sheathing paper, nonbituminous saturated felt, or other approved material over which a roof covering, with a slope of 2 to 12 (17-percent slope) or greater, is applied.

❖ Underlayment for roof covering is the initial layer of weather protection installed over a roof deck. The underlayment may consist of single or multiple layers of material and is to be used on roofs having a minimum slope of 2 to 12. Examples of underlayment include felt, sheathing paper and nonbituminous saturated felt.

UNUSUALLY TIGHT CONSTRUCTION. Construction meeting the following requirements:

1. Walls comprising the building thermal envelope have a continuous water vapor retarder with a rating of 1 perm [57.4 ng/(s · m^2 · Pa)] or less with openings therein gasketed or sealed.

2. Doors and openable windows meet the air leakage requirements of IECC Section 502.1.4.1; and

3. Caulking or sealants are applied to areas such as joints around window and door frames between sole plates and floors, between wall-ceiling joints, between wall panels, at penetrations for plumbing, electrical and gas lines, and at other openings.

❖ Due to energy conservation efforts, buildings are being constructed to limit the infiltration of outdoor air into the building. Buildings of such construction cannot provide combustion air by infiltration alone.

VACUUM BREAKERS. A device which prevents backsiphonage of water by admitting atmospheric pressure through ports to the discharge side of the device.

❖ A vacuum breaker is a device used as a backflow preventer. The device neutralizes (breaks) the negative pressures (vacuum) in closed systems by allowing atmospheric pressure to enter the system.

VAPOR PERMEABLE MEMBRANE. A material or covering having a permeance rating of 5 perms (52.9 · 10^{-10} kg/Pa · s · m^2) or greater, when tested in accordance with the desiccant method using Procedure A of ASTM E96. A vapor permeable material permits the passage of moisture vapor.

❖ A vapor permeable membrane can consist of breather paper or other materials that allow the passage of moisture vapor. An acceptable vapor permeable mem-

DEFINITIONS

R202

brane can be any material that has a vapor permeance of no less than 5 perm (52.9 · 10^{-10} kg/Pa · s · m^2) when tested in accordance with the desiccant method using Procedure A of ASTM E 96.

[B] VAPOR RETARDER. A vapor resistant material, membrane or covering such as foil, plastic sheeting, or insulation facing having a permeance rating of 1 perm (5.7 · 10^{-11} kg/Pa · s · m^2) or less, when tested in accordance with the desiccant method using Procedure A of ASTM E 96. Vapor retarders limit the amount of moisture vapor that passes through a material or wall assembly.

❖ An acceptable vapor retarder can be any material that has been tested in accordance with the desiccant method using Procedure A of ASTM E 96 that does not have a permeance rating greater than 1.0. Permeance is a measure of water vapor transmission (wvt) rate of grains per hour per square foot divided by the vapor pressure difference in inches of mercury. Materials tested to the standard are papers, plastic films, asphalt-saturated construction papers and sheet materials. Vapor retarders are installed on the warm-in-winter side of walls, floors and ceilings of the building's thermal insulation.

VENT. A passageway for conveying flue gases from fuel-fired appliances, or their vent connectors, to the outside atmosphere.

❖ In code terminology, vents are distinguished from chimneys and usually are constructed of factory-made listed and labeled components intended to function as a system. Type B and BW vents are constructed of galvanized steel and aluminum sheet metal and are double-wall and air insulated. Such vents are designed to vent gas-fired appliances and equipment that are equipped with draft hoods or specifically listed (labeled) for use with Type B or BW vents. Type L vents are typically constructed of sheet steel and stainless steel. They are double wall, air insulated and designed to vent gas and oil-fired appliances and equipment. Mid- to high-efficiency appliances are designed for use with corrosion-resistant vents, such as those made of plastic pipe and special alloys of stainless steel. Pellet vents (PL) are specialized vents similar in design and construction to Type L vents. Pellet vents must not be used with any appliances other than pellet-burning appliances.

VENT COLLAR. See "Flue collar."

VENT CONNECTOR. That portion of a venting system which connects the flue collar or draft hood of an appliance to a vent.

❖ In most cases, appliances are not located directly in line with the vertically rising chimney or vent; therefore, a vent connector is necessary to connect the appliance flue outlet with the vent. Vent connectors can be single- or double-wall pipes and are usually made from galvanized steel, stainless steel or aluminum sheet metal. In many installations, the vent connectors must be constructed of the same material as the vent, as is typically done with Type B vent systems.

VENT DAMPER DEVICE, AUTOMATIC. A device intended for installation in the venting system, in the outlet of an individual, automatically operated fuel burning appliance and that is designed to open the venting system automatically when the appliance is in operation and to close off the venting system automatically when the appliance is in a standby or shutdown condition.

❖ The purpose of this device is the conservation of energy by preventing the cooling of the combustion chamber of the appliance. This is accomplished by preventing the flow of air (by means of connection) through the combustion chamber when the appliance is not in operation.

VENT GASES. Products of combustion from fuel-burning appliances, plus excess air and dilution air, in the venting system above the draft hood or draft regulator.

❖ These products of combustion are the result of burning fuel gas.

VENT STACK. A vertical vent pipe installed to provide circulation of air to and from the drainage system and which extends through one or more stories.

❖ This is a dry vent that extends from the base of a waste or soil stack and typically runs parallel and adjacent to the waste or soil stack it serves. The vent stack can connect to the horizontal drain downstream of the drain stack or directly to the drainage stack.

VENT SYSTEM. Piping installed to equalize pneumatic pressure in a drainage system to prevent trap seal loss or blow-back due to siphonage or back pressure.

❖ This system is a piping arrangement designed to maintain pressure fluctuation at fixture traps to within plus or minus 2 inches of water column (249 Pa).

[B] VENTILATION. The natural or mechanical process of supplying conditioned or unconditioned air to, or removing such air from, any space.

❖ Ventilation can be used for comfort cooling, the control of air contaminants, equipment cooling and replenishing oxygen levels.

VENTING. Removal of combustion products to the outdoors.

❖ This is the desired result of the venting system.

VENTING SYSTEM. A continuous open passageway from the flue collar of an appliance to the outside atmosphere for the purpose of removing flue or vent gases. A venting system is usually composed of a vent or a chimney and vent connector, if used, assembled to form the open passageway.

❖ Venting systems operate by gravity (natural draft) or mechanical means. Venting systems include both vents and chimneys.

VERTICAL PIPE. Any pipe or fitting that makes an angle of 45 degrees (0.79 rad) or more with the horizontal.

❖ Vertical pipes are angled 45 degrees (0.79 rad) or more above the horizontal.

[B] WALLS. Walls shall be defined as follows:

Load-bearing wall is a wall supporting any vertical load in addition to its own weight.

Nonbearing wall is a wall which does not support vertical loads other than its own weight.

❖ Walls are either load-bearing or nonbearing. A load-bearing wall supports not only its own weight but also an additional vertical load. An example is a wood stud wall supporting roof rafters and ceiling joists. A nonbearing wall does not support any vertical load other than its own weight.

WASTE. Liquid-borne waste that is free of fecal matter.

❖ Waste is any plumbing fixture discharge other than the discharge from a water closet or urinal. This material does not contain human fecal matter.

WASTE PIPE OR STACK. Piping that conveys only liquid sewage not containing fecal material.

❖ A waste pipe or stack is any waste-conducting pipe not falling under the definition of soil pipe. Waste pipes do not convey human fecal matter.

WATER-DISTRIBUTION SYSTEM. Piping which conveys water from the service to the plumbing fixtures, appliances, appurtenances, equipment, devices or other systems served, including fittings and control valves.

❖ This is all the water piping within a building from the point of connection to the water service to the various connections of the fixture supplies or outlets.

WATER HEATER. Any heating appliance or equipment that heats potable water and supplies such water to the potable hot water distribution system.

❖ The types of water heaters include storage (tank), circulating and instantaneous. Point-of-use water heaters typically supply hot water to a single fixture or outlet and are located in close proximity to the fixture or outlet.

WATER MAIN. A water-supply pipe for public use.

❖ A water main is owned and operated by municipalities, rural water districts, privately owned water purveyors or other such entities.

WATER OUTLET. A valved discharge opening, including a hose bibb, through which water is removed from the potable water system supplying water to a plumbing fixture or plumbing appliance that requires either an air gap or backflow prevention device for protection of the supply system.

❖ A water outlet is a discharge opening through which water is supplied to a fixture, into the atmosphere (except into an open tank that is part of the water supply system), to a boiler or heating system or to any device or equipment requiring water to operate but that is not part of the plumbing system.

WATER-SERVICE PIPE. The outside pipe from the water main or other source of potable water supply to the water-distribution system inside the building, terminating at the service valve.

❖ Water-service pipe is any acceptable piping that conveys potable water.

WATER-SUPPLY SYSTEM. The water-service pipe, the water-distributing pipes and the necessary connecting pipes, fittings, control valves and all appurtenances in or adjacent to the building or premises.

❖ The water supply system includes all piping and components that convey potable water from the public main or other source to the points of water usage.

WET VENT. A vent that also receives the discharge of wastes from other fixtures.

❖ While Section P3108 describes "Wet Venting" systems, the term "wet vent" applies to any pipe that is used concurrently as both a drain and a vent. This arrangement can be found in the following systems: Common Vent (Section P3107), Wet Venting (Section P3108), Waste Stack Vent (Section P3109), Circuit Venting (Section P3110), and Combination Waste and Vent System (Section P3111).

WIND BORNE DEBRIS REGION. Areas within hurricane-prone regions within one mile of the coastal mean high water line where the basic wind speed is 110 miles per hour (177 km/h) or greater; or where the basic wind speed is equal to or greater than 120 miles per hour (193 km/h); or Hawaii.

❖ The IRC regulates the exterior openings of a building where there is a concern about the extensive damage that can be caused by debris carried through the air by extremely high winds. Wind-borne debris regions are those areas where the potential for such damage is high. The code identifies three such areas: 1) hurricane-prone regions within one mile of the coastal mean high water line, provided the wind speed is at least 110 mph (48 m/s); 2) those locations where the wind speed is at least 120 mph (153 m/s); and 3) the islands of Hawaii. See also the definition for "Hurricane-prone Region."

WINDER. A tread with non-parallel edges.

❖ Winders are treads in a winding stairway. They are used as components of stairs that change direction, just as straight treads are components in straight

DEFINITIONS

stairs. A winder serves as a tread, but its shape allows the additional function of a gradual turning of the stairway direction. The tread depth of a winder at the walk line and the minimum tread depth at the narrow end can control the turn made by each winder.

[B] WOOD STRUCTURAL PANEL. A panel manufactured from veneers; or wood strands or wafers; bonded together with waterproof synthetic resins or other suitable bonding systems. Examples of wood structural panels are plywood, OSB or composite panels.

❖ Wood structural panels are sheet products commonly known as plywood, wafer board, oriented strand board (OSB), or chip board manufactured to meet the United States Department of Commerce (DOC) standards for Construction and Industrial Plywood, PS 1-95, and Performance Standards for Wood-Based Structural-Use Panels, PS 2-93.

[B] YARD. An open space, other than a court, unobstructed from the ground to the sky, except where specifically provided by this code, on the lot on which a building is situated.

❖ Used throughout the code to describe an open space at the exterior of a building, a yard must be unobstructed from the ground to the sky and located on the same lot on which the building is situated. The openness of a yard allows it to be used to supply buildings with natural light and ventilation. A court, which is bounded on three or more sides by the exterior walls of the building, is not a yard.

Bibliography

The following resource materials are referenced in this chapter or are relevant to the subject matter addressed in this chapter.

ANSI Z21.83—1998, *Fuel Cell Power Plants*. New York: American National Standards Institute, 1998.

ASTM C 55–01, *Standard Specification for Concrete Brick*. West Conshohocken, PA: ASTM International, 2001.

ASTM C 62–01, *Standard Specification for Building Brick (Solid Masonry Units Made from Clay or Shale)*. West Conshohocken, PA: ASTM International, 2001.

ASTM C 73–99a, *Standard Specification for Calcium Silicate Face Brick (Sand Lime Brick)*. West Conshohocken, PA: ASTM International, 1999.

ASTM C 90–01, *Standard Specification for Load–Bearing Concrete Masonry Units*. West Conshohocken, PA: ASTM International, 2001.

ASTM C 216–01a, *Standard Specification for Facing Brick (Solid Masonry Units Made from Clay or Shale)*. West Conshohocken, PA: ASTM International, 2001.

ASTM E 84–01, *Standard Test Method for Surface Burning Characteristics of Building Materials*. West Conshohocken, PA: ASTM International, 2001.

ASTM E 96–00, *Standard Test Method for Water Vapor Transmission of Materials*. West Conshohocken, PA: ASTM International, 2000.

ASTM E 136–99, *Standard Test Method for Behavior of Materials in a Vertical Tube Furnace at 750°C*. West Conshohocken, PA: ASTM International, 1999.

DOC PS 1—95, *Construction and Industrial Plywood*. Gaithersburg, MD: U.S. Department of Commerce, 1995.

DOC PS 2—92, *Performance Standard for Wood-Based Structural-Use Panels*. Gaithersburg, MD: U.S. Department of Commerce, 1992.

IBC-2003, *International Building Code*. Falls Church, VA: International Code Council, Inc., 2003.

IFC-2003, *International Fire Code*. Falls Church, VA: International Code Council, Inc., 2003.

IFGC-2003, *International Fuel Gas Code*. Falls Church, VA: International Code Council, Inc. 2003.

IMC-2003, *International Mechanical Code*. Falls Church, VA: International Code Council, Inc., 2003.

IPC-2003, *International Plumbing Code*. Falls Church, VA: International Code Council, Inc., 2003.

NFPA 501—99, *Standard on Manufactured Housing*. Quincy, MA: National Fire Protection Association, 1999.

NFPA 853—00, *Installation of Stationary Fuel Cell Power Plants*, Quincy, MA: National Fire Protection Association, 2000.

UL 103–98, *Chimneys, Factory–Built, Residential Type and Building Heating Appliance—with Revision through March 1999*. Northbrook, IL: Underwriters Laboratories, 1998.

UL 641–95, *Type L, Low-Temperature Venting Systems—with Revisions through April 1999*. Northbrook, IL: Underwriters Laboratories, 1999.

Part III — Building Planning and Construction

Chapter 3: Building Planning

General Comments

Chapter 3 is a compilation of the code requirements specific to the building planning sector of the design and construction process. The provisions address a wide variety of issues important to designing a building that is both safe and usable. The limitations placed on the materials and methods of construction can lead to structural integrity. Snow, wind, and seismic loading are regulated, as are the expected live and dead loads.

Fire-resistance-rated assemblies are necessary under two different conditions: 1) where a building is situated in very close proximity to a property line, the code addresses the concern for radiant heat exposure under fire conditions; 2) where two or more dwelling units are provided in a single structure, the code mandates a minimum level of fire separation between all units. Other concerns related to fire conditions include the limitations on wall and ceiling finishes, the requirement for emergency escape and rescue openings, the required installation of smoke alarms throughout the dwelling unit and limitations on the use of foam plastics and other insulation materials.

This chapter sets forth traditional code requirements dealing with light, ventilation, sanitation, room size, ceiling height and environmental comfort. Life-safety provisions include limitations on glazing used in hazardous areas, specifications on the use of guards at elevated surfaces and basic rules for the exiting system. This chapter also contains most of the regulations found in the code that deal with the planning and design of dwelling units.

Section R301 establishes the design criteria, including dead loads, live loads, roof loads, floor loads, snow loads, wind loads and seismic loads. Section R302 identifies the requirements for the building based on its location on the lot. Section R303 establishes the light, ventilation and heating requirements for dwelling units. Section R304 establishes the minimum requirements for rooms in dwelling units. Section R305 establishes the ceiling height requirements for dwelling units. Section R306 contains requirements for sanitation. Section R307 contains requirements for toilets, bath and shower spaces. Section R308 contains requirements for glazing, hazardous locations of glazing, site built windows and skylights. Section R309 contains provisions for garages and carports. Section R310 contains provisions for emergency escape and rescue openings. Section R311 establishes the means of egress requirements including provisions for exterior balconies; hallways; doors including their landings; stairways including their width, treads, risers, landings, and handrails; special stairways and ramps including their landings and handrails. Section R312 addresses guards. Section R313 contains the requirements for smoke alarms. Section R314 addresses the use of foam plastic. Section R315 addresses flame spread and smoke density requirements for walls and ceilings. Section R316 contains provisions for insulation. Section R317 addresses the separation of dwelling units. Section R318 addresses moisture vapor retarders. Section R319 contains requirements for decay protection. Section R320 contains requirements for termite protection. Section R321 provides the requirements for premise identification (site address). Section R322 contains the scoping requirements for accessibility, and Section R323 establishes the flood-resistant construction provisions.

Purpose

The purpose of Chapter 3 is to provide guidelines for a minimum level of structural integrity, life safety, fire safety and livability for inhabitants of dwelling units regulated by the code. The chapter sets forth the requirements that affect the most basic planning and design aspects of dwelling construction. It identifies the various structural loads that are imposed on a building, and it establishes criteria that address each of the imposed loads. In the design of residential structures scoped by the IRC, there are many climatic and geographical issues that must be considered. This chapter provides guidance in the determination of all appropriate design criteria. In addition, it sets forth the limiting conditions under which a building may be designed and constructed using the IRC.

Fundamental issues of livability and sanitation are satisfied through the regulation of minimum room sizes and ceiling heights, as well as basic requirements for toilet rooms and kitchens. Life-safety concerns are addressed in a number of areas, including provisions regulating emergency escape and rescue openings, glazing in areas subject to human impact and exiting. The chapter establishes minimum specifications for a number of different building components, including stairways, ramps, landings, handrails and guards. It deals with fire-safety issues such as early fire detection by smoke alarms, exterior wall protection for proximity to property lines, separation of dwelling units in multiple-family buildings and control of fire spread across wall and ceiling finishes. Property protection is also a concern, with provisions established for protection against decay and termites.

SECTION R301
DESIGN CRITERIA

R301.1 Design. Buildings and structures, and all parts thereof, shall be constructed to safely support all loads, including dead loads, live loads, roof loads, flood loads, snow loads, wind loads and seismic loads as prescribed by this code. The construction of buildings and structures shall result in a system that provides a complete load path capable of transferring all loads from their point of origin through the load-resisting elements to the foundation.

❖ This section specifies the minimum design loads required for structures built in accordance with the *International Residential Code*. In structural design, loads are generally divided into two categories: gravity loads, which act vertically, and lateral loads, which act horizontally. Lateral loads typically result from either wind (see Section R301.2.1) or earthquakes (see Section R301.2.2). While wind and earthquake design may concern themselves with lateral loads, there are also vertical force components that should be considered.

All structures must be designed to support these loads and provide a complete load path capable of transferring these loads from their point of origin through the appropriate load-resisting elements and foundation and ultimately to the supporting soil. A load path that is either incomplete or inadequate will expose the structure to damage just as surely as an undersized structural member will. The concept of a complete load path is a fundamental principle in structural engineering, and the code makes it clear that a complete load path must be provided.

R301.1.1 Alternative provisions. As an alternative to the requirements in Section R301.1 the following standards are permitted subject to the limitations of this code and the limitations therein. Where engineered design is used in conjunction with these standards the design shall comply with the *International Building Code*.

1. American Forest and Paper Association (AF&PA) *Wood Frame Construction Manual* (WFCM).
2. American Iron and Steel Institute (AISI), *Standard for Cold-Formed Steel Framing—Prescriptive Method for One- and Two-family Dwellings* (COFS/PM).

❖ This section permits the use of alternate prescriptive framing methods. Wood framing is permitted to comply with the provisions of the American Forest and Paper Association, *Wood Frame Construction Manual for One-and Two-Family Dwellings*. Cold-formed steel framing is permitted to comply with American Iron and Steel Institute, *Standard for Cold-Formed Steel Framing-Prescriptive Method for One-and Two-Family Dwellings*. Engineered design in accordance with the *International Building Code* is required when a building exceeds the applicability limits of either standard.

R301.1.2 Construction systems. The requirements of this code are based on platform and balloon-frame construction for light-frame buildings. The requirements for concrete and masonry buildings are based on a balloon framing system. Other framing systems must have equivalent detailing to ensure force transfer, continuity and compatible deformations.

❖ The requirements of the *International Residential Code* are based on platform or balloon-frame construction for light-frame buildings (see the definitions of "Platform construction" and "Light-framed construction" in Chapter 2) and on a balloon-framing system for concrete and masonry buildings.

R301.1.3 Engineered design. When a building of otherwise conventional construction contains structural elements exceeding the limits of Section R301 or otherwise, not conforming to this code, these elements shall be designed in accordance with accepted engineering practice. The extent of such design need only demonstrate compliance of nonconventional elements with other applicable provisions and shall be compatible with the performance of the conventional framed system. Engineered design in accordance with the *International Building Code* is permitted for all buildings and structures, and parts thereof, included in the scope of this code.

❖ Generally, proper application of the *International Residential Code* requires a clear understanding of and adherence to its prescriptive limitations, which are based on conventional construction. However, a building may contain structural elements that are either unconventional or exceed the prescriptive limitations of the code. This is acceptable, provided these elements are designed in accordance with accepted engineering practice.

[B] R301.2 Climatic and geographic design criteria. Buildings shall be constructed in accordance with the provisions of this code as limited by the provisions of this section. Additional criteria shall be established by the local jurisdiction and set forth in Table R301.2(1).

❖ This section establishes the design criteria that vary based on location and/or climate. Some of the criteria reflect loading such as earthquake and wind, while others reflect susceptibility to damage from hazards such as weather exposure or termites. Additional criteria may be established by local jurisdictions as necessary. These would include, for example, whether a site is within a wind borne debris region as described in Section R301.2.1.2 of this commentary. Table R301.2(1) lists the criteria that need to be established within each jurisdiction for any project constructed under the IRC. Note that some of these criteria (e.g., wind exposure category) can vary within a given jurisdiction and may need to be established on a site-by-site (or project-by-project) basis. The table serves as a useful reminder for code enforcement personnel, builders, designers, and owners. Verifying this information up front aids compliance with the IRC.

Commentary Figure R301.2(9) shows guidelines for frost-line depths. This figure should only be used when existing conditions are unknown, because frost penetration is dependent on soil conditions and elevations. In some areas of the United States, this depth may vary up to 18 inches (457 mm) within a two- to three-mile (3.22 to 4.83 km) area.

TABLE R301.2(1)
CLIMATIC AND GEOGRAPHIC DESIGN CRITERIA

GROUND SNOW LOAD	WIND SPEED[e] (mph)	SEISMIC DESIGN CATEGORY[g]	SUBJECT TO DAMAGE FROM				WINTER DESIGN TEMP[f]	ICE SHIELD UNDERLAYMENT REQUIRED[i]	FLOOD HAZARDS[h]	AIR FREEZING INDEX[j]	MEAN ANNUAL TEMP[k]
			Weathering[a]	Frost line depth[b]	Termite[c]	Decay[d]					

For SI: 1 pound per square foot = 0.0479 kN/m^2, 1 mile per hour = 1.609 km/h.

a. Weathering may require a higher strength concrete or grade of masonry than necessary to satisfy the structural requirements of this code. The weathering column shall be filled in with the weathering index (i.e., "negligible," "moderate" or "severe") for concrete as determined from the Weathering Probability Map [Figure R301.2(3)]. The grade of masonry units shall be determined from ASTM C 34, C 55, C 62, C 73, C 90, C 129, C 145, C 216 or C 652.
b. The frost line depth may require deeper footings than indicated in Figure R403.1(1). The jurisdiction shall fill in the frost line depth column with the minimum depth of footing below finish grade.
c. The jurisdiction shall fill in this part of the table with "very heavy," "moderate to heavy," "slight to moderate," or "none to slight" in accordance with Figure R301.2(6) depending on whether there has been a history of local damage.
d. The jurisdiction shall fill in this part of the table with "moderate to severe," "slight to moderate," or "none to slight" in accordance with Figure R301.2(7) depending on whether there has been a history of local damage.
e. The jurisdiction shall fill in this part of the table with the wind speed from the basic wind speed map [Figure R301.2(4)]. Wind exposure category shall be determined on a site-specific basis in accordance with Section R301.2.1.4.
f. The outdoor design dry-bulb temperature shall be selected from the columns of 97^1/$_2$-percent values for winter from Appendix D of the *International Plumbing Code*. Deviations from the Appendix D temperatures shall be permitted to reflect local climates or local weather experience as determined by the building official.
g. The jurisdiction shall fill in this part of the table with the Seismic Design Category determined from Section R301.2.2.1.
h. The jurisdiction shall fill in this part of the table with (a) the date of the jurisdiction's entry into the National Flood Insurance Program (date of adoption of the first code or ordinance for management of flood hazard areas), (b) the date(s) of the currently effective FIRM and FBFM, or other flood hazard map adopted by the community, as may be amended.
i. In accordance with Sections R905.2.7.1, R905.4.3, R905.5.3, R905.6.3, R905.7.3 and R905.8.3, for areas where the average daily temperature in January is 25°F (-4°C) or less, or where there has been a history of local damage from the effects of ice damming, the jurisdiction shall fill in this part of the table with "YES." Otherwise, the jurisdiction shall fill in this part of the table with "NO."
j. The jurisdiction shall fill in this part of the table with the 100-year return period air freezing index (BF-days) from Figure R403.3(2) or from the 100-year (99%) value on the National Climatic Data Center data table "Air Freezing Index- USA Method (Base 32° Fahrenheit)" at www.ncdc.noaa.gov/fpsf.html.
k. The jurisdiction shall fill in this part of the table with the mean annual temperature from the National Climatic Data Center data table "Air Freezing Index- USA Method (Base 32° Fahrenheit)" at www.ncdc.noaa.gov/fpsf.html.

❖ Table R301.2(1) is designed so that jurisdictions recognize certain climatic and geographic design criteria that vary from location to location. Communities are directed to complete the table with a variety of factors. See the table footnotes for the sources of the information to be determined by the local jurisdiction so that it can complete the table.

The required information includes the date of entry into the National Flood Insurance Program (NFIP), which is the date of adoption of the first code or ordinance for management of flood hazard areas. With respect to the official map that shows flood hazard areas, the community inserts the date of the currently effective Flood Insurance Rate Map (FIRM) (and Flood Boundary and Floodway Map, if applicable) or the date of other maps that are adopted. Another flood hazard map may be specified if it shows flood hazard areas that are greater than those shown on the FIRM, as may be the case if a community elects to define its floodplains based on higher standards such as the "ultimate development" or "no-rise" rules to define the floodway.

TABLE R301.2(2)
COMPONENT AND CLADDING LOADS FOR A BUILDING WITH A MEAN ROOF HEIGHT OF 30 FEET LOCATED IN EXPOSURE B (psf)

	ZONE	EFFECTIVE WIND AREA (feet^2)	\\multicolumn{26}{c}{BASIC WIND SPEED (mph—3-second gust)}																							
			\\multicolumn{2}{c}{85}	\\multicolumn{2}{c}{90}	\\multicolumn{2}{c}{100}	\\multicolumn{2}{c}{105}	\\multicolumn{2}{c}{110}	\\multicolumn{2}{c}{120}	\\multicolumn{2}{c}{125}	\\multicolumn{2}{c}{130}	\\multicolumn{2}{c}{140}	\\multicolumn{2}{c}{145}	\\multicolumn{2}{c}{150}	\\multicolumn{2}{c}{170}				
Roof > 0 to 10 degrees	1	10	10.0	-13.0	10.0	-14.6	10.0	-18.0	10.0	-19.8	10.0	-21.8	10.5	-25.9	11.4	-28.1	12.4	-30.4	14.3	-35.3	15.4	-37.8	16.5	-40.5	21.1	-52.0
	1	20	10.0	-12.7	10.0	-14.2	10.0	-17.5	10.0	-19.3	10.0	-21.2	10.0	-25.2	10.7	-27.4	11.6	-29.6	13.4	-34.4	14.4	-36.9	15.4	-39.4	19.8	-50.7
	1	50	10.0	-12.2	10.0	-13.7	10.0	-16.9	10.0	-18.7	10.0	-20.5	10.0	-24.4	10.0	-26.4	10.6	-28.6	12.3	-33.2	13.1	-35.6	14.1	-38.1	18.1	-48.9
	1	100	10.0	-11.9	10.0	-13.3	10.0	-18.5	10.0	-18.2	10.0	-19.9	10.0	-23.7	10.0	-25.7	10.0	-27.8	11.4	-32.3	12.2	-34.6	13.0	-37.0	16.7	-47.6
	2	10	10.0	-21.8	10.0	-24.4	10.0	-30.2	10.0	-33.3	10.0	-36.5	10.5	-43.5	11.4	-47.2	12.4	-51.0	14.3	-59.2	15.4	-63.5	16.5	-67.9	21.1	-87.2
	2	20	10.0	-19.5	10.0	-21.8	10.0	-27.0	10.0	-29.7	10.0	-32.6	10.0	-38.8	10.7	-42.1	11.6	-45.6	13.4	-52.9	14.4	-56.7	15.4	-60.7	19.8	-78.0
	2	50	10.0	-16.4	10.0	-18.4	10.0	-22.7	10.0	-25.1	10.0	-27.5	10.0	-32.7	10.0	-35.5	10.6	-38.4	12.3	-44.5	13.1	-47.8	14.1	-51.1	18.1	-65.7
	2	100	10.0	-14.1	10.0	-15.8	10.0	-19.5	10.0	-21.5	10.0	-23.6	10.0	-28.1	10.0	-30.5	10.0	-33.0	11.4	-38.2	12.2	-41.0	13.0	-43.9	16.7	-56.4
	3	10	10.0	-32.8	10.0	-36.8	10.0	-45.4	10.0	-50.1	10.0	-55.0	10.5	-65.4	11.4	-71.0	12.4	-76.8	14.3	-89.0	15.4	-95.5	16.5	-102.2	21.1	-131.3
	3	20	10.0	-27.2	10.0	-30.5	10.0	-37.6	10.0	-41.5	10.0	-45.5	10.0	-54.2	10.7	-58.8	11.6	-63.6	13.4	-73.8	14.4	-79.1	15.4	-84.7	19.8	-108.7
	3	50	10.0	-19.7	10.0	-22.1	10.0	-27.3	10.0	-30.1	10.0	-33.1	10.0	-39.3	10.0	-42.7	10.6	-46.2	12.3	-53.5	13.1	-57.4	14.1	-61.5	18.1	-78.9
	3	100	10.0	-14.1	10.0	-15.8	10.0	-19.5	10.0	-21.5	10.0	-23.6	10.0	-28.1	10.0	-30.5	10.0	-33.0	11.4	-38.2	12.2	-41.0	13.0	-43.9	16.7	-56.4
Roof > 10 to 30 degrees	1	10	10.0	-11.9	10.0	-13.3	10.4	-16.5	11.4	-18.2	12.5	-19.9	14.9	-23.7	16.2	-25.7	17.5	-27.8	20.3	-32.3	21.8	-34.6	23.3	-37.0	30.0	-47.6
	1	20	10.0	-11.6	10.0	-13.0	10.0	-16.0	10.4	-17.6	11.4	-19.4	13.6	-23.0	14.8	-25.0	16.0	-27.0	18.5	-31.4	19.9	-33.7	21.3	-36.0	27.3	-46.3
	1	50	10.0	-11.1	10.0	-12.5	10.0	-15.4	10.0	-17.0	10.0	-18.6	11.9	-22.2	12.9	-24.1	13.9	-26.0	16.1	-30.2	17.3	-32.4	18.5	-34.6	23.8	-44.5
	1	100	10.0	-10.8	10.0	-12.1	10.0	-14.9	10.0	-16.5	10.0	-18.1	10.5	-21.5	11.4	-23.3	12.4	-25.2	14.3	-29.3	15.4	-31.4	16.5	-33.6	21.1	-43.2
	2	10	10.0	-25.1	10.0	-28.2	10.4	-34.8	11.4	-38.3	12.5	-42.1	14.9	-50.1	16.2	-54.3	17.5	-58.7	20.3	-68.1	21.8	-73.1	23.3	-78.2	30.0	-100.5
	2	20	10.0	-22.8	10.0	-25.6	10.0	-31.5	10.4	-34.8	11.4	-38.2	13.6	-45.4	14.8	-49.3	16.0	-53.3	18.5	-61.8	19.9	-66.3	21.3	-71.0	27.3	-91.2
	2	50	10.0	-19.7	10.0	-22.1	10.0	-27.3	10.0	-30.1	10.0	-33.0	11.9	-39.3	12.9	-42.7	13.9	-46.1	16.1	-53.5	17.3	-57.4	18.5	-61.4	23.8	-78.9
	2	100	10.0	-17.4	10.0	-19.5	10.0	-24.1	10.0	-26.6	10.0	-29.1	10.5	-34.7	11.4	-37.6	12.4	-40.7	14.3	-47.2	15.4	-50.6	16.5	-54.2	21.1	-69.6
	3	10	10.0	-25.1	10.0	-28.2	10.4	-34.8	11.4	-38.3	12.5	-42.1	14.9	-50.1	16.2	-54.3	17.5	-58.7	20.3	-68.1	21.8	-73.1	23.3	-78.2	30.0	-100.5
	3	20	10.0	-22.8	10.0	-25.6	10.0	-31.5	10.4	-34.8	11.4	-38.2	13.6	-45.4	14.8	-49.3	16.0	-53.3	18.5	-61.8	19.9	-66.3	21.3	-71.0	27.3	-91.2
	3	50	10.0	-19.7	10.0	-22.1	10.0	-27.3	10.0	-30.1	10.0	-33.0	11.9	-39.3	12.9	-42.7	13.9	-46.1	16.1	-53.5	17.3	-57.4	18.5	-61.4	23.8	-78.9
	3	100	10.0	-17.4	10.0	-19.5	10.0	-24.1	10.0	-26.6	10.0	-29.1	10.5	-34.7	11.4	-37.6	12.4	-40.7	14.3	-47.2	15.4	-50.6	16.5	-54.2	21.1	-69.6
Roof > 30 to 45 degrees	1	10	11.9	-13.0	13.3	-14.6	16.5	-18.0	18.2	-19.8	19.9	-21.8	23.7	-25.9	25.7	-28.1	27.8	-30.4	32.3	-35.3	34.6	-37.8	37.0	-40.5	47.6	-52.0
	1	20	11.6	-12.3	13.0	-13.8	16.0	-17.1	17.6	-18.8	19.4	-20.7	23.0	-24.6	25.0	-26.7	27.0	-28.9	31.4	-33.5	33.7	-35.9	36.0	-38.4	46.3	-49.3
	1	50	11.1	-11.5	12.5	-12.8	15.4	-15.9	17.0	-17.5	18.6	-19.2	22.2	-22.8	24.1	-24.8	26.0	-25.8	30.2	-31.1	32.4	-33.3	34.6	-35.7	44.5	-45.8
	1	100	10.8	-10.8	12.1	-12.1	14.9	-14.9	16.5	-16.5	18.1	-18.1	21.5	-21.5	23.3	-23.3	25.2	-25.2	29.3	-29.3	31.4	-31.4	33.6	-33.6	43.2	-43.2
	2	10	11.9	-15.2	13.3	-17.0	16.5	-21.0	18.2	-23.2	19.9	-25.5	23.7	-30.3	25.7	-32.9	27.8	-35.6	32.3	-41.2	34.6	-44.2	37.0	-47.3	47.6	-60.8
	2	20	11.6	-14.5	13.0	-16.3	16.0	-20.1	17.6	-22.2	19.4	-24.3	23.0	-29.0	25.0	-31.4	27.0	-34.0	31.4	-39.4	33.7	-42.3	36.0	-45.3	46.3	-58.1
	2	50	11.1	-13.7	12.5	-15.3	15.4	-18.9	17.0	-20.8	18.6	-22.9	22.2	-27.2	24.1	-29.5	26.0	-32.0	30.2	-37.1	32.4	-39.8	34.6	-42.5	44.5	-54.6
	2	100	10.8	-13.0	12.1	-14.6	14.9	-18.0	16.5	-19.8	18.1	-21.8	21.5	-25.9	23.3	-28.1	25.2	-30.4	29.3	-35.3	31.4	-37.8	33.6	-40.5	43.2	-52.0
	3	10	11.9	-15.2	13.3	-17.0	16.5	-21.0	18.2	-23.2	19.9	-25.5	23.7	-30.3	25.7	-32.9	27.8	-35.6	32.3	-41.2	34.6	-44.2	37.0	-47.3	47.6	-60.8
	3	20	11.6	-14.5	13.0	-16.3	16.0	-20.1	17.6	-22.2	19.4	-24.3	23.0	-29.0	25.0	-31.4	27.0	-34.0	31.4	-39.4	33.7	-42.3	36.0	-45.3	46.3	-58.1
	3	50	11.1	-13.7	12.5	-15.3	15.4	-18.9	17.0	-20.8	18.6	-22.9	22.2	-27.2	24.1	-29.5	26.0	-32.0	30.2	-37.1	32.4	-39.8	34.6	-42.5	44.5	-54.5
	3	100	10.8	-13.0	12.1	-14.6	14.9	-18.0	16.5	-19.8	18.1	-21.8	21.5	-25.9	23.3	-28.1	25.2	-30.4	29.3	-35.3	31.4	-37.8	33.6	-40.5	43.2	-52.0
Wall	4	10	13.0	-14.1	14.6	-15.8	18.0	-19.5	19.8	-21.5	21.8	-23.6	25.9	-28.1	28.1	-30.5	30.4	-33.0	35.3	-38.2	37.8	-41.0	40.5	-43.9	52.0	-56.4
	4	20	12.4	-13.5	13.9	-15.1	17.2	-18.7	18.9	-20.6	20.8	-22.6	24.7	-26.9	26.8	-29.2	29.0	-31.6	33.7	-36.7	36.1	-39.3	38.7	-42.1	49.6	-54.1
	4	50	11.6	-12.7	13.0	-14.3	16.1	-17.6	17.8	-19.4	19.5	-21.3	23.2	-25.4	25.2	-27.5	27.2	-29.8	31.6	-34.6	33.9	-37.1	36.2	-39.7	46.6	-51.0
	4	100	11.1	-12.2	12.4	-13.6	15.3	-16.8	16.9	-18.5	18.5	-20.4	22.0	-24.2	23.9	-26.3	25.9	-28.4	30.0	-33.0	32.2	-35.4	34.4	-37.8	44.2	-48.6
	5	10	13.0	-17.4	14.6	-19.5	18.0	-24.1	19.8	-26.6	21.8	-29.1	25.9	-34.7	28.1	-37.6	30.4	-40.7	35.3	-47.2	37.8	-50.6	40.5	-54.2	52.0	-69.6
	5	20	12.4	-16.2	13.9	-18.2	17.2	-22.5	18.9	-24.8	20.8	-27.2	24.7	-32.4	26.8	-35.1	29.0	-38.0	33.7	-44.0	36.1	-47.2	38.7	-50.5	49.6	-64.9
	5	50	11.6	-14.7	13.0	-16.5	16.1	-20.3	17.8	-22.4	19.5	-24.6	23.2	-29.3	25.2	-31.8	27.2	-34.3	31.6	-39.8	33.9	-42.7	36.2	-45.7	46.6	-58.7
	5	100	11.1	-13.5	12.4	-15.1	15.3	-18.7	16.9	-20.6	18.5	-22.6	22.0	-26.9	23.9	-29.2	25.9	-31.6	30.0	-36.7	32.2	-39.3	34.4	-42.1	44.2	-54.1

For SI: 1 foot = 304.8 mm, 1 square foot = 0.0929 m^2, 1 mile per hour = 1.609 km/h.
NOTES: For effective areas between those given above the load may be interpolated, otherwise use the load associated with the lower effective area.
Table values shall be adjusted for height and exposure by multiplying by the adjustment coefficient in Table R301.2(3).
See Figure R301.2(8) for location of zones.
Plus and minus signs signify pressures acting toward and away from the building surfaces.

❖ This table lists wind pressures for components or cladding building elements. The term cladding is defined in Section R202. While the basic wind speed from Figure R301.2(4) is applicable to the structure as a whole, this table applies to elements of the building that typically are not part of the wind-force-resisting system. Wind pressures on these elements vary depending on the tributary area, the location/orientation (i.e. a wall that is typically vertical versus a roof and its slope), the region or zone, as well as the basic wind speed. "Zones" refer to various portions of a structure as illustrated in Figure R301.2(8). Some zones, such as at roof edges and the ends of a wall, represent discontinuities in the exterior building surfaces. Discontinuities interrupt the flow of wind and result in higher wind pressures at these locations. For each combination of these various factors there are two entries in the table, one positive and one negative. The positive number denotes the pressure acting inward or toward the exterior surface of the building; the nega-

BUILDING PLANNING

tive number denotes the pressure acting outward or away from the building surface. The latter is of interest in designing roof elements and their hold-downs that must resist wind uplift.

Unlike the *International Building Code,* the *International Residential Code* does not elaborate on the distinctions between wind loads on the main wind-force-resisting system versus components and cladding wind loads. Instead, compliance with the components and cladding pressures is only necessary where code provisions explicitly refer to this table. This occurs in the establishment of design wind loads for skylights, exterior windows and doors (Sections R301.2.1 and R613.2) and setting the threshold for requiring rafter ties (Sections R611.9, R802.11, and R804.4).

TABLE R301.2(3)
HEIGHT AND EXPOSURE ADJUSTMENT COEFFICIENTS FOR TABLE R301.2(2)

MEAN ROOF HEIGHT	EXPOSURE		
	B	C	D
15	1.00	1.21	1.47
20	1.00	1.29	1.55
25	1.00	1.35	1.61
30	1.00	1.40	1.66
35	1.05	1.45	1.70
40	1.09	1.49	1.74
45	1.12	1.53	1.78
50	1.16	1.56	1.81
55	1.19	1.59	1.84
60	1.22	1.62	1.87

❖ This table provides adjustment factors for the components and cladding wind pressures of Table R301.2(2), which are based on a mean roof height of 30 feet (9144 mm) and wind exposure category B (see the commentary on Section R301.2.1.4 for a discussion of exposure). For mean roof heights and exposure categories that vary from those assumed in Table R301.2(2), the resulting pressures will change accordingly. Thus, adjustments to the pressures are required by the factors given in this table.

FIGURE R301.2(1)   **BUILDING PLANNING**

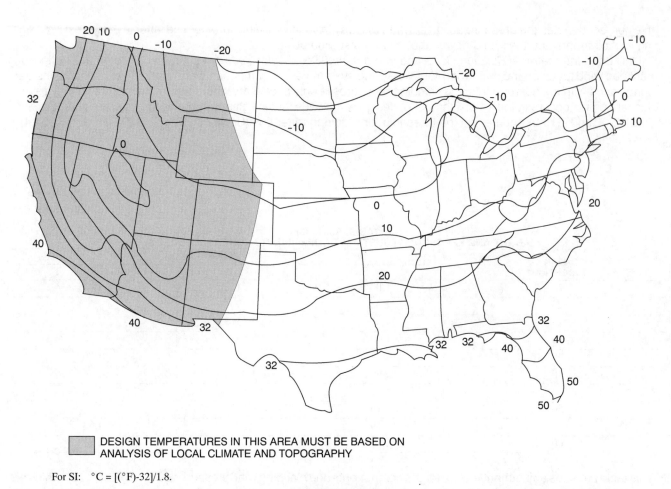

▨ DESIGN TEMPERATURES IN THIS AREA MUST BE BASED ON ANALYSIS OF LOCAL CLIMATE AND TOPOGRAPHY

For SI: °C = [(°F)-32]/1.8.

FIGURE R301.2(1)
ISOLINES OF THE 97$^{1}/_{2}$ PERCENT WINTER (DECEMBER, JANUARY AND FEBRUARY) DESIGN TEMPERATURES (°F)

❖ This figure establishes the winter design temperature, which is a criteria for determining the need for dwelling unit heating (R303.8) as well as determining the need for freeze protection of piping (M2301.2.5, P2603.6, and P3001.2).

BUILDING PLANNING FIGURE R301.2(2)

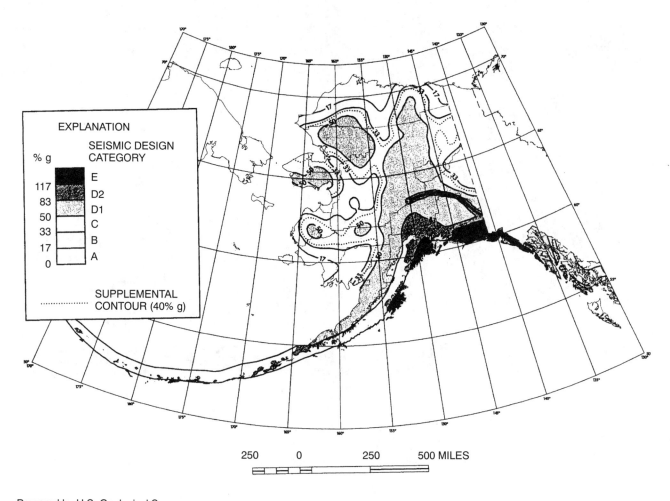

Prepared by U.S. Geological Survey

Scale 1:17,000,000

REFERENCES

Digital data prepared with ARC/INFO 7.1.1
running under Solaris 2.5 on a UNIX workstation

U.S. Geological Survey National Seismic-Hazard Mapping Project,
1998 Alaska Seismic-Hazard Maps: Documentation: U.S. Geological
Survey Open-File Report, in progress.

Albers Equal-Area Conic Projection
Standard parallels 55°N and 65°N
Central Meridian 160°W

U.S. Geological Survey National Seismic-Hazard Mapping Project,
1998, Seismic-Hazard Maps of Alaska: U.S. Geological Survey Open-File
Report, 6 sheet, scale 1:5,000,000, in progress.

For SI: 1 mile = 1.61 km.

**[B] FIGURE R301.2(2)
SEISMIC DESIGN CATEGORIES—SITE CLASS D**
(continued)

FIGURE R301.2(2)

BUILDING PLANNING

SCALE 1:15,000,000

Digital data prepared with ARC/INFO 7.1.1
running under Solaris 2.5 on a UNIX workstation

Albers Equal-Area Conic Projection
Standard Parallels 29.5°N and 45.5°N
Central Meridian 95°W

Prepared by U.S. Geological Survey

For SI: 1 mile = 1.61 km.

REFERENCES

Frankel, A. Mueller, C., Barnhard, T., Perkins, D., Leyendecker, E. V., Dickman, N., Hanson, S., and Hopper, M., 1996, National Seismic-Hazard Maps: Documentation June 1996: U.S. Geological Survey Open-File Report 96-532, 110 p.

Frankel, A., Muller, C., Barnhard, T., Perkins, D., Leyendecker, E. V., Dickman, N., Hanson, S., and Spectral Response Acceleration for 0.2 Second Period with 2% Probability of Exceedance in 50 Years: U. S. Geological Survey Open-File Report 97-131-F, scale 1:7,000,000.

Petersen, M., Bryant, W., Cramer, C., Cao, T., Reichle, M., Frankel, A., Lienkaemper, J., McCrory, P., and Schwartz, D., 1996, Probabilitic Seismic Hazard Assessment for the State of California: California Division of Mines and Geology Open-File Report 96-08, 66 p., and U.S. Geological Survey Open-File Report 96-706, 66 p.

**[B] FIGURE R301.2(2)—continued
SEISMIC DESIGN CATEGORIES—SITE CLASS D**
(continued)

BUILDING PLANNING FIGURE R301.2(2)

Prepared by U.S. Geological Survey

SCALE 1:3,500,000

Digital data prepared with ARC/INFO 7.1.1
running under Solaris 2.5 on a UNIX workstation

Albers Equal-Area Conic Projection
Standard parallels 8°N and 18°N
Central Meridian 157.5°W

For SI: 1 mile = 1.61 km.

REFERENCES

U.S. Geological Survey National Seismic-Hazard Mapping Project, 1998
Hawaii Seismic-Hazard Maps:
Documentation: U.S. Geological Survey Open-File Report, in progress.

U.S. Geological Survey National Seismic-Hazard Mapping Project, 1998,
Seismic-Hazard Maps of Hawaii: U.S. Geological Survey Open-File
Report, 6 sheet, scale 1:2,000,000, in progress.

**[B] FIGURE R301.2(2)—continued
SEISMIC DESIGN CATEGORIES—SITE CLASS D**

❖ This figure establishes the seismic design category for a site as presented in Section R301.2.2.1. It provides the most direct determination of seismic design category in the code. The earthquake-related provisions of the code state requirements as a function of the seismic design category.

For over fifty years, the United States Geological Survey (USGS) has prepared seismic hazard maps that have been used to establish building code provisions for earthquake-resistant construction. The seismic provisions incorporated into the *International Building Code* and *International Residential Code* are largely based upon the 1997 National Earthquake Hazards Reduction Program Provisions. The seismic hazard maps from these provisions are the latest USGS ground-motion spectral-response maps. The map incorporated into the IRC is essentially a plot of Short

Perlod Design Spectral Response, S_{DS}, assuming a site class D. In the IRC, the seismic design category is dependent only on the value of S_{DS}. This allows the seismic design category to be read directly from the map in the same manner as the seismic zone was read off a single map in prior building codes.

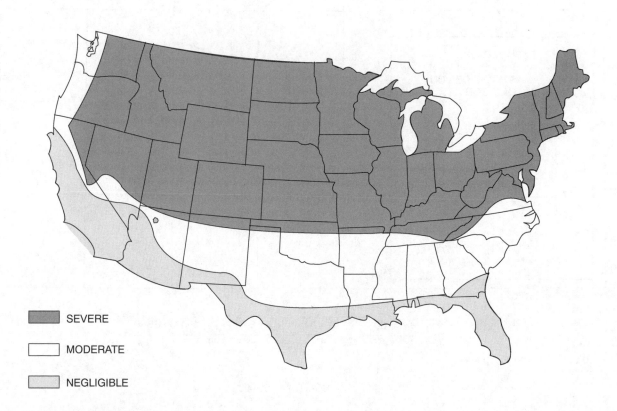

a. Alaska and Hawaii are classified as severe and negligible, respectively.
b. Lines defining areas are approximate only. Local conditions may be more or less severe than indicated by region classification. A severe classification is where weather conditions result in significant snowfall combined with extended periods during which there is little or no natural thawing causing deicing salts to be used extensively.

**[B] FIGURE R301.2(3)
WEATHERING PROBABILITY MAP FOR CONCRETE**

❖ The weathering probability map for concrete in Figure R301.2(3) classifies regions as severe, moderate, or negligible. These classifications are based on climatic data, surveys of state transportation departments, and field surveys from Portland Cement Association regional offices. The map provides a guideline for the severity of weathering that can be expected for concrete surfaces exposed to weather or de-icing chemicals. As indicated in Note "b," the map boundaries are approximate, and local conditions may warrant a classification that is different from that provided by the map.

The weathering index from this map indicates the need to increase the strength of concrete, to add air-entrainment agents to concrete (see Section R402.2), or both. In masonry construction, the index determines the grade of masonry units required in accordance with the ASTM standards listed in Note "a" to Table R301.2(1).

BUILDING PLANNING FIGURE R301.2(4)

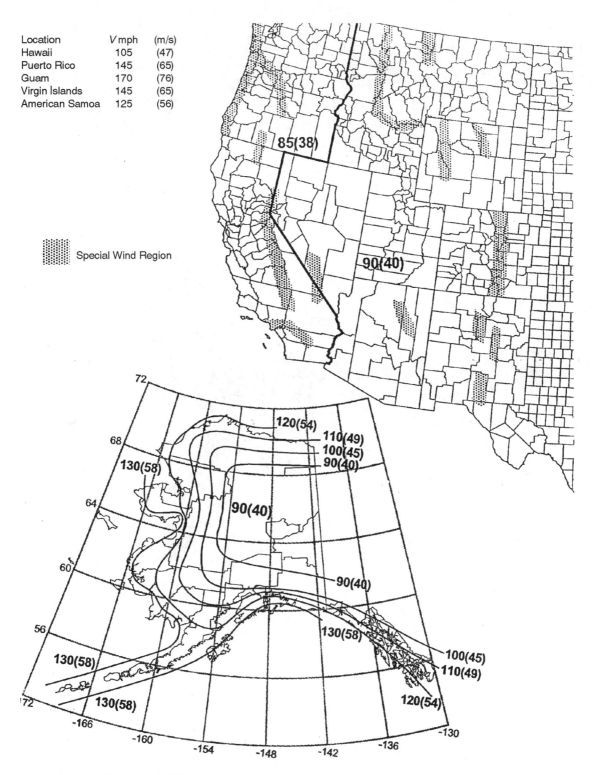

Location	V mph	(m/s)
Hawaii	105	(47)
Puerto Rico	145	(65)
Guam	170	(76)
Virgin Islands	145	(65)
American Samoa	125	(56)

For SI: 1 foot = 304.8 mm, 1 mile per hour = 0.447 m/s.

a. Values are nominal design 3-second gust wind speeds in miles per hour at 33 feet above ground for Exposure C category.
b. Linear interpolation between wind contours is permitted.
c. Islands and coastal areas outside the last contour shall use the last wind speed contour of the coastal area.
d. Mountainous terrain, gorges, ocean promontories, and special wind regions shall be examined for unusual wind conditions.

[B] FIGURE R301.2(4)
BASIC WIND SPEEDS FOR 50-YEAR MEAN RECURRENCE INTERVAL
(continued)

FIGURE R301.2(4)

BUILDING PLANNING

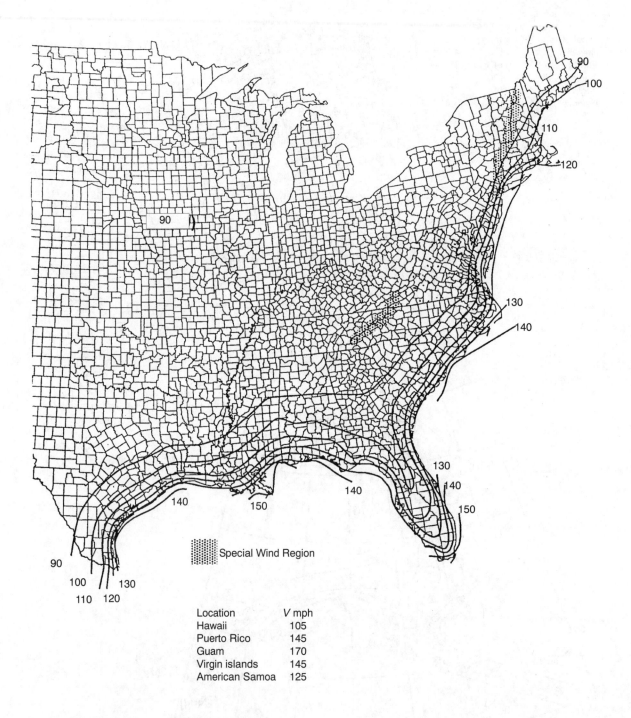

Location	V mph
Hawaii	105
Puerto Rico	145
Guam	170
Virgin islands	145
American Samoa	125

For SI: 1 foot = 304.8 mm, 1 mile per hour = 0.447 m/s.

a. Values are nominal design 3-second gust wind speeds in miles per hour at 33 feet above ground for Exposure C category.
b. Linear interpolation between wind contours is permitted.
c. Islands and coastal areas outside the last contour shall use the last wind speed contour of the coastal area.
d. Mountainous terrain, gorges, ocean promontories, and special wind regions shall be examined for unusual wind conditions.
e. Enlarged views of Eastern and Southern seaboards are on the following pages.

[B] FIGURE R301.2(4)—continued
BASIC WIND SPEEDS FOR 50-YEAR MEAN RECURRENCE INTERVAL
(continued)

BUILDING PLANNING FIGURE R301.2(4)

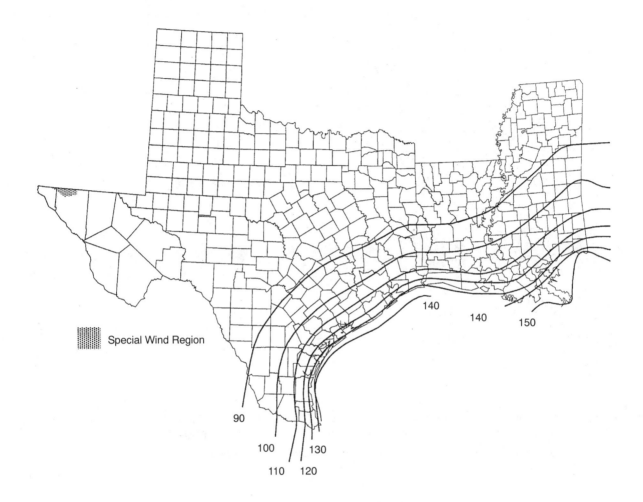

For SI: 1 foot = 304.8 mm, 1 mile per hour = 0.447 m/s.
a. Values are nominal design 3-second gust wind speeds in miles per hour at 33 feet above ground for Exposure C category.
b. Linear interpolation between wind contours is permitted.
c. Islands and coastal areas outside the last contour shall use the last wind speed contour of the coastal area.
d. Mountainous terrain, gorges, ocean promontories, and special wind regions shall be examined for unusual wind conditions.

**[B] FIGURE R301.2(4)—continued
BASIC WIND SPEEDS FOR 50-YEAR MEAN RECURRENCE INTERVAL**
(continued)

FIGURE R301.2(4)

BUILDING PLANNING

For SI: 1 foot = 304.8 mm, 1 mile per hour = 0.447 m/s.

a. Values are nominal design 3-second gust wind speeds in miles per hour at 33 feet above ground for Exposure C category.
b. Linear interpolation between wind contours is permitted.
c. Islands and coastal areas outside the last contour shall use the last wind speed contour of the coastal area.
d. Mountainous terrain, gorges, ocean promontories, and special wind regions shall be examined for unusual wind conditions.

**[B] FIGURE R301.2(4)—continued
BASIC WIND SPEEDS FOR 50-YEAR MEAN RECURRENCE INTERVAL**
(continued)

BUILDING PLANNING

FIGURE R301.2(4)

For SI: 1 foot = 304.8 mm, 1 mile per hour = 0.447 m/s.

a. Values are nominal design 3-second gust wind speeds in miles per hour at 33 feet above ground for Exposure C category.
b. Linear interpolation between wind contours is permitted.
c. Islands and coastal areas outside the last contour shall use the last wind speed contour of the coastal area.
d. Mountainous terrain, gorges, ocean promontories, and special wind regions shall be examined for unusual wind conditions.

[B] FIGURE R301.2(4)—continued
BASIC WIND SPEEDS FOR 50-YEAR MEAN RECURRENCE INTERVAL

❖ The basic wind-speed map is taken from the ASCE 7 wind provisions and is used to establish the basic wind speed of any particular site. This latest map depicts 3-second gust velocities having a 50-year mean recurrence interval, or a 2-percent annual chance of being exceeded. Because the National Weather Service has phased out the measurement of fastest-mile wind speeds, the wind speed used in design has been revised to a measure of wind speed consistent with current weather data and wind speeds reported by the media.

While the code does not provide any specific guidance on the topic, one should take note of the areas identified as "special wind regions" as well as other areas described in the footnote where wind anomalies are observed. As the footnote explains, these areas must be examined for unusual wind conditions. ASCE 7 discusses the use of regional climatic data in these areas.

[B] FIGURE R301.2(5)
GROUND SNOW LOADS, P_g, FOR THE UNITED STATES (lb/ft²)

❖ The ground snow-load map is taken from the ASCE 7 snow-load provisions and is based on statistical analysis of ground snow data. This map provides ground snow loads for direct use in the prescriptive provisions and tables of the code.

BUILDING PLANNING FIGURE R301.2(6)

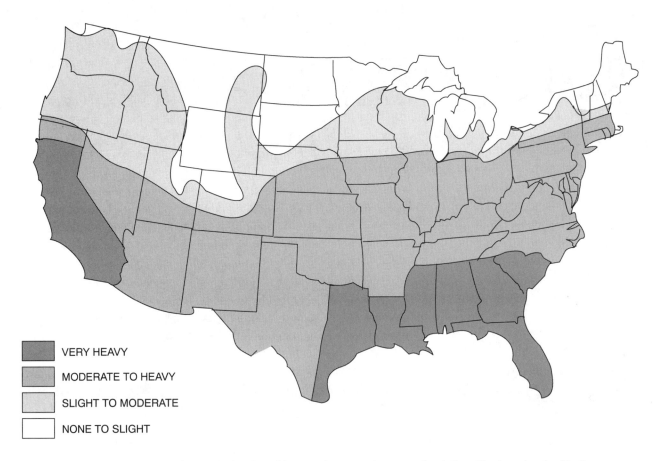

Note: Lines defining areas are approximate only. Local conditions may be more or less severe than indicated by the region classification.

**FIGURE R301.2(6)
TERMITE INFESTATION PROBABILITY MAP**

❖ This figure establishes the potential for termite damage in areas of the continental United States. The boundaries for the various classifications are approximate, and local experience should be relied on either to affirm or to modify the classification determined from this map.

FIGURE R301.2(7) BUILDING PLANNING

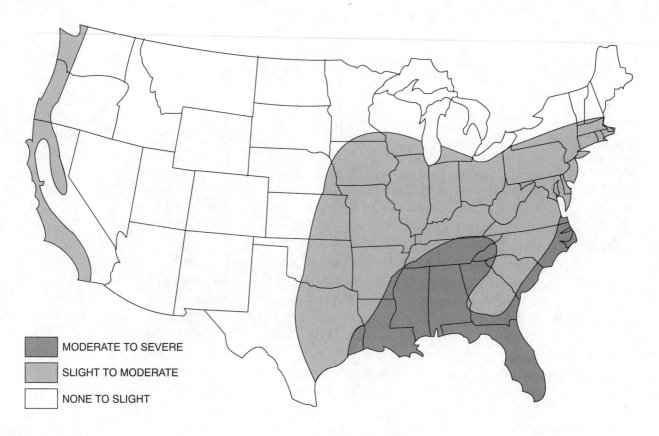

Notes: Lines defining areas are approximate only. Local conditions may be more or less severe than indicated by the region classification.

**FIGURE R301.2(7)
DECAY PROBABILITY MAP**

❖ This figure serves as a guideline for establishing the susceptibility to decay damage for areas in the continental United States. The boundaries for the various classifications are approximate, and local experience should be relied on to either affirm or to modify the classification determined from this map.

BUILDING PLANNING

For SI: 1 foot = 304.8 mm, 1 degree = 0.009 rad.

Note: a = 4 feet in all cases

**FIGURE R301.2(8)
COMPONENT AND CLADDING PRESSURE ZONES**

❖ This figure depicts the zones for the components and cladding wind pressures that are given in Table R301.2(2). See the commentary to this table for a discussion of components and cladding.

R301.2.1 Wind limitations. Buildings and portions thereof shall be limited by wind speed, as defined in Table R301.2(1), and construction methods in accordance with this code. Basic wind speeds shall be determined from Figure R301.2(4). Where different construction methods and structural materials are used for various portions of a building, the applicable requirements of this section for each portion shall apply. Where loads for windows, skylights and exterior doors are not otherwise specified, the loads listed in Table R301.2(2) adjusted for height and exposure per Table R301.2(3), shall be used to determine design load performance requirements for windows and doors.

❖ As described above, wind loads are a major consideration in designing a structure's lateral force-resisting system. Commentary Figure R301.2.1 is a schematic representation of the lateral component of wind loading on a building. Wind loads affect more than the lateral load system, as evidenced by provisions such as roof tie-downs in Section R802.11.

Buildings must be capable of withstanding the wind loads specified in this section. Wind speeds used for design are determined from Figure R301.2(4). For the most part, no further calculation is required, since the prescriptive tables of the code are based on wind speed and exposure.

Windows, skylights, and exterior doors must be capable of withstanding the component and cladding wind pressures of Table R301.2(2) adjusted by the height and exposure coefficients given in Table R301.2(3).

NOTE: FROST PENETRATION IN INCHES

SOURCE: U.S. WEATHER BUREAU

For SI: 1 inch = 25.4 mm.

**Figure R301.2(9)
FROST LINE DEPTH**

**Figure R301.2.1
WIND FORCES ACTING ON AREA**

R301.2.1.1 Design criteria. Construction in regions where the basic wind speeds from Figure R301.2(4) equal or exceed 110 miles per hour (177.1 km/h) shall be designed in accordance with one of the following:

1. American Forest and Paper Association (AF&PA) *Wood Frame Construction Manual for One- and Two-Family Dwellings* (WFCM); or
2. *Southern Building Code Congress International Standard for Hurricane Resistant Residential Construction* (SSTD 10); or

BUILDING PLANNING

3. *Minimum Design Loads for Buildings and Other Structures* (ASCE-7); or
4. American Iron and Steel Institute (AISI), *Standard for Cold-Formed Steel Framing—Prescriptive Method for One- and Two-family Dwellings* (COFS/PM).
5. Concrete construction shall be designed in accordance with the provisions of this code.

❖ With the exception of concrete construction, the prescriptive provisions of the IRC apply only where the basic wind speed is less than 110 miles per hour (177 km/h). Concrete provisions are applicable in regions with wind speeds up to 150 mph (241 km/h), where the site is classified as Exposure B, C or D (see the commentary on Section R611.2). Where the basic wind speed exceeds the limitations given in the code, structures must be designed for wind loads. For this purpose, the IRC expressly allows the use of AF&PA WFCM: *Wood Frame Construction Manual for One- and Two-Family Dwellings*; SBCCI SSTD 10: *Standard for Hurricane Resistant Residential Construction*; ASCE 7: *Minimum Design Loads for Buildings and Other Structures*; or AISI: *Standard for Cold-Formed Steel Framing-Prescriptive Method for One-and Two-Family Dwellings*.

R301.2.1.2 Internal pressure. Windows in buildings located in wind borne debris regions shall have glazed openings protected from windborne debris or the building shall be designed as a partially enclosed building in accordance with the *International Building Code*. Glazed opening protection for windborne debris shall meet the requirements of the Large Missile Test of ASTM E 1996 and of ASTM E 1886 referenced therein.

> **Exception:** Wood structural panels with a minimum thickness of $^7/_{16}$ inch (11.1 mm) and a maximum span of 8 feet (2438 mm) shall be permitted for opening protection in one- and two-story buildings. Panels shall be precut to cover the glazed openings with attachment hardware provided. Attachments shall be provided in accordance with Table R301.2.1.2 or shall be designed to resist the components and cladding loads determined in accordance with the provisions of the *International Building Code*.

❖ Windows in buildings located in wind borne debris regions must have glazed openings that provide protection from wind borne debris. The opening protection, usually in the form of permanent shutters or laminated glass, must meet the requirements of the Large Missile Test of ASTM E 1996 and ASTM E 1886, listed in Chapter 43. The exception provides a prescriptive and more economical approach for one- and two-story buildings. This method of opening protection uses wood structural panels that are limited to a maximum span of 8 feet (2338 mm), with fastening as specified in Table R301.2.1.2. The builder must precut these to fit each glazed opening and provide the necessary attachment hardware.

To determine where this requirement applies, refer to the definition in Chapter 2, which states that wind borne debris regions include areas in hurricane prone regions within one mile of the coastal mean high water line where the basic wind speed [from Figure R301.2(4)] is 110 mph (177 km/h) or greater; areas where the basic wind speed is at least 120 mph (193 km/h); or Hawaii. For the definition of a hurricane-prone region, see Section R202.

If protection of glazed openings is not provided in one of the manners specified, the building must be designed as a partially enclosed" building in accordance with the *International Building Code*, which in turn requires design in accordance with ASCE 7. This approach accounts for the increased internal pressure that occurs as a result of a breach in the building envelope.

TABLE R301.2.1.2
WINDBORNE DEBRIS PROTECTION FASTENING SCHEDULE FOR WOOD STRUCTURAL PANELS[a,b,c]

FASTENER TYPE	FASTENER SPACING		
	Panel span ≤ 4 foot	4 foot < panel span ≤ 6 foot	6 foot < panel span ≤ 8 foot
$2^1/_2$" #6 Wood screws	16"	12"	9"
$2^1/_2$" #8 Wood screws	16"	16"	12"

For SI: 1 inch = 25.4 mm, 1 foot = 304.8 mm, 1 pound = 0.454 kg, 1 mile per hour = 0.447 m/s.

a. This table is based on 130 mph wind speeds and a 33-foot mean roof height.
b. Fasteners shall be installed at opposing ends of the wood structural panel.
c. Where screws are attached to masonry or masonry/stucco, they shall be attached utilizing vibration-resistant anchors having a minimum ultimate withdrawal capacity of 490 pounds.

❖ The fastening schedule provided in this table allows selection of the appropriate fastener size and spacing when the prescriptive alternative is applied to a tested assembly as required by Section R301.2.1.2. Installing the specified fasteners at both ends of the wood structural panel provides the necessary panel support.

R301.2.1.3 Wind speed conversion. When referenced documents are based on fastest mile wind speeds, the three second gust wind velocities of Figure R301.2(4) shall be converted to fastest mile wind velocities using Table R301.2.1.3.

❖ Prior to publication of the latest edition of ASCE 7, most wind-related code provisions were based on fastest-mile wind velocity. Because many of the documents referenced in the code still use this criterion as a basis for wind design rather than the three-second gust wind velocities of Figure R301.2(4), Table R301.2.1.3 provides the necessary conversion to fastest-mile wind velocities used in those documents. For example, the AF&PA *Wood Frame Construction Manual* and the SSTD 10 standard that are referenced in Section R301.2.1.1 are based on fastest-mile wind velocity.

TABLE R301.2.1.3
EQUIVALENT BASIC WIND SPEEDS[a]

3-second gust	85	90	100	105	110	120	125	130	140	145	150	160	170
Fastest mile	70	75	80	85	90	100	105	110	120	125	130	140	150

For SI: 1 mile per hour = 1.609 km/h.

a. Linear interpolation is permitted.

❖ When referenced documents are based on fastest-mile wind velocities, the appropriate wind speed can be selected directly from this table using the basic wind speed determined from Figure R301.2(4). For example, if the basic wind speed for a site is determined to be 110 mph (177 km/h), and it is necessary to design using one of the referenced standards in Section R301.2.1.1, the corresponding fastest-mile wind velocity would be 90 mph (145 km/h).

R301.2.1.4 Exposure category. For each wind direction considered, an exposure category that adequately reflects the characteristics of ground surface irregularities shall be determined for the site at which the building or structure is to be constructed. For a site located in the transition zone between categories, the category resulting in the largest wind forces shall apply. Account shall be taken of variations in ground surface roughness that arise from natural topography and vegetation as well as from constructed features. For any given wind direction, the exposure in which a specific building or other structure is sited shall be assessed as being one of the following categories:

1. Exposure A. Large city centers with at least 50 percent of the buildings having a height in excess of 70 feet (21 336 mm). Use of this exposure category shall be limited to those areas for which terrain representative of Exposure A prevails in the upwind direction for a distance of at least 0.5 mile (0.8 km) or 10 times the height of the building or other structure, whichever is greater. Possible channeling effects or increased velocity pressures due to the building or structure being located in the wake of adjacent buildings shall be taken into account.

2. Exposure B. Urban and suburban areas, wooded areas, or other terrain with numerous closely spaced obstructions having the size of single-family dwellings or larger. Exposure B shall be assumed unless the site meets the definition of another type exposure.

3. Exposure C. Open terrain with scattered obstructions, including surface undulations or other irregularities, having heights generally less than 30 feet (9144 mm) extending more than 1,500 feet (457 m) from the building site in any quadrant. This exposure shall also apply to any building located within Exposure B type terrain where the building is directly adjacent to open areas of Exposure C type terrain in any quadrant for a distance of more than 600 feet (183 m). This category includes flat open country, grasslands and shorelines in hurricane prone regions.

4. Exposure D. Flat, unobstructed areas exposed to wind flowing over open water (excluding shorelines in hurricane prone regions) for a distance of at least 1 mile (1.61 km). Shorelines in Exposure D include inland waterways, the Great Lakes and coastal areas of California, Oregon, Washington and Alaska. This exposure shall apply only to those buildings and other structures exposed to the wind coming from over the water. Exposure D extends inland from the shoreline a distance of 1,500 feet (457 m) or 10 times the height of the building or structure, whichever is greater.

❖ Wind loading on structures is a function of the site's exposure category, which reflects the characteristics of ground surface irregularities and accounts for variations in ground surface roughness that arise from natural topography and vegetation as well as from constructed features. These categories range from Exposure A for minimal exposure to Exposure D for the most severe exposure category. Exposure B should be used unless the site meets the definition for another category of exposure. Exposure B includes urban and suburban areas, wooded areas and other terrain with numerous, closely spaced obstructions having the size of single-family dwellings or larger. A detailed discussion of wind exposure categories can be found in the commentary of ASCE 7.

R301.2.2 Seismic provisions. The seismic provisions of this code shall apply to buildings constructed in Seismic Design Categories C, D_1, and D_2, as determined in accordance with this section. Buildings in Seismic Design Category E shall be designed in accordance with the *International Building Code,* except when the Seismic Design Category is reclassified to a lower Seismic Design Category in accordance with Section R301.2.2.1.

Exception: Detached one- and two-family dwellings located in Seismic Design Category C are exempt from the seismic requirements of this code.

The weight and irregularity limitations of Section R301.2.2.2 shall apply to buildings in all Seismic Design Categories regulated by the seismic provision of this code. Buildings in Seismic Design Category C shall be constructed in accordance with the additional requirements of Sections R301.2.2.3. Buildings in Seismic Design Categories D_1 and D_2 shall be constructed in accordance with the additional requirements of Section R301.2.2.4.

❖ Earthquakes generate internal forces in a structure due to inertia; this is depicted in Commentary Figure R301.2.2. These forces can cause a building to be distorted and severely damaged. The objective of earthquake-resistant construction is to resist these forces and the resulting distortions.

Providing for the adequate earthquake performance of any structure begins with assessing the degree of hazard at the proposed site. As in the *International*

Building Code, this is accomplished in the IRC by assigning a seismic design category (see Section R301.2.2.1 and the definition in Chapter 2). These range from Seismic Design Category A, representing the lowest level of seismic hazard, to Seismic Design Category E, representing the highest hazard. Buildings in Seismic Design Category E must be designed in accordance with the *International Building Code*, except as allowed in Section R301.2.2.1. Under the IRC, the seismic design category depends on a building's location according to Figure R301.2(2), unless one of the options permitting a more precise determination is exercised.

The IRC seismic provisions apply to buildings classified as Seismic Design Categories C, D_1 or D_2. Detached one- and two-family dwellings need to comply with only the seismic requirements of this code when classified as Seismic Design Category D_1 or D_2. In other words, only townhouses constructed under the IRC's Section R101.2 need to comply with Seismic Design Category C requirements. This indicates that all buildings classified as Seismic Design Category A or B as well as detached one- and two-family dwellings classified as Seismic Design Category C will perform satisfactorily when constructed in accordance with the basic prescriptive requirements of the IRC.

Values for seismic loads on structures are based on a percentage of the dead load, and the IRC prescriptive provisions for structural elements that resist seismic forces are based on an assumed dead load. Generally, the building dead loads as described in Section R301.4 are reflected in the prescriptive design tables throughout the code. The weight and irregularity limitations of Section R301.2.2.2 apply to buildings in the seismic design categories specifically regulated by the code (i.e., SDC C, D_1 and D_2).

An excellent source of additional information on seismic-resistant construction, particularly as it pertains to residential buildings, can be found in FEMA 232: *Home Builder's Guide to Seismic-Resistant Construction*.

R301.2.2.1 Determination of seismic design category. Buildings shall be assigned a Seismic Design Category in accordance with Figure 301.2(2).

❖ Unlike the *International Building Code*, the *International Residential Code* does not require a calculation in order to determine the seismic design category of a structure. Instead, the seismic design category is taken directly from Figure R301.2(2). The seismic design categories shown in Figure R301.2(2) are based solely on the Short Period Design Spectral Response Accelerations, S_{DS}, assuming soil Site Class D, as defined in Section 1615.1 of the IBC.

A convenient alternative to the maps of Figure R301.2(2) and the calculations described in Section R301.2.2.1.1 is the Seismic Design Parameters program prepared by USGS and available on CD-ROM from any of the model code organizations. Another option is the USGS National Seismic Hazard Mapping Project web site (http://geohazards.cr.usgs.gov/eq/index.html). Either of these allows a computer user to

Figure R301.2.2
SEISMIC FORCE

view the seismic hazard map of the IRC in more detail. The Seismic Design Parameters program also provides determination of a site's seismic design category using either the zip code or the longitude and latitude of the site. Since use of longitude and latitude is more precise, this approach is preferable.

R301.2.2.1.1 Alternate determination of seismic design category. The Seismic Design Categories and corresponding Short Period Design Spectral Response Accelerations, S_{DS} shown in Figure R301.2(2) are based on soil Site Class D, as defined in Section 1615.1.1 of the *International Building Code*. If soil conditions are other than Site Class D, the Short Period Design Spectral Response Acceleration, S_{DS}, for a site can be determined according to Section 1615.1 of the *International Building Code*. The value of S_{DS} determined according to Section 1615.1 of the *International Building Code* is permitted to be used to set the Seismic Design Category according to Table R301.2.2.1.1, and to interpolate be-tween values in Tables R602.10.1, R603.7, and other seismic design requirements of this code.

❖ In recognition that there may be an advantage to determining a building's seismic design category using the more detailed procedure of the *International Building Code*, the IRC allows this as an option. If soil conditions are other than Site Class D, for instance, the Short Period Design Spectral Response Acceleration (S_{DS}) for a site may be determined using Section 1615.1 of the IBC. This value of S_{DS} can then be used to establish the seismic design category according to Table R301.2.2.1.1. Determining S_{DS} in this manner, while requiring more work, will provide the benefit of reducing the amount of wall bracing that would otherwise be required by the provisions in Chapter 6, based solely on seismic design category. This is accomplished, in Table R602.10.1 for instance, by interpolating between the tabulated values using the values of S_{DS} determined as described above.

TABLE R301.2.2.1.1
SEISMIC DESIGN CATEGORY DETERMINATION

CALCULATED S_{DS}	SEISMIC DESIGN CATEGORY
$S_{DS} \leq 0.17g$	A
$0.17g < S_{DS} \leq 0.33g$	B
$0.33g < S_{DS} \leq 0.50g$	C
$0.50g < S_{DS} \leq 0.83g$	D_1
$0.83g < S_{DS} \leq 1.17g$	D_2
$1.17g < S_{DS}$	E

❖ This table lists the seismic design category for ranges of Short Period Design Spectral Response Acceleration, S_{DS}. These numbers correspond exactly to the legend in the Seismic Design Category Map, Figure R301.2(2). It is necessary to use this table only if the alternative seismic design category determination is employed.

R301.2.2.1.2 Alternative determination of Seismic Design Category E. Buildings located in Seismic Design Category E in accordance with Figure R301.2(2) are permitted to be reclassified as being in Seismic Design Category D2 provided one of the following is done:

1. A more detailed evaluation of the Seismic Design Category is made in accordance with the provisions and maps of the *International Building Code*. Buildings located in Seismic Design Category E per Table R301.2.2.1.1, but located in Seismic Design Category D per the *International Building Code*, may be designed using the Seismic Design Category D2 requirements of this code.

2. Buildings located in Seismic Design Category E that conform to the following additional restrictions are permitted to be constructed in accordance with the provisions for Seismic Design Category D_2 of this code:

   2.1. All exterior shear wall lines or braced wall panels are in one plane vertically from the foundation to the uppermost story.

   2.2. Floors shall not cantilever past the exterior walls.

   2.3. The building is within all of the requirements of Section R301.2.2.2.2 for being considered as regular.

❖ The IRC permits buildings classified as Seismic Design Category E in Figure R301.2(2) to be constructed in accordance with the provisions for Seismic Design Category D_2, provided the building has no irregular features as described in Section R301.2.2.2.2. In addition, out-of-plane offsets in the exterior shear-wall lines or braced wall panels are not allowed, and floors are not permitted to cantilever beyond the exterior walls. In other words, this reclassification to a lower seismic design category is limited to very regular, boxlike structures.

Also, a building classified as Seismic Design Category E using Figure R301.2(2) may possibly be reclassified to Seismic Design Category D_2 if a more detailed evaluation in accordance with the provisions of the *International Building Code* is undertaken. If application of the IBC provisions results in a classification of Seismic Design Category D, then the building may be constructed as Seismic Design Category D_2 under the IRC.

R301.2.2.2 Seismic limitations. The following limitations apply to buildings in all Seismic Design Categories regulated by the seismic provisions of this code.

❖ This section establishes the limitations for weights, irregularity and structures in Seismic Design Categories C, D_1 and D_2.

R301.2.2.2.1 Weights of materials. Average dead loads shall not exceed 15 psf (0.72 kN/m^2) for roofs/ceiling assemblies or 10 psf (0.48 kN/m^2) for floor assemblies, except as further limited by Section R301.2.2. Dead loads for walls above grade shall not exceed:

1. Fifteen psf (0.72 kN/m^2) for exterior light-frame wood walls.

2. Fourteen psf (0.67 kN/m^2) for exterior light-frame cold-formed steel walls.
3. Ten psf (0.48 kN/m^2) for interior light-frame wood walls.
4. Five psf (0.24 kN/m^2) for interior light-frame cold-formed steel walls.
5. Eighty psf (3.83 kN/m^2) for 8-inch-thick (203 mm) masonry walls.
6. Eighty-five psf (4.07 kN/m^2) for 6-inch-thick (152 mm) concrete walls.

Exception: Roof/ceiling dead loads not exceeding 25 psf (1.19 kN/m^2) shall be permitted provided the wall bracing amounts in Chapter 6 are increased in accordance with Table R301.2.2.2.1.

❖ As stated in Section R301.2.2, the dead loads of structures in seismic design categories regulated by the code must be within the limits listed in this section. Exceeding these limits would necessitate a design of the lateral-force-resisting system as described in Section R301.1.2. In some cases, these dead load limits are more restrictive than the dead load allowances in the prescriptive tables of the code [e.g. Table R802.5(1) for rafters and Table R502.3.1(1) for floor joists].

The exception permits the use of roof/ceiling assemblies that exceed the 15 psf (0.72 kN/m^2) dead load limit, provided additional wall bracing is installed as determined using the adjustment factors of Table R301.2.2.2.1. This would permit the use of heavier roof coverings such as tile. The additional wall bracing compensates for the additional seismic force that would be attributed to the added dead load.

TABLE R301.2.2.2.1
WALL BRACING ADJUSTMENT FACTORS BY ROOF COVERING DEAD LOAD^a

WALL SUPPORTING	ROOF/CEILING DEAD LOAD	ROOF/CEILING DEAD LOAD
	15 psf or less	25 psf
Roof only	1.0	1.2
Roof plus one story	1.0	1.1

For SI: 1 pound per square foot = 0.049 kN/m^2.
a. Linear interpolation shall be permitted.

❖ The adjustment factors in this table must be used to increase the amount of wall bracing if the exception is used from Section R301.2.2.2.1, which allows the installation of a roof/ceiling assembly exceeding the 15 psf (0.72 kN/m^2) dead load limit. The resulting additional wall bracing accounts for the additional seismic forces in order to avoid overloading the lateral force resisting system.

R301.2.2.2.2 Irregular buildings. Concrete construction complying with Section R611 or R612 and conventional light-frame construction shall not be used in irregular portions of structures in Seismic Design Categories C, D$_1$ and D$_2$. Only such irregular portions of structures shall be designed in accordance with accepted engineering practice to the extent such irregular features affect the performance of the conventional framing system. A portion of a building shall be considered to be irregular when one or more of the following conditions occur:

1. When exterior shear wall lines or braced wall panels are not in one plane vertically from the foundation to the uppermost story in which they are required.

   Exception: For wood light-frame construction, floors with cantilevers or setbacks not exceeding four times the nominal depth of the wood floor joists are permitted to support braced wall panels that are out of plane with braced wall panels below provided that:

   1. Floor joists are nominal 2 inches by 10 inches (51 mm by 254 mm) or larger and spaced not more than 16 inches (406 mm) on center.
   2. The ratio of the back span to the cantilever is at least 2 to 1.
   3. Floor joists at ends of braced wall panels are doubled.
   4. For wood-frame construction, a continuous rim joist is connected to ends of all cantilever joists. When spliced, the rim joists shall be spliced us-ing a galvanized metal tie not less than 0.058 inch (1.47mm) (16 gage) and $1^1/_2$ inches (38 mm) wide fastened with six 16d nails on each side of the splice or a block of the same size as the rim joist of sufficient length to fit securely between the joist space at which the splice occurs fastened with eight 16d nails on each side of the splice; and
   5. Gravity loads carried at the end of cantilevered joists are limited to uniform wall and roof load and the reactions from headers having span of 8 feet (2438 mm) or less.

2. When a section of floor or roof is not laterally supported by shear walls or braced wall lines on all edges.

   Exception: Portions of floors that do not support shear walls or braced wall panels above, or roofs, shall be permitted to extend no more than 6 feet (1829 mm) beyond a shear wall or braced wall line.

3. When the end of a braced wall panel occurs over an opening in the wall below and ends at a horizontal distance greater than 1 foot (305 mm) from the edge of the opening. This provision is applicable to shear walls and braced wall panels offset in plane and to braced wall panels offset out of plane as permitted by the exception to Item 1 above.

   Exception: For wood light-frame wall construction, one end of a braced wall panel shall be permitted to extend more than 1 foot (305 mm) over an opening of not more than eight feet (2438 mm) in width in the wall below provided that the opening includes a header in accordance with the following:

   1. The building width, loading condition, and member species limitations of Table R502.5(1) shall apply and
   2. Not less than 1-2x12 or 2-2x10 for an opening not more than 4 feet in width or

3. Not less than 2-2x12 or 3-2x10 for an opening not more than 6 feet in width or
4. Not less than 3-2x12 or 4-2x10 for an opening not more than 8 feet in width and
5. The entire length of the braced wall panel shall not occur over an opening in the wall below.

4. When an opening in a floor or roof exceeds the lesser of 12 feet (3657 mm) or 50 percent of the least floor or roof dimension.

5. When portions of a floor level are vertically offset.

Exceptions:
1. Framing supported directly by continuous foundations at the perimeter of the building.
2. For wood light-frame construction, floors shall be permitted to be vertically offset when the floor framing is lapped or tied together as required by Section R502.6.1.

6. When shear walls and braced wall lines do not occur in two perpendicular directions.

7. When stories above grade partially or completely braced by wood wall framing in accordance with Section R602 or steel wall framing in accordance with Section R603 include masonry or concrete construction.

   Exception: Fireplaces, chimneys, and masonry veneer as permitted by this code.

When this irregularity applies, the entire story shall be designed in accordance with accepted engineering practice.

❖ Conventional light-frame construction typically allows cantilevers, offsets, etc. within certain limits in order to accommodate common design features and options. These features are not well suited to resisting earthquake forces. This becomes more of a concern in areas of higher seismic hazard where it is preferable that the building, or more importantly the lateral-force-resisting system, be more box-like or regular.

Accordingly, this section identifies building features that are irregular and are not permitted under the concrete or conventional light-frame construction provisions in Seismic Design Categories C, D_1 and D_2. These irregular portions of structures must be designed in accordance with accepted engineering practice. A portion of a building is classified as "irregular" when one of the several conditions described in the code occurs.

Out-of-plane offsets in exterior braced wall lines have been recognized as a factor in buildings damaged by earthquakes. Where an offset occurs, the earthquake forces must be transferred through the floor framing. Thus, item 1 stipulates that exterior shear-wall lines or braced wall panels that are not in one plane vertically from the foundation to the uppermost story in which they are required constitute an irregularity as illustrated in Commentary Figure R301.2.2.2.2(1). The exception, illustrated in Commentary Figure R301.2.2.2.2(2), allows out-of-plane offsets not exceeding four times the nominal depth of the wood floor joists supporting the braced wall line above. There are five additional limitations that must be met to assure the adequacy of the supporting floor framing. Use of this exception is limited to light-frame wood construction.

Typically, braced wall lines are located around the perimeter of floors and walls where forces are greatest. As item 2 indicates, an irregularity occurs where a portion of a floor or roof is not laterally supported by shear walls or braced wall lines on all edges. This is shown in Commentary Figure R301.2.2.2.2(3). The exception permits a floor or roof to extend beyond a braced wall line as illustrated in Commentary Figure R301.2.2.2.2(4).

Where the end of a braced wall panel occurs over an opening in the wall below, an irregularity will be present if the panel extends a horizontal distance greater than 1 foot (305 mm) from the edge of the opening. As Commentary Figure R301.2.2.2.2(5) illustrates, this applies to braced wall panels offset out of plane as well as in plane. An exception for wood light-framed construction, permits a braced wall panel to extend more than 1 foot (305 mm) over an opening in the wall below provided the opening includes a header in accordance with this exception, and the entire length of the braced wall panel does not occur over the opening.

Openings in floors are required in most buildings to provide for stairways, equipment chases, etc. If the

**Figure R301.2.2.2.2(1)
BRACED WALL PANELS OUT OF PLANE**

BUILDING PLANNING

size of the opening is large compared to the size of the floor diaphragm, earthquake forces may overstress portions of the diaphragm. Since this could result in premature failure of the diaphragm and thus constitute an incomplete load path, the code limits the size of such openings allowed under conventional light-frame construction. Commentary Figure R301.2.2.2.2(6) illustrates the irregularity created by an opening in a floor or roof that exceeds the lesser of 12 feet (3657 mm) or 50 percent of the least floor or roof dimension.

Commentary Figure R301.2.2.2.2(7) illustrates item 5. An irregularity results where portions of a floor level are vertically offset. This limitation does not apply to framing that is supported directly by continuous foundations at the perimeter of the building. Also, in light-frame wood construction, floors may be vertically offset provided the floor framing is lapped or tied together as required by Section R502.6.1.

More often than not, conventional light-frame buildings are regular in shape with wall lines perpendicular to one another. Commentary Figure R301.2.2.2.2(8) depicts a non-orthogonal system. This type of irregularity occurs where shear walls or braced wall lines are not oriented perpendicular to each other.

In addition, an irregularity occurs if shear walls or braced wall lines are constructed of dissimilar bracing systems on any story level above grade. This requirement applies to the underlying framing system material such as steel framed shear walls or wood framed braced wall panels combined with masonry or concrete construction in the same story. Where this condition occurs, the entire story must be designed in accordance with accepted engineering practice.

Masonry or concrete fireplaces, chimneys, and masonry veneer are permitted to be combined with steel or wood framing and are not considered to be irregular.

For SI: 1 inch = 25.4 mm, 1 foot = 304.8 mm.

Figure R301.2.2.2.2(2)
BRACED WALL PANELS SUPPORTED BY CANTILEVER OR SET BACK

Figure R301.2.2.2.2(3)
FLOOR OR ROOF NOT SUPPORTED ALL EDGES

FIGURE R301.2.2.2.2(4) - FIGURE R301.2.2.2.2(6) BUILDING PLANNING

Figure R301.2.2.2.2(4)
ROOF OR FLOOR EXTENSION BEYOND BRACED WALL LINE

Figure R301.2.2.2.2(5)
BRACED WALL PANEL EXTENSION OVER OPENING

Figure R301.2.2.2.2(6)
OPENING LIMITATIONS FOR FLOOR AND ROOF DIAPHRAGMS

BUILDING PLANNING

Figure R301.2.2.2.2(7)
PORTIONS OF FLOOR LEVEL OFFSET VERTICALLY

Figure R301.2.2.2.2(8)
BRACED WALL LINES NOT PERPENDICULAR

R301.2.2.3 Seismic Design Category C. Structures assigned to Seismic Design Category C shall conform to the requirements of this section.

❖ This section establishes the limitations for structures in Seismic Design Category C.

R301.2.2.3.1 Anchored stone and masonry veneer. Anchored stone and masonry veneer shall be limited to the first story above grade and shall not exceed 5 inches (127 mm) in thickness.

Exception: In Seismic Design Category C, anchored stone and masonry veneer not exceeding 5 inches (127 mm) in thickness shall be permitted to the height allowed in Section R703.7. In other than the topmost story, the length of wall bracing shall be 1.5 times the length otherwise required in Table R602.10.1.

❖ This section limits the height of anchored stone and masonry veneer so the bracing capacity will not be exceeded during an earthquake. This section references Section R703.7 and duplicates some of the wording. Generally, anchored stone and masonry veneer is limited to the first story above grade and not more than 5 inches (127 mm) in thickness. However, the exceptions to Section R703.7 permit a height of up to 30 feet (9144 mm) above a noncombustible foundation in Seismic Design Categories A, B and C. In order to accommodate this additional height in Seismic Design Category C, the length of wall bracing required by Chapter 6 must be increased by 50 percent, except on the topmost story of the building.

R301.2.2.3.2 Masonry construction. Masonry construction shall comply with the requirements of Section R606.11.2.

❖ This is a cross reference to the appropriate masonry section for seismic requirements.

R301.2.2.3.3 Concrete construction. Concrete construction shall comply with the requirements of Section R611.7.1.4.

❖ This is a cross reference to the appropriate concrete section for seismic requirements.

R301.2.2.4 Seismic Design Categories D_1 and D_2. Structures assigned to Seismic Design Categories D_1 and D_2 shall conform to the requirements for Seismic Design Category C and the additional requirements of this section.

❖ Seismic requirements are cumulative as the seismic design category becomes higher. A structure that is assigned to Seismic Design Categories D_1 or D_2 must comply with all the seismic requirements for Seismic Design Category C structures in addition to those for Seismic Design Categories D_1 or D_2.

R301.2.2.4.1 Height limitations. Wood framed buildings shall be limited to three stories above grade or the limits given in Table R602.10.1. Cold-formed steel framed buildings shall be limited to two stories above grade in accordance with COFS/PM. Mezzanines as defined in Section 202 shall not be considered as stories.

❖ This section limits the building height so the lateral-force-resisting elements will not be overloaded during an earthquake. It provides cross references to other sections that limit the height of structures. Wood framed buildings are generally limited to three stories except in Seismic Design Category D_2, where the limit is two stories in accordance with Table R602.10.1. Floor levels meeting the limitations for mezzanines (see the definition in Chapter 2) need not be counted as a story.

R301.2.2.4.2 Anchored stone and masonry veneer. Buildings with anchored stone and masonry veneer shall be designed in accordance with accepted engineering practice.

Exceptions:

1. In Seismic Design Category D_1, exterior masonry veneer with a maximum nominal thickness of 4 inches (102 mm) is permitted in accordance with Section R703.7, Exception 3.
2. In Seismic Design Category D_2, exterior masonry veneer with a maximum actual thickness of 3 inches (76 mm) is permitted in accordance with Section R703.7, Exception 4.

❖ An engineered design is required for anchored stone and masonry veneer. For detached one- and two-family dwellings, the exceptions permit the prescriptive requirements of Section R703.7. See the commentary for Section R703.7.

R301.2.2.4.3 Masonry construction. Masonry construction in Seismic Design Category D_1 shall comply with the requirements of Section R606.11.3. Masonry construction in Seismic Design Category D_2 shall comply with the requirements of Section R606.11.4.

❖ This section provides cross references to the appropriate masonry sections for the seismic requirements.

R301.2.2.4.4 Concrete construction. Buildings with above-grade concrete walls shall be in accordance with Section R611, R612, or designed in accordance with accepted engineering practice

❖ This section provides cross references to the prescriptive provisions for concrete construction in Seismic Design Categories D_1 and D_2. As an alternative, concrete construction may be designed in accordance with accepted engineering practice.

R301.2.2.4.5 Cold-formed steel framing in Seismic Design Category D_1 and D_2. In Seismic Design Category D_1 and D_2 in addition to the requirements of this code, cold-formed steel framing shall comply with the requirements of COFS/PM.

❖ This section contains the prescriptive requirements for cold-formed steel framing in Seismic Design Category C. In addition to the seismic requirements in this code, the cold-formed steel framing must comply with AISI, *Standard for Cold-Formed Steel Framing—Prescriptive Method for One- and Two-family Dwellings* (COFS/PM).

R301.2.3 Snow loads. Wood framed construction, cold-formed steel framed construction and masonry and concrete construction in regions with ground snow loads 70 psf (3.35 kN/m^2) or less, shall be in accordance with Chapters 5, 6 and 8. Buildings in regions with ground snow loads greater than 70 psf (3.35 kN/m^2) shall be designed in accordance with accepted engineering practice.

❖ This section specifies the maximum ground snow load where the prescriptive provisions in Chapters 5, 6, and 8 apply.

The prescriptive provisions for floors, walls, and roofs in Chapters 5, 6, and 8 apply in regions with ground snow loads of 70 psf (3.35 kN/m^2) or less. Structures in regions with ground snow loads exceeding 70 psf (3.35 kN/m^2) are beyond the limitations of these prescriptive provisions and, as stated in Section R301.1, a design would be required for all elements that carry snow loads. The ground snow load is to be determined by Figure R301.2(5) or by site-specific case studies as discussed in Section 7.0 of ASCE 7.

R301.2.4 Floodplain construction. Buildings and struc-tures constructed in flood hazard areas (including A or V Zones) as established in Table R301.2(1) shall be designed and constructed in accordance with Section R323.

Exception: All buildings and structures in identified floodways as established in Table R301.2(1) shall be designed and constructed as stipulated in the *International Building Code*.

❖ See IBC Section 1612.2 for the definitions of the flood terminology used in this section. The flood hazard areas shown on maps prepared by FEMA are determined using the base flood, which is defined as having

BUILDING PLANNING

a 1-percent chance (one chance in 100) of occurring in any given year. The maps do not show the worst case flood, nor the "flood of record," which usually refers to the most severe flood in the history of the community. Although a 1-percent chance seems fairly remote, larger floods occur regularly throughout the United States. Application of the flood-resistant provisions of the code can not prevent or eliminate all future flood damage. These provisions represent a reasonable balance of the knowledge and awareness of flood hazards, methods to guide development to less hazard-prone locations, methods of design and construction intended to resist flood damage and each community's and landowner's reasonable expectations to use the land.

FEMA prepares flood maps and studies, and they are to be used as a community's official map unless the community chooses to adopt a map that shows more extensive flood hazard areas. From time to time, FEMA's floodplain maps and studies are revised and republished. When this occurs, FEMA involves the community and provides a formal opportunity to review the revised documents. Once the revisions are finalized, FEMA requires specific adoption of the new maps by the community. Many communities minimize the effort of having to adopt each revision separately by referencing the date of the original map or study, and referencing as well any future revisions that may be issued. This is a method by which subsequent revisions to flood maps and studies may be adopted administratively without requiring legislative action on the part of the community.

FEMA uses multiple designations for the flood hazard areas shown on each Flood Insurance Rate Map (FIRM), including AO, AH, A1-30, AE, A99, AR, AR/A1-30, AR/AE, AR/AO, AR/AH, AR/A, VO, V1-30, VE and V. Along many open coasts and lake shores, where wind-driven waves are predicted, the flood hazard area is commonly referred to as the "V Zone." Flood hazard areas that are inland of coastal high-hazard areas subject to high velocity wave action plus flood hazard areas along rivers and streams are commonly referred to as "A Zones." Due to waves, the flood loads in areas subject to high velocity wave action differ from those in other flood hazard areas.

Some coastal communities have FIRMs that show areas that are designated as units of the Coastal Barrier Resource System (CBRS) established by the Coastal Barrier Resource Act (CoBRA) of 1982 and subsequent amendments. Within these designated areas, the National Flood Insurance Program (NFIP) is prohibited from offering flood insurance on new or substantially improved buildings. The community is responsible for application of the provisions of the IRC concerning flood resistance in all designated flood hazard areas, whether or not flood insurance is made available.

An exception states explicitly that buildings in floodways are not to be approved under the IRC and that the provisions of the *International Building Code* apply. Because floodways typically are characterized by deeper and faster moving water, it is appropriate that flood loads be computed for foundation design.

R301.3 Story height. Buildings constructed in accordance with these provisions shall be limited to story heights of not more than the following:

1. For wood wall framing, the laterally unsupported bearing wall stud height permitted by Table R602.3(5) plus a height of floor framing not to exceed sixteen inches.

   Exception: For wood framed wall buildings with bracing in accordance with Table R602.10.1, the wall stud clear height used to determine the maximum permitted story height may be increased to 12 feet without requiring an engineered design for the building wind and seismic force resisting systems provided that the length of bracing required by Table R602.10.1 is increased by multiplying by a factor of 1.20. Wall studs are still subject to the requirements of this section.

2. For steel wall framing, a stud height of 10 feet, plus a height of floor framing not to exceed 16 inches.

3. For masonry walls, a maximum bearing wall clear height of 12 feet plus a height of floor framing not to exceed 16 inches.

   Exception: An additional 8 feet is permitted for gable end walls.

4. For insulating concrete form walls, the maximum bearing wall height per story as permitted by Section 611 tables plus a height of floor framing not to exceed 16 inches.

Individual walls or walls studs shall be permitted to exceed these limits as permitted by Chapter 6 provisions, provided story heights are not exceeded. An engineered design shall be provided for the wall or wall framing members when they exceed the limits of Chapter 6. Where the story height limits are exceeded, an engineered design shall be provided in accordance with the *International Building Code* the overall wind and seismic force resisting systems.

❖ This section defines story height and wall height limits. Story heights are the basis for the wind and seismic force resisting system. Wall heights are based on the materials used for the wall construction. Wall-height limits are prescribed for wood, steel, masonry and concrete. The story height permitted is the wall height plus the height of floor framing not to exceed 16 inches (406 mm). Story height may vary depending upon the material used for the wall construction. Chapter 6 permits wall heights to exceed the limits of this section under certain circumstances. However, if the story heights of this section are exceeded an engineered design in accordance with the *International Building Code* must be provided for the structure for the overall wind and seismic lateral forces.

R301.4 Dead load. The actual weights of materials and construction shall be used for determining dead load with consideration for the dead load of fixed service equipment.

❖ The actual weights of materials of construction must be used to determine the dead loads. The code requires the weight of fixed service equipment (see the definition in Chapter 2) to be included as well. It is important to verify that the actual dead loads of proposed buildings are within the limits of the dead load allowance in the prescriptive design tables of subsequent chapters. Otherwise, in accordance with Section R301.1, a design is required. Also, the dead load limits of Section R301.2.2.2 must be met where required by a structure's seismic design category as explained in the commentary for Section R301.2.2.

The dead load of a building or other structure is the weight of all permanent construction, such as floors, roofs, permanent partitions, stairways, and walls. The actual weights of materials and construction should be considered along with the dead load of fixed service equipment in building design. Commentary Table R301.4 provides the weight of typical residential building components. Sources of information for weights of fixed service equipment include manufacturer's literature and trade association publications.

R301.5 Live load. The minimum uniformly distributed live load shall be as provided in Table R301.5.

❖ Table R301.5 lists the minimum uniformly distributed live loads (see the definition in Chapter 2) required for design of various portions of a residence. These loads are the basis for the prescriptive tables for floor systems in Chapter 5.

R301.6 Roof load. Roof shall be designed for the live load indicated in Table R301.6 or the snow load indicated in Table R301.2(1), whichever is greater.

❖ The basic 20-psf (0.96 kN/m^2) roof live load is a severe enough loading condition on the roof to support live loads created by maintenance workers, including their equipment and materials. The code permits reduction of this basic live load based on the tributary area supported by any structural member of the roof. The rationale for this reduction of roof live load is that it is highly improbable that structural members with large tributary areas would be loaded over the entire area with the full live load. In addition, the reduction of the basic roof live load for roofs is also a function of the slope of the roof, because it becomes less probable that the loads on the roof members would be at maximum levels as the roof slope increases.

The code requires that the roof be designed to resist snow loading and does not permit a reduction of snow loads based on the tributary area, as it does for the roof live load. It is cautioned, however, that snow on the roof rarely accumulates evenly. The design should account for unbalanced snow loading. One case would be the loading of one slope of gable roof with snow while the other slope is unloaded. Many roofs fail from accumulation of snow at valleys, parapets, roof structures, and offsets in roofs. However, the code does not specify criteria for the determination of how these potential accumulations are to be handled. Snow loading provisions in the *International Building Code* may be consulted for these loading scenarios.

R301.7 Deflection. The allowable deflection of any structural member under the live load listed in Sections R301.5 and R301.6 shall not exceed the values in Table R301.7.

❖ The allowable deflection of structural members from the design live load must not exceed the values in Table R301.7. These limits are expressed in terms of the span length. Protection of brittle finishes, such as plaster ceilings and exterior stucco walls, is achieved by limiting the deflection of those elements.

Commentary Figure R301.7 shows a simply supported beam with dead load deflection before the live load has been applied. It also shows the same beam after the live load has been applied. The vertical distance that the center of the beam has moved from the initial dead load deflection position is called the live load deflection. If a finish material is attached to this beam, the finish material is subject to distortion in proportion to this deflection.

R301.8 Nominal sizes. For the purposes of this code, where dimensions of lumber are specified, they shall be deemed to be nominal dimensions unless specifically designated as actual dimensions.

❖ Because sawn lumber sizes are normally referred to using the nominal lumber dimensions, this clarification explains that any code references to lumber dimensions are to be taken as nominal dimensions unless explicitly stated otherwise.

BUILDING PLANNING

TABLE R301.4

TABLE R301.4
MINIMUM DESIGN DEAD LOADS FOR TYPICAL RESIDENTIAL COMPONENTS

COMPONENT	LOAD (psf)
CEILINGS	
Acoustical fiber tile	1
Gypsum board (per $1/8$ inch thickness)	0.55
Mechanical duct allowance	4
Plaster on tile or concrete	5
Plaster on wood lath	8
Suspended steel channel system	2
Suspended metal lath and cement plaster	15
Suspended metal lath and gypsum plaster	10
Wood-furring suspension system	2.5
COVERINGS, ROOF AND WALL	
Asbestos-cement shingles	4
Asphalt shingles	2
Cement tile	16
Clay tile (for mortar add 10 lb.)	
Book tile, 2-inch	12
Book tile, 3-inch	20
Ludowici	10
Roman	12
Spanish	19
Composition:	
Three-ply ready roofing	1
Four-ply felt and gravel	5.5
Five-ply felt and gravel	6
Copper or tin	1
Corrugated asbestos-cement roofing	4
Deck, metal, 20 gage	2.5
Deck, metal, 18 gage	3
Decking, 2-inch wood (Douglas fir)	5
Decking, 3-inch wood (Douglas fir)	8
Fiberboard, 2-inch	0.75
Gypsum sheathing, 2-inch	2
Insulation, roof boards (per inch thickness)	
Cellular glass	0.7
Fibrous glass	1.1
Fiberboard	1.5
Perlite	0.8
Polystyrene foam	0.2
Urethane foam with skin	0.5
Plywood (per $1/8$-inch thickness)	0.4
Rigid insulation, 2-inch	0.75
Skylight, metal frame, $3/8$-inch wired glass	8
Slate, $3/16$-inch	7
Slate, $1/4$-inch	10
Waterproofing membranes	
Bituminous, gravel-covered	5.5
Bituminous, smooth surface	1.5
Liquid applied	1.0
Single-ply, sheet	0.7
Wood sheathing (per 1-inch thickness)	3
Wood shingles	3
FLOORS AND FLOOR FINISHES	
Asphalt block (2-inch), 2-inch mortar	30
Cement finish (1-inch) on stone-concrete fill	32

(continued)

TABLE R301.4 — (continued)

COMPONENT	LOAD (psf)
FLOORS AND FLOOR FINISHES—(continued)	
Ceramic or quarry tile ($3/4$-inch) on 2-inch mortar bed	16
Ceramic or quarry tile ($3/4$-inch) on 1-inch mortar bed	23
Concrete fill finish (per inch thickness)	12
Hardwood flooring, $7/8$-inch	4
Linoleum or asphalt tile, $1/4$-inch	1
Marble and mortar on stone-concrete fill	33
Slate (per inch thickness)	15
Solid flat tile on 1-inch mortar base	23
Subflooring, $3/4$-inch	3
Terrazzo (12-inch) directly on slab	19
Terrazzo (1-inch) on stone-concrete fill	32
Terrazzo (1-inch) 2-inch stone concrete	32
Wood block (3-inch) on mastic, no fill	10
Wood block (3-inch) on 2-inch mortar base	16

FLOORS, WOOD-JOIST (NO PLASTER) DOUBLE WOOD FLOOR

Joint Sizes (inches)	12-in. Spacing (lb./ft.^{2})	16-in. Spacing (lb./ft.^{2})	24-in. Spacing (lb./ft.^{2})
2 × 6	6	5	5
2 × 8	6	6	5
2 × 10	7	6	6
2 × 12	8	7	6

COMPONENT	LOAD (psf)
FRAME PARTITIONS	
Moveable steel partitions	4
Wood or steel studs, 2-inch gypsum board each side	8
Wood studs, 2 × 4, unplastered	4
Wood studs, 2 × 4, plastered one side	12
Wood studs, 2 × 4, plastered two sides	20
FRAME WALLS	
Exterior stud walls:	
2 × 4 @ 16 inches, $5/8$-inch gypsum insulated, $3/8$-inch siding	11
2 × 6 @ 16 inches, $5/8$-inch gypsum insulated, $3/8$-inch siding	12
Exterior stud walls, with brick veneer	48
Windows, glass, frame and sash	8
MASONRY PARTITIONS AND WALLS	
Clay tile:	
4-inch	18
6-inch	24
8-inch	24
Concrete block, heavy aggregate:	
4-inch	30
6-inch	42
8-inch	55
12-inch	85
Concrete block, light aggregate:	
4-inch	20
6-inch	28
8-inch	38
12-inch	55

For SI: 1 psf = 0.0479 kN/m^2, 1 inch = 25.4 mm, 1 pound = 0.4536 kg, 1 psf = 4.882 kg/m^2.

NOTE: Weights of masonry include mortar but not plaster. For plaster add 5 lbs./ft.^{2} for each inch of plaster. Values given represent averages. In some cases there is a considerable range of weight for the same construction.

TABLE R301.5
MINIMUM UNIFORMLY DISTRIBUTED LIVE LOADS
(in pounds per square foot)

USE	LIVE LOAD
Attics with storage[b]	20
Attics without storage[b]	10
Decks[e]	40
Exterior balconies	60
Fire escapes	40
Guardrails and handrails[d]	200
Guardrails in-fill components[f]	50
Passenger vehicle garages[a]	50[a]
Rooms other than sleeping rooms	40
Sleeping rooms	30
Stairs	40[c]

For SI: 1 pound per square foot = 0.0479 kN/m^2, 1 square inch = 645 mm^2, 1 pound = 4.45 N.

a. Elevated garage floors shall be capable of supporting a 2,000-pound load applied over a 20-square-inch area.
b. No storage with roof slope not over 3 units in 12 units.
c. Individual stair treads shall be designed for the uniformly distributed live load or a 300-pound concentrated load acting over an area of 4 square inches, whichever produces the greater stresses.
d. A single concentrated load applied in any direction at any point along the top.
e. See Section R502.2.1 for decks attached to exterior walls.
f. Guard in-fill components (all those except the handrail), balusters and panel fillers shall be designed to withstand a horizontally applied normal load of 50 pounds on an area equal to 1 square foot. This load need not be assumed to act concurrently with any other live load requirement.

❖ The uniform live loads of this table are consistent with Table 1607.1 of the *International Building Code,* which in turn is similar to ASCE 7. The table provides the minimum loads based on the use of a particular area or portion of a structure that must be considered for the design of corresponding structural elements of any residence constructed under the code. For instance, bedrooms (sleeping rooms) require use of a 30-psf (1.48 kN/m^2) live load, while all other rooms must be designed for 40-psf (1.96 kN/m^2) uniform live load.

Attics with storage must be designed for 20-psf (0.96 kN/m^2) live loads, while a 10-psf live load (0.48 kN/m^2) is required without storage. Note "b" explains that storage must be considered when the roof slope is greater than 3:12. With a roof slope of 3:12 or less, there is insufficient clearance or headroom to accumulate significant storage.

TABLE R301.6
MINIMUM ROOF LIVE LOADS IN POUNDS-FORCE PER SQUARE FOOT OF HORIZONTAL PROJECTION

ROOF SLOPE	TRIBUTARY LOADED AREA IN SQUARE FEET FOR ANY STRUCTURAL MEMBER		
	0 to 200	201 to 600	Over 600
Flat or rise less than 4 inches per foot (1:3)	20	16	12
Rise 4 inches per foot (1:3) to less than 12 inches per foot (1:1)	16	14	12
Rise 12 inches per foot (1:1) and greater	12	12	12

For SI: 1 square foot = 0.0929 m^2, 1 pound per square foot = 0.0479 kN/m^2, 1 inch per foot = 0.0833 mm/m.

❖ See the commentary for Section R301.6.

**Figure R301.7
BEAM DEFLECTION**

BUILDING PLANNING

TABLE R301.7
ALLOWABLE DEFLECTION OF STRUCTURAL MEMBERS[a,b,c]

STRUCTURAL MEMBER	ALLOWABLE DEFLECTION
Rafters having slopes greater than 3/12 with no finished ceiling attached to rafters	L/180
Interior walls and partitions	H/180
Floors and plastered ceilings	L/360
All other structural members	L/240
Exterior walls with plaster or stucco finish	H/360
Exterior walls—wind loads[a] with brittle finishes	L/240
Exterior walls—wind loads[a] with flexible finishes	L/120

Note: L = span length, H = span height.

a. The wind load shall be permitted to be taken as 0.7 times the Component and Cladding loads for the purpose of the determining deflection limits herein.
b. For cantilever members, L shall be taken as twice the length of the cantilever.
c. For aluminum structural members or panels used in roofs or walls of sunroom additions or patio covers, not supporting edge of glass or sandwich panels, the total load deflection shall not exceed L/60. For sandwich panels used in roofs or walls of sunroom additions or patio covers, the total load deflection shall not exceed L/120.

❖ Generally, it is not necessary to calculate the deflections indicated in Table R301.7, since the deflection limit is typically accounted for in the prescriptive design tables of the code. Since this is not the case for elements that require design in accordance with Section R301.1.2, the following example illustrates this requirement.

Example:
What is the maximum allowable deflection of a 20'-0" (6096 mm) span for the following conditions?
  a) Rafter with no ceiling load having a slope of 5 units vertical in 12 units horizontal (42% slope)
  b) Rafter with no ceiling load having a slope of 2 units vertical in 12 units horizontal (17% slope)
  c) A floor joist supporting a finished floor

Solution:
Length = Span = 20'-0" (6096 mm)
  a) D MAX = L/180 = 20' x 12"/180 = 1.33" or approx. 1 3/8" (35 mm)
  b) D MAX = L/240 = 20' x 12"/240 = 1.00" or approx. 1" (25.4 mm)
  c) D MAX = L/360 = 20' x 12"/360 = 0.67" or approx. 5/8" (15.9 mm)

SECTION R302
LOCATION ON LOT

R302.1 Exterior walls. Exterior walls with a fire separation distance less than 3 feet (914 mm) shall have not less than a one-hour fire-resistive rating with exposure from both sides. Projections shall not extend to a point closer than 2 feet (610 mm) from the line used to determine the fire separation distance.

Exception: Detached garages accessory to a dwelling located within 2 feet of a lot line shall be permitted to have roof eave projections not exceeding 4 inches.

Projections extending into the fire separation distance shall have not less than one-hour fire-resistive construction on the underside. The above provisions shall not apply to walls which are perpendicular to the line used to determine the fire separation distance.

Exception: Tool and storage sheds, playhouses and similar structures exempted from permits by R105.2 are not required to provide wall protection based on location on the lot. Projections beyond the exterior wall shall not extend over the lot line.

❖ This section provides details for issues related to the building's location on the property, including the fire rating of exterior walls, permitted openings and protected openings.

Concerning exterior wall protection, the IRC assumes that an owner has no control over an adjoining property. Thus, the location of buildings on the owner's property requires regulation relative to the property line.

The property line concept provides a convenient means of protecting one building from another as far as exposure is concerned. Exposure is the potential for heat to be transmitted from one building to another during a fire in the exposing building. Radiation is the primary means of heat transfer.

Section R302.1 begins with a requirement that walls less than 3 feet (914 mm) from the property line be of one-hour fire-resistive construction. The fire-resistive rating also requires the rating exposure to be for both sides. Specific provisions are not found in the IRC for the fire-resistance ratings of walls or other elements, so one may use either the *International Building Code* or another manual of fire-resistant assemblies. Section R302.1 also regulates projections into the area where openings are not permitted. Projections should not extend more than 12 inches (305 mm) into that area. Projections extending into an area where walls are required to be of one-hour fire-resistive construction must be protected on the underside with one-hour fire-resistive construction. See Commentary Figure R302.1(1).

The fire separation distance is measured at right angles to the property line. See the definition of Fire Separation Distance in Chapter 2. The requirements for protection do not apply to walls that are at a right angle (perpendicular to) the property line. See Commentary Figures R302.1(2) and (3).

Tool and storage sheds, playhouses, and similar structures that are exempt from permits are not required to comply with the fire-separation distance and fire-resistance requirement. Projections from these structures, however, are not permitted to extend over the property line.

R302.2 Openings. Openings shall not be permitted in the exterior wall of a dwelling or accessory building with a fire separation distance less than 3 feet (914 mm). This distance shall be measured perpendicular to the line used to determine the fire separation distance.

Exceptions:

1. Openings shall be permitted in walls that are perpendicular to the line used to determine the fire separation distance.
2. Foundation vents installed in compliance with this code are permitted.

❖ Unlike the *International Building Code*, the IRC does not set a distance from the property line at which openings must be protected. Openings are not permitted in exterior walls where the exterior wall has a fire-separation distance of less than 3 feet (714 mm) from the property line. See Commentary Figures R302.1(2) and (3). Openings that are 90 degrees (perpendicular) to the property line are not prohibited. Foundation vents installed in compliance with the IRC are permitted in areas where openings are otherwise prohibited.

R302.3 Penetrations. Penetrations located in the exterior wall of a dwelling with a fire separation distance less than 3 feet (914 mm) shall be protected in accordance with Section R317.3.

Exception: Penetrations shall be permitted in walls that are perpendicular to the line used to determine the fire separation distance.

❖ Section R317.3 describes through penetrations and membrane penetrations in detail. Only penetrations of exterior walls that are located closer than 3 feet (914 mm) to the property line require protection. Penetration protection is not required for walls that are perpendicular to the property line. See Commentary Figures R302.1(2) and (3).

For SI: 1 foot = 304.8 mm.

**Figure R302.1(1)
EXTERIOR WALLS AND PROJECTIONS**

BUILDING PLANNING

FIGURE R302.1(2) - R303.1

**Figure R302.1(2)
EXTERIOR WALL FIRE-RESISTANCE-RATED REQUIREMENTS AND PROHIBITED OPENINGS**

SECTION R303
LIGHT, VENTILATION AND HEATING

R303.1 Habitable rooms. All habitable rooms shall be provided with aggregate glazing area of not less than 8 percent of the floor area of such rooms. Natural ventilation shall be through windows, doors, louvers or other approved openings to the outdoor air. Such openings shall be provided with ready access or shall otherwise be readily controllable by the building occupants. The minimum openable area to the outdoors shall be 4 percent of the floor area being ventilated.

Exceptions:

1. The glazed areas need not be openable where the opening is not required by Section R310 and an approved mechanical ventilation system is provided capable of producing 0.35 air change per hour in the room or a whole-house mechanical ventilation system is installed capable of supplying outdoor ventilation air of 15 cubic feet per minute (cfm) (7.08 L/s) per occupant computed on the basis of two occupants for the first bedroom and one occupant for each additional bedroom.

2. The glazed areas need not be provided in rooms where Exception 1 above is satisfied and artificial light is provided capable of producing an average illumination of 6 footcandles (6.46 lux) over the area of the room at a height of 30 inches (762 mm) above the floor level.

❖ All habitable rooms are to be provided with an aggregate glazing area of 8 percent of the floor area, with a minimal openable area of 4 percent of the floor area being ventilated. The purpose of the glazing and the portion that is openable is to provide both natural light and ventilation. The requirement for adequate light and ventilation dates back to at least 1619 when King James I issued a proclamation known as the Birth of Building Act.

Two exceptions modify the requirement of Section R303.1. Exception 1 permits the glazed area to be fixed or "not openable" when an approved mechanical

Figure R302.1(3)
EXTERIOR WALL FIRE-RESISTANCE REQUIREMENTS AND PROHIBITED OPENINGS

ventilation system or whole-house mechanical ventilation system is installed. The mechanical ventilation system must provide a 35-percent air change per hour in each habitable room. As an example, if a bedroom is 10 feet by 12 feet with an eight foot ceiling height, the air change per hour would be as follows:

10 x 12 x 8 = 960 cubic feet
960 x .35 = 333 cubic feet/hour
or 5.55 cubic feet/minute

The whole-house mechanical ventilation system must provide outdoor ventilation at the rate of 15 cubic feet per minute (7.08 L/s) per occupant, based on two occupants in the first bedroom and one occupant in each additional bedroom. Thus, a three bedroom dwelling unit would require the whole-house ventilation to provide outdoor ventilation at the rate of 60 cfm. This is calculated as follows:

First Bedroom	-	2 Occupants
Second Bedroom	-	1 Occupant
Third Bedroom	-	1 Occupant
Total Occupants	-	4 Occupants

4 x 15 cfm = 60 cfm (28.32 L/s)

Exception 2 to Section R301.1 allows glazing to be completely deleted from habitable rooms where artificial light is available that is capable of producing an average illumination of 6 foot candles (6.46 lux) over the room at a height of 30 inches above the floor level and where a mechanical or whole-house mechanical ventilation system complying with Exception 1 to Section R303.1 is provided.

Calculations should be submitted for review for both exceptions. Various handbooks are available for calculating the required illumination from various lighting manufacturers.

R303.2 Adjoining rooms. For the purpose of determining light and ventilation requirements, any room shall be considered as a portion of an adjoining room when at least one-half of the area of the common wall is open and unobstructed and provides an opening of not less than one-tenth of the floor area of the interior room but not less than 25 square feet (2.32 m^2).

Exception: Openings required for light and/or ventilation shall be permitted to open into a thermally isolated sunroom addition or patio cover, provided that there is an openable area between the adjoining room and the sunroom addition or patio cover of not less than one-tenth of the floor area of the interior room but not less than 20 square feet (1.86 m^2). The minimum openable area to the outdoors shall be based upon the total floor area being ventilated.

❖ Where rooms do not have access to an exterior wall for light and ventilation, they may be considered as part of

BUILDING PLANNING

an adjoining room provided they are open to that room as depicted in Commentary Figure R303.2.

The exception deals with a very common circumstance, especially in residential construction. As long as the sunroom addition is large enough and is thermally isolated, the building owner need not move openings for lighting and/or ventilation when installing an addition that meets the definition of Sunroom Addition.

R303.3 Bathrooms. Bathrooms, water closet compartments and other similar rooms shall be provided with aggregate glazing area in windows of not less than 3 square feet (0.279 m²), one-half of which must be openable.

Exception: The glazed areas shall not be required where artificial light and a mechanical ventilation system are provided. The minimum ventilation rates shall be 50 cfm (23.6 L/s) for intermittent ventilation or 20 cfm (9.4 L/s) for continuous ventilation. Ventilation air from the space shall be exhausted directly to the outside.

❖ Like habitable rooms, bathrooms are required to be provided with natural light and ventilation or adequate artificial light and mechanical ventilation. For natural light, exterior glazing of at least 3 square feet (0.279 m²) must be provided. To meet the natural ventilation requirement, at least 1.5 square feet (0.140 m²) must be openable.

For mechanical ventilation, the ventilation rates must be either 50 cfm (23.6 L/s) for intermittent ventilation or 20 cfm (9.4 L/s) for continuous ventilation. Intermittent ventilation would be controlled by the occupant on an as-needed basis. Continuous ventilation is on at all times. In either case, the ventilation must be to the exterior of the building to the outside atmosphere.

R303.4 Opening location. Outdoor intake and exhaust openings shall be located in accordance with Sections R303.4.1 and R303.4.2.

❖ This section specifies the locations of outdoor intake and exhaust openings. These locations are intended to prevent the introduction of contaminants into the

For SI: 1 square foot = 0.0929 m².

**Figure R303.2
NATURAL LIGHT**

ventilation air of a building or to avoid the exhaust of contaminants into other buildings or areas that may be occupied.

R303.4.1 Intake openings. Mechanical and gravity outdoor air intake openings shall be located a minimum of 10 feet (3048 mm) from any hazardous or noxious contaminant, such as vents, chimneys, plumbing vents, streets, alleys, parking lots and loading docks, except as otherwise specified in this code. Where a source of contaminant is located within 10 feet (3048 mm) of an intake opening, such opening shall be located a minimum of 2 feet (610 mm) below the contaminant source.

For the purpose of this section, the exhaust from dwelling unit toilet rooms, bathrooms and kitchens shall not be considered as hazardous or noxious.

❖ In the context of this section, intake openings include windows, doors, gravity air intakes, soffit vents, combustion air intake openings, outside air intakes for air handlers, makeup air intakes and similar openings that naturally or mechanically draw in air from the building exterior. This section identifies specific locations that are known to generate or emit noxious contaminants and requires both mechanical and gravity air intake openings be located a minimum of 10 feet (3048 mm) from such hazards to avoid introducing contaminants into the building. As an alternative, mechanical and gravity air intakes can be located within 10 feet (3048 mm) of such sources of contamination if the intakes are at an elevation of at least 2 feet (610 mm) below the contaminant source. A 2-foot (610 mm) vertical separation will allow the noxious gases and contaminants to disperse into the atmosphere before they can be drawn into an air intake opening. Placing the source of contamination above an air intake takes advantage of the fact that normally encountered sources of contamination are lighter (less dense) than the surrounding air and therefore will rise above the vicinity of an air intake located below. Commentary Figure R303.4.1 shows an example of the relative locations for intake air openings for a building where sources of contaminants are present.

Particular types of exhausts may have more specific restrictions on their location that would supercede this section (see Section M1602.2, for example).

The air exhausts discharging from a dwelling unit (clothes dryer, kitchen and bathroom) are not considered to be significantly hazardous or noxious and are of low volume. In these situations, the code official must determine an appropriate distance or location for the relative placement of intake and exhaust openings. In evaluating each installation, consideration should be given to the orientation of the exhaust or intake louver and its spatial relationship to any source of contaminant or adjacent intake opening, as well as to the direction of the prevailing winds at the location.

R303.4.2 Exhaust openings. Outside exhaust openings shall be located so as not to create a nuisance. Exhaust air shall not be directed onto walkways.

❖ Mechanical exhaust air openings must not be directed onto walkways or be located where they create a nuisance. A "nuisance" is defined in Chapter 2 of the *International Plumbing Code* as "whatever is dangerous to human life or detrimental to health; whatever structure or premises is not sufficiently ventilated, sewered, drained, cleaned or lighted, with respect to its intended occupancy; and whatever renders the air, or human food, drink or water supply unwholesome." A nuisance is much worse than simply "bothersome." This definition is rather broad and allows the code official to decide what may or may not constitute a nuisance. Unfortunately, it is not an easy task to determine whether or not a nuisance will be present because the conditions under which an exhaust system performs vary considerably with the change of seasons, the ambient temperatures, and the prevailing winds. The code official should gather as much information regarding the installation as possible to evaluate the hypothetical worst-case scenario. This would include the characteristics and geometry of the installation as well as the local ambient conditions, so that an educated guess may be made to determine the "nuisance effect" of the exhaust outlet.

For obvious health reasons, exhaust air cannot be directed onto walkways in such a manner that the users of the walkway are subjected to the exhaust air stream.

R303.5 Outside opening protection. Air exhaust and intake openings that terminate outdoors shall be protected with corrosion-resistant screens, louvers or grilles having a minimum opening size of $^1/_4$ inch (6.4 mm) and a maximum opening size of $^1/_2$ inch (12.7 mm), in any dimension. Openings shall be protected against local weather conditions. Outdoor air exhaust and intake openings shall meet the provisions for exterior wall opening protectives in accordance with this code.

❖ Outside air exhaust and intake openings must be provided with corrosion-resistant screens, grilles, or louvers to prevent foreign objects (such as insects or debris) from entering the system or building. Also, such openings must be protected against the entry of falling or wind-driven water, snow, and ice. Exhaust systems sometimes incorporate rotating hoods over the opening to prevent high winds from restricting the flow of exhaust gases out of the system. The hoods align themselves with the direction of the wind to allow the unimpeded, and actually induced, discharge from the exhaust outlet. Rotating turbines are also employed to serve as both weather protection and as a means of inducing airflow.

The opening sizes for louvers and grilles and the mesh sizes for screens must be within the specified range as indicated. The opening size must be large enough to inhibit blockage by debris to prevent signifi-

BUILDING PLANNING

cant resistance to airflow and yet must be small enough to keep out what is intended to be kept out. A screen of such mesh size would restrict the passage of rodents and large insects and would be resistant to blockage by lint, debris and plant fibers.

R303.6 Stairway illumination. All interior and exterior stairways shall be provided with a means to illuminate the stairs, including the landings and treads. Interior stairways shall be provided with an artificial light source located in the immediate vicinity of each landing of the stairway. For interior stairs the artificial light sources shall be capable of illuminating treads and landings to levels not less than 1 foot-candles (11 lux) measured at the center of treads and landings. Exterior stairways shall be provided with an artificial light source located in the immediate vicinity of the top landing of the stairway. Exterior stairways providing access to a basement from the outside grade level shall be provided with an artificial light source located in the immediate vicinity of the bottom landing of the stairway.

Exception: An artificial light source is not required at the top and bottom landing, provided an artificial light source is located directly over each stairway section.

❖ Stairway illumination for interior and exterior stairs may be accomplished in two ways. The first option is to provide artificial light in the vicinity of each landing. This would include top, intermediate and bottom landings. For interior stairs, the artificial light must be capable of illuminating treads and landings to not less than 1 foot-candle (11 lux). The measurement of 1 foot-candle is to be taken at the center of landings and treads. Exterior stairs require illumination only at the top landing. See Commentary Figure R303.6.

Exterior stairs to a basement must be provided with artificial illumination near the bottom landing.

The exception allows the light source to be provided over each individual stair section, thus eliminating the lighting over the landings.

R303.6.1 Light activation. The control for activation of the required interior stairway lighting shall be accessible at the top

**Figure R303.4.1
AIR INTAKE OPENING LOCATIONS**

and bottom of each stairway without traversing any steps. The illumination of exterior stairways shall be controlled from inside the dwelling unit.

Exception: Lights that are continuously illuminated or automatically controlled.

❖ The location of the light activation is based on the need to illuminate an area before it is used. Interior stairway lighting control is required at both the top and the bottom of the stair unless the lights are continuously illuminated or automatically activated. When manual switches are provided, they must be accessible without the switch operator stepping onto the stair. Exterior stairway light control must be from the interior of the structure.

R303.7 Required glazed openings. Required glazed openings shall open directly onto a street or public alley, or a yard or court located on the same lot as the building.

❖ The placement of glazed openings is to be such that natural light will be available even after future adjacent construction occurs. Where glazing is required by the code, it must open onto a street, public alley, yard or court of the lot on which the building is located. Glazed openings are not permitted in walls that are required to be of one-hour fire-resistant construction in accordance with Section R302.1.

R303.7.1 Roofed porches. Required glazed openings may face into a roofed porch where the porch abuts a street, yard or court and the longer side of the porch is at least 65 percent open and unobstructed and the ceiling height is not less than 7 feet (2134 mm).

❖ The openness of the porch is required so that natural light will be available. Glazed openings required by the code are permitted to open onto a porch provided it abuts a yard, street, or court and the longer side is open at least 65 percent. The porch ceiling height must be at least 7 feet (2134 mm).

R303.8 Required heating. When the winter design temperature in Table R301.2(1) is below 60°F (16°C), every dwelling unit shall be provided with heating facilities capable of maintaining a minimum room temperature of 68°F (20°C) at a point 3 feet (914 mm) above the floor and 2 feet (610 mm) from exterior walls in all habitable rooms at the design temperature. The installation of one or more portable space heaters shall not be used to achieve compliance with this section.

❖ Minimum heating requirements are for health reasons. In areas where the design temperature is based on Table R301.2(1) and is below 60°F (16°C), dwelling units are required to be provided with heating facilities capable of maintaining a minimum room temperature of 68°F (20°C). The primary need here is that of human comfort. The minimum temperature is measured at 3 feet (914 mm) above the floor and 2 feet (610 mm) from the exterior walls of all habitable rooms. See Commentary Figure R303.8. Portable space heaters cannot be used to achieve compliance with this section.

Figure R303.6
STAIRWAY ILLUMINATION

BUILDING PLANNING

FIGURE R303.8 - R305.1

**Figure R303.8
REQUIRED HEATING**

For SI: 1 foot = 304.8 mm, C = [(°F)-32]/1.8

SECTION R304
MINIMUM ROOM AREAS

R304.1 Minimum area. Every dwelling unit shall have at least one habitable room that shall have not less than 120 square feet (11.2 m²) of gross floor area.

❖ The interior living environment is affected by a number of issues. Among these are the size of the room, tightness of construction, ceiling height, number of occupants and ventilation. These all interact and impact the interior living conditions including odors, moisture, and disease transmission. The IRC regulates room sizes to assist in maintaining a safe and comfortable interior environment. See Commentary Figure R304.1. At least one habitable room must be at least 120 square feet (11.2 m²) of gross floor area. Since the definition of habitable space in Section R202 includes rooms used for living, sleeping, eating, or cooking, any one of these rooms can be used to meet the requirement.

R304.2 Other rooms. Other habitable rooms shall have a floor area of not less than 70 square feet (6.5 m²).

Exception: Kitchens.

❖ One habitable room must comply with the provisions of R304.1. The remainder of the habitable rooms except kitchens are required to have a floor area of 70 square feet (6.5 m²). Kitchens are exempt from the minimum floor area requirement.

R304.3 Minimum dimensions. Habitable rooms shall not be less than 7 feet (2134 mm) in any horizontal dimension.

Exception: Kitchens.

❖ Except for kitchens, all habitable rooms are required to have a minimum horizontal dimension in any direction of at least 7 feet (2134 mm).

R304.4 Height effect on room area. Portions of a room with a sloping ceiling measuring less than 5 feet (1524 mm) or a furred ceiling measuring less than 7 feet (2134 mm) from the finished floor to the finished ceiling shall not be considered as contributing to the minimum required habitable area for that room.

❖ In a room with a sloping ceiling, any portion of the room with a vertical ceiling height of less than 5 feet (1524 mm) from the finished floor does not provide the minimum required floor area for habitation purposes. Likewise, the area under a furred ceiling with a vertical height of less than 7 feet (2134 mm) from the finished floor is not part of the habitable area. An example of the first case would be an A-frame structure, which consists of a sloping roof and no or minimal exterior walls. This condition could also exist in any room that has a sloping ceiling. The low height makes those portions of the room generally unusable for adults. See Commentary Figure R304.4.

SECTION R305
CEILING HEIGHT

R305.1 Minimum height. Habitable rooms, hallways, corridors, bathrooms, toilet rooms, laundry rooms and basements shall have a ceiling height of not less than 7 feet (2134 mm). The required height shall be measured from the finish floor to the lowest projection from the ceiling.

Exceptions:

1. Beams and girders spaced not less than 4 feet (1219 mm) on center may project not more than 6 inches (152 mm) below the required ceiling height.

2. Ceilings in basements without habitable spaces may project to within 6 feet, 8 inches (2032 mm) of the fin-

ished floor; and beams, girders, ducts or other obstructions may project to within 6 feet, 4 inches (1931 mm) of the finished floor.

3. Not more than 50 percent of the required floor area of a room or space is permitted to have a sloped ceiling less than 7 feet (2134 mm) in height with no portion of the required floor area less than 5 feet (1524 mm) in height.

For SI: 1 foot = 304.8 mm. 1 square foot = 0.0929 m^2.

**Figure R304.1
FLOOR AREA**

THE CROSS HATCHED PORTIONS OF EACH ROOM WOULD NOT BE CONSIDERED IN DETERMINING THE MINIMUM REQUIRED FLOOR AREA.

For SI: 1 inch = 25.4 mm, 1 foot = 304.8 mm.

**Figure R304.4
HEIGHT EFFECT ON ROOM AREA**

BUILDING PLANNING

4. Bathrooms shall have a minimum ceiling height of 6 feet 8 inches (2036 mm) over the fixture and at the front clearance area for fixtures as shown in Figure R307.2. A shower or tub equipped with a showerhead shall have a minimum ceiling height of 6 feet 8 inches (2036 mm) above a minimum area 30 inches (762 mm) by 30 inches (762 mm) at the showerhead.

❖ Minimum ceiling heights are required for habitable rooms, hallways, corridors, bathrooms, toilet rooms, laundry rooms and basements. The minimum required height of 7 feet (2134 mm) helps maintain a healthy interior environment. The dimension must be measured to the lowest projection of the ceiling.

For ceilings having exposed beams that project down from the ceiling surface, the ceiling beam members may project no more than 6 inches (152 mm) below the required ceiling height, provided the beams or girders are opened at no less than 4 feet (1219 mm) on center. Basements need have a ceiling height of only 6 feet, 8 inches (2033 mm) or more, with at least 6 feet, 4 inches (1932 mm) of clear height under beams, girders, ducts and similar obstructions.

For rooms with sloped ceilings, the code requires only that the prescribed ceiling height be maintained in one-half the area of the room. However, no portion of the room that has a ceiling height of less than 5 feet (1524 mm) must be used in the computations for minimum floor area.

Exception 4 defines the required minimum ceiling height over toilet, bath and shower fixtures. This exception allows a sloping ceiling over toilet, bath or shower fixtures provided the minimum ceiling height of 6 feet, 8 inches (2036 mm) is maintained over the fixture and at the front clearance area.

SECTION R306
SANITATION

R306.1 Toilet facilities. Every dwelling unit shall be provided with a water closet, lavatory, and a bathtub or shower.

❖ Dwelling units must be provided with at least one each of the fixtures indicated in the code, and the fixtures must be connected to an approved sanitary sewer or private sewage disposal system. A water closet, lavatory and bathtub or shower are the minimum fixtures needed to maintain the occupant's health and cleanliness.

R306.2 Kitchen. Each dwelling unit shall be provided with a kitchen area and every kitchen area shall be provided with a sink.

❖ Dwelling units must have a kitchen area with a sink. This area is for the basic preparation of food.

R306.3 Sewage disposal. All plumbing fixtures shall be connected to a sanitary sewer or to an approved private sewage disposal system.

❖ In order to maintain sanitary conditions in the dwelling, all plumbing fixtures must be attached to either a sanitary sewer or approved private sewage disposal system.

R306.4 Water supply to fixtures. All plumbing fixtures shall be connected to an approved water supply. Kitchen sinks, lavatories, bathtubs, showers, bidets, laundry tubs and washing machine outlets shall be provided with hot and cold water.

❖ In order to provide proper sanitation for occupants of dwelling units, each plumbing fixture must be connected to an approved water supply. Additionally, specific fixtures must be provided with both a hot and cold water supply.

SECTION R307
TOILET, BATH AND SHOWER SPACES

R307.1 Space required. Fixtures shall be spaced as per Figure R307.2.

❖ Fixtures require certain clearances to be accessible and usable. Figure R307.2 depicts the minimum fixture clearances.

R307.2 Bathtub and shower spaces. Bathtub and shower floors and walls above bathtubs with installed shower heads and in shower compartments shall be finished with a nonabsorbent surface. Such wall surfaces shall extend to a height of not less than 6 feet (1829 mm) above the floor.

❖ Wall surfaces subject to water spray by showerheads must be protected with a nonabsorbent surface to a height of at least 6 feet (1829 mm) above the floor of a bathtub or shower.

SECTION R308
GLAZING

[B] R308.1 Identification. Except as indicated in Section R308.1.1, each pane of glazing installed in hazardous locations as defined in Section R308.4 shall be provided with a manufacturer's or installer's label, designating the type and thickness of glass and the safety glazing standard with which it complies, which is visible in the final installation. The label shall be acid etched, sandblasted, ceramic-fired, embossed mark, or shall be of a type which once applied cannot be removed without being destroyed.

Exceptions:

1. For other than tempered glass, labels may be omitted provided the building official approves the use of a certificate, affidavit or other evidence confirming compliance with this code.

2. Tempered spandrel glass may be identified by the manufacturer with a removable paper label.

❖ Once glass is installed in a window frame by a manufacturer, it is not easily determined whether a particular piece of glass is safety glazing or not. In theory, this can only be established by breaking the particular piece of glass, in which case the glass is no longer useful. Thus, the code requires that safety glazing be labeled so that the label is visible during the final building inspection. Except for tempered glass labels, labels may be omitted where approved by the building official and an affidavit, certificate or other evidence is submitted indicating compliance with the code. A manufacturer can identify safety glazing with a removable paper label, provided the label is destroyed during its removal. This ensures that the label will not be applied to a non-complying piece of glass.

R308.1.1 Identification of multipane assemblies. Multipane assemblies having individual panes not exceeding 1 square foot (0.09 m^2) in exposed area shall have at least one pane in the assembly identified in accordance with Section R308.1. All other panes in the assembly shall be labeled "16 CFR 1201."

❖ Multipane assemblies of glass need identification for the same reasons noted in Section R308.1. This provision allows labeling of only one pane of glass per Section R308.1.1, when the exposed area of each pane is 1 square foot (0.09 m^2) or less. All other panes must be labeled "16 CFR 1201."

R308.2 Louvered windows or jalousies. Regular, float, wired or patterned glass in jalousies and louvered windows shall be no thinner than nominal $^3/_{16}$ inch (4.76 mm) and no longer than 48 inches (1219 mm). Exposed glass edges shall be smooth.

❖ The requirements for louvered windows exist because there is no edge support on the longitudinal edges of these panes. The code requires that the exposed edges be smooth for safety. The minimum thickness and maximum span are specified so that the glass has sufficient resistance to human impact loads.

For SI: 1 inch = 25.4 mm.

**FIGURE R307.2
MINIMUM FIXTURE CLEARANCES**

BUILDING PLANNING

R308.2.1 Wired glass prohibited. Wired glass with wire exposed on longitudinal edges shall not be used in jalousies or louvered windows.

❖ Wired glass is not permitted if the wire is exposed on the longitudinal edge, since it would be a hazard.

[B] R308.3 Human impact loads. Individual glazed areas includ-ing glass mirrors in hazardous locations such as those indicated as defined in Section R308.4 shall pass the test requirements of CPSC 16 CFR, Part 1201. Glazing shall comply with the CPSC 16 CFR, Part 1201 criteria for Category I or Category II as indicated in Table R308.3.

Exceptions:

1. Polished wired glass for use in fire doors and other fire resistant locations shall comply with ANSI Z97.1.
2. Louvered windows and jalousies shall comply with Section R308.2.

❖ The code requires that glazing in hazardous locations subject to human impact pass the test requirements of CPSC 16 CFR, Part 1201. This is the standard developed by the Consumer Product Safety Commission (CPSC) when it was determined that former standards (such as ANSI Z97.1) were inadequate to properly protect the public from injury due to accidental impact with glazing. Exception 1 of the provision exempts the installation of polished wire glass in fire doors or as fire windows and view panels in fire resistant walls, if it complies with ANSI Z97.1. The problem is that wired glass cannot pass the test requirements of the CPSC standard, but its use is necessary in fire assemblies. Except for wired glass, other glazing material used in fire assemblies installed in hazardous locations must comply with CPSC 16 CFR 1201. Additional exceptions are provided for louvered windows and jalousies complying with Section R308.2. See the categories listed in Table 308.3.

[B] R308.4 Hazardous locations. The following shall be considered specific hazardous locations for the purposes of glazing:

1. Glazing in swinging doors except jalousies.
2. Glazing in fixed and sliding panels of sliding door assemblies and panels in sliding and bifold closet door assemblies.
3. Glazing in storm doors.
4. Glazing in all unframed swinging doors.
5. Glazing in doors and enclosures for hot tubs, whirlpools, saunas, steam rooms, bathtubs and showers. Glazing in any part of a building wall enclosing these compartments where the bottom exposed edge of the glazing is less than 60 inches (1524 mm) measured vertically above any standing or walking surface.
6. Glazing, in an individual fixed or operable panel adjacent to a door where the nearest vertical edge is within a 24-inch (610 mm) arc of the door in a closed position and whose bottom edge is less than 60 inches (1524 mm) above the floor or walking surface.
7. Glazing in an individual fixed or operable panel, other than those locations described in Items 5 and 6 above, that meets all of the following conditions:
    7.1. Exposed area of an individual pane greater than 9 square feet (0.836 m^2).
    7.2. Bottom edge less than 18 inches (457 mm) above the floor.
    7.3. Top edge greater than 36 inches (914 mm) above the floor.
    7.4. One or more walking surfaces within 36 inches (914 mm) horizontally of the glazing.
8. All glazing in railings regardless of an area or height above a walking surface. Included are structural baluster panels and nonstructural in-fill panels.
9. Glazing in walls and fences enclosing indoor and outdoor swimming pools, hot tubs and spas where the bottom edge of the glazing is less than 60 inches (1524 mm) above a walking surface and within 60 inches (1524 mm) horizontally of the water's edge. This shall apply to single glazing and all panes in multiple glazing.
10. Glazing adjacent to stairways, landings and ramps within 36 inches (914 mm) horizontally of a walking surface when the exposed surface of the glass is less than 60 inches (1524 mm) above the plane of the adjacent walking surface.

TABLE R308.3
MINIMUM CATEGORY CLASSIFICATION OF GLAZING

EXPOSED SURFACE AREA OF ONE SIDE OF ONE LITE	GLAZING IN STORM OR COMBINATION DOORS (Category Class)	GLAZING IN DOORS (Category Class)	GLAZED PANELS REGULATED BY ITEM 7 OF SECTION R308.4 (Category Class)	GLAZED PANELS REGULATED BY ITEM 6 OF SECTION R308.4 (Category Class)	GLAZING IN DOORS AND ENCLOSURES REGULATED BY ITEM 5 OF SECTION R308.4 (Category Class)	SLIDING GLASS DOORS PATIO TYPE (Category Class)
9 sq. ft. or less	I	I	NR[a]	I	II	II
More than 9 sq. ft	II	II	II	II	II	II

For SI: 1 square foot = 0.0929 m^2.
[a] NR means "No Requirement."

❖ See the commentary for Section R308.3.

11. Glazing adjacent to stairways within 60 inches (1524 mm) horizontally of the bottom tread of a stairway in any direction when the exposed surface of the glass is less than 60 inches (1524 mm) above the nose of the tread.

Exception: The following products, materials and uses are exempt from the above hazardous locations:

1. Openings in doors through which a 3-inch (76 mm) sphere is unable to pass.
2. Decorative glass in Items 1, 6 or 7.
3. Glazing in Section R308.4, Item 6, when there is an intervening wall or other permanent barrier between the door and the glazing.
4. Glazing in Section R308.4, Item 6, in walls perpendicular to the plane of the door in a closed position or where access through the door is to a closet or storage area 3 feet (914 mm) or less in depth. Glazing in these applications shall comply with Section R308.4, Item 7.
5. Glazing in Section R308.4, Items 7 and 10, when a protective bar is installed on the accessible side(s) of the glazing 36 inches ± 2 inches (914 mm ± 51 mm) above the floor. The bar shall be capable of withstanding a horizontal load of 50 pounds per linear foot (74.5 kg/m) without contacting the glass and be a minimum of $1^1/_2$ inches (38 mm) in height.
6. Outboard panes in insulating glass units and other multiple glazed panels in Section R308.4, Item 7, when the bottom edge of the glass is 25 feet (7620 mm) or more above grade, a roof, walking surface, or other horizontal [within 45 degrees (0.79 rad) of horizontal] surface adjacent to the glass exterior.
7. Louvered windows and jalousies complying with the requirements of Section R308.2.
8. Mirrors and other glass panels mounted or hung on a surface that provides a continuous backing support.
9. Safety glazing in Section R308.4, Items 10 and 11 is not required where:
   9.1. The side of a stairway, landing or ramp has a guardrail or handrail, including balusters or in-fill panels, complying with the provisions of Sections 1012 and 1607.7 of the *International Building Code;* and
   9.2. The plane of the glass is greater than 18 inches (457 mm) from the railing.

❖ This section lists ten specific hazardous locations where safety glazing is required. Some of these locations are depicted in Commentary Figures R308.4(1) through R308.4(9). In addition to the hazardous locations shown in the nine drawings, safety glazing is also required in a number of other locations, including fixed and sliding panels of sliding door assemblies, storm doors and glass railings.

Commentary Figure R308.4(1) illustrates several locations where safety glazing may or may not be required. To facilitate discussion, each glazed panel has been numbered. Panel 1 is not required to have a safety glazing because a protective bar has been installed in compliance with Exception 1 to item 7, the details of which are illustrated in Commentary Figure R308.4(2). Panels 4 and 7 require safety glazing because they are door sidelights. The exception mentioned above does not apply to panels adjacent to a door, so even though Panel 7 has been provided with a protective bar, safety glazing is still required. Commentary Figures R308.4(3) and (4) illustrate where safety glazing is required for panels adjacent to a door. This requirement applies to both fixed and operable panels. Where there is an intervening wall or permanent barrier, as shown in Commentary Figure R308.4(5), safety glazing would not be required. Commentary Figure R308.4(6) illustrates an exception found in Section 2406 of the *International Building Code* applicable to one- and two-family dwellings and within dwelling units in Group R, Division 2 uses. This exception in the IBC could be used in structures regulated by the IRC by applying the provisions of Section R104.11.

Panels 8 and 9, as well as Panels 2 and 3, fall under item 6 of Section R308.4. Under this item all four stated conditions must occur before safety glazing is required. These conditions are as follows:

6.1. The area of an individual pane must be greater than 9 square feet (0.84 m^2);

6.2. The bottom edge must be less than 18 inches (457 mm) above the floor:

6.3. The top edge must be more than 36 inches (914 mm) above the floor; and

6.4. One or more walking surfaces must be within 36 inches (914 mm), measured horizontally from the glazed panel.

Panels 2 and 3 do not require safety glazing because their bottom edges are not less than 18 inches (457 mm) from the floor.

If Panels 8 and 9 have a walking surface within 36 inches (914 mm) of the interior, safety glazing would be required. From the exterior side, as shown in Commentary Figure R308.4(1), the bottom of the panel appears to be greater than 18 inches (457 mm) above the exterior walking surface, so the exterior condition would have no bearing on the determination. Panels 5 and 6 are glass doors, which require safety glazing based on the provisions of Items 1 or 4. Most ingress and egress doors (except jalousies), unframed swinging doors, and glazing in storm doors require safety glazing, but there are several exceptions. If openings in a door will not pass a 3-inch diameter (76 mm) sphere, the glazing is exempt, as are assemblies of leaded, faceted or carved glass used for decorative purposes. The latter exception not only applies to doors but also to sidelights and other glazed panels covered by items 5 and 6.

BUILDING PLANNING

FIGURE R308.4(1) - FIGURE R308.4(2)

**Figure R308.4(1)
HAZARDOUS LOCATIONS**

**Figure R308.4(2)
PROTECTIVE BAR ALTERNATIVE**

**Figure R308.4(3)
GLASS IN SIDELIGHTS—ELEVATION**

**Figure R308.4(4)
GLASS IN SIDELIGHTS—PLAN**

Commentary Figure R308.4(7) illustrates the condition where a window occurs within a shower enclosure. If this window is less than 60 inches (1524 mm) above a standing surface, then safety glazing would be required. This same requirement applies not only to showers but also to windows installed adjacent to hot tubs, whirlpools, saunas, steam rooms and bathtubs. Because of the presence of moisture, all of these locations represent slip hazards and need safety glazing to prevent injury in case of a fall.

Glass in railings, balusters panels and in-fill panels, regardless of their height above a walking surface, require safety glazing. Due to the high probability that people will impact guards, it is critical that an increased level of protection be provided.

Commentary Figure R308.4(8) illustrates the requirements of item 9. This provision applies to walls and fences used as barriers for indoor or outdoor swimming pools and spas. Before safety glazing is required, the glazed panels must be within 5 feet (1524 mm) above the decking for a pool or spa. Items 10 and 11 address the hazardous locations to be considered for stairways, landing and ramps. Stairways and ramps present users with a greater risk for injury caused by falling than a flat surface. Not only is the risk of falling greater when using a stair, but the injuries are generally more severe. Unlike falling on a flat surface where the floor will, for the most part, break a person's fall, there is nothing to stop someone from continuing to fall until he or she reaches the bottom of the stair. The increased risks inherent in stairways, as well as attempting to be consistent with other chapters in the code that mandate more restrictive requirements when addressing safety issues involving stairs and ramps, account for the more restrictive requirements for glazing in and around stairways and ramps. Item 10 includes any glazing within 36 inches (914 mm) horizontally of any walking surface when the exposed surface of that glazing is within 60 inches (1524 mm) of the walking surface. The walking surface in question would be part of a stair or ramp itself, including top, bottom and intermediate landings. It does not include adjacent floors or other walking surfaces.

In item 11, the concern is any glass that may be located within 60 inches (1524 mm) from the bottom tread in a run of stairs and within 60 inches (1524 mm) vertically of the walking surface of a stair. The code does not distinguish between a bottom tread at the primary floor level or at an intermediate landing. The last tread in a run of steps is the bottom tread. The 60-inch (1524 mm) dimension is from any point on the bottom tread, horizontally in any direction to any surface of any glazing within that range. Commentary Figure R308.4(9) illustrates the applications of item 11. Safety glazing is not required for items 10 and 11 where protected by a railing or guard at least 18 inches (457 mm) from the glass.

[B] R308.5 Site built windows. Site built windows shall comply with Section 2404 of the *International Building Code*.

❖ Since site built windows are not constructed in a manufacturing facility that follows industry standards, they must be constructed in accordance with Section 2404 of the *International Building Code*, which sets forth the wind, snow and dead loads on glass.

[B] R308.6 Skylights and sloped glazing. Skylights and sloped glazing shall comply with the following sections.

❖ Sloped glazing and skylights consist of glazing installed in roofs or walls that are on a slope of more than 15 degrees from the vertical. The provisions of the IRC address loads normally attributed to roofs. The provisions also enhance the protection of the occupants of a building from the possibility of falling glazing materials.

Figure R308.4(5)
BARRIER BETWEEN GLAZING AND DOOR

Figure R308.4(6)
GLASS IN SIDELIGHTS—DWELLINGS

Figure R308.4(7)
GLAZING WITHIN A SHOWER ENCLOSURE

R308.6.1 Definitions.

❖ This section contains definitions of terms that are associated with the subject matter of this section. These terms are not exclusively related to this section but are applicable everywhere the term is used in the code.

SKYLIGHTS AND SLOPED GLAZING. Glass or other transparent or translucent glazing material installed at a slope of more than 15 degrees (0.26 rad) from vertical. Glazing materials in skylights, including unit skylights, solariums, sunrooms, roofs and sloped walls are included in this definition.

❖ The failure of skylights and sloped glazing could result in injury and building damage. This definition establishes the criterion to which the requirements of Section R308.6 apply.

UNIT SKYLIGHT. A factory assembled, glazed fenestration unit, containing one panel of glazing material, that allows for natural daylighting through an opening in the roof assembly while preserving the weather resistant barrier of the roof.

❖ Unit skylights are a specific type of factory-assembled sloped glazing assembly. The *International Building Code* and the *International Residential Code* contain specific provisions that are appropriate for this type of building component. Factory-assembled units, as opposed to site-built skylights, can be designed, tested and rated as one component that incorporates both glazing and framing, if applicable. The individual components of site-built glazing must be designed to resist the design loads of the codes individually and are not usually rated as an assembly.

R308.6.2 Permitted materials. The following types of glazing may be used:

1. Laminated glass with a minimum 0.015-inch (0.38 mm) polyvinyl butyral interlayer for glass panes 16 square feet (1.5 m^2) or less in area located such that the highest point of the glass is not more than 12 feet (3658 mm) above a walking surface or other accessible area; for higher or larger sizes, the minimum interlayer thickness shall be 0.030 inch (0.76 mm).
2. Fully tempered glass.
3. Heat-strengthened glass.
4. Wired glass.
5. Approved rigid plastics.

❖ The provisions of this section limit glazing materials in skylights and sloped glazing to those specified, and they outline glazing materials and protective measures for sloped glazing and skylights. The materials and their characteristics and limitations are as follows:

Laminated glass. Laminated glass is usually furnished with an inner layer of polyvinyl butyral, which has a minimum thickness of 30 mil (0.76 mm). Such glass is highly resistant to impact and as a result requires no further protection below. When used within

BUILDING PLANNING

dwelling units, laminated glass is permitted to have a 15-mil (0.38 mm) polyvinyl butyral inner layer, provided each pane of glass is 16 square feet (1.5 m²) or less in area, and the highest point of the glass is no more than 12 feet (3658 mm) above a walking surface or other accessible area.

Fully tempered glass. Tempered glass is glass that has been specifically heat-treated or chemically

Figure R308.4(8)
GLAZING ADJACENT TO SWIMMING POOL OR SPA

Figure R308.4(9)
GLAZING ADJACENT TO STAIRWAYS

treated to provide high strength. When broken, the entire piece of glass immediately breaks into numerous small granular pieces. Because of its high strength and manner of breakage, tempered glass has been considered in the past to be a desirable glazing material for skylights that have no protective screens. However, as a result of studies by the industry that show that tempered glass is subject to spontaneous breakage such that large chunks of glass may fall under this condition, the IRC requires screen protection below tempered glass.

Heat-strengthened glass. Heat-strengthened glass is glass that has been reheated to just below its melting point and then cooled. This process forms a compression on the outer surface and increases the strength of the glass. However, heat-strengthened glass requires screen protection below the skylight to protect the occupants from falling shards.

Wired glass. Wired glass is resistant to impact and when used as a single-layer glazing requires no additional protection below.

Approved rigid plastics. Rigid plastics are fairly durable as a glazing material.

Annealed glass. Annealed glass is not allowed since it is subject to breakage by impact and has very low strength. Annealed glass is also unsatisfactory for use as a skylight since it breaks up under impact into large sharp shards, which, when they fall, are hazardous to occupants of a building.

R308.6.3 Screens, general. For fully tempered or heat-strengthened glass, a retaining screen meeting the requirements of Section R308.6.7 shall be installed below the glass, except for fully tempered glass that meets either condition listed in Section R308.6.5.

❖ As a general rule, single-layer glazing of heat-strengthened glass and fully tempered glass must be provided with screens below the glazing material.

R308.6.4 Screens with multiple glazing. When the inboard pane is fully tempered, heat-strengthened, or wired glass, a retaining screen meeting the requirements of Section R308.6.7 shall be installed below the glass, except for either condition listed in Section R308.6.5. All other panes in the multiple glazing may be of any type listed in Section R308.6.2.

❖ As does Section R308.6.3, this section states that screens are required for the inbound plane of glazing when such glazing is fully tempered, heat-strengthened, or wired glass. The screen must comply with Section R308.6.7. Screens are not required for either approved laminated glass or approved rigid plastics.

R308.6.5 Screens not required. Screens shall not be required when fully tempered glass is used as single glazing or the inboard pane in multiple glazing and either of the following conditions are met:

1. Glass area 16 square feet (1.49 m^2) or less. Highest point of glass not more than 12 feet (3658 mm) above a walking surface or other accessible area, nominal glass thickness not more than $^3/_{16}$ inch (4.76 mm), and (for multiple glazing only) the other pane or panes fully tempered, laminated or wired glass.
2. Glass area greater than 16 square feet (1.49 m^2). Glass sloped 30 degrees (0.52 rad) or less from vertical, and highest point of glass not more than 10 feet (3048 mm) above a walking surface or other accessible area.

❖ Section R308.6.5 provides two exceptions to the provisions of Section R308.6.3 and Section R308.6.4. The first exception applies to glazing that is no greater in area than 16 square feet (1.49 m^2) and is no more than 12 feet (3658 mm) above a walking surface or other accessible area. The second exception applies to sloped glazing with a maximum slope of 30 degrees (0.52 rad) from vertical, is not greater than 16 square feet (1.49 m^2) in area, and is no greater than 10 feet (3048 mm) above a walking surface or other accessible area. Generally, installed skylights and sloped glazing will meet one of these exceptions, so screens are not required.

R308.6.6 Glass in greenhouses. Any glazing material is permitted to be installed without screening in the sloped areas of greenhouses, provided the greenhouse height at the ridge does not exceed 20 feet (6096 mm) above grade.

❖ The glazing regulations for greenhouses are less stringent since greenhouses are seldom occupied during storms that might break the glass. These provisions also explain an exception to the screening provisions of Section R308.6 and R308.6.4, specifically for the sloped glazing areas within greenhouses. Screens are not required for sloped areas of greenhouses provided the ridge of the greenhouse is not more than 20 feet (6096 mm) above grade.

R308.6.7 Screen characteristics. The screen and its fastenings shall be capable of supporting twice the weight of the glazing, be firmly and substantially fastened to the framing members, and have a mesh opening of no more than 1 inch by 1 inch (25.4 mm by 25.4 mm).

❖ It is critical that screens are installed in a manner that will adequately support the weight of the glass. In utilizing a safety factor of two, the screen and its fastenings must be capable of supporting twice the weight of the glazing. In order to accomplish this, the screen is to be fastened firmly to the framing members.

R308.6.8 Curbs for skylights. All unit skylights installed in a roof with a pitch flatter than three units vertical in 12 units horizontal (25-percent slope) shall be mounted on a curb extending at least 4 inches (102 mm) above the plane of the roof unless otherwise specified in the manufacturer's installation instructions.

❖ Skylights installed on low-sloped roofs are more susceptible to leaking than those on higher pitched roofs,

BUILDING PLANNING

because water does not drain as quickly on low-sloped roofs.

In order that these skylights be properly flashed to prevent leakage, they must be placed on a 4-inch high curb unless the manufacturers' installation instructions indicate otherwise.

R308.6.9 Testing and labeling. Unit skylights shall be tested by an approved independent laboratory, and bear a label identifying manufacturer, performance grade rating, and approved inspection agency to indicate compliance with the requirements of AAMA/WDMA 101/I.S.2/NAFS.

❖ The referenced standard, AAMA/WDMA 101/I.S.2/NAFS: *Voluntary Performance Specification for Windows, Skylights and Glass Doors*, includes a separate rating system for positive and negative pressure on skylights that allows the manufacturer to design and fabricate products that are best suited for the climate in which they will be used. 101/I.S.2/NAFS establishes the performance requirements for skylights based on the desired performance grade rating, which includes minimum requirements for resistance to air leakage, water infiltration and the design load pressures. The resulting performance grade rating states the design load pressure used to rate the product, but it also includes consideration of these additional performance characteristics. For skylights certified for only one performance grade, the rating is based on the minimum requirements met for both positive and negative design pressure. Skylights certified for two performance grades are rated separately for positive and negative design pressure.

Skylights must be capable of withstanding the component and cladding wind pressures of Table R301.2(2) adjusted by the height and exposure coefficients given in Table R301.2(3).

The most critical load on a skylight is determined by the climate in which it is installed. In a colder climate with heavier snow loads and moderate design wind speeds, the positive load on a skylight from the combined snow and dead load will be more critical than the negative load from wind uplift. The opposite will be the case in warmer, coastal climates with higher design wind speeds and little or no snow load.

SECTION R309
GARAGES AND CARPORTS

R309.1 Opening protection. Openings from a private garage directly into a room used for sleeping purposes shall not be permitted. Other openings between the garage and residence shall be equipped with solid wood doors not less than $1^3/_8$ inches (35 mm) in thickness, solid or honeycomb core steel doors not less than $1^3/_8$ inches (35 mm) thick, or 20-minute fire-rated doors.

❖ Openings to sleeping rooms from garages are not allowed, since a person might not wake up in time if there was a hazard from carbon monoxide fumes or smoke from the garage. Openings from the garage are only permitted into rooms that are not used for sleeping purposes. Those openings must be protected by the installation of a door complying with the provisions of this section. $1^3/_8$-inch (35 mm) thick solid wood doors, solid or honeycomb steel doors and 20-minute fire rated doors are the only acceptable doors for use in the opening between the garage and the dwelling unit.

R309.1.1 Duct penetration. Ducts in the garage and ducts penetrating the walls or ceilings separating the dwelling from the garage shall be constructed of a minimum No. 26 gage (0.48 mm) sheet steel or other approved material and shall have no openings into the garage.

❖ Ducts are permitted to penetrate the required separation (see Section R309.2) between the garage and the dwelling unit when the ducts within the garage and the portion of the duct penetrating the wall are of No. 26 gage (0.48 mm) sheet steel or other materials acceptable to the building official. Steel ducts are required to help prevent the passage of an undetected fire within the garage to the dwelling unit. See Commentary Figure R309.1.1. The opening limitation in the garage is to limit the path for smoke to enter the dwelling unit.

R309.2 Separation required. The garage shall be separated from the residence and its attic area by not less than $1/_2$-inch (12.7 mm) gypsum board applied to the garage side. Garages beneath habitable rooms shall be separated from all habitable rooms above by not less than $5/_8$-inch (15.9 mm) Type X gypsum board or equivalent. Where the separation is a floor-ceiling assembly, the structure supporting the separation shall also be protected by not less than $1/_2$-inch (12.7 mm) gypsum board or equivalent.

❖ Numerous potential hazards exist within garages because occupants of dwelling units tend to store a variety of hazardous materials there. Along with this and the potential for carbon monoxide build-up within the garage, the IRC requires that the garage be separated from the dwelling unit and the attic with at least $1/_2$-inch (12.7 mm) gypsum board or other equivalent material. If a habitable room is above the garage, the separation must be at least $5/_8$-inch (15.9 mm) Type X gypsum board or equivalent.

There are two primary reasons for providing an enhanced fire endurance for a garage ceiling located beneath a habitable room. First, a fire occurring in a garage may well go undetected for an extended period prior to activation of a detector or other visual alerting.

Second, the inherent fire load and hazardous household activities associated with a garage necessitate this additional level of protection if fire suppression forces are to have a reasonable opportunity to contain a garage fire to the area of origin.

The single layer of $5/_8$-inch Type X gypsum at the garage ceiling increases the fire endurance of the assembly considerably, from 15 minutes for a $1/_2$-inch layer, to at least 40 minutes, or a 167-percent increase in endurance. When added to the rating for floor joists and certain subflooring combinations, the final endurance is close to 1 hour.

Additionally, the exterior walls of the garage are required to have $1/2$-inch gypsum board on the interior face where they support floors separating all or part of a dwelling unit above the garage. Commentary Figure R309.2 shows two locations of gypsum wallboard; each achieves the protection required by the code.

R309.3 Floor surface. Garage floor surfaces shall be of approved noncombustible material.

The area of floor used for parking of automobiles or other vehicles shall be sloped to facilitate the movement of liquids to a drain or toward the main vehicle entry doorway.

❖ Garage floor surfaces must be of an approved noncombustible material such as concrete. Additionally, the floor surface must either slope toward the garage

Figure R309.1.1
GARAGE/DWELLING SEPARATION

BUILDING PLANNING

door opening or slope to an approved drain. This allows grease, flammable liquids, or other hazardous materials that might drain from an automobile to be drained from the garage.

R309.4 Carports. Carports shall be open on at least two sides. Carport floor surfaces shall be of approved noncombustible material. Carports not open on at least two sides shall be considered a garage and shall comply with the provisions of this section for garages.

Exception: Asphalt surfaces shall be permitted at ground level in carports.

The area of floor used for parking of automobiles or other vehicles shall be sloped to facilitate the movement of liquids to a drain or toward the main vehicle entry doorway.

❖ Carports must have at least two sides open to outside air. If two sides are not open, the structure is a garage and comes under the provisions of R309.1, R309.2, R309.3 and R309.5.

R309.5 Flood hazard areas. For buildings located in flood hazard areas as established by Table R301.2(1), garage floors shall be:

1. Elevated to or above the design flood elevation as determined in Section R323; or

2. Located below the design flood elevation provided they are at or above grade on all sides, are used solely for parking, building access, or storage, meet the requirements of Section R323, and are otherwise constructed in accordance with this code.

❖ Garage floors of buildings in flood hazard areas must meet one of two requirements. The first option is to simply construct the garage floor above the design flood elevation. The second option allows the floor of the garage to be below the design flood elevation provided construction is compliant with the applicable pro-

For SI: 1 inch = 25.4 mm.

Figure R309.2
SEPARATION BETWEEN ATTACHED GARAGES AND DWELLING UNITS

visions of R323, including: 1) the floor is at or above finished exterior grade (not a basement); 2) it complies with provisions for enclosed areas below the design flood elevation (R323.2.2); 3) materials and finishes below the design flood elevation are flood-resistant (R323.1.7) and 4) service equipment and systems comply with R323.1.5. If a garage below the design flood elevation does not meet the requirements for enclosures, then the garage floor is the "lowest floor," and the building is non-compliant. Non-compliance means that although the living area is elevated, flood insurance will be rated as though the building is not.

R309.6 Automatic garage door openers. Automatic garage door openers, if provided, shall be listed in accordance with UL 325.

❖ This code does not require an automatic garage door opener. However, if one is installed it must be listed in accordance with UL 325. Federal law requires automatic residential garage door openers to conform to the entrapment protection requirements of UL 325.

SECTION R310
EMERGENCY ESCAPE AND RESCUE OPENINGS

R310.1 Emergency escape and rescue required. Basements with habitable space and every sleeping room shall have at least one openable emergency escape and rescue opening. Where basements contain one or more sleeping rooms, emergency egress and rescue openings shall be required in each sleeping room, but shall not be required in adjoining areas of the basement. Where emergency escape and rescue openings are provided they shall have a sill height of not more than 44 inches (1118 mm) above the floor. Where a door opening having a threshold below the adjacent ground elevation serves as an emergency escape and rescue opening and is provided with a bulkhead enclosure, the bulkhead enclosure shall comply with Section 310.3. The net clear opening dimensions required by this section shall be obtained by the normal operation of the emergency escape and rescue opening from the inside. Emergency escape and rescue openings with a finished sill height below the adjacent ground elevation shall be provided with a window well in accordance with Section R310.2.

❖ Because so many fire deaths occur as the result of occupants of residential buildings being asleep during a fire, the IRC requires that basements with habitable spaces and all sleeping rooms have windows or doors that may be used for emergency escape or rescue. The requirement for emergency escape and egress openings in sleeping rooms exists because a fire will usually have spread before the occupants are aware of the problem, and the normal exit channels may be blocked. The requirement for basements exists because they are so often used as sleeping rooms.

Openings required for emergency escape or rescue must be located on the exterior of the building so that rescue can be performed from the exterior. Alternatively, occupants may escape from that opening to the exterior of the building without having to travel through the building itself. Therefore, where openings are required, they should open directly into a public street, public alley, yard or court. After the occupants pass through the emergency escape and rescue opening, their continued egress is essential. Where a basement contains sleeping rooms and a habitable space, an emergency escape and rescue opening is required in each sleeping room but is not required in adjoining areas of the basement.

The dimensions prescribed in the code, and as illustrated in Commentary Figure R310.1 for exterior wall openings used for emergency egress and rescue, are based in part on extensive testing by the San Diego Building and Fire Departments to determine the proper relationships of the height and width of window openings to adequately serve for both rescue and escape. The minimum of 20 inches (508 mm) for the width is based on two criteria: the width necessary to place a ladder within the window opening and the width necessary to admit a fire fighter with full rescue equipment. The minimum 24-inch (610 mm) height is based on the minimum size necessary to admit a fire fighter with full rescue equipment. By requiring a minimum net clear opening size of the least 5.7 square feet (0.53 m^2), the code provides for an opening of adequate dimensions. Where the opening occurs at grade level, the opening need be only 5 square feet (0.46 m^2) due to the increased ease of access from the exterior.

In order to be accessible from the interior of the sleeping room or basement, the emergency escape and rescue opening cannot be located more than 44 inches (1118 mm) above the floor. The measurement is to be taken from the floor to the bottom of the clear opening.

The required opening dimensions must be achieved by the normal operation of the emergency escape and rescue opening from the inside. Where a window is used as the emergency escape and rescue opening, then it must be the usual double-hung, horizontal sliding, or casement window operated by the turn of a crank. It is impractical to assume that all occupants can operate a window that requires a special sequence of operations to achieve the required opening size. While most occupants are familiar with the normal operation by which to open the window, children and guests are frequently unfamiliar with special procedures necessary to remove the sashes. The time spent comprehending the special operation unnecessarily delays egress from the bedroom and could lead to panic and further confusion. Thus, windows that achieve the required opening dimensions only by having a special sequence of operations performed upon them, such as the removal of sashes or mullions, are not permitted.

BUILDING PLANNING

FIGURE R310.1 - R310.2

**Figure R310.1
EMERGENCY ESCAPE AND RESCUE WINDOW**

R310.1.1 Minimum opening area. All emergency escape and rescue openings shall have a minimum net clear opening of 5.7 square feet (0.530 m²).

Exception: Grade floor openings shall have a minimum net clear opening of 5 square feet (0.465 m²).

❖ The minimum net clear opening area of 5 square feet (0.465 m²) is necessary so that fire fighters in full gear can enter through the opening.

R310.1.2 Minimum opening height. The minimum net clear opening height shall be 24 inches (610 mm).

❖ The minimum opening height for emergency space and rescue opening is 24 inches (610 mm), based on the minimum dimension of a fire fighter with full rescue equipment.

R310.1.3 Minimum opening width. The minimum net clear opening width shall be 20 inches (508 mm).

❖ This section establishes a minimum width of 20 inches (508 mm) for emergency space and rescue openings, based on the minimum dimension of a fire fighter with full rescue equipment.

R310.1.4 Operational constraints. Emergency escape and rescue openings shall be operational from the inside of the room without the use of keys or tools.

❖ Openings for emergency escape and rescue must be operational from the inside. Keys or special tools must not be needed to operate these openings. If keys or tools were necessary, they might not be readily available in an emergency or panic situation, and an individual might not be able to use them, so the opening would be unusable. Section R310.1 also requires the opening size to be obtained by the normal operation of the window. See the commentary for Section R310.1.

R310.2 Window wells. The minimum horizontal area of the window well shall be 9 square feet (0.84 m²), with a minimum horizontal projection and width of 36 inches (914 mm). The area of the window well shall allow the emergency escape and rescue opening to be fully opened.

Exception: The ladder or steps required by Section R310.2.1 shall be permitted to encroach a maximum of 6 inches (152 mm) into the required dimensions of the window well.

❖ Window wells in front of emergency escape and rescue openings also have minimum size requirements. These provisions address those emergency escape windows that occur below grade. Just applying the standard emergency escape window criteria to these windows will result in an opening that occupants can get through, but the window well may actually trap the occupants against the building without providing for their escape from the window well or providing for a fire fighter to enter the residence.

The minimum size requirements in cross-section are similar to the emergency escape and opening criteria; that is, they are sufficient to provide a nominal size to allow for the escape of occupants or the entry of fire fighters. See Commentary Figure R310.2. The ladder or steps requirement is the main difference.

R310.2.1 Ladder and steps. Window wells with a vertical depth greater than 44 inches (1118 mm) shall be equipped with a permanently affixed ladder or steps usable with the window in the fully open position. Ladders or steps required by this section shall not be required to comply with Sections R311.5 and R311.6. Ladders or rungs shall have an inside width of at least 12 inches (305 mm), shall project at least 3 inches (76 mm) from the wall and shall be spaced not more than 18 inches (457 mm) on center vertically for the full height of the window well.

❖ When the depth of a window well exceeds 44 inches (1118 mm), a ladder or steps from the window is required. The details for construction of steps are not identified in the provisions; however, the design of the ladder is specifically addressed. Since ladders and steps in window wells are provided for emergency use only, they are not required to comply with the provisions for stairways found in Sections R311.5 and R311.6.

R310.3 Bulkhead enclosures. Bulkhead enclosures shall provide direct access to the basement. The bulkhead enclosure with the door panels in the fully open position shall provide the minimum net clear opening required by Section R310.1.1. Bulkhead enclosures shall also comply with Section R311.5.8.2.

❖ Bulkhead enclosures, when provided for access to below-grade openings used for emergency escape and rescue, must meet the net openable provisions of Section R310.1.1. Also, bulkhead enclosures must comply with Section R311.5.8.2.

R310.4 Bars, grills, covers and screens. Bars, grills, covers, screens or similar devices are permitted to be placed over emergency escape and rescue openings, bulkhead enclosures, or window wells that serve such openings, provided the minimum net clear opening size complies with Sections R310.1.1 to R310.1.3, and such devices shall be releasable or removable from the inside without the use of a key, tool or force greater than that which is required for normal operation of the escape and rescue opening.

❖ The ever-increasing concern for security, particularly in residential buildings, has created a fairly large demand for security devices such as grilles, bars, and steel shutters. Unless properly designed and constructed, the security devices over bedroom win-

For SI: 1 inch = 25.4 mm, 1 square foot = 0.0929 m^2.

**Figure R310.2
WINDOW WELLS**

dows can completely defeat the purpose of the emergency escape and rescue opening. Therefore, the IRC makes provisions for security devices provided the release mechanism has been approved and is operable from the inside without the use of a key, tool or force greater than that which is required for normal operation of the escape and rescue opening.

The essence of the requirement for emergency escape openings is that a person must be able to effect escape or be rescued in a short period of time, because the fire might have spread to a point where all other exit routes are blocked. Thus, time cannot be wasted in figuring out means of opening rescue windows or obtaining egress through them. Any impediment to escape or rescue caused by security devices, inadequate window size or difficult operating mechanisms is not permitted by the code.

SECTION R311
MEANS OF EGRESS

R311.1 General. Stairways, ramps, exterior exit balconies, hallways and doors shall comply with this section.

❖ Sections R311.2 through R311.6 contains the requirements for the exit and the means of egress components.

R311.2 Construction.

❖ Sections R311.2.1 and R311.2.2 contain the provisions for attachment to the main structure of means of egress components and for fire protection of understair enclosures.

R311.2.1 Attachment. Required exterior exit balconies, stairs and similar exit facilities shall be positively anchored to the primary structure to resist both vertical and lateral forces. Such attachment shall not be accomplished by use of toenails or nails subject to withdrawal.

❖ It is necessary to attach exterior exit balconies, stairs, and similar exit facilities to the primary structure so that vertical and lateral forces will be resisted in order to avoid separation of the exit facilities from the structure. This is reiterating the requirement for a complete load path in Section R301.1. The reasons for doing so are 1) the need to maintain key elements of the egress system for emergency evacuations, and 2) the increased possibility of overlooking such connections. "Exterior exit balconies" and "exit facilities" are not defined or otherwise described in this code. The requirement for positive anchor to the primary structure applies to all exterior means of egress components used as part of an egress system whether part of the required exit or not.

R311.2.2 Under stair protection. Enclosed accessible space under stairs shall have walls, under stair surface and any soffits protected on the enclosed side with $^1/_2$-inch (12.7 mm) gypsum board.

❖ Often times the space under a stairway is used for storage, since this space is often of little use for other purposes. The code permits the use of an open space beneath a stair without the need for any additional protection. Additionally, if the space is walled off and there is no access to the area, then the code is also not concerned. If, however, the area beneath the stairway is enclosed and any type of access is provided into the space, then the walls, soffits and ceilings of the enclosed space must be protected on the enclosed side with at least $^1/_2$-inch (12.7 mm) gypsum board.

R311.3 Hallways. The minimum width of a hallway shall be not less than 3 feet (914 mm).

❖ Hallways must be a minimum of 3 feet (914 mm) in width to accommodate moving furniture into rooms off the hallway and for the safe egress of people from the structure.

R311.4 Doors.

❖ Sections R311.4.1 through R311.4.4 contain the requirements for doors.

R311.4.1 Exit door required. Not less than one exit door conforming to this section shall be provided for each dwelling unit. The required exit door shall provide for direct access from the habitable portions of the dwelling to the exterior without requiring travel through a garage. Access to habitable levels not having an exit in accordance with this section shall be by a ramp in accordance with Section R311.6 or a stairway in accordance with Section R311.5.

❖ All dwelling units are required to have at least one exit door that complies with the provisions of Sections R311.4.2, R311.4.3 and R311.4.4. This door must have access to the exterior without the dwelling's occupants traveling through a garage where hazards could prevent a suitable means of egress. A ramp or stairway is required for access to habitable areas not having an exit.

R311.4.2 Door type and size. The required exit door shall be a side-hinged door not less than 3 feet (914 mm) in width and 6 feet 8 inches (2032 mm) in height. Other doors shall not be required to comply with these minimum dimensions.

❖ This provision applies only to a single exit door required by Section R311.4.1. Other exterior doors are not required to comply with provisions of this section.

R311.4.3 Landings at doors. There shall be a floor or landing on each side of each exterior door.

Exception: Where a stairway of two or fewer risers is located on the exterior side of a door, other than the required exit door, a landing is not required for the exterior side of the door.

The floor or landing at the exit door required by Section R311.4.1 shall not be more than 1.5 inches (38 mm) lower than the top of the threshold. The floor or landing at exterior doors other than the exit door required by Section R311.4.1 shall not be required to comply with this requirement but shall have a rise no greater than that permitted in Section R311.5.3.

Exception: The landing at an exterior doorway shall not be more than $7^3/_4$ inches (196 mm) below the top of the threshold, provided the door, other than an exterior storm or screen door does not swing over the landing.

The width of each landing shall not be less than the door served. Every landing shall have a minimum dimension of 36 inches (914 mm) measured in the direction of travel.

❖ Landings are required for exterior doors and must be provided on both the exterior and interior sides of the door. An exception allows the deletion of the landing if two or fewer stair risers are on the exterior side of a door if the door is other than an exit door. Thresholds must not be higher than 1.5 inches (38 mm) above floor level. See Commentary Figure R311.5.4.

Additionally, an exception permits the exterior landing of an exterior door be $7^3/_4$ inches (196 mm) below the top of the threshold. See Commentary Figure R311.4.3.

Landings must be the same width of the door they serve and must be at least 36 inches (914 mm) in length.

R311.4.4 Type of lock or latch. All egress doors shall be readily openable from the side from which egress is to be made without the use of a key or special knowledge or effort.

❖ This section mandates that the required exit door is under the control of and operable by the person seeking egress, thus the statement that the egress door be operable from the side from which egress is sought, without the need of a key or any special knowledge or effort. This section applies to all egress doors.

R311.5 Stairways.

❖ The requirements for stairways are contained in Sections R311.5.1 through R311.5.8.

R311.5.1 Width. Stairways shall not be less than 36 inches (914 mm) in clear width at all points above the permitted handrail height and below the required headroom height. Handrails shall not project more than 4.5 inches (114 mm) on either side of the stairway and the minimum clear width of the stairway at and below the handrail height, including treads and landings, shall not be less than 31.5 inches (787 mm) where a handrail is installed on one side and 27 (698 mm) where handrails are provided on both sides.

Exception: The width of spiral stairways shall be in accordance with Section R311.5.8.

❖ Section R311.5 provides the code requirements for regulating stairways in an effort to make them usable and also to help in reducing the number of injuries caused by stairway accidents.

Section R311.5.1 provides for a minimum stairway width of 36 inches (914 mm). Generally, when the code specifies a required width of a component in the egress system, the width will be the clear, net, usable, unobstructed width. In this case, however, the width is specified as applying only to the area "above the permitted handrail height and below the required headroom height."

At and below the handrail height, the required width for the stairway including treads and landings is 27 inches (686 mm) if handrails are provided on each side and $31^1/_2$ inches (800 mm) if there is a handrail installed on only one side. In essence, the code is not concerned about elements such as trim, stringers or other items that may be found below the level of the handrail, as long as they do not exceed the handrail's projection. This reduced width below the handrail is based on a body's movements as a person walks on a stair or other surface. See Commentary Figure

For SI: 1 inch = 25.4 mm.

Figure R311.4.3
THRESHOLD HEIGHTS

BUILDING PLANNING

R311.5.1. The exception and the provisions of Section R311.5.8 will permit a minimum width of 26 inches (660 mm) for spiral stairways.

R311.5.2 Headroom. The minimum headroom in all parts of the stairway shall not be less than 6 feet 8 inches (2036 mm) measured vertically from the sloped plane adjoining the tread nosing or from the floor surface of the landing or platform.

❖ A minimum headroom clearance of 6 feet, 8 inches (2 032 mm) is required in connection with every stairway. This includes not only the above-the-tread portion but also above any landings serving the stairway. The clearance is to be measured vertically above a plane that connects the stair nosings and also vertically above any landing or floor surface that is a part of the stairway. See Commentary Figure R311.5.2. This specific height requirement overrides the general ceiling height limitations of Section R305 and is modified for spiral stairways by Section R311.5.8.1.

R311.5.3 Stair treads and risers.

❖ The riser height, tread depth, and profile requirements for stairways are specified in Sections R311.5.3.1 through R311.5.3.3.

R311.5.3.1 Riser height. The maximum riser height shall be $7^3/_4$ inches (196 mm). The riser shall be measured vertically between leading edges of the adjacent treads. The greatest riser height within any flight of stairs shall not exceed the smallest by more than $^3/_8$ inch (9.5 mm).

❖ The code establishes that the maximum riser height is $7^3/_4$ inches (197 mm). The *International Residential Code* does not provide a minimum riser height as does the *International Building Code*, where a 4-inch (102 mm) limit is specified. The provisions specify how the riser height is to be measured. See Commentary Figure R311.5.3.1(1). A significant safety factor relative to stairways is the uniformity of risers and treads in any flight of stairs. The section of a stairway leading from one landing to the next is defined as a flight of stairs. It is very important that any variation that would interfere with the rhythm of the stair user be avoided. While it is true that adequate attention to the use of the stair can compensate for substantial variations in risers and treads, too frequently the stair user does not give the necessary attention.

To obtain the best uniformity possible in a flight of stairs, the maximum variation between the highest and lowest risers is limited to $^3/_8$ inch (9.5 mm). This tolerance is not to be used as a design variation, but its inclusion is in recognition that construction practices make it difficult to get exactly identical riser heights and tread dimensions in constructing a stairway in the field. Therefore, the code allows the variation indicated in Commentary Figure 311.5.3.1(2).

R311.5.3.2 Tread depth. The minimum tread depth shall be 10 inches (254 mm). The tread depth shall be measured horizontally between the vertical planes of the foremost projection of adjacent treads and at a right angle to the tread's leading edge. The greatest tread depth within any flight of stairs shall not exceed the smallest by more than $^3/_8$ inch (9.5 mm). Winder treads

Figure R311.5.1
STAIRWAY CLEARANCES

shall have a minimum tread depth of 10 inches (254 mm) measured as above at a point 12 inches (305) mm from the side where the treads are narrower. Winder treads shall have a minimum tread depth of 6 inches (152 mm) at any point. Within any flight of stairs, the greatest winder tread depth at the 12 inch (305 mm) walk line shall not exceed the smallest by more than $^3/_8$ inch (9.5 mm).

❖ The code establishes that the minimum tread depth is 10 inches (102 mm). The provisions specify how the tread depth is to be measured. See Commentary Figure R311.5.3.1(1). To obtain the best uniformity possible in a flight of stairs, the maximum variation between the greatest and smallest tread depth is limited to $^3/_8$ inch (9.5 mm). See the commentary for Section R311.5.3.1 for a discussion on uniformity. The same criteria apply to winder treads. However, the depth is to be measured at the "walk line." The "walk line" is at a point 12 inches (305 mm) from the side where the treads are narrower.

A stairway may consist of straight treads, or it may be constructed using winders. If winders are used, they can either be used for an entire flight of a stairway,

For SI: 1 inch = 25.4 mm, 1 foot = 304.8 mm.

Figure R311.5.2
MINIMUM HEADROOM

For SI: 1 inch = 25.4 mm.

Figure R311.5.3.1(1)
CONVENTIONAL STAIRWAY

BUILDING PLANNING

as a portion of a stairway such as at a change of direction, or to form a circular stairway. Winders consist of tapered treads that are narrow on one end and widen out, pie-shaped, toward the opposite side of the stairs. Since they are primarily used to change the direction of the stair, and they create a change in the rhythm for the stair users, it is important that winders comply with the specified dimensional criteria. Winder treads must have a minimum depth of 6 inches (152 mm) at any point. See Commentary Figure R311.5.3.2(1) for examples of winders used as a portion of a stairway at a change of direction. See Commentary Figure R311.5.3.2(2) for an example of winders used to form a circular stairway.

R311.5.3.3 Profile. The radius of curvature at the leading edge of the tread shall be no greater than $9/16$ inch (14.3 mm). A nosing not less than $3/4$ inch (19 mm) but not more than $1 1/4$ inch (32 mm) shall be provided on stairways with solid risers. The greatest nosing projection shall not exceed the smallest nosing projection by more than $3/8$ inch (9.5 mm) between two stories, including the nosing at the level of floors and landings. Beveling of nosing shall not exceed $1/2$ inch (12.7 mm). Risers shall

Figure R311.5.3.1(2)
STAIR TOLERANCES

Figure R311.5.3.2(1)
WINDERS

be vertical or sloped from the underside of the leading edge of the tread above at an angle not more than 30 (0.51 rad) degrees from the vertical. Open risers are permitted, provided that the opening between treads does not permit the passage of a 4-inch diameter (102 mm) sphere.

Exceptions:
1. A nosing is not required where the tread depth is a minimum of 11 inches (279 mm).
2. The opening between adjacent treads is not limited on stairs with a total rise of 30 inches (762 mm) or less.

❖ The leading edge or nosing of the treads is limited to a maximum radius of $9/16$ inch (14.3 mm). This varies from the $1/2$ inch (12.7 mm) maximum used in the *International Building Code*. These limitations apply to not only the nosings on the stair treads, but also to nosings at the level of floors or landings that are a part of the stairway. The *International Residential Code* places a requirement for a minimum nosing of $3/4$ inch (19.1 mm) when the stair has a solid riser. This also varies from the *International Building Code*, which does not specify a minimum nosing requirement. However, the first exception will eliminate the nosing requirement if the stair tread is at least 11 inches (279 mm) in depth. Tread nosings are limited to a maximum projection of $1^1/4$ inches (32 mm). The nosing projection is to be consistent from one story level to the next, instead of simply within each flight as is required for the treads and risers. The greatest variation permitted for the nosing is $3/8$ inch (9.5 mm), which is a construction tolerance and not a design variation. See Commentary Figure R311.5.3.3 for examples of the profile provisions. The code does not require solid risers, but where the height of the stairway exceeds 30 inches (762 mm), either solid risers or another method to limit the opening between adjacent treads is needed. As with the guard provisions of Section R312, a 4-inch (102 mm) sphere is used to determine compliance for any type of open riser system.

R311.5.4 Landings for stairways. There shall be a floor or landing at the top and bottom of each stairway.

Exception: A floor or landing is not required at the top of an interior flight of stairs, provided a door does not swing over the stairs.

A flight of stairs shall not have a vertical rise greater than 12 feet (3658 mm) between floor levels or landings.

The width of each landing shall not be less than the stairway served. Every landing shall have a minimum dimension of 36 inches (914 mm) measured in the direction of travel.

❖ A landing is required at the top and bottom of each stairway; however, a landing is not required at the top of interior stairways provided a door does not swing over the stairway. See Commentary Figure R311.5.4.

R311.5.5 Stairway walking surface. The walking surface of treads and landings of stairways shall be sloped no steeper than one unit vertical in 48 inches horizontal (2-percent slope).

❖ The slope of the walking surfaces must provide drainage to stairs and landings that may be subjected to accumulation of liquids such as water, rain or melting snow. The section also provides for a safe limit of the slope so as not to yield an unsafe walking surface. This requirement applies to all stairs and landings, both exterior and interior.

Figure R311.5.3.2(2)
WINDERS USED FOR CIRCULAR STAIRWAY

BUILDING PLANNING

Figure R311.5.3.3
TREAD PROFILE

Figure R311.5.4
LANDINGS AT DOORS

R311.5.6 Handrails. Handrails shall be provided on at least one side of each continuous run of treads or flight with four or more risers.

❖ One of the best means of creating safer stairs and assisting stairway users is to provide a graspable handrail. This can be used for support and guidance as people travel from one level to the next or to assist in arresting a fall should they slip while they are on the stairs. Other than controlling the people who use stairs by making them pay attention and not carry things while on the stairs, the handrail probably will provide the greatest benefit in increased safety for the least amount of cost. It will never be known how many missteps, accidents, injuries or even fatalities have been prevented by having a properly installed, sturdy handrail.

The *International Residential Code* requires that a handrail be provided on at least one side of any stairway that has four or more risers. The code does not provide any exemption for the elimination of handrails if the stairway has four or more risers.

R311.5.6.1 Height. Handrail height, measured vertically from the sloped plane adjoining the tread nosing, or finish surface of ramp slope, shall be not less than 34 inches (864 mm) and not more than 38 inches (965 mm).

❖ Where handrails are required, they must be installed at a height of at least 34 inches (864 mm) and not more than 38 inches (965 mm), measured vertically from the nosing of the stair treads. This height should be measured to the top of the handrail at the point that it is directly above the nosing.

R311.5.6.2 Continuity. Handrails for stairways shall be continuous for the full length of the flight, from a point directly above the top riser of the flight to a point directly above lowest riser of the flight. Handrail ends shall be returned or shall terminate in newel posts or safety terminals. Handrails adjacent to a wall shall have a space of not less than $1^{1}/_{2}$ inch (38 mm) between the wall and the handrails.

Exceptions:
1. Handrails shall be permitted to be interrupted by a newel post at the turn.
2. The use of a volute, turnout, starting easing or starting newel shall be allowed over the lowest tread.

❖ This required handrail is to be "continuous" for the length of the stairs. See Commentary Figure R311.5.6.2. The term "continuous" means not only that a single handrail must run from the top riser to the bottom riser, but it also indicates that users should be able to grasp the handrail and maintain their grasp without having to release the rail where it is supported. There is no requirement within the code for a second handrail to be installed, but depending on the design and the placement of the required handrail, Section R316 and the requirement for a guard should be reviewed. The two exceptions to this section create situations where the graspable portion of the handrail may not end up being completely continuous from the top riser to the bottom riser. These traditional situations have routinely been accepted in the past and therefore are permitted by the code.

In view of the desire to make the handrail graspable, and considering the requirement that the handrail be continuous, it is necessary to provide a clear space of at least $1^{1}/_{2}$ inches (38 mm) between the handrail and any abutting wall. Under normal handrail use this distance will permit the fingers to slide past any adjacent rough surface that may cause injury, and it will provide an adequate distance so that the handrail may be quickly grabbed to help arrest a fall.

One important aspect of this section is the requirement for the ends of handrails to be returned to the wall or floor or to terminate in some type of end that will not catch clothing or limbs. It is still common to find many locations where handrails into unfinished basements or from raised deck areas do not have the ends properly protected.

R311.5.6.3 Handrail grip size. All required handrails shall be of one of the following types or provide equivalent graspability.

1. Type I. Handrails with a circular cross section shall have an outside diameter of at least $1^{1}/_{4}$ inches (32 mm) and not greater than 2 inches (51 mm). If the handrail is not circular it shall have a perimeter dimension of at least 4 inches (102 mm) and not greater than $6^{1}/_{4}$ inches (160 mm) with a maximum cross section of dimension of $2^{1}/_{4}$ inches (57 mm).

2. Type II. Handrails with a perimeter greater than $6^{1}/_{4}$ inches (160 mm) shall provide a graspable finger recess area on both sides of the profile. The finger recess shall begin within a distance of $^{3}/_{4}$ inch (19 mm) measured vertically from the tallest portion of the profile and achieve a depth of at least $^{5}/_{16}$ inch (8 mm) within $^{7}/_{8}$ inch (22 mm) below the widest portion of the profile. This required depth shall continue for at least $^{3}/_{8}$ inch (10 mm) to a level that is not less than $1^{3}/_{4}$ inches (45 mm) below the tallest portion of the profile. The minimum width of the handrail above the recess shall be $1^{1}/_{4}$ inches (32 mm) to a maximum of $2^{3}/_{4}$ inches (70 mm). Edges shall have a minimum radius of 0.01 inches (0.25 mm).

❖ To be effective, a handrail must be of a size that can easily be grasped by the vast majority of users. If it is too large, then it is difficult for a user to get a strong enough grip to provide the needed support. Due to the variations in people's sizes and physical conditions, it is often very difficult to determine which types of commonly used handrails fall into this acceptable range.

The code specifies that the handrail be either a Type I or Type II. A Type I can be either circular or noncircular in shape. See Commentary Figure R311.5.6.3(1) for examples of Type I handrails.

BUILDING PLANNING

A Type II handrail has a perimeter greater than $6^{1}/_{4}$ inches (160 mm) with graspable finger recess area on both sides of the profile. See Commentary Figure R311.6.6.3(2) for the limitations of a Type II handrail.

R311.5.7 Illumination. All stairs shall be provided with illumination in accordance with Section R303.6.

❖ This section provides a reference to the illumination provisions of Section R303.6. The proper illumination of stairways is an important aspect for stairway safety. This lighting can assist users by making sure the level changes do not occur in areas with shadows or in area with contrasting light, which would make the changes difficult to perceive. See the discussion at Section R303.6 for additional information.

R311.5.8 Special stairways. Circular stairways, spiral stairways, winders and bulkhead enclosure stairways shall comply with all requirements of Section R311.5 except as specified below.

❖ Sections R311.5.8.1 and R311.5.8.2 provide exceptions to the general requirements for stairways as prescribed in Section R311.5.

R311.5.8.1 Spiral stairways. Spiral stairways are permitted, provided the minimum width shall be 26 inches (660 mm) with each tread having a $7^{1}/_{2}$-inches (190 mm) minimum tread depth at 12 inches from the narrower edge. All treads shall be identical, and the rise shall be no more than $9^{1}/_{2}$ inches (241

Figure R311.5.6.2
HANDRAILS

For SI: 1 inch = 25.4 mm.

Figure R311.5.6.3(1)
TYPE I HANDRAIL

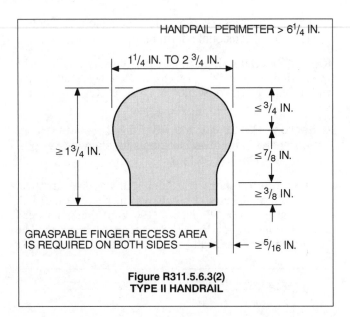

Figure R311.5.6.3(2)
TYPE II HANDRAIL

mm). A minimum headroom of 6 feet 6 inches (1982 mm) shall be provided.

❖ A spiral stairway is one of several types of special stairs that the code permits. Although a spiral stair may be difficult to use for moving furniture from one level to another, the code places no limitations on its use within the egress system provided it meets the size requirements of this section. A spiral stairway that meets these requirements may provide the only means of egress from a level regardless of the occupant load or size of area served.

A spiral stairway is one in which the treads radiate from a central pole. Such a stair must provide a clear width of at least 26 inches (660 mm). Each tread must be identical and have a minimum dimension of $7^1/_2$ inches (191 mm) at a point 12 inches (305 mm) from its narrow end. The stair must have at least 6 feet, 6 inches (1981 mm) of headroom measured vertically from the leading edge of the tread. The rise between treads can be as much as but not more than $9^1/_2$ inches (241 mm). Commentary Figure R311.5.8.1 depicts the required dimensions of a spiral stairway.

R311.5.8.2 Bulkhead enclosure stairways. Stairways serving bulkhead enclosures, not part of the required building egress, providing access from the outside grade level to the basement shall be exempt from the requirements of Sections R311.4.3 and R311.5 where the maximum height from the basement finished floor level to grade adjacent to the stairway does not exceed 8 feet (2438 mm), and the grade level opening to the stairway is covered by a bulkhead enclosure with hinged doors or other approved means.

❖ This section exempts exterior bulkhead enclosure stairways from the landing, stairway, and handrail requirements in Chapter 3, and it therefore permits a situation that has been fairly common in some areas. See Commentary Figure R311.5.8.2 for an illustration of the requirements. Since these stairways are not a part of the building's egress system and serve only as a convenient way to access the basement from the exterior, the code exemption will not greatly affect the occupants' safety. Through this exemption, the size of the enclosure that is needed to provide weather protection for the stairway is greatly reduced.

R311.6 Ramps.

❖ Section R311.6 provides the code requirements for ramps, which are defined by this code as being a walking surfaces that have a running slope steeper than 1 unit vertical in 20 units horizontal (5-percent slope).

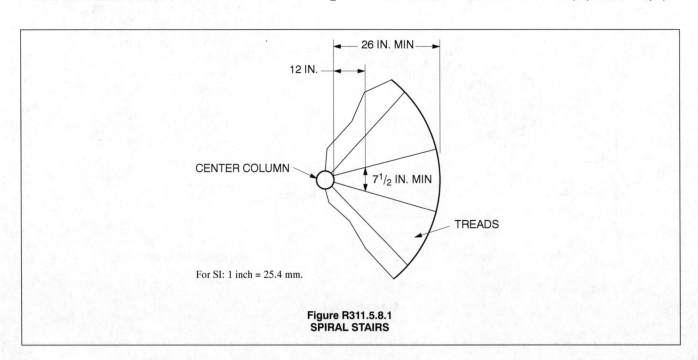

Figure R311.5.8.1
SPIRAL STAIRS

R311.6.1 Maximum slope. Ramps shall have a maximum slope of one unit vertical in eight units horizontal (12.5-percent slope).

❖ Section R311.6.1 places a maximum slope of 1 unit vertical in 8 units horizontal (12.5-percent slope) on ramps. This requirement applies to all ramps, including those on circulation routes and those leading to an exit. This maximum slope matches what is permitted by the *International Building Code* for ramps that are not a part of the means of egress. Egress ramps under the IBC have a maximum slope limit of 1:12 due to accessibility requirements.

R311.6.2 Landings required. A minimum 3-foot-by-3-foot (914 mm by 914 mm) landing shall be provided:

1. At the top and bottom of ramps,
2. Where doors open onto ramps,
3. Where ramps change direction.

❖ The code requires a minimum 3-foot-by-3-foot (914 mm by 914 mm) landing at three specific locations related to the ramps. These dimensions are not tied to the actual width of the ramp. Item 2, dealing with doors that open onto ramps, calls for a larger size landing if it is also required by Section R311.4.3. The specified landing dimensions coordinate with the requirements for nonaccessible dwelling units, which are found in exceptions in the *International Building Code*.

R311.6.3 Handrails required. Handrails shall be provided on at least one side of all ramps exceeding a slope of one unit vertical in 12 units horizontal (8.33-percent slope).

❖ Where a ramp exceeds a slope of 1 unit vertical in 12 units horizontal (8.33-percent slope) the code requires that a handrail be provided on at least one side to assist ramp users. This provision differs from that in the *International Building Code*, where a slope of 1 unit

Figure R311.5.8.2
BULKHEAD ENCLOSURE STAIRWAY

vertical in 20 units horizontal (5-percent slope) and a ramp rise of 6 inches (152 mm) establish the limits.

R311.6.3.1 Height. Handrail height, measured above the finished surface of the ramp slope, shall be not less than 34 inches (864 mm) and not more than 38 inches (965 mm).

❖ Where handrails are required, they must be installed at a height of at least 34 inches (864 mm) and not more than 38 inches (965 mm), measured vertically from the finished surface of the ramp slope. This height should be measured to the top of the handrail.

R311.6.3.2 Handrail grip size. Handrails on ramps shall comply with Section R311.5.6.3.

❖ See the commentary for Section R311.5.6.3.

R311.6.3.3 Continuity. Handrails where required on ramps shall be continuous for the full length of the ramp. Handrail ends shall be returned or shall terminate in newel posts or safety terminals. Handrails adjacent to a wall shall have a space of not less than 1.5 inches (38 mm) between the wall and the handrails.

❖ The continuity requirement for the ramp handrail is similar to the continuity requirement for the stair handrail. See the commentary for Section R311.5.6.2

SECTION R312
GUARDS

R312.1 Guards required. Porches, balconies or raised floor surfaces located more than 30 inches (762 mm) above the floor or grade below shall have guards not less than 36 inches (914 mm) in height. Open sides of stairs with a total rise of more than 30 inches (762 mm) above the floor or grade below shall have guards not less than 34 inches (864 mm) in height measured vertically from the nosing of the treads.

Porches and decks which are enclosed with insect screening shall be provided with guards where the walking surface is located more than 30 inches (762 mm) above the floor or grade below.

❖ The guard provisions of this code address the issue of providing protection for occupants from falling off of any type of elevated walking surface. The provisions in Section R312 provide the scoping requirements as well as the general construction requirements for the guards. Code users should be aware that Section R301.5 as well as this section contain the design load criteria for guards.

Section R312.1 establishes the requirement for and the minimum height requirements for guards. The code provides for guard protection at open sides along raised floor or walking surfaces such as those at balconies, mezzanines, stairways, ramps, porches, and landings that are more than 30 inches (762 mm) above the grade or floor surface below. The requirements for guards on stairs are different from other guard requirements in two ways. The first is the scoping requirements that establish the need for the guard, and the second is the required height of the guard. The scoping requirement for guards along open sides of stairs not only applies to the portion of a stairway that is more than 30 inches (762 mm) above the adjacent floor, but it also applies to the entire open side of the stair, including the parts that are less than 30 inches (762 mm) above the floor. This requirement applies to the entire open side of the stairway if any point of the open side is greater than 30 inches (762 mm) in height. See Commentary Figures R312.1(1) and (2) for examples of how this provision is applied.

Where a guard is required, it must be of an adequate height to prevent someone from falling over the edge of the protected areas. It must as well be designed to prevent someone, including small children, from falling through or under the top rail. Therefore, the code establishes 36 inches (914 mm) as the minimum acceptable height for guard protection. While this height matches what has commonly been accepted within residential occupancies, it is less than the 42-inch (1067 mm) height that is generally required by the *International Building Code* and that places the top of the rail above the center of gravity for the vast majority of the general population. The height of a guard located on the open side of a stair is permitted to be a minimum of 34 inches (864 mm) when measured above the nosing of the tread. See Commentary Figure R312.1(3). As mentioned in the discussion of Section R311.5.6.1, this height should be measured to the top of the guard. This 34-inch height (864 mm) is acceptable on the stairs because the handrail height on a stairway is required to be between 34 and 38 inches (864 and 965 mm). Since reduced-height rails have been used on the open sides of stairs for years, the code has continued this practice, so a single element can serve as both a handrail and as a guard.

Insect screening is not of sufficient strength to prevent someone from falling under a top rail. For this reason a guard is required for porches and decks enclosed with insect screening where the walking surface is located more than 30 inches above a floor or grade below.

R312.2 Guard opening limitations. Required guards on open sides of stairways, raised floor areas, balconies and porches shall have intermediate rails or ornamental closures which do not allow passage of a sphere 4 inches (102mm) or more in diameter.

Exceptions:

1. The triangular openings formed by the riser, tread and bottom rail of a guard at the open side of a stairway are permitted to be of such a size that a sphere 6 inches (152 mm) cannot pass through.

BUILDING PLANNING

2. Openings for required guards on the sides of stair treads shall not allow a sphere $4^3/_8$ inches (107 mm) to pass through.

❖ Guards must be constructed so that they not only prevent people from falling over them but also prevent smaller occupants such as children from falling through them. To prevent people from slipping through a guard, any required guard would need to have supports, spindles, intermediate rails, or some type of ornamental pattern so that a 4-inch (102 mm) sphere cannot pass through it. This spacing was chosen after many years of discussion and because information was submitted to show that based on the size of a child's head, very few children would be able to crawl or walk through a 4-inch opening. The code does provide two exceptions for this spacing requirement, permitting the use of a 6-inch (152 mm) sphere for the triangular area formed by the riser, tread and bottom rail of a guard along the open side of a stair and 4 3/8 inch (111 mm) sphere for the guard on open side of stair treads. See Commentary Figure R312.2 for an illustration of the guard requirements.

SECTION R313
SMOKE ALARMS

[F] R313.1 Smoke alarms. Smoke alarms shall be installed in the following locations:

1. In each sleeping room.
2. Outside each separate sleeping area in the immediate vicinity of the bedrooms.
3. On each additional story of the dwelling, including basements but not including crawl spaces and uninhabitable attics. In dwellings or dwelling units with split levels and without an intervening door between the adjacent levels, a smoke alarm installed on the upper level shall suffice for the adjacent lower level provided that the lower level is less than one full story below the upper level.

When more than one smoke alarm is required to be installed within an individual dwelling unit the alarm devices shall be interconnected in such a manner that the actuation of one alarm will activate all of the alarms in the individual unit. The alarm shall be clearly audible in all bedrooms over background noise levels with all intervening doors closed.

All smoke alarms shall be listed and installed in accordance with the provisions of this code and the household fire warning equipment provisions of NFPA 72.

❖ Section R313 provides the details of smoke detection and notification to alert occupants of potential problems. When asleep, the occupants of residential buildings will usually be unaware of a fire, and it will have an opportunity to spread before being detected. As a result, a majority of fire deaths occurring in residential buildings have occurred because of this delay in detection. It is for this reason that the IRC requires smoke alarms. This alarm system provides early notification to occupants of the building in the event of a fire, thereby providing a greater opportunity for everyone in the building to evacuate or relocate to a safe area. Of all of the provisions for safety features that have been placed within the code over the past few decades, the provisions for these detection and alarm devices have probably provided the greatest benefit in increased safety when compared to their minor expense.

For SI: 1 inch = 25.4 mm.

**Figure R312.1(1)
STAIRWAY GUARD**

Figure R312.1(2)
STAIRWAY GUARD

Figure R312.1(3)
STAIRWAY GUARD

BUILDING PLANNING

**Figure R312.2
GUARD REQUIREMENTS**

So that all areas have at least some level of protection, and so that sleeping areas are adequately protected, Section R313.1 specifies where the devices are to be installed. The code requires that alarms be located within each sleeping room and outside each separate sleeping area but in the immediate vicinity of the bedrooms. The device within the bedroom will provide protection should the fire begin within that sleeping room, while the device outside of the room will provide early notification and protection should a problem develop in the area that generally will serve as the egress path for the bedroom. In addition, Item 3 will require at least one smoke alarm to be installed on each story of the dwelling, including basements. The code does not require the installation of alarms within crawl spaces or within attics that are not habitable. This provides detection and notification within the areas of general occupancy but ignores spaces that are not occupied. See Commentary Figure R313.1(1) for an illustration of the required alarm locations.

Where split levels occur in a dwelling and the adjacent levels are openly communicating with each other, then the alarm may be placed on the upper portion of the split level provided it is not more than one full story different in elevation. Commentary Figure R313.1(2) provides an example of this provision. This requirement is based on the fact that any fire initiating on the lower portion of the level will send products of combustion up to the upper portion and that a detector there will provide a quick response and early warning.

In order to assure that alarm notification is provided and audible in order to alert the occupants of any problem within the unit, the code requires that two or more smoke alarms be interconnected so that if one device is activated, all alarms within the dwelling unit will be activated. This interconnection is required so that no matter where the smoke first develops or is detected, occupants throughout the unit will be made aware of the situation. Since one of the main concerns of the code is occupants who may be asleep and unaware of any developing problem, the code requires that the alarm signal be "clearly audible" in the bedroom area even when the doors to the rooms are closed. The code does not specify what sound level is required, so the building official will need to make this determination. If the smoke alarms are being installed in an existing building, see the commentary for Section R313.1.1, which contains an exception to the requirement for interconnection.

[EB] R313.1.1 Alterations, repairs and additions. When interior alterations, repairs or additions requiring a permit occur, or when one or more sleeping rooms are added or created in existing dwellings, the individual dwelling unit shall be provided with smoke alarms located as required for new dwellings; the smoke alarms shall be interconnected and hard wired.

Exceptions:

1. Smoke alarms in existing areas shall not be required to be interconnected and hard wired where the alterations or repairs do not result in the removal of interior wall or ceiling finishes exposing the structure, unless there is an attic, crawl space, or basement available which could provide access for hard wiring and interconnection without the removal of interior finishes.

2. Repairs to the exterior surfaces of dwellings are exempt from the requirements of this section.

❖ This section contains a unique provision in the code, applying the smoke alarm provisions to existing buildings when an addition, alteration, or repair is made that will require a permit, or if any sleeping rooms are added or created. See the discussion at Section R105.2 regarding what types of repair or alterations require a permit. The installation of the smoke alarms in these existing buildings is to be done in the same manner as

Figure R313.1(1)
LOCATION OF SMOKE DETECTORS

Figure R313.1(2)
SMOKE ALARM LOCATION

required for new dwellings. This would not only require their installation in the same locations within the dwelling but also that they be interconnected and receive their power from the building wiring. The commentary for Section R313.2 provides more discussion of the power source.

Two exceptions provide relief from the normal smoke alarm requirements in existing buildings that undergo some type of alteration, repair, or addition. The first exception negates the requirement for hardwired and interconnected alarms in areas where interior finishes are not removed, making the placement of wiring difficult. In these cases it is acceptable to use a single-station battery operated alarm to provide the protection. This exception will not apply if there is an attic, crawl space, or unfinished basement available that would permit the wiring to be run on that side of the existing finish materials.

The second exception exempts "exterior surface" repairs from initiating the requirement for smoke alarms being placed in an existing dwelling. This exception exempts work that is done on the exterior only, and it does require that access be provided to the interior of the dwelling. The final determination of what type of work is included is left to the building official, but this would generally be viewed as covering reroofing, siding repairs or siding replacement and could possibly include some window replacement systems.

[F] R313.2 Power source. In new construction, the required smoke alarms shall receive their primary power from the building wiring when such wiring is served from a commercial source, and when primary power is interrupted, shall receive power from a battery. Wiring shall be permanent and without a disconnecting switch other than those required for overcurrent protection. Smoke alarms shall be permitted to be battery operated when installed in buildings without commercial power or in buildings that undergo alterations, repairs or additions regulated by Section R313.1.1.

❖ Smoke alarms are required to use AC power as their primary source and battery power as a secondary source in order to enhance their reliability. For example, during a power outage, the probability of fire is increased because of the use of candles or lanterns for temporary light. Required backup battery power provides for continued performance of the smoke alarms. Smoke alarms are commonly designed to emit a recurring signal when batteries are low and need to be replaced. The last part of the section acknowledges that the code does not require smoke alarms in all existing buildings be served from a commercial power source. Battery-operated smoke alarms may be the only power source when a commercial power source is not available or when extensive alterations or repairs are not being conducted. Where permanent building wiring can be installed without the removal of interior finishes (see Section R313.1.1, Exception 1), this section

recognizes the increased reliability that the "hard wired" commercial power source with battery back-up can provide. Therefore, where feasible, permanent wiring is to be provided. It is also for the reliability issue that the code does not permit the alarms to be on any type of circuit that could be disconnected or turned off, such as a lighting circuit with a switch. The only way to disconnect power to the smoke-alarms should be through the electrical panel box by either flipping a circuit breaker or removing the circuit's fuse.

SECTION R314
FOAM PLASTIC

R314.1 General. The provisions of this section shall govern the requirements and uses of foam plastic insulation.

❖ Section R314 covers several topics related to the use and installation of various types of foam plastic materials used for insulation, trim and finishes. These requirements provide for the acceptable uses of this combustible product and the associated protection needed to use it in building construction.

Section R314.1 provides the two basic issues that serve as the basis for the foam plastic insulation requirements: the flame-spread rating of the material and the separation of the foam plastic insulation from the interior of the building.

R314.1.1 Surface burning characteristics. Except where otherwise noted in Section R314.2, all foam plastic or foam plastic cores in manufactured assemblies used in building construction shall have a flame-spread rating of not more than 75 and shall have a smoke-developed rating of not more than 450 when tested in the maximum thickness intended for use in accordance with ASTM E 84.

❖ Any foam plastic insulation or foam plastic core material found in manufactured assemblies must be limited in flame spread and smoke development, so the occupants will not be quickly overcome by smoke and fire. In this section, the code limits foam plastic materials to a flame-spread index of 75 and a smoke-developed index of 450 where tested at the maximum intended thickness of use. The section does provide a reference to Section R314.2, which provides various exceptions for interior trim, roofing, foam-filled doors and other items with specific approval based on the reference to Section R314.3.

R314.1.2 Thermal barrier. Foam plastic, except where otherwise noted, shall be separated from the interior of a building by minimum $1/2$-inch (12.7 mm) gypsum board or an approved finish material equivalent to a thermal barrier to limit the average temperature rise of the unexposed surface to no more than 250°F (121°C) after 15 minutes of fire expo-sure to the ASTM E 119 standard time temperature curve. The gypsum board shall be installed using a mechanical fastening system in accordance with Section R702.3.5. Reliance on adhesives to ensure that the gypsum board will remain in place when exposed to fire shall be prohibited.

❖ Because of potential hazards involved in foam plastic igniting, it must be separated from the interior of a building by an approved thermal barrier. The thermal barrier's job is to limit the temperature rise that the foam plastic will be exposed to. Gypsum wallboard at least $1/2$ inch (12.7 mm) thick satisfies this requirement, as does any other material that can provide a thermal barrier complying with the criteria of this section. The barrier must remain in place for the required 15-minute time period. Therefore the code requires some method of mechanical fastening, referencing Section R702.3.5 for the gypsum board and prohibiting adhesive attachment.

R314.2 Specific requirements. The following requirements shall apply to all uses of foam plastic unless specifically approved in accordance with Section R314.3 or by other sections of the code.

❖ This section provides specific requirements regulating the use of foam plastic in the listed locations. The wording in Section R314.1.2 and especially Section R314.1.1 provides a reference to this section, and therefore the provisions of Section R314.2 can remove some of the requirements for both the surface burning characteristics as well as the required thermal barrier. In addition, this section provides a reference that connects to the specific approval provisions of Section R314.3. This reference provides another method that will affect the general requirements from Section R314.1. See the discussion of Section R314.3.

R314.2.1 Masonry or concrete construction. Foam plastics may be used without the thermal barrier described in Section R314.1 when the foam plastic is protected by a minimum 1-inch (25.4 mm) thickness of masonry or concrete.

❖ See Commentary Figure R314.2.1. When foam plastics are encapsulated within concrete or masonry walls, floors or roof systems, the code does not require a thermal barrier as long as the foam plastic is covered by a minimum of 1-inch (25.4 mm) thickness of masonry or concrete. This exemption is reasonable based on the separation and insulating value that these two materials provide.

R314.2.2 Roofing. Foam plastic may be used in a roof-covering assembly without the thermal barrier when the foam is separated from the interior of the building by wood structural panel sheathing in accordance with Section R803, not less than $15/32$ inch (11.9 mm) in thickness bonded with exterior glue and identified as Exposure 1, with edge supported by blocking or tongue-and-groove joints. The smoke-developed rating shall not be limited.

❖ This section will permit the installation of foam plastic as a part of the roof covering assembly where the wood roof sheathing provides an adequate separation. This section removes the smoke-developed rating from Section R314.1.1 and eliminates the thermal barrier requirements from Section R314.1.2, provided that the foam is separated from the interior of the building by a minimum $15/32$-inch (11.9 mm) wood structural-panel sheathing. These wood structural panels are required to be bonded with exterior glue and must carry an "Exposure 1" identification. The sheathing in this case must have either tongue-and groove joints or edges that are supported by blocking where they do not occur over framing members. This issue of joint treatment may not be a typical method of construction and should be inspected for compliance. This section does not eliminate the flame-spread rating specified in Section R314.1.1.

R314.2.3 Attics and crawlspaces. Within attics and crawlspaces where entry is made only for service of utilities, foam plastics shall be protected against ignition by $1^{1}/_{2}$-inch-thick (38 mm) mineral fiber insulation, $1/4$-inch-thick (6.4 mm) wood structural panels, $3/8$-inch (9.5 mm) particleboard,

For SI: 1 inch = 25.4 mm.

**Figure R314.2.1
ENCAPSULATED FOAM PLASTIC**

BUILDING PLANNING

$^1/_4$-inch (6.4 mm) hardboard, $^3/_8$-inch (9.5 mm) gypsum board, or corrosion-resistant steel having a base metal thick-ness of 0.016 inch (0.406 mm).

❖ See Commentary Figure R314.2.3. This section describes specific methods used to protect foam plastics located within attics and crawl spaces in lieu of the normal thermal barrier specified in Section R314.1.2. These described methods of protection are acceptable where entry is provided to the attic or crawl space only for the service of utilities. The phrase "where entry is made only for service of utilities" restricts these reduced requirements for a thermal barrier to those areas that contain mechanical equipment, electrical wiring, fans, plumbing, etc. that might require service and the areas have no other uses, such as storage. Thus, the reduced provisions provide a barrier whose only purpose is to prevent the direct impingement of flame on the foam plastic.

R314.2.4 Foam-filled doors. Foam-filled doors are exempt from the requirements of Section R314.1.

❖ This section exempts any door that encapsulates a foam plastic material from both the surface-burning characteristics as well as the thermal barrier requirements of Section R314.1. These requirements differ from the provisions contained in the *International Building Code,* which address doors in three specific situations. The IRC provisions exempt all foam-filled doors regardless of their location.

R314.2.5 Siding backer board. Foam plastic board of not more than $^1/_2$-inch (12.7 mm) thickness may be used as siding backer board when separated from interior spaces by not less than 2 inches (51 mm) of mineral fiber insulation or $^1/_2$-inch (12.7 mm) gypsum wallboard or installed over existing exterior wall finish in conjunction with re-siding, providing the plastic board does not have a potential heat of more than 2,000 Btu per square foot (22 720 kJ/m^2) when tested in accordance with NFPA 259.

❖ The code permits up to a $^1/_2$-inch (12.7 mm) thickness of foam plastic to be used as a backer board to insulate exterior walls under exterior siding, provided the insulation has a potential heat of not more than 2,000 Btu per square foot (22 720 kJ/m^2). The thermal barrier is not required under these circumstances as long as the siding backer board is separated from the interior of the building by not less than 2 inches (51 mm) of mineral-fiber insulation or $^1/_2$-inch (12.7 mm) gypsum wallboard.

The code also permits the siding backer board to be installed without a thermal barrier when the siding is applied as part of a re-siding over existing wall construction. This is reasonable considering the separation provided by the existing construction and the limitations on the potential heat imposed by the code.

R314.2.6 Interior trim. Foam plastic trim defined as picture molds, chair rails, baseboards, handrails, ceiling beams, door trim and window trim may be installed, provided:

1. The minimum density is 20 pounds per cubic foot (3.14 kg/m^3).

Figure R314.2.3
FOAM PLASTIC, ATTIC AND CRAWL SPACES

2. The maximum thickness of the trim is 0.5 inch (12.7 mm) and the maximum width is 4 inches (102 mm).
3. The trim constitutes no more than 10 percent of the area of any wall or ceiling.
4. The flame-spread rating does not exceed 75 when tested per ASTM E 84. The smoke-developed rating is not limited.

❖ This section provides four limitations on foam plastic used as interior trim. The code does not permit the use of foam plastic as a base or finish material in a large area. These provisions are only applicable when the foam plastic is used as a trim or decorative item in the examples listed in the code. The hazard level created by the exposure of foam plastics is low because the code limits the density, thickness, size, wall and ceiling area, and flame spread of the foam plastic materials. Foam plastic trim that is accepted by this section does not have to meet the smoke-developed rating of 450 that is normally required by Section R314.

R314.2.7 Sill plates and headers. Foam plastic shall be permitted to be spray applied to a sill plate and header without thermal barrier subject to all of the following:

1. The maximum thickness of the foam plastic shall be $3^1/_4$ inches (82.6 mm).
2. The density of the foam plastic shall be in the range of 1.5 to 2.0 pcf (24 to 32 kg/m^3).
3. The foam plastic shall have a flame spread index of 25 or less and an accompanying smoke developed index of 450 or less when tested in accordance with ASTM E84.

❖ See Commentary Figure R314.2.7. Foam plastic spray applied to the sill plate and joist header does not require the thermal barrier described in Section R314.1.2 when the conditions of Section R314.2.7 are complied with. The prescriptive installation methods described in these sections must be strictly followed in order for foam plastic to serve as an alternative to the thermal barrier.

R314.3 Specific approval. Plastic foam not meeting the requirements of Sections R314.1 and R314.2 may be specifically approved on the basis of one of the following approved tests: ASTM E 84, FM 4880, UL 1040, NFPA 286, ASTM E 152, or UL 1715, or fire tests related to actual end-use configurations. The specific approval may be based on the end use, quantity, location and similar considerations where such tests would not be applicable or practical.

❖ In this section, the code provides for those cases where foam plastic products and protective coverings do not comply with the specific requirements of Section R314. The specific approvals are based on testing that is related to the actual end use of the products. The code refers to a number of test standards for determining specific approvals, and there are others that

Figure R314.2.7
FOAM PLASTIC SPRAY APPLIED TO SILL PLATE AND HEADER

BUILDING PLANNING

use some variation of the room test and are designed for testing exterior wall applications.

R314.4 Interior finish. Foam plastics that are used as interior finish shall also meet the flame-spread requirements for interior finish.

❖ This section serves as a reminder regarding the applicability of the provisions found within Section R315 when foam plastic material is used as an interior finish. Foam-plastic materials that are used as a finish must meet the general requirements of Section R319 and must also comply with the specific provisions found in Section R314 for the particular use.

R314.5 Termite damage. The use of foam plastics in areas of "very heavy" termite infestation probability shall be in accordance with Section R320.4.

❖ This section serves as a warning flag and provides a reference to another section of the code. Because of the type of material and the way that termites are able to tunnel and stay alive in foam plastic insulation, the code provides this reference to the requirements found in Section R320.4. See the discussion at that section for details regarding the protection of the foam plastic insulation.

[B] SECTION R315
FLAME SPREAD AND SMOKE DENSITY

R315.1 Wall and ceiling. Wall and ceiling finishes shall have a flame-spread classification of not greater than 200.

> **Exception:** Flame-spread requirements for finishes shall not apply to trim defined as picture molds, chair rails, baseboards and handrails; to doors and windows or their frames; or to materials that are less than $1/28$ inch (0.907 mm) in thickness cemented to the surface of walls or ceilings if these materials have a flame-spread characteristic no greater than paper of this thickness cemented to a noncombustible backing.

❖ Section R315 contains the requirements for controlling fire growth within buildings by restricting interior finish materials. The control of interior finishes is an important aspect of fire protection. The dangers of unregulated interior finish include both the rapid spread of fire and the contribution of additional fuel to the fire. The rapid spread of fire presents a threat to the occupants of a building by limiting or denying their use of exitways within and outside the building. This can be caused by the rapid spread of the fire itself or by the production of large quantities of dense, black smoke, which obscures the exit path or makes movement difficult. Unregulated finish materials also have the potential for adding fuel to the fire, thus increasing its intensity and shortening the time available to the occupants to exit safely. However, based on the test standard that is used, the *International Residential Code* does not regulate interior finish materials in regard to the fuel-contribution basis.

The IRC regulates interior finish materials on both walls and ceilings. These provisions do not address the floor or any coverings applied to the floor. The level of performance the code establishes for finish materials is a flame-spread classification of 200 or less. The flame-spread classification for a material is based on reviewing its performance under the test standard specified in Section R315.3. The limitation of 200 for the flame-spread classification matches a "Class C" material in the *International Building Code*. Therefore, any material that has a flame-spread rating of less than 200 may be used as a finish material on both walls and ceilings.

The exception will permit the installation of materials that will not significantly contribute to fire conditions. This includes various types of trim such as chair rails, door and window frames, and baseboards, which due to their quantities and locations do not cause great concern. The actual size and quantity of such trim is not specified in the code since it has not been a concern. However, if some type of foam plastic is being used instead of other types of combustible trim, then Section R314.2.6 does limit those situations. The exception also does not regulate thin materials such as wallpaper, which are less than $1/28$ inch (0.907 mm) thick when they are properly installed. These thin materials, when cemented to the surface of the wall or ceiling, behave essentially as the backing to which they are applied, and as a result are not regulated.

R315.2 Smoke-developed index. Wall and ceiling finishes shall have a smoke-developed index of not greater than 450.

❖ The development of smoke affects the occupants' safety. This section places a limitation of 450 on the density of smoke that is allowed from the wall and ceiling finishes. A product's smoke-developed index is established for the finish materials when they are tested under the standard specified in Section R315.3. The test only measures the obscurity caused by the smoke and does not consider the toxic content within the smoke.

R315.3 Testing. Tests shall be made in accordance with ASTM E 84.

❖ This section establishes that the standard test for flame spread and smoke-development characteristics is ASTM E 84, commonly known as the Steiner Tunnel Test. ASTM E 84 determines the relative burning behavior of materials on exposed surfaces, such as ceilings and walls, by visually observing the flame spread along the test specimen. Flame spread and smoke density are reported. The test method renders measurements of surface flame spread and smoke density in comparison with test results obtained by using select red oak and asbestos-cement board as control materials. Red oak is used for the furnace calibration because it is a fairly uniform grade of lumber that is readily available nationally, is uniform in thickness and moisture content and generally gives consistent and reproducible results. Asbestos-cement board has a flame spread of zero, while the red oak is assigned a flame-spread rating of 100. All other materials are then given a classification based on a comparison with these two materials. Therefore, the flame spread rating of 200 that is permitted by Section R315.1 essentially means that the maximum flame-spread for any finish material is twice that of the sample specimen of red oak.

R315.4 Alternate test method. As an alternate to having a flame-spread classification of not greater than 200 and a smoke developed index of not greater than 450 when tested in accordance with ASTM E 84, wall and ceiling finishes, other than textiles, shall be permitted to be tested in accordance with NFPA 286. Materials tested in accordance with NFPA 286 shall meet the following criteria:

During the 40 kW exposure, the interior finish shall comply with Item 1. During the 160 kW exposure, the interior finish shall comply with Item 2. During the entire test, the interior finish shall comply with Item 3.

1. During the 40 kW exposure, flames shall not spread to the ceiling.
2. During the 160 kW exposure, the interior finish shall comply with the following:
    2.1. Flame shall not spread to the outer extremity of the sample on any wall or ceiling.
    2.2. Flashover, as defined in NFPA 286, shall not occur.
3. The total smoke released throughout the NFPA 286 test shall not exceed 1,000 m^2.

❖ This section allows the use of NFPA 286 in lieu of ASTM E84 for testing of wall and ceiling finishes other than textiles. NFPA 286 is known as a "room corner" fire test. In this test, a fire source consisting of a wood crib is placed in the corner of a compartment. The materials tested are then placed upon the walls of the compartment. See Commentary Figure R315.4. This generally provides a more realistic understanding of the hazards involved with the materials.

There are two levels of exposure during an NFPA 286 fire test in order to better represent a growing fire. The first is 40 kw fire size for 5 minutes and then 160 kw for 10 minutes. The 40 kw exposure represents the beginning of a fire where the initial spread is critical. Therefore the stated criteria is that the fire can not spread to the ceiling. The 160 kw exposure is obviously a more intense fire, and the criteria relates to preventing flashover (as defined by NFPA 286) and the extent of flame spread throughout the entire test assembly. There is also a total smoke production limit of 1000 m^2 for both levels of exposure.

The flashover criteria for NFPA 286 are:

Heat release exceeds 1 MW,
Heat flux at the floor exceeds 20 kw/m^2,
Average upper layer temperature exceeds 600°C (1112°F),
Flames exit the doorway,
Auto ignition of paper target on the floor occurs.

[B] SECTION R316
INSULATION

R316.1 Insulation. Insulation materials, including facings, such as vapor retarders or vapor permeable membranes installed within floor-ceiling assemblies, roof-ceiling assemblies, wall assemblies, crawl spaces and attics shall have a flame-spread index not to exceed 25 with an accompanying smoke-developed index not to exceed 450 when tested in accordance with ASTM E 84.

Exceptions:
1. When such materials are installed in concealed spaces, the flame-spread and smoke-developed limitations do not apply to the facings, provided that the facing is installed in substantial contact with the unexposed surface of the ceiling, floor or wall finish.
2. Cellulose loose-fill insulation, which is not spray applied, complying with the requirements of Section R316.3, shall only be required to meet the smoke-developed index of not more than 450.

❖ Section R316 addresses insulating materials installed in building spaces. Insulating materials can affect fire development and fire-spread and therefore are regulated. Section R316.1 addresses the various insulating materials that may be installed in building spaces, including insulating batts, blankets, fills (including vapor barriers and vapor permeable membranes), and other coverings. Exposed insulating materials represent the same fire exposure hazard as any other exposed material, such as an interior finish. The provisions of Sections R316.2, R316.3, R316.4, as well as the foam plastic provisions of Section R314 should also be reviewed based on the actual type of insulation and how it is installed. As a general requirement, insulation, including facings used as vapor retarders or as breather papers, must have a flame-spread index not in excess of 25 and a smoke-developed index not in excess of 450. These values limit the contribution of the insulation to a fire. The flame-spread requirement of 25 for the insulation will be more limiting than the 200, which is accepted for interior finishes by Section R315. The test method used to establish these limits is the ASTM E 84 standard. See the discussion of Section R315.3 for additional information.

The two exceptions address situations where, because of the way the material is installed or because of other imposed regulations, the material does not make any significant contribution to a fire. Exception 1 eliminates the flame-spread and smoke-developed rating for the facing portion of the insulation if it is installed "in

BUILDING PLANNING

substantial contact" with the unexposed surface of the ceiling, floor, or wall finish. For example, when paper-backed insulation is placed directly on top of a ceiling, the paper facing is not required to meet the smoke and flame-spread limits. If the same material is applied to the underside of a roof deck and the paper facing is exposed to the attic space, the paper facing must then meet the general criteria. The potential for flame spread is greatly diminished when the facings are installed in direct contact with the finish material because of the lack of airspace to support a fire if the facing were to be exposed to a source of ignition. See Commentary Figure R316.1 for an example of the various facing provisions.

Exception 2 addresses the fact that cellulose loose-fill insulation is federally regulated by the Consumer Product Safety Commission (CPSC). Parts 1209 and 1404 of CPSC 16 CFR contain various requirements that regulate the product in order to avoid excessive flammability or significant fire hazards. The smoke-developed index for cellulose loose-fill insulation must be determined by the ASTM E 84 test and must be 450 or less. The statement regarding the smoke-developed index requires that the procedure is done by the ASTM E 84 test and not by the test procedures specified in Section R316.2.

R316.2 Loose-fill insulation. Loose-fill insulation materials that cannot be mounted in the ASTM E 84 apparatus without a screen or artificial supports shall have a flame-spread rating not to exceed 25 with an accompanying smoke-developed factor not to exceed 450 when tested in accordance with CAN/ULC-S102.2.

Exception: Cellulose loose-fill insulation shall not be required to comply with this test method provided that such insulation complies with the requirements of Section R316.3.

❖ The main provision establishes that CAN/ULC-S102.2 is used as the test standard to determine the flame-spread and smoke-developed ratings for loose-fill insulation materials that cannot be tested using the normal ASTM E 84 test method. The exception makes a distinction between cellulose insulation, which is spray applied using a water-mist spray applicator, and cellulose loose-fill insulation, which is poured or blown into place. Spray-applied cellulose insulation can be exposed on vertical and horizontal ceiling-type surfaces, so it is treated as any other insulating material in regard to testing. Cellulose loose-fill insulation, which is poured or blown in, is regulated by the Consumer Product Safety Commission requirements (see discussion at Section R316.1, Exception 2), and is exempt from the test procedure described in this section.

Figure R315.4
INTERIOR FIRE TEST ROOM DIMENSIONS

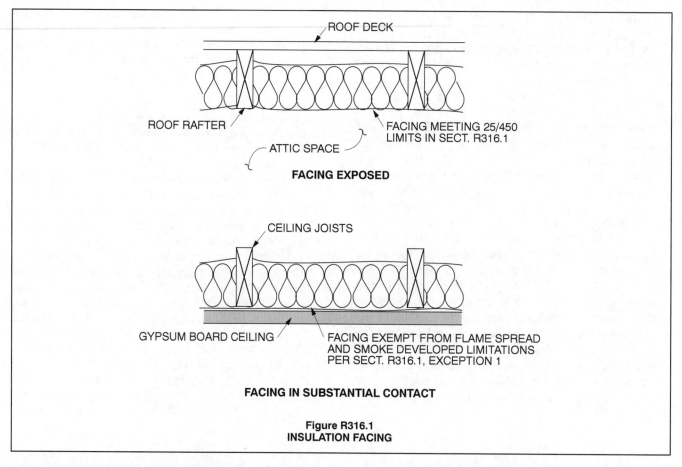

Figure R316.1
INSULATION FACING

R316.3 Cellulose loose-fill insulation. Cellulose loose-fill insulation shall comply with CPSC 16 CFR, Parts 1209 and 1404. Each package of such insulating material shall be clearly labeled in accordance with CPSC 16 CFR, Parts 1209 and 1404.

❖ Because cellulose loose-fill insulation is federally regulated, this section provides the reference to the various federal regulations that are to be used. Because the federal regulations have precedence, the IRC cannot impose other requirements on this product. Therefore, this section and the reference to it from Section R316.1, Exception 2 establish the requirements for this product and provide the exemption from the normal ASTM E 84 test standard.

R316.4 Exposed attic insulation. All exposed insulation materials installed on attic floors shall have a critical radiant flux not less than 0.12 watt per square centimeter.

❖ This section provides the performance requirements for the test exposure that insulation must meet when it is exposed on the floor of an attic. It is tied to the testing provisions found in Section R316.5, which specifies that the ASTM E 970 test is to be used for determining the critical radiant flux. See the commentary for Section R316.5 regarding the application of this requirement to cellulose loose-fill insulation.

R316.5 Testing. Tests for critical radiant flux shall be made in accordance with ASTM E 970.

❖ ASTM E 970 is a test method developed by the insulation industry to evaluate the fire hazard of exposed attic insulation and is referenced in the material standards for insulation. Cellulose loose-fill insulation must comply with CPSC 16 CFR, Part 1209 (see Section R316.3), which requires testing by this standard. Spray-applied cellulose insulation that is not subject to the CPSC standard (see R316.1 and Exception 2 in that section) is also subject to the ASTM E 970 testing through Section R316.4.

SECTION R317
DWELLING UNIT SEPARATION

R317.1 Two-family dwellings. Dwelling units in two-family dwellings shall be separated from each other by wall and/or floor assemblies having not less than 1-hour fire-resistance rating when tested in accordance with ASTM E 119. Fire-resistance-rated floor-ceiling and wall assemblies shall extend to and be tight against the exterior wall, and wall assemblies shall extend to the underside of the roof sheathing.

Exception: A fire resistance rating of $1/2$ hour shall be permitted in buildings equipped throughout with an automatic sprinkler system installed in accordance with NFPA 13.

❖ Most of the nation's fires occur in residential buildings, particularly one- and two-family dwellings. These fires

BUILDING PLANNING

account for more than 80 percent of all deaths from fire in residential uses (including hotels, apartments, dormitories, etc.) and about two-thirds of all fire fatalities in any type of building. One- and two-family dwellings also account for more than 80 percent of residential property losses and more than one-half of all property losses from fire. Despite this poor fire record, there is wide-spread resistance to mandating much in the way of fire protection systems or methods due to our society's viewpoint that people's homes are their castles. This viewpoint has limited the types of protection that are imposed on these private homes to the installation of smoke-alarms and the more recent requirement of dwelling unit separation. Section R317 provides a separation for protection of the occupants of one dwelling unit from the actions of their neighbor. The requirements of this section pertain to any structure regulated by this code other than a single-family dwelling unit. To accomplish this protection, the code addresses separation between the units, structural support, and any openings or penetrations of the separation.

Depending on the layout of the various dwelling units, Section R317.1 requires that the walls and/or floor assemblies that divide one dwelling unit from the adjacent unit be of at least one-hour fire-resistant construction. See Commentary Figure R317.1 for examples of the separation. The separation rating is to be determined by ASTM E 119: Standard Test Method for Fire Tests of Building Construction and Materials, which is the normal test used for determining fire-resistance. Many tested assemblies are available for use in these locations. The provisions of the section also address the continuity of the separation, so that one dwelling unit is completely divided from the other. The horizontal aspect of the separation, which requires the assemblies to extend to and be tight against the exterior wall, is not difficult to comply with. It is most likely the vertical aspect (continuing a wall assembly to the underside of the roof sheathing) that will require some detailed planning, careful construction, and careful inspection in order for the units to be separated.

The exception grants a reduction in the required separation for those cases in which the building is equipped with an automatic sprinkler system. In these cases, a rating of $1/2$ hour is permitted versus a one-hour fire-resistance rating. The sprinkler system is required to be "installed in accordance with NFPA 13," and it is required to be installed "throughout" the building. The type of sprinkler system used must meet the NFPA 13 standard and may not be installed to either the NFPA 13R or 13D standards, even though those

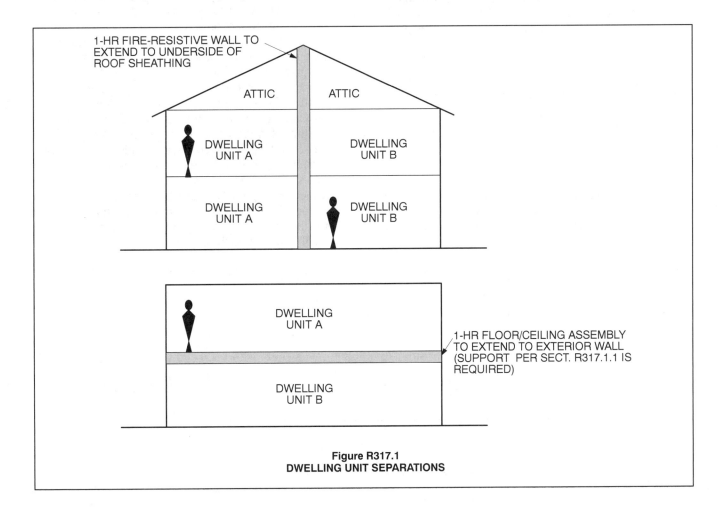

Figure R317.1
DWELLING UNIT SEPARATIONS

two standards do address certain types of residential uses. The word "throughout" requires that the sprinkler system be installed in all portions of both dwelling units and any common spaces. The provisions of the NFPA 13 standard that permit sprinklers to be omitted in certain areas such as small concealed spaces are applicable. Therefore, the provision requires a complying sprinkler system "throughout" the building (that is, in all areas of the building that are required to be protected according to the standard), and it does not accept any partial system such as one installed in only one dwelling unit or only in the basement level of both units.

R317.1.1 Supporting construction. When floor assemblies are required to be fire-resistance-rated by Section R317.1, the supporting construction of such assemblies shall have an equal or greater fire-resistive rating.

❖ This provision applies to only the floor assemblies that form the separation between different dwelling units. When either all or portions of a dwelling unit separation is provided by a floor assembly, the code requires that the structural supports for the separation have a rating equal to or higher than the floor. This is conceptually similar to the garage separation of Section R309.2. Without the supporting construction being protected, a fire on the lower level could lead to an early failure of the dwelling unit separation. See Commentary Figure R317.1.1.

R317.2 Townhouses. Each townhouse shall be considered a separate building and shall be separated by fire-resistance-rated wall assemblies meeting the requirements of Section R302 for exterior walls.

Exception: A common 2-hour fire-resistance-rated wall is permitted for townhouses if such walls do not contain plumbing or mechanical equipment, ducts or vents in the cavity of the common wall. Electrical installations shall be installed in accordance with Chapters 33 through 42. Penetrations of electrical outlet boxes shall be in accordance with Section R317.3.

❖ The application of this section has its basis in the exterior wall requirements found in Section R302 that deal with the building's location on the lot. The definition of a townhouse in Section R202 should be reviewed as well as the requirement for structural independence in Section R317.2.4. In general, since the "exterior wall" of the townhouse is essentially being constructed with no fire separation distance where one townhouse adjoins another, the code requires by Section R302 that the wall have not less than a one-hour fire-resistive rating. The adjacent townhouse would have the same requirement. Therefore, the general requirement at this location (based on Sections R302.1 and R317.1) would be that each townhouse have its own "exterior wall." This would result in the construction of two separate one-hour walls located side by side where one townhouse adjoins another.

Because of the difficulties of constructing and the potential for unnecessary duplication, the exception will provide an alternative to the two separate one-hour walls by permitting the construction of a single or "common" two-hour rated wall between the townhouses. See Commentary Figure R317.2 for an illustration of the separate and the common walls. This exception has its basis in the actions of many building officials who permit this type of common wall as an alternative method of construction using provisions similar to those found in Section R104.11. Because the common wall has the potential to create an interconnection between the adjacent dwelling units and reduce the clear separation that would exist if two separate walls were constructed, the code places limits on services being located within the wall. This exception does not permit the inclusion of any type of plumbing, mechanical equipment, ducts or vents within the cavity of the common wall. This prohibition is applicable even if the penetrations or openings are protected by the penetration provisions of Section R317.3 or if a damper is installed in the duct or vent. The prohibition on plumbing includes all types of plumbing materials and systems as well as water supply and drainage piping of either combustible or noncombustible materials. The exception

**Figure R317.1.1
SUPPORT OF DWELLING UNIT SEPARATION**

BUILDING PLANNING

permits the cavity of the wall to be used for electrical installations provided they comply with the electrical provisions of the code and that the penetrations are properly protected.

R317.2.1 Continuity. The common wall for townhouses shall be continuous from the foundation to the underside of the roof sheathing, deck or slab and shall extend the full length of the common wall including walls extending through and separating attached accessory structures.

❖ This section addresses the continuity of the common wall permitted between townhouses by the exception in Section R317.2. The requirements are conceptually similar to the continuity issues that exist in Section R317.1. These provisions, by regulating the extension and termination of the wall, can make possible the separation of dwelling units from each other. The provisions do not address or apply to the condition where two separate one-hour fire-resistive "exterior" walls complying with Section R302 exist between the dwelling units.

R317.2.2 Parapets. Parapets constructed in accordance with Section R317.2.3 shall be provided for townhouses as an extension of exterior walls or common walls in accordance with the following:

1. Where roof surfaces adjacent to the wall or walls are at the same elevation, the parapet shall extend not less than 30 inches (762 mm) above the roof surfaces.

2. Where roof surfaces adjacent to the wall or walls are at different elevations and the higher roof is not more than 30 inches (762 mm) above the lower roof, the parapet shall extend not less than 30 inches (762 mm) above the lower roof surface.

   Exception: A parapet is not required in the two cases above when the roof is covered with a minimum class C roof covering, and the roof decking or sheathing is of noncombustible materials or approved fire-retardant-treated wood for a distance of 4 feet (1219 mm) on each side of the wall or walls, or one layer of $^5/_8$-inch (15.9 mm) Type X gypsum board is installed directly beneath the roof decking or sheathing for a distance of 4 feet (1219 mm) on each side of the wall or walls.

3. A parapet is not required where roof surfaces adjacent to the wall or walls are at different elevations and the higher roof is more than 30 inches (762 mm) above the lower roof. The common wall construction from the lower roof to the underside of the higher roof deck shall not have less than a 1-hour fire-resistive rating. The wall shall be rated for exposure from both sides.

❖ This section provides for the continuation of the dwelling-unit separation by requiring that a parapet be constructed above the common wall or the two exterior walls that occur between adjacent townhouses. This parapet requirement is only applicable to the wall or walls between townhouses and does not apply to the separation in a two-family dwelling, nor does it apply to other exterior walls that occur on either townhouses or dwellings.

The code provides three requirements that address the details of the parapet depending on the height of the adjacent roofs. The code also provides an exception that can be used to eliminate the need for the parapet in the two conditions that require them. In general, parapets are required to extend at least 30 inches (762

Figure R317.2
DWELLING UNIT SEPARATION FOR TOWNHOUSES

mm) above the roof surfaces of the adjacent townhouses. This standard requirement is found in Item 1 and is applicable where the roof surfaces of the adjacent dwelling units are at the same level. The second item addresses the requirement for a parapet where the roofs are at different levels, but the difference is less that 30 inches (762 mm). Under this condition, the parapet height is still required to be 30 inches (762 mm), but it is measured only from the roof surface of the lower roof. The third item requires a parapet, but it may not be apparent to an observer since there is no requirement for anything to extend above the roof surfaces. In this case, the roofs must be more than 30 inches (762 mm) different in their heights, and the code requires the common wall between the units to be rated to the height of the upper roof deck. The rating of this portion of the wall is a one-hour fire-resistive rating and not the typical two-hour (or two separate one-hour walls) that separate the actual dwelling units. See Commentary Figure R317.2.2(1) for an illustration of the three requirements.

The exception found in this section applies only to the first two items of the section. See Commentary Figure R317.2.2(2) for a depiction of the exception. It does not apply where the roofs are more than 30 inches (762 mm) different in height and the provisions of Item 3 have been applied.

Figure R317.2.2(1)
PARAPET REQUIREMENTS FOR WALL BETWEEN TOWNHOUSES

BUILDING PLANNING

**Figure R317.2.2(2)
EXCEPTION TO ELIMINATE PARAPET**

R317.2.3 Parapet construction. Parapets shall have the same fire-resistance rating as that required for the supporting wall or walls. On any side adjacent to a roof surface, the parapet shall have noncombustible faces for the uppermost 18 inches (457 mm), to include counterflashing and coping materials. Where the roof slopes toward a parapet at slopes greater than two units vertical in 12 units horizontal (16.7-percent slope), the parapet shall extend to the same height as any portion of the roof within a distance of 3 feet (914 mm), but in no case shall the height be less than 30 inches (762 mm).

❖ In addition to having the same degree of fire resistance as required for the wall or walls below, the surface of the parapet that faces the roof must be of noncombustible materials for the upper 18 inches (457 mm). Thus, a fire that might travel along the roof and reach the parapet would not be able to continue upward along the face of the parapet and over the top, exposing the adjacent building. The requirement applies to the upper 18 inches (457 mm) of the parapet to allow for extending the roof covering up the base of the parapet so that it can be effectively flashed. The 18-inch (457 mm) figure is based on a parapet height of at least 30 inches (762 mm) and would not be applicable to the face of the parapet that was toward the higher roof deck as constructed using Section R317.2.2, Item 2, nor would it be applicable where the parapet is not required based on the exception in Section R317.2.2 or on Item 3 of that section.

The 30-inch (762 mm) requirement is measured "above the roof surface" as stated in Section R317.2.2. When a cricket or other element is installed adjacent to the parapet, the 30-inch (762 mm) dimension must be taken from the top of the roof surface on the cricket so that the parapet truly extends 30 inches (762 mm) above the roof.

In those cases where the roof slopes upward away from the parapet and the slope exceeds 2 units vertical in 12 units horizontal (16.7 percent slope), the parapet must be extended to the same height as any portion of the roof that is within a distance of three feet. However, in no case can the height of the parapet be less than 30 inches (762 mm). See Commentary Figure R317.2.3 for an illustration of this requirement.

R317.2.4 Structural independence. Each individual townhouse shall be structurally independent.

Exceptions:
1. Foundations supporting exterior walls or common walls.
2. Structural roof and wall sheathing from each unit may fasten to the common wall framing.
3. Nonstructural wall coverings.
4. Flashing at termination of roof covering over common wall.
5. Townhouses separated by a common 2-hour fire-resistance-rated wall as provided in Section R317.2.

❖ Each townhouse is to be structurally independent and capable of being removed without affecting the adjacent dwelling unit. This provision is only applicable to townhouses, not two-family dwellings. This independence is useful not only in the event of a fire within one unit, but also during any remodeling or alteration. The objective of this structural independence is that a complete burnout could occur on one side of the wall without causing the collapse of the adjacent townhouse. This condition occurs rarely. The provision also helps if there ever is a fire or other problem by creating a clear separation between the units. With separate ownership and each owner having a different insurance company, the ability to gain access or get repairs made can be difficult and time consuming. By having clearly separated units, it is much easier to determine who is responsible and to make any needed repairs.

The code provides five exceptions that waive the structural independence. A quick review of the exceptions shows that these exceptions generally deal with items that will not structurally affect the townhouses should a problem develop in the adjacent dwelling unit. Exception 1 is based on the norm within the industry for foundation construction. Section R402 only lists wood and concrete within the foundation materials section, although Section R404 accepts masonry foundation walls. In general, concrete and masonry are the most common types of foundations; wood foundations are viewed as unique. Given the performance of both masonry and concrete and the fact that these foundation systems must sustain loads from both the structure and the adjacent soils, it is reasonable to assume that the foundation will not be the item that fails in most situations. Permitting a common foundation also helps solve other problems that would arise if the structural independence issue were taken as an absolute. An example where requiring separate foundations would probably create more problems or difficulty is in the damp-proofing or water-proofing of below-grade foundation walls.

If a wood foundation is used between adjacent units, what is the level of fire-protection that may be needed? Since concrete and masonry foundations are the norm, it would be easy to forget or overlook protecting the foundation when it is constructed of wood. In these cases, it would seem appropriate to deal with the foundation as any other wall, and provide it with protection on any exposed side. The level of fire-resistance should be equal to that of the wall or walls that the foundation supports.

R317.3 Rated penetrations. Penetrations of wall or floor/ceiling assemblies required to be fire-resistance-rated in accordance with Section R317.1 or R317.2 shall be protected in accordance with this section.

❖ This section addresses the specific requirements for maintaining the integrity of fire-resistance rated assemblies at penetrations. If the penetration of a rated assembly is not properly constructed, then the assembly itself is jeopardized and may not perform as intended. The provisions of this section apply to penetrations of fire-resistance rated walls and floor/ceiling assemblies that are a part of the dwelling unit separation in either two-family dwellings or in townhouses. Penetrations of the rated assemblies range from combustible pipe and tubing, to noncombustible wiring with combustible covering to noncombustible items such as pipe, tube, conduit and ductwork.

Each type of penetration requires a specific method of protection, which is based on the type of fire-resist-

For SI: 1 inch = 25.4 mm, 1 foot = 304.8 mm.

**Figure R317.2.3
PARAPET REQUIREMENTS**

BUILDING PLANNING

ance rated assembly penetrated and the size and type of the penetrating item. The first step in determining the type of penetration protection required is to identify whether a wall or floor/ceiling assembly is being penetrated. The next step is to determine the type of penetrating item and whether it is a membrane or through-penetration. Once these factors are known, then the applicable section must be applied and the applicable method of protection must be decided upon.

R317.3.1 Through penetrations. Through penetrations of fire-resistance-rated wall or floor assemblies shall comply with Section R317.3.1.1 or R317.3.1.2.

Exception: Where the penetrating items are steel, ferrous or copper pipes or steel conduits, the annular space shall be permitted to be protected as follows:

1. In concrete or masonry wall or floor assemblies where the penetrating item is a maximum 6 inches (152 mm) nominal diameter and the opening is a maximum 144 square inches (92 900 mm^2), con-crete, grout or mortar shall be permitted where installed to the full thickness of the wall or floor assembly or the thickness required to maintain the fire-resistance rating.

2. The material used to fill the annular space shall prevent the passage of flame and hot gases sufficient to ignite cotton waste where subjected to ASTM E 119 time temperature fire conditions under a minimum positive pressure differential of 0.01 inch of water (3 Pa) at the location of the penetration for the time period equivalent to the fire-resistance rating of the construction penetrated.

❖ This section provides the general requirements for through penetrations, which is an opening that passes through an entire assembly. A through penetration is in contrast to a membrane penetration, which creates an opening through only one side of an assembly. Membrane penetrations are addressed later in Section R317.3.2. See Commentary Figure R317.3.1 for an illustration of these two types of penetrations.

Through penetrations must be protected in order to maintain the fire-resistance of the penetrated assembly. The code provides two methods, found in Sections R317.3.1.1 and R317.3.1.2, that can be used to assure the adequacy of the penetration protection. The difference between these two is the test methodology used, but they both provide essentially the same results. The commentary for those sections provides additional discussion of the differences.

Based on the history of these provisions and on the wealth of fire test data that exists concerning items such as conduit, water piping and other similar penetrations, the code provides two exceptions that permit protection by methods other than those generally required. The first permits the use of concrete, grout or mortar to provide the protection for certain penetrations of concrete and masonry wall or floor assemblies. The concrete, grout or mortar must be provided for the full thickness of the assembly, unless evidence can be produced demonstrating that the required fire-resistance rating can be achieved with a lesser depth. Concrete, grout and mortar have traditionally been used as protection for the annular space in penetrations of concrete and masonry assemblies. Experience has shown this form of protection to be viable. However, caution must be used any time something such as a water pipe or conduit is placed in concrete or masonry. Sections P2603.3 and P2603.5 provide examples that relate to the protection of plumbing systems.

Exception 2 addresses the space between the penetrating item and the original assembly construction. This gap is called the annular space, and this exception provides a method to simply evaluate the performance of the material used to fill that space. It is often mistakenly believed that this exception permits a variety of untested items, but as can be seen from the provision itself, the materials need to meet a specific performance level. This exception requires that the material be prequalified as to its ability to prevent the passage of flame and hot gasses sufficient to ignite cotton when subjected to the time—temperature criteria of the ASTM E 119 test standard. This requirement is similar to provisions found within both the ASTM E 119 and ASTM E 814 standards, the standards used to evaluate fire-resistive assemblies and penetration protection. Since it is very likely that the penetration in the actual fire will be exposed to a positive pressure, this section specifies that the test-fire exposure include a positive pressure of 0.01 inch (0.25 mm) of water column as a further means to verify the performance of this protection method. Thus the protection will not be blown out or moved from its place during a fire.

R317.3.1.1 Fire-resistance-rated assembly. Penetrations shall be installed as tested in the approved fire-resistance-rated assembly.

❖ This section addresses situations where the penetration is tested as a part of the regular full-scale test for the wall or floor/ceiling assembly. The penetration and proposed type of protection are evaluated as a part of the regular ASTM E 119 test, which evaluates the wall or floor/ceiling's rating. This section and the option it provides are not used frequently because of the cost of conducting such full-scale tests and the limitations placed on the tested assembly relative to its application. Because of these issues, penetration protection is most often provided in accordance with one of the exceptions in Section R317.3.1 or based on the provisions of Section R317.3.1.2.

R317.3.1.2 Penetration firestop system. Penetrations shall be protected by an approved penetration firestop system installed as tested in accordance with ASTM E 814 or UL 1479, with a minimum positive pressure differential of 0.01 inch of water (3 Pa) and shall have an F rating of not less than the required fire-resistance rating of the wall or floor/ceiling assembly penetrated.

❖ Through-penetration firestop systems consist of specific materials or an assembly of materials that are designed to restrict the passage of fire and hot gasses for a prescribed period of time through openings made in fire-resistance-rated assemblies. In order to determine the effectiveness of a through-penetration firestop system in restricting the passage of fire, and in order to determine that the penetration has not jeopardized the original fire-resistive assembly, firestop systems are required to be subjected to fire testing by the ASTM E 814 or UL 1479 test standard. This is a small-scale test method developed specifically for the evaluation of a firestop system's ability to resist the passage of flame and hot gases, withstand thermal stresses, and restrict the transfer of heat through the penetrated assembly. There are hundreds if not thousands of tested through-penetration firestop systems available today. The actual type of system used will depend on the type and construction of the assembly being penetrated, the material makeup and size of the penetrating item, and the size of the annular space that exists between the penetrating item and the original assembly. Because there are a multitude of products available, and since there is no "one size fits all" system available, it is helpful if the methods of protection are included on the construction documents as covered by Section R106.1.1.

The actual rating of the through-penetration firestop system is generated from the results of the testing and are reported as an "F" (flame) rating and a "T" (temperature) rating. The IRC requires only an "F" rating. The "F" rating indicates the period of time, in hours, that the TPFS tested remained in place without allowing the passage of fire during the fire exposure test, or the passage of water during the hose stream portion of the test. The required "F" rating of a TPFS must be equal to the fire-resistance rating of the wall or floor/ceiling assembly that is being penetrated. This means either a one- or two-hour rating depending on the dwelling unit separation.

Two of the most common materials used in through-penetration firestop systems are intumescent and endothermic materials. Intumescent materials expand approximately 8 to 10 times their original volume when exposed to temperatures exceeding 250°F (121°C). The expansion of the material fills the voids or openings within the penetration to resist the passage of flame, while the outer layer of the expanded intumescent material forms an insulating charred layer that assists in limiting the transfer of heat. The expansion properties of intumescent materials allow them to seal openings left by combustible penetrating items that burn away during a fire, but they do not retard heat as well as endothermic materials. Intumescent materials

Figure R317.3.1
TYPES OF PENETRATIONS

BUILDING PLANNING

are typically used with combustible penetrating items or where a higher "T" rating is not required.

Endothermic materials provide protection through chemically-bound water released in the form of steam when exposed to temperatures exceeding 600°F (316°C). This released water cools the penetration and retards heat transfer through the penetration. Endothermic materials tend to be superior in heat-transfer resistance and have higher "T" ratings, but they do not expand to fill voids left by combustible penetrating items that burn away during a fire. Therefore, endothermic materials are typically used with noncombustible penetrating items and where a higher "T" rating is required.

R317.3.2 Membrane penetrations. Membrane penetrations shall comply with Section R317.3.1. Where walls are required to have a minimum 1-hour fire resistance rating, recessed light fixtures shall be so installed such that the required fire resistance will not be reduced.

Exceptions:

1. Steel electrical boxes that do not exceed 16 square inches (0.0103m^2) in area provided the total area of such openings does not exceed 100 square inches (0.0645 m^2 for any 100 square feet (9.29 m^2) of wall area. Outlet boxes on opposite sides of the wall shall be separated as follows:

   1.1. By a horizontal distance of not less than 24 inches (610 mm);

   1.2. By a horizontal distance of not less than the depth of the wall cavity when the wall cavity is filled with cellulose loose-fill, rockwool or slag mineral wool insulation;

   1.3. By solid fire blocking in accordance with Section R602.8.1;

   1.4. By protecting both outlet boxes by listed putty pads; or

   1.5. By other listed materials and methods.

2. Membrane penetrations for listed electrical outlet boxes of any materials are permitted provided such boxes have been tested for use in fire resistance-rated assemblies and are installed in accordance with the instructions included in the listing. Outlet boxes on opposite sides of the wall shall be separated as follows:

   2.1. By a horizontal distance of not less than 24 inches (610 mm);

   2.2. By solid fire-blocking in accordance with Section R602.8;

   2.3. By protecting both outlet boxes by listed putty pads; or

   2.4. By other listed materials and methods.

3. The annular space created by the penetration of a fire sprinkler provided it is covered by a metal escutcheon plate.

❖ This section deals with instances where only a single side of the fire-resistance rated assembly is penetrated. This would be the situation for items such as electrical outlet boxes or plumbing fixtures located on one side of the wall only. Commentary Figure R317.3.1 shows this type of penetration. For the most part, a membrane penetration is to be protected by one of the previously described methods established for through penetrations. However, there are some penetrations that are allowed without a specific firestopping material in the annular space around such penetrations. These are addressed by the exceptions. This section also deals with the installation of recessed light fixtures in fire-resistance rated assemblies and states that their installation may not reduce the assembly's protection. Although these fixtures are common, they do represent a penetration of the assembly's protection and as such need to be installed so that the assembly is not compromised.

Exception 1 allows penetrations of steel electrical outlet boxes under certain conditions. The criteria of this section limit the size of the box to 16 square inches (0.0103 m^2) or less in area and to an aggregate area not to exceed 100 square inches (64 500 mm^2) in 100 square feet (9.3 m^2). Commentary Figure R317.3.2(1) shows some of the requirements of this section. The

For SI: 1 inch = 25.4 mm, 1 square inch = 645.16 mm^2, 1 square foot = 0.093m^2.

**Figure R317.3.2(1)
MEMBRANE PENETRATION BY OUTLET BOX**

area limitations are consistent with the criteria from fire tests, which have shown that within these limitations, these penetrations will not adversely affect the fire-resistance rating of the assembly. However, the boxes are assumed to be installed as they were during the fire tests. In general the test requirements match the limitations shown by the code regarding their size and the need to be offset. An additional requirement, one that does not appear in the code, regulates the size of the annular space created around the outlet boxes. Both the Underwriter's Laboratory's Fire Resistance Directory and the Gypsum Association's Fire-Resistance Design Manual specify a maximum over-cut of $^1/_8$ inch (3 mm) for the annular space around the outlet boxes. Additionally, Article 373 of the National Electrical Code (NEC) includes the size limitation of the over-cut. Therefore, the exception only applies when the boxes are installed as they were during the original fire tests, including the limited size annular space. Since outlet boxes on both sides of a wall create penetrations of both layers of a wall assembly's protection, the code provides five methods to address this problem. This gives code users several options and does not limit them to the usual 24-inch (610 mm) offset.

Exception 2 permits outlet boxes of nonmetallic materials to be used if they have been specifically tested. Since there are many different types of nonmetallic boxes available, it is important to determine that the boxes being used in the rated dwelling unit separation have been tested. Although the exception applies to nonmetallic electrical outlet boxes, the same concept would be apply to steel boxes that exceed the sizes specified in Exception 1.

Exception 3 provides an alternative to the annular space protection provisions for a fire sprinkler that penetrates a single membrane. This exception is available provided the annular space around the sprinkler is completely covered by an escutcheon plate of noncombustible material. The nature of the hazard posed by single-membrane penetrations of the sprinkler is limited due to the size of the opening, the potential amount of openings present, and the presence of a sprinkler system. The installation of a noncombustible escutcheon provides protection against free passage of fire through the annular space and allows for the movement of the sprinkler piping without breaking during a seismic event. See Commentary Figure R317.3.2(2).

**Figure R317.3.2(2)
EXCEPTION TO ANNULAR SPACE PROTECTION**

SECTION R318
MOISTURE VAPOR RETARDERS

R318.1 Moisture control. In all framed walls, floors and roof/ceilings comprising elements of the building thermal envelope, a vapor retarder shall be installed on the warm-in-winter side of the insulation.

Exceptions:
1. In construction where moisture or freezing will not damage the materials.
2. Where the framed cavity or space is ventilated to allow moisture to escape.
3. In counties identified with footnote a in Table N1101.2.

❖ This section addresses the use of a vapor retarder for moisture control purposes at the exterior thermal envelope; it is applicable to walls, floors, and roof/ceiling assemblies of frame construction.

Water condensation in an exterior framed element can diminish the insulation value and lead to early structural failure. The code takes steps to eliminate these detrimental effects by requiring a vapor retarder, also referred to as a vapor barrier. As attic and under-floor spaces generally are provided with adequate means of ventilation, the exterior wall cavities are the locations where a vapor retarder is necessary. Water vapor is part of the atmosphere and as such travels with the air flow. However, as does independent gas, it migrates by diffusion through air and materials according to its own pressure differentials. Once inside the wall cavity, the vapor can encounter surfaces that are below its dew point and condense.

The location of the vapor barrier is important, but to understand why we must examine how moisture enters the wall cavity. Moisture can enter in at least two ways, through a hole with direct air migration or by diffusion. Much more moisture can enter and be deposited in a wall cavity through air currents than by diffusion alone. The first step to effective moisture control is construction of a tight envelope.

Assuming a tight envelope is present, moisture flows through a wall by the process of diffusion. Diffusion is driven by the water-vapor pressure differences between indoor air and outdoor air. For most of the year, in a large part of the United States, the vapor pressure is greater indoors than outdoors. Under these conditions, the moisture diffuses through the wall from inside to outside. Hence, the code provides for the vapor barrier on the warm-in-winter side of the wall, as shown in Commentary Figure R318.1.

Moisture migration is more complicated in air-conditioned buildings. A portion of the southern United States (along the Gulf of Mexico and the lower Atlantic

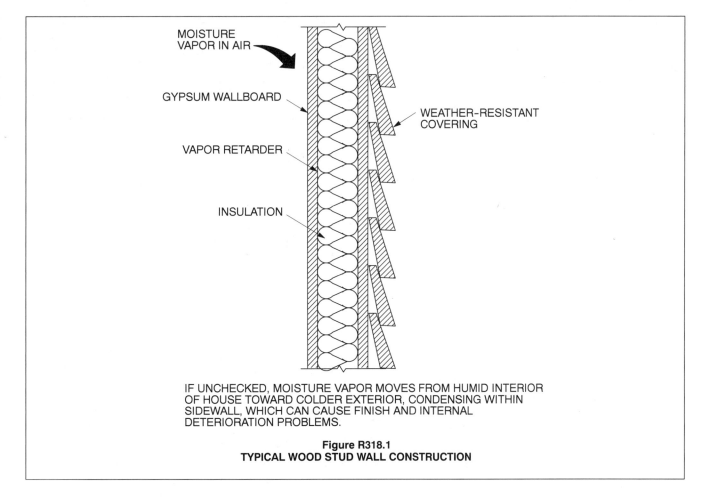

Figure R318.1
TYPICAL WOOD STUD WALL CONSTRUCTION

coast) is a humid climate. In these locations, identified in Note "a" in Table N1101.2, no vapor retarder is required, as the exterior weather membrane provides the necessary protection.

SECTION R319
PROTECTION AGAINST DECAY

R319.1 Location required. In areas subject to decay damage as established by Table R301.2(1), the following locations shall require the use of an approved species and grade of lumber, pressure treated in accordance with AWPA C1, C2, C3, C4, C9, C15, C18, C22, C23, C24, C28, C31, C33, P1, P2 and P3, or decay-resistant heartwood of redwood, black locust, or cedars.

1. Wood joists or the bottom of a wood structural floor when closer than 18 inches (457 mm) or wood girders when closer than 12 inches (305 mm) to the exposed ground in crawl spaces or unexcavated area located within the periphery of the building foundation.
2. All wood framing members that rest on concrete or masonry exterior foundation walls and are less than 8 inches (203 mm) from the exposed ground.
3. Sills and sleepers on a concrete or masonry slab that is in direct contact with the ground unless separated from such slab by an impervious moisture barrier.
4. The ends of wood girders entering exterior masonry or concrete walls having clearances of less than 0.5 inch (12.7 mm) on tops, sides and ends.
5. Wood siding, sheathing and wall framing on the exterior of a building having a clearance of less than 6 inches (152 mm) from the ground.
6. Wood structural members supporting moisture-permeable floors or roofs that are exposed to the weather, such as concrete or masonry slabs, unless separated from such floors or roofs by an impervious moisture barrier.
7. Wood furring strips or other wood framing members attached directly to the interior of exterior masonry walls or concrete walls below grade except where an approved vapor retarder is applied between the wall and the furring strips or framing members.

❖ This section addresses the need for minimum protection against decay damage for wood members located in certain locations.

For those portions of a wood-framed structure that are subject to damage by decay, the code mandates that the lumber be pressure-preservatively treated or be of a species of wood having a natural resistance to decay.

Crawl spaces and unexcavated areas under a building usually contain moisture-laden air. These spaces are required to be ventilated in accordance with Section R408 to remove as much moisture as possible before it causes decay. Wood placed a minimum specified distance above grade in unexcavated under-floor areas or crawl spaces as shown in Commentary Figure R319.1(1) is not required to be either treated wood or wood that is naturally decay resistant. These clearances below floor joists and beams are deemed to be the minimum necessary to allow adequate circulation and removal of moisture from the air and from the wood framing members. Such clearances apply within the exterior wall line of the building foundation.

Foundation walls will absorb moisture from the ground and by capillary action move it to framing mem-

Figure R319.1(1)
MINIMUM SEPARATION OF NONTREATED WOOD FROM GROUND

bers that are in contact with the foundation. Unless a minimum clearance of 8 inches (203 mm) is maintained from the finished grade to wood sills resting on concrete or masonry exterior foundation walls, decay-resistant or preservatively-treated wood as shown in Commentary Figure R319.1(2) must be used. The 8-inch (203 mm) clearance specified in this section has been determined to be great enough to prevent wetting of wood framing members under most circumstances.

Concrete and masonry slabs that are in direct contact with the earth are very susceptible to moisture because of absorption of ground water. This can occur on interior slabs as well as at the perimeter. In the case of wood sills or sleepers placed on concrete or masonry slabs, decay-resistant wood or pressure-treated wood is required where such slabs are in direct contact with the ground, as illustrated in Commentary Figure R319.1(3). Concrete that is fully separated from the ground by a vapor barrier is not in direct contact with earth.

A minimum $^1/_2$-inch (12.7 mm) clearance along the top, sides, and ends of wood members projecting into exterior masonry or concrete walls must be maintained as illustrated in Commentary Figure R319.1(4), unless the wood is treated or is of a species that is naturally decay resistant.

Experience has shown that wood siding may extend below the sill plate to within 6 inches (152 mm) of the earth without decaying. Commentary Figure R319.1(5) depicts the required minimum 6-inch (152 mm) clearance from the ground for wood siding,

Figure R319.1(2)
WOOD SILLS ON FOUNDATION WALLS

Figure R319.1(3)
WOOD SILLS AND SLEEPERS

sheathing, and wall framing on the exterior of a building. It should not be in direct contact with the foundation wall.

R319.1.1 Ground contact. All wood in contact with the ground and that supports permanent structures intended for human occupancy shall be approved pressure preservatively treated wood suitable for ground contact use, except un-treated wood may be used where entirely below groundwater level or continuously submerged in fresh water.

❖ Wood members that are designed to be in contact with the ground must be suitable for ground contact use. This provision applies to all wood members that support permanent structures designed for human occupancy. Untreated wood is permitted only where the wood members will be located below the groundwater

**Figure R319.1(4)
NONTREATED WOOD PROJECTING INTO WALL**

**Figure R319.1(5)
NONTREATED WOOD SIDING OR SHEATHING**

BUILDING PLANNING

level or where the members are continuously submerged in fresh water, since fungus that decays wood and termites cannot survive in a water-only environment with no oxygen.

R319.1.2 Geographical areas. In geographical areas where experience has demonstrated a specific need, approved naturally durable or pressure preservatively treated wood shall be used for those portions of wood members that form the structural supports of buildings, balconies, porches or simi-lar permanent building appurtenances when such members are exposed to the weather without adequate protection from a roof, eave, overhang or other covering that would prevent moisture or water accumulation on the surface or at joints between members. Depending on local experience, such members may include:

1. Horizontal members such as girders, joists and decking.
2. Vertical members such as posts, poles and columns.
3. Both horizontal and vertical members.

❖ This section provides the local jurisdiction with the authority to require additional protection when the need can be documented by experience. Such experience may include high water tables, extended high-humidity conditions, etc.

Guidance is provided in this section for the protection of wood members other than those specifically identified elsewhere in this section that are exposed to the weather without benefit of protective elements. Included in the list of possible wood members required to be pressure-preservatively treated or naturally durable wood are girders, joists, decking, columns and similar structural members of porches, balconies and other exterior building elements.

R319.1.3 Posts, poles and columns. Posts, poles and col-umns supporting permanent structures that are embedded in concrete in direct contact with the ground or embedded in concrete exposed to the weather shall be approved pressure preservatively treated wood suitable for ground contact use.

❖ Commentary Figure R319.1.3 illustrates post, pole, and column embedment of wood structural members within concrete in direct contact with the ground. This section contains provisions that will prevent decay in a situation that could result in allowing wetting of the wood. Wood posts or columns that are embedded in concrete are exposed to the wetting potential of the concrete on many surfaces, and they have no opportunity to be dried directly by circulating air. As such, they are susceptible to decay and rot such that naturally durable woods are not considered adequate. Only wood that has been preservative treated in accordance with the more stringent treatment methods is permitted for such applications.

R319.1.4 Wood columns. Wood columns shall be approved wood of natural decay resistance or approved pressure preservatively treated wood.

Exceptions:

1. Posts or columns which are either exposed to the weather or located in basements or cellars, supported by piers or metal pedestals projecting 1 inch (25.4 mm) above the floor or finished grade and 6 inches (152 mm) above exposed earth, and are separated therefrom by an approved impervious moisture barrier.
2. Posts or columns in enclosed crawl spaces or unexcavated areas located within the periphery of the building, supported by a concrete pier or metal pedestal at a height greater than 8 inches (203mm) from exposed ground, are separated therefrom by an impervious moisture barrier.

❖ Commentary Figure R319.1.4 shows where protection against decay is required for wood structural members on permeable surfaces in direct contact with the ground. Concrete or masonry that is in direct contact with the ground is susceptible to moisture because

Figure R319.1.3
WOOD POST EMBEDDED IN CONCRETE

of absorption of ground water. The exceptions permit the use of common framing lumber when the post or column is isolated from the earth or concrete. Generally, structural elements in basements or cellars are easier to inspect, which explains the lower clearances. Posts or columns in crawl spaces must have a greater separation distance to safeguard against termites between inspections. In both exceptions, an impervious moisture barrier must be provided to insure the post or column does not absorb water from the earth or concrete pier.

R319.2 Quality mark. Lumber and plywood required to be pressure preservatively treated in accordance with Section R319.1 shall bear the quality mark of an approved inspection agency that maintains continuing supervision, testing and inspection over the quality of the product and that has been approved by an accreditation body that complies with the requirements of the American Lumber Standard Committee treated wood program.

❖ The quality of lumber or plywood that has been pressure-preservatively treated in accordance with the code is identifiable through a quality mark, such as shown in Commentary Figure R319.2.1. Without the proper identifying mark, it would be impossible to ascertain in the field that preservative-treated wood conforms to the applicable standards. The identifying mark must be from an approved inspection agency that has continuous follow-up services. Additionally, the inspection agency must be certified as being competent by an approved organization. The required identifying mark is in addition to the requirements for grade marks to identify the species and grade as specified elsewhere in the code.

R319.2.1 Required information. The required quality mark on each piece of pressure preservatively treated lumber or plywood shall contain the following information:

1. Identification of the treating plant.
2. Type of preservative.
3. The minimum preservative retention.
4. End use for which the product was treated.
5. Standard to which the product was treated.
6. Identity of the approved inspection agency.
7. The designation "Dry," if applicable.

Exception: Quality marks on lumber less than 1 inch (25.4 mm) nominal thickness, or lumber less than nominal 1 inch by 5 inches (25.4 mm by 127 mm) or 2 inches by 4 inches (51 mm by 102 mm) or lumber 36 inches (914 mm) or less in length shall be applied by stamping the faces of exterior pieces or by end labeling not less than 25 percent of the pieces of a bundled unit.

❖ This section provides the information required to be shown on a quality mark for pressure-preservatively treated lumber or plywood. Commentary Figure R319.2.1 illustrates how this information may be shown on a quality label.

R319.3 Fasteners. Fasteners for pressure preservative and fire-retardant-treated wood shall be of hot-dipped galvanized steel, stainless steel, silicon bronze or copper.

Exception: One-half-inch (12.7 mm) diameter or greater steel bolts.

❖ Except for large steel bolts, all fasteners used in conjunction with pressure-preservatively treated wood must be corrosion resistant. Some chemicals used to preservative-treat or fire-retardant-treat wood may have a corrosive effect on fasteners when installed in locations that are high in moisture. Acceptable materials include stainless steel, silicon bronze, copper or

Figure R319.1.4
WOOD POST ON PERMEABLE PAD

hot-dipped galvanized. Such fasteners are also mandated for use with fire-retardant-treated wood members. Use of durable fasteners is required to decrease the likelihood that the load-carrying capacity of fasteners will be reduced by corrosion.

SECTION R320
PROTECTION AGAINST TERMITES

R320.1 Subterranean termite control. In areas favorable to termite damage as established by Table R301.2(1), methods of protection shall be by chemical soil treatment, pressure preservatively treated wood in accordance with the AWPA standards listed in Section R319.1, naturally termite-resistant wood or physical barriers (such as metal or plastic termite shields), or any combination of these methods.

❖ This section establishes the rules for protection of structures from damage due to termites. The methods of protection address not only wood members but also foam plastic materials.

Figure R301.2(6) illustrates those geographical areas where termite damage is probable. In those areas, the structure must be protected from termite damage in an appropriate manner. There are a number of methods permitted by the code to provide the necessary protection against termite damage. The most common method of termite control is soil poisoning. Alternatives include the use of pressure-preservatively treated, naturally termite-resistant wood and termite shields over perimeter walls. Often a combination of these methods is necessary to establish the required level of protection.

R320.1.1 Quality mark. Lumber and plywood required to be pressure preservatively treated in accordance with Section R320.1 shall bear the quality mark of an approved inspection agency which maintains continuing supervision, testing and inspection over the quality of the product and which has been approved by an accreditation body which complies with the requirements of the American Lumber Standard Committee treated wood program.

❖ See the commentary for Section R319.2.

R320.2 Chemical soil treatment. The concentration, rate of application and treatment method of the termiticide shall be consistent with and never less than the termiticide label.

❖ Where using a chemical poisoning of the soil for termite control, the chemical's instructions must be followed in detail. All treatment must be consistent with the recommendations of the chemical manufacturer.

R320.3 Pressure preservatively treated and naturally resistant wood. Heartwood of redwood and eastern red cedar shall be considered termite resistant. Pressure preservatively treated wood and naturally termite-resistant wood shall not be used as a physical barrier unless a barrier can be inspected for any termite shelter tubes around the inside and outside edges and joints of a barrier.

❖ Only the heartwood of redwood and eastern red cedar are naturally resistant to termite damage. Where such naturally termite-resistant wood or pressure-treated wood is used as a physical barrier, it must be located where it can be occasionally inspected for signs of termite activity.

R320.3.1 Field treatment. Field cut ends, notches and drilled holes of pressure preservatively treated wood shall be retreated in the field in accordance with AWPA M4.

❖ When pressure-preservatively treated wood is used for protection against termites, any cuts, notches, or

```
                    20_-20_ [3]
              GROUND CONTACT [6]
ABC [1]                 .40 [5]
                  AWPA—STDS [2]
XXX [7]
                  PRESERVATIVE [4]
                         KDAT [8]
                         X-XX [9]
```

1- THE IDENTIFYING SYMBOL, LOGO OR NAME OF THE ACCREDITED AGENCY.

2- THE APPLICABLE AMERICAN WOOD PRESERVER'S ASSOCIATION (AWPA) COMMODITY STANDARD.

3- THE YEAR OF TREATMENT IF REQUIRED BY AWPA STANDARD.

4- THE PRESERVATIVE USED, WHICH MAY BE ABBREVIATED.

5- THE PRESERVATIVE RETENTION.

6- THE EXPOSURE CATEGORY (E.G., ABOVE GROUND, GROUND CONTACT, ETC.).

7- THE PLANT NAME AND LOCATION; PLANT NAME AND NUMBER, OR PLANT NUMBER.

8- IF APPLICABLE, MOISTURE CONTENT AFTER TREATMENT.

9- IF APPLICABLE, LENGTH, AND/OR CLASS.

**Figure R319.2.1
TYPICAL LABEL FOR PRESSURE-TREATED WOOD**

bored holes done in the field must be retreated. The retreatment methods and materials are to be accomplished in accordance with AWPA M4, which regulates the care of preservative-treated wood products.

R320.4 Foam plastic protection. In areas where the probability of termite infestation is "very heavy" as indicated in Figure R301.2(6), extruded and expanded polystyrene, polyisocyanurate and other foam plastics shall not be installed on the exterior face or under interior or exterior foundation walls or slab foundations located below grade. The clearance between foam plastics installed above grade and exposed earth shall be at least 6 inches (152 mm).

Exceptions:

1. Buildings where the structural members of walls, floors, ceilings and roofs are entirely of noncombustible materials or pressure preservatively treated wood.

2. When in addition to the requirements of R320.1, an approved method of protecting the foam plastic and structure from subterranean termite damage is provided.

3. On the interior side of basement walls.

❖ In those geographical areas where there is a very high risk of termite infestation, the code prohibits the use of foam plastic materials below or adjacent to ground level. This prohibition includes foam plastics installed on the exterior face of foundation walls or slab foundations, under exterior or interior foundation walls or slab foundations below grade, or where located within 6 inches (152 mm) of exposed earth. Foam plastics are permitted where they are adequately protected or where the structural members are either noncombustible or of pressure-preservatively-treated wood.

[F] SECTION R321
SITE ADDRESS

R321.1 Premises identification. Approved numbers or addresses shall be provided for all new buildings in such a position as to be plainly visible and legible from the street or road fronting the property.

❖ The code provides that buildings have plainly visible and legible address numbers posted on the building or in such a place on the property that the building may be identified by emergency services such as fire, medical, and police. The primary concern is that emergency forces should be able to locate the building without going through a lengthy search procedure. In furthering the concept, the code states that the approved street numbers be placed in a location readily visible from the street or roadway fronting the property if a sign on the building would not be visible from the street.

[B] SECTION R322
ACCESSIBILITY

R322.1 Scope. Where there are four or more dwelling units or sleeping units in a single structure, the provisions of Chapter 11 of the *International Building Code* for Group R-3 shall apply.

❖ Multiple single-family dwellings (townhouses) constructed in a group of four or more attached units must meet the accessibility requirements of Chapter 11 of the *International Building Code*, where townhouses are occupancy classification Group R3.

Section 1107.6.3 of the *International Building Code* requires that for Group R3 with four or more dwelling units in a single structure every dwelling unit must be a Type B dwelling unit. To determine the number of units in a single structure, the Fair Housing Act (FHA) considers all the units built under the same roof. The FHA does not recognize separating units with fire walls to create separate buildings. All dwelling units built in a single structure are counted together. If four or more units are built as a single structure, all units are required to be Type B dwelling units. The exception to Section 1107.6.3 permits the number of Type B units to be reduced; see the commentary for Section 1107.6.3 in the 2003 International Building Code *Commentary*.

SECTION R323
FLOOD-RESISTANT CONSTRUCTION

R323.1 General. Buildings and structures constructed in flood hazard areas (including A or V Zones) as established in Table R301.2 (1) shall be designed and constructed in accordance with the provisions contained in this section.

Exception: All buildings and structures in identified floodways as established in Table R301.2 (1) shall be designed and constructed as stipulated in the *International Building Code*.

❖ This section addresses additional requirements for all buildings and structures proposed to be located in areas designated as flood hazard areas. These areas are commonly referred to as "floodplains" and are shown on a community's Flood Insurance Rate Map (FIRM) prepared by the Federal Emergency Management Agency (FEMA) or other adopted flood hazard map. Through the adoption of the IRC, communities meet some of the requirements necessary to participate in the National Flood Insurance Program (NFIP). To learn more about how communities may meet the requirements by adoption of the *International Residential Code* and the *International Building Code* (with Appendix G), see *Reducing Flood Losses Through the International Code Series: Meeting the Requirements of the National Flood Insurance Program.*

The NFIP was established to reduce flood losses, to better indemnify individuals from flood losses, and to reduce federal expenditures for disaster assistance. A community that has flood hazard areas participates in the NFIP to protect health, safety, and property, and so that its citizens can purchase federally-backed flood insurance. FEMA administers the NFIP and monitors community compliance with the floodplain management requirements of the NFIP.

Many states require that communities regulate floodplain development to a higher standard than the minimum standards established by the NFIP. Communities considering using the IRC and other I-Codes to meet the floodplain management requirements of the NFIP are advised to consult with their state NFIP coordinator or the appropriate FEMA regional office.

R323.1.1 Structural systems. All structural systems of all buildings and structures shall be designed, connected and anchored to resist flotation, collapse or permanent lateral movement due to structural loads and stresses from flooding equal to the design flood elevation.

❖ New buildings, new structures, and substantial improvements to existing buildings and structures in flood hazard areas are to be designed and constructed to resist flood forces to minimize damage. Buildings are to be connected to structural foundation systems that are capable of resisting flood forces, including flotation, lateral (hydrostatic) pressures, moving water (hydrodynamic) pressures, wave impact, and debris impact.

R323.1.2 Flood-resistant construction. All buildings and structures erected in areas prone to flooding shall be constructed by methods and practices that minimize flood damage.

❖ In addition to the structural requirement of Section R323.1.1, this section establishes a broad requirement that techniques used for construction in flood hazard areas are to contribute to the flood resistance by minimizing flood damage.

R323.1.3 Establishing the design flood elevation. The design flood elevation shall be used to define areas prone to flooding, and shall describe, at a minimum, the base flood elevation at the depth of peak elevation of flooding (includ-ing wave height) which has a 1 percent (100-year flood) or greater chance of being equaled or exceeded in any given year.

❖ This section defines the "flood hazard area," which is the area predicted to be inundated during a flood that has a 1-in-100 (or 1-percent) chance of occurring in any given year. At a minimum, the floodplain is the Special Flood Hazard Area shown on the community's Flood Insurance Rate Map prepared by FEMA. This section also defines the term "design flood elevation," which is used throughout the code. The design flood elevation is the greater of the base flood elevation shown on the FIRM or other flood hazard area map identified in Table R301.2(1).

R323.1.4 Lowest floor. The lowest floor shall be the floor of the lowest enclosed area, including basement, but excluding any unfinished flood-resistant enclosure that is useable solely for vehicle parking, building access or limited storage provided that such enclosure is not built so as to render the building or structure in violation of this section.

❖ The lowest floor is the most important reference point when designing and constructing a building or structure in a flood hazard area. The term is specifically defined to include basements, which are any areas that are below grade on all sides. This definition differs from the one in Chapter 2, which includes areas that are partly below grade.

It is important to understand the distinction between the lowest floor of a building or structure and how it relates to enclosures. The National Flood Insurance Program recognizes that elevated buildings may reasonably have certain enclosures that will be subject to flooding. Enclosures that meet certain provisions are deemed not to be the lowest floor, and flood insurance premium rates are based on the elevated portion of the building. For flood hazard areas known as "A zones," enclosure provisions are set forth in R323.2.2. For coastal high hazard areas including "V zones," provisions for enclosures are set forth in R323.3.3 and R323.3.4.

R323.1.5 Protection of mechanical and electrical systems. Electrical systems, equipment and components, and heating, ventilating, air conditioning and plumbing appliances, plumbing fixtures, duct systems, and other service equipment shall be located at or above the design flood elevation. If replaced as part of a substantial improvement, electrical systems, equipment and components, and heating, ventilating, air conditioning, and plumbing appliances, plumbing fixtures, duct systems, and other service equipment shall meet the requirements of this section. Systems, fixtures, and equipment and components shall not be mounted on or penetrate through walls intended to break away under flood loads.

Exception: Electrical systems, equipment and components, and heating, ventilating, air conditioning and plumbing appliances, plumbing fixtures, duct systems, and other service equipment are permitted to be located below the design flood elevation provided that they are designed and installed to prevent water from entering or accumulating within the components and to resist hydrostatic and hydrodynamic loads and stresses, including the effects of buoyancy, during the occurrence of flooding to the design flood elevation in compliance with the flood-resistant construction requirements of the *International Building Code*. Electrical wiring systems are permitted to be located below the design flood elevation provided they conform to the provisions of the electrical part of this code for wet locations.

❖ This section sets a broad requirement that electrical and mechanical system elements, including electrical, heating, ventilating, air conditioning, plumbing, and other service equipment are to be elevated above the anticipated flood level. The same requirement applies to replacement of electrical and mechanical system elements that are included as part of substantial improvements to existing buildings. However, if new or replacement system elements are permitted below the design flood elevations, they must be protected against flood damage. If a building's floor is at the design flood elevation (the minimum requirement), then anything installed below the floor is automatically below the minimum elevation and therefore not in compliance with this section. For additional guidance, refer to FEMA 348: Protecting Building Utilities from Flood Damage.

R323.1.6 Protection of water supply and sanitary sewage systems. New and replacement water supply systems shall be designed to minimize or eliminate infiltration of flood waters into the systems in accordance with the plumbing provisions of this code. New and replacement sanitary sewage systems shall be designed to minimize or eliminate infiltration of floodwaters into systems and discharges from systems into floodwaters in accordance with the plumbing provisions of this code and Chapter 3 of the *International Private Sewage Disposal Code*.

❖ Health concerns arise when water supply systems are exposed to flood waters, and contamination from flooded sewage systems can pose health and environmental risks. To reduce these risks, water supply systems and sanitary sewage systems are to be designed to minimize or eliminate infiltration and discharges under flood conditions.

R323.1.7 Flood-resistant materials. Building materials used below the design flood elevation shall comply with the following:

1. All wood, including floor sheathing, shall be pressure preservatively treated in accordance with AWPA C1, C2, C3, C4, C9, C15, C18, C22, C23, C24, C28, P1, P2 and P3 or decay-resistant heartwood or redwood, black locust, or cedars.

2. Materials and installation methods used for flooring and interior and exterior walls and wall coverings shall conform to the provisions of FEMA/FIA-TB-2.

❖ To minimize flood damage, foundation materials used below the design flood elevation are to resist damage due to flood waters. While certain enclosures may be allowed under otherwise properly elevated buildings, the materials used in those enclosures are to be resistant to water damage to minimize flood losses. For further guidance, refer to FEMA TB-2: *Flood Resistant Material Requirements* and FEMA FIA-TB #8: *Corrosion Protection for Metal Connectors In Coastal Areas for Structures Located in Special Flood Hazard Areas*.

R323.1.8 Manufactured housing. New or replacement manufactured housing shall be elevated in accordance with Section R323.2 and the anchor and tie-down requirements of Sections AE604 and AE605 of Appendix E shall apply. The foundation and anchorage of manufactured housing to be located in identified flood ways as established in Table R301.2(1) shall be designed and constructed in accordance with the applicable provisions in the *International Building Code*.

❖ The placement of new manufactured-housing units is subject to the same elevation requirements as other new construction. Replacement of existing manufactured homes is considered "new construction" and is to comply with the provisions of this section. Manufactured units are to be elevated so that their lowest floors are at or above the design flood elevation. In part, this provision exceeds the minimum requirements of the NFIP, which allows replacement manufactured homes to be located below the DFE in some limited situations in existing manufactured home parks and subdivisions. Because of anticipated flood forces, units are to be anchored and tied down in accordance with Appendix E (AE 604 and AE 605). Within floodways, where water depths and velocities are anticipated to be greatest, foundation and anchorage are to be designed in accordance with the *International Building Code*.

R323.1.9 As-built elevation documentation. A registered design professional shall prepare and seal documentation of the elevations specified in Section R323.2 or R323.3.

❖ Documentation of the "as-built" lowest floor elevation is to be submitted to verify compliance. Most communities require or request submission of FEMA's Elevation Certificate (FEMA Form 81-31, available on-line at www.fema.gov/nfip/forms.htm), which is designed to obtain the information necessary to show compliance. Building owners need elevation information to obtain flood insurance, and insurance agents use the Elevation Certificates to compute the proper flood insurance premium rates.

Ideally, the elevation information should be obtained when the floor elevation is set and before further significant vertical construction occurs. If an error is discovered, it will be significantly easier and less expensive to correct at this stage. The building official must obtain the lowest floor elevation when the building is completed and before the Certificate of Occupancy is issued to determine that the building is in compliance with the flood hazard provisions of the code. The building official must maintain copies of the elevation information on file and make elevation and related building information available for insurance rating and other purposes.

R323.2 Flood hazard areas (including A Zones). All areas that have been determined to be prone to flooding but not subject to high velocity wave action shall be designated as flood hazard areas. All buildings and structures erected in flood hazard areas shall be designed and constructed in accordance with Sections R323.2.1 through R323.2.3.

❖ Flood hazard areas often referred to as A Zones are found along non-tidal waterways and lakes, and include the inland portions of coastal floodplains that are not subject to high velocity wave action. The zone designation does not indicate anticipated velocities of river and stream flooding. Especially in areas where the slopes of streambeds are relatively steep, it is recommended that designers check the Flood Insurance Study to see whether velocity information is available.

R323.2.1 Elevation requirements.

1. Buildings and structures shall have the lowest floors elevated to or above the design flood elevation.
2. In areas of shallow flooding (AO Zones), buildings and structures shall have the lowest floor (including basement) elevated at least as high above the highest adjacent grade as the depth number specified in feet (mm) on the FIRM, or at least 2 feet (610 mm) if a depth number is not specified.
3. Basement floors that are below grade on all sides shall be elevated to or above the design flood elevation.

Exception: Enclosed areas below the design flood elevation, including basements whose floors are not below grade on all sides, shall meet the requirements of Section R323.2.2.

❖ This section presents the required minimum floor elevations for all buildings and structures proposed within flood hazard areas known as A zones and AO zones. The minimum requirement is that lowest floors are to be elevated to or above the design flood elevation. AO zones are areas of shallow flooding where FEMA has specified a "depth" number rather than a flood elevation that is referenced to a datum. Buildings in AO zones are to be elevated so the lowest floor, including the basement, is at least as high above grade as the depth number, or at least 2 feet if a depth number is not specified. The point where this depth is measured is at the highest natural grade adjacent to the proposed building.

This section emphasizes that basements that are below grade on all sides are to be elevated to or above the design flood elevation. The NFIP defines a basement not by its common use, but by whether it is below grade on all sides. Even if below grade on all sides by only an inch, such an area is, technically, a basement. The specification in Item 3 means that basements excavated into fill within a flood plain cannot be excavated below the design flood elevation. The exception allows enclosures that, if designed and constructed to meet the flood opening specifications of R323.2.2, are not the lowest floor.

R323.2.2 Enclosed area below design flood elevation. Enclosed areas, including crawl spaces, that are below the design flood elevation shall:

1. Be used solely for parking of vehicles, building access or storage.
2. Be provided with flood openings which shall meet the following criteria:
    2.1. There shall be a minimum of two openings on different sides of each enclosed area; if a building has more than one enclosed area below the design flood elevation, each area shall have openings on exterior walls.
    2.2. The total net area of all openings shall be at least 1 square inch for each square foot (275 mm for each square meter) of enclosed area.
    2.3. The bottom of each opening shall be 1 foot (305 mm) or less above the adjacent ground-level.
    2.4. Openings shall be at least 3 inches (76 mm) in diameter.
    2.5. Any louvers, screens or other opening covers shall allow the automatic flow of floodwaters into and out of the enclosed area.
    2.6. Openings installed in doors and windows, that meet requirements 2.1 through 2.5, are accept-able; however, doors and windows without installed openings do not meet the requirements of this section.

❖ This section contains the use limitations and flood opening requirements that apply to enclosures below otherwise properly elevated buildings. If enclosures meet these provisions they are not considered the lowest floor for insurance purposes. The use limitation states that enclosures, including crawl spaces, are to be used only for parking of vehicles, building access or storage. Although the NFIP regulations do not define storage, most states interpret it to mean "limited" storage. Some states specify that hazardous materials must not be stored in enclosures. Experience has suggested that two design factors influence the use of enclosures and may discourage conversions by owners: the size of the enclosed area and utility service. Plumbing service is not appropriate for the allowable uses, and minimal electrical service can be provided (a light switch for a stairwell, for example, or electrical outlets dropped from the ceiling).

R323.2.3 Foundation design and construction. Foundation walls for all buildings and structures erected in flood hazard areas shall meet the requirements of Chapter 4.

Exception: Unless designed in accordance with Section 404:

1. The unsupported height of 6 inches (152 mm) plain masonry walls shall be no greater than 3 feet (914 mm).
2. The unsupported height of 8 inches (203 mm) plain masonry walls shall be no greater than 4 feet (1219 mm).
3. The unsupported height of 8 inches (203 mm) reinforced masonry walls shall be no greater than 8 feet (2438 mm).

For the purpose of this exception, unsupported height is the distance from the finished grade of the under-floor space and the top of the wall.

❖ Flood conditions impose additional loads on foundation walls, including hydrostatic and hydrodynamic loads. Under a cooperative agreement with FEMA, the American Society of Civil Engineers (ASCE) performed an analysis of flood loads and flood-load combinations to examine the structural adequacy of the foundation walls prescribed in Chapter 4. A range of flood depths and velocities was used to determine if there were any height limitations for certain types of walls. The results suggested that unless the walls are designed in accordance with Section R404, it is appropriate to limit the unsupported height of certain masonry walls based on wall thickness and whether they are plain or reinforced.

R323.3 Coastal high-hazard areas (including V Zones). Areas that have been determined to be subject to wave heights in excess of 3 feet (914 mm) or subject to high velocity wave action or wave-induced erosion shall be designated as coastal high-hazard areas. All buildings and structures erected in coastal high-hazard areas shall be designed and constructed in accordance with Sections R323.3.1 through R323.3.6.

❖ Buildings and structures located in flood hazard areas known as V zones, or coastal high hazard areas, are exposed to flooding and wind-driven waves in excess of 3 feet. Waves impose additional loads on foundations and can cause erosion.

R323.3.1 Location and site preparation.

1. Buildings and structures shall be located landward of the reach of mean high tide.
2. For any alteration of sand dunes and mangrove stands the building official shall require submission of an engineering analysis which demonstrates that the proposed alteration will not increase the potential for flood damage.

❖ Buildings and structures are to be sited such that they are not regularly affected by normal (mean) high tides. Many coastal locations benefit from natural storm protection due to the presence of sand dunes and mangrove stands. Those natural features are to remain intact. However, alterations may be considered if specific analyses indicate that the alterations will not increase the potential for flood damage, including increasing the exposure of the site to greater flood depths and higher wave heights.

R323.3.2 Elevation requirements.

1. All buildings and structures erected within coastal high hazard areas shall be elevated so that the lowest portion of all structural members supporting the lowest floor, with the exception of mat or raft foundations, piling, pile caps, columns, grade beams and bracing, is located at or above the design flood elevation.
2. Basement floors that are below grade on all sides are prohibited.
3. The use of fill for structural support is prohibited.
4. The placement of fill beneath buildings and structures is prohibited.

Exception: Walls and partitions enclosing areas below the design flood elevation shall meet the requirements of Sections R323.3.4 and R323.3.5.

❖ Buildings and structures to be located in coastal high hazard areas (V zones) are to be elevated so that the bottom of the lowest horizontal structural member will be at or above the design flood elevation. Elevation is one of the most important requirements to provide resistance to flood damage. It is also the main factor used in determining flood insurance premium rates for buildings and structures. In coastal high-hazard areas, the use of fill to provide structural support is prohibited because areas subject to wave action are more likely to experience flood-related erosion, which may lead to failure of foundations on fill. The NFIP does allow certain enclosures under buildings that otherwise meet the elevation requirements, and the exception notes that the requirements for such enclosures are found in R323.3.3 and R323.3.5. Additional guidance about design and construction in coastal high hazard areas can be found in FEMA FIA-TB 5: *Free-of-Obstruction Requirements for Buildings Located in Coastal High Hazard Areas* and FEMA 55: *Coastal Construction Manual*.

R323.3.3 Foundations. All buildings and structures erected in coastal high hazard areas shall be supported on pilings or columns and shall be adequately anchored to such pilings or columns. Piling shall have adequate soil penetrations to re-sist the combined wave and wind loads (lateral and uplift). Water loading values used shall be those associated with the design flood. Wind loading values shall be those required by this code. Pile embedment shall include consideration of decreased resistance capacity caused by scour of soil strata surrounding the piling. Pile systems design and installation shall be certified in accordance with Section R323.3.6. Mat, raft or other foundations that support columns shall not be permitted where soil investigations that are required in accordance with Section R401.4 indicate that soil material under the mat, raft or other foundation is subject to scour or erosion from wave-velocity flow conditions.

❖ Several provisions applicable to foundations for buildings and structures in coastal high hazard areas (V zones) are in this section. During coastal storms, both wind and water loads are significant, and buildings are to be designed and constructed to withstand both loads. Buildings are to be anchored to their foundations, and the foundations are to withstand the anticipated forces. A limitation is outlined on the use of mat, raft, and other foundations that support columns in areas where soil investigations indicate that scour or erosion is likely.

R323.3.4 Walls below design flood elevation. Walls and partitions are permitted below the elevated floor, provided that such walls and partitions are not part of the structural support of the building or structure and:

1. Electrical, mechanical, and plumbing system components are not to be mounted on or penetrate through walls that are designed to break away under flood loads, and
2. Are constructed with insect screening or open lattice, or;
3. Are designed to break away or collapse without causing collapse, displacement or other structural damage to the elevated portion of the building or supporting foundation

system. Such walls, framing and connections shall have a design safe loading resistance of not less than 10 (0.48 kN/m^2) and no more than 20 pounds per square foot (0.96 kN/m^2); or

4. Where wind loading values of this code exceed 20 pounds per square foot (0.96 kN/m^2), the construction documents shall include documentation prepared and sealed by a registered design professional that:

    4.1. The walls and partitions below the design flood elevation have been designed to collapse from a water load less than that which would occur during the design flood.

    4.2. The elevated portion of the building and supporting foundation system have been designed to withstand the effects of wind and flood loads acting simultaneously on all building components (structural and nonstructural). Water loading values used shall be those associated with the design flood. Wind loading values shall be those required by this code.

❖ All buildings and structures in coastal high-hazard areas are to be elevated to or above the design flood elevation. However, certain enclosures are allowed below otherwise elevated buildings provided they meet certain specifications. These specifications allow the walls surrounding the enclosures to fail or break away under certain loads without causing damage to the foundation. Post-flood investigations indicate that breakaway walls do not perform as intended if wires, pipes, and other utility components are mounted on or penetrate through the walls. Walls other than lattice or insect screening are to meet specific loading resistance; they are to be designed to fail under loads of not less than 10 (0.48kN/m^2) and no more than 20 pounds per square foot (0.96kN/m^2). If anticipated loads exceed 20 pounds per square foot (0.96kN/m^2), the wall design is to be certified as meeting certain specifications. For additional guidance, refer to NFIP Technical Bulletin #9: *Design and Construction Guidance for Breakaway Walls below Elevated Buildings* and FEMA 55: *Coastal Construction Manual*.

R323.3.5 Enclosed areas below design flood elevation. Enclosed areas below the design flood elevation shall be used solely for parking of vehicles, building access or storage.

❖ The limitations on how enclosures below elevated buildings are used that are outlined in R323.2.2 are repeated here. They apply to enclosures below elevated buildings in coastal high hazard areas. Enclosures are to be used only for parking, building access and storage. To minimize the likelihood of conversion to other uses, some communities limit the size of enclosures. Another factor that influences enclosure use is the availability of utility services. Plumbing service is not associated with the allowable uses and in accordance with R323.1.5 must not be located below the design flood elevation. Minimal electrical service can be provided (a light switch for a stairwell, for example, or electrical outlets that are dropped from the ceiling).

R323.3.6 Construction documents. The construction documents shall include documentation that is prepared and sealed by a registered design professional that the design and methods of construction to be used meet the applicable criteria of this section.

❖ Documentation is to be submitted to demonstrate that the design and proposed construction methods address the provisions of the code. This requirement recognizes that coastal high hazard areas subject to waves and erosion are dynamic environments that impose significant loads on buildings. Documentation of the elevation of the bottom of the lowest horizontal structural member is addressed in R323.1.9. The documentation provided by the registered design professional is applicable to the design criteria for Section R323.3 for coastal high-hazard areas in addition to the as-built elevations as required by Section R323.1.9. See the commentary for Section R323.1.9.

Bibliography

The following resource materials are referenced in this chapter or are relevant to the subject matter addressed in this chapter.

16 CFR Part 1201-(1977), *Safety Standard for Architectural Glazing*. Bethesda, MD: Consumer Product Safety Commission, 1977.

16 CFR Part 1404-(1979), *Cellulose Insulation*. Bethesda, MD: Consumer Product Safety Commission, 1979.

AAMA 101/I.S2/NAFS.02, *Voluntary Performance Specification for Windows, Skylights and Glass Doors*. Schaumburg, IL: American Architectural Manufacturers Association, 2002.

AFPA WFCM-2001, *Wood Frame Construction Manual for One-and Two-family Dwellings*. Washington, DC: American Forest and Paper Association, 2001.

AISI PM-2001, *Standard for Cold-formed Steel Framing-Prescriptive Method for One- and Two-family Dwellings*. Washington, DC: American Iron and Steel Institute, 2001.

An Architect's Compendium to the UBC. Whittier, CA: International Conference of Building Officials, 2000.

ANSI Z97.1-84 (R1994), *Safety Glazing Materials Used in Buildings—Safety Performance Specifications and Methods of Test (Reaffirmed 1994)*. New York: American National Standards Institute, 1984.

APA PAS #473, *Subdivision Design in Flood Hazard Areas*. Washington, DC: American Planning Association, 1997.

ASCE 7-02, *Minimum Design Loads for Buildings and Other Structures*. Reston, VA: American Society of Civil Engineers, 2002.

ASCE 24-98, *Flood Resistant Design and Construction*. Reston, VA: American Society of Civil Engineers, 1998.

ASTM C 34-96 (2001), *Specification for Structural Clay Load-Bearing Wall Tile*. West Conshohocken, PA: ASTM International, 1996.

ASTM C 55-01, *Specification for Concrete Brick*. West Conshohocken, PA: ASTM International, 2001.

ASTM C 67-02, *Test Methods of Sampling and Testing Brick and Structural Clay Tile*. West Conshohocken, PA: ASTM International, 2002.

ASTM C 129-97, *Standard Specification for Nonload-Bearing Concrete Masonry Units*. West Conshohocken, PA: ASTM International, 1997.

ASTM C 216-98, *Standard Specification for Facing Brick (Solid Masonry Units Made from Clay or Shale)*. West Conshohocken, PA: ASTM International, 1998.

ASTM C 652-07, *Standard Specification for Hollow Brick (Hollow Masonry Units Made from Clay or Shale)*. West Conshohocken, PA: ASTM International, 1997.

ASTM C 1996-99, *Specification for Performance of Exterior Windows, Glazed Curtain Walls, Doors and Storm Shutters Impacted by Windborne Debris in Hurricanes*. West Conshohocken, PA: ASTM International, 1999.

ASTM C 73-99a, *Specification for Calcium Silicate Face Brick (Sand Lime Brick)*. West Conshohocken, PA: ASTM International, 1999.

ASTM C 90-01, *Specification for Load-bearing Concrete Masonry Units*. West Conshohocken, PA: ASTM International, 2001.

ASTM C 129-01, *Specification for Nonload-bearing Concrete Masonry Units and Related Units*. West Conshohocken, PA: ASTM International, 2001.

ASTM C 216-01a, *Specification for Facing Brick (Solid Masonry Units Made from Clay or Shale)*. West Conshohocken, PA: ASTM International, 2001.

ASTM C 652-01a, *Specification for Hollow Brick (Hollow Masonry Units Made from Clay or Shale)*. West Conshohocken, PA: ASTM International, 2001.

ASTM E 84-01, *Test Method for Surface Burning Characteristics of Building Materials*. West Conshohocken, PA: ASTM International, 2001.

ASTM E 119-00a, *Test Method for Fire Tests of Building Construction and Materials*. West Conshohocken, PA: ASTM International, 2000.

ASTM E 152-95, *Method of Fire Tests for Door Assemblies*. West Conshohocken, PA: ASTM International, 1995.

ASTM E 814-00, *Test Method for Fire Tests of Through-Penetration Fire Stops*. West Conshohocken, PA: ASTM International, 2000.

ASTM E 970-00, *Test Method for Critical Radiant Flux of Exposed Attic Floor Insulation Using a Radiant Heat Energy Source*. West Conshohocken, PA: ASTM International, 2000.

ASTM E 1886-97, *Test Method for Performance of Exterior Windows, Curtain Walls, Doors and Storm Shutters Impacted by Missile(s) and Exposed to Cyclic Pressure Differentials*. West Conshohocken, PA: ASTM International, 1997.

ASTM E 1996-01, *Specification for Performance of Exterior Windows, Curtain Walls, Doors and Storm Shutters Impacted by Windborne Debris in Hurricanes*. West Conshohocken, PA: ASTM International, 2001.

AWPA C1-00, *All Timber Products — Preservative Treatment by Pressure Processes*. Granbury, TX: American Wood-Preservers' Association, 2000.

AWPA C2-01, *Lumber, Timbers, Bridge Ties and Mine Ties — Preservative Treatment by Pressure Processes*. Granbury, TX: American Wood-Preservers' Association, 2001.

AWPA C3-99, *Piles — Preservative Treatment by Pressure Processes*. Granbury, TX: American Wood-Preservers' Association, 1999.

AWPA C4-99, *Poles — Preservative Treatment by Pressure Processes*. Granbury, TX: American Wood-Preservers' Association, 1999.

AWPA C9-00, *Plywood — Preservative Treatment by Pressure Processes*. Granbury, TX: American Wood-Preservers' Association, 2000.

AWPA C15-00, *Wood for Commercial-Residential Construction Preservative Treatment by Pressure Processes*. Granbury, TX: American Wood-Preservers' Association, 2000.

AWPA C18-99, *Standard for Pressure Treated Material in Marine Construction*. Granbury, TX: American Wood-Preservers' Association, 1999.

AWPA C22-96, *Lumber and Plywood for Permanent Wood Foundations — Preservative Treatment by Pressure Processes*. Granbury, TX: American Wood-Preservers' Association, 1996.

AWPA C23-00, *Round Poles and Posts Used in Building Construction — Preservative Treatment by Pressure Processes*. Granbury, TX: American Wood-Preservers' Association, 2000.

AWPA C24-96, *Sawn Timber Used to Support Residential and Commercial Structures*. Granbury, TX: American Wood-Preservers' Association, 1996.

AWPA C28-99, *Standard for Preservative Treatment by Pressure Process of Structural Glued Laminated Members and Laminations Before Gluing*. Granbury, TX: American Wood-Preservers' Association, 1999.

AWPA M4-01, *Standard for the Care of Preservative-Treated Wood Products*. Granbury, TX: American Wood-Preservers' Association, 2001.

BUILDING PLANNING

AWPA P1/P13-01, *Standard for Creosote Preservative*. Granbury, TX: American Wood-Preservers' Association, 2001.

AWPA P2-01, *Standard for Creosote Solutions*. Granbury, TX: American Wood-Preservers' Association, 2001.

AWPA P3-01, *Standard for Creosote — Petroleum Solutions*. Granbury, TX: American Wood-Preservers' Association, 2001.

FEMA 232, *Home Builder's Guide to Seismic Resistant Construction.* Washington, DC: Federal Emergency Management Agency, 1998.

FEMA 303, *National Earthquake Hazards Reduction Program Recommended Provisions for Seismic Regulations for New Buildings and Other Structures.* 1997 Edition. Part 2: Commentary. Washington, DC: Building Seismic Safety Council, 1997.

Fire Resistance Directory. Northbrook, IL: Underwriters Laboratories, Inc., 2003.

FM 4880-(2001), *American National Standard for Evaluating Insulated Wall or Wall and Roof/Ceiling Assemblies, Plastic Interior Finish Materials, Plastic Exterior Building Panels, Wall/Ceiling Coating Systems, Interior or Exterior Finish Systems*. Norwood, MA; Factory Mutual Global Research, 2001.

GA 600-00, *Fire-resistance Design Manual.* 16th Edition. Washington, DC: Gypsum Association, 2000.

Handbook to the International Building Code. Whittier, CA: International Conference of Building Officials, 2000.

IBC-2003, *International Building Code*. Falls Church, VA: International Code Council, 2003.

ICC SBCCI-10-99, *Standard for Hurricane Resistant Construction.* Falls Church, VA: International Code Council, 1999.

Interior Finish to Room Fire Growth. Quincy, MA: National Fire Protection Association, 2000.

IPC-2003, *International Plumbing Code*. Falls Church, VA: International Code Council, 2003.

NFPA 13-99, *Installation of Sprinkler Systems*. Quincy, MA: National Fire Protection Association, 1999.

NFPA 70-02, *National Electrical Code*. Quincy, MA: National Fire Protection Association, 2002.

NFPA 72-99, *National Fire Alarm Code*. Quincy, MA: National Fire Protection Association, 1999.

NFPA 13-96, *Installation of Sprinkler Systems*. Quincy, MA: National Fire Protection Association, 1996.

NFPA 259-98, *Standard Test Method for Potential Heat of Building Materials*. Quincy, MA: National Fire Protection Association, 1998.

NFPA 286-00, *Standard Methods of Fire Tests for Evaluating Contribution of Wall and Ceiling*

UL 325-95, *Door, Drapery, Louver and Window Operations and Systems—with Revisions through June 2001*. Northbrook, IL: Underwriters Laboratories, 1995.

UL 1040-96, *Fire Test of Insulated Wall Construction—with Revisions through April 2001*. Northbrook, IL: Underwriters Laboratories, 1996.

UL 1479-94, *Fire Tests of Through-Penetration Firestops*. Northbrook, IL: Underwriters Laboratories, 1994.

UL 1715-97, *Fire Test of Interior Finish Material*. Northbrook, IL: Underwriters Laboratories, 1997.

ULC S 102-M88, *Surface Burning Characteristics of Building Materials and Assemblies*. Scarborough, Ontario: Underwriters' Laboratories of Canada, 1988.

The following publications and forms may be obtained at no cost from the Federal Emergency Management Agency (FEMA). These and other materials may also be available online at http://www.fema.gov/library/lib06.htm, or ordered from:

Federal Emergency Management Agency
P.O. Box 2012
Jessup, Maryland 20794-2012
800-480-2520

FEMA 44 CFR Parts 59-73: *National Flood Insurance Program* (NFIP).

FEMA 55CD (third edition): *Coastal Construction Manual.*

FEMA 213: *Answers to Questions about Substantially Damaged Buildings.*

FEMA 232: *Home Builder's Guide to Seismic Resistant Construction.*

FEMA 259: *Engineering Principles and Practices for Retrofitting Flood Prone Residential Buildings.*

FEMA 265: *Managing Floodplain Development in Approximate Zone A Areas: A Guide for Obtaining and Developing Base (100-Year) Flood Elevations.*

FEMA 311: *Guidance on Estimating Substantial Damage Using the NFIP Residential Substantial Damage Estimator.*

FEMA 348: *Protecting Building Utilities From Flood Damage: Principles and Practices for the Design and Construction of Flood Resistant Building Utility Systems.*

FEMA Form 81-31: *Elevation Certificate.* Available at http://www.fema.gov/library/elvcert.pdf

NFIP Technical Bulletin Series. Available at http://www.fema.gov/mit/techbul.htm

FEMA FIA-TB #1: *Openings in Foundation Walls for Buildings Located in Special Flood Hazard Areas.*

FEMA FIA-TB #2: *Flood-Resistant Material Requirements for Buildings Located in Special Flood Hazard Areas.*

FEMA FIA-TB #3: *Non-Residential Floodproofing Requirements and Certification for Buildings Located in Special Flood Hazard Areas.* (Note: applies to non-residential only.)

FEMA FIA-TB #4: *Elevator Installation for Buildings Located in Special Flood Hazard Areas.*

FEMA FIA-TB #5: *Free-of-Obstruction Requirements for Buildings Located in Coastal High Hazard Areas.*

FEMA FIA-TB #6: *Below-Grade Parking Requirements for Buildings Located in Special Flood Hazard Areas.*

FEMA FIA-TB #7: *Wet Floodproofing Requirements for Structures Located in Special Flood Hazard Areas.* (Note: applies to non-residential only.)

FEMA FIA-TB #8: *Corrosion Protection for Metal Connectors In Coastal Areas for Structures Located in Special Flood Hazard Areas.*

FEMA FIA-TB #9: *Design and Construction Guidance for Breakaway Walls Below Elevated Coastal Buildings.*

FEMA/FIA-TB-2-93: *Flood-Resistant Materials Requirements for Buildings Located in Special Flood Hazard Areas in Accordance with the National Flood Insurance Program.*

FM 4880-94: *Approval Standard for Class 1: A) Insulated Wall or Wall & Roof/Ceiling Panels, B) Plastic Interior Finish Materials, C) Plastic Exterior Building Panels, D) Wall/Ceiling coating Systems and E) Interior or Exterior Finish Systems.*

Reducing Flood Losses Through the International Code Series: Meeting the Requirements of National Flood Insurance Program.

Chapter 4: Foundations

General Comments

Section R401 establishes the scope and applicability of the chapter. The criteria provide the details necessary for designing, building, and inspecting most common foundation systems for homes. The selection of a particular type of foundation is based on many variables, and it should result in the system that best suits the needs of a particular project. Also, the lot itself must be evaluated to determine the type of soil and the drainage pattern.

Section R402 specifies material requirements for wood and concrete foundations. Section R403 provides minimum requirements for concrete and masonry footings as well as wood foundations. Section R404 regulates foundation walls of concrete, masonry, and wood as well as insulating concrete form foundation walls. Section R405 has important criteria for maintaining the drainage of water away from foundations. The requirements for waterproofing and dampproofing in Section R406 protect below-grade habitable spaces from moisture. Section R407 provides requirements for underfloor areas and for protecting columns. Section R408 includes provisions for proper ventilation, adequate access, and the removal of debris.

Purpose

Chapter 4 regulates foundations designed and constructed under the *International Residential Code*. A foundation system consists of two interdependent components: the foundation structure itself and the supporting soil. Proper design and construction of a foundation system is vital to the satisfactory performance of the entire structure that it supports.

The prescriptive provisions of this chapter provide requirements for constructing foundation walls and footings of concrete, masonry, or wood. In addition to a foundation's ability to support the required design loads, this chapter addresses several other factors that can affect foundation performance. These include controlling surface water and subsurface drainage, requiring soil tests where conditions warrant, and evaluating proximity to slopes and minimum depth requirements. The chapter also provides requirements to minimize adverse effects of moisture, decay, and pests in basements and crawl spaces.

SECTION R401
GENERAL

R401.1 Application. The provisions of this chapter shall control the design and construction of the foundation and foundation spaces for all buildings. Wood foundations shall be designed and installed in accordance with AF&PA Report No. 7.

Exceptions:
1. The provisions of this chapter shall be permitted to be used for wood foundations only in the following situations:
   1.1. In buildings that have no more than two floors and a roof.
   1.2. When interior basement and foundation walls are provided at intervals not exceeding 50 feet.
2. In addition to the provisions of this chapter, the design and construction of foundations in areas prone to flooding as established by Table R301.2(1) shall meet the provisions of Section R323.

Wood foundations in Seismic Design Categories D_1 and D_2 shall be designed in accordance with accepted engineering practice.

❖ This chapter contains all requirements relating to the design and construction of foundations and underfloor spaces. The second exception provides an explicit link to the flood-resistant construction provisions of Section R323. Thus, foundations for buildings to be located in areas prone to flooding are to meet the provisions of Section R323 as well.

This section also specifically allows a wood foundation system designed and installed according to the American Forest & Paper Association manual Basic Requirements for Permanent Wood Foundation Systems, Technical Report No. 7. When this system is installed under the prescriptive code requirements of this chapter, the code limits its use to structures where the loads exerted on the foundation are relatively light and to areas of low or moderate seismic hazard. This type of foundation construction is essentially a below-grade, loadbearing, wood-frame system capable of providing support for light frame structures. It generally would consist of the following components:
1. Walls consisting of plywood fastened to wood studs (Section R404.2).
2. A composite footing made up of a continuous wood plate set on a bed of granular materials, such as sand, gravel, or crushed stone, which in turn supports the foundation walls and transmits their loads to the bearing soil below (Section R403.2).

3. Polyethylene film that serves as a vapor barrier and covers the exterior side of the plywood foundation walls from grade level down to the footing plate (Section R406.3).
4. Caulking compounds used for sealing the joints in the plywood walls, bonding agents for attaching the polyethylene film to the plywood, and for the film joints (Section R406.3).
5. Metal fasteners made of silicon bronze, copper, or stainless steel or hot-dipped zinc-coated steel nails or staples (Section R402.1).
6. Pressure-treated plywood and lumber to protect the foundation material against decay, termites, and other insects (Section R402.1).

R401.2 Requirements. Foundation construction shall be capable of accommodating all loads according to Section R301 and of transmitting the resulting loads to the supporting soil. Fill soils that support footings and foundations shall be designed, installed and tested in accordance with accepted engineering practice. Gravel fill used as footings for wood and precast concrete foundations shall comply with Section R403.

❖ In order to fulfill its role in the complete load path, the foundation must support the required design loads and transmit these to the soil. The phrase "accepted engineering practice" in this section means common practice that is acceptable to the code official of the jurisdiction.

R401.3 Drainage. Surface drainage shall be diverted to a storm sewer conveyance or other approved point of collection so as to not create a hazard. Lots shall be graded so as to drain surface water away from foundation walls. The grade away from foundation walls shall fall a minimum of 6 inches (152 mm) within the first 10 feet (3048 mm).

Exception: Where lot lines, walls, slopes or other physical barriers prohibit 6 inches (152 mm) of fall within 10 feet (3048 mm), drains or swales shall be provided to ensure drainage away from the structure.

❖ Along with the proper support for a structure through the foundation system, adequate preparation of the building site is necessary to keep water drainage away from the supporting foundations. Proper site drainage is an important element in preventing wet basements, damp crawl spaces, eroded banks, and possible failure of a foundation system.

One of the most important considerations is the arrangement of structures on a building site in a manner that retains natural drainage patterns and minimizes the alteration or disturbance to existing grades. If the designer keeps such factors in mind, the result will be a reduction of ground surface stabilization problems and opportunities for differential settlement through the reduction in the use of fills. A detailed treatment of drainage design is beyond the scope of this document; therefore, only rough guidelines can be provided for areas where a more comprehensive set of grading regulations does not exist.

As illustrated in Commentary Figures R401.3(1) and R401.3(2), drainage patterns should result in adequate slopes to approved drainage devices that are capable of carrying concentrated runoff. In some cases, control of concentrated roof runoff by gutters and downspouts may be needed, and if gutters and downspouts are used, provisions should be made to discharge runoff in order to prevent soil erosion. Refer also to Section R801.3.

GRADING METHOD FOR LOT WHERE SLOPE IS FROM REAR TO FRONT LOT. DRAINAGE SWALES ARE LOCATED AT REAR AND SIDES OF DWELLING.

Figure R401.3(1)
LOT GRADING EXAMPLE

Figure R401.3(2)
LOT GRADING EXAMPLE

❖ Cross-lot drainage and drainage over graded slopes should generally be avoided. Slopes should be designed with as moderate a grade as possible to minimize slope instability and erosion. The minimum slope gradients which should be used are a function of the combined ground frost and moisture conditions, soil type, geological features, and geographic condition. The slope away from the building is required to be 6 inches (152 mm) within the first 10 feet (3048 mm). The exception in this section allows for use of drains and swales if lot lines or physical barriers, such as a steeply sloping lot, do not allow for a 6-inch (152 mm) slope.

Commentary Figure R401.3(3) provides guidance for determining minimum slope gradients for swales based on combined ground frost and moisture conditions only. The permeability of soil and the type of drainage devices used will also affect the desired slope gradient; therefore, the gradients stated in Commentary Figure R401.3(3) should not be considered absolutes.

R401.4 Soil tests. In areas likely to have expansive, compressible, shifting or other unknown soil characteristics, the building official shall determine whether to require a soil test to determine the soil's characteristics at a particular location. This test shall be made by an approved agency using an approved method.

❖ Loading conditions beyond the normal live, dead, and wind or earthquake loading generally include the types of loading attributable to special regional conditions. Such conditions might include soil instability, forces generated on foundations by expansive soils, and increased lateral pressures due to a high water table or surcharge loads from adjacent structures. Accounting for such conditions in the structure is generally considered to be beyond the scope of the conventional methods of foundation construction specified within this chapter. In some cases, a soil test may be required to evaluate the existing conditions.

Where the bearing capacity of the soil has not been determined by geotechnical evaluation such as borings, field load tests, laboratory tests, and engineering analysis, it is a common practice to use presumptive bearing values for the design of the foundation system. Section R401.4.1 requires that the presumptive bearing values in Table R401.4.1 be used if a geotechnical evaluation has not been performed.

R401.4.1 Geotechnical evaluation. In lieu of a complete geotechnical evaluation, the load-bearing values in Table R401.4.1 shall be assumed.

❖ See the commentary for Section R401.4.

TABLE R401.4.1
PRESUMPTIVE LOAD-BEARING VALUES OF
FOUNDATION MATERIALS[a]

CLASS OF MATERIAL	LOAD-BEARING PRESSURE (pounds per square foot)
Crystalline bedrock	12,000
Sedimentary and foliated rock	4,000
Sandy gravel and/or gravel (GW and GP)	3,000
Sand, silty sand, clayey sand, silty gravel and clayey gravel (SW, SP, SM, SC, GM and GC)	2,000
Clay, sandy clay, silty clay, clayey silt, silt and sandy silt (CI, ML, MH and CH)	1,500[b]

For SI: 1 pound per square foot = 0.0479 kN/m^2.

a. When soil tests are required by Section R401.4, the allowable bearing capacities of the soil shall be part of the recommendations.

b. Where the building official determines that in–place soils with an allowable bearing capacity of less than 1,500 psf are likely to be present at the site, the allowable bearing capacity shall be determined by a soils investigation.

❖ As explained in the commentary for Section R401.4, where a soils investigation is not required by the code, the load-bearing values of Table R401.4.1 are to be used. The values vary by soil classifications, which are based on the Unified Soil Classification System. Table R405.1 also gives these soil descriptions in more detail. The allowable soil bearing values listed in the table are based on lengthy experience with the behavior of these materials in supporting loads from all types of structures.

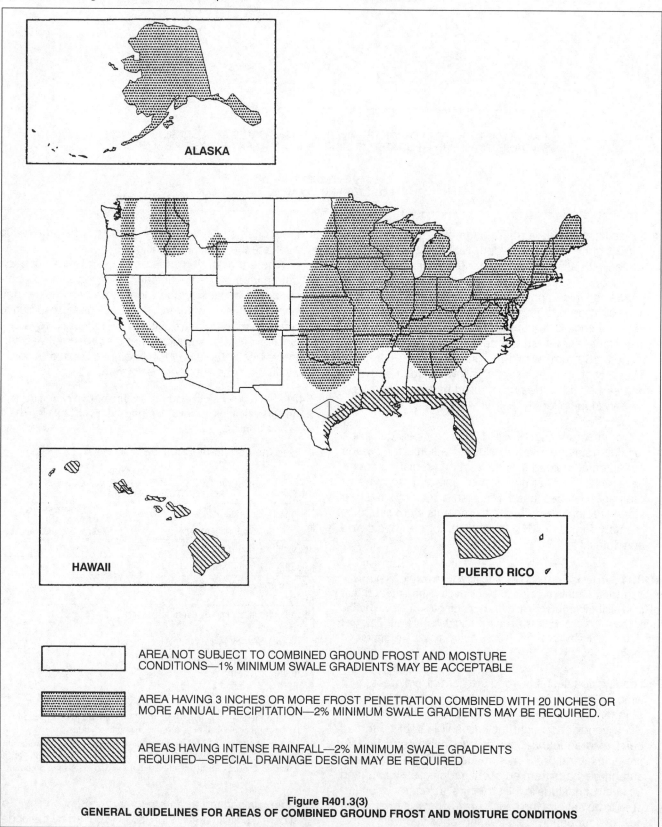

Figure R401.3(3)
GENERAL GUIDELINES FOR AREAS OF COMBINED GROUND FROST AND MOISTURE CONDITIONS

R401.5 Compressible or shifting soil. When top or subsoils are compressible or shifting, such soils shall be removed to a depth and width sufficient to assure stable moisture content in each active zone and shall not be used as fill or stabilized within each active zone by chemical, dewatering, or presaturation.

❖ A geotechnical engineer should be consulted by the owner or the owner's builder if there is a concern that a compressible or shifting soil condition exists. Where these conditions are present, one of two methods must be used. The first of these deals with removal and replacement of the soil, thus reducing the forces associated with soil movement. The second approach is soil stabilization by chemical means, dewatering, or presaturation.

SECTION R402
MATERIALS

R402.1 Wood foundations. Wood foundation systems shall be designed and installed in accordance with the provisions of this code.

❖ The wood foundation system requirements are from the American Forest & Paper Association manual *Basic Requirements for the Permanent Wood Foundation System,* Technical Report No. 7. Figures R403.1(2) and R403.1(3) of the code illustrate some typical details of this system. Also refer to the commentary for Section R401.1.

R402.1.1 Fasteners. Fasteners used below grade to attach plywood to the exterior side of exterior basement or crawlspace wall studs, or fasteners used in knee wall construction, shall be of Type 304 or 316 stainless steel. Fasteners used above grade to attach plywood and all lumber-to-lumber fasteners except those used in knee wall construction shall be of Type 304 or 316 stainless steel, silicon bronze, copper, hot-dipped galvanized (zinc coated) steel nails, or hot-tumbled galvanized (zinc coated) steel nails. Electrogalvanized steel nails and galvanized (zinc coated) steel staples shall not be permitted.

❖ Appropriate fasteners must be used in wood foundation construction because the presence of any moisture in combination with the preservative treatment can corrode incompatible fasteners.

R402.1.2 Wood treatment. All lumber and plywood shall be treated in accordance with AWPA C22, and shall bear the label of an accredited agency showing 0.60 retention. Where lumber and/or plywood is cut or drilled after treatment, the treated surface shall be field treated with Copper Napthenate, the concentration of which shall contain a minimum of 2 percent copper metal, by repeated brushing, dipping or soaking until the wood absorbs no more preservative.

❖ Performance of the wood foundation system is dependent on the use of properly treated materials; thus, the code provision emphasizes the use of properly treated lumber and plywood. Verification of the proper materials is provided by identification showing the approval of an accredited inspection agency. An example of such identification is shown in Commentary Figure R402.1.2. The label must reflect the required retention of 0.60. Where treated lumber or plywood is field cut (exposing wood that is untreated), the code specifies the method of treating the cut surface.

R402.2 Concrete. Concrete shall have a minimum specified compressive strength as shown in Table R402.2. Concrete sub-

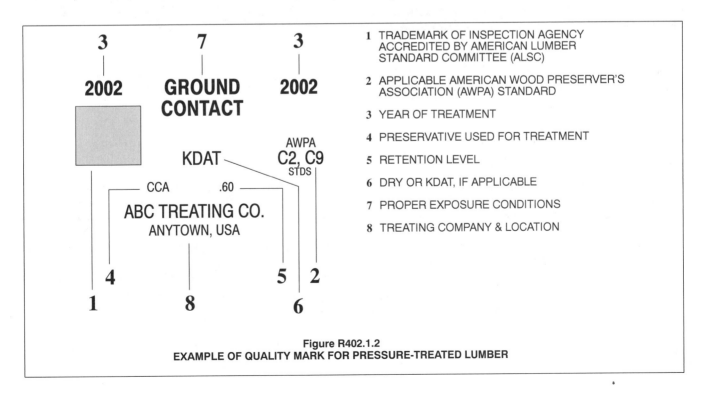

Figure R402.1.2
EXAMPLE OF QUALITY MARK FOR PRESSURE-TREATED LUMBER

ject to weathering as indicated in Table R301.2(1) shall be air entrained as specified in Table R402.2. The maximum weight of fly ash, other pozzolans, silica fume, or slag that is included in concrete mixtures for garage floor slabs and for exterior porches, carport slabs, and steps that will be exposed to deicing chemicals shall not exceed the percentages of the total weight of cementitious materials specified in ACI 318. Materials used to produce concrete and testing thereof shall comply with the applicable standards listed in ACI 318. In addition to the cements permitted by ACI 318, cement complying with ASTM C 1157 is permitted.

❖ The code specifies minimum concrete compressive strengths ranging from 2,500 to 3,000 psi (17.2 to 24.1 Mpa). Table R402.2 specifies the required compressive strength of concrete based on a locale's weathering potential and a building element's exposure. Section R301 uses the Weathering Probability Map for Concrete to classify an area's weathering potential as negligible, moderate, or severe. For concrete that will be subject to freezing and thawing in a moist condition (i.e., weathering) or subject to direct or indirect application of deicing chemicals, Table R402.2 requires the use of higher-strength, air-entrained concrete. Freezing and thawing cycles can be the most destructive weathering factors for concrete when it is wet. Water freezing in the cement matrix, in the aggregate, or both causes deterioration of non-air-entrained concrete. Studies have documented that concrete provided with proper air entrainment is highly resistant to this deterioration.

All materials and testing must be in accordance with the American Concrete Institute's ACI 318, *Building Code Requirements for Structural Concrete*. Cementitious materials include portland cement (ASTM C150), blended hydraulic cements (ASTM C595 and ASTM C1157), and expansive cement (ASTM C845). They can also include pozzolanic materials such as fly ash and other raw or calcined natural pozzolan (ASTM C618), ground granulated blast-furnace slag (ASTM C989), and/or silica fume (ASTM C1240) when used in combination with the cements. Most concrete contains admixtures such as fly ash or other pozzolans that are added as a replacement for a portion of the cement. Although such substitutions can be for economic reasons, in some cases they can also be employed to improve certain concrete properties.

R402.3 Precast concrete. Approved precast concrete foundations shall be designed and installed in accordance with the provisions of this code and the manufacturer's installation instructions.

❖ This section permits the design and installation of precast concrete foundations. All requirements in this code for concrete foundations apply to precast concrete as well. Also, any manufacturer's instructions must be complied with.

TABLE R402.2
MINIMUM SPECIFIED COMPRESSIVE STRENGTH OF CONCRETE

TYPE OR LOCATIONS OF CONCRETE CONSTRUCTION	MINIMUM SPECIFIED COMPRESSIVE STRENGTH[a] (f'_c)		
	Weathering potential[b]		
	Negligible	Moderate	Severe
Basement walls, foundations and other concrete not exposed to the weather	2,500	2,500	2,500[c]
Basement slabs and interior slabs on grade, except garage floor slabs	2,500	2,500	2,500[c]
Basement walls, foundation walls, exterior walls and other vertical concrete work exposed to the weather	2,500	3,000[d]	3,000[d]
Porches, carport slabs and steps exposed to the weather, and garage floor slabs	2,500	3,000[d,e]	3,500[d,e]

For SI: 1 pound per square inch = 6.895 kPa.
a. At 28 days psi.
b. See Table R301.2(1) for weathering potential.
c. Concrete in these locations that may be subject to freezing and thawing during construction shall be air-entrained concrete in accordance with Footnote d.
d. Concrete shall be air entrained. Total air content (percent by volume of concrete) shall not be less than 5 percent or more than 7 percent.
e. See Section R402.2 for minimum cement content.

❖ The code requires the use of higher-strength, air-entrained concrete for enclosed garage floor slabs in moderate and severe weathering regions. Even though these slabs may be enclosed and not subject to freezing and thawing conditions, they are likely to be subject to the deteriorating effects of deicing chemicals that drip from vehicles. The replacement of portland cement with other cementitious materials can be detrimental to the durability of the concrete if the concrete will be exposed to deicing chemicals. Thus, the code limits the total amount of these replacement materials to 50 percent of the total weight of all cementitious materials.

Footnote "c" to Table R402.2 requires air-entrained concrete for concrete elements not exposed to freezing and thawing conditions when the building is completed if it is likely that these elements will be subject to freezing and thawing during construction. Also see the commentary for Section R404.2.

FOUNDATIONS

SECTION R403
FOOTINGS

R403.1 General. All exterior walls shall be supported on continuous solid or fully grouted masonry or concrete footings, wood foundations, or other approved structural systems which shall be of sufficient design to accommodate all loads according to Section R301 and to transmit the resulting loads to the soil within the limitations as determined from the character of the soil. Footings shall be supported on undisturbed natural soils or engineered fill.

❖ The provisions of this section include the general statement, also contained in Section R401.2, that a footing must be capable of supporting the required design loads. The section expands on this by referencing the character of the soil and the minimum extension below the frost line. The code mandates that footings be supported on undisturbed natural soil or engineered fill.

TABLE R403.1
MINIMUM WIDTH OF CONCRETE OR MASONRY FOOTINGS
(inches)[a]

	LOAD-BEARING VALUE OF SOIL (psf)			
	1,500	2,000	3,000	≥ 4,000
Conventional light-frame construction				
1-story	12	12	12	12
2-story	15	12	12	12
3-story	23	17	12	12
4-inch brick veneer over light frame or 8-inch hollow concrete masonry				
1-story	12	12	12	12
2-story	21	16	12	12
3-story	32	24	16	12
8-inch solid or fully grouted masonry				
1-story	16	12	12	12
2-story	29	21	14	12
3-story	42	32	21	16

For SI: 1 inch = 25.4 mm, 1 pound per square foot = 0.0479 kN/m^2.

a. Where minimum footing width is 12 inches, a single wythe of solid or fully grouted 12-inch nominal concrete masonry units is permitted to be used.

❖ Table R403.1 specifies minimum footing widths where normal soil conditions are encountered. The minimum footing widths are based on the soil load-bearing values in Table R401.4.1. The minimum values were calculated based on 20 feet of tributary roof area, 16 feet of tributary floor areas, 10-foot-high first story walls, and 8-foot-high second and third story walls. A 50 psf snow load was used.

A minimum footing width of 12 inches is required so that footings can span over weak locations in the soil and to allow for minor misalignment of foundations. The footnote permits solid or fully grouted 12-inch-nominal concrete masonry block to be used for this minimum width of footing. Section R403.1.1 specifies the minimum thickness of footings. These provisions provide for the transfer of loads to the supporting soil without exceeding the capacity of the materials used to construct the footings.

If any of the following conditions exist in the area of the foundation, conventional spread footings should not be used, and a designed foundation may be necessary:

- Filled ground, except when properly compacted.
- Foundation soils subject to subsidence.
- Expansive soils such as those having a plasticity index greater than (see Section R403.1.8.1)
- Highly compressive clays.
- Unconfined sands and silts.

In cold weather, care must be taken that foundations are not placed on frozen soil, nor should such foundations be placed in freezing weather unless a method of ensuring that the underlying soil is free of frost is employed and the foundations are properly protected against the weather.

R403.1.1 Minimum size. Minimum sizes for concrete and masonry footings shall be as set forth in Table R403.1 and Figure R403.1(1). The footing width, W, shall be based on the load-bearing value of the soil in accordance with Table R401.4.1. Spread footings shall be at least 6 inches (152 mm) in thickness. Footing projections, P, shall be at least 2 inches (51 mm) and shall not exceed the thickness of the footing. The size of footings supporting piers and columns shall be based on the tributary load and allowable soil pressure in accordance with Table R401.4.1. Footings for wood foundations shall be in accordance with the details set forth in Section R403.2, and Figures R403.1(2) and R403.1(3).

❖ Table R403.1 specifies minimum footing widths based on soil pressure and the number of stories supported. Figure R403.1(1) illustrates masonry and concrete footings. To avoid construction that might overstress the footing, this section also specifies the minimum footing thickness and places a limitation on the maximum projection of the footing beyond the face of the foundation wall as shown in Figure R403.1(1). The size of isolated footings for piers and columns should be determined from the tributary (design) load supported using the presumptive soil values of Table R401.4.1.

The limitation on the maximum projection of the footing beyond the face of the foundation wall is particularly important for 6-inch-thick (152 mm) footings of plain concrete subjected to soil-bearing pressures on the order of 2,000 pounds per square foot (95.8 kPa). Projections in excess of the maximum allowable could result in the footing being cracked in a plane coinciding with the face of the foundation wall. A crack could occur if the allowable flexural tension stress of the concrete is exceeded.

FIGURE R403.1(1)

For SI: 1 inch = 25.4 mm.

FIGURE R403.1(1)
CONCRETE AND MASONRY FOUNDATION DETAILS

❖ See the commentary for Section R403.1.

FOUNDATIONS

**FIGURE R403.1(2)
PERMANENT WOOD FOUNDATION BASEMENT WALL SECTION**

❖ See the commentary for Section R403.1.

FIGURE R403.1(3)
PERMANENT WOOD FOUNDATION CRAWL SPACE SECTION

❖ See the commentary for Section R403.1.

R403.1.2 Continuous footings in Seismic Design Categories D_1 and D_2. The braced wall panels at exterior walls of all buildings located in Seismic Design Categories D_1 and D_2 shall be supported by continuous footings. All required interior braced wall panels in buildings with plan dimensions greater than 50 feet (15 240 mm) shall also be supported by continuous footings.

❖ The code requires continuous footings for support of braced wall panels in exterior walls, a critical component of a structure's lateral force resisting system (see Chapter 6). For a structure classified as Seismic Design Category D_1 or D_2, the code requires continuous footings at all exterior walls as well as at interior braced wall lines in buildings with plan dimensions greater than 50 feet (15,240 mm).

R403.1.3 Seismic reinforcing. Concrete footings located in Seismic Design Categories D_1 and D_2, as established in Table R301.2(1), shall have minimum reinforcement. Bottom reinforcement shall be located a minimum of 3 inches (76 mm) clear from the bottom of the footing.

In Seismic Design Categories D_1 and D_2 where a construction joint is created between a concrete footing and stem wall, a minimum of one No. 4 bar shall be provided at not more than 4 feet (1219 mm) on center. The vertical bar shall extend to 3 inches (76 mm) clear of the bottom of the footing, have a standard hook and extend a minimum of 14 inches (357 mm) into the stem wall.

In Seismic Design Categories D_1 and D_2 where a grouted masonry stem wall is supported on a concrete footing and stem wall, a minimum of one No. 4 bar shall be provided at not more than four feet on center. The vertical bar shall extend to 3 inches (76 mm) clear of the bottom of the footing and have a standard hook.

In Seismic Design Categories D_1 and D_2 masonry stem walls without solid grout and vertical reinforcing shall not be permitted.

FOUNDATIONS

Exception: In detached one and two family dwellings which are three stories or less in height and constructed with stud bearing walls, plain concrete footings without longitudinal reinforcement supporting walls and isolated plain concrete footings supporting columns or pedestals are permitted.

❖ This section specifies minimal reinforcement for buildings classified as Seismic Design Category D_1 or D_2. Interconnection of stem walls and supporting foundations is necessary to resist the tendency of the wall to slip in an earthquake. In grouted masonry stem walls supported on concrete footings and at construction joints between concrete stem walls and their supporting footing, a minimum of one vertical No. 4 bar at 48 inches (1219 mm) spacing must be provided. A monolithically cast foundation and stem wall do not require this vertical reinforcement. Masonry stem walls must be grouted and provided with vertical reinforcement. However, plain concrete footings are permitted for detached one- and two-family dwellings. If an alternate braced wall panel is used in accordance with Section R602.10.6, that section contains more specific foundation requirements that also must be met.

R403.1.3.1 Foundations with stemwalls. Foundations with stemwalls shall be provided with a minimum of one No. 4 bar at the top of the wall and one No. 4 bar at the bottom of the footing.

❖ Where a stem wall exists, longitudinal reinforcement consisting of one No. 4 bar at the top of the wall and one No. 4 bar at the bottom of the footing must be provided for footings in Seismic Design Category D_1 or D_2.

R403.1.3.2 Slabs-on-ground with turned-down footings. Slabs-on-ground with turned down footings shall have a minimum of one No. 4 bar at the top and bottom of the footing.

Exception: For slabs-on-ground cast monolithically with a footing, one No. 5 bar or two No. 4 bars shall be located in the middle third of the footing depth.

❖ For footings in Seismic Design Category D_1 or D_2, a slab-on-ground with a turned down footing requires a No. 4 bar at the top and bottom of the footing. If cast monolithically with the slab, either one No. 5 or two No. 4 bars should be placed in the middle third of the footing depth.

R403.1.4 Minimum depth. All exterior footings shall be placed at least 12 inches (305 mm) below the undisturbed ground support. Where applicable, the depth of footings shall also conform to Sections R403.1.4.1 through R403.1.4.2.

❖ Footings are required to extend below the ground surface a minimum of 12 inches (305 mm). This is considered a minimum depth to protect the footing from movement of the soil caused by freezing and thawing in mild climate areas. See Section R403.1.4.1 for general frost protection requirements.

Footing depths may also be influenced by adjacent footings at different elevations; the footing at a higher elevation imposes surcharge pressures on lower adjacent footings. One method of alleviating the surcharge pressures on lower adjacent footings is the use of the guidelines illustrated in Commentary Figure R403.1.4. Other methods based on a soil and foundation analysis are also acceptable.

R403.1.4.1 Frost protection. Except where otherwise protected from frost, foundation walls, piers and other permanent supports of buildings and structures shall be protected from frost by one or more of the following methods:

1. Extending below the frost line specified in Table R301.2(1);
2. Constructing in accordance with Section R403.3;
3. Constructing in accordance with ASCE 32-01; and
4. Erected on solid rock.

Exceptions:
1. Freestanding accessory structures with an area of 400 square feet (37 m^2) or less and an eave height of 10 feet (3048 mm) or less shall not be required to be protected.
2. Decks not supported by a dwelling need not be provided with footings that extend below the frost line.

Footings shall not bear on frozen soil unless such frozen condition is of a permanent character.

❖ Buildings must not be founded on or within frozen ground. This provision prevents damage to the exterior walls and other walls bearing on the frozen soil. The volume changes (frost heave) that take place during freezing and thawing produce excessive stresses in the foundations and, as a result, may create extensive damage to the walls that are supported.

Section R403.1.4.1, Frost Protection, contains the three acceptable ways of protecting the footings and foundations from frost: 1) extend below the frost line; 2) insulate the foundation according to the IRC prescriptive frost protected shallow foundation requirements or the ASCE 32-01, *Design and Construction of Frost Protected Shallow Foundations* performance standard; and 3) erect the foundation on solid rock.

The frost-line depth is established by the authority having jurisdiction based on field experience and must be listed along with other design criteria in Table R301.2(1). Exterior foundations must extend below the frost line and a minimum of 12 inches (305 mm) below undisturbed soil. Foundations supported on rock or that are frost-protected in accordance with Section R403.3 are exempt from the extension below the frost line, but the 12-inch (305 mm) minimum still applies.

ASCE 32 contains requirements for:
1. a simplified method for heated buildings
2. a more complicated performance method that permits trading off insulation for frost-susceptible fill or use of varying foundation depths

3. requirements for slabs-on-ground, crawlspaces (without outside vents), walk out basements, semi-heated buildings
4. requirements for unheated buildings that conserve heat from the ground with thick layers of insulation under the entire foundation and beyond it several feet; continuous walls and columns
5. special design conditions.

All the ASCE standard requirements are consistent with theory, widespread practice in Scandinavia, and the successful construction of thousands of buildings in the U.S.

Foundations are not to be placed on frozen soil because when the ground thaws, uneven settlement of the structure is apt to occur, thereby causing structural damage. This section does, however, permit footings to be constructed on permanently frozen soil. In permafrost areas, special precautions are necessary to prevent heat from the structure from thawing the soil beneath the foundation.

R403.1.4.2 Seismic conditions. In Seismic Design Categories D_1 and D_2, interior footings supporting bearing or bracing walls and cast monolithically with a slab on grade shall extend to a depth of not less than 18 inches (457 mm) below the top of slab.

❖ In buildings classified as Seismic Design Category D_1 or D_2, the code establishes the minimum depth below the top of slab for interior footings that support bearing or bracing walls and are poured monolithically with the slab. This allows for the installation of anchor bolts, tie-downs, etc.

R403.1.5 Slope. The top surface of footings shall be level. The bottom surface of footings shall not have a slope exceeding one unit vertical in 10 units horizontal (10-percent slope). Footings shall be stepped where it is necessary to change the elevation of the top surface of the footings or where the slope of the bottom surface of the footings will exceed one unit vertical in ten units horizontal (10-percent slope).

❖ The code requires that the top surface of footings for buildings be essentially level. However, a slope of 1 unit vertical in 10 units horizontal (10-percent slope) is permitted for the bottom bearing surface. If the slope is steeper, the foundation is required to be stepped. Commentary Figure R403.1.5 schematically shows a stepped foundation. Although the code places no restriction on a stepped foundation, the figure shows a recommended overlap of the top of the foundation wall beyond the step in the foundation. It is larger than the vertical step in the foundation wall at that point. This recommendation is based on possible crack propagation at an angle of 45 degrees.

For SI: 1 foot = 304.8 mm, 1 degree = 0.01745 rad.

Figure R403.1.4
RECOMMENDATION FOR ADJACENT FOOTINGS AT DIFFERENT ELEVATIONS

R403.1.6 Foundation anchorage. When braced wall panels are supported directly on continuous foundations, the wall wood sill plate or cold-formed steel bottom track shall be anchored to the foundation in accordance with this section.

The wood sole plate at exterior walls on monolithic slabs and wood sill plate shall be anchored to the foundation with anchor bolts spaced a maximum of 6 feet (1829 mm) on center. There shall be a minimum of two bolts per plate section with one bolt located not more than 12 inches (305 mm) or less than seven bolt diameters from each end of the plate section. In Seismic Design Categories D_1 and D_2, anchor bolts shall also be spaced at 6 feet (1829 mm) on center and located within 12 inches (305 mm) from the ends of each plate section at interior braced wall lines when required by Section R602.10.9 to be supported on a continuous foundation. Bolts shall be at least $1/2$ inch (12.7 mm) in diameter and shall extend a minimum of 7 inches (178 mm) into masonry or concrete. Interior bearing wall sole plates on monolithic slab foundations shall be positively anchored with approved fasteners. A nut and washer shall be tightened on each bolt to the plate. Sills and sole plates shall be protected against decay and termites where required by Sections R318 and R319. Cold-formed steel framing systems shall be fastened to the wood sill plates or anchored directly to the foundation as required in Section R505.3.1 or R603.1.1.

Exception: Foundation anchor straps, spaced as required to provide equivalent anchorage to $1/2$-inch-diameter (12.7 mm) anchor bolts.

❖ To prevent walls and floors from shifting under lateral loads, the code requires anchorage to the supporting foundation. Anchor bolts installed as specified in this section supply the minimum required capacity. Commentary Figure R403.1.6 illustrates anchorage of wood sill plates. This anchorage applies at all exterior walls in any seismic design category. An exception explicitly allows foundation anchor straps provided they

Figure R403.1.5
RECOMMENDATION FOR STEPPED WALL FOOTINGS

Figure R403.1.6
WOOD SILL PLATE ANCHORAGE TO FOUNDATIONS

are spaced to provide anchorage capacity equivalent to that of the $^{1}/_{2}$-inch (12.7 mm) diameter anchor bolts specified. In this case it would be necessary to use manufacturers' data, such as evaluation reports, to document the anchor straps shear and tension capacities.

R403.1.6.1 Foundation anchorage in Seismic Design Categories C, D_1 and D_2. In addition to the requirements of Section R403.1.6, the following requirements shall apply to wood light-frame structures in Seismic Design Categories D_1 and D_2 and wood light-frame townhouses in Seismic Design Category C.

1. Plate washers conforming to Section R602.11.1 shall be used on each bolt.
2. Interior braced wall plates shall have anchor bolts spaced at not more than 6 feet (1829 mm) on center and located within 12 inches (305 mm) from the ends of each plate section when supported on a continuous foundation.
3. Interior bearing wall sole plates shall have anchor bolts spaced at not more than 6 feet (1829 mm) on center and located within 12 inches (305 mm) from the ends of each plate section when supported on a continuous foundation.
4. The maximum anchor bolt spacing shall be 4 feet (1219 mm) for buildings over two stories in height.
5. Stepped cripple walls shall conform to Section R602.11.3.
6. Where continuous wood foundations in accordance with Section R404.2 are used, the force transfer shall have a capacity equal to or greater than the connections required by Section R602.11.1 or the braced wall panel shall be connected to the wood foundations in accordance with the braced wall panel-to-floor fastening requirements of Table 602.3(1).

❖ The additional anchorage requirements of this section apply to light-frame wood structures classified as Seismic Design Category D_1 or D_2 and wood light-frame townhouses in Seismic Design Category C. These provisions were added following observations of past earthquakes, especially the 1994 Northridge earthquake in California. Considerable longitudinal splitting of sill plates occurred in that earthquake, resulting in substantial damage to the plates and subsequent loss of lateral load capacity. Plate washers are required on each bolt. Refer to the commentary for Section R602.11.1.

R403.1.7 Footings on or adjacent to slopes. The placement of buildings and structures on or adjacent to slopes steeper than 1 unit vertical in 3 units horizontal (33.3-percent slope) shall conform to Sections R403.1.7.1 through R403.1.7.4.

❖ The provisions of the referenced sections apply to buildings placed on or adjacent to slopes steeper than 1 unit vertical in 3 units horizontal (33.3-percent slope) only.

R403.1.7.1 Building clearances from ascending slopes. In general, buildings below slopes shall be set a sufficient distance from the slope to provide protection from slope drainage, erosion and shallow failures. Except as provided in Section R403.1.7.4 and Figure R403.1.7.1, the following criteria will be assumed to provide this protection. Where the existing slope is steeper than one unit vertical in one unit horizontal (100-percent slope), the toe of the slope shall be assumed to be at the intersection of a horizontal plane drawn from the top of the foundation and a plane drawn tangent to the slope at an angle of 45 degrees (0.79 rad) to the horizontal. Where a retaining wall is constructed at the toe of the slope, the height of the slope shall be measured from the top of the wall to the top of the slope.

❖ Figure R403.1.7.1 provides the criteria for the location of foundations adjacent to the toe of ascending slopes not exceeding 1 unit vertical in 1 unit horizontal. Commentary Figure R403.1.7.2 illustrates the criteria for the determination of the location of the toe of the slope where the slope exceeds 1 unit vertical in 1 unit horizontal.

R403.1.7.2 Footing setback from descending slope surfaces. Footings on or adjacent to slope surfaces shall be founded in material with an embedment and setback from the slope surface sufficient to provide vertical and lateral support for the footing without detrimental settlement. Except as provided for in Section R403.1.7.4 and Figure R403.1.7.1, the following setback is deemed adequate to meet the criteria. Where the slope is

FOUNDATIONS

**FIGURE R403.1.7.1
FOUNDATION CLEARANCE FROM SLOPES**

❖ The setback required by the code provides protection to the structure from shallow failures (sometimes referred to as sloughing) and protection from erosion and slope drainage. Furthermore, the space provided by the setback provides access around the building and helps to create a light and open-air environment. The dimension from the toe of the slope to the face of the structure need not exceed 15 feet. Also refer to the commentary for Section R403.1.7.1.

**Figure R403.1.7.2
BUILDINGS ADJACENT TO ASCENDING SLOPE EXCEEDING 1 TO 1**

steeper than one unit vertical in one unit horizontal (100-percent slope), the required setback shall be measured from an imaginary plane 45 degrees (0.79 rad) to the horizontal, projected upward from the toe of the slope.

❖ In this section, the code restricts the placement of footings adjacent to or on descending slopes so that both vertical and lateral support are provided. The criteria for this condition are as shown in Figure R403.1.7.1. It is possible to locate buildings closer to the slope than the indicated setback in Figure R403.1.7.1, and in fact it is possible to locate the footing of the structure on the slope itself. In these two cases, it will be necessary to provide an adequate depth of embedment of the footing so that the face of the footing at the bearing plane is set back from the edge of the slope at least the distance required by the code (H/3, but need not exceed 40 feet).

Commentary Figure R403.1.7.3 depicts the condition where the descending slope is steeper than 1 unit vertical in 1 unit horizontal and shows the proper location of the top of the slope as required by the code. The setback at the top of descending slopes will primarily provide lateral support for the foundations. The area so allocated also provides space for lot drainage away from the slope without creating too steep a drainage profile, which could create erosion problems. Furthermore, this space also provides for access around the building.

R403.1.7.3 Foundation elevation. On graded sites, the top of any exterior foundation shall extend above the elevation of the street gutter at point of discharge or the inlet of an approved drainage device a minimum of 12 inches (305 mm) plus 2 percent. Alternate elevations are permitted subject to the approval of the building official, provided it can be demonstrated that required drainage to the point of discharge and away from the structure is provided at all locations on the site.

❖ Commentary Figure R403.1.7.4 depicts the requirements of the code in this item for the elevation for exterior foundations with respect to the street, gutter, or point of inlet of a drainage device. The elevation of the street or gutter shown is that point at which drainage from the site reaches the street or gutter.

This requirement protects the building from water encroachment in case of unusually heavy rains and may be modified on the approval of the building official if the building official finds that positive drainage slopes are provided to drain water away from the building and that the drainage pattern is not subject to temporary flooding due to landscaping or other impediments to drainage.

R403.1.7.4 Alternate setback and clearances. Alternate setbacks and clearances are permitted, subject to the approval of the building official. The building official is permitted to require an investigation and recommendation of a qualified engineer to demonstrate that the intent of this section has been satisfied. Such an investigation shall include consideration of material, height of slope, slope gradient, load intensity and erosion characteristics of slope material.

❖ This item provides that the building official may approve alternate setbacks and clearances from slopes provided the building official is satisfied that the intent of this section has been met. The code gives the building official authority to require a foundation investigation by a qualified geotechnical engineer. This section also specifies which parameters must be considered by the geotechnical engineer in such an investigation.

R403.1.8 Foundations on expansive soils. Foundation and floor slabs for buildings located on expansive soils shall be designed in accordance with Section 1805.8 of the *International Building Code*.

Exception: Slab-on-ground and other foundation systems which have performed adequately in soil conditions similar to those encountered at the building site are permitted subject to the approval of the building official.

Figure R403.1.7.3
BUILDINGS ADJACENT TO DESCENDING SLOPE EXCEEDING 1 TO 1

Figure R403.1.7.4
DETERMINATION OF FOUNDATION ELEVATION ON GRADED SITES

FOUNDATIONS

❖ Expansive soils are those that shrink and swell due to changes in moisture content. The resulting movements can be damaging to residential structures. The amount and depth of potential swelling that can occur in a clay material are to some extent functions of the cyclical moisture content in the soil. In dryer climates where the moisture content in the soil near the ground surface is low because of evaporation, there is a greater potential for extensive swelling than in the same soil in wetter climates where the variations of moisture content are not as severe.

When foundations or floor slabs are supported on soils determined to be expansive, the code defers to the *International Building Code* provisions that address the design of foundations and slabs on expansive soil. The exception permits the use of systems that have demonstrated adequate performance with the approval of the building official.

R403.1.8.1 Expansive soils classifications. Soils meeting all four of the following provisions shall be considered expansive, except that tests to show compliance with Items 1, 2 and 3 shall not be required if the test prescribed in Item 4 is conducted:

1. Plasticity Index (PI) of 15 or greater, determined in accordance with ASTM D 4318.
2. More than 10 percent of the soil particles pass a No. 200 sieve (75 mm), determined in accordance with ASTM D 422.
3. More than 10 percent of the soil particles are less than 5 micrometers in size, determined in accordance with ASTM D 422.
4. Expansion Index greater than 20, determined in accordance with ASTM D4829.

❖ This section defines "expansive soil" as any plastic material with a PI of 15 or greater with more than 10 percent of the soil particles passing a No. 200 sieve and less than 5 micrometers in size. As an alternative, tests in accordance with ASTM D4829 can be used to determine if a soil is expansive. The expansion index is a measure of the swelling potential of the soil.

R403.2 Footings for wood foundations. Footings for wood foundations shall be in accordance with Figures R403.1(2) and R403.1(3). Gravel shall be washed and well graded. The maximum size stone shall not exceed $^3/_4$ inch (19.1 mm). Gravel shall be free from organic, clayey or silty soils. Sand shall be coarse, not smaller than $^1/_{16}$-inch (1.6 mm) grains and shall be free from organic, clayey or silty soils. Crushed stone shall have a maximum size of $^1/_2$ inch (12.7 mm).

❖ See Figures R403.1(2) and R403.1(3) and refer to the commentary for Section R401.1.

R403.3 Frost protected shallow foundations. For buildings where the monthly mean temperature of the building is maintained at a minimum of 64°F (18°C), footings are not required to extend below the frost line when protected from frost by insulation in accordance with Figure R403.3(1) and Table R403.3. Foundations protected from frost in accordance with Figure R403.3(1) and Table R403.3 shall not be used for unheated spaces such as porches, utility rooms, garages and carports, and shall not be attached to basements or crawl spaces that are not maintained at a minimum monthly mean temperature of 64°F (18°C).

Materials used below grade for the purpose of insulating footings against frost shall be labeled as complying with ASTM C 578.

❖ This section provides an alternative method of protecting foundations against frost heave, thus allowing foundations to be constructed above the frost line. Frost-protected foundations use insulation to reduce the heat loss at the slab edge. By holding heat from the dwelling in the ground under the foundation, the insulation, in effect, raises the frost line around the foundation.

The provisions require that the building be heated and that insulation be installed in accordance with the criteria specified in Figure R403.3(1) and Table R403.3 based on the air-freezing index established by Figure R403.3(2). As Figure R403.3(1) illustrates, these provisions apply only to slab-on-ground floors. Frost Protected Shallow Foundation (FPSF) designs for heated buildings are not to be used for unheated spaces. The provisions also prohibit the attachment of additions with FPSF to basements or crawl spaces that are not heated because, although unlikely, in very cold climates frost could penetrate under FPSFs from unheated basements or ventilated crawl spaces. The insulation values specified in this section are only for the purposes of exercising this option, and the Chapter 11 energy conservation provisions could result in slab edge insulation with a greater degree of thermal resistance. Conversely, if slab-edge insulation is used to comply with Chapter 11, a builder may take advantage of that insulation to reduce the foundation depth in accordance with this section.

R403.3.1 Foundations adjoining frost protected shallow foundations. Foundations that adjoin frost protected shallow foundations shall be protected from frost in accordance with Section R403.1.4.

❖ This section requires that foundations that adjoin FPSFs be protected from frost according to the current requirements of Section R403.1.4, which requires that foundations extend below the frost line, be built on solid rock, or use FPSF designs. Each method prevents frost from heaving foundations. This language prevents the attachment of FPSFs to reinforced floating garage foundations, which may be permitted by some local codes to be placed at shallow depths without insulating for frost protection (i.e., the foundation moves up and down with frost).

R403.3.1.1 Attachment to unheated slab-on-ground structure. Vertical wall insulation and horizontal insulation of frost protected shallow foundations that adjoin a slab-on-ground

FIGURE R403.3(1)

foundation that does not have a monthly mean temperature maintained at a minimum of 64°F (18°C), shall be in accordance with Figure R403.3(3) and Table R403.3. Vertical wall insulation shall extend between the frost protected shallow foundation and the adjoining slab foundation. Required horizontal insulation shall be continuous under the adjoining slab foundation and through any foundation walls adjoining the frost protected shallow foundation. Where insulation passes through a foundation wall, it shall either be of a type complying with this section and having bearing capacity equal to or greater than the structural loads imposed by the building, or the building shall be designed and constructed using beams, lintels, cantilevers or other means of transferring building loads such that the structural loads of the building do not bear on the insulation.

❖ This section requires that a building with a FPSF be insulated to current FPSF requirements even where an unheated garage or other slab-on-ground is attached to it. The vertical and horizontal insulation must be in accordance with Figure 403.3(3) and Table R403.3 and be continuous through the foundation of the unheated structure. The vertical wall insulation must extend between the FPSF and the adjoining slab foundation, and any required horizontal insulation must be continuous through any foundation walls adjoining the FPSF. No frost penetration or differential movement occurred with this construction in a HUD demonstration in Fargo, North Dakota. Deep foundations of unheated garages have been attached to hundreds of

For SI: 1 inch = 25.4 mm.

a. See Table R403.3 for required dimensions and *R*-values for vertical and horizontal insulation.

**FIGURE R403.3(1)
INSULATION PLACEMENT FOR FROST-PROTECTED FOOTINGS IN HEATED BUILDINGS**

❖ See the commentary for Section R403.3.

FOUNDATIONS

other buildings built on FPSFs in the U.S. In most cases the FPSF insulation has been placed continuously between the garage floor and the house FPSF; however, the insulation has not been continuous through the foundation walls of the unheated structures. In these buildings, no differential movement between the deep and shallow foundations has been reported.

Insulation passing through a foundation wall must comply with the requirements of Section R403.3, i.e., it must meet ASTM C578 and the requirements of Table R403.3, and the insulation must be of a type that has sufficient compressive strength to carry the load of the building. Alternatively, the building must be designed to carry those loads over the insulation and not bear on it.

TABLE R403.3
MINIMUM INSULATION REQUIREMENTS FOR FROST-PROTECTED FOOTINGS IN HEATED BUILDINGS[a]

AIR FREEZING INDEX (°F-days)[b]	VERTICAL INSULATION R-VALUE[c,d]	HORIZONTAL INSULATION R-VALUE[c,e]		HORIZONTAL INSULATION DIMENSIONS PER FIGURE R403.3(1) (inches)		
		Along walls	At corners	A	B	C
1,500 or less	4.5	Not required	Not required	Not required	Not required	Not required
2,000	5.6	Not required	Not required	Not required	Not required	Not required
2,500	6.7	1.7	4.9	12	24	40
3,000	7.8	6.5	8.6	12	24	40
3,500	9.0	8.0	11.2	24	30	60
4,000	10.1	10.5	13.1	24	36	60

a. Insulation requirements are for protection against frost damage in heated buildings. Greater values may be required to meet energy conservation standards. Interpolation between values is permissible.

b. See Figure R403.3(2) for Air Freezing Index values.

c. Insulation materials shall provide the stated minimum R-values under long-term exposure to moist, below-ground conditions in freezing climates. The following R-values shall be used to determine insulation thicknesses required for this application: Type II expanded polystyrene—2.4R per inch; Type IV extruded polystyrene—4.5R per inch; Type VI extruded polystyrene—4.5R per inch; Type IX expanded polystyrene—3.2R per inch; Type X extruded polystyrene—4.5R per inch.

d. Vertical insulation shall be expanded polystyrene insulation or extruded polystyrene insulation.

e. Horizontal insulation shall be extruded polystyrene insulation.

❖ This table provides the required R-value of horizontal and vertical insulation based on the air-freezing index. In addition, it gives the plan dimensions required for the horizontal insulation as illustrated in Figure R403.3(1).

R403.3.1.2 Attachment to heated structure. Where a frost protected shallow foundation abuts a structure that has a monthly mean temperature maintained at a minimum of 64°F (18°C), horizontal insulation and vertical wall insulation shall not be required between the frost protected shallow foundation and the adjoining structure. Where the frost protected shallow foundation abuts the heated structure, the horizontal insulation and vertical wall insulation shall extend along the adjoining foundation in accordance with Figure R403.3(4) a distance not of not less than Dimension A in Table R403.3.

Exception: Where the frost protected shallow foundation abuts the heated structure to form an inside corner, vertical insulation extending along the adjoining foundation is not required.

❖ This section stipulates that additions built on FPSFs are not required to be insulated where they join a heated structure because heat from both structures keeps the ground from freezing. However, where the FPSF abuts the heated building, the horizontal and vertical insulation are required to extend along the adjoining foundation to prevent cold from intruding under the slab through the foundation wall. The dimension of the insulation is related to the width of the horizontal insulation. This technique was shown to be effective by the U.S. Army Corps of Engineers Cold Regions Research and Engineering Laboratory (CRREL) on a FPSF addition to an airport control tower with a deep foundation in a severe Alaskan climate with a 13-foot frost line. An abutment at an inside corner is an exception. In this condition, both the building and the addition heat the ground, and extra insulation is not needed.

R403.3.2 Protection of horizontal insulation below gound. Horizontal insulation placed less than 12 inches (305 mm) below the ground surface or that portion of horizontal insulation extending outward more than 24 inches (610 mm) from the foundation edge shall be protected against damage by use of a concrete slab or asphalt paving on the ground surface directly above the insulation or by cementitious board, plywood rated for below-ground use, or other approved materials placed below ground, directly above the top surface of the insulation.

❖ This protection for the insulation prevents damage due to excavating (e.g., for landscaping purposes).

FIGURE R403.3(2)

For SI: °C = [(°F)-32]/1.8.

Note: The air-freezing index is defined as cumulative degree days below 32°F. It is used as a measure of the combined magnitude and duration of air temperature below freezing. The index was computed over a 12-month period (July-June) for each of the 3,044 stations used in the above analysis. Data from the 1951-80 period were fitted to a Weibull probability distribution to produce an estimate of the 100-year return period.

**FIGURE R403.3(2)
AIR-FREEZING INDEX
AN ESTIMATE OF THE 100-YEAR RETURN PERIOD**

❖ This figure provides the criteria necessary to apply the minimum insulation requirements of Table R404.3. The note explains the term air-freezing index.

FOUNDATIONS

FIGURE R403.3(3)

INSULATION DETAIL

HORIZONTAL INSULATION PLAN

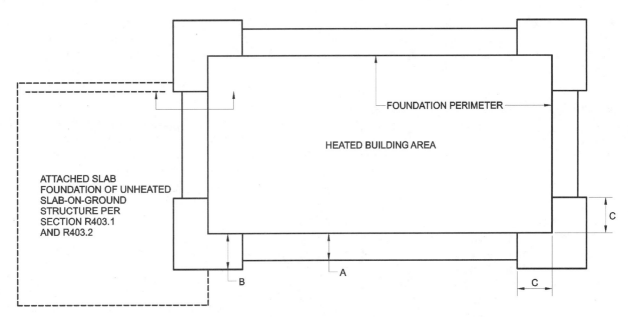

For SI: 1 inch = 25.4 mm.

a. See Table R403.3 for required dimensions and R-values for vertical and horizontal insulation.

**FIGURE R403.3(3)
INSULATION PLACEMENT FOR FROST-PROTECTED FOOTINGS
ADJACENT TO UNHEATED SLAB-ON-GROUND STRUCTURE**

❖ This figure of insulation detail shows 1) the maximum vertical distance between the top of the FPSF and the top of slab of adjacent unheated slab-on-ground structure and 2) the minimum vertical distance from the top of the slab and the bottom of the insulation. The horizontal insulation plan shows placement of a FPSF where it meets an unheated structure.

FIGURE R403.3(4)
INSULATION PLACEMENT FOR FROST-PROTECTED FOOTINGS ADJACENT TO HEATED STRUCTURE

❖ This figure shows where extra insulation is required when a FPSF abuts a heated structure.

R403.3.3 Drainage. Final grade shall be sloped in accordance with Section R401.3. In other than Group I Soils, as detailed in Table R405.1, gravel or crushed stone beneath horizontal insulation below ground shall drain to daylight or into an approved sewer system.

❖ See Figure R403.3(1) and the commentary for Section R401.3.

R403.3.4 Termite damage. The use of foam plastic in areas of "very heavy" termite infestation probability shall be in accordance with Section R320.4.

❖ Studies have shown that rigid board insulation installed below grade, particularly in areas where the hazard associated with termite infestation is very heavy, creates a pathway for termites that cannot be blocked with currently available termiticide treatments. Because it provides a place for termites to burrow in areas of very heavy termite infestation probability, the foam plastic insulation must be protected per Section R320.4.

SECTION R404
FOUNDATION WALLS

R404.1 Concrete and masonry foundation walls. Concrete and masonry foundation walls shall be selected and constructed in accordance with the provisions of this section or in accordance with ACI 318, NCMA TR68-A or ACI 530/ASCE 5/TMS 402 or other approved structural standards. When ACI 318 or ACI 530/ASCE 5/TMS 402 or the provisions of this section are used to design concrete or masonry foundation walls, project drawings, typical details and specifications are not required to bear the seal of the architect or engineer responsible for design, unless otherwise required by the state law of the jurisdiction having authority.

❖ These provisions are primarily for masonry or concrete basement walls. Foundation walls should be in accordance with this section or designed under one of the following design documents:
  • ACI 318: *Building Code Requirements for Structural Concrete*
  • NCMA TR68-A: *Concrete Masonry Foundation Walls*
  • ACI 530/ASCE 5/TMS 402: *Building Code Requirements for Masonry Structures.*

Foundation walls are usually designed and constructed to carry the vertical loads from the structure above, resist wind and any lateral forces transmitted to the foundations, and sustain earth pressures exerted against the walls, including any forces that may be imposed by frost action.

Most states do not require the seal of an architect or engineer for small residential buildings. Check with the state authorities in the location of the project for local regulations.

R404.1.1 Masonry foundation walls. Concrete masonry and clay masonry foundation walls shall be constructed as set forth in Tables R404.1.1(1), R404.1.1(2), R404.1.1(3) and

FOUNDATIONS

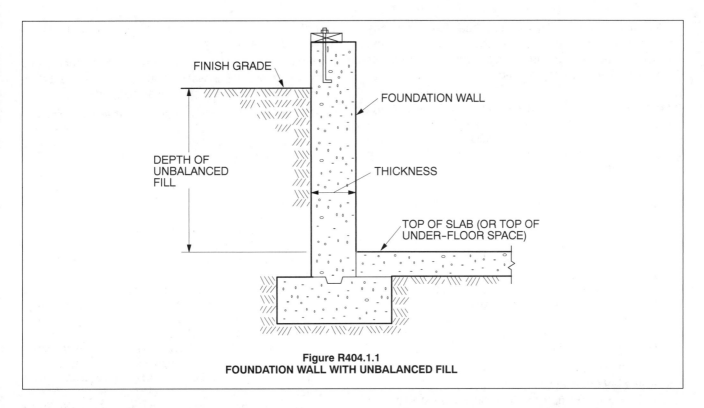

Figure R404.1.1
FOUNDATION WALL WITH UNBALANCED FILL

R404.1.1(4) and shall also comply with the provisions of this section and the applicable provisions of Sections R606, R607 and R608. In Seismic Design Categories D_1 and D_2, concrete masonry and clay masonry foundation walls shall comply with Section R404.1.4. Rubble stone masonry foundation walls shall be constructed in accordance with Sections R404.1.8 and R606.2.2. Rubble stone masonry walls shall not be used in Seismic Design Categories D_1 and D_2.

❖ Minimum thickness of concrete masonry and clay masonry foundation walls listed in Table R404.1.1(1) are predicated on the type of soil (see Table R405.1), the maximum heights of wall and unbalanced backfill (see Commentary Figure R404.1.1), and the type of wall system that is to be supported. The tabulated values are based on the assumption that the lateral loads will be carried by the walls through bending in the vertical direction. The actual thickness may be as much as $^1/_2$ inch (12.7 mm) less than the required nominal thickness specified in the table.

Example:
    From Table R404.1.1(1): an 8-foot (2438 mm) high foundation wall constructed of hollow-unit masonry in a gravel-sand-clay mixture (SM) with a backfill height above the basement floor of 5 feet (1524 mm) would have to be minimum nominal thickness of 10 inches (254 mm). The actual wall thickness could be a minimum of $9^1/_2$ inches (241 mm).

The reinforcement specified in Tables R404.1.1(2) through R404.1.1(4) of the code are based on the nominal wall thickness, type of soil, the height of the wall, and the height of unbalanced fill. The height of unbalanced fill is measured as shown in Commentary Figure R404.1.1. The tabulated values are based on the assumption that lateral loads will be carried by the walls through bending, primarily in the vertical direction.

Additional (general) requirements for masonry construction found in Section R606 through R608 must be followed also. See Section R404.1.4 for buildings in Seismic Design Category D_1 or D_2, and for rubble stone masonry, see the commentary for Section R404.1.8.

R404.1.2 Concrete foundation walls. Concrete foundation walls shall be constructed as set forth in Tables R404.1.1(1), R404.1.1(2), R404.1.1(3) and R404.1.1(4), and shall also comply with the provisions of this section and the applicable provisions of Section R402.2. In Seismic Design Categories D_1 and D_2, concrete foundation walls shall comply with Section R404.1.4.

❖ See the commentary for Section R404.1.1.

R404.1.3 Design required. A design in accordance with accepted engineering practice shall be provided for concrete or masonry foundation walls when any of the following conditions exist:

1. Walls are subject to hydrostatic pressure from groundwater.
2. Walls supporting more than 48 inches (1219 mm) of unbalanced backfill that do not have permanent lateral support at the top and bottom.

❖ Retaining walls and walls subjected to hydrostatic pressure from groundwater are not specifically included in the code. This section specifies conditions under which foundation walls require design. Section R404.1 cites the standards that are commonly used for "accepted engineering practice."

R404.1.4 Seismic Design Categories D_1 and D_2. In addition to the requirements of Table R404.1.1(1), plain concrete and plain masonry foundation walls located in Seismic Design Categories D_1 and D_2, as established in Table R301.2 (1), shall comply with the following:

1. Minimum reinforcement shall consist of one No. 4 (No. 13) horizontal bar located in the upper 12 inches (305 mm) of the wall,
2. Wall height shall not exceed 8 feet (2438 mm),
3. Height of unbalanced backfill shall not exceed 4 feet (1219 mm), and
4. A minimum thickness of 7.5 inches (191 mm) is required for plain concrete foundation walls except that a minimum thickness of 6 inches (152 mm) shall be permitted for plain concrete foundation walls with a maximum height of 4 feet, 6 inches (1372 mm).
5. Plain masonry foundation walls shall be a minimum of 8 inches (203 mm) thick.

Vertical reinforcement for masonry stem walls shall be tied to the horizontal reinforcement in the footings. Masonry stem walls located in Seismic Design Categories D_1 and D_2 shall have a minimum vertical reinforcement of one No. 3 bar located a maximum of 4 feet (1220 mm) on center in grouted cells.

Foundation walls located in Seismic Design Categories D_1 and D_2, as established in Table R301.2(1), supporting more than 4 feet (1219 mm) of unbalanced backfill or exceeding 8 feet (2438 mm) in height shall be constructed in accordance with Table R404.1.1(2), R404.1.1(3) or R404.1.1(4) and shall have two No. 4 (No. 13) horizontal bars located in the upper 12 inches (305 mm) of the wall.

❖ This section places five restrictions on the requirements in Table R404.1.1(1) for plain concrete and masonry foundation walls in Seismic Design Category D_1 or D_2.

The code requires minimum vertical reinforcement of masonry stem walls in order for them to resist lateral earthquake forces. Any wall over 8 feet (2338 mm) or supporting more than 4 feet (1219 mm) of unbalanced backfill must be constructed with minimum vertical reinforcement in accordance with Tables R404.1.1(2) through R404.1.1(4) and have the specified longitudinal reinforcement.

R404.1.5 Foundation wall thickness based on walls supported. The thickness of concrete and masonry foundation walls shall not be less than the thickness of the wall supported, except that foundation walls of at least 8-inch (203 mm) nominal thickness shall be permitted under brick-veneered frame walls and under 10-inch-wide (254 mm) cavity walls where the total height of the wall supported, including gables, is not more than 20 feet (6096 mm), provided the requirements of Sections R404.1.1 and R404.1.2 are met.

❖ Other than the two exceptions noted, concrete and masonry foundation walls should be equal to or greater than the thickness of the wall supported.

R404.1.5.1 Pier and curtain wall foundations. Pier and curtain wall foundations shall be permitted to be used to support light-frame construction not more than two stories in height, provided the following requirements are met:
1. All load-bearing walls shall be placed on continuous concrete footings placed integrally with the exterior wall footings.
2. The minimum actual thickness of a load-bearing masonry wall shall be not less than 4 inches (102 mm) nominal or $3^3/_8$ inches (92 mm) actual thickness, and shall be bonded integrally with piers spaced in accordance with Section R606.8.
3. Piers shall be constructed in accordance with Section R606.5 and Section R606.5.1, and shall be bonded into the load-bearing masonry wall in accordance with Section R608.1.1 or Section R608.1.1.2.
4. The maximum height of a 4-inch (102 mm) load-bearing masonry foundation wall supporting wood framed walls and floors shall not be more than 4 feet (1219 mm) in height.
5. Anchorage shall be in accordance with R403.1.6, Figure R404.1.5(1), or as specified by engineered design accepted by the building official.
6. The unbalanced fill for 4-inch (102 mm) foundation walls shall not exceed 24 inches (610 mm) for solid masonry or 12 inches (305 mm) for hollow masonry.
7. In Seismic Design Categories D_1 and D_2 prescriptive reinforcement shall be provided in the horizontal and vertical directions. Provide minimum horizontal joint reinforcement of two No. 9 gage wires spaced not less than 6 inches (152 mm) or one $^1/_4$ inch (6.4 mm) diameter wire at 10 inches (254 mm) on center vertically. Provide minimum vertical reinforcement of one #4 bar at 48 inches (1220 mm) on center horizontally grouted in place.

❖ Pier and curtain wall foundations may be used to support light-frame construction for buildings in Seismic Design Category A, B, C, D_1 or D_2, provided several conditions are met. A curtain wall typically refers to a nonload-bearing wall, but within the limits established in this section, it may be used to support up to two stories of light-frame construction. 4-inch (102 mm) nominal masonry walls are limited to 4 feet (1219 mm) in height so that a thin wall does not fail laterally due to the vertical loads on the wall. In addition, no more than 2 feet (610 mm) of unbalanced fill is permitted for solid masonry and 1 foot (305 mm) of unbalanced fill for hollow masonry to prevent failure due to the pressures created by unbalanced fill. Additional requirements are shown in Figure R404.1.5(1). This type of foundation wall must be reinforced when located in Seismic Design Category D_1 or D_2.

R404.1.6 Height above finished grade. Concrete and masonry foundation walls shall extend above the finished grade adja-

FOUNDATIONS

cent to the foundation at all points a minimum of 4 inches (102 mm) where masonry veneer is used and a minimum of 6 inches (152 mm) elsewhere.

❖ The minimum distance above adjacent grade to which the foundation must be extended provides termite protection and minimizes the chance of decay resulting from moisture migrating to the wood framing. A reduced foundation extension is permitted when masonry veneer is used.

TABLE R404.1.1(1)
PLAIN CONCRETE AND PLAIN MASONRY FOUNDATION WALLS

MAXIMUM WALL HEIGHT (feet)	MAXIMUM UNBALANCED BACKFILL HEIGHT[c] (feet)	PLAIN CONCRETE MINIMUM NOMINAL WALL THICKNESS (inches)			PLAIN MASONRY[a] MINIMUM NOMINAL WALL THICKNESS (inches)		
		GW, GP, SW and SP	GM, GC, SM, SM-SC and ML	SC, MH, ML-CL and inorganic CL	GW, GP, SW and SP	GM, GC, SM, SM-SC and ML	SC, MH, ML-CL and inorganic CL
5	4	6	6	6	6 solid[d] or 8	6 solid[d] or 8	6 solid[d] or 8
	5	6	6	6	6 solid[d] or 8	8	10
6	4	6	6	6	6 solid[d] or 8	6 solid[d] or 8	6 solid[d] or 8
	5	6	6	6	6 solid[d] or 8	8	10
	6	6	8[g]	8[g]	8	10	12
7	4	6	6	6	6 solid[d] or 8	8	8
	5	6	6	8[g]	6 solid[d] or 8	10	10
	6	6	8	8	10	12	10 solid[d]
	7	8	8	10	12	10 solid[d]	12 solid[d]
8	4	6	6	6	6 solid[d] or 8	6 solid[d] or 8	8
	5	6	6	8	6 solid[d] or 8	10	12
	6	8[g]	8	10	10	12	12 solid[d]
	7	8	10	10	12	12 solid[d]	Footnote e
	8	10	10	12	10 solid[d]	12 solid[d]	Footnote e
9	4	6	6	6	6 solid[d] or 8	6 solid[d] or 8	8
	5	6	8[g]	8	8	10	12
	6	8	8	10	10	12	12 solid[d]
	7	8	10	10	12	12 solid[d]	Footnote e
	8	10	10	12	12 solid[d]	Footnote e	Footnote e
	9	10	12	Footnote f	Footnote e	Footnote e	Footnote e

For SI: 1 inch = 25.4 mm, 1 foot = 304.8 mm, 1 pound per square inch = 6.895 Pa.

a. Mortar shall be Type M or S and masonry shall be laid in running bond. Ungrouted hollow masonry units are permitted except where otherwise indicated.

b. Soil classes are in accordance with the Unified Soil Classification System. Refer to Table R405.1.

c. Unbalanced backfill height is the difference in height of the exterior and interior finish ground levels. Where an interior concrete slab is provided, the unbalanced backfill height shall be measured from the exterior finish ground level to the top of the interior concrete slab.

d. Solid grouted hollow units or solid masonry units.

e. Wall construction shall be in accordance with Table R404.1.1(2) or a design shall be provided.

f. A design is required.

g. Thickness may be 6 inches, provided minimum specified compressive strength of concrete, f_c, is 4,000 psi.

❖ See Section R404.1.1 and example.

TABLE R404.1.1(2)
REINFORCED CONCRETE AND MASONRY[a] FOUNDATION WALLS

MAXIMUM WALL HEIGHT (feet)	MAXIMUM UNBALANCED BACKFILL HEIGHT[e] (feet)	MINIMUM VERTICAL REINFORCEMENT SIZE AND SPACING[b, c] FOR 8-INCH NOMINAL WALL THICKNESS		
		Soil classes[d]		
		GW, GP, SW and SP soils	GM, GC, SM, SM-SC and ML soils	SC, MH, ML-CL and inorganic CL soils
6	5	#4 at 48" o.c.	#4 at 48" o.c.	#4 at 48" o.c.
	6	#4 at 48" o.c.	#4 at 40" o.c.	#5 at 48" o.c.
7	4	#4 at 48" o.c.	#4 at 48" o.c.	#4 at 48" o.c.
	5	#4 at 48" o.c.	#4 at 48" o.c.	#4 at 40" o.c.
	6	#4 at 48" o.c.	#5 at 48" o.c.	#5 at 40" o.c.
	7	#4 at 40" o.c.	#5 at 40" o.c.	#6 at 48" o.c.
8	5	#4 at 48" o.c.	#4 at 48" o.c.	#4 at 40" o.c.
	6	#4 at 48" o.c.	#5 at 48" o.c.	#5 at 40" o.c.
	7	#5 at 48" o.c.	#6 at 48" o.c.	#6 at 40" o.c.
	8	#5 at 40" o.c.	#6 at 40" o.c.	#6 at 24" o.c.
9	5	#4 at 48" o.c.	#4 at 48" o.c.	#5 at 48" o.c.
	6	#4 at 48" o.c.	#5 at 48" o.c.	#6 at 48" o.c.
	7	#5 at 48" o.c.	#6 at 48" o.c.	#6 at 32" o.c.
	8	#5 at 40" o.c.	#6 at 32" o.c.	#6 at 24" o.c.
	9	#6 at 40" o.c.	#6 at 24" o.c.	#6 at 16" o.c.

For SI: 1 inch = 25.4 mm, 1 foot = 304.8 mm.

a. Mortar shall be Type M or S and masonry shall be laid in running bond.
b. Alternative reinforcing bar sizes and spacings having an equivalent cross-sectional area of reinforcement per lineal foot of wall shall be permitted provided the spacing of the reinforcement does not exceed 72 inches.
c. Vertical reinforcement shall be Grade 60 minimum. The distance from the face of the soil side of the wall to the center of vertical reinforcement shall be at least 5 inches.
d. Soil classes are in accordance with the Unified Soil Classification System. Refer to Table R405.1.
e. Unbalanced backfill height is the difference in height of the exterior and interior finish ground levels. Where an interior concrete slab is provided, the unbalanced backfill height shall be measured from the exterior finish ground level to the top of the interior concrete slab.

❖ See the commentary for Section R404.1.1.

TABLE R404.1.1(3)
REINFORCED CONCRETE AND MASONRY[a] FOUNDATION WALLS

MAXIMUM WALL HEIGHT (feet)	MAXIMUM UNBALANCED BACKFILL HEIGHT[e] (feet)	VERTICAL REINFORCEMENT SIZE AND SPACING[b,c] FOR 12-INCH NOMINAL WALL THICKNESS		
		Soil classes[d]		
		GW, GP, SW and SP soils	GM, GC, SM, SM-SC and ML soils	SC, MH, ML-CL and inorganic CL soils
7	4	#4 at 72" o.c.	#4 at 72" o.c.	#4 at 72" o.c.
	5	#4 at 72" o.c.	#4 at 72" o.c.	#4 at 72" o.c.
	6	#4 at 72" o.c.	#4 at 64" o.c.	#4 at 48" o.c.
	7	#4 at 72" o.c.	#4 at 48" o.c.	#5 at 56" o.c.
8	5	#4 at 72" o.c.	#4 at 72" o.c.	#4 at 72" o.c.
	6	#4 at 72" o.c.	#4 at 56" o.c.	#5 at 72" o.c.
	7	#4 at 64" o.c.	#5 at 64" o.c.	#4 at 32" o.c.
	8	#4 at 48" o.c.	#4 at 32" o.c.	#5 at 40" o.c.
9	5	#4 at 72" o.c.	#4 at 72" o.c.	#4 at 72" o.c.
	6	#4 at 72" o.c.	#4 at 56" o.c.	#5 at 64" o.c.
	7	#4 at 56" o.c.	#4 at 40" o.c.	#6 at 64" o.c.
	8	#4 at 64" o.c.	#6 at 64" o.c.	#6 at 48" o.c.
	9	#5 at 56" o.c.	#7 at 72" o.c.	#6 at 40" o.c.

For SI: For SI: 1 inch = 25.4 mm, 1 foot = 304.8 mm.

a. Mortar shall be Type M or S and masonry shall be laid in running bond.
b. Alternative reinforcing bar sizes and spacing having an equivalent cross-sectional area of reinforcement per lineal foot of wall shall be permitted provided the spacing of the reinforcement does not exceed 72 inches.
c. Vertical reinforcement shall be Grade 60 minimum. The distance from the face of the soil side of the wall to the center of vertical reinforcement shall be at least 8.75 inches.
d. Soil classes are in accordance with the Unified Soil Classification System. Refer to Table R405.1.
e. Unbalanced backfill height is the difference in height of the exterior and interior finish ground levels. Where an interior concrete slab is provided, the unbalanced backfill height shall be measured from the exterior finish ground level to the top of the interior concrete slab.

❖ See the commentary for Section R404.1.1.

TABLE R404.1.1(4)
REINFORCED CONCRETE AND MASONRY[a] FOUNDATION WALLS

MAXIMUM WALL HEIGHT (feet)	MAXIMUM UNBALANCED BACKFILL HEIGHT[e] (feet)	MINIMUM VERTICAL REINFORCEMENT SIZE AND SPACING[b,c] FOR 10-INCH NOMINAL WALL THICKNESS		
		Soil Classes[d]		
		GW, GP, SW and SP soils	GM, GC, SM, SM-SC and ML soils	SC, MH, ML-CL and inorganic CL soils
7	4	#4 at 56" o.c.	#4 at 56" o.c.	#4 at 56" o.c
	5	#4 at 56" o.c	#4 at 56" o.c.	#4 at 56" o.c.
	6	#4 at 56" o.c.	#4 at 48" o.c	#4 at 40" o.c.
	7	#4 at 56" o.c.	#5 at 56" o.c.	#5 at 40" o.c.
8	5	#4 at 56" o.c.	#4 at 56" o.c.	#4 at 48" o.c.
	6	#4 at 56" o.c	#4 at 48" o.c	#5 at 56" o.c
	7	#4 at 48" o.c.	#4 at 32" o.c.	#6 at 56" o.c.
	8	#5 at 56" o.c.	#5 at 40" o.c.	#7 at 56" o.c.
9	5	#4 at 56" o.c.	#4 at 56" o.c.	#4 at 48" o.c.
	6	#4 at 56" o.c	#4 at 40" o.c	#4 at 32" o.c
	7	#4 at 56" o.c.	#5 at 48" o.c.	#6 at 48" o.c.
	8	#4 at 32" o.c.	#6 at 48" o.c.	#4 at 16" o.c.
	9	#5 at 40" o.c.	#6 at 40" o.c.	#7 at 40" o.c.

For SI: 1 inch = 25.4 mm, 1 foot = 304.8 mm.

a. Mortar shall be Type M or S and masonry shall be laid in running bond.
b. Alternative reinforcing bar sizes and spacings having an equivalent cross-sectional area of reinforcement per lineal foot of wall shall be permitted provided the spacing of the reinforcement does not exceed 72 inches.
c. Vertical reinforcement shall be Grade 60 minimum. The distance from the face of the soil side of the wall to the center of vertical reinforcement shall be at least 6.75 inches.
d. Soil classes are in accordance with the Unified Soil Classification System. Refer to Table R405.1.
e. Unbalanced backfill height is the difference in height of the exterior and interior finish ground levels. Where an interior concrete slab is provided, the unbalanced backfill height shall be measured from the exterior finish ground level to the top of the interior concrete slab.

❖ See the commentary for Section R404.1.1.

R404.1.7 Backfill placement. Backfill shall not be placed against the wall until the wall has sufficient strength and has been anchored to the floor above, or has been sufficiently braced to prevent damage by the backfill.

Exception: Such bracing is not required for walls supporting less than 4 feet (1219 mm) of unbalanced backfill.

❖ Backfilling should not begin until a foundation wall is in suitable condition to resist lateral earth pressure. The wall must have sufficient strength, and the floor framing system should be in place and anchored to the foundation wall as shown in Commentary Figure R404.1.7 to minimize the chance of damage to the foundation wall resulting from placement of backfill against the foundation wall.

R404.1.8 Rubble stone masonry. Rubble stone masonry foundation walls shall have a minimum thickness of 16 inches (406 mm), shall not support an unbalanced backfill exceeding 8 feet (2438 mm) in height, shall not support a soil pressure greater than 30 psf (481 kg/m^2), and shall not be constructed in Seismic Design Categories D_1 or D_2 as established in Figure R301.2(2).

❖ Rubble stone masonry is masonry walls constructed of roughly shaped stones. See the definitions in the *International Building Code*. Rubble stone masonry is almost never used for foundation walls. This type of wall construction is permitted only in Seismic Design Category A, B, or C. Furthermore, it is restricted to a maximum unbalanced backfill height of 8 feet (2438 mm) and must not be subjected to lateral soil pressures more than 30 psf (481 kg/m^2).

R404.2 Wood foundation walls. Wood foundation walls shall be constructed in accordance with the provisions of Sections R404.2.1 through R404.2.6 and with the details shown in Figures R403.1(2) and R403.1(3).

❖ This section covers the design and installation of a multi-component wood foundation system. The construction is essentially a below-grade load-bearing wood frame system that serves as the enclosure for basements and crawl spaces and also serves as the structural foundation for the support of light-frame structures. See Figures R403.1(2) and R403.1(3) for typical installation details for wood foundation walls. Also see the commentary for Section R401.1.

Certain cold-weather precautions should be taken in the installation of the wood foundation system. The composite footing consisting of a wood plate supported on a bed of stone or sand fill should not be placed on frozen ground. While the bottom of the wood plate footing would normally be placed below the frost line (i.e., basement construction) under certain drainage conditions, AFPA TR7-87 permits the wood plate to be set above the frost level. The code official should ascertain that such an alternative design satisfies the intent of the code. Also important is the use of proper sealants during very cold weather. All manufacturers of sealants and bonding agents impose temperature restrictions on the use of their products. Only sealants and bonding agents specifically produced for cold weather conditions should be used.

FIGURE R404.1.5(1)

For SI: 1 inch = 25.4 mm, 1 foot = 304.8 mm, 1 degree = $^{0.79}/_{45}$ rad.

**FIGURE R404.1.5(1)
FOUNDATION WALL CLAY MASONRY CURTAIN WALL WITH CONCRETE MASONRY PIERS**

❖ This figure illustrates several requirements for pier and curtain wall foundations described in Section R404.1.5.1. The pier spacing of not more than 6 feet, 0 inches (1829 mm) on center establishes minimum lateral (horizontal) wall support similar to that of Section R606.8. Also see the commentary for Section R404.1.5.1.

FOUNDATIONS

Figure R404.1.7
BEARING FOUNDATION WALLS AGAINST LATERAL EARTH PRESSURE

R404.2.1 Identification. All load-bearing lumber shall be identified by the grade mark of a lumber grading or inspection agency which has been approved by an accreditation body that complies with DOC PS 20. In lieu of a grade mark, a certificate of inspection issued by a lumber grading or inspection agency meeting the requirements of this section shall be accepted. Wood structural panels shall conform to DOC PS 1 or DOC PS 2 and shall be identified by a grade mark or certificate of inspection issued by an approved agency.

❖ See the commentary for Sections R502.1 and R503.2.1 as well as Figures R502.1 and R503.2.1.

R404.2.2 Stud size. The studs used in foundation walls shall be 2-inch-by-6-inch (51 mm by 152 mm) members. When spaced 16 inches (406 mm) on center, a wood species with an F_b value of not less than 1,250 (8612 kPa) as listed in AF&PA/NDS shall be used. When spaced 12 inches (305 mm) on center, an F_b of not less than 875 (6029 kPa) shall be required.

❖ This section specifies the minimum bending stress, F_b, for studs based on their spacing in a foundation wall. The appropriate wood species that meet this requirement can be found in the supplement to the AF&PA/NDS: *National Design Specification for Wood Construction.*

R404.2.3 Height of backfill. For wood foundations that are not designed and installed in accordance with AF&PA Report No.7, the height of backfill against a foundation wall shall not exceed 4 feet (1219 mm). When the height of fill is more than 12 inches (305 mm) above the interior grade of a crawl space or floor of a basement, the thickness of the plywood sheathing shall meet the requirements of Table R404.2.3.

❖ Table R404.2.3 specifies requirements for plywood based on height of backfill and plywood span (stud spacing) where the height of retained earth is more than 12 inches (305 mm). See Figure R403.1(2). The backfill limits of Table R404.2.3 do not apply to wood foundations designed and installed in accordance with AF&PA report No. 7.

R404.2.4 Backfilling. Wood foundation walls shall not be backfilled until the basement floor and first floor have been constructed or the walls have been braced. For crawl space construction, backfill or bracing shall be installed on the interior of the walls prior to placing backfill on the exterior.

❖ These provisions require that lateral support be in place prior to backfilling on the exterior.

R404.2.5 Drainage and dampproofing. Wood foundation basements shall be drained and dampproofed in accordance with Sections R405 and R406, respectively.

❖ See Sections R405.2 and R406.3 of this commentary.

R404.2.6 Fastening. Wood structural panel foundation wall sheathing shall be attached to framing in accordance with Table R602.3(1) and Section R402.1.1.

❖ The fastening schedule must be followed for wall sheathing attachment. Also see the commentary for Section R402.1.1.

R404.3 Wood sill plates. Wood sill plates shall be a minimum of 2-inch by 4-inch (51 mm by 102 mm) nominal lumber. Sill plate anchorage shall be in accordance with Sections R403.1.6 and R602.11.

❖ Minimum 2 x 4 (51 mm x 102 mm) nominal sill plates are required, and they should be anchored to the foundation as required by Section R403.1.6.

R404.4 Insulating concrete form foundation walls. Insulating concrete form (ICF) foundation walls shall be designed and constructed in accordance with the provisions of this section or in accordance with the provisions of ACI 318. When ACI 318 or the provisions of this section are used to design insulating concrete form foundation walls, project drawings, typical details and specifications are not required to bear the seal of the architect or engineer responsible for design unless otherwise required by the state law of the jurisdiction having authority.

❖ This concrete forming system utilizes stay-in-place, rigid foam plastic forms for basement wall construction. The three systems covered in the code are flat wall, waffle-grid, and screen-grid. This section provides prescriptive requirements, and ACI 318 may also be used for design and construction of insulating concrete form foundation walls.

R404.4.1 Applicability limits. The provisions of this section shall apply to the construction of insulating concrete form foundation walls for buildings not greater than 60 feet (18 288 mm) in plan dimensions, and floors not greater than 32 feet (9754 mm) or roofs not greater than 40 feet (12 192 mm) in clear span. Buildings shall not exceed two stories in height above-grade with each story not greater than 10 feet (3048 mm) high. Foundation walls constructed in accordance with the provisions of this section shall be limited to buildings subjected to a maximum ground snow load of 70 psf (3.35 kN/m^2) and located in Seismic Design Category A, B or C.

❖ Figure R611.2 illustrates the applicability limits for buildings using insulating concrete form wall construction. In Chapter 6, the use of insulating concrete form walls is also limited to wind speeds of 150 mph (241 km/h) or less.

Section R611.2 limits the use of ICF walls to Seismic Design Categories A, B, C, D_1, and D_2. However, ICF foundation walls are limited to Seismic Design Categories A, B, and C. Foundation walls located in Seismic Design Categories D_1 and D_2 require conventionally formed concrete foundation walls with flat surfaces in accordance with the provisions of this code or in accordance with the provisions of ACI 318.

R404.4.2 Flat insulating concrete form wall systems. Flat ICF wall systems shall comply with Figure R611.3, shall have a minimum concrete thickness of 5.5 inches (140 mm), and shall have reinforcement in accordance with Table R404.4(1), R404.4(2) or R404.4(3).

❖ Figure R611.3 illustrates flat-insulating concrete form walls. They must be in accordance with the limits and requirements of Tables R404.4(1) through R404.4(3). These tables are based on 2500 psi (17,235 kPa) concrete and reinforcing steel with a minimum yield strength of 40,000 psi (275,760 kPa). When reinforcing steel with a minimum yield strength or 60,000 psi (413,640 kPa) is used, the spacing of the reinforcement can be increased as permitted by Footnote "a." Manufacturers of insulating concrete form systems may have systems that vary from these requirements; these systems are permitted, provided appropriate engineering data (e.g. evaluation reports) are furnished.

R404.4.3 Waffle grid insulating concrete form wall systems. Waffle-grid wall systems shall have a minimum nominal concrete thickness of 6 inches (152 mm) for the horizontal and vertical concrete members (cores) and shall be reinforced in accordance with Table R404.4(4). The minimum core dimension shall comply with Table R611.4 (2) and Figure R611.4.

❖ Figure R611.4 illustrates waffle-grid wall systems. They must be constructed in accordance with Table R404.4(4).

R404.4.4 Screen grid insulating concrete form wall systems. Screen-grid ICF wall systems shall have a minimum nominal concrete thickness of 6 inches (152 mm) for the horizontal and vertical concrete members (cores). The minimum core dimensions shall comply with Table R611.4(2) and Figure R611.5. Walls shall have reinforcement in accordance with Table R404.4(5).

❖ Figure R611.5 illustrates screen-grid wall systems. They must be constructed in accordance with Table R404.4(5).

TABLE R404.2.3[d]
PLYWOOD GRADE AND THICKNESS FOR WOOD FOUNDATION CONSTRUCTION
(30 pcf equivalent-fluid weight soil pressure)

HEIGHT OF FILL (inches)	STUD SPACING (inches)	FACE GRAIN ACROSS STUDS			FACE GRAIN PARALLEL TO STUDS		
		Grade[a]	Minimum thickness (inches)	Span rating	Grade[a]	Minimum thickness (inches)[b,c]	Span rating
24	12	B	$15/32$	32/16	A	$15/32$	32/16
					B	$15/32^c$	32/16
	16	B	$15/32$	32/16	A	$15/32^c$	32/16
					B	$19/32^c$ (4, 5 ply)	40/20
36	12	B	$15/32$	32/16	A	$15/32$	32/16
					B	$15/32^c$ (4, 5 ply)	32/16
					B	$19/32$ (4, 5 ply)	40/20
	16	B	$15/32^c$	32/16	A	$19/32$	40/20
					B	$23/32$	48/24
48	12	B	$15/32$	32/16	A	$15/32^c$	32/16
					B	$19/32^c$ (4, 5 ply)	40/20
	16	B	$19/32$	40/20	A	$19/32^c$	40/20
					A	$23/32$	48/24

For SI: 1 inch = 25.4 mm, 1 foot = 304.8 mm, 1 pound per cubic foot = 0.1572 kN/m^3.

a. Plywood shall be of the following minimum grades in accordance with DOC PS 1 or DOC PS 2:
   1. DOC PS 1 Plywood grades marked:
      1.1. Structural I C-D (Exposure 1)
      1.2. C-D (Exposure 1)
   2. DOC PS 2 Plywood grades marked:
      2.1. Structural I Sheathing (Exposure 1)
      2.2. Sheathing (Exposure 1)
   3. Where a major portion of the wall is exposed above ground and a better appearance is desired, the following plywood grades marked exterior are suitable:
      3.1. Structural I A-C, Structural I B-C or Structural I C-C (Plugged) in accordance with DOC PS 1
      3.2. A-C Group 1, B-C Group 1, C-C (Plugged) Group 1 or MDO Group 1 in accordance with DOC PS 1
      3.3. Single Floor in accordance with DOC PS 1 or DOC PS 2

b. Minimum thickness $15/32$ inch, except crawl space sheathing may be $3/8$ inch for face grain across studs 16 inches on center and maximum 2-foot depth of unequal fill.

c. For this fill height, thickness and grade combination, panels that are continuous over less than three spans (across less than three stud spacings) require blocking 16 inches above the bottom plate. Offset adjacent blocks and fasten through studs with two 16d corrosion-resistant nails at each end.

d. Fastening shall be in accordance with Section R319.3.

❖ This table specifies the minimum grade and thickness of plywood based on the height of retained earth (fill) and the plywood span.

TABLE R404.4(1)
5.5-INCH THICK FLAT ICF FOUNDATION WALLS[a, b, c, d]

HEIGHT OF BASEMENT WALL (feet)	MAXIMUM UNBALANCED BACKFILL HEIGHT[e] (feet)	MINIMUM VERTICAL REINFORCEMENT SIZE AND SPACING[f]		
		Soil group I[f]	Soil group II[f]	Soil group III[f]
8	4	#4@48″	#4@48″	#4@48″
8	5	#4@48″	#3@12″; #4@22″; #5@32″	#3@8″; #4@14″; #5@20″; #6@26″
8	6	#3@12″; #4@22″; #5@30″;	#3@8″; #4@14″; #5@20″; #6@24″	#3@6″; #4@10″: #5@14″; #6@20″
8	7	#3@8″; #4@14″; #5@22″; #6@26″	#3@5″; #4@10″; #5@14″; #6@18″	#3@4″; #4@6″; #5@10″; #6@14″
9	4	#4@48″	#4@48″	#4@48″
9	5	#4@48″	#3@12″; #4@20″; #5@28″; #6@36″	#3@8″; #4@14″; #5@20″; #6@22″
9	6	#3@10″; #4@20″; #5@28″; #6@34″	#3@6″; #4@12″; #5@18″; #6@20″	#4@8″; #5@14″; #6@16″
9	7	#3@8″; #4@14″; #5@20″; #6@22″	#4@8″; #5@12″; #6@16″	#4@6″; #5@10″; #6@12″
9	8	#3@6″; #4@10″; #5@14″; #6@16″	#4@6″; #5@10″; #6@12″	#4@4″; #5@6″; #6@8″
10	4	#4@48″	#4@48″	#4@48″
10	5	#4@48″	#3@10″; #4@18″; #5@26″; #6@30″	#3@6″; #4@14″; #5@18″; #6@20″
10	6	#3@10″; #4@18″; #5@24″; #6@30″	#3@6″; #4@12″; #5@16″; #6@18″	#3@4″; #4@8″; #5@12″; #6@14″
10	7	#3@6″; #4@12″; #5@16″; #6@18″	#3@4″; #4@8″; #5@12″	#4@6″; #5@8″; #6@10″
10	8	#4@8″; #5@12″; #6@14″	#4@6″; #5@8″; #6@12″	#4@4″; #5@6″; #6@8″
10	9	#4@6″; #5@10″; #6@12″	#4@4″; #5@6″; #6@8″	#5@4″; #6@6″

For SI: 1 inch = 25.4 mm, 1 foot = 304.8 mm, 1 pound per square inch = 6.895 kPa.

a. This table is based on concrete with a minimum specified concrete strength of 2500 psi, reinforcing steel with a minimum yield strength of 40,000 psi. When reinforcing steel with a minimum yield strength of 60,000 psi is used, the spacing of the reinforcement shall be increased to 1.5 times the spacing value in the table but in no case greater than 48 inches on center.

b. This table is not intended to prohibit the use of an ICF manufacturer's tables based on engineering analysis in accordance with ACI 318.

c. Deflection criteria: $L/240$.

d. Interpolation between rebar sizes and spacing is not permitted.

e. Unbalanced backfill height is the difference in height of the exterior and interior finished ground. Where an interior concrete slab is provided, the unbalanced backfill height shall be measured from the exterior finished ground level to the top of the interior concrete slab.

f. Soil classes are in accordance with the Unified Soil Classification System. Refer to Table R405.1.

❖ Tables R404.4(1) through R404.4(3) specify vertical reinforcing requirements for flat insulating concrete form walls based on the basement wall height, unbalanced backfill height and the soil group as determined in accordance with Table R405.1.

TABLE R404.4(2)
7.5-INCH-THICK FLAT ICF FOUNDATION WALLS[a, b, c, d, e]

HEIGHT OF BASEMENT WALL (feet)	MAXIMUM UNBALANCED BACKFILL HEIGHT[f] (feet)	MINIMUM VERTICAL REINFORCEMENT SIZE AND SPACING[g]		
		Soil group I[g]	Soil group II[g]	Soil group III[g]
8	6	N/R	N/R	N/R
	7	N/R	#3@8"; #4@14"; #5@20"; #6@28"	#3@6"; #4@10"; #5@16"; #6@20"
9	6	N/R	N/R	#3@8"; #4@14"; #5@20"; #6@28"
	7	N/R	#3@6"; #4@12"; #5@18"; #6@26"	#3@4"; #4@8"; #5@14"; #6@18"
	8	#3@8"; #4@14"; #5@22"; #6@28"	#3@4"; #4@8"; #5@14"; #6@18"	#3@4"; #4@6"; #5@10"; #6@14"
10	6	N/R	N/R	#3@6"; #4@12"; #5@18"; #6@26"
	7	N/R	#3@6"; #4@12"; #5@18"; #6@24"	#3@4"; #4@8"; #5@12"; #6@18"
	8	#3@6"; #4@12"; #5@20"; #6@26"	#3@4"; #4@8"; #5@12"; #6@16"	#3@4"; #4@6"; #5@8"; #6@12"
	9	#3@6"; #4@10"; #5@14"; #6@20"	#3@4"; #4@6"; #5@10"; #6@12"	#4@4"; #5@6"; #6@10"

For SI: 1 inch = 25.4 mm, 1 foot = 304.8 mm, 1 pound per square inch = 6.895 kPa.

a. This table is based on concrete with a minimum specified concrete strength of 2500 psi, reinforcing steel with a minimum yield strength of 40,000 psi. When reinforcing steel with a minimum yield strength of 60,000 psi is used, the spacing of the reinforcement shall be increased to 1.5 times the spacing value in the table.
b. This table is not intended to prohibit the use of an ICF manufacturer's tables based on engineering analysis in accordance with ACI 318.
c. N/R denotes "not required."
d. Deflection criteria: $L/240$.
e. Interpolation between rebar sizes and spacing is not permitted.
f. Unbalanced backfill height is the difference in height of the exterior and interior finished ground. Where an interior concrete slab is provided, the unbalanced backfill height shall be measured from the exterior finished ground level to the top of the interior concrete slab.
g. Soil classes are in accordance with the Unified Soil Classification System. Refer to Table R405.1.

❖ See the commentary for Table R404.4(1).

TABLE R404.4(3)
9.5-INCH-THICK FLAT ICF FOUNDATION WALLS[a, b, c, d, e]

HEIGHT OF BASEMENT WALL (feet)	MAXIMUM UNBALANCED BACKFILL HEIGHT[f] (feet)	MINIMUM VERTICAL REINFORCEMENT SIZE AND SPACING[g]		
		Soil I[g]	Soil II[g]	Soil III[g]
8	7	N/R	N/R	N/R
9	6	N/R	N/R	N/R
	7	N/R	N/R	#3@6"; #4@12"; #5@18"; #6@26"
	8	N/R	#3@6"; #4@12"; #5@18"; #6@26"	#3@4"; #4@8"; #5@14"; #6@18"
10	5	N/R	N/R	N/R
	6	N/R	N/R	#3@10"; #4@18"; #5@26"; #6@36"
	7	N/R	N/R	#3@6"; #4@10"; #5@18"; #6@24"
	8	N/R	#3@6"; #4@12"; #5@16"; #6@24"	#3@4"; #4@8"; #5@12"; #6@16"
	9	N/R	#3@4"; #4@8"; #5@12"; #6@18"	#3@4"; #4@6"; #5@10"; #6@12"

For SI: 1 inch = 25.4 mm, 1 foot = 304.8 mm, 1 pound per square inch = 6.895 kPa.

a. This table is based on concrete with a minimum specified concrete strength of 2500 psi, reinforcing steel with a minimum yield strength of 40,000 psi. When reinforcing steel with a minimum yield strength of 60,000 psi is used, the spacing of the reinforcement shall be increased to 1.5 times the spacing value in the table.
b. This table is not intended to prohibit the use of an ICF manufacturer's tables based on engineering analysis in accordance with ACI 318.
c. N/R denotes "not required."
d. Deflection criteria: $L/240$.
e. Interpolation between rebar sizes and spacing is not permitted.
f. Unbalanced backfill height is the difference in height of the exterior and interior finished ground. Where an interior concrete slab is provided, the unbalanced backfill height shall be measured from the exterior finished ground level to the top of the interior concrete slab.
g. Soil classes are in accordance with the Unified Soil Classification System. Refer to Table R405.1.

❖ See the commentary for Table R404.4(1).

TABLE R404.4(4)
WAFFLE GRID ICF FOUNDATION WALLS[a,b,c,d,e]

MINIMUM NOMINAL WALL THICKNESS[f] (inches)	HEIGHT OF BASEMENT WALL (feet)	MAXIMUM UNBALANCED BACKFILL HEIGHT[g] (feet)	MINIMUM VERTICAL REINFORCEMENT SIZE AND SPACING[h]		
			Soil group I[h]	Soil group II[h]	Soil group III[h]
6	8	4	#4@48"	#3@12"; #4@24"	#3@12"
		5	#3@12"; #5@24"	#4@12"	#7@12"
		6	#4@12"	Design required	Design required
		7	#7@12"	Design required	Design required
	9	4	#4@48"	#3@12"; #5@24"	#3@12"
		5	#3@12"	#4@12"	Design required
		6	#5@12"	Design required	Design required
		7	Design required	Design required	Design required
	10	4	#4@48"	#4@12"	#5@12"
		5	#3@12"	Design required	Design required
		6	Design required	Design required	Design required
		7	Design required	Design required	Design required
8	8	4	N/R	N/R	N/R
		5	N/R	#3@12"; #4@24"; #5@36"	#3@12"; #5@24"
		6	#3@12"; #4@24"; #5@36"	#4@12"; #5@24"	#4@12"
		7	#3@12"; #6@24"	#4@12"	#5@12"
	9	4	N/R	N/R	N/R
		5	N/R	#3@12"; #5@24"	#3@12"; #5@24"
		6	#3@12"; #4@24"	#4@12"	#4@12"
		7	#4@12"; #5@24"	#5@12"	#5@12"
		8	#4@12"	#5@12"	#8@12"
	10	4	N/R	#3@12"; #4@24"; #6@36"	#3@12"; #5@24"
		5	N/R	#3@12"; #4@24"; #6@36"	#4@12"; #5@24"
		6	#3@12"; #5@24"	#4@12"	#5@12"
		7	#4@12"	#5@12"	#6@12"
		8	#4@12"	#6@12"	Design required
		9	#5@12"	Design required	Design required

For SI: 1 inch = 25.4 mm, 1 foot = 304.8 mm, 1 pound per square foot = 0.0479 kN/m².

a. This table is based on concrete with a minimum specified concrete strength of 2500 psi, reinforcing steel with a minimum yield strength of 40,000 psi. When reinforcing steel with a minimum yield strength of 60,000 psi is used, the spacing of the reinforcement shall be increased 12 inches but in no case greater than 48 inches on center.
b. This table is not intended to prohibit the use of an ICF manufacturer's tables based on engineering analysis in accordance with ACI 318.
c. N/R denotes "not required."
d. Deflection criteria: $L/240$.
e. Interpolation between rebar sizes and spacing is not permitted.
f. Refer to Table R611.4(2) for wall dimensions.
g. Unbalanced backfill height is the difference in height of the exterior and interior finished ground. Where an interior concrete slab is provided, the unbalanced backfill height shall be measured from the exterior finished ground level to the top of the interior concrete slab.
h. Soil classes are in accordance with the Unified Soil Classification System. Refer to Table R405.1.

❖ This table specifies verticle reinforcement requirements for waffle-grid basement walls based on the height and thickness of the wall, the height of unbalanced backfill and the soil group as determined in accordance with Table R405.1. Several table entries note that a design is required.

TABLE R404.4(5)
SCREEN-GRID ICF FOUNDATION WALLS[a,b,c,d,e]

MINIMUM NOMINAL WALL THICKNESS[f] (inches)	HEIGHT OF BASEMENT WALL (feet)	MAXIMUM UNBALANCED BACKFILL HEIGHT[g] (feet)	MINIMUM VERTICAL REINFORCEMENT SIZE AND SPACING Soil classes		
			Soil group I[h]	Soil group II[h]	Soil group III[h]
6	8	4	#4@48"	#3@12"; #4@24"; #5@36"	#3@12"; #5@24"
		5	#3@12"; #4@24"	#3@12"	#4@12"
		6	#4@12"	#5@12"	Design required
		7	#4@12"	Design required	Design required
	9	4	#4@48"	#3@12"; #4@24"	#3@12"; #6@24"
		5	#3@12"; #5@24"	#4@12"	#7@12"
		6	#4@12"	Design required	Design required
		7	Design required	Design required	Design required
		8	Design required	Design required	Design required
	10	4	#4@48"	#3@12"; #5@24"	#3@12"
		5	#3@12"	#4@12"	#7@12"
		6	#4@12"	Design required	Design required
		7	Design required	Design required	Design required
		8	Design required	Design required	Design required

For SI: 1 inch = 25.4 mm, 1 foot = 304.8 mm, 1 pound per square foot = 0.0479 kN/m^2.

a. This table is based on concrete with a minimum specified concrete strength of 2500 psi, reinforcing steel with a minimum yield strength of 40,000 psi. When reinforcing steel with a minimum yield strength of 60,000 psi is used, the spacing of the reinforcement in the shaded cells shall be increased 12 inches.
b. This table is not intended to prohibit the use of an ICF manufacturer's tables based on engineering analysis in accordance with ACI 318.
c. N/R denotes "not required."
d. Deflection criteria: $L/240$.
e. Interpolation between rebar sizes and spacing is not permitted.
f. Refer to Table R611.4(2) for wall dimensions.
g. Unbalanced backfill height is the difference in height of the exterior and interior finished ground. Where an interior concrete slab is provided, the unbalanced backfill height shall be measured from the exterior finished ground level to the top of the interior concrete slab.
h. Soil classes are in accordance with the Unified Soil Classification System. Refer to Table R405.1.

❖ This table provides insulating concrete form screen-grid wall requirements and is similar to Table R404.4(4).

R404.4.5 Concrete material. Ready-mixed concrete for insulating concrete form walls shall be in accordance with Section R402.2. Maximum slump shall not be greater than 6 inches (152 mm) as determined in accordance with ASTM C 143. Maximum aggregate size shall not be larger than $^3/_4$ inch (19.1 mm).

Exception: Concrete mixes conforming to the ICF manufacturer's recommendations.

❖ The requirements for concrete in Section R402.2 must be followed for insulating concrete form walls. In addition, this section specifies maximum slump and aggregate size. Insulating concrete form manufacturer's recommendations for concrete mixes are also acceptable.

R404.4.6 Reinforcing steel.

R404.4.6.1 General. Reinforcing steel shall meet the requirements of ASTM A 615, A 706 or A 996. The minimum yield strength of reinforcing steel shall be 40,000 psi (Grade 40) (276 MPa). Vertical and horizontal wall reinforcements shall be placed no closer to the outside face of the wall than one-half the wall thickness. Steel reinforcement for foundation walls shall have concrete cover in accordance with ACI 318.

Exception: Where insulated concrete forms are used and the form remains in place as cover for the concrete, the minimum concrete cover for the reinforcing steel is permitted to be reduced to $^3/_4$ inch (19.1 mm).

❖ This section specifies the minimum grades of reinforcing steel that may be used. The cover must be in accordance with ACI 318, and the cover on the exterior wall face must be at least one-half of the wall thickness.

R404.4.6.2 Horizontal reinforcement. When vertical reinforcement is required, ICF foundation walls shall have horizontal reinforcement in accordance with this section. ICF foundation walls up to 8 feet (2438 mm) in height shall have a minimum of one continuous No. 4 horizontal reinforcing bar placed at 48 inches (1219 mm) on center with one bar located within 12 inches (305 mm) of the top of the wall story. ICF Foundation walls greater than 8 feet (2438 mm) in height shall have a minimum of one continuous No. 4 horizontal reinforcing bar placed at 36 inches (914 mm) on center with one bar located within 12 inches (305 mm) of the top of the wall story.

❖ Where the insulating concrete form wall tables indicate a requirement for vertical reinforcement, horizontal reinforcement must also be provided in accordance with this section.

R404.4.6.3 Wall openings. Vertical wall reinforcement required by Section R404.4.2, R404.4.3 or R404.4.4 that is interrupted by wall openings shall have additional vertical reinforcement of the same size placed within 12 inches (305 mm) of each side of the opening.

❖ If a wall opening interrupts reinforcement, additional reinforcement of equivalent area should be provided as illustrated in Figure R404.4.6. While the code doesn't elaborate, a good practice is to distribute this equivalent reinforcing symmetrically on either side of the opening.

R404.4.7 Foam plastic insulation. Foam plastic insulation in insulating concrete foam construction shall comply with this section.

❖ Foam plastic insulation must be installed in accordance with Section R314.1.

R404.4.7.1 Material. Insulating concrete form material shall meet the surface burning characteristics of Section R314.1.1. A thermal barrier shall be provided on the building interior in accordance with Section R314.1.2.

❖ See the commentary for Section R314.1.

R404.4.7.2 Termite hazards. In areas where hazard of termite damage is very heavy in accordance with Figure R301.2(6), foam plastic insulation shall be permitted below grade on foundation walls in accordance with one of the following conditions:

1. When in addition to the requirements in Section R320.1, an approved method of protecting the foam plastic and structure from subterranean termite damage is provided.
2. The structural members of walls, floors, ceilings and roofs are entirely of noncombustible materials or pressure preservatively treated wood.
3. On the interior side of basement walls.

❖ Where the termite hazard is very heavy, foam plastic insulation may be installed below grade only if it installed on the interior or if all walls, floors, and ceilings are noncombustible material or treated wood or in accordance with Section R320.1. See the commentary for Section R404.3.3.

R404.4.8 Foundation wall thickness based on walls supported. The thickness of ICF foundation walls shall not be less than the thickness of the wall supported above.

❖ The thickness of insulating concrete form foundation walls must not be less then that of the wall supported.

R404.4.9 Height above finished ground. ICF foundation walls shall extend above the finished ground adjacent to the foundation at all points a minimum of 4 inches (102 mm) where masonry veneer is used and a minimum of 6 inches (152 mm) elsewhere.

❖ See the commentary for Section R404.1.6.

R404.4.10 Backfill placement. Backfill shall be placed in accordance with Section R404.1.7.

❖ See the commentary for Section R404.1.7.

R404.4.11 Drainage and dampproofing/waterproofing. ICF foundation basements shall be drained and damp-proofed/ waterproofed in accordance with Sections R405 and R406.

❖ See the requirements for concrete walls in Sections R405 and R406 of this commentary.

SECTION R405
FOUNDATION DRAINAGE

R405.1 Concrete or masonry foundations. Drains shall be provided around all concrete or masonry foundations that retain earth and enclose habitable or usable spaces located below grade. Drainage tiles, gravel or crushed stone drains, perforated pipe or other approved systems or materials shall be installed at or below the area to be protected and shall discharge by gravity or mechanical means into an approved drainage system. Gravel or crushed stone drains shall extend at least 1 foot (305 mm) beyond the outside edge of the footing and 6 inches (153 mm) above the top of the footing and be covered with an approved filter membrane material. The top of open joints of drain tiles shall be protected with strips of building paper, and the drainage tiles or perforated pipe shall be placed on a minimum of 2 inches (51 mm) of washed gravel or crushed rock at least one sieve size larger than the tile joint opening or perforation and covered with not less than 6 inches (153 mm) of the same material.

Exception: A drainage system is not required when the foundation is installed on well-drained ground or sand-gravel mixture soils according to the Unified Soil Classification System, Group I Soils, as detailed in Table R405.1.

❖ To allow free groundwater that may be present adjacent to the foundation wall to be drained away, drains are usually placed around houses to remove water and prevent leakage into habitable or usable spaces below grade. The top of the gravel or crushed stone drain, with or without drainage tile or perforated pipe, must be covered with an approved filter membrane. Drainage tiles are extremely important in areas having moderate to heavy rainfall and in soils having a low percolation rate. The filter membrane allows water to pass through the perimeter drain tiles or pipes without allowing, or at least greatly reducing, the possibility of fine soil materials entering the drainage system. A few examples of such systems are illustrated in Commentary Figures R405.1(1) and R405.1(2). The exception allows omission of the drainage system in Group I soils, which by definition exhibit good drainage characteristics as shown in Table R405.1.

R405.2 Wood foundations. Wood foundations enclosing habitable or usable spaces located below grade shall be adequately drained in accordance with Sections R405.2.1 through R405.2.3.

❖ Similar to Section R405.1, this section gives specific drainage requirements for wood foundations. Also refer to the commentary for Section R401.1.

R405.2.1 Base. A porous layer of gravel, crushed stone or coarse sand shall be placed to a minimum thickness of 4 inches (102 mm) under the basement floor. Provision shall be made for automatic draining of this layer and the gravel or crushed stone wall footings.

❖ See Figure R403.1(2). This requirement for drainage under the basement floor is similar to those in Sections R504.2.1 (wood floor) and R406.2.2 (concrete floor).

R405.2.2 Moisture barrier. A 6-mil-thick (0.15 mm) polyethylene moisture barrier shall be applied over the porous layer with the basement floor constructed over the polyethylene.

❖ See Figure R403.1(2). Also see the commentary for Section R506.2.3.

R405.2.3 Drainage system. In other than Group I soils, a sump shall be provided to drain the porous layer and footings. The sump shall be at least 24 inches (610 mm) in diameter or 20 inches square (0.0129 m^2), shall extend at least 24 inches (610 mm) below the bottom of the basement floor and shall be capable of positive gravity or mechanical drainage to remove any accumulated water. The drainage system shall discharge into an approved sewer system or to daylight.

❖ Except for Group I soils, which exhibit good drainage (see Table R405.1), a sump is necessary to drain the footing (Section R403.2) and porous layer (Section R405.2.1). Also see Figure R403.1(2).

TABLE R405.1
PROPERTIES OF SOILS CLASSIFIED ACCORDING TO THE UNIFIED SOIL CLASSIFICATION SYSTEM

SOIL GROUP	UNIFIED SOIL CLASSIFICATION SYSTEM SYMBOL	SOIL DESCRIPTION	DRAINAGE CHARACTERISTICS[a]	FROST HEAVE POTENTIAL	VOLUME CHANGE POTENTIAL EXPANSION[b]
Group I	GW	Well-graded gravels, gravel sand mixtures, little or no fines.	Good	Low	Low
	GP	Poorly graded gravels or gravel sand mixtures, little or no fines.	Good	Low	Low
	SW	Well-graded sands, gravelly sands, little or no fines.	Good	Low	Low
	SP	Poorly graded sands or gravelly sands, little or no fines.	Good	Low	Low
	GM	Silty gravels, gravel-sand-silt mixtures.	Good	Medium	Low
	SM	Silty sand, sand-silt mixtures.	Good	Medium	Low
Group II	GC	Clayey gravels, gravel-sand-clay mixtures.	Medium	Medium	Low
	SC	Clayey sands, sand-clay mixture.	Medium	Medium	Low
	ML	Inorganic silts and very fine sands, rock flour, silty or clayey fine sands or clayey silts with slight plasticity.	Medium	High	Low
	CL	Inorganic clays of low to medium plasticity, gravelly clays, sandy clays, silty clays, lean clays.	Medium	Medium	Medium to Low
Group III	CH	Inorganic clays of high plasticity, fat clays.	Poor	Medium	High
	MH	Inorganic silts, micaceous or diatomaceous fine sandy or silty soils, elastic silts.	Poor	High	High
Group IV	OL	Organic silts and organic silty clays of low plasticity.	Poor	Medium	Medium
	OH	Organic clays of medium to high plasticity, organic silts.	Unsatisfactory	Medium	High
	Pt	Peat and other highly organic soils.	Unsatisfactory	Medium	High

For SI: 1 inch = 25.4 mm.

a. The percolation rate for good drainage is over 4 inches per hour, medium drainage is 2 inches to 4 inches per hour, and poor is less than 2 inches per hour.

b. Soils with a low potential expansion typically have a plasticity index (PI) of 0 to 15, soils with a medium potential expansion have a PI of 10 to 35 and soils with a high potential expansion have a PI greater than 20.

❖ This table provides soil properties based on the soil's classification in accordance with the Unified Soil Classification System. This method of identifying soil is determined in ASTM 2487, *Standard Classification of Soils for Engineering Purposes*.

FOUNDATIONS

**Figure R405.1(1)
FOUNDATION DRAINAGE FOR HABITABLE SPACE BELOW GRADE**

SECTION R406
FOUNDATION WATERPROOFING AND DAMPPROOFING

R406.1 Concrete and masonry foundation dampproofing. Except where required to be waterproofed by Section R406.2, foundation walls that retain earth and enclose habitable or usable spaces located below grade shall be dampproofed from the top of the footing to the finished grade. Masonry walls shall have not less than $^3/_8$ inch (9.5 mm) portland cement parging applied to the exterior of the wall. The parging shall be dampproofed with a bituminous coating, 3 pounds per square yard (1.63 kg/m^2) of acrylic modified cement, $^1/_8$-inch (3.2 mm) coat of surface-bonding mortar complying with ASTM C 887 or any material permitted for waterproofing in Section R406.2. Concrete walls shall be dampproofed by applying any one of the above listed dampproofing materials or any one of the waterproofing materials listed in Section R406.2 to the exterior of the wall.

❖ To minimize moisture in the form of water vapor from entering below-ground spaces from the outside, dampproofing of the exterior foundation walls is necessary, unless waterproofing is required by Section R406.2. While the terms "waterproofing" and "dampproofing" both relate to moisture protection, they are sometimes misapplied in the sense that the "waterproofing" is used when "dampproofing" is really meant. Dampproofing does not give the same degree of moisture protection as does waterproofing. Methods of dampproofing for masonry and concrete foundation walls are shown in Commentary Figures R406.1(1) and R406.1(2).

Dampproofing installations generally consist of the application of one or more coatings of impervious compounds that are intended to prevent the passage of water vapor through walls or other building elements. Dampproofing may also restrict the flow of water under slight pressure.

FIGURE R405.1(2)

Figure R405.1(2)
OTHER SUGGESTED METHODS OF FOUNDATION DRAINAGE

For SI: 1 inch = 25.4 mm.

FOUNDATIONS

Figure R406.1(1)
DAMPPROOFING OF MASONRY FOUNDATION WALL
[SEE ALSO FIGURE R405.1(1)]

For SI: 1 inch = 25.4 mm.

Figure R406.1(2)
DAMPPROOFING OF CONCRETE FOUNDATION WALL
[SEE ALSO FIGURE R405.1.(1)]

R406.2 Concrete and masonry foundation waterproofing. In areas where a high water table or other severe soil-water conditions are known to exist, exterior foundation walls that retain earth and enclose habitable or usable spaces located below grade shall be waterproofed with a membrane extending from the top of the footing to the finished grade. The membrane shall consist of 2-ply hot-mopped felts, 55 pound (25 kg) roll roofing, 6-mil (0.15 mm) polyvinyl chloride, 6-mil (0.15 mm) polyethylene or 40-mil (1 mm) polymer-modified asphalt. The joints in the membrane shall be lapped and sealed with an adhesive compatible with the waterproofing membrane.

Exception: Organic solvent based products such as hydrocarbons, chlorinated hydrocarbons, ketones and esters shall not be used for ICF walls with expanded polystyrene form material. Plastic roofing cements, acrylic coatings, latex coatings, mortars and pargings are permitted to be used to seal ICF walls. Cold setting asphalt or hot asphalt shall conform to type C of ASTM D 449. Hot asphalt shall be applied at a temperature of less than 200 degrees.

❖ If the foundation walls enclosing habitable or storage spaces extend below groundwater level (seasonal or otherwise), a positive means of preventing moisture migration must be employed. Waterproofing installations consist of the application of a combination of sealing materials and impervious coatings used on walls or other building elements to prevent the passage of moisture in either a vapor or liquid form under conditions of significant hydrostatic pressure. Methods for waterproofing concrete and masonry foundation walls are shown in Commentary Figure R406.2.

R406.3 Dampproofing for wood foundations. Wood foundations enclosing habitable or usable spaces located below grade shall be dampproofed in accordance with Sections R406.3.1 through R406.3.4.

❖ The requirements for dampproofing wood foundation walls are shown in Figure R403.1(2). These requirements for sealing the plywood panel joints, installing a moisture barrier, and installing porous fill can minimize the intrusion of water into a basement or usable space. Also refer to the commentary for Section R401.1.

R406.3.1 Panel joint sealed. Plywood panel joints in the foundation walls shall be sealed full length with a caulking compound capable of producing a moisture-proof seal under the conditions of temperature and moisture content at which it will be applied and used.

❖ See the commentary for Section R406.3.

R406.3.2 Below grade moisture barrier. A 6-mil-thick (0.15 mm) polyethylene film shall be applied over the below-grade portion of exterior foundation walls prior to backfilling. Joints in the polyethylene film shall be lapped 6 inches (152 mm) and sealed with adhesive. The top edge of the polyethylene film shall be bonded to the sheathing to form a seal. Film areas at grade level shall be protected from mechanical damage and exposure by a pressure preservatively treated lumber or plywood strip attached to the wall several inches above finish grade level and extending approximately 9 inches (229 mm) below grade. The joint between the strip and the wall shall be caulked full length prior to fastening the strip to the wall. Other coverings appropriate to the architectural treatment may also be used. The polyethylene film shall extend down to the bottom of the wood footing plate but shall not overlap or extend into the gravel or crushed stone footing.

❖ See the commentary for Section R406.3.

R406.3.3 Porous fill. The space between the excavation and the foundation wall shall be backfilled with the same material used for footings, up to a height of 1 foot (305 mm) above the footing for well-drained sites, or one-half the total back-fill height for poorly drained sites. The porous fill shall be covered with strips of 30-pound (13.6 kg) asphalt paper or 6-mil (0.15 mm) polyethylene to permit water seepage while avoiding infiltration of fine soils.

❖ See the commentary for Section R406.3.

R406.3.4 Backfill. The remainder of the excavated area shall be backfilled with the same type of soil as was removed during the excavation.

❖ See the commentary for Sections R404.2.4 and R406.3.

Figure R406.2
METHODS OF WATERPROOFING BASEMENT WALLS
[SEE ALSO FIGURES R405.1(1) AND R405.1(2) FOR TYPICAL DRAINAGE DETAILS.]

FOUNDATIONS

SECTION R407
COLUMNS

R407.1 Wood column protection. Wood columns shall be protected against decay as set forth in Section R319.

❖ See the commentary for Section R319.

R407.2 Steel column protection. All surfaces (inside and outside) of steel columns shall be given a shop coat of rust-inhibitive paint, except for corrosion-resistant steel and steel treated with coatings to provide corrosion resistance.

❖ The requirements of this section protect steel columns from the adverse affects of corrosion.

R407.3 Structural requirements. The columns shall be restrained to prevent lateral displacement at the bottom end. Wood columns shall not be less in nominal size than 4 inches by 4 inches (102 mm by 102 mm) and steel columns shall not be less than 3-inch-diameter (76 mm) standard pipe or approved equivalent.

Exception: In Seismic Design Categories A, B and C columns no more than 48 inches (1219 mm) in height on a pier or footing are exempt from the bottom end lateral displacement requirement within underfloor areas enclosed by a continuous foundation.

❖ Minimum sizes for wood columns and steel columns are specified to reduce concerns about the structural capacity of slender columns.

To minimize the chance of accidentally displacing columns supporting beams or girders, a means of mechanically anchoring a column is required. Accepted construction details that have performed adequately in the field are typically employed in lieu of a structural analysis and design. See Commentary Figures R407.3(1), R407.3(2), and R407.3(3). If prefabricated metal-wood connectors are used, they should be installed in accordance with the manufacturer's installation instructions.

SECTION R408
UNDER-FLOOR SPACE

R408.1 Ventilation. The under-floor space between the bottom of the floor joists and the earth under any building (except space occupied by a basement or cellar) shall be provided with ventilation openings through foundation walls or exterior walls. The minimum net area of ventilation openings shall not be less than 1 square foot for each 150 square feet (0.67 m^2 for each 100 m^2) of under-floor space area. One such ventilating opening shall be within 3 feet (914 mm) of each corner of said building.

❖ Raised floor construction results in an under-floor space, commonly referred to as a crawl space. To control condensation within crawl space areas and thus reduce the chance of dry rot, natural ventilation of such spaces by reasonably distributed openings through foundation walls or exterior walls is required. Condensation is a function of the geographical location

**Figure R407.3(1)
COLUMN ANCHORAGE**

**Figure R407.3(2)
COLUMN ANCHORAGE**

**Figure R407.3(3)
COLUMN ANCHORAGE**

and the climatic conditions and, thus, the dependence upon ventilating openings through the foundation wall or exterior wall may run counter to energy-conservation measures. Commentary Figure R408.1 illustrates the use of openings through the foundation walls.

R408.2 Openings for under-floor ventilation. The minimum net area of ventilation openings shall not be less than 1 square foot (0.0929 m^2) for each 150 square feet (100 m^2) of under-floor space area. One such ventilating opening shall be within 3 feet (914 mm) of each corner of the building. Ventilation openings shall be covered for their height and width with any of the

FOUNDATIONS

FIGURE R408.1

For SI: 1 inch = 25.4 mm, 1 foot = 304.8 mm.

SCREENED OPENINGS (MESH OPENING ≥ 1/8 IN.) THROUGH FOUNDATION WALL TO CRAWL SPACE
TOTAL NET CLEAR AREA OF OPENING EQUALS THE CRAWL SPACE AREA DIVIDED BY 150

**Figure R408.1
CRAWL-SPACE VENTILATION**

following materials provided that the least dimension of the covering shall not exceed $1/4$ inch (6.4 mm):

1. Perforated sheet metal plates not less than 0.070 inch (1.8 mm) thick.
2. Expanded sheet metal plates not less than 0.047 inch (1.2 mm) thick.
3. Cast iron grills or grating.
4. Extruded load-bearing brick vents.
5. Hardware cloth of 0.035 inch (0.89 mm) wire or heavier.
6. Corrosion-resistant wire mesh, with the least dimension being $1/8$ inch (3.2 mm).

Exceptions:

1. Where warranted by climatic conditions, ventilation openings to the outdoors are not required if ventilation openings to the interior are provided.
2. The total area of ventilation openings may be reduced to $1/1500$ of the under-floor area where the ground surface is treated with an approved vapor retarder material and the required openings are placed so as to provide cross-ventilation of the space. The installation of operable louvers shall not be prohibited.
3. Under-floor spaces used as supply plenums for distribution of heated and cooled air shall comply with the requirements of Section M1601.4
4. Ventilation openings are not required where continuously operated mechanical ventilation is provided at a rate of 1.0 cfm (10 m^2) for each 50 square feet (1.02 L/s) of underfloor space floor area and ground surface is covered with an approved vapor retarder material.
5. Ventilation openings are not required when the ground surface is covered with an approved vapor retarder material, the space is supplied with conditioned air and the perimeter walls are insulated in accordance with Section N1102.1.7.

❖ Installing a covering material over the ventilation opening keeps animals such as rodents or vermin from entering the crawl space. There are several options provided. The area of the ventilation opening accounts for net free area of the covering material used to protect the ventilation opening. For some covering materials the net free area can be up to 50 percent less than the gross area of the ventilation opening.

A number of exceptions are provided to address a variety of conditions under which ventilation openings are either reduced or eliminated.

Example 1:

A house having a crawl space area of 1,300 square feet (121 m^2) would require a total net clear area of openings not less than 8.7 square feet (0.805 m^2). With ten openings, an opening size of 8 inches by 16 inches (203 mm x 406 mm) would be acceptable.

Example 2:
In Example 1, the total area of openings to the exterior from the crawl space could be reduced to 0.87 ft.^{2} (80,826 mm^2) using the second exception to Section R408.2, if an approved vapor retarder is placed over the ground surface of the crawl space. Note that opening placement must provide cross ventilation.

R408.3 Access. Access shall be provided to all under-floor spaces. Access openings through the floor shall be a minimum of 18 inches by 24 inches (457 mm x 610 mm). Openings through a perimeter wall shall be 16 inches by 24 inches (407 mm x 610 mm). When any portion of the through wall access is below grade, an areaway of not less than 16 inches by 24 inches shall be provided. The bottom of the areaway shall be below the threshold of the access opening. Through wall access openings shall not be located under a door to the residence. See M1305.1.4 for access requirements where mechanical equipment is located under floors.

❖ The provisions of this section require access to be provided to all under-floor spaces. Access is required for continuing maintenance of the building and for inspection and repair of such items as plumbing, mechanical systems, or electrical system runs within the crawl space. This section addresses the different conditions encountered in accessing an under floor space through the floor system or the foundation wall. Where the access opening is provided through the floor, an opening of not less than 18 inches by 24 inches is required. Where the access opening is through the foundation wall, an opening of not less than 16 inches by 24 inches is required. See Figure R408.1. The 16-and 24-inch dimensions are to work with standard CMU coursing. Where mechanical equipment is located under the floor, the minimum access opening must comply with Section M1305.1.4.

R408.4 Removal of debris. The under-floor grade shall be cleaned of all vegetation and organic material. All wood forms used for placing concrete shall be removed before a building is occupied or used for any purpose. All construction materials shall be removed before a building is occupied or used for any purpose.

❖ Vegetation, stumps, roots, and other matter left in an excavation around a building are major causes of termite infestation and moisture problems. As such material decays, the ground settles, negating the original drainage plan. Even before decay, the material provides pockets for water accumulation, which can have subsequent destructive impact on the structure. To eliminate a natural attraction to termites, insects, or animals, all vegetation and organic material must be cleared.

R408.5 Finished grade. The finished grade of under-floor surface may be located at the bottom of the footings; however, where there is evidence that the groundwater table can rise to within 6 inches (152 mm) of the finished floor at the building perimeter or where there is evidence that the surface water does not readily drain from the building site, the grade in the under-floor space shall be as high as the outside finished grade, unless an approved drainage system is provided.

❖ To circumvent moisture accumulation in the crawl space area, the grade in the under-floor space is required to be at the same elevation as grade outside of the building if the groundwater table can rise to within 6 inches (152 mm) of the finished floor elevation or the surface water does not readily drain from the site. An alternative would be approved drainage system.

R408.6 Flood resistance. For buildings located in areas prone to flooding as established in Table R301.2(1):

1. Walls enclosing the underfloor space shall be provided with flood openings in accordance with Section R323.2.2.
2. The finished ground level of the underfloor space shall be equal to or higher than the outside finished ground level.

Exception: Underfloor spaces that meet the requirements of FEMA/FIA TB 11-1.

❖ To minimize hydrostatic loads by allowing the free inflow and outflow of floodwaters, buildings in flood hazard areas (A zones) that have walls enclosing underfloor spaces are to have flood openings in those walls. This provision alerts the designer that flood openings are to be provided in addition to standard ventilation openings. This requirement applies only to buildings in flood hazard areas not subject to high velocity wave action. Buildings in coastal high hazard areas subject to high velocity wave action may have enclosed areas below the design flood elevation, but only if those walls are intended to break away as specified in Section R323.3.3.

The provisions of this section require the elevation of the finished ground level of the underfloor space to be at or above the elevation of the outside finish ground level. The exception permits the use of FEMA/FIA TB 11-01, which permits the elevation of the underfloor space to be below the elevation of the outside finish ground level. Use of FEMA/TIA TB 11-01 will result in flood insurance rates higher than for buildings that have the finished ground level of the under floor space at or above the elevation of the outside finish ground level.

Bibliography

The following resource materials are referenced in this chapter or are relevant to the subject matter addressed in this chapter.

ACI 318-02, *Building Code Requirements for Structural Concrete*. Farmington Hills, MI: American Concrete Institute, 2002.

FOUNDATIONS

ACI 530-02/ASCE 5-02/TMS 402-02, *Building Code Requirements for Masonry Structures*. Farmington Hills, MI: American Concrete Institute, 2002.

AFPA NDS-01, *National Design Specification (NDS) for Wood Construction—with 2001 Supplement*. Washington, D.C.: American Forest and Paper Association, 2001.

AFPA TR7-87, *Basic Requirements for Permanent Wood Foundation System*. Washington, D.C.: American Forest and Paper Association, 1987.

ASCE 32-01, *Design and Construction of Frost Protected Shallow Foundations*. Reston, VA: American Society of Civil Engineers, 2001.

ASTM A 615/A 615M-01b, *Specification for Deformed and Plain Billet Steel bars for Concrete Reinforcement*. West Conshohocken, PA: ASTM International, 2001.

ASTM A 706M-00, *Specification for Low-Alloy Steel Deformed Plain Bars for Concrete Reinforcement*. West Conshohocken, PA: ASTM International, 2000.

ASTM C 143/C 0143M-00, *Test Method for Slump or Hydraulic Cement Concrete*. West Conshohocken, PA: ASTM International, 2000.

ASTM C578-01, *Specification for Rigid, Cellular Polystyrene Thermal Insulation*. West Conshohocken, PA: ASTM International, 2001.

ASTM C887-79a(96) e[01], *Specification for Packaged, Dry, Combined Materials for Surface Bonding Mortar*. West Conshohocken, PA: ASTM International, 1996.

ASTM C 1157-00a, *Performance Specification for Hydraulic Cements*. West Conshohocken, PA: ASTM International, 2000.

ASTM D422-63(1998), *Test Method for Particle-Size Analysis of Soils*. West Conshohocken, PA: ASTM International, 1998.

ASTM D449-89(1999)e[01], *Specification for Asphalt Used in Dampproofing and Waterproofing*. West Conshohocken, PA: ASTM International, 1999.

ASTM D2487-93, *Standard Practice for Classification of Soils for Engineering Purposes*. West Conshohocken, PA: ASTM International, 1993.

ASTM D4318-00, *Test Method for Liquid Limit, Plastic Limit and Plasticity Index of Soils*. West Conshohocken, PA: ASTM International, 2000.

ASTM D4829-95, *Test Method for Expansion Index of Soils*. West Conshohocken, PA: ASTM International, 1995.

AWPA C22-96, *Lumber and Plywood for Permanent Wood Foundation-Preservative Treatment by Pressure Process*. Granbury, TX: American Wood-Preservers' Association, 1996.

DOC PS1-95, Construction and Industrial Plywood. Gaithersburg, MD: United States Department of Commerce, National Institute of Standards and Technology, 1995.

DOC PS 2-92, Performance Standard for Wood-Based Structural-Use Panels. Gaithersburg, MD: United States Department of Commerce, National Institute of Standards and Technology, 1992.

DOC PS 20-99, *American Softwood Lumber Standard*. Gaithersburg, MD: United States Department of Commerce, National Institute of Standards and Technology, 1999.

FEMA 303, *Recommended Provisions for Seismic Regulations for New Buildings and Other Structures*. Part 2: Commentary. Washington, DC: Building Seismic Safety Council, 1997.

FEMA TB-11-01, *Crawlspace Construction for Buildings Located in Special Flood Hazard Area*. Washington, DC: Federal Emergency Management Agency, 2001.

NCMA TR 68-A-75, *Design and Construction of Plain and Reinforced Concrete Masonry and Basement and Foundation Walls*. Herndon, VA: National Concrete Masonry Association, 1975.

Portland Cement Association. *Prescriptive Method for Insulating Concrete Forms in Residential Construction*. Skokie, IL: Portland Cement Association, 1998.

Sculling, Michael. *Excavation and Grading Code Administration, Inspection and Enforcement*. Englewood Cliffs, NJ: Prentice Hall, 1990.

Chapter 5: Floors

General Comments

Section R501 provides the scope of the chapter and states the general performance requirements for floor systems. The floor systems covered in the *International Residential Code*® (IRC®) consist of four different types: wood floor framing, wood floors on the ground, steel floor framing and concrete slabs on the ground. Section R502 addresses wood floor framing criteria and provides prescriptive tables from which the size of the joists can be determined based on the span and species of wood.

Section R503 specifies floor-sheathing requirements. Section R504 governs wood floors constructed on ground that would serve as part of a wood foundation system described in Chapter 4. This system relies on pressure-preservatively-treated wood to resist the damage from exposure to moisture from contact with the ground. Section R505 provides steel floor framing provisions, which include prescriptive criteria that greatly simplify the use of steel framing in a floor system. Section R506 contains requirements for concrete slab-on-ground floors. Concrete-slab-on-grade construction requires a minimum 3.5-inch (89 mm) thick slab and a vapor retarder in most cases.

The primary consideration of this chapter is the structural integrity of the floor system. In a two-family dwelling, a floor that separates dwelling units must be a fire-resistance-rated assembly in accordance with Section R317.1. Such a floor must conform to the conditions of the floor assembly's listing in addition to the requirements of this chapter.

Purpose

Chapter 5 provisions govern the design and construction of floor systems that will be capable of supporting minimum required design loads. The allowable-span tables greatly simplify the determination of joist, girder and sheathing sizes for raised floor systems.

The floor system must also serve as a diaphragm; therefore, it plays a key role in resisting lateral loads from earthquakes and/or wind. Although this chapter does not base floor system requirements on these loads, floor systems constructed as required in this chapter will perform this function. Chapter 6 of the *International Residential Code* accounts for these loads by varying the spacing of the supporting braced wall lines or shear walls based on seismic design category and basic wind speed or, in selected cases, specifying additional diaphragm requirements.

SECTION R501 GENERAL

R501.1 Application. The provisions of this chapter shall control the design and construction of the floors for all buildings including the floors of attic spaces used to house mechanical and/or plumbing fixtures and equipment.

❖ All floors (including attic floors) that house mechanical equipment or plumbing fixtures (see the definition in Chapter 2) must comply with this chapter.

R501.2 Requirements. Floor construction shall be capable of accommodating all loads according to Section R301 and of transmitting the resulting loads to the supporting structural elements.

❖ This is a general performance statement that requires the floor system to support the design loads and provide the necessary load path to supporting elements.

SECTION R502 WOOD FLOOR FRAMING

R502.1 Identification. Load-bearing dimension lumber for joists, beams and girders shall be identified by a grade mark of a lumber grading or inspection agency that has been approved by an accreditation body that complies with DOC PS 20. In lieu of a grade mark, a certificate of inspection issued by a lumber grading or inspection agency meeting the requirements of this section shall be accepted.

❖ Load-carrying wood members must be identified in order to verify their conformance to a minimum quality control standard. The required grade mark must reflect the following information:
  1. The species.
  2. The grade.
  3. The moisture content (MC) at time of grading: B S-Dry or S-Gm, MC15, KD15, or KD19. S-Grn indicates that the moisture content was in excess of 19 percent at the time of grading, and KD indicates kiln dried.
  4. The grading agency.
  5. The mill name or mill number.
  6. In the case of independent agencies, the rules used for grading.

For examples of grade marks, see Commentary Figure R502.1.

Figure R502.1
GRADE MARK EXAMPLES

❖ Traditionally, industry practice places only one grade mark on each piece. Therefore, the grade mark can be removed from precut or remanufactured lumber. It has also been industry practice not to grade mark material having a thickness greater than 3 inches (76 mm) or rough-sawn (unfinished) lumber where a grade mark would be illegible. In these cases, the code permits the substitution of a certificate of inspection for the grade mark from an approved grading agency. This certificate should be filed with the permanent records of the building structure.

R502.1.1 Preservatively treated lumber. Preservatively treated dimension lumber shall also be identified as required by Section R319.1.

❖ Grade marks as described in Section R502.1 must also include the quality mark of an agency accredited in the ALSC treated wood program. See Commentary Figure R502.1.1 for an example of a preservatively-treated-lumber grade mark.

R502.1.2 Blocking and subflooring. Blocking shall be a minimum of utility grade lumber. Subflooring may be a minimum of utility grade lumber or No. 4 common grade boards.

❖ Allowing the use of lower lumber grades in blocking and subflooring in these instances provides possible economies.

R502.1.3 End-jointed lumber. Approved end-jointed lumber identified by a grade mark conforming to Section R501.2 may be used interchangeably with solid-sawn members of the same species and grade.

❖ End-jointed lumber is generally considered equivalent to solid sawn lumber, provided it is properly identified by a grade mark. The key is verifying proper end use. For instance, grade marks including the phrase "STUD USE ONLY" denote finger-jointed lumber that is limited to stud use and/or bending or tension stresses of short duration loads such as wind and seismic. Structural fingerjoints, on the other hand, allow horizontal as well as vertical applications.

Figure R502.1.1
TREATED LUMBER GRADE MARK

R502.1.4 Prefabricated wood I-joists. Structural capacities and design provisions for prefabricated wood I-joists shall be established and monitored in accordance with ASTM D 5055.

❖ Prefabricated wood I-joists are some of the many engineered wood products currently available. These structural members consist of sawn lumber of structural composite lumber flanges bonded to wood structural panel webs, forming an I-shaped cross section. Since these products are widely used in residential floor construction ASTM D 5055, *Standard Specification for Establishing and Monitoring Structural Capacities of Prefabricated Wood I-joists* is referenced as the appropriate standard.

This standard does not specify the makeup of the component materials or how they must be assembled. Each manufacturer must develop product specifications and construction details for its products. These are subject to qualification tests, which are conducted or witnessed by qualified, independent agencies. The best results are used to establish the following design values:
- Allowable bending moment.
- Allowable shear.
- Member stiffness (EI).
- Sheer deflection factor (K).
- Allowable reactions.
- Bearing length requirements.
- Web stiffener requirements.
- A chart detailing the size, location, shape and spacing of any holes permitted in the web.

R502.1.5 Structural glued laminated timbers. Glued laminated timbers shall be manufactured and identified as required in AITC A190.1 and ASTM D3737.

❖ This section requires glulam timbers to be manufactured following ANSI/AITC 90.1 and ASTM D 3737. Knowing these standards makes it easier to determine that the product found in the field will meet the design requirements.

R502.2 Design and construction. Floors shall be designed and constructed in accordance with the provisions of this chapter, Figure R502.2 and Sections R319 and R320 or in accordance with AF&PA/NDS.

❖ Design and construction of wood floors must be in accordance with the prescriptive requirements of this section or where required in accordance with AF&PA/NDS, *National Design Specification for Wood Construction*. Figure R502.2 provides section references to provisions that apply to typical elements of floor systems.

R502.2.1 Decks. Where supported by attachment to an exterior wall, decks shall be positively anchored to the primary structure and designed for both vertical and lateral loads as applicable. Such attachment shall not be accomplished by the use of toenails or nails subject to withdrawal. Where positive connection to the primary building structure cannot be verified during inspection, decks shall be self-supporting. For decks with cantilevered framing members, connections to exterior walls or other framing members, shall be designed and constructed to resist uplift resulting from the full live load specified in Table R301.5 acting on the cantilevered portion of the deck.

❖ If an exterior wall is used to support a deck, the deck framing must be positively attached to the building structure. This connection design must include a consideration of both vertical and lateral loads, and the connection must be available for inspection. If it is not, this method of support is not permitted and the deck must be self-supporting.

If a deck has cantilevered framing, the framing must have a connection to its support that is designed to resist any uplift resulting from the full live load acting on the cantilevered span only. This load condition will produce maximum uplift at the support opposite the cantilevered end.

R502.3 Allowable joist spans. Spans for floor joists shall be in accordance with Tables R502.3.1(1) and R502.3.1(2). For other grades and species and for other loading conditions, refer to the AF&PA Span Tables for Joists and Rafters.

❖ Tables R502.3.1(1) and R502.3.1(2) provide allowable floor joist spans for common lumber species and grades based on design loads and joist spacing. The referenced standard may be used for grades and species of lumber not included in these tables.

Table R502.3.1(1) covers sleeping rooms where the design live load is 30 pounds per square foot (psf) (1.44 K N/m^2) per Table R301.5. This includes attics that are provided with fixed stair access. It is important to identify the intended use because an attic space used as a bonus room, for example, could require use of a 40-pounds-per-square-foot (1.92 Kn/m^2) live load. Table 502.3.1(2) covers rooms other than sleeping rooms where the design load is 40 pounds per square foot (1.92 K N/m^2) per Table R301.5.

Example:
A floor has a design live load of 40 psf (1.92 kN/m^2) and a dead load of 10 psf (0.48 kN/m^2). The joists span 17 feet, 6 inches (5334 mm). Find the required joist size using Douglas-Fir Larch #2 lumber and a joist spacing of 16 inches (406 mm).

Solution:
In Table R502.3.1(2), which is based on LL=40 psf (1.92 kN/m^2), find 16-inch (406 mm) joist spacing in the left hand column. Locate Douglas Fir-Larch #2 and under DL = 10 psf (0.48 kN/m^2) look for an allowable span of 17 feet, 6 inches (5334 mm) or greater. Note that the allowable span for 2 inch by 12 inch (51 mm by 305 mm) is 17 feet, 10 inches (5436 mm). Therefore, use 2-inch by 12-inch (51 mm by 305 mm) joists at 16-inch (406 mm) spacing.

FIGURE R502.2

For SI: 1 inch = 25.4 mm, 1 foot = 304.8 mm.

**FIGURE R502.2
FLOOR CONSTRUCTION**

❖ This figure is a schematic plan of typical wood floor framing. It serves as a key, providing references to the applicable provisions of the code.

The span tables account for a uniform load condition. They will also permit isolated concentrated loads such as nonbearing partitions offset from a support by a distance less than or equal to the joist depth. They may not support large concentrated loads such as ones that result from an entire kitchen utility wall or bathtubs parallel to joists. In such instances, additional joists and other adequate supports must be provided.

R502.3.1 Sleeping areas and attic joists. Table R502.3.1(1) shall be utilized to determine the maximum allowable span of floor joists that support sleeping areas and attics that are accessed by means of a fixed stairway provided that the design live load does not exceed 30 psf (1.44 kN/m^2) and the design dead load does not exceed 10 psf (0.48 kN/m^2). The allowable span of ceiling joists that support attics utilized for limited storage or no storage shall be determined in accordance with Section R802.4.

❖ See the commentary for Section R502.3.

TABLE R502.3.1(1)
FLOOR JOIST SPANS FOR COMMON LUMBER SPECIES
(Residential sleeping areas, live load=30 psf, L/Δ=360)

JOIST SPACING (inches)	SPECIE AND GRADE		DEAD LOAD = 10 psf				DEAD LOAD = 20 psf			
			2x6	2x8	2x10	2x12	2x6	2x8	2x10	2x12
			(ft.- in.)	(ft.- in.)	(ft.- in.)	(ft.- in.)	(ft.- in.)	(ft.- in.)	(ft.- in.)	(ft.- in.)
12	Douglas fir-larch	SS	12- 6	16- 6	21- 0	25- 7	12- 6	16- 6	21- 0	25- 7
	Douglas fir-larch	#1	12- 0	15-10	20- 3	24- 8	12- 0	15- 7	19- 0	22- 0
	Douglas fir-larch	#2	11-10	15- 7	19-10	23- 0	11- 6	14- 7	17- 9	20- 7
	Douglas fir-larch	#3	9- 8	12- 4	15- 0	17- 5	8- 8	11- 0	13- 5	15- 7
	Hem-fir	SS	11-10	15- 7	19-10	24- 2	11-10	15- 7	19-10	24- 2
	Hem-fir	#1	11- 7	15- 3	19- 5	23- 7	11- 7	15- 2	18- 6	21- 6
	Hem-fir	#2	11- 0	14- 6	18- 6	22- 6	11- 0	14- 4	17- 6	20- 4
	Hem-fir	#3	9- 8	12- 4	15- 0	17- 5	8- 8	11- 0	13- 5	15- 7
	Southern pine	SS	12- 3	16- 2	20- 8	25- 1	12- 3	16- 2	20- 8	25- 1
	Southern pine	#1	12- 0	15-10	20- 3	24- 8	12- 0	15-10	20- 3	24- 8
	Southern pine	#2	11-10	15- 7	19-10	24- 2	11-10	15- 7	18- 7	21- 9
	Southern pine	#3	10- 5	13- 3	15- 8	18- 8	9- 4	11-11	14- 0	16- 8
	Spruce-pine-fir	SS	11- 7	15- 3	19- 5	23- 7	11- 7	15- 3	19- 5	23- 7
	Spruce-pine-fir	#1	11- 3	14-11	19- 0	23- 0	11- 3	14- 7	17- 9	20- 7
	Spruce-pine-fir	#2	11- 3	14-11	19- 0	23- 0	11- 3	14- 7	17- 9	20- 7
	Spruce-pine-fir	#3	9- 8	12- 4	15- 0	17- 5	8- 8	11- 0	13- 5	15- 7
16	Douglas fir-larch	SS	11- 4	15- 0	19- 1	23- 3	11- 4	15- 0	19- 1	23- 0
	Douglas fir-larch	#1	10-11	14- 5	18- 5	21- 4	10- 8	13- 6	16- 5	19- 1
	Douglas fir-larch	#2	10- 9	14- 1	17- 2	19-11	9-11	12- 7	15- 5	17-10
	Douglas fir-larch	#3	8- 5	10- 8	13- 0	15- 1	7- 6	9- 6	11- 8	13- 6
	Hem-fir	SS	10- 9	14- 2	18- 0	21-11	10- 9	14- 2	18- 0	21-11
	Hem-fir	#1	10- 6	13-10	17- 8	20- 9	10- 4	13- 1	16- 0	18- 7
	Hem-fir	#2	10- 0	13- 2	16-10	19- 8	9-10	12- 5	15- 2	17- 7
	Hem-fir	#3	8- 5	10- 8	13- 0	15- 1	7- 6	9- 6	11- 8	13- 6
	Southern pine	SS	11- 2	14- 8	18- 9	22-10	11- 2	14- 8	18- 9	22-10
	Southern pine	#1	10-11	14- 5	18- 5	22- 5	10-11	14- 5	17-11	21- 4
	Southern pine	#2	10- 9	14- 2	18- 0	21- 1	10- 5	13- 6	16- 1	18-10
	Southern pine	#3	9- 0	11- 6	13- 7	16- 2	8- 1	10- 3	12- 2	14- 6
	Spruce-pine-fir	SS	10- 6	13-10	17- 8	21- 6	10- 6	13-10	17- 8	21- 4
	Spruce-pine-fir	#1	10- 3	13- 6	17- 2	19-11	9-11	12- 7	15- 5	17-10
	Spruce-pine-fir	#2	10- 3	13- 6	17- 2	19-11	9-11	12- 7	15- 5	17-10
	Spruce-pine-fir	#3	8- 5	10- 8	13- 0	15- 1	7- 6	9- 6	11- 8	13- 6
19.2	Douglas fir-larch	SS	10- 8	14- 1	18- 0	21-10	10- 8	14- 1	18- 0	21- 0
	Douglas fir-larch	#1	10- 4	13- 7	16- 9	19- 6	9- 8	12- 4	15- 0	17- 5
	Douglas fir-larch	#2	10- 1	12-10	15- 8	18- 3	9- 1	11- 6	14- 1	16- 3
	Douglas fir-larch	#3	7- 8	9- 9	11-10	13- 9	6-10	8- 8	10- 7	12- 4
	Hem-fir	SS	10- 1	13- 4	17- 0	20- 8	10- 1	13- 4	17- 0	20- 7
	Hem-fir	#1	9-10	13- 0	16- 4	19- 0	9- 6	12- 0	14- 8	17- 0
	Hem-fir	#2	9- 5	12- 5	15- 6	17-1	8-11	11- 4	13-10	16- 1
	Hem-fir	#3	7- 8	9- 9	11- 10	13- 9	6-10	8- 8	10- 7	12- 4
	Southern pine	SS	10- 6	13-10	17- 8	21- 6	10- 6	13-10	17- 8	21- 6
	Southern pine	#1	10- 4	13- 7	17- 4	21- 1	10- 4	13- 7	16- 4	19- 6
	Southern pine	#2	10- 1	13- 4	16- 5	19- 3	9- 6	12- 4	14- 8	17- 2
	Southern pine	#3	8- 3	10- 6	12- 5	14- 9	7- 4	9- 5	11- 1	13- 2
	Spruce-pine-fir	SS	9-10	13- 0	16- 7	20- 2	9-10	13- 0	16- 7	19- 6
	Spruce-pine-fir	#1	9- 8	12- 9	15- 8	18- 3	9- 1	11- 6	14- 1	16- 3
	Spruce-pine-fir	#2	9- 8	12- 9	15- 8	18- 3	9- 1	11- 6	14- 1	16- 3
	Spruce-pine-fir	#3	7- 8	9- 9	11-10	13- 9	6-10	8- 8	10- 7	12- 4

(continued)

TABLE R502.3.1(1)—continued
FLOOR JOIST SPANS FOR COMMON LUMBER SPECIES
(Residential sleeping areas, live load=30 psf, L/Δ=360)

JOIST SPACING (inches)	SPECIE AND GRADE		DEAD LOAD = 10 psf				DEAD LOAD = 20 psf			
			2x6	2x8	2x10	2x12	2x6	2x8	2x10	2x12
			(ft.- in.)	(ft.- in.)	(ft.- in.)	(ft.- in.)	(ft.- in.)	(ft.- in.)	(ft.- in.)	(ft.- in.)
24	Douglas fir-larch	SS	9-11	13-1	16-8	20-3	9-11	13-1	16-2	18-9
	Douglas fir-larch	#1	9-7	12-4	15-0	17-5	8-8	11-0	13-5	15-7
	Douglas fir-larch	#2	9-1	11-6	14-1	16-3	8-1	10-3	12-7	14-7
	Douglas fir-larch	#3	6-10	8-8	10-7	12-4	6-2	7-9	9-6	11-0
	Hem-fir	SS	9-4	12-4	15-9	19-2	9-4	12-4	15-9	18-5
	Hem-fir	#1	9-2	12-0	14-8	17-0	8-6	10-9	13-1	15-2
	Hem-fir	#2	8-9	11-4	13-10	16-1	8-0	10-2	12-5	14-4
	Hem-fir	#3	6-10	8-8	10-7	12-4	6-2	7-9	9-6	11-0
	Southern pine	SS	9-9	12-10	16-5	19-11	9-9	12-10	16-5	19-11
	Southern pine	#1	9-7	12-7	16-1	19-6	9-7	12-4	14-7	17-5
	Southern pine	#2	9-4	12-4	14-8	17-2	8-6	11-0	13-1	15-5
	Southern pine	#3	7-4	9-5	11-1	13-2	6-7	8-5	9-11	11-10
	Spruce-pine-fir	SS	9-2	12-1	15-5	18-9	9-2	12-1	15-0	17-5
	Spruce-pine-fir	#1	8-11	11-6	14-1	16-3	8-1	10-3	12-7	14-7
	Spruce-pine-fir	#2	8-11	11-6	14-1	16-3	8-1	10-3	12-7	14-7
	Spruce-pine-fir	#3	6-10	8-8	10-7	12-4	6-2	7-9	9-6	11-0

For SI: 1 inch = 25.4 mm, 1 foot = 304.8 mm, 1 pound per square foot = 0.0479 kN/m^2.

NOTE: Check sources for availability of lumber in lengths greater than 20 feet.

❖ This table provides allowable joist spans for common lumber species, grades and joist sizes. The spans are applicable to residential sleeping areas. Use of this table is similar to the use described in the example in Section R502.3.

TABLE R502.3.1(2)
FLOOR JOIST SPANS FOR COMMON LUMBER SPECIES (Residential living areas, live load=40 psf, L/Δ=360)

JOIST SPACING (inches)	SPECIE AND GRADE		DEAD LOAD = 10 psf				DEAD LOAD = 20 psf			
			2x6	2x8	2x10	2x12	2x6	2x8	2x10	2x12
			(ft.- in.)	(ft.- in.)	(ft.- in.)	(ft.- in.)	(ft.- in.)	(ft.- in.)	(ft.- in.)	(ft.- in.)
12	Douglas fir-larch	SS	11-4	15-0	19-1	23-3	11-4	15-0	19-1	23-3
	Douglas fir-larch	#1	10-11	14-5	18-5	22-0	10-11	14-2	17-4	20-1
	Douglas fir-larch	#2	10-9	14-2	17-9	20-7	10-6	13-3	16-3	18-10
	Douglas fir-larch	#3	8-8	11-0	13-5	15-7	7-11	10-0	12-3	14-3
	Hem-fir	SS	10-9	14-2	18-0	21-11	10-9	14-2	18-0	21-11
	Hem-fir	#1	10-6	13-10	17-8	21-6	10-6	13-10	16-11	19-7
	Hem-fir	#2	10-0	13-2	16-10	20-4	10-0	13-1	16-0	18-6
	Hem-fir	#3	8-8	11-0	13-5	15-7	7-11	10-0	12-3	14-3
	Southern pine	SS	11-2	14-8	18-9	22-10	11-2	14-8	18-9	22-10
	Southern pine	#1	10-11	14-5	18-5	22-5	10-11	14-5	18-5	22-5
	Southern pine	#2	10-9	14-2	18-0	21-9	10-9	14-2	16-11	19-10
	Southern pine	#3	9-4	11-11	14-0	16-8	8-6	10-10	12-10	15-3
	Spruce-pine-fir	SS	10-6	13-10	17-8	21-6	10-6	13-10	17-8	21-6
	Spruce-pine-fir	#1	10-3	13-6	17-3	20-7	10-3	13-3	16-3	18-10
	Spruce-pine-fir	#2	10-3	13-6	17-3	20-7	10-3	13-3	16-3	18-10
	Spruce-pine-fir	#3	8-8	11-0	13-5	15-7	7-11	10-0	12-3	14-3
16	Douglas fir-larch	SS	10-4	13-7	17-4	21-1	10-4	13-7	17-4	21-0
	Douglas fir-larch	#1	9-11	13-1	16-5	19-1	9-8	12-4	15-0	17-5
	Douglas fir-larch	#2	9-9	12-7	15-5	17-10	9-1	11-6	14-1	16-3
	Douglas fir-larch	#3	7-6	9-6	11-8	13-6	6-10	8-8	10-7	12-4
	Hem-fir	SS	9-9	12-10	16-5	19-11	9-9	12-10	16-5	19-11
	Hem-fir	#1	9-6	12-7	16-0	18-7	9-6	12-0	14-8	17-0
	Hem-fir	#2	9-1	12-0	15-2	17-7	8-11	11-4	13-10	16-1
	Hem-fir	#3	7-6	9-6	11-8	13-6	6-10	8-8	10-7	12-4
	Southern pine	SS	10-2	13-4	17-0	20-9	10-2	13-4	17-0	20-9
	Southern pine	#1	9-11	13-1	16-9	20-4	9-11	13-1	16-4	19-6
	Southern pine	#2	9-9	12-10	16-1	18-10	9-6	12-4	14-8	17-2
	Southern pine	#3	8-1	10-3	12-2	14-6	7-4	9-5	11-1	13-2
	Spruce-pine-fir	SS	9-6	12-7	16-0	19-6	9-6	12-7	16-0	19-6
	Spruce-pine-fir	#1	9-4	12-3	15-5	17-10	9-1	11-6	14-1	16-3
	Spruce-pine-fir	#2	9-4	12-3	15-5	17-10	9-1	11-6	14-1	16-3
	Spruce-pine-fir	#3	7-6	9-6	11-8	13-6	6-10	8-8	10-7	12-4

(continued)

FLOORS

TABLE R502.3.1(2)—continued
FLOOR JOIST SPANS FOR COMMON LUMBER SPECIES (Residential living areas, live load=40 psf, L/Δ=360)

JOIST SPACING (inches)	SPECIE AND GRADE		DEAD LOAD = 10 psf				DEAD LOAD = 20 psf			
			2x6	2x8	2x10	2x12	2x6	2x8	2x10	2x12
			\multicolumn{8}{c}{Maximum floor joist spans}							
			(ft.- in.)	(ft.- in.)	(ft.- in.)	(ft.- in.)	(ft.- in.)	(ft.- in.)	(ft.- in.)	(ft.- in.)
19.2	Douglas fir-larch	SS	9- 8	12-10	16- 4	19-10	9- 8	12-10	16- 4	19- 2
	Douglas fir-larch	#1	9- 4	12- 4	15- 0	17- 5	8-10	11- 3	13- 8	15-11
	Douglas fir-larch	#2	9- 1	11- 6	14- 1	16- 3	8- 3	10- 6	12-10	14-10
	Douglas fir-larch	#3	6-10	8- 8	10- 7	12- 4	6- 3	7-11	9- 8	11- 3
	Hem-fir	SS	9- 2	12- 1	15- 5	18- 9	9- 2	12- 1	15- 5	18- 9
	Hem-fir	#1	9- 0	11-10	14- 8	17- 0	8- 8	10-11	13- 4	15- 6
	Hem-fir	#2	8- 7	11- 3	13-10	16- 1	8- 2	10- 4	12- 8	14- 8
	Hem-fir	#3	6-10	8- 8	10- 7	12- 4	6- 3	7-11	9- 8	11- 3
	Southern pine	SS	9- 6	12- 7	16- 0	19- 6	9- 6	12- 7	16- 0	19- 6
	Southern pine	#1	9- 4	12- 4	15- 9	19- 2	9- 4	12- 4	14-11	17- 9
	Southern pine	#2	9- 2	12- 1	14- 8	17- 2	8- 8	11- 3	13- 5	15- 8
	Southern pine	#3	7- 4	9- 5	11- 1	13- 2	6- 9	8- 7	10- 1	12- 1
	Spruce-pine-fir	SS	9- 0	11-10	15- 1	18- 4	9- 0	11-10	15- 1	17- 9
	Spruce-pine-fir	#1	8- 9	11- 6	14- 1	16- 3	8- 3	10- 6	12-10	14-10
	Spruce-pine-fir	#2	8- 9	11- 6	14- 1	16- 3	8- 3	10- 6	12-10	14-10
	Spruce-pine-fir	#3	6-10	8- 8	10- 7	12- 4	6- 3	7-11	9- 8	11- 3
24	Douglas fir-larch	SS	9- 0	11-11	15- 2	18- 5	9- 0	11-11	14- 9	17- 1
	Douglas fir-larch	#1	8- 8	11- 0	13- 5	15- 7	7-11	10- 0	12- 3	14- 3
	Douglas fir-larch	#2	8- 1	10- 3	12- 7	14- 7	7- 5	9- 5	11- 6	13- 4
	Douglas fir-larch	#3	6- 2	7- 9	9- 6	11- 0	5- 7	7- 1	8- 8	10- 1
	Hem-fir	SS	8- 6	11- 3	14- 4	17- 5	8- 6	11- 3	14- 4	16-10[a]
	Hem-fir	#1	8- 4	10- 9	13- 1	15- 2	7- 9	9- 9	11-11	13-10
	Hem-fir	#2	7-11	10- 2	12- 5	14- 4	7- 4	9- 3	11- 4	13- 1
	Hem-fir	#3	6- 2	7- 9	9- 6	11- 0	5- 7	7- 1	8- 8	10- 1
	Southern pine	SS	8-10	11- 8	14-11	18- 1	8-10	11- 8	14-11	18- 1
	Southern pine	#1	8- 8	11- 5	14- 7	17- 5	8- 8	11- 3	13- 4	15-11
	Southern pine	#2	8- 6	11- 0	13- 1	15- 5	7- 9	10- 0	12- 0	14- 0
	Southern pine	#3	6- 7	8- 5	9-11	11-10	6- 0	7- 8	9- 1	10- 9
	Spruce-pine-fir	SS	8- 4	11- 0	14- 0	17- 0	8- 4	11- 0	13- 8	15-11
	Spruce-pine-fir	#1	8- 1	10- 3	12- 7	14- 7	7- 5	9- 5	11- 6	13- 4
	Spruce-pine-fir	#2	8- 1	10- 3	12- 7	14- 7	7- 5	9- 5	11- 6	13- 4
	Spruce-pine-fir	#3	6- 2	7- 9	9- 6	11- 0	5- 7	7- 1	8- 8	10- 1

Check sources for availability of lumber in lengths greater than 20 feet.
For SI: 1 inch = 25.4 mm, 1 foot = 308.4 mm, 1 pound per square foot = 0.0479 kN/m.
a. End bearing length shall be increased to 2 inches.

❖ See the commentary for Section R502.3.

R502.3.2 Other floor joists. Table R502.3.1(2) shall be utilized to determine the maximum allowable span of floor joists that support all areas of the building, other than sleeping and attics, provided that the design live load does not exceed 40 psf (1.92 kN/m^2) and the design dead does not exceed 10 psf (0.48 kN/m^2).

❖ See the commentary for Section R502.3.

R502.3.3 Floor cantilevers. Floor cantilever spans shall not exceed the nominal depth of the wood floor joist. Floor cantilevers constructed in accordance with Table R502.3.3(1) shall be permitted when supporting a light-frame bearing wall and roof only. Floor cantilevers supporting an exterior balcony are permitted to be constructed in accordance with Table R502.3.3(2).

❖ The cantilever span is permitted to be equal to the nominal depth of the joist without additional limitations. This provides for load transfer to the support by direct bearing so that shear and bending of the joist is not a concern.

Larger cantilevers are permitted in accordance with the limitations of the appropriate table for floor joists supporting an exterior balcony or supporting a light-frame bearing wall and roof.

R502.4 Joists under bearing partitions. Joists under parallel bearing partitions shall be of adequate size to support the load. Double joists, sized to adequately support the load, that are separated to permit the installation of piping or vents shall be full depth solid blocked with lumber not less than 2 inches (51 mm) in nominal thickness spaced not more than 4 feet (1219 mm) on center. Bearing partitions perpendicular to joists shall not be offset from supporting girders, walls or partitions more than the joist depth unless such joists are of sufficient size to carry the additional load.

❖ Where floor joists support bearing partitions, the joists must be of adequate size. Commentary Figure R502.4(1) shows an example of using adequate sized double joists. This provides added support for the additional load from the bearing partition above. Alternatively, a beam of adequate size may also be substituted for the double joist.

Where piping or vents must penetrate the floor system, the double joists must be separated to allow the pipe or vent to pass through. In this case, full depth solid blocking spaced a maximum of 4 feet (1219 mm) on center, as shown in Commentary Figure R502.4(2), is required so that the two joists will function as a combined member without twisting.

Bearing partitions oriented perpendicular to joists cannot be offset from supporting girders, walls or partitions more than the depth of the joists unless the joists are of sufficient size to carry the additional load.

TABLE R502.3.3(1)
CANTILEVER SPANS FOR FLOOR JOISTS SUPPORTING LIGHT-FRAME EXTERIOR BEARING WALL AND ROOF ONLY[a, b, c, f, g, h]
(Floor Live Load ≤ 40 psf, Roof Live Load ≤ 20 psf)

Member & Spacing	Maximum Cantilever Span (Uplift Force at Backspan Support in Lbs.)[d, e]											
	Ground Snow Load											
	≤ 20 psf			30 psf			50 psf			70 psf		
	Roof Width			Roof Width			Roof Width			Roof Width		
	24 ft.	32 ft.	40 ft.	24 ft.	32 ft.	40 ft.	24 ft.	32 ft.	40 ft.	24 ft.	32 ft.	40 ft.
2 × 8 @ 12″	20″ (177)	15″ (227)	—	18″ (209)	—	—	—	—	—	—	—	—
2 × 10 @ 16″	29″ (228)	21″ (297)	16″ (364)	26″ (271)	18″ (354)	—	20″ (375)	—	—	—	—	—
2 × 10 @ 12″	36″ (166)	26″ (219)	20″ (270)	34″ (198)	22″ (263)	16″ (324)	26″ (277)	—	—	19″ (356)	—	—
2 × 12 @ 16″	—	32″ (287)	25″ (356)	36″ (263)	29″ (345)	21″ (428)	29″ (367)	20″ (484)	—	23″ (471)	—	—
2 × 12 @ 12″	—	42″ (209)	31″ (263)	—	37″ (253)	27″ (317)	36″ (271)	27″ (358)	17″ (447)	31″ (348)	19″ (462)	—
2 × 12 @ 8″	—	48″ (136)	45″ (169)	—	48″ (164)	38″ (206)	—	40″ (233)	26″ (294)	36″ (230)	29″ (304)	18″ (379)

For SI: 1 inch = 25.4 mm, 1 pound per square foot = 0.0479 kN/m^2.

a. Tabulated values are for clear-span roof supported solely by exterior bearing walls.
b. Spans are based on No. 2 Grade lumber of Douglas fir-larch, hem-fir, southern pine, and spruce-pine-fir for repetitive (3 or more) members.
c. Ratio of backspan to cantilever span shall be at least 3:1.
d. Connections capable of resisting the indicated uplift force shall be provided at the backspan support.
e. Uplift force is for a backspan to cantilever span ratio of 3:1. Tabulated uplift values are permitted to be reduced by multiplying by a factor equal to 3 divided by the actual backspan ratio provided (3/backspan ratio).
f. See Section R301.2.2.7.1 for additional limitations on cantilevered floor joists for detached one- and two-family dwellings in Seismic Design Categories D_1 and D_2 and townhouses in Seismic Design Categories C, D_1, and D_2.
g. A full-depth rim joist shall be provided at the cantilevered end of the joists. Solid blocking shall be provided at the cantilever support.
h. Linear interpolation shall be permitted for building widths and ground snow loads other than shown.

❖ This table provides the prescriptive span and uplift requirements for cantilever wood floor joists supporting a light-frame bearing wall and roof only. The permitted cantilever span ranges from 15 to 48 inches (331 to 1219 mm) based on the limitations prescribed in the table and in the footnotes.

TABLE R502.3.3(2)
CANTILEVER SPANS FOR FLOOR JOISTS SUPPORTING EXTERIOR BALCONY[a, b, e, f]

Member Size	Spacing	Maximum Cantilever Span (Uplift Force at Backspan Support in Lbs.)[c, d]		
		Ground Snow Load		
		≤30 psf	50 psf	70 psf
2 × 8	12″	42″ (139)	39″ (156)	34″ (165)
2 × 8	16″	36″ (151)	34″ (171)	29″ (180)
2 × 10	12″	61″ (164)	57″ (189)	49″ (201)
2 × 10	16″	53″ (180)	49″ (208)	42″ (220)
2 × 10	24″	43″ (212)	40″ (241)	34″ (255)
2 × 12	16″	72″ (228)	67″ (260)	57″ (268)
2 × 12	24″	58″ (279)	54″ (319)	47″ (330)

For SI: 1 inch = 25.4 mm, 1 pound per square foot = 0.0479 kN/m^2.

a. Spans are based on No. 2 Grade lumber of Douglas fir-larch, hem-fir, southern pine, and spruce-pine-fir for repetitive (3 or more) members.
b. Ratio of backspan to cantilever span shall be at least 2:1.
c. Connections capable of resisting the indicated uplift force shall be provided at the backspan support.
d. Uplift force is for a backspan to cantilever span ratio of 2:1. Tabulated uplift values are permitted to be reduced by multiplying by a factor equal to 2 divided by the actual backspan ratio provided (2/backspan ratio).
e. A full-depth rim joist shall be provided at the cantilevered end of the joists. Solid blocking shall be provided at the cantilevered support.
f. Linear interpolation shall be permitted for ground snow loads other than shown.

❖ This table provides the prescriptive span and uplift requirements for cantilever wood floor joists supporting an exterior balcony. The table was developed based on an engineered design consistent with the requirements of the IBC. A live load of 60 psf (293 kg/m^2) is used for the cantilevered portion of the floor joists and is combined with anticipated dead loads and snow loads. Snow loads vary up to 70 psf (342 ks/m^2) and include consideration of drifted snow from an adjacent roof. Deflections are limited to the values specified in IBC Table 1604.3. As indicated in Note b, the ratio of backspan to cantilever span is limited to a minimum of 2:1.

Figure R502.4(1)
DOUBLE JOIST UNDER BEARING PARTITION

Figure R502.4(2)
SPACED JOISTS UNDER BEARING PARTITION WITH BLOCKING AT PENETRATION

For SI: 1 foot = 304.8 mm.

R502.5 Allowable girder spans. The allowable spans of girders fabricated of dimension lumber shall not exceed the values set forth in Tables R502.5(1) and R502.5(2).

❖ Tables R502.5(1) and R502.5(2) list allowable spans for girders. These tables also apply to header spans and are referenced in Section R602.7 for that purpose. See Commentary Figure R502.5(1) and the following example for girder span-spacing relationships.
  Example:
     Assuming two equal floor joist spans of 10 feet (3048 mm) as indicated in Figure R502.5(1) (only one joist span and the two supporting girders are shown for clarity), determine the maximum span for a double 2-inch by 10-inch (51 mm by 254 mm) interior girder supporting one floor only.
  Solution:
     Table R502.5(2) contains the information for an interior girder supporting one floor only. For a double 2-inch by 10-inch (51 mm by 254 mm) girder and a building width of 20 feet (6096 mm), the maximum allowable girder span is read directly from Table R502.5(2) as 7 feet (2134 mm).

FLOORS TABLE R502.5(1)

TABLE R502.5(1)
GIRDER SPANS[a] AND HEADER SPANS[a] FOR EXTERIOR BEARING WALLS
(Maximum spans for Douglas fir-larch, hem-fir, southern pine and spruce-pine-fir[b] and required number of jack studs)

GIRDERS AND HEADERS SUPPORTING	SIZE	GROUND SNOW LOAD (psf)[e]											
		30						50					
		Building width[c] (feet)											
		20		28		36		20		28		36	
		Span	NJ[d]	Span	NJ[d]	Span	NJ[d]	Span	NJ[d]	Span	NJ[d]	Span	NJ[d]
Roof and ceiling	2-2×4	3-6	1	3-2	1	2-10	1	3-2	1	2-9	1	2-6	1
	2-2×6	5-5	1	4-8	1	4-2	1	4-8	1	4-1	1	3-8	2
	2-2×8	6-10	1	5-11	2	5-4	2	5-11	2	5-2	2	4-7	2
	2-2×10	8-5	2	7-3	2	6-6	2	7-3	2	6-3	2	5-7	2
	2-2×12	9-9	2	8-5	2	7-6	2	8-5	2	7-3	2	6-6	2
	3-2×8	8-4	1	7-5	1	6-8	1	7-5	1	6-5	2	5-9	2
	3-2×10	10-6	1	9-1	2	8-2	2	9-1	2	7-10	2	7-0	2
	3-2×12	12-2	2	10-7	2	9-5	2	10-7	2	9-2	2	8-2	2
	4-2×8	9-2	1	8-4	1	7-8	1	8-4	1	7-5	1	6-8	1
	4-2×10	11-8	1	10-6	1	9-5	2	10-6	1	9-1	2	8-2	2
	4-2×12	14-1	1	12-2	2	10-11	2	12-2	2	10-7	2	9-5	2
Roof, ceiling and one center-bearing floor	2-2×4	3-1	1	2-9	1	2-5	1	2-9	1	2-5	1	2-2	1
	2-2×6	4-6	1	4-0	1	3-7	2	4-1	1	3-7	2	3-3	2
	2-2×8	5-9	2	5-0	2	4-6	2	5-2	2	4-6	2	4-1	2
	2-2×10	7-0	2	6-2	2	5-6	2	6-4	2	5-6	2	5-0	2
	2-2×12	8-1	2	7-1	2	6-5	2	7-4	2	6-5	2	5-9	3
	3-2×8	7-2	1	6-3	2	5-8	2	6-5	2	5-8	2	5-1	2
	3-2×10	8-9	2	7-8	2	6-11	2	7-11	2	6-11	2	6-3	2
	3-2×12	10-2	2	8-11	2	8-0	2	9-2	2	8-0	2	7-3	2
	4-2×8	8-1	1	7-3	1	6-7	1	7-5	1	6-6	1	5-11	2
	4-2×10	10-1	1	8-10	2	8-0	2	9-1	2	8-0	2	7-2	2
	4-2×12	11-9	2	10-3	2	9-3	2	10-7	2	9-3	2	8-4	2
Roof, ceiling and one clear span floor	2-2×4	2-8	1	2-4	1	2-1	1	2-7	1	2-3	1	2-0	1
	2-2×6	3-11	1	3-5	2	3-0	2	3-10	2	3-4	2	3-0	2
	2-2×8	5-0	2	4-4	2	3-10	2	4-10	2	4-2	2	3-9	2
	2-2×10	6-1	2	5-3	2	4-8	2	5-11	2	5-1	2	4-7	3
	2-2×12	7-1	2	6-1	3	5-5	3	6-10	2	5-11	3	5-4	3
	3-2×8	6-3	2	5-5	2	4-10	2	6-1	2	5-3	2	4-8	2
	3-2×10	7-7	2	6-7	2	5-11	2	7-5	2	6-5	2	5-9	2
	3-2×12	8-10	2	7-8	2	6-10	2	8-7	2	7-5	2	6-8	2
	4-2×8	7-2	1	6-3	2	5-7	2	7-0	1	6-1	2	5-5	2
	4-2×10	8-9	2	7-7	2	6-10	2	8-7	2	7-5	2	6-7	2
	4-2×12	10-2	2	8-10	2	7-11	2	9-11	2	8-7	2	7-8	2
Roof, ceiling and two center-bearing floors	2-2×4	2-7	1	2-3	1	2-0	1	2-6	1	2-2	1	1-11	1
	2-2×6	3-9	2	3-3	2	2-11	2	3-8	2	3-2	2	2-10	2
	2-2×8	4-9	2	4-2	2	3-9	2	4-7	2	4-0	2	3-8	2
	2-2×10	5-9	2	5-1	2	4-7	3	5-8	2	4-11	2	4-5	3
	2-2×12	6-8	2	5-10	3	5-3	3	6-6	2	5-9	3	5-2	3
	3-2×8	5-11	2	5-2	2	4-8	2	5-9	2	5-1	2	4-7	2
	3-2×10	7-3	2	6-4	2	5-8	2	7-1	2	6-2	2	5-7	2
	3-2×12	8-5	2	7-4	2	6-7	2	8-2	2	7-2	2	6-5	3
	4-2×8	6-10	1	6-0	2	5-5	2	6-8	1	5-10	2	5-3	2
	4-2×10	8-4	2	7-4	2	6-7	2	8-2	2	7-2	2	6-5	2
	4-2×12	9-8	2	8-6	2	7-8	2	9-5	2	8-3	2	7-5	2

(continued)

TABLE R502.5(1)—continued
GIRDER SPANS[a] AND HEADER SPANS[a] FOR EXTERIOR BEARING WALLS
(Maximum spans for Douglas fir-larch, hem-fir, southern pine and spruce-pine-fir[b] and required number of jack studs)

GIRDERS AND HEADERS SUPPORTING	SIZE	GROUND SNOW LOAD (psf)[e]											
		30						50					
		Building width[c] (feet)											
		20		28		36		20		28		36	
		Span	NJ[d]	Span	NJ[d]	Span	NJ[d]	Span	NJ[d]	Span	NJ[d]	Span	NJ[d]
Roof, ceiling and two clear span floor	2-2×4	2-1	1	1-8	1	1-6	2	2-0	1	1-8	1	1-5	2
	2-2×6	3-1	2	2-8	2	2-4	2	3-0	2	2-7	2	2-3	2
	2-2×8	3-10	2	3-4	3	3-0	3	3-10	2	3-4	2	2-11	3
	2-2×10	4-9	2	4-1	3	3-8	3	4-8	2	4-0	3	3-7	3
	2-2×12	5-6	3	4-9	3	4-3	3	5-5	3	4-8	3	4-2	3
	3-2×8	4-10	2	4-2	2	3-9	2	4-9	2	4-1	2	3-8	2
	3-2×10	5-11	2	5-1	2	4-7	3	5-10	2	5-0	2	4-6	3
	3-2×12	6-10	2	5-11	3	5-4	3	6-9	2	5-10	3	5-3	3
	4-2×8	5-7	2	4-10	2	4-4	2	5-6	2	4-9	2	4-3	2
	4-2×10	6-10	2	5-11	2	5-3	2	6-9	2	5-10	2	5-2	2
	4-2×12	7-11	2	6-10	2	6-2	3	7-9	2	6-9	2	6-0	3

For SI: 1 inch = 25.4 mm, 1 pound per square foot = 0.0479 kN/m^2.

a. Spans are given in feet and inches.
b. Tabulated values assume #2 grade lumber.
c. Building width is measured perpendicular to the ridge. For widths between those shown, spans are permitted to be interpolated.
d. NJ - Number of jack studs required to support each end. Where the number of required jack studs equals one, the header is permitted to be supported by an approved framing anchor attached to the full-height wall stud and to the header.
e. Use 30 psf ground snow load for cases in which ground snow load is less than 30 psf and the roof live load is equal to or less than 20 psf.

❖ In addition to providing the allowable spans for girders and headers, this table lists the number of jack studs required to support the girder or header. The header studs (jack studs) on which the headers rest should be continuous from the header to the sill plate of the wall. Cutting the header stud to support a sill is not allowed. Headers should be adequately nailed together and to the wall studs. The table is broken down into five loading conditions, illustrated in Commentary Figure R502.5(2).

FLOORS

TABLE R502.5(2)
GIRDER SPANS[a] AND HEADER SPANS[a] FOR INTERIOR BEARING WALLS
(Maximum spans for Douglas fir-larch, hem-fir, southern pine and spruce-pine-fir[b] and required number of jack studs)

HEADERS AND GIRDERS SUPPORTING	SIZE	BUILDING WIDTH[c] (feet)					
		20		28		36	
		Span	NJ[d]	Span	NJ[d]	Span	NJ[d]
One floor only	2-2×4	3-1	1	2-8	1	2-5	1
	2-2×6	4-6	1	3-11	1	3-6	1
	2-2×8	5-9	1	5-0	2	4-5	2
	2-2×10	7-0	2	6-1	2	5-5	2
	2-2×12	8-1	2	7-0	2	6-3	2
	3-2×8	7-2	1	6-3	1	5-7	2
	3-2×10	8-9	1	7-7	2	6-9	2
	3-2×12	10-2	2	8-10	2	7-10	2
	4-2×8	5-10	1	5-1	2	4-6	2
	4-2×10	10-1	1	8-9	1	7-10	2
	4-2×12	11-9	1	10-2	2	9-1	2
Two floors	2-2×4	2-2	1	1-10	1	1-7	1
	2-2×6	3-2	2	2-9	2	2-5	2
	2-2×8	4-1	2	3-6	2	3-2	2
	2-2×10	4-11	2	4-3	2	3-10	3
	2-2×12	5-9	2	5-0	3	4-5	3
	3-2×8	5-1	2	4-5	2	3-11	2
	3-2×10	6-2	2	5-4	2	4-10	2
	3-2×12	7-2	2	6-3	2	5-7	3
	4-2×8	4-2	2	3-7	2	3-2	2
	4-2×10	7-2	2	6-2	2	5-6	2
	4-2×12	8-4	2	7-2	2	6-5	2

For SI: 1 inch = 25.4 mm, 1 foot = 304.8 mm.

a. Spans are given in feet and inches.
b. Tabulated values assume #2 grade lumber.
c. Building width is measured perpendicular to the ridge. For widths between those shown, spans are permitted to be interpolated.
d. NJ - Number of jack studs required to support each end. Where the number of required jack studs equals one, the headers are permitted to be supported by an approved framing anchor attached to the full-height wall stud and to the header.

❖ In addition to providing the allowable spans for girders and headers, this table lists the number of jack studs required based on the bearing necessary for the girder or header. The table is broken down into two loading conditions, illustrated in Commentary Figure R502.5(3).

Figure R502.5(1)
GIRDER SPAN-SPACING RELATIONSHIP

SOURCE: AF&PA WOOD FRAME CONSTRUCTION MANUAL FOR ONE- AND TWO-FAMILY DWELLINGS, 1995 SBC HIGH WIND EDITION

Figure R502.5(2)
HEADER SPANS FOR EXTERIOR LOAD BEARING WALLS

Figure R502.5(3)
HEADER SPANS FOR INTERIOR LOAD BEARING WALLS

R502.6 Bearing. The ends of each joist, beam or girder shall have not less than 1.5 inches (38 mm) of bearing on wood or metal and not less than 3 inches (76 mm) on masonry or concrete except where supported on a 1-inch-by-4-inch (25.4 mm by 102 mm) ribbon strip and nailed to the adjacent stud or by the use of approved joist hangers.

❖ This section establishes minimum lengths of bearing for several alternative support systems to provide for the transfer of floor loads to supporting elements. These minimum requirements are based on anticipated loads and the allowable compressive stresses perpendicular to the grain for beam sizes and grades typical for wood-frame construction, in addition to consideration of shear failure of the masonry. For joists, beams or girders bearing on wood or metal, the minimum bearing shown in Commentary Figure R502.6(1) must be provided.

As an alternative, the members may be supported by a 1-inch by 4-inch (25 mm by 102 mm) ribbon strip when joists are nailed to adjacent studs as shown in Commentary Figure R502.6(3) or by the use of approved joist hangers as shown in Commentary Figure R502.6(4).

When such joists, beams or girders bear on masonry, a minimum bearing of 3 inches (76 mm) is required, as shown in Commentary Figure R502.6(2).

R502.6.1 Floor systems. Joists framing from opposite sides over a bearing support shall lap a minimum of 3 inches (76 mm) and shall be nailed together with a minimum three 10d face nails. A wood or metal splice with strength equal to or greater than that provided by the nailed lap is permitted.

❖ To provide a concentric application of load from the joist to supporting beams or girders, joists framing from opposite sides of a beam or girder are required to lap at least 3 inches (76 mm) or the opposing joists are to be tied together with a wood or metal splice. See Commentary Figure R502.6.1.

R502.6.2 Joist framing. Joists framing into the side of a wood girder shall be supported by approved framing anchors or on ledger strips not less than nominal 2 inches by 2 inches (51 mm by 51 mm).

❖ For joists framed into the side of a wood beam or girder, the joists must be supported by approved framing anchors, as shown in Commentary Figure R502.6.2(1), or by ledger strips having a minimum nominal dimension of 2 inches (51 mm), as shown in Commentary Figure R502.6.2(2).

R502.7 Lateral restraint at supports. Joists shall be supported laterally at the ends by full-depth solid blocking not less than 2 inches (51 mm) nominal in thickness; or by attachment to a header, band, or rim joist, or to an adjoining stud; or shall be otherwise provided with lateral support to prevent rotation.

Exception: In Seismic Design Categories D_1 and D_2, lateral restraint shall also be provided at each intermediate support.

❖ Bridging, blocking or some other acceptable means of holding a joist in place is required so the floor joists do not twist out of the plane of the applied load. Lateral support at the ends of joists provide an additional function by transferring lateral loads to the supporting elements. Lateral support at ends may be provided by full-depth solid blocking not less than 2 inches (51 mm) in thickness, or the ends of joists may be nailed or bolted to a header, band or rim joist or to an adjoining stud as shown in Commentary Figure R502.7.

R502.7.1 Bridging. Joists exceeding a nominal 2 inches by 12 inches (51 mm by 305 mm) shall be supported laterally by solid blocking, diagonal bridging (wood or metal), or a continuous 1-inch-by-3-inch (25.4 mm by 76 mm) strip nailed across the bottom of joists perpendicular to joists at intervals not exceeding 8 feet (2438 mm).

❖ In addition to the lateral support at the ends, joists are required to have intermediate lateral support at intervals not exceeding 8 feet (2438 mm). Intermediate blocking is not required for joists 2 inches by 12 inches (51 mm by 305 mm) or smaller. The intermediate lateral support may be provided by solid blocking, diagonal bridging or wood bridging not less than 1 inch by 3 inches (25 mm by 76 mm) nominal, nailed to the bot-

Figure R502.6(1)
FLOOR JOIST BEARING ON WOOD

tom of the joist. Commentary Figure R502.7.1 illustrates the various alternatives for intermediate lateral support.

R502.8 Drilling and notching. Structural floor members shall not be cut, bored or notched in excess of the limitations specified in this section. See Figure R502.8.

❖ Some designs and installation practices require that limited notching and cutting occur. Notching should be avoided when possible, and holes bored in beams and joists create the same problems as notches. When necessary, the holes should be located in areas with the least stress concentration, generally along the neutral axis of the joist. Limitations on the allowable cutting and notching of wood floor joists are meant to retain structural or functional integrity.

R502.8.1 Sawn lumber. Notches in solid lumber joists, rafters and beams shall not exceed one-sixth of the depth of the member, shall not be longer than one-third of the depth of the member and shall not be located in the middle one-third of the span. Notches at the ends of the member shall not exceed one-fourth the depth of the member. The tension side of members 4 inches (102 mm) or greater in nominal thickness shall not be notched except at the ends of the members. The diameter of holes bored or cut into members shall not exceed one-third the depth of the member. Holes shall not be closer than 2 inches (51 mm) to the top or bottom of the member, or to any other hole located in the member. Where the member is also notched, the hole shall not be closer than 2 inches (51 mm) to the notch.

❖ Cutting and notching limitations are illustrated in Figure R502.8. Additionally, the tension side of members with a thickness of 4 inches (102 mm) or more can be notched at the ends only. See the commentary for Section R502.8.

R502.8.2 Engineered wood products. Cuts, notches and holes bored in trusses, laminated veneer lumber, glue-laminated members or I-joists are not permitted unless the effects of such penetrations are specifically considered in the design of the member.

Figure R502.6(2)
FLOOR JOIST BEARING ON MASONRY

Figure R502.6(3)
FLOOR JOIST BEARING ON RIBBON STRIP

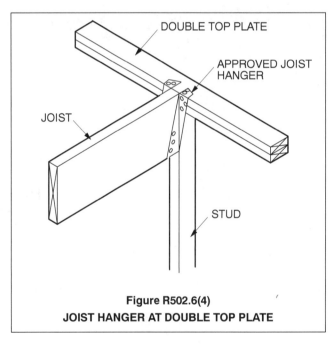

Figure R502.6(4)
JOIST HANGER AT DOUBLE TOP PLATE

wood products without documentation that the design has taken this possibility into account.

R502.9 Fastening. Floor framing shall be nailed in accordance with Table R602.3(1). Where posts and beam or girder construction is used to support floor framing, positive connections shall be provided to ensure against uplift and lateral displacement.

❖ Cutting and notching limitations for sawn lumber do not apply to engineered wood products. The code does not allow drilling and notching of engineered

❖ Commentary Figure R502.9 shows various methods of accomplishing the mandatory positive connection between post and beam or girder construction.

For SI: 1 inch = 25.4 mm.

Figure R502.6.1
JOIST AT GIRDER

Figure R502.6.2(1)
JOIST HANGER AT GIRDER

For SI: 1 inch = 25.4 mm.

Figure R502.6.2(2)
LEDGER STRIPS AT BEAM OR GIRDER

Figure R502.7
BLOCKING OF JOISTS

Figure R502.7.1
INTERMEDIATE LATERAL SUPPORT

FLOORS FIGURE R502.8

FIGURE R502.8
CUTTING, NOTCHING AND DRILLING

For SI: 1 inch = 25.4 mm.

❖ See the commentary to Section R502.8.

Figure R502.9
POSITIVE CONNECTION BETWEEN POSTS AND GIRDERS

R502.10 Framing of openings. Openings in floor framing shall be framed with a header and trimmer joists. When the header joist span does not exceed 4 feet (1219 mm), the header joist may be a single member the same size as the floor joist. Single trimmer joists may be used to carry a single header joist that is located within 3 feet (914 mm) of the trimmer joist bearing. When the header joist span exceeds 4 feet (1219 mm), the trimmer joists and the header joist shall be doubled and of sufficient cross section to support the floor joists framing into the header. Approved hangers shall be used for the header joist to trimmer joist connections when the header joist span exceeds 6 feet (1829 mm). Tail joists over 12 feet (3658 mm) long shall be supported at the header by framing anchors or on ledger strips not less than 2 inches by 2 inches (51 mm by 51 mm).

❖ Where larger floor openings are necessary, adequate load-carrying transfer capability must be provided.

Header joists the same size as floor joists may be used for spans not exceeding 4 feet (1219 mm), as shown in Commentary Figure R502.10(1). Trimmers must also be doubled to carry the additional load, unless the header joist is located within 3 feet (914 mm) of the trimmer joist bearing.

Where the header joist span exceeds 4 feet (1219 mm), a larger load-carrying capability is required. The header and trimmer joists must be doubled and be of sufficient size to support floor joist framing as illustrated in Commentary Figure R502.10(2).

In some cases, nailing would be insufficient to transfer vertical loads. Therefore, positive connections as shown in Commentary Figure R502.10(3) must be employed. These hangers are required when the header joist span exceeds 6 feet (1829 mm).

Figure R502.10(1)
FLOOR FRAMING FOR MAXIMUM 4-FOOT OPENINGS

Figure R502.10(2)
FLOOR FRAMING FOR GREATER THAN 4-FOOT OPENING

Figure R502.10(3)
HANGERS FOR JOIST-HEADER CONNECTION

R502.11 Wood trusses.

R502.11.1 Design. Wood trusses shall be designed in accordance with approved engineering practice. The design and manufacture of metal plate connected wood trusses shall comply with ANSI/TPI 1. The truss design drawings shall be prepared by a registered professional where required by the statutes of the jurisdiction in which the project is to be constructed in accordance with Section R106.1.

❖ The code contains no prescriptive provisions for the design and installation of wood trusses. A design is required in accordance with accepted engineering practice. Recognizing the extensive use of trusses in residential construction, the code references ANSI/TPI 1, *National Design Standard for Metal-Plate-Connected Wood Truss Construction*. The standard provides regulations for the design and installation of metal-plate-connected wood trusses, including the procedures for full-scale tests and testing methods for evaluating metal plate connectors. In addition to adequate design, it is important that the handling and erection of the trusses are performed properly so that the performance capability of the trusses is not compromised. A truss member should never be cut without approval from the design engineer.

Usually trusses are delivered in bundles, which reduces the potential for damage. Sufficient slings and spreader bars should be used to reduce stresses caused by sway and bending.

Shop drawings showing the lumber schedule, design loads and panel point details (size, location and attachment of plates) should be filed with the building permit application and should be available at the time of inspection.

R502.11.2 Bracing. Trusses shall be braced to prevent rotation and provide lateral stability in accordance with the requirements specified in the construction documents for the building and on the individual truss design drawings. In the absence of specific bracing requirements, trusses shall be braced in accordance with the TPI, HIB.

❖ To prevent collapse during construction and until permanent bracing is installed, trusses should be adequately braced temporarily. When braced for use, trusses should be positioned as vertical as possible; tilted trusses will not perform as required.

R502.11.3 Alterations to trusses. Truss members and components shall not be cut, notched, spliced or otherwise altered in any way without the approval of a registered design professional. Alterations resulting in the addition of load (e.g., HVAC equipment, water heater, etc.), that exceed the design load for the truss, shall not be permitted without verification that the truss is capable of supporting the additional loading.

❖ Addition of loads in excess of the design load is only allowed if it can be verified that the truss has additional capacity. Also see the commentary for Section R502.8.2.

R502.11.4 Truss design drawings. Truss design drawings, prepared in compliance with Section R502.11.1, shall be provided to the building official and approved prior to installation. Truss design drawing shall be provided with the shipment of trusses delivered to the job site. Truss design drawings shall include, at a minimum, the information specified below:

1. Slope or depth, span, and spacing.
2. Location of all joints.
3. Required bearing widths.
4. Design loads as applicable.
    4.1. Top chord live load (including snow loads).
    4.2. Top chord dead load.
    4.3. Bottom chord live load.
    4.4. Bottom chord dead load.
    4.5. Concentrated loads and their points of application.
    4.6. Controlling wind and earthquake loads.
5. Adjustments to lumber and joint connector design values for conditions of use.

FLOORS

6. Each reaction force and direction.
7. Joint connector type and description (e.g., size, thickness or gauge); and the dimensioned location of each joint connector except where symmetrically located relative to the joint interface.
8. Lumber size, species and grade for each member.
9. Connection requirements for:
    9.1. Truss-to-truss girder.
    9.2. Truss ply-to-ply.
    9.3. Field splices.
10. Calculated deflection ratio and/or maximum description for live and total load.
11. Maximum axial compression forces in the truss members to enable the building designer to design the size, connections and anchorage of the permanent continuous lateral bracing. Forces shall be shown on the truss drawing or on supplemental documents.
12. Required permanent truss member bracing location.

❖ See the definition in Chapter 2 and the commentary for Section R502.11.1.

R502.12 Draftstopping required. When there is usable space both above and below the concealed space of a floor/ceiling assembly, draftstops shall be installed so that the area of the concealed space does not exceed 1,000 square feet (92.9 m^2). Draftstopping shall divide the concealed space into approximately equal areas. Where the assembly is enclosed by a floor membrane above and a ceiling membrane below draftstopping shall be provided in floor/ceiling assemblies under the following circumstances:

1. Ceiling is suspended under the floor framing.
2. Floor framing is constructed of truss-type open-web or perforated members.

❖ Draftstopping is required to limit the spread of fire through combustible spaces in floor/ceiling assemblies when such spaces create a connected area beyond the normal joist cavity. For example, draftstopping is required when the ceiling is suspended under the floor framing, as illustrated in Commentary Figure R502.12(1), or when the floor framing is constructed of truss-type or open-web perforated members, as illustrated in Commentary Figure R502.12(2). When such space exceeds 1,000 square feet (92.9 m^2) in area, the code requires the space between the ceiling membrane and the floor to be divided into two approximately equal areas with no area greater than 1,000 square feet (92.9 m^2). A floor/ceiling assembly having a space exceeding 2,000 square feet (185.8 m^2), but not exceeding 3,000 square feet (278.7 m^2), would have to be divided into a minimum of three approximately equal areas.

Figure R502.12(1)
DRAFTSTOPPING OF CEILING SUSPENDED UNDER FLOOR FRAMING

Figure R502.12(2)
DRAFTSTOPPING OF TRUSS-TYPE FLOOR FRAMING

R502.12.1 Materials. Draftstopping materials shall not be less than $^1/_2$-inch (12.7 mm) gypsum board, $^3/_8$-inch (9.5 mm) wood structural panels, $^3/_8$-inch (9.5 mm) Type 2-M-W particleboard or other approved materials adequately supported. Draftstopping shall be installed parallel to the floor framing members unless otherwise approved by the building official. The integrity of all draftstops shall be maintained.

❖ Materials used to meet the draftstopping requirements are typically $^1/_2$-inch (12.7 mm) gypsum board, $^3/_8$-inch (9.5 mm) plywood or $^3/_8$-inch (9.5 mm) Type 2-M-W particle board adequately attached to the supporting members. Other materials may be used for draftstopping, provided that such materials produce an equivalent barrier.

The code places an additional restriction on the specific location of the draftstopping, mandating that the draftstopping be placed parallel to the main framing members. Commentary Figures R502.12(1) and (2) illustrate two examples of draftstopping orientation placed as required.

When a penetration through the draftstopping is necessary, the integrity of the draftstop must be maintained in a manner similar to firestopping penetrations.

R502.13 Fireblocking required. Fireblocking shall be provided in wood-frame floor construction and floor-ceiling assemblies in accordance with Section R602.8.

❖ See the commentary for Section R602.8.

SECTION R503
FLOOR SHEATHING

R503.1 Lumber sheathing. Maximum allowable spans for lumber used as floor sheathing shall conform to Tables R503.1, R503.2.1.1(1) and R503.2.1.1(2).

❖ Table R503.1 of the code sets forth the required thickness of lumber used as floor sheathing. The allowable spans are based on the floor sheathing thickness, joist spacing and the orientation of the sheathing with respect to the joist. Commentary Figures R503.1(1) and R503.1(2) show floor sheathing applications that are perpendicular and diagonal to the joist.

**TABLE R503.1
MINIMUM THICKNESS OF LUMBER FLOOR SHEATHING**

JOIST OR BEAM SPACING (inches)	MINIMUM NET THICKNESS	
	Perpendicular to joist	Diagonal to joist
24	$^{11}/_{16}$	$^{3}/_{4}$
16	$^{5}/_{8}$	$^{5}/_{8}$
48[a]	$1^{1}/_{2}$ T & G	N/A
54[b]		
60[c]		

For SI: 1 inch = 25.4 mm, 1 pound per square inch = 6.895 kPa.

a. For this support spacing, lumber sheathing shall have a minimum F_b of 675 and minimum E of 1,100,000 (see AF&PA/NDS).
b. For this support spacing, lumber sheathing shall have a minimum F_b of 765 and minimum E of 1,400,000 (see AF&PA/NDS).
c. For this support spacing, lumber sheathing shall have a minimum F_b of 855 and minimum E of 1,700,000 (see AF&PA/NDS).

❖ See the commentary for Section R503.1.

R503.1.1 End joints. End joints in lumber used as subflooring shall occur over supports unless end-matched lumber is used, in which case each piece shall bear on at least two joists. Subflooring may be omitted when joist spacing does not exceed 16 inches (406 mm) and a 1-inch (25.4 mm) nominal tongue-and-groove wood strip flooring is applied perpendicular to the joists.

❖ When lumber is to be used as subflooring, adequate load-carrying capability and continuity must be provided through either the proper placement of end joints over supports or the use of end-matched lumber as illustrated in Commentary Figure R503.1.1.

R503.2 Wood structural panel sheathing.

❖ Wood structural panels are manufactured with fully waterproof adhesive and include plywood, oriented strand board (OSB) and composite panels made up of a combination of wood veneers and reconstructed wood layers. Additional information on wood structural panels is available from the APA—The Engineered Wood Association (formerly the American Plywood Association), other trade groups and manufacturers.

Plywood is a wood structural panel manufactured by gluing three or more cross-laminated wood layers together. Plywood may be manufactured in accordance with PS 1 or PS 2. It is inspected, and if it is certified, the trademark of an approved testing and grading agency is stamped on it.

OSB panels are fabricated out of multiple layers of wood flakes or strands called "furnish." Like the layers in plywood, these layers of furnish are oriented 90 degrees (1.57 rad) to each other. This gives OSB panels properties that are very similar to those of plywood.

Composite panels are structural panels made up of a combination of wood veneers and wood-based materials. The wood veneer usually forms the two outer layers and may also be used in the core of the panel.

**Figure R503.1(1)
FLOOR SHEATHING AS SUBFLOORING**

Figure R503.1(2)
SUBFLOOR (FLOOR SHEATHING) ORIENTATION

These layers of veneer and furnish may be cross-laminated.

Both OSB and composite panels are manufactured in accordance with PS 2. They are inspected, and if they are certified, the trademark of an approved testing and grading agency is stamped on them. The grade stamps of plywood, OSB and composite panels are almost identical in appearance, and in most cases, wood structural panels can be specified without regard to specific panel type.

Wood structural panels can be manufactured with Exterior, Exposure 1, Exposure 2 and Interior exposure durability classifications. Exterior and Exposure 1 are by far the most common classifications and both are manufactured with the same class of fully waterproof adhesives. Exterior panels are intended for applications subject to permanent exposure to the weather or moisture. Exposure 1 panels are intended for applications where long construction delays may be expected prior to protection being installed. Exposure 2 panels may be used for protected applications that are not continuously exposed to high humidity conditions. Interior panels may be used only for permanently protected interior applications. Exposure 2 and Interior panels are not readily available.

The intended end use of wood structural panels is designated in their grade-marking stamp.

R503.2.1 Identification and grade. Wood structural panel sheathing used for structural purposes shall conform to DOC PS 1, DOC PS 2 or, when manufactured in Canada, CSA 0437 or CSA 0325. All panels shall be identified by a grade mark of certificate of inspection issued by an approved agency.

❖ Plywood that performs a load-carrying function in floor construction must conform to a known quality control

Figure R503.1.1
SUBFLOOR END-JOINT REQUIREMENTS

FLOORS

standard; therefore, a method of identifying the limitations under which the plywood may be used is needed. This is accomplished by a grade mark or certificate of inspection issued by an approved agency. Examples of plywood grade marks are shown in Commentary Figure R503.2.1.

R503.2.1.1 Subfloor and combined subfloor underlayment. Where used as subflooring or combination subfloor underlayment, wood structural panels shall be of one of the grades specified in Table R503.2.1.1(1). When sanded plywood is used as combination subfloor underlayment, the grade shall be as specified in Table R503.2.1.1(2).

❖ This section specifies grades of wood structural panels when they are used as subflooring or combination subflooring underlayment. It also addresses the use of sanded plywood as a combination subfloor underlayment.

A floor system could consist of three elements: (1) a subfloor, (2) an underlayment or a combination subfloor/underlayment system and (3) a finished floor surface material. A finished surfacing material may be wood-strip flooring; tongue-and-groove flooring or various types of resilient flooring coverings such as vinyl, tile or carpeting.

Wood structural panels are manufactured for use as either structural subfloor or combination subfloor underlayment. The allowable spans for structural subflooring and combination subfloor underlayment are based on the wood structural panels' face grain strength axis parallel to its supporting member, or they are based on the panels being continuous over two or more spans, with a face grain placed perpendicular to the supports. These qualifications are critical in determining the permissible spans. Most wood structural panels are considerably stronger when their face grain is parallel to the supports and continuous over two or more spans. Panels with multiple spans have greater capacity than those that are simply supported between the two joists.

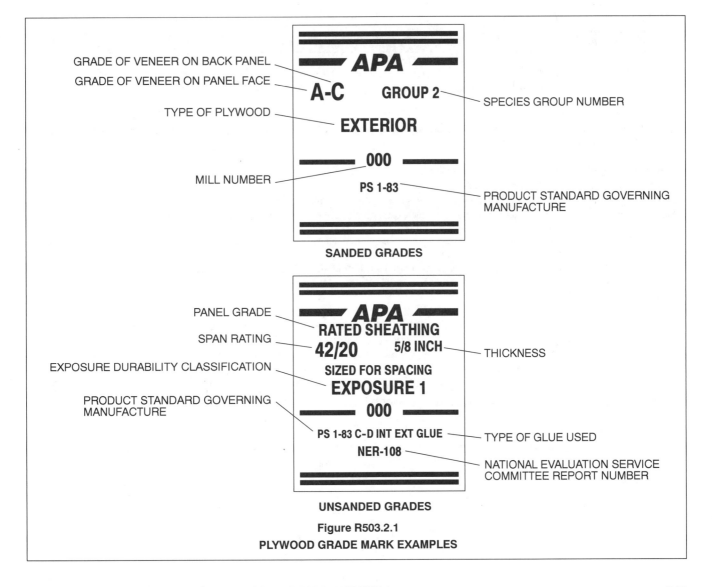

Figure R503.2.1
PLYWOOD GRADE MARK EXAMPLES

TABLE R503.2.1.1(1)
ALLOWABLE SPANS AND LOADS FOR WOOD STRUCTURAL PANELS FOR ROOF AND SUBFLOOR SHEATHING AND COMBINATION SUBFLOOR UNDERLAYMENT[a,b,c]

SPAN RATING	MINIMUM NOMINAL PANEL THICKNESS (inch)	MAXIMUM SPAN (inches)[d]		LOAD (pounds per square foot, at maximum span)		MAXIMUM SPAN (inches)
		With edge support	Without edge support	Total load	Live load	
Sheathing[e]		Roof[f]				Subfloor[j]
12/0	5/16	12	12	40	30	0
16/0	5/16	16	16	40	30	0
20/0	5/16	20	20	40	30	0
24/0	3/8	24	20[g]	40	30	0
24/16	7/16	24	24	50	40	16
32/16	15/32, 1/2	32	28	40	30	16[h]
40/20	19/32, 5/8	40	32	40	30	20[h,i]
48/24	23/32, 3/4	48	36	45	35	24
60/32	7/8	60	48	45	35	32
Underlayment, C-C plugged, single floor[e]		Roof[f]				Combination subfloor underlayment[k]
16 o.c.	19/32, 5/8	24	24	50	40	16[i]
20 o.c.	19/32, 5/8	32	32	40	30	20[i,j]
24 o.c.	23/32, 3/4	48	36	35	25	24
32 o.c.	7/8	48	40	50	40	32
48 o.c.	1 3/32, 1 1/8	60	48	50	40	48

For SI: 1 inch = 25.4 mm, 1 pound per square foot = 0.0479 kN/m^2.

a. The allowable total loads were determined using a dead load of 10 psf. If the dead load exceeds 10 psf, then the live load shall be reduced accordingly.
b. Panels continuous over two or more spans with long dimension perpendicular to supports. Spans shall be limited to values shown because of possible effect of concentrated loads.
c. Applies to panels 24 inches or wider.
d. Lumber blocking, panel edge clips (one midway between each support, except two equally spaced between supports when span is 48 inches), tongue-and-groove panel edges, or other approved type of edge support.
e. Includes Structural 1 panels in these grades.
f. Uniform load deflection limitation: 1/180 of span under live load plus dead load, 1/240 of span under live load only.
g. Maximum span 24 inches for 15/32- and 1/2-inch panels.
h. Maximum span 24 inches where 3/4-inch wood finish flooring is installed at right angles to joists.
i. Maximum span 24 inches where 1.5 inches of lightweight concrete or approved cellular concrete is placed over the subfloor.
j. Unsupported edges shall have tongue-and-groove joints or shall be supported with blocking unless minimum nominal 1/4-inch thick underlayment with end and edge joints offset at least 2 inches or 1.5 inches of lightweight concrete or approved cellular concrete is placed over the subfloor, or 3/4-inch wood finish flooring is installed at right angles to the supports. Allowable uniform live load at maximum span, based on deflection of 1/360 of span, is 100 psf.
k. Unsupported edges shall have tongue-and-groove joints or shall be supported by blocking unless nominal 1/4-inch-thick underlayment with end and edge joints offset at least 2 inches or 3/4-inch wood finish flooring is installed at right angles to the supports. Allowable uniform live load at maximum span, based on deflection of 1/360 of span, is 100 psf, except panels with a span rating of 48 on center are limited to 65 psf total uniform load at maximum span.

❖ The maximum spans for wood structural panel floor and roof sheathing are limited by the stresses and deflection imposed by the design live loads. For convenience, the trademarks of the inspection agencies include a span rating, which appears as two numbers separated by a slash (32/16 or 48/24, for example). The first number represents the maximum recommended span for roof sheathing when the panels are applied with the long dimension across three or more supports and the edges are blocked or when other support required by the table is provided. The second number indicates the maximum recommended span when the panel is used for structural floor sheathing with the panels applied with the long dimension across three or more supports. Single panels intended for single floor applications will be marked with a single number representing this span.

The edges of the wood structural panels between floor supports are prevented from moving relative to each other by tongue and groove panel edges, the addition of wood blocking or by the use of an approved underlayment or structural finished floor system.

TABLE R503.2.1.1(2)
ALLOWABLE SPANS FOR SANDED PLYWOOD COMBINATION SUBFLOOR UNDERLAYMENT[a]

IDENTIFICATION	SPACING OF JOISTS (inches)		
	16	20	24
Species group[b]	—	—	—
1	$^1/_2$	$^5/_8$	$^3/_4$
2, 3	$^5/_8$	$^3/_4$	$^7/_8$
4	$^3/_4$	$^7/_8$	1

For SI: 1 inch = 25.4 mm, 1 pound per square foot = 0.0479 kN/m^2.

a. Plywood continuous over two or more spans and face grain perpendicular to supports. Unsupported edges shall be tongue-and-groove or blocked except where nominal $^1/_4$-inch-thick underlayment or $^3/_4$-inch wood finish floor is used. Allowable uniform live load at maximum span based on deflection of $^1/_{360}$ of span is 100 psf.

b. Applicable to all grades of sanded exterior-type plywood.

❖ For exterior-type sanded plywood grades, the thickness required for a specific span is related to the species grouping of the panel used. For example, species Group 1 may be $^1/_2$ inches (12.7 mm) for 16-inch (406 mm) joist spacing, while Group 4 will require a thickness of $^3/_4$ inch (19.1 mm) for the same span.

When the panels are used as a combination subfloor/underlayment, extra precautions should be observed in attaching the panels to the floor framing. Joints in adjacent panels should not be continuous and should not occur at locations where the orientation of the joist supports is different.

R503.2.2 Allowable spans. The maximum allowable span for wood structural panels used as subfloor or combination subfloor underlayment shall be as set forth in Table R503.2.1.1(1). The maximum span for sanded plywood combination subfloor underlayment shall be as set forth in Table R503.2.1.1(2).

❖ Table R503.2.1.1(1) indicates spans for plywood and wood structural panels used as subflooring. This table also covers roof sheathing and is referenced in Section R803.2.2 for that purpose. The span limitations are predicated on the grade of plywood used. In the case of rated sheathing used as structural subflooring, the maximum span is easily identified through the panel span rating (identification index). Where the panel span rating is stamped on the sheet of plywood, the denominator represents the allowable span of the plywood floor sheathing. See Commentary Figure R503.2.2(1). An illustration of the limitations imposed by Notes j and k to Table R503.2.1.1(1) on combination subfloor underlayment applications is illustrated in Commentary Figure R503.2.2.(2). Allowable increases in the maximum span for certain span ratings are permitted as described in Notes h and i to Table R503.2.1.1(1).

When plywood combination subfloor underlayment is to be used, the allowable span as listed in Table R503.2.1.1(2) is applicable to underlayment grade C-C plugged and sanded exterior-type plywood of specific species.

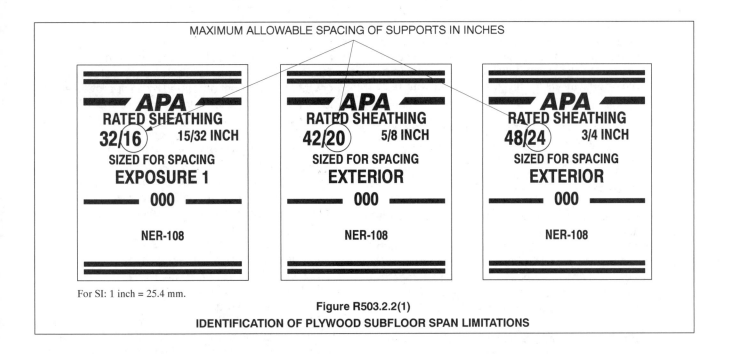

For SI: 1 inch = 25.4 mm.

Figure R503.2.2(1)
IDENTIFICATION OF PLYWOOD SUBFLOOR SPAN LIMITATIONS

Figure R503.2.2(2)
COMBINATION SUBFLOOR UNDERLAYMENT ALTERNATIVES

R503.2.3 Installation. Wood structural panels used as subfloor or combination subfloor underlayment shall be attached to wood framing in accordance with Table R602.3(1) and shall be attached to cold-formed steel framing in accordance with Table R505.3.1(2).

❖ References are provided to the wood framing fastener schedule in Chapter 6 and the cold-formed steel floor-fastening schedule.

R503.3 Particleboard.

❖ Particleboard is a generic term given to panels manufactured from cellulosic materials, usually wood in the form of discrete pieces and particles rather than fibers. The particles and pieces are combined with synthetic resins and other binders and bonded together under conditions of heat and pressure. Particleboard must be in accordance with ANSI A208.1.

R503.3.1 Identification and grade. Particleboard shall conform to ANSI A208.1 and shall be so identified by a grade mark or certificate of inspection issued by an approved agency.

❖ Particleboard used in floor construction that performs a load-carrying function must conform to a known quality control standard. To this end, a grade mark or certificate of inspection issued by an approved agency identifies the limitations under which the particleboard may be used. Commentary Figure R503.3.1 shows an example of a particleboard grade mark.

The particleboard grades in ANSI A208.1 are identified by a letter followed by a hyphen and a digit or another letter. A letter, number or term following the grade designation identifies special performance characteristics. The first letter has the following meaning:

H: High density (above 50 lbs./f^3),
M: Medium density (40—50 lbs./f^3),
LD: Low density (below 40 lbs./f^3),
D: Manufactured home decking
PBU: Underlayment.

The digit or letter following the hyphen indicates the grade identification within a particular description. For example, M-2 indicates medium density particleboard, Grade 2.

Figure R503.3.1
PARTICLEBOARD GRADE MARK EXAMPLE

R503.3.2 Floor underlayment. Particleboard floor underlayment shall conform to Type PBU and shall not be less than $^1/_4$ inch (6.4 mm) in thickness.

❖ Particleboard is often used over a structural subfloor to provide a smooth, even surface under textile or resilient type finish floors. Particleboard is required to conform to Type PBU with a minimum thickness of $^1/_4$ inch (6.4 mm).

R503.3.3 Installation. Particleboard underlayment shall be installed in accordance with the recommendations of the manufacturer and attached to framing in accordance with Table R602.3(1).

❖ Particleboard used as underlayment only is not subject to span limitations since the plywood or lumber subflooring provides the load-carrying capability for the floor system; the particleboard has a nonstructural role. However, joints in particleboard should not occur over joints in the subfloor; the joints should be staggered.

SECTION R504
PRESSURE PRESERVATIVELY TREATED-WOOD FLOORS (ON GROUND)

R504.1 General. Pressure preservatively treated-wood basement floors and floors on ground shall be designed to withstand axial forces and bending moments resulting from lateral soil pressures at the base of the exterior walls and floor live and dead loads. Floor framing shall be designed to meet joist deflection requirements in accordance with Section R301.

❖ As indicated in the commentary for Section R404.1.1, lateral soil pressure loads are carried by the foundation walls through bending in the vertical direction. The resulting forces (reactions) are resisted by the basement floor and the first floor above. This section stipulates the requirements for basement wood-floor framing in order for it to resist the applied loads at the base of the foundation wall. It is implicit that this type of floor would be used in conjunction with wood foundations described in Sections R401.1.

R504.1.1 Unbalanced soil loads. Unless special provision is made to resist sliding caused by unbalanced lateral soil loads, wood basement floors shall be limited to applications where the differential depth of fill on opposite exterior foundation walls is 2 feet (610 mm) or less.

❖ The requirements of foundation walls permit up to 4 feet (1219 mm) of unbalanced fill (see Table R404.2.3). When exterior foundation walls on opposite sides of a building support an equal (or nearly equal) height of unbalanced fill, sliding is not a concern. When there is a difference in the amount of fill supported, however, there is a net horizontal force that must be resisted, otherwise sliding may occur. To limit this unbalanced force, the height of differential fill is accordingly limited to 2 feet (610 mm) for this floor system.

R504.1.2 Construction. Joists in wood basement floors shall bear tightly against the narrow face of studs in the foundation wall or directly against a band joist that bears on the studs. Plywood subfloor shall be continuous over lapped joists or over butt joints between in-line joists. Sufficient blocking shall be provided between joists to transfer lateral forces at the base of the end walls into the floor system.

❖ These construction requirements provide a load path for the horizontal force described in the previous section.

R504.1.3 Uplift and buckling. Where required, resistance to uplift or restraint against buckling shall be provided by interior bearing walls or properly designed stub walls anchored in the supporting soil below.

❖ This is a general caution that the effects of uplift forces on the floor system, which could result from fluctuations in the water table, should be considered.

R504.2 Site preparation. The area within the foundation walls shall have all vegetation, topsoil and foreign material removed, and any fill material that is added shall be free of vegetation and foreign material. The fill shall be compacted to assure uniform support of the pressure preservatively treated-wood floor sleepers.

❖ This requirement is similar to Section R506.2. See the commentary for Section R506.2.2.

R504.2.1 Base. A minimum 4-inch-thick (102 mm) granular base of gravel having a maximum size of $^3/_4$ inch (19.1 mm) or crushed stone having a maximum size of $^1/_2$ inch (12.7 mm) shall be placed over the compacted earth.

❖ This requirement is similar to Section R506.2.2. See the commentary for Section R506.2.2.

R504.2.2 Moisture barrier. Polyethylene sheeting of minimum 6-mil (0.15 mm) thickness shall be placed over the granular base. Joints shall be lapped 6 inches (152 mm) and left unsealed. The polyethylene membrane shall be placed over the pressure preservatively treated-wood sleepers and shall not extend beneath the footing plates of the exterior walls.

❖ See the commentary for Section R506.2.2 concerning application of the moisture barrier.

R504.3 Materials. All framing materials, including sleepers, joists, blocking and plywood subflooring, shall be pressure preservatively treated and dried after treatment in accordance with AWPA C22.

❖ To resist potential decay, wood floor material is to be treated in accordance with AW&PA C22, *Lumber and Plywood for Permanent Wood Foundations—Preservative Treatment by Pressure Process.*

SECTION R505
STEEL FLOOR FRAMING

R505.1 Cold-formed steel floor framing. Elements shall be straight and free of any defects that would significantly affect structural performance. Cold-formed steel floor framing members shall comply with the requirements of this section.

❖ The provisions of this section apply to the construction of floor systems using cold-formed steel framing, which is a type of construction made up in part or entirely of steel structural members cold formed to shape from sheet or strip steel. The general shape of the framing members is in the form of the letter "C" (C-shape) and is described in further detail in Section R505.2. The use of cold-formed steel-framed designs not consistent with these provisions is beyond the scope of this code, and an analysis by a design professional is recommended.

R505.1.1 Applicability limits. The provisions of this section shall control the construction of steel floor framing for buildings not greater than 60 feet (18 288 mm) in length perpendicular to the joist span, not greater than 36 feet (10 973 mm) in width parallel to the joist span, and not greater than two stories in height with each story not greater than 10 feet (3048 mm) high. Steel floor framing constructed in accordance with the provisions of this section shall be limited to sites subjected to a maximum design wind speed of 110 miles per hour Exposure A, B or C and a maximum ground snow load of 70 pounds per square foot (3.35 kN/m²).

❖ Commentary Figure R505.1.1 depicts building limitations. These limits are repeated in Section R603.1.1 for walls and in Section R804.1.1 for roofs. Steel floor framing constructed in accordance with the provisions of this section must be limited to sites with a maximum design wind speed of 110 miles per hour (209 km/h); Exposure A, B, or C and a maximum ground snow load of 70 pounds per square foot (3.35 kN/m²).

R505.1.2 In-line framing. When supported by steel-framed walls in accordance with Section R603, steel floor framing shall be constructed with floor joists located directly in-line with load-bearing studs located below the joists with a maximum tolerance of $^3/_4$ inch (19.1 mm) between the center lines of the joist and the stud.

❖ In-line framing is the preferred framing method because it provides a direct load path for the transfer of forces from joists to studs. In-line framing maximizes framing alignment in order to minimize secondary moments on the framing members, taking into account that the track cannot function as a load-transfer member. The ¾-inch (19 mm) alignment value is accepted industry practice and addresses those conditions where two framing members (e.g. roof rafter and ceiling joist) are to be aligned over a wall stud. In the absence of in-line framing, a load distribution member, such as a structural track, may be required for this force transfer.

R505.2 Structural framing. Load-bearing floor framing members shall comply with Figure R505.2(1) and with the dimensional and minimum thickness requirements specified in Tables R505.2(1) and R505.2(2). Tracks shall comply with Figure R505.2(2) and shall have a minimum flange width of $1^1/_4$ inches (32 mm). The maximum inside bend radius for members shall be the greater of $^3/_{32}$ inch (2.4 mm) or twice the uncoated steel thickness. Holes in joist webs shall conform to Figure R505.2(3) and to the dimensional requirements specified in Table R505.2(3). Holes shall be permitted only along the centerline of the web of the framing member. Holes for 800S162-33, 1000S162-43, 1200S162-43 and 1200S162-54 nominal joist sizes located less than 10 inches (254 mm) from the edge of load-bearing surface shall be patched in accordance with Section R505.3.6.

❖ Joists must comply with the dimensional and minimum thickness requirements specified in Tables R505.2(1) and R505.2(2) respectively. Note "a" to Table R505.2(1) explains the meaning of the alphanumeric member designation used for all steel framing members in the IRC. This system replaces the varied designation approaches that were produced by each individual manufacturer. In addition, the designation is used to identify not only a specific steel framing member but also the section properties of that same member through the use of the product technical information document.

Steel thickness is expressed in mils, which pertains to the base metal thickness measured prior to painting or the application of metallic (corrosion resistant) coatings. The base metal thickness is typically stamped or embossed on the member by the manufacturer. Table R505.2(2) also references the equivalent thickness in inches and provides a further reference to the gage number. The use of gage numbers to define the thickness (a specific mean decimal thickness for which rolling tolerances were established) of steel sheet is a

FLOORS

Figure R505.1.1
STEEL FRAMING APPLICABILITY LIMITS

practice that has been discontinued by the steel industry.

To allow for routing utilities through steel-framed floors, holes (also referred to as "penetrations," "utility holes" and "punchouts") are permitted in the webs of floor joists. To avoid adversely affecting the strength of floor joists, the web holes are limited to the locations along the centerline of the web of the framing member as illustrated in Figure R505.2(3) and Table R505.2(3). To avoid overstressing the joist web through high shear, these web holes must occur more than 10 inches (254 mm) from the edge of a load-bearing surface. Patching of web holes must extend at least 1 inch (25.4 mm) beyond all edges of the hole, and it must be fastened to the web with No. 8 screws having a minimum edge distance of $^1/_2$ inch (12.7 mm).

Steel track sections are to serve in nonload-bearing applications only and are shown in Figure R505.2(2). They must have a minimum flange width of $1^1/_4$ inches (32 mm).

R505.2.1 Material. Load-bearing members utilized in steel floor construction shall be cold formed to shape from structural quality sheet steel complying with the requirements of one of the following:

1. ASTM A 653: Grades 33, 37, 40 and 50 (Class 1 and 3).
2. ASTM A 792: Grades 33, 37, 40 and 50A.
3. ASTM A 875: Grades 33, 37, 40 and 50 (Class 1 and 3).
4. Steels that comply with ASTM A 653, except for tensile and elongation, shall be permitted provided the ratio of tensile strength to yield point is at least 1.08 and the total elongation is at least 10 percent for a 2-inch (51 mm) gage length or 7 percent for an 8-inch (203 mm) gage length.

❖ Load bearing steel framing members must have a legible label, stencil, stamp, or embossment. This identification allows for verification that materials installed are consistent with the design and meet the provisions of the code.

**FIGURE R505.2(1)
C-SECTION**

**FIGURE R505.2(2)
TRACK SECTION**

For SI: 1 inch = 25.4 mm.

**FIGURE R505.2(3)
FLOOR JOIST WEB HOLES**

❖ See the commentary for Section R505.2 for Figures R505.2 (1) through (3).

R505.2.2 Identification. Load-bearing steel framing members shall have a legible label, stencil, stamp or embossment with the following information as a minimum:

1. Manufacturer's identification.
2. Minimum uncoated steel thickness in inches (mm).
3. Minimum coating designation.
4. Minimum yield strength, in kips per square inch (ksi) (kPa).

❖ Load-bearing steel framing members must have a legible label, stencil, stamp or embossment. This allows for verification that the materials installed are consistent with the design and meet the code requirements.

R505.2.3 Corrosion protection. Load-bearing steel framing shall have a metallic coating complying with one of the following:

1. A minimum of G 60 in accordance with ASTM A 653.
2. A minimum of AZ 50 in accordance with ASTM A 792.
3. A minimum of GF 60 in accordance with ASTM A 875.

❖ The specified metallic coatings correspond to requirements in the referenced material standards. The minimum coating designations assume normal exposure conditions that are best defined as having the framing members enclosed within a building envelope or wall assembly within a controlled environment. When more severe exposure conditions are probable, such as industrial or marine atmospheres, consideration should be given to specifying a heavier coating.

TABLE R505.2(1)
COLD-FORMED STEEL JOIST SIZES

MEMBER DESIGNATION[a]	WEB DEPTH (inches)	MINIMUM FLANGE WIDTH (inches)	MAXIMUM FLANGE WIDTH (inches)	MINIMUM LIP SIZE (inches)
550S162-t	5.5	1.625	2	0.5
800S162-t	8	1.625	2	0.5
1000S162-t	10	1.625	2	0.5
1200S162-t	12	1.625	2	0.5

For SI: 1 inch = 25.4 mm.

a. The member designation is defined by the first number representing the member depth in $1/100$ inches, the letter "S" representing a stud or joist member, the second number representing the flange width in $1/100$ inches, and the letter "t" shall be a number representing the minimum base metal thickness in mils [See Table R505.2(2)].

❖ See the commentary for Section R505.2.

TABLE R505.2(2)
MINIMUM THICKNESS OF COLD-FORMED STEEL MEMBERS

DESIGNATION (mils)	MINIMUM UNCOATED THICKNESS (inches)	REFERENCE GAGE NUMBER
33	0.033	20
43	0.043	18
54	0.054	16
68	0.068	14

For SI: 1 inch = 25.4 mm, 1 mil = 0.0254 mm.

❖ See the commentary for Section R505.2.

TABLE R505.2(3)
MAXIMUM HOLE DIMENSIONS AND SPACING IN JOIST WEBS

NOMINAL MEMBER SIZE	MAXIMUM HOLE DEPTH[a] (inches)	MAXIMUM HOLE LENGTH[b] (inches)	MINIMUM HOLE SPACING (inches)	MINIMUM HOLE EDGE DISTANCE[c] (inches)
550S162-33	2	5.25	16.5	10
550S162-43	2	5.25	16.5	10
550S162-54	2	5.25	16.5	10
550S162-68	2	5.25	16.5	10
800S162-33	1.5	4	24	10
800S162-43	3	6	24	10
800S162-54	3	6	24	10
800S162-68	3	6	24	10
1000S162-43	1.5	4	24	10
1000S162-54	4	6	24	10
1000S162-68	4	6	24	10
1200S162-43	1.5	4	24	10
1200S162-54	1.5	4	24	10
1200S162-68	4.75	6	24	10

For SI: 1 inch = 25.4 mm.

a. The dimension of the hole measured across the depth of the joist web.
b. The dimension of the hole measured along the length of the joist.
c. Edge distance is measured from the edge of the hole to the edge of bearing support.

❖ This table lists the maximum dimensions of holes permitted in the web of steel joists as illustrated in Figure R505.2(3).

R505.2.4 Fastening requirements. Screws for steel-to-steel connections shall be installed with a minimum edge distance and center-to-center spacing of 0.5 inch (12.7 mm), shall be self-drilling tapping, and shall conform to SAE J78. Floor sheathing shall be attached to steel joists with minimum No. 8 self-drilling tapping screws that conform to SAE J78. Screws attaching floor-sheathing-to-steel joists shall have a minimum head diameter of 0.292 inch (7.4 mm) with countersunk heads and shall be installed with a minimum edge distance of 0.375 inch (9.5 mm). Gypsum board ceilings shall be attached to steel joists with minimum No. 6 screws conforming to ASTM C 954 and shall be installed in accordance with Section R702. For all connections, screws shall extend through the steel a minimum of three exposed threads. All self-drilling tapping screws conforming to SAE J78 shall have a Type II coating in accordance with ASTM B 633.

Where No. 8 screws are specified in a steel to steel connection the required number of screws in the connection is permitted to be reduced in accordance with the reduction factors in Table R505.2.4 when larger screws are used or when one of the sheets of steel being connected is thicker than 33 mils (0.84 mm). When applying the reduction factor the resulting number of screws shall be rounded up.

❖ Fasteners meeting the referenced standards and installed in accordance with this section provide the necessary load capacity consistent with these prescriptive provisions. This section specifies self-drilling tapping screws conforming to SAE J78, *Steel Self-Drilling Tapping Screws* for all steel-to-steel connections as well as for fastening floor sheathing to steel joists. They must have a Type II coating in accordance with ASTM B 633F, *Standard Specification for Electrodeposited Coatings of Zinc on Iron and Steel*. Gypsum board must be attached to steel joists with No. 6 screws (minimum) conforming to ASTM C 954, *Standard Specification for Steel Drill Screws for the Application of Gypsum Panel Products or Metal Plaster Bases to Steel Studs from 0.033 in. (0.84 mm) to 0.112 in. (2.84 mm) in Thickness*. See Section R702 for installation. For all connections, screws must extend through the steel a minimum of three exposed threads.

TABLE R505.2.4
SCREW SUBSTITUTION FACTOR

SCREW SIZE	THINNEST CONNECTED STEEL SHEET (mils)	
	33	43
#8	1.0	0.67
#10	0.93	0.62
#12	0.86	0.56

For SI: 1 mil = 0.0254 mm.

❖ In providing prescriptive connection requirements, the code typically specifies the quantity of No. 8 screws necessary to resist the required design forces transferred through a given connection. If the quantity of screws required for a particular connection is large, it may be desirable to reduce the quantity of screws by substituting a larger screw size. Where No. 8 screws are specified for a steel-to-steel connection, the required quantity of screws may be reduced if a larger screw size is substituted in accordance with the reduction factors of Table R505.2.4. Likewise, if the thinnest of the connected sheets is at least 43 mils (1.1 mm), the required quantity of screws may be reduced.

R505.3 Floor construction. Cold-formed steel floors shall be constructed in accordance with this section and Figure R505.3.

❖ Load-bearing steel-framing members must be cold formed to shape from structural quality sheet steel complying with appropriate material standards. The steel grades specified have the ductility and strength to met these provisions. For more detailed information refer to the AISI *Specification for Design of Cold-Formed Steel Structural Members Commentary*.

R505.3.1 Floor to foundation or bearing wall connections. Cold-formed steel floors shall be anchored to foundations, wood sills or load-bearing walls in accordance with Table R505.3.1(1) and Figure R505.3.1(1), R505.3.1(2), R505.3.1(3), R505.3.1(4), R505.3.1(5) or R505.3.1(6). Continuous steel joists supported by interior load-bearing walls shall be constructed in accordance with Figure R505.3.1(7). Lapped steel joists shall be constructed in accordance with Figure R505.3.1(8). Fastening of steel joists to other framing members shall be in accordance with Table R505.3.1(2).

❖ This section provides minimum fastening requirements for attachment of steel floor framing to foundations, wood sills and both exterior and interior load-bearing walls. Figures R505.3.1(1) through R505.3.1(6) illustrate exterior wall conditions; thus, the corresponding connection requirements in Table R505.3.1(1) vary based on a site's basic wind speed and seismic design category. In the event these criteria give rise to different requirements from the table, the more restrictive would govern. Figures R505.3.1(7) and R505.3.1(8) illustrate the minimum connection requirements for floor joists supported by interior load-bearing walls in conjunction with the first line of Table R505.3.1(2).

FLOORS FIGURE R505.3

For SI: 1 inch = 25.4 mm, 1 foot = 304.8 mm.

**FIGURE R505.3
STEEL FLOOR CONSTRUCTION**
(continued)

FIGURE R505.3

BUILT-UP HEADER OR TRIMMER JOIST

JOIST TO HEADER JOIST

X-BRACING   BLOCKING OPTIONS   SOLID BLOCKING

For SI: 1 inch = 25.4 mm, 1 mil = 0.0254 mm.

**FIGURE R505.3—continued
STEEL FLOOR CONSTRUCTION**

❖ This figure is a schematic of a typical steel-framed floor. It illustrates several of the detailed provisions for steel floor framing.

**TABLE R505.3.1(1)
FLOOR TO FOUNDATION OR BEARING WALL CONNECTION REQUIREMENTS a,b**

FRAMING CONDITION	WIND SPEED (mph) AND EXPOSURE	
	Up to 110 A/B or 85 C or Seismic Design Categories A, B, C	Up to 110 C
Floor joist to wall track of exterior steel load-bearing wall per Figure R505.3.1(1)	2-No. 8 screws	3-No. 8 screws
Floor joist track to wood sill per Figure R505.3.1(2)	Steel plate spaced at 3' o.c., with 4-No. 8 screws and 4-10d or 6-8d common nails	Steel plate, spaced at 2' o.c., with 4-No. 8 screws and 4-10d or 6-8d common nails
Floor joist track to foundation per Figure R505.3.1(3)	$1/2''$ minimum diameter anchor bolt and clip angle spaced at 6' o.c. with 8-No. 8 screws	$1/2''$ minimum diameter anchor bolt and clip angle spaced at 4' o.c. with 8-No. 8 screws
Joist cantilever to wall track per Figure R505.3.1(4)	2-No. 8 screws per stiffener or bent plate	3-No. 8 screws per stiffener or bent plate
Joist cantilever to wood sill per Figure R505.3.1(5)	Steel plate spaced at 3' o.c., with 4-No. 8 screws and 4-10d or 6-8d common nails	Steel plate spaced at 2' o.c., with 4-No. 8 screws and 4-10d or 6-8d common nails
Joist cantilever to foundation per Figure R505.3.1(6)	$1/2''$ minimum diameter anchor bolt and clip angle spaced at 6' o.c. with 8-No. 8 screws	$1/2''$ minimum diameter anchor bolt and clip angle spaced at 4' o.c. with 8-No. 8 screws

For SI: 1 inch = 25.4 mm, 1 foot = 304.8 mm, 1 mile per hour = 1.609 km/h.

a. Anchor bolts shall be located not more than 12 inches from corners or the termination of bottom tracks (e.g., at door openings). Bolts shall extend a minimum of 15 inches into masonry or 7 inches into concrete.

b. All screw sizes shown are minimum.

❖ See the commentary for Section R505.3.1 as well as Figures R505.3.1(1) through R505.3.1(6).

**TABLE R505.3.1(2)
FLOOR FASTENING SCHEDULE a**

DESCRIPTION OF BUILDING ELEMENTS	NUMBER AND SIZE OF FASTENERS	SPACING OF FASTENERS
Floor joist to track of an interior load-bearing wall per Figures R505.3.1(7) and R505.3.1(8)	2 No. 8 screws	Each joist
Floor joist to track at end of joist	2 No. 8 screws	One per flange or two per bearing stiffener
Subfloor to floor joists	No. 8 screws	6" o.c. on edges and 10" o.c. at intermediate supports

For SI: 1 inch = 25.4 mm.

a. All screw sizes shown are minimum.

❖ In addition to the other connections covered under Section R505.3.1, Table R505.3.1(2) provides the required fastening requirements for tracks at the end of floor joists and for subfloor-to-floor joists.

For SI: 1 inch = 25.4 mm, 1 mil = 0.0254 mm.

**FIGURE R505.3.1(1)
FLOOR TO LOAD-BEARING WALL STUD CONNECTION**

❖ Figure R505.3.1(1) depicts connection requirements for steel floor joists at exterior steel bearing walls. The minimum number of screws through either the joist flange, clip angle or bent stiffener should be selected from the first row of Table R505.3.1(1) based on the seismic design category or wind speed and exposure. Similarly, if these joists cantilever beyond the wall as illustrated in Figure R505.3.1(4), the required number of screws should be selected from the fourth row of Table R505.3.1(1).

FLOORS FIGURE R505.3.1(2)

For SI: 1 inch = 25.4 mm, 1 mil = 0.0254 mm.

**FIGURE R505.3.1(2)
FLOOR TO WOOD SILL CONNECTION**

❖ Figures R505.3.1(2) and R505.3.1(5) depict the connection requirements of steel floor framing to a wood sill, which, in turn, is anchored to the foundation. Details such as the spacing of the clip angle and anchor bolt should be selected from the appropriate row of Table R505.3.1(1). See the commentary for Figure R505.3.1(1).

FIGURE R505.3.1(3)

For SI: 1 inch = 25.4 mm, 1 mil = 0.0254 mm.

**FIGURE R505.3.1(3)
FLOOR TO FOUNDATION CONNECTION**

❖ Figures R505.3.1(3) and R505.3.1(6) depict requirements for anchoring steel floor framing directly to concrete or masonry foundations. Details such as the spacing of the clip angle and anchor bolt should be selected from the appropriate row of Table R505.3.1(1). See the commentary for Figure R505.3.1(1).

FLOORS

**FIGURE R505.3.1(4)
FLOOR CANTILEVER TO LOAD-BEARING WALL CONNECTION**

❖ See the commentary for Figure R505.3.1(1).

For SI: 1 inch = 25.4 mm, 1 mil = 0.0254 mm.

**FIGURE R505.3.1(5)
FLOOR CANTILEVER TO WOOD SILL CONNECTION**

❖ See the commentary for Figures R505.3.1(1) and R505.3.1(2).

FIGURE R505.3.1(6)

For SI: 1 inch = 25.4 mm, 1 mil = 0.0254 mm.

**FIGURE R505.3.1(6)
FLOOR CANTILEVER TO FOUNDATION CONNECTION**

❖ See the commentary for Figures R505.3.1(1) and R505.3.1(3).

**FIGURE R505.3.1(7)
CONTINUOUS JOIST SPAN SUPPORTED ON STUD**

❖ See the commentary for Section R505.3.1.

**FIGURE R505.3.1(8)
LAPPED JOISTS SUPPORTED ON STUD**

❖ See the commentary for Section R505.3.1.

R505.3.2 Allowable joist spans. The clear span of cold-formed steel floor joists shall not exceed the limits set forth in Table R505.3.2. Floor joists shall have a minimum bearing length of 1.5 inches (38 mm). When continuous joists are used the interior bearing supports shall be located within 2 feet (610 mm) of mid span of the steel joists, and the individual spans shall not exceed the spans in Table R505.3.2. Bearing stiffeners shall be installed at each bearing location in accordance with Section R505.3.4 and as shown in Figure R505.3.

❖ Table R505.3.2 lists the allowable clear spans for steel floor joists. The tabulated spans are based on dead loads of 10 psf (0.48 kN/m^2) and design live loads of 30 psf (1.44 kN/m^2) or 40 psf (1.92 kN/m^2). The following example illustrates use of the tables.

Example:
Steel floor joists spaced at 16 inches (406 mm) on center support a floor live load of 40 psf (1.92 kN/m^2), a dead load of 10 psf (0.48 kN/m^2) and span 17 feet, 6 inches (5334 mm). Determine the required joist size.

Solution:
In Table R505.3.2 under 40 psf (1.92 kN/m^2), find the live load column headed 16 inches (406 mm). Read down this column to find the listed span of 17 feet, 6 inches (5334 mm) or greater. The required joist is 1000S 162-54.

When joists are continuous over interior bearing supports, the spans must be arranged as illustrated in Commentary Figure R505.3.2.

TABLE R505.3.2
ALLOWABLE SPANS FOR COLD-FORMED STEEL JOISTS[a,b]

NOMINAL JOIST SIZE	30 PSF LIVE LOAD Spacing (inches)		40 PSF LIVE LOAD Spacing (inches)	
	16	24	16	24
550S162-33	10'-7"	9'-1"	9'-7"	8'-1"
550S162-43	11'-6"	10'-0"	10'-5"	9'-1"
550S162-54	12'-4"	10'-9"	11'-2"	9'-9"
550S162-68	13'-2"	11'-6"	12'-0"	10'-6"
800S162-33	13'-3"	8'-10"	10'-7"	7'-1"
800S162-43	15'-6"	13'-7"	14'-1"	12'-3"
800S162-54	16'-8"	14'-7"	15'-2"	13'-3"
800S162-68	17'-11"	15'-7"	16'-3"	14'-2"
1000S162-43	18'-8"	15'-3"	16'-8"	13'-1"
1000S162-54	20'-1"	17'-6"	18'-3"	15'-11"
1000S162-68	21'-6"	18'-10"	19'-7"	17'-1"
1200S162-43	20'-3"	14'-1"	16'-10"	11'-3"
1200S162-54	23'-4"	19'-7"	21'-3"	17'-6"
1200S162-68	25'-1"	21'-11"	22'-10"	19'-11"

For SI: 1 inch = 25.4 mm, 1 foot = 304.8 mm, 1 pound per square foot = 0.0479 kN/m^2.
a. Deflection criteria: $L/480$ for live loads, $L/360$ for total loads.
b. Floor dead load = 10 psf.

❖ See the commentary and example for Section R505.3.2.

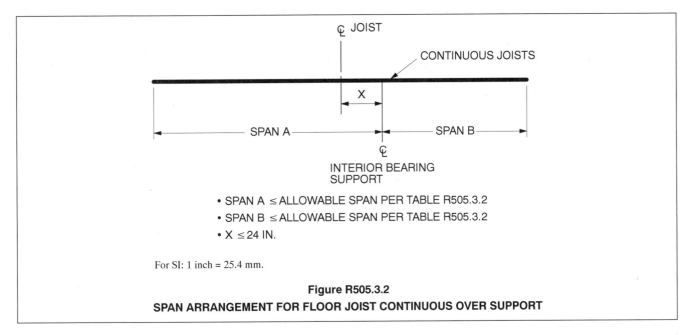

Figure R505.3.2
SPAN ARRANGEMENT FOR FLOOR JOIST CONTINUOUS OVER SUPPORT

R505.3.3 Joist bracing. The top flanges of steel joists shall be laterally braced by the application of floor sheathing fastened to the joists in accordance with Table R505.3.1(2). Floor joists with spans that exceed 12 feet (3658 mm) shall have the bottom flanges laterally braced in accordance with one of the following:

1. Gypsum board installed with minimum No. 6 screws in accordance with Section R702.
2. Continuous steel strapping installed in accordance with Figure R505.3. Steel straps shall be at least 1.5 inches (38 mm) in width and 33 mils (0.84 mm) in thickness. Straps shall be fastened to the bottom flange at each joist with at least one No. 8 screw and shall be fastened to blocking with at least two No. 8 screws. Blocking or bridging (X-bracing) shall be installed between joists in-line with straps at a maximum spacing of 12 feet (3658 mm) measured perpendicular to the joist run and at the termination of all straps.

❖ The application of floor sheathing to steel floor joists in accordance with Table R505.3.1(2) of the code allows the transfer of loads to the joists and provides the necessary lateral bracing for the top flanges of those joists. Steel floor joists with spans greater than 12 feet (3658 mm) are more prone to overturning and therefore must also have the bottom flanges laterally braced either with gypsum board or continuous steel strapping as illustrated in Figure R505.3 of the code. Strap bracing (see Figure R505.3) requires the installation of solid blocking or bridging (X-bracing) in line with the straps at the termination of all straps and at a maximum spacing of 12 feet (3658 mm) measured perpendicular to the joist run.

R505.3.4 Bearing stiffeners. Bearing stiffeners shall be installed at all bearing locations for steel floor joists. A bearing stiffener shall be fabricated from a minimum 33 mil (0.84 mm) C-section or 43 mil (1.09 mm) track section. Each stiffener shall be fastened to the web of the joist with a minimum of four No. 8 screws equally spaced as shown in Figure R505.3.4. Stiffeners shall extend across the full depth of the web and shall be installed on either side of the web.

❖ Stiffeners (also referred to as "transverse stiffeners" or "web stiffeners") at points of bearing provide resistance to web crippling in the joist. These stiffeners are illustrated in Figure R505.3.4, but a stiffener can be installed on either side of the web of a joist as depicted in Figures R505.3.1(1) through R505.3.1(8).

R505.3.5 Cutting and notching. Flanges and lips of load-bearing steel floor framing members shall not be cut or notched.

❖ Flanges and lips of load-bearing steel floor framing members must not be cut or notched because this would affect their structural integrity. The only holes permitted in steel floor joists are those that conform to Section R505.2.

R505.3.6 Hole patching. Web holes for 800S162-33, 1000S162-43, 1200S162-43 and 1200S162-54 nominal joist sizes with dimensions conforming to Section R505.2 that are closer than 10 inches (305 mm) from the edge of the hole to the edge of the bearing surface shall be patched with a solid steel plate, C-section or track section in accordance with Figure R505.3.6. The steel patch shall be of a minimum thickness as the receiving member and shall extend at least 1 inch (25.4 mm) beyond all edges of the hole. The steel patch shall be fastened to the web with No. 8 screws (minimum) spaced no greater than 1 inch (25.4 mm) center-to-center along the edges of the patch, with a minimum edge distance of 0.5 inch (12.7 mm).

❖ See the commentary for Section R505.2.

R505.3.7 Floor cantilevers. Floor cantilevers shall not exceed 24 inches (610 mm) as illustrated in Figure R505.3. The cantilever back-span shall extend a minimum of 6 feet (1830 mm)

**FIGURE R505.3.4
BEARING STIFFENER**

❖ See the commentary for Section R505.3.4.

within the building, and shall be fastened to a bearing condition in accordance with Section R505.3.1. Floor cantilevers shall be permitted only on the second floor of a two-story building or the first floor of a one-story building. Floor framing that is cantilevered and supports the cantilevered floor only shall consist of single joist members in accordance with Section R505.3.2. Floor framing that is cantilevered and supports the cantilevered floor and the roof framing load above shall consist of double joist members of the same size and material thickness as that for single joist members in accordance with Section R505.3.2, and shall be fastened web-to-web with minimum No. 8 screws at 24 inches (610 mm) maximum on-center spacing top and bottom. Built-up floor framing consisting of a C-section inside a track section, fastened at the top and bottom flanges by minimum No. 8 screws at 24 inches (610 mm) maximum on center spacing, is permitted in lieu of the web-to-web double joist method.

❖ Figure R505.3 illustrates the maximum permitted cantilever. Floor cantilevers are permitted on the second floor of two-story buildings or the first floor of one-story buildings.

Floor framing that is cantilevered but supports only the cantilevered floor may be a single joist conforming to the allowable span requirements of Section R505.3.2. Cantilevered floor framing that also supports the roof framing above must consist of double joist members of the same size and material thickness as that for single joist members in accordance with the allowable span requirements of Section R505.3.2. Built-up floor framing consisting of a C-section inside a track section is permitted as an option to the web-to-web double joist member.

R505.3.8 Splicing. Joists and other structural members shall not be spliced. Splicing of tracks shall conform with Figure R505.3.8.

❖ Splicing of structural members is not permitted except when lapped joists occur at interior bearing points.

Splicing of tracks is allowed since they are not permitted to act as load-carrying members (see Figure R505.3.8). The exception to this would be under lateral loads from wind and earthquakes. In this case, the floor system must act as a diaphragm, with tracks serving as diaphragm chords. In order to provide continuity and carry the anticipated chord force, the splice must be as shown in Figure R505.3.8.

FLOORS

**FIGURE R505.3.6
HOLE PATCH**

❖ See the commentary for Section R505.2.

**FIGURE R505.3.8
TRACK SPLICE**

❖ See the commentary for Section R505.3.8.

R505.3.9 Framing of openings. Openings in floor framing shall be framed with header and trimmer joists. Header joist spans shall not exceed 8 feet (2438 mm). Header and trimmer joists shall be fabricated from joist and track sections, which shall be of a minimum size and thickness as the adjacent floor joists and shall be installed in accordance with Figure R505.3. Each header joist shall be connected to trimmer joists with a minimum of four 2-inch-by-2-inch (51 mm by 51 mm) clip angles. Each clip angle shall be fastened to both the header and trimmer joists with four No. 8 screws, evenly spaced, through each leg of the clip angle. The clip angles shall have a steel thickness not less than that of the floor joist.

❖ Recognizing that floor openings are necessary for such things as stairways and utility chases, the code provides prescriptive criteria for framing these openings. These requirements for openings in floors are illustrated in Figure R505.3.

SECTION R506
CONCRETE FLOORS (ON GROUND)

R506.1 General. Concrete slab-on-ground floors shall be a minimum 3.5 inches (89 mm) thick (for expansive soils, see Section R403.1.8). The specified compressive strength of concrete shall be as set forth in Section R402.2.

❖ For weathering requirements, see the commentary for Sections R301.2 and R402.2. Commentary Figure R506.1 shows the requirements of this section.

R506.2 Site preparation. The area within the foundation walls shall have all vegetation, top soil and foreign material removed.

❖ Removal of construction debris and foreign materials, such as lumber formwork, stakes, tree stumps and other vegetation, limits the attraction of termites, insects and vermin. Top soil and soil vegetation should also be removed, since such top soil is generally loosely compacted or full of vegetation to such an extent as to allow soil settlement upon decay of the vegetation. For concrete slabs placed on uncompacted fill or on large quantities of foreign materials, differential settlement may take place due to subsequent compaction of the soil, which can result in cracking of the floor slab and the interior wall/ceiling finishes.

R506.2.1 Fill. Fill material shall be free of vegetation and foreign material. The fill shall be compacted to assure uniform support of the slab, and except where approved, the fill depths shall not exceed 24 inches (610 mm) for clean sand or gravel and 8 inches (203 mm) for earth.

❖ To minimize differential settlement due to consolidation of uncompacted fill and the problems associated with such differential settlement, any fill beneath a concrete slab is required to be compacted. Properly compacted fill, besides minimizing settlement, increases the soil load-bearing characteristics and soil stability and reduces water penetration. The amount of compaction for fill is not specifically stipulated. Generally, fill soils should be compacted to 95-percent maximum density as determined by a Standard Proctor Test (ASTM D-698). Compaction requirements for expansive soils should be determined from an engineering analysis.

Compaction may be performed using equipment appropriate to the type of material being compacted, with lifts not exceeding 8 inches (203 mm). In general, thinner layers produce better compaction of fill, regardless of the type of soil being compacted.

R506.2.2 Base. A 4-inch-thick (102 mm) base course consisting of clean graded sand, gravel, crushed stone or crushed blast-furnace slag passing a 2-inch (51 mm) sieve shall be placed on the prepared subgrade when the slab is below grade.

Exception: A base course is not required when the concrete slab is installed on well-drained or sand-gravel mixture soils classified as Group I according to the United Soil Classification System in accordance with Table R405.1.

❖ Slabs need to be protected from the penetration of water and water vapor from below to prevent damage to interior finish material. Additionally, the effectiveness of thermal insulation at the slab may be adversely affected because of moisture. A 4-inch (102 mm) minimum granular base course is placed over the fill or undisturbed soil for slabs below grade to provide a capillary stop for water rising through the soil and into the slab. A base course is especially necessary when the site soil is other than gravel or clean sand. Both surface and groundwater problems must be addressed. Proper site preparation, selection of adequate fill and base course materials, and the installation of a vapor barrier minimize potential moisture problems associated with slab-on-grade construction.

R506.2.3 Vapor retarder. A 6 mil (0.006 inch; 152 μm) polyethylene or approved vapor retarder with joints lapped not less

Figure R506.1
CONCRETE SLAB-ON-GROUND REQUIREMENTS

than 6 inches (152 mm) shall be placed between the concrete floor slab and the base course or the prepared subgrade where no base course exists.

Exception: The vapor retarder may be omitted:
1. From garages, utility buildings and other unheated accessory structures.
2. From driveways, walks, patios and other flatwork not likely to be enclosed and heated at a later date.
3. Where approved by the building official, based on local site conditions.

❖ While good quality, uncracked concrete is practically impermeable to the passage of water (unless the water is under a considerable pressure), concrete is not impervious to the passage of water vapors. If the surface of the slab is not sealed, water vapor will pass through the slab. If a floor finish such as linoleum, vinyl tile, wood flooring, or any type of covering is placed on top of the slab, the moisture will be trapped in the slab. If the floor finish is adhered to the concrete, it may eventually loosen and buckle or blister.

Many of the moisture problems associated with enclosed slabs-on-ground can be minimized by installing a vapor retarder. When required, vapor retarders of either single-layer membranes or multiple-layer membranes should be acceptable provided that the vapor retarder is properly installed with lapped joints and the barrier is not punctured during construction.

Bibliography

The following resource materials are referenced in this chapter or are relevant to the subject matter addressed in this chapter.

AISI (1996), *Specification for Design of Cold-formed Steel Structural Members*. Washington, DC: American Iron and Steel Institute, 1996.

AFPA B 93, *Span Tables for Joists and Rafters*. Washington, DC: American Forest and Paper Association, 1993.

AFPA NDS 01, *National Design Specification for Wood Construction with 2001 Supplement*. Washington, DC: American Forest and Paper Association, 2001.

ANSI A208.1-99, *Particleboard*. New York: American National Standards Institute, 1999.

ASTM A 653/A 653M-00, *Standard Specification for Steel Sheet, Zinc-Coated Galvanized or Zinc-Iron Alloy-Coated (Galvannealed) by the Hot-Dip Process*. West Conshohocken, PA: ASTM International, 2000.

ASTM A 792M-99, *Standard Specification for Steel Sheet, 55% Aluminum-Zinc Alloy-Coated by the Hot-Dip Process*. West Conshohocken, PA: ASTM International, 1999.

ASTM A 875/875M-01a, *Standard Specification for Steel Sheet, Zinc-5% Aluminum Alloy-Coated by the Hot-Dip Process*. West Conshohocken, PA: ASTM International, 2001.

ASTM B 633-98e^{01}, *Standard Specification for Electrodeposited Coatings of Zinc on Iron and Steel*. West Conshohocken, PA: ASTM International, 1998.

ASTM C 954-00, *Standard Specification for Steel Drill Screws for the Application of Gypsum Panel Products or Metal Plaster Bases to Steel Studs from 0.033 in. (0.84 mm) to 0.112 in. (2.84 mm) in Thickness*. West Conshohocken, PA: ASTM International, 2000.

ASTM D 5055-00, *Standard Specification for Establishing and Monitoring Structural Capacities of Prefabricated Wood I-Joists*. West Conshohocken, PA: ASTM International, 2000.

AWPA C22-96, *Lumber and Plywood for Permanent Wood Foundations — Preservative Treatment by Pressure Processes*. Granbury, TX: American Wood-Preservers' Association, 1996.

CSA 0325.0-92, *Construction Sheathing*. Rexdale, Ontario, Canada: Canadian Standards Association, 1992.

DOC PS1-95, *Construction and Industrial Plywood*. Gaithersburg, MD: United States Department of Commerce, National Institute of Standards and Technology, 1995.

DOC PS2-92, *Performance Standard for Wood-Based Structural-Use Panels*. Gaithersburg, MD: United States Department of Commerce, National Institute of Standards and Technology, 1992.

DOC PS20-99, *American Softwood Lumber Standard*. Gaithersburg, MD: United States Department of Commerce, National Institute of Standards and Technology, 1999.

FEMA FIA-TB #2—93, *Flood Resistant Material Requirements*. Washington, DC: Federal Emergency Management Agency, 1993.

SAE J78-1998, *Steel Self-Drilling Tapping Screws*. Warrendale, PA: Society of Automotive Engineers, 1998.

TPI 1—2002, *National Design Standard for Metal-Plate-Connected Wood Truss Construction*. Madison, WI: Truss Plate Institute, 2002.

The following and other materials are available online at http://www.fema.gov/mit/techbul.htm

FEMA. NFIP Technical Bulletin Series. Washington, DC: National Flood Insurance Program.

Chapter 6:
Wall Construction

General Comments

Chapter 6 contains provisions that regulate the design and construction of walls. The primary concern of this chapter is the structural integrity of wall construction.

Section R601 contains the scope as well as the performance expectations for walls. Section R602 addresses wood wall framing. Section R603 addresses steel wall framing. Section R604 specifies requirements for wood-structural-panel wall sheathing. Section R605 provides requirements for particleboard used as wall sheathing. Section R606 states the general requirements pertaining to masonry wall construction. Section R607 addresses unit masonry walls. Section R608 covers walls of multiple-wythe masonry. Section R609 provides grouted masonry provisions. Section R610 specifies requirements for wall panels of glass unit masonry. Section R611 addresses prescriptive requirements for the Insulated Concrete Form system, while Section R612 contains provisions for concrete wall construction, including conventionally formed concrete. Section R613 contains the criteria for the performance of exterior windows and doors.

In certain instances a wall must have a fire-resistance rating. Where this is necessary, the wall system must be a tested assembly, and any conditions specific to the installation of said assembly must apply as well. Examples of these provisions include:

- Exterior walls based on location on property (Section R302).
- Walls serving as dwelling unit separations in two-family dwellings (Section R317.1).
- Common wall between townhouses (Section R317.2, Exception).

Purpose

Chapter 6 provisions for the design and construction of walls and partitions contain prescriptive requirements that provide for the support of the required dead, live and snow loads. In addition, these walls are a key component in resisting the lateral forces attributable to earthquakes and wind. Where the hazards are greater, the requirements for braced wall lines and/or shear walls are correspondingly more stringent.

So that windows and glass doors do not compromise the integrity of the building envelope, this chapter also includes performance expectations for these elements. The proper testing and labeling of these elements are key to verifying their performance.

SECTION R601
GENERAL

R601.1 Application. The provisions of this chapter shall control the design and construction of all walls and partitions for all buildings.

❖ This chapter concerns wall and partition construction.

R601.2 Requirements. Wall construction shall be capable of accommodating all loads imposed according to Section R301 and of transmitting the resulting loads to the supporting structural elements.

❖ This is a general performance statement, requiring that walls support the required design loads and provide an adequate load path to supporting elements.

R601.2.1 Compressible floor-covering materials. Compressible floor-covering materials that compress more than $1/32$ inch (0.794 mm) when subjected to 50 pounds (23 kg) applied over 1 inch square (645 mm) of material and are greater than $1/8$ inch (3.2 mm) in thickness in the uncompressed state shall not extend beneath walls, partitions or columns, which are fastened to the floor.

❖ While it is preferable to fasten walls directly to the supporting structure, this provision allows them to be installed over finish floor materials that meet the specified criteria.

SECTION R602
WOOD WALL FRAMING

R602.1 Identification. Load-bearing dimension lumber for studs, plates and headers shall be identified by a grade mark of a lumber grading or inspection agency that has been approved by an accreditation body that complies with DOC PS 20. In lieu of a grade mark, a certification of inspection issued by a lumber grading or inspection agency meeting the requirements of this section shall be accepted.

❖ Wood materials performing a load-carrying function must conform to a minimum quality control standard. In order to verify this, all load-bearing lumber must be properly identified to indicate that the grades and species meet the minimum requirements specified in this chapter. Commentary Figure R602.1 shows the information contained in a grade mark. For more examples, see Commentary Figure R502.1. For examples of marks for wood structural panels, see Commentary Figure R503.2.1.

```
                    INTERPRETING GRADE STAMPS              A.  THE TRADEMARK INDICATES AGENCY QUALITY
                                                              SUPERVISION.
          MOST GRADE STAMPS, EXCEPT
          THOSE FOR ROUGH LUMBER OR                        B.  MILL IDENTIFICATION – FIRM NAME, BRAND OR
          HEAVY TIMBERS, CONTAIN FIVE                          ASSIGNED MILL NUMBER.
          BASIC ELEMENTS:
                                                           C.  GRADE DESIGNATION – GRADE NAME, NUMBER OR
                                                               ABBREVIATION.
              (B) 1 2  STAND (C)
                                                           D.  SPECIAL IDENTIFICATION – INDICATES SPECIES
               A B C   S-DRY                                   INDIVIDUALLY OR IN COMBINATION.
                        (E)    /D\ (D)
                (A)           /FIR\                        E.  CONDITION OF SEASONING AT TIME OF SURFACING:
                                                               S-DRY  – 19% MAX.
                                                                        MOISTURE CONTENT
                                                               MC 15   – 15% MAX.
                                                                        MOISTURE CONTENT
                                                               S-GRN   – OVER 19%
                                                                        MOISTURE CONTENT (UNSEASONED)
```

**Figure R602.1
GRADE STAMP EXAMPLE FOR DIMENSIONAL LUMBER**

R602.1.1 End-jointed lumber. Approved end-jointed lumber identified by a grade mark conforming to Section R602.1 may be used interchangeably with solid-sawn members of the same species and grade.

❖ End-jointed lumber that has been fabricated in accordance with nationally recognized standards is permitted to be used interchangeably with solid-sawn members, provided that the species and grade of the wood are comparable. Commentary Figure 602.1.1 shows end-jointed lumber, which is also referred to as "finger-jointed." Also see the commentary for Section R502.1.3.

R602.1.2 Structural glued laminated timbers. Glued laminated timbers shall be manufactured and identified as required in AITC A190.1 and ASTM D3737.

❖ Glued laminated (gluelam) timbers are engineered wood elements consisting of layers of plywood plies with varying grades of plywood through the thickness of the member. The strength of the members and the use of the members (beams, studs, etc.) are based on different combinations of plywood veneer grades, called layups. The design, testing, and manufacture of different layups for different applications are standardized in the referenced standards. The standards also address the identification of these timbers, which is useful at the construction site during inspection.

R602.2 Grade. Studs shall be a minimum No. 3, standard or stud grade lumber.

Exception: Bearing studs not supporting floors and nonbearing studs may be utility grade lumber, provided the studs are spaced in accordance with Table R602.3(5).

❖ A minimum grade of No. 3, standard, or stud grade lumber is specified for studs in conventional wood-frame walls in order to maintain minimum load-carrying capabilities. The maximum size of standard grade lumber is limited to 2 to 4 inches (51 to 102 mm) in thickness and 4 inches (102 mm) in width. Thus, a 4-inch by 6-inch (102 mm by 152 mm) piece of dimensional lumber would not be available in the standard grade.

**Figure R602.1.1
END-JOINTED LUMBER—FINGER JOINT**

WALL CONSTRUCTION

To allow more economical use of lumber for bearing studs not supporting floors and nonbearing studs, the exception permits the use of utility grade lumber, provided that the studs are spaced according to the limits established in Table R602.3(5). Note "a" of the table places a more restrictive limit on utility grade stud spacing where permitted in a load-bearing application.

R602.3 Design and construction. Exterior walls of wood-frame construction shall be designed and constructed in accordance with the provisions of this chapter and Figures R602.3(1) and R602.3(2) or in accordance with AF&PA's NDS. Components of exterior walls shall be fastened in accordance with Table R602.3(1) through R602.3(4). Exterior walls covered with foam plastic sheathing shall be braced in accordance with Section R602.10. Structural sheathing shall be fastened directly to structural framing members.

❖ Figures R602.3(1) and R602.3(2) in conjunction with the minimum nailing requirements of Tables R602.3(1) and R602.3(2) provide typical wall-framing details for construction of exterior walls. Walls must be in accordance with this section or American Forest & Paper Association's *National Design Specification* (NDS).

Typically, foam-plastic material used for insulation is nailed to the outside face of walls as sheathing, but the materials are usually not structural sheathing and therefore provide no lateral bracing for the building. Thus, the code calls out requirements for structural sheathing to remind the user that foam-plastic sheathing should not be confused with elements that provide sufficient lateral bracing.

TABLE R602.3(1)
FASTENER SCHEDULE FOR STRUCTURAL MEMBERS

DESCRIPTION OF BUILDING ELEMENTS	NUMBER AND TYPE OF FASTENER[a,b,c,d]	SPACING OF FASTENERS
Joist to sill or girder, toe nail	3-8d	—
1" × 6" subfloor or less to each joist, face nail	2-8d 2 staples, $1^3/_4$"	— —
2" subfloor to joist or girder, blind and face nail	2-16d	—
Sole plate to joist or blocking, face nail	16d	16" o.c.
Top or sole plate to stud, end nail	2-16d	—
Stud to sole plate, toe nail	3-8d or 2-16d	—
Double studs, face nail	10d	24" o.c.
Double top plates, face nail	10d	24" o.c.
Sole plate to joist or blocking at braced wall panels	3-16d	16" o.c.
Double top plates, minimum 24-inch offset of end joints, face nail in lapped area	8-16d	—
Blocking between joists or rafters to top plate, toe nail	3-8d	—
Rim joist to top plate, toe nail	8d	6" o.c.
Top plates, laps at corners and intersections, face nail	2-10d	—
Built-up header, two pieces with $1/_2$" spacer	16d	16" o.c. along each edge
Continued header, two pieces	16d	16" o.c. along each edge
Ceiling joists to plate, toe nail	3-8d	—
Continuous header to stud, toe nail	4-8d	—
Ceiling joist, laps over partitions, face nail	3-10d	—
Ceiling joist to parallel rafters, face nail	3-10d	—
Rafter to plate, toe nail	2-16d	—
1" brace to each stud and plate, face nail	2-8d 2 staples, $1^3/_4$"	— —
1" × 6" sheathing to each bearing, face nail	2-8d 2 staples, $1^3/_4$"	— —
1" × 8" sheathing to each bearing, face nail	2-8d 3 staples, $1^3/_4$"	— —
Wider than 1" × 8" sheathing to each bearing, face nail	3-8d 4 staples, $1^3/_4$"	— —

(continued)

TABLE R602.3(1)—continued
FASTENER SCHEDULE FOR STRUCTURAL MEMBERS

Built-up corner studs		10d	24″ o.c.
Built-up girders and beams, 2-inch lumber layers		10d	Nail each layer as follows: 32″ o.c. at top and bottom and staggered. Two nails at ends and at each splice.
2″ planks		2-16d	At each bearing
Roof rafters to ridge, valley or hip rafters: toe nail face nail		4-16d 3-16d	— —
Rafter ties to rafters, face		3-8d	—
Wood structural panels, subfloor, roof and wall sheathing to framing, and particleboard wall sheathing to framing			
$5/16″$-$1/2″$	6d common nail (subfloor, wall) 8d common nail (roof)[f]	6	12[g]
$19/32″$-$1″$	8d common nail	6	12[g]
$1 1/8″$-$1 1/4″$	10d common nail or 8d deformed nail	6	12

DESCRIPTION OF BUILDING MATERIALS	DESCRIPTION OF FASTENER[b,c,d,e]	SPACING OF FASTENERS	
		Edges (inches)[i]	Intermediate supports[c,e] (inches)
Other wall sheathing[h]			
$1/2″$ regular cellulosic fiberboard sheathing	$1 1/2″$ galvanized roofing nail 6d common nail staple 16 ga., $1 1/2$ long	3	6
$1/2″$ structural cellulosic fiberboard sheathing	$1 1/2″$ galvanized roofing nail 8d common nail staple 16 ga., $1 1/2$ long	3	6
$25/32″$ structural cellulosic fiberboard sheathing	$1 3/4″$ galvanized roofing nail 8d common nail staple 16 ga., $1 3/4$ long	3	6
$1/2″$ gypsum sheathing	$1 1/2″$ galvanized roofing nail; 6d common nail; staple galvanized, $1 1/2″$ long; $1 1/4″$ screws, Type W or S	4	8
$5/8″$ gypsum sheathing	$1 3/4″$ galvanized roofing nail; 8d common nail; staple galvanized, $1 5/8″$ long; $1 5/8″$ screws, Type W or S	4	8
Wood structural panels, combination subfloor underlayment to framing			
$3/4″$ and less	6d deformed nail or 8d common nail	6	12
$7/8″$-$1″$	8d common nail or 8d deformed nail	6	12
$1 1/8″$-$1 1/4″$	10d common nail or 8d deformed nail	6	12

For SI: 1 inch = 25.4 mm, 1 foot = 304.8 mm, 1 mile per hour = 1.609 km/h.

a. All nails are smooth-common, box or deformed shanks except where otherwise stated. Nails used for framing and sheathing connections shall have minimum average bending yield strengths as shown: 80 ksi (551 MPa) for shank diameter of 0.192 inch (20d common nail), 90 ksi (620 MPa) for shank diameters larger than 0.142 inch but not larger than 0.177 inch, and 100 ksi (689 MPa) for shank diameters of 0.142 inch or less.

b. Staples are 16 gage wire and have a minimum $7/16$-inch on diameter crown width.

c. Nails shall be spaced at not more than 6 inches on center at all supports where spans are 48 inches or greater.

d. Four-foot-by-8-foot or 4-foot-by-9-foot panels shall be applied vertically.

e. Spacing of fasteners not included in this table shall be based on Table R602.3(1).

f. For regions having basic wind speed of 110 mph or greater, 8d deformed nails shall be used for attaching plywood and wood structural panel roof sheathing to framing within minimum 48-inch distance from gable end walls, if mean roof height is more than 25 feet, up to 35 feet maximum.

g. For regions having basic wind speed of 100 mph or less, nails for attaching wood structural panel roof sheathing to gable end wall framing shall be spaced 6 inches on center. When basic wind speed is greater than 100 mph, nails for attaching panel roof sheathing to intermediate supports shall be spaced 6 inches on center for minimum 48-inch distance from ridges, eaves and gable end walls; and 4 inches on center to gable end wall framing.

h. Gypsum sheathing shall conform to ASTM C 79 and shall be installed in accordance with GA 253. Fiberboard sheathing shall conform to either AHA 194.1 or ASTM C 208.

i. Spacing of fasteners on floor sheathing panel edges applies to panel edges supported by framing members and at all floor perimeters only. Spacing of fasteners on roof sheathing panel edges applies to panel edges supported by framing members and at all roof plane perimeters. Blocking of roof or floor sheathing panel edges perpendicular to the framing members shall not be required except at intersection of adjacent roof planes. Floor and roof perimeter shall be supported by framing members or solid blocking.

❖ The fastener schedule provides minimum nailing requirements (i.e., size, spacing) for connecting building elements used in wood frame construction. For wood structural panels, both edge nailing and intermediate (field) nailing is specified. In addition to the nailing for wood structural panels, fasteners are specified for gypsum wall sheathing, cellulosic fiberboard wall sheathing and combination subfloor underlayment.

WALL CONSTRUCTION

TABLE R602.3(2)
ALTERNATE ATTACHMENTS

NOMINAL MATERIAL THICKNESS (inches)	DESCRIPTION[a,b] OF FASTENER AND LENGTH (inches)	SPACING[c] OF FASTENERS Edges (inches)	Intermediate supports (inches)
Wood structural panels subfloor, roof and wall sheathing to framing and particleboard wall sheathing to framing[f]			
5/16	0.097 - 0.099 Nail 1 1/2 Staple 15 ga. 1 3/8 Staple 16 ga. 1 3/4	6	12
3/8	Staple 15 ga. 1 3/8	6	12
	0.097 - 0.099 Nail 1 1/2	4	10
	Staple 16 ga. 1 3/4	6	12
15/32 and 1/2	Staple 15 ga. 1 1/2	6	12
	0.097 - 0.099 Nail 1 5/8	3	6
	Staple 16 ga. 1 3/4	6	12
19/32 and 5/8	0.113 Nail 1 7/8 Staple 15 and 16 ga. 1 5/8	6	12
	0.097 - 0.099 Nail 1 3/4	3	6
23/32 and 3/4	Staple 14 ga. 1 3/4	6	12
	Staple 15 ga. 1 3/4	5	10
	0.097 - 0.099 Nail 1 7/8	3	6
	Staple 16 ga. 2	4	8
1	Staple 14 ga. 2	5	10
	0.113 Nail 2 1/4, Staple 15 ga. 2	4	8
	0.097 - 0.099 Nail 2 1/8	3	6

NOMINAL MATERIAL THICKNESS (inches)	DESCRIPTION[a,b] OF FASTENER AND LENGTH	SPACING[c] OF FASTENERS Edges (inches)	Body of panel[d] (inches)
Floor underlayment; plywood-hardboard-particleboard[f]			
Plywood			
1/4 and 5/16	1 1/4 ring or screw shank nail—minimum 12 1/2 ga. (0.099") shank diameter	3	6
	Staple 18 ga., 7/8, 3/16 crown width	2	5
11/32, 3/8, 15/32 and 1/2	1 1/4 ring or screw shank nail—minimum 12 1/2 ga. (0.099) shank diameter	6	8[e]
19/32, 5/8, 23/32 and 3/4	1 1/2 ring or screw shank nail—minimum 12 1/2 ga. (0.099) shank diameter	6	12
	Staple 16 ga. 1 1/4	6	8
Hardboard[f]			
0.200	1 1/2 long ring-grooved underlayment nail	6	6
	4d cement-coated sinker nail	6	6
	Staple 18 ga., 7/8 long (plastic coated)	3	6
Particleboard			
1/4	4d ring-grooved underlayment nail	3	6
	Staple 18 ga., 7/8 long, 3/16 crown	3	6
3/8	6d ring-grooved underlayment nail	6	10
	Staple 16 ga., 1 1/8 long, 3/8 crown	3	6
1/2, 5/8	6d ring-grooved underlayment nail	6	10
	Staple 16 ga., 1 5/8 long, 3/8 crown	3	6

For SI: 1 inch = 25.4 mm.
a. Nail is a general description and may be T-head, modified round head or round head.
b. Staples shall have a minimum crown width of 7/16-inch on diameter except as noted.
c. Nails or staples shall be spaced at not more than 6 inches on center at all supports where spans are 48 inches or greater. Nails or staples shall be spaced at not more than 12 inches on center at intermediate supports for floors.
d. Fasteners shall be placed in a grid pattern throughout the body of the panel.
e. For 5-ply panels, intermediate nails shall be spaced not more than 12 inches on center each way.
f. Hardboard underlayment shall conform to ANSI/AHA A135.4.

❖ This table offers alternatives to the nailing specified for wood structural panels in Table R602.3(1).

TABLE R602.3(3)
ALLOWABLE STUD SPACING FOR WOOD STRUCTURAL PANEL WALL SHEATHING

PANEL SPAN RATING	PANEL NOMINAL THICKNESS (inch)	MAXIMUM STUD SPACING (inches) Siding nailed to:[a]	
		Stud	Sheathing
12/0, 16/0, 20/0, or wall —16 o.c.	$5/16$, $3/8$	16	16[b]
24/0, 24/16, 32/16 or wall—24 o.c.	$3/8$, $7/16$, $15/32$, $1/2$	24	24[c]

For SI: 1 inch = 25.4 mm.

a. Blocking of horizontal joints shall not be required.
b. Plywood sheathing $3/8$-inch thick or less shall be applied with long dimension across studs.
c. Three-ply plywood panels shall be applied with long dimension across studs.

❖ Table R602.3(3) gives maximum permissable stud spacing based on the thicknesses of wood structural panels used as sheathing. Although the span ratings are listed, they are not applicable for determining the spacing of vertical framing members. Some wood structural panels are manufactured for use as wall sheathing only. The maximum stud spacing for these specialty panels is specified on the trademark. All wood structural panels may be installed with the face veneers either parallel or perpendicular to the framing, except where thinner panels are used as a nailing base for cladding, as indicated in Notes "b" and "c." Face veneers must be perpendicular to studs.

TABLE R602.3(4)
ALLOWABLE SPANS FOR PARTICLEBOARD WALL SHEATHING[a]

THICKNESS (inch)	GRADE	STUD SPACING (inches)	
		When siding is nailed to studs	When siding is nailed to sheathing
$3/8$	M-1 Exterior glue	16	—
$1/2$	M-2 Exterior glue	16	16

For SI: 1 inch = 25.4 mm.

a. Wall sheathing not exposed to the weather. If the panels are applied horizontally, the end joints of the panel shall be offset so that four panels corners will not meet. All panel edges must be supported. Leave a $1/16$-inch gap between panels and nail no closer than $3/8$ inch from panel edges.

❖ The allowable spans for particleboard wall sheathing listed in Table R602.3(4) are for wall sheathing that is not exposed to weather. The panels may be applied with the long dimension parallel or perpendicular to the studs.

TABLE R602.3(5)
SIZE, HEIGHT AND SPACING OF WOOD STUDS[a]

STUD SIZE (inches)	BEARING WALLS					NONBEARING WALLS	
	Laterally unsupported stud height[a] (feet)	Maximum spacing when supporting roof and ceiling only (inches)	Maximum spacing when supporting one floor, roof and ceiling (inches)	Maximum spacing when supporting two floors, roof and ceiling (inches)	Maximum spacing when supporting one floor only (inches)	Laterally unsupported stud height[a] (feet)	Maximum spacing (inches)
2 × 3[b]	—	—	—	—	—	10	16
2 × 4	10	24	16	—	24	14	24
3 × 4	10	24	24	16	24	14	24
2 × 5	10	24	24	—	24	16	24
2 × 6	10	24	24	16	24	20	24

For SI: 1 inch = 25.4 mm.

a. Listed heights are distances between points of lateral support placed perpendicular to the plane of the wall. Increases in unsupported height are permitted where justified by analysis.
b. Shall not be used in exterior walls.

❖ This table provides the allowed stud spacing in bearing walls up to 10 feet high as stated in Section R602.3.1. As described in Note "a," increases in unsupported height is permitted where justified by an engineering analysis. The "unsupported height" of a stud is really the distance between points of lateral support, which generally occurs at each floor level.

WALL CONSTRUCTION

FIGURE R602.3(1)

For SI: 1 inch = 25.4 mm.

FIGURE R602.3(1)
TYPICAL WALL, FLOOR AND ROOF FRAMING

❖ This is a schematic wood-frame building section and depicts code provisions for wood floor systems as well as wall construction.

2003 INTERNATIONAL RESIDENTIAL CODE® COMMENTARY—VOLUME 1   6-7

FIGURE R602.3(2)

FIGURE R602.3(2)
FRAMING DETAILS

❖ This figure provides an elevation of typical wood-frame wall construction and depicts several code provisions as well as illustrating features such as cripple walls.

WALL CONSTRUCTION

R602.3.1 Stud size, height and spacing. The size, height and spacing of studs shall be in accordance with Table R602.3.(5).

Exceptions:

1. Utility grade studs shall not be spaced more than 16 inches (406 mm) on center, shall not support more than a roof and ceiling, and shall not exceed 8 feet (2438 mm) in height for exterior walls and load-bearing walls or 10 feet (3048 mm) for interior nonload-bearing walls.

2. Studs more than 10 feet (3048 mm) in height which are in accordance with Table R602.3.1.

❖ Table R602.3(5) lists required stud spacing based on the stud size and loading condition.

Exception 1 addresses the fact that utility grade lumber is lower in strength and quality and therefore requires tighter spacing and shorter lengths in exterior walls. Note that the minimum grade lumber for application of Table R602.3(5) is, by implication, stud grade or standard grade.

As noted in Exception 2, if the bearing wall height exceeds 10 feet (3048 mm), stud size is determined using Table 602.3.1. A design is required if the limits given in the table's title are exceeded or if Note "a" applies.

TABLE R602.3.1
MAXIMUM ALLOWABLE LENGTH OF WOOD WALL STUDS EXPOSED TO WIND SPEEDS OF 100 MPH OR LESS IN SEISMIC DESIGN CATEGORIES A, B, C and D_1 [b,c]

HEIGHT (feet)	ON-CENTER SPACING (inches)			
	24	16	12	8
Supporting a roof only				
>10	2×4	2×4	2×4	2×4
12	2×6	2×4	2×4	2×4
14	2×6	2×6	2×6	2×4
16	2×6	2×6	2×6	2×4
18	NA[a]	2×6	2×6	2×6
20	NA[a]	NA[a]	2×6	2×6
24	NA[a]	NA[a]	NA[a]	2×6
Supporting one floor and a roof				
>10	2×6	2×4	2×4	2×4
12	2×6	2×6	2×6	2×4
14	2×6	2×6	2×6	2×6
16	NA[a]	2×6	2×6	2×6
18	NA[a]	2×6	2×6	2×6
20	NA[a]	NA[a]	2×6	2×6
24	NA[a]	NA[a]	NA[a]	2×6
Supporting two floors and a roof				
>10	2×6	2×6	2×4	2×4
12	2×6	2×6	2×6	2×6
14	2×6	2×6	2×6	2×6
16	NA[a]	NA[a]	2×6	2×6
18	NA[a]	NA[a]	2×6	2×6
20	NA[a]	NA[a]	NA[a]	2×6
22	NA[a]	NA[a]	NA[a]	NA[a]
24	NA[a]	NA[a]	NA[a]	NA[a]

For SI: 1 inch = 25.4 mm, 1 foot = 304.8 mm, 1 pound per square foot = 0.0479 kN/m², 1 pound per square inch = 6.895 kPa, 1 mile per hour = 1.609 km/h.

a. Design required.

b. Applicability of this table assumes the following: Snow load not exceeding 25 psf, f_b not less than 1310 psi determined by multiplying the AF&PA NDS tabular base design value by the repetitive use factor, and by the size factor for all species except southern pine, E not less than 1.6 by 10^6 psi, tributary dimensions for floors and roofs not exceeding 6 feet, maximum span for floors and roof not exceeding 12 feet, eaves not greater than 2 feet in dimension and exterior sheathing. Where the conditions are not within these parameters, design is required.

c. Utility, standard, stud and No. 3 grade lumber of any species are not permitted.

(continued)

TABLE R602.3.1—continued
MAXIMUM ALLOWABLE LENGTH OF WOOD WALL STUDS EXPOSED TO WIND SPEEDS OF 100 MPH OR LESS IN SEISMIC DESIGN CATEGORIES A, B, C and D_1

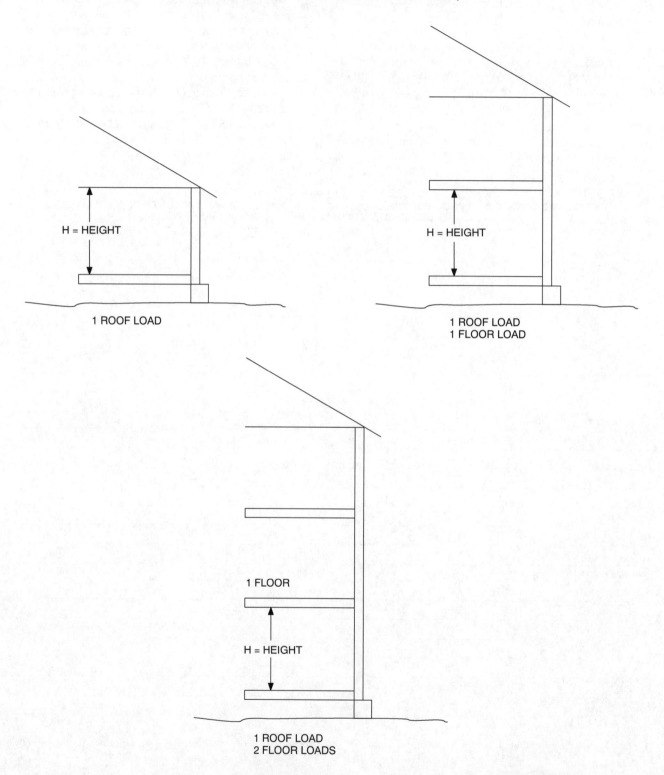

❖ This table provides the stud spacing limits in bearing walls over 10 feet (3048 mm) high as stated in Section R602.3.1. Notes "b" and "c" make it clear that the lumber grades specified in Section R602.2 are not applicable, but instead must be determined by using AF&PA NDS and applying the limitations of Note "b." Conditions noted as "NA" in the table require a design per Note "a."

R602.3.2 Top plate. Wood stud walls shall be capped with a double top plate installed to provide overlapping at corners and intersections with bearing partitions. End joints in top plates shall be offset at least 24 inches (610 mm). Plates shall be a nominal 2 inches in depth (51 mm) and have a width at least equal to the width of the studs.

Exception: A single top plate may be installed in stud walls, provided the plate is adequately tied at joints, corners and intersecting walls by a minimum 3-inch-by-6-inch by a 0.036-inch-thick (76 mm by 152 mm by 0.914 mm) galvanized steel plate that is nailed to each wall or segment of wall by six 8d nails on each side, provided the rafters or joists are centered over the studs with a tolerance of no more than 1 inch (25.4 mm). The top plate may be omitted over lintels that are adequately tied to adjacent wall sections with steel plates or equivalent as previously described.

❖ Plates are the horizontal elements of walls capping the top of walls and framing the bottom of stud walls. They are called plates because they are laid flat, with the deeper dimension horizontal. In building construction applications, top plates must be a nominal 2-inch (51 mm) thickness.

Double top plates serve three major functions:
1. They overlap at corners and interior wall intersections, providing a means of tying the building together.
2. They serve as beams to support joists and rafters that are not located directly over the studs.
3. They serve as chords for floor and roof diaphragms.

Along with provisions for stud size and spacing limitations, this section requires the installation of double top plates to provide a continuous tie along the tops of the walls. With the advent of wider wall framing to accommodate increased thickness of insulation, a desire to save on material costs led to the allowance of a single top plate alternative. The exception permits the use of a single top plate in bearing and exterior walls as long as adequate top-plate ties are provided. In addition, joists or rafters framing into the wall must be placed more closely to the vertical stud below. This is necessary to limit the bending stress in the top plate. The single top plate exception is illustrated in Commentary Figure R602.3.2.

R602.3.3 Bearing studs. Where joists, trusses or rafters are spaced more than 16 inches (406 mm) on center and the bearing studs below are spaced 24 inches (610 mm) on center, such members shall bear within 5 inches (127 mm) of the studs beneath.

Exceptions:
1. The top plates are two 2-inch by 6-inch (38 mm by 140 mm) or two 3-inch by 4-inch (64 mm by 89 mm) members.
2. A third top plate is installed.
3. Solid blocking equal in size to the studs is installed to reinforce the double top plate.

❖ The bottom plate serves to anchor the wall to the floor, and studs are attached to it by end nailing or toe-nailing. The basic requirements of this section are shown in Commentary Figures R602.3.3(1) and R602.3.3(3). The code permits three options (exceptions) to account for loading that must be resisted by the double top plate when the stud spacing is 24 inches (610 mm)

For SI: 1 inch = 25.4 mm.

Figure R602.3.2
SINGLE TOP-PLATE SPLICE

on center. The first option is an increase in the minimum top plate size. The second option is to provide a third plate. The third option is solid blocking used as reinforcement for a double plate, installed so that the load from the framing members above is transferred to the supporting studs. Commentary Figures R602.3.3(2) and R602.3.3(4) illustrate the first and third options.

R602.3.4 Bottom (sole) plate. Studs shall have full bearing on a nominal 2 by (38 mm) or larger plate or sill having a width at least equal to the width of the studs.

❖ This provision requires that studs bear on an adequately sized bottom plate. This allows the bottom plate to serve as a nailing surface for the wall sheathing while controlling the compression-perpendicular-to-grain on the sill. If too great, this stress could result in crushing.

R602.4 Interior load-bearing walls. Interior load-bearing walls shall be constructed, framed and fireblocked as specified for exterior walls.

❖ Interior load-bearing walls must be designed and constructed using the criteria for exterior walls. Thus, Table R602.3(5) should be used to establish stud spacing of walls up to 10 feet (3048 mm) high, and Table R602.3.1 applies to walls over 10 feet (3048 mm).

R602.5 Interior nonbearing walls. Interior nonbearing walls shall be permitted to be constructed with 2-inch-by-3-inch (51 mm by 76 mm) studs spaced 24 inches (610 mm) on center or, when not part of a braced wall line, 2-inch-by-4-inch (51 mm

For SI: 1 inch = 25.4 mm.

Figure R602.3.3(1)
TOP PLATE WITH 16-INCH STUD SPACING

For SI: 1 inch = 25.4 mm.

Figure R602.3.3(2)
TOP PLATE WITH 24-INCH STUD SPACING

WALL CONSTRUCTION

by 102 mm) flat studs spaced at 16 inches (406 mm) on center. Interior nonbearing walls shall be capped with at least a single top plate. Interior nonbearing walls shall be fireblocked in accordance with Section R602.8.

❖ The code permits interior nonbearing partitions to have reduced stud sizes and increased stud spacing and be capped with a single top plate. These allowances are based on minimal superimposed vertical loads.

R602.6 Drilling and notching—studs. Any stud in an exterior wall or bearing partition may be cut or notched to a depth not exceeding 25 percent of its width. Studs in nonbearing partitions may be notched to a depth not to exceed 40 percent of a single stud width. Any stud may be bored or drilled, provided that the diameter of the resulting hole is no greater than 40 percent of the stud width, the edge of the hole is no closer than $^5/_8$ inch (15.9 mm) to the edge of the stud, and the hole is not located in the same section as a cut or notch. See Figures R602.6(1) and R602.6(2).

Exceptions:
1. A stud may be bored to a diameter not exceeding 60 percent of its width, provided that such studs located in exterior walls or bearing partitions are doubled and that not more than two successive studs are bored.
2. Approved stud shoes may be used when installed in accordance with the manufacturer's recommendation.

❖ This section addresses the allowable drilling and notching of studs used to frame partitions. See Figures R602.6(1) and R602.6(2) for examples of permitted drilling and notching of studs in exterior walls, bearing walls and nonbearing walls. These limitations retain the structural integrity of the studs.

R602.6.1 Drilling and notching of top plate. When piping or ductwork is placed in or partly in an exterior wall or interior load-bearing wall, necessitating cutting, drilling or notching of

Figure R602.3.3(3)
TOP PLATE WITH 24-INCH STUD SPACING AND BEARING POINT LIMITATIONS

NOTE: NO LIMIT ON FLOOR FRAMING OR ROOF TRUSS CHORD LOCATION

For SI: 1 inch = 25.4 mm.

Figure R602.3.3(4)
BLOCKED TOP PLATE WITH 24-INCH STUD SPACING

FIGURE R602.6(1)
NOTCHING AND BORED HOLE LIMITATIONS FOR EXTERIOR WALLS AND BEARING WALLS

❖ See the commentary for Section R602.6.

the top plate by more than 50 percent of its width, a galvanized metal tie of not less than 0.054 inches thick (1.37mm) (16ga) and $1^1/_2$ inches (38mm) wide shall be fastened to each plate across and to each side of the opening with not less than eight 16d nails at each side or equivalent. See Figure R602.6.1.

Exception: When the entire side of the wall with the notch or cut is covered by wood structural panel sheathing.

❖ In many cases, drilling or notching of the top plate is necessary to allow plumbing, heating or other pipes to be placed within the exterior walls and load-bearing interior walls. When drilling or notching the top plate by more than one-half its width is necessary, strapping across the plates as illustrated in Figure R602.6.1 is required to provide top plate continuity and to retain the structural integrity of the wall system as a whole. In the case of a double top plate, only the upper of the two plates needs to have a strap installed as described.

R602.7 Headers. For header spans see Tables R502.5(1) and R502.5(2).

❖ At a wall opening, headers are required to transfer loads received from the wall and floors/roof to the foundation. Framing over openings such as doors or

WALL CONSTRUCTION

FIGURE R602.6(2)
NOTCHING AND BORED HOLE LIMITATIONS FOR INTERIOR NONBEARING WALLS

❖ See the commentary for Section R602.6.

windows within bearing walls must be accomplished through the installation of headers of sufficient size to span the opening and transfer loads to jamb studs (header studs). Commentary Figure R602.7 illustrates a double header over a door opening. Tables R502.5(1) and R502.5(2) of the code are used to determine allowable spans for headers as well as girders.

Example:
You wish to size a header in the exterior wall of a one-story house that is 30 feet (9144 mm) in width. The opening size is 8 feet (2338 mm). No. 2 grade southern pine is common for the area, and the ground-snow loading is 20 pounds per square foot (958 Pa).

Solution:
Table R502.5(1) lists, for a 30 psf (1436 Pa) snow load and a building width of 36 feet (10,973 mm), a triple 2 x 10 (51 mm x 254 mm) header; it will span 8 feet, 2 inches (2489 mm).

R602.7.1 Wood structural panel box headers. Wood structural panel box headers shall be constructed in accordance with Figure R602.7.2 and Table R602.7.2.

❖ Wood structural panel box headers are used in some construction and are an efficient way of carrying load over a opening. The box header must be built in accordance with the criteria specified in Figures R602.7.2 and Table R602.7.2. Spans of box headers are estab-

**FIGURE R602.6.1
TOP PLATE FRAMING TO ACCOMMODATE PIPING**

❖ See the commentary for Section R602.6.1.

lished based on the construction of the header, the header depth, and the depth of the house. Note "a" to the table explains the depth criteria, which is based on a single story with a clear-span trussed roof or a two-story in which part of the load from the floors and roofs is supported by interior bearing walls. House depth is based on the span from the header to an interior bearing wall or the opposing exterior wall.

R602.7.2 Nonbearing walls. Load-bearing headers are not required in interior or exterior nonbearing walls. A single flat 2-inch-by-4-inch (51 mm by 102 mm) member may be used as a header in interior or exterior nonbearing walls for openings up to 8 feet (2438 mm) in width if the vertical distance to the parallel nailing surface above is not more than 24 inches (610 mm). For such nonbearing headers, no cripples or blocking are required above the header.

❖ In walls that are not supporting significant loads, the code provides a simple (more economical) option for constructing headers.

**TABLE R602.7.2
MAXIMUM SPANS FOR WOOD STRUCTURAL PANEL BOX HEADERS[a]**

HEADER CONSTRUCTION[b]	HEADER DEPTH (inches)	HOUSE DEPTH (feet)				
		24	26	28	30	32
Wood structural panel—one side	9	4	4	3	3	—
	15	5	5	4	3	3
Wood structural panel—both sides	9	7	5	5	4	3
	15	8	8	7	7	6

For SI: 1 inch = 25.4 mm, 1 foot = 304.8 mm.
a. Spans are based on single story with clear-span trussed roof or two-story with floor and roof supported by interior-bearing walls.
b. See Figure R602.7.2 for construction details.

❖ See the commentary for Section R602.7.1.

WALL CONSTRUCTION

FIGURE R602.7 - R602.8

For SI: 1 inch = 25.4 mm.

Figure R602.7
HEADER DETAIL

R602.8 Fireblocking required. Fireblocking shall be provided to cut off all concealed draft openings (both vertical and horizontal) and to form an effective fire barrier between stories, and between a top story and the roof space. Fireblocking shall be provided in wood-frame construction in the following locations.

1. In concealed spaces of stud walls and partitions, including furred spaces and parallel rows of studs or staggered studs; as follows:

    1.1. Vertically at the ceiling and floor levels.

    1.2. Horizontally at intervals not exceeding 10 feet (3048 mm).

2. At all interconnections between concealed vertical and horizontal spaces such as occur at soffits, drop ceilings and cove ceilings.

3. In concealed spaces between stair stringers at the top and bottom of the run. Enclosed spaces under stairs shall comply with Section R311.2.2.

4. At openings around vents, pipes, and ducts at ceiling and floor level, with an approved material to resist the free passage of flame and products of combustion.

5. For the fireblocking of chimneys and fireplaces, see Section R1001.16.

6. Fireblocking of cornices of a two-family dwelling is required at the line of dwelling unit separation.

❖ To restrict the movement of flame and gasses to other areas of a building through concealed passages in building components such as floors, walls and stairs, fireblocking of these concealed combustible spaces is required to form a barrier between stories and between a top story and the roof space. For example, the following locations must be firestopped in wood-frame construction:

- In concealed spaces of stud walls and partitions, including spaces at the ceiling and floor levels. See Commentary Figures R602.8(1) and R602.8(2).

- At all interconnections between concealed vertical and horizontal spaces such as soffits [Commentary Figure R602.8(3)], dropped ceilings [Commentary Figure R602.8(4)], and cove ceilings [Commentary Figure R602.8(5)]. Interconnections shown in Commentary Figure R602.8(6) for a bathtub installation must also be firestopped.

- In concealed spaces between stair stringers at the top and bottom of the run; see Commentary Figure R602.8(7).

- At openings around vent pipes, ducts, chimneys, and fireplaces at ceiling and floor levels with noncombustible materials; see Commentary Fig-

**FIGURE R602.7.2
TYPICAL WOOD STRUCTURAL PANEL BOX HEADER CONSTRUCTION**

❖ See the commentary for Section R602.7.2.

ures R602.8(8) and R602.8(9). Commentary Figure R602.8(9) illustrates fireblocking at chimneys and fireplaces. The fireblocking at ductwork would be similar.

R602.8.1 Materials. Except as provided in Section R602.8, Item 4, fireblocking shall consist of 2-inch (51 mm) nominal lumber, or two thicknesses of 1-inch (25.4 mm) nominal lumber with broken lap joints, or one thickness of $^{23}/_{32}$-inch (19.8 mm) wood structural panels with joints backed by $^{23}/_{32}$-inch (19.8 mm) wood structural panels or one thickness of $^{3}/_{4}$-inch (19.1 mm) particleboard with joints backed by $^{3}/_{4}$-inch (19.1 mm) particleboard, $^{1}/_{2}$-inch (12.7 mm) gypsum board, or $^{1}/_{4}$-inch (6.4 mm) cement-based millboard. Batts or blankets of mineral wool or glass fiber or other approved materials installed in such a manner as to be securely retained in place shall be permitted as an acceptable fire block. Batts or blankets of mineral or glass fiber or other approved non-rigid materials shall be permitted for compliance with the 10 foot horizontal fireblocking in walls constructed using parallel rows of studs or staggered studs. Loose-fill insulation material shall not be used as a fire block unless specifically tested in the form and manner intended for use to demonstrate its ability to remain in place and to retard the spread of fire and hot gases.

❖ This section specifies the material required for fireblocking. Fireblocking around chimneys and fireplaces is required to be noncombustible according to Section R1001.6 as specified in item 5 of Section R602.8. The fireblocking at openings around vents, pipes and ducts at the ceiling and floor level is required only to be adequate to resist the free passage of flames and smoke according to item 4 of Section R602.8. So that it is not accidentally displaced, all fireblocking material must be securely fastened in place.

R602.8.1.1 Unfaced fiberglass. Unfaced fiberglass batt insulation used as fireblocking shall fill the entire cross section of the wall cavity to a minimum height of 16 inches (406 mm) measured vertically. When piping, conduit or similar obstructions are encountered, the insulation shall be packed tightly around the obstruction.

❖ Fiberglass batts may serve as fireblocking if installed per this section.

WALL CONSTRUCTION

Figure R602.8(1)
FIRESTOPPING—BALLOON FRAMING

R602.8.1.2 Fireblocking integrity. The integrity of all fireblocks shall be maintained.

❖ Piping, ducts, or other similar items that pass through firestops must be installed so that the integrity of the firestop is maintained. This may be accomplished by packing an oversized hole with an acceptable fireblocking material.

R602.9 Cripple walls. Foundation cripple walls shall be framed of studs not less in size than the studding above. When exceeding 4 feet (1219 mm) in height, such walls shall be framed of studs having the size required for an additional story.

Cripple walls with a stud height less than 14 inches (356 mm) shall be sheathed on at least one side with a wood structural panel that is fastened to both the top and bottom plates in accordance with Table R602.3(1), or the cripple walls shall be constructed of solid blocking. Cripple walls shall be supported on continuous foundations.

❖ A cripple wall is a framed stud extending from the top of foundation to the underside of the floor framing above the foundation. Should these foundation studs exceed 4 feet (1219 mm) in height, the partial-height wall must, for purposes of stud sizing, be sized as if the partial-height wall were an additional story. Such a condition may occur in buildings of a split-level type with floor levels partially below grade or in buildings with underfloor crawl spaces larger than the minimum specified. See Figure R602.3(2). The rationale for the wall being considered an additional story is based on the load capacity of studs of a particular size changing as the unsupported length is increased.

The minimum length of 14 inches (356 mm) for cripple wall studs provides sufficient clear space in which required nailing of the framing may be accomplished. If the wood-framed foundation wall does not permit the installation of foundation studs of such a length, wood structural panel sheathing must be applied, or a solid-blocking method of construction may be used.

R602.10 Wall bracing. All exterior walls shall be braced in accordance with this section. In addition, interior braced wall lines shall be provided in accordance with Section R602.10.1.1. For buildings in Seismic Design Categories D_1 and D_2, walls shall be constructed in accordance with the additional requirements of R602.10.9, R602.10.11 and R602.11.

❖ Wall construction must include bracing to resist imposed lateral loads resulting from wind or seismic loading and to provide stability to the structure. When subjected to wind loads, the upper portion of the struc-

Figure R602.8(2)
FIRESTOPPING—PLATFORM FRAMING

ture moves horizontally while the lower portion is restrained at ground level. During an earthquake, the ground motion displaces the foundation while the top portion of the structure tries to remain stationary. In both of these cases the bracing resists the differential movement and thus prevents or limits damage to the building.

R602.10.1 Braced wall lines. Braced wall lines shall consist of braced wall panel construction methods in accordance with Section R602.10.3. The amount and location of bracing shall be in accordance with Table R602.10.1 and the amount of bracing shall be the greater of that required by the Seismic Design Category or the design wind speed. Braced wall panels shall begin no more than 12.5 feet (3810 mm) from each end of a braced wall line. Braced wall panels that are counted as part of a braced wall line shall be in line, except that offsets out-of-plane of up to 4 feet (1219 mm) shall be permitted provided that the total out-to-out offset dimension in any braced wall line is not more than 8 feet (2438 mm).

A designed collector shall be provided if the bracing begins more than 12 feet (3658 mm) from each end of a braced wall line.

❖ Wall bracing in wood wall construction must consist of braced wall panels constructed by the methods specified in Section R602.10.3. The amount of wall bracing is determined from Table R602.10.1. The percentage and spacing of wall bracing will vary based on wind speed and seismic design category (use the larger value determined), location within the structure and the method of wall panel construction. These are illustrated in Figure R602.10.1. Let-in braces (Method 1) have very limited applicability.

Ideally, the braced wall panels forming a braced wall line should be in the same plane, but the code does allow out-of-plane offsets of braced wall panels of up to 4 feet (1219 mm) on either side of the braced wall line.

While Table R602.10.1 seems to require bracing to start at the end of each braced wall line, this section clarifies that the first panel may be located up to $12^1/_2$ feet (3810 mm) from the end of the braced wall line. This is exactly one-half of the typical center-to-center spacing of panels. Also see Section R602.10.11 for further modifications and limitations in Seismic Design Categories D_1 and D_2.

Figure R602.8(3)
FIRESTOPPING—FURRED SOFFIT

Figure R602.8(4)
FIRESTOPPING—DROPPED CEILING

Figure R602.8(5)
FIRESTOPPING—COVE CEILING

Figure R602.8(6)
FIRESTOPPING—AT TUB

Figure R602.8(7)
FIRESTOPPING—AT STAIRWAYS

Figure R602.8(8)
FIRESTOPPING—AROUND PIPING

Figure R602.8(9)
FIRESTOPPING—AROUND CHIMNEYS AND FIREPLACES

TABLE R602.10.1
WALL BRACING

SEISMIC DESIGN CATEGORY OR WIND SPEED	CONDITION	TYPE OF BRACE[b,c]	AMOUNT OF BRACING[a,d,e]
Categories A and B ($S_s \leq 0.35g$ and $S_{ds} \leq 0.33g$) or 100 mph and less	One story Top of two or three story	Methods 1, 2, 3, 4, 5, 6, 7 or 8	Located at each end and at least every 25 feet on center but not less than 16% of braced wall line.
	First story of two story Second story of three story	Methods 1, 2, 3, 4, 5, 6, 7 or 8	Located at each end and at least every 25 feet on center but not less than 16% of braced wall line for Method 3 and 25% of braced wall line for Methods 2, 4, 5, 6, 7 or 8.
	First story of three story	Methods 2, 3, 4, 5, 6, 7 or 8	Minimum 48-inch-wide panels located at each end and at least every 25 feet on center but not less than 25% of braced wall line for method 3 and 35% of braced wall line for Methods 2, 4, 5, 6, 7 or 8.
Category C ($S_s \leq 0.6g$ and $S_{ds} \leq 0.53g$) or less than 110 mph	One story Top of two or three story	Methods 1, 2, 3, 4, 5, 6, 7 or 8	Located at each end and at least every 25 feet on center but not less than 16% of braced wall line for Method 3 and 25% of braced wall line for Methods 2, 4, 5, 6, 7 or 8.
	First story of two story Second story of three story	Methods 2, 3, 4, 5, 6, 7 or 8	Located at each end and at least every 25 feet on center but not less than 30% of braced wall line for Method 3 and 45% of braced wall line for Methods 2, 4, 5, 6, 7 or 8.
	First story of three story	Methods 2, 3, 4, 5, 6, 7 or 8	Located at each end and at least every 25 feet on center but not less than 45% of braced wall line for Method 3 and 60% of braced wall line for Methods 2, 4, 5, 6, 7 or 8.
Category D_1 ($S_s \leq 1.25g$ and $S_{ds} \leq 0.83g$) or less than 110 mph	One story Top of two or three story	Methods 2, 3, 4, 5, 6, 7 or 8	Located at each end and at least every 25 feet on center but not less than 20% of braced wall line for Method 3 and 30% of braced wall line for Methods 2, 4, 5, 6, 7 or 8.
	First story of two story Second story of three story	Methods 2, 3, 4, 5, 6, 7 or 8	Located at each end and not more than 25 feet on center but not less than 45% of braced wall line for Method 3 and 60% of braced wall line for Methods 2, 4, 5, 6, 7 or 8.
	First story of three story	Methods 2, 3, 4, 5, 6, 7 or 8	Located at each end and not more than 25 feet on center but not less than 60% of braced wall line for Method 3 and 85% of braced wall line for Method 2, 4, 5, 6, 7 or 8.
Category D_2 or less than 110 mph	One story Top of two story	Methods 2, 3, 4, 5, 6, 7 or 8	Located at each end and at least every 25 feet on center but not less than 25% of braced wall line for Method 3 and 40% of braced wall line for Methods 2, 4, 5, 6, 7 or 8.
	First story of two story	Methods 2, 3, 4, 5, 6, 7 or 8	Located at each end and not more than 25 feet on center but not less than 55% of braced wall line for Method 3 and 75% of braced wall line for Methods 2, 4, 5, 6, 7 or 8.
	Cripple walls	Method 3	Located at each end and not more than 25 feet on center but not less than 75% of braced wall line.

For SI: 1 inch = 25.4 mm, 1 foot = 304.8 mm, 1 pound per square foot = 0.0479 kN/m^2, 1 mile per hour = 1.609 km/h.

a. Wall bracing amounts are based on a soil site class "D." Interpolation of bracing amounts between the S_{ds} values associated with the Seismic Design Categories shall be permitted when a site specific S_{ds} value is determined in accordance with Section 1615 of the *International Building Code*.

b. Foundation cripple wall panels shall be braced in accordance with Section R602.10.2.

c. Methods of bracing shall be as described in Section R602.10.3. The alternate braced wall panels described in Section R602.10.6 shall also be permitted.

d. The bracing amounts for Seismic Design Categories are based on a 15 psf wall dead load. For walls with a dead load of 8 psf or less, the bracing amounts shall be permitted to be multiplied by 0.85 provided that the adjusted bracing amount is not less than that required for the site's wind speed. The minimum length of braced panel shall not be less than required by Section R602.10.3.

e. When the dead load of the roof/ceiling exceeds 15 psf, the bracing amounts shall be increased in accordance with Section R301.2.2.4. Bracing required for a site's wind speed shall not be adjusted.

❖ See the commentary for Section R602.10.1 and Commentary Figure R602.10.1. The amount of wall bracing required for buildings is a function of basic wind speed or seismic design category. This must be based on the most restrictive of these criteria. For example, a building classified as Seismic Design Category A but having a basic wind speed of 105 mph (169 km/h) would need to comply with the bracing amounts listed under Seismic Design Category C or less than 110 mph (177 km/h).

Notes "a" and "d" allow decreases to the tabulated amounts of bracing based on seismic design categories, while Note "e" requires an increase if a roof/ceiling dead load exceeds 15 psf (0.72 kN/m^2). In these cases, wind speed criteria remain unchanged.

FIGURE R602.10.1 — WALL CONSTRUCTION

Figure R602.10.1
REQUIRED WALL BRACING, PERCENT OF BRACED WALL LINE LENGTH

R602.10.1.1 Spacing. Spacing of braced wall lines shall not exceed 35 feet (10,668 mm) on center in both the longitudinal and transverse directions in each story.

> **Exception:** Spacing of braced wall lines not exceeding 50 feet shall be permitted where:
>
> 1. The wall bracing provided equals or exceeds the amount of bracing required by Table R602.10.1 multiplied by a factor equal to the braced wall line spacing divided by 35 feet, and
> 2. The length-to-width ratio for the floor/wall diaphragm does not exceed 3:1.

❖ Buildings that are not in Seismic Design Categories D_1 or D_2 must be braced in accordance with this section. The requirements for wall bracing vary, depending upon the level of seismic activity for the region in which the building is located and whether the story in question is supporting other stories. The bracing requirements increase for higher seismic design categories and for walls that support the weight of additional stories above. Various types of structural sheathing materials are permitted, although not every material is permitted in every instance.

The spacing of the braced wall lines is permitted to be increased provided both of the items in the exception are met. The concept is that the spacing is allowed to be increased where additional length of wall bracing is provided versus the minimum length required by the code. This increased spacing is only allowed for residential buildings that are not long in relation to their width.

Nailing requirements are critical to the performance of wall bracing. To insure that the bracing material is effective in resisting racking of the wall, it must be installed in the manner prescribed. See Commentary Figure 602.10.1.1.

R602.10.2 Cripple wall bracing.

❖ Cripple walls participate in the load path necessary to resist earthquake and wind loads. See the commentaries for Section R602.9 and Figure R602.3(2). This section outlines wall bracing requirements for buildings with cripple walls based on the seismic design category.

R602.10.2.1 Seismic Design Categories Other than D_2. In Seismic Design Categories other than D_2, cripple walls shall be braced with an amount and type of bracing as required for the wall above in accordance with Table R602.10.1 with the following modifications for cripple wall bracing:

1. The percent bracing amount as determined from Table R602.10.1 shall be increased by 15 percent, and

2. The wall panel spacing shall be decreased to 18 feet (5486 mm) instead of 25 feet (7620 mm).

❖ The bracing requirement for the wall (story) above is increased 15 percent for the supporting cripple wall, and the braced wall panel spacing is reduced from 25 feet to 18 feet (7620 mm to 5486 mm). A cripple wall, in effect, has the impact of being an additional story in seismic performance of a building. The intent of the provision is to provide additional lateral stability to a building with a cripple wall in order to compensate for the possible overstress that could occur in this situation.

R602.10.2.2 Seismic Design Category D_2. In Seismic Design Category D_2, cripple walls shall be braced in accordance with Table R602.10.1.

❖ Table R602.10.1 contains specific requirements for bracing of cripple walls in Seismic Design Category D_2. This is done to ensure stability for the structure in a serious seismic event.

R602.10.2.3 Redesignation of cripple walls. In any Seismic Design Category, cripple walls are permitted to be redesignated as the first story walls for purposes of determining wall bracing requirements. If the cripple walls are redesignated, the stories above the redesignated story shall be counted as the second and third stories respectively.

❖ The provisions of R602.10.2.1 and R602.10.2.2 are predicated on the fact that the designer is not treating the cripple wall as a story. If the cripple wall is designed as a first story in the considerations for bracing (with the first story being counted as the second story, the second story as the third story, etc.), then the concerns regarding less stability related to cripple walls becomes moot.

R602.10.3 Braced wall panel construction methods. The construction of braced wall panels shall be in accordance with one of the following methods:

1. Nominal 1-inch-by-4-inch (25.4 mm by 102 mm) continuous diagonal braces let in to the top and bottom plates and the intervening studs or approved metal strap devices installed in accordance with the manufacturer's specifications. The let-in bracing shall be placed at an angle not more than 60 degrees (1.06 rad) or less than 45 degrees (0.79 rad) from the horizontal.

2. Wood boards of $^5/_8$ inch (15.9 mm) net minimum thickness applied diagonally on studs spaced a maximum of 24 inches (610 mm). Diagonal boards shall be attached to studs in accordance with Table R602.3(1).

Figure R602.10.1.1
ACCEPTABLE TWO-STORY BRACING EXAMPLE

WALL CONSTRUCTION

3. Wood structural panel sheathing with a thickness not less than $5/16$ inch (7.9 mm) for 16-inch (406 mm) stud spacing and not less than $3/8$ inch (9.5 mm) for 24-inch (610 mm) stud spacing. Wood structural panels shall be installed in accordance with Table R602.3(3).
4. One-half-inch (12.7 mm) or $25/32$-inch (19.8 mm) thick structural fiberboard sheathing applied vertically or horizontally on studs spaced a maximum of 16 inches (406 mm) on center. Structural fiberboard sheathing shall be installed in accordance with Table R602.3(1).
5. Gypsum board with minimum $1/2$-inch (12.7 mm) thickness placed on studs spaced a maximum of 24 inches (610 mm) on center and fastened at 7 inches (178 mm) on center with the size nails specified in Table R602.3(1) for sheathing and Table R702.3.5 for interior gypsum board.
6. Particleboard wall sheathing panels installed in accordance with Table R602.3(4)
7. Portland cement plaster on studs spaced a maximum of 16 inches (406 mm) on center and installed in accordance with Section R703.6.
8. Hardboard panel siding when installed in accordance with Table R703.4.

Exception: Alternate braced wall panels constructed in accordance with Section R602.10.6 shall be permitted to replace any of the above methods of braced wall panels.

❖ Braced wall panels must be constructed according to one of these methods and in the amounts required in Table R602.10.1. In accordance with the exception, an alternative braced panel may be substituted where allowed by Section R602.10.6.

R602.10.4 Length of braced panels. For Methods 2, 3, 4, 6, 7 and 8 above, each braced wall panel shall be at least 48 inches (1219 mm) in length, covering a minimum of three stud spaces where studs are spaced 16 inches (406 mm) on center and covering a minimum of two stud spaces where studs are spaced 24 inches (610 mm) on center. For Method 5 above, each braced wall panel shall be at least 96 inches (2438 mm) in length where applied to one face of a braced wall panel and at least 48 inches (1219 mm) where applied to both faces.

Exceptions:
1. Lengths of braced wall panels for continuous wood structural panel sheathing shall be in accordance with Section R602.10.5.
2. Lengths of alternate braced wall panels shall be in accordance with Section R602.10.6.

❖ For the bracing methods specified, braced wall panels of structural sheathing are required to be at least 48 inches (1219 mm) in width, covering three 16-inch (406 mm) stud spaces or two 24-inch (610 mm) stud spaces. Structural sheathing fastening should include edge nailing and field nailing (i.e., at intermediate supports) in accordance with Tables 602.3(1) or R602.3(2).

R602.10.5 Continuous structural panel sheathing. When continuous wood structural panel sheathing is provided in accordance with Method 3 of R602.10.3 on all sheathable areas of all exterior walls, and interior braced wall lines, where required, including areas above and below openings, braced wall panel lengths shall be in accordance with Table R602.10.5. Wood structural panel sheathing shall be installed at corners in accordance with Figure R602.10.5. The bracing amounts in Table R602.10.1 for Method 3 shall be permitted to be multiplied by a factor of 0.9 for walls with a maximum opening height that does not exceed 85 percent of the wall height or a factor of 0.8 for walls with a maximum opening height that does not exceed 67 percent of the wall height.

❖ This section allows two reductions to the requirements for braced wall lines that use wood structural panel sheathing installed continuously in accordance with method 3 of Section R602.10.3. The first is an exception to the 48-inch (1219 mm) length required by Section R602.10.4. Figure R602.10.5 illustrates the reduced wall panel lengths of Table R602.10.5. Where the opening is 100 percent of the wall height, no reduction is permitted. In addition, the total required length of braced wall panels from Table R602.10.1 might be reduced by the appropriate multipliers shown in Commentary Figure R602.10.1.

R602.10.6 Alternate braced wall panels. Alternate braced wall lines constructed in accordance with one of the following provisions shall be permitted to replace each 4 feet (1219 mm) of braced wall panel as required by Section R602.10.4:

1. In one-story buildings, each panel shall have a length of not less than 2 feet, 8 inches (813 mm) and a height of not more than 10 feet (3048 mm). Each panel shall be sheathed on one face with $3/8$-inch-minimum-thickness (9.5 mm) wood structural panel sheathing nailed with 8d common or galvanized box nails in accordance with Table R602.3(1) and blocked at all wood structural panel sheathing edges. Two anchor bolts installed in accordance with Figure R403.1(1) shall be provided in each panel. Anchor bolts shall be placed at panel quarter points. Each panel end stud shall have a tie-down device fastened to the foundation, capable of providing an uplift capacity of at least 1,800 pounds (816.5 kg). The tie-down device shall be installed in accordance with the manufacturer's recommendations. The panels shall be supported directly on a foundation or on floor framing supported directly on a foundation which is continuous across the entire length of the braced wall line. This foundation shall be reinforced with not less than one No. 4 bar top and bottom. When the continuous foundation is required to have a depth greater than 12 inches (305 mm), a minimum 12-inch-by-12-inch (305 mm by 305 mm) continuous footing or turned down slab edge is permitted at door openings in the braced wall line. This continuous footing or turned down slab edge shall be reinforced with not less than one No. 4 bar top and bottom. This reinforcement shall be lapped 15 inches (381 mm) with the reinforcement required in the continuous foundation located directly under the braced wall line.
2. In the first story of two-story buildings, each braced wall panel shall be in accordance with Item 1 above, except that

FIGURE R602.10.5   **WALL CONSTRUCTION**

(a) OUTSIDE CORNER DETAIL

(b) INSIDE CORNER DETAIL

For SI: 1 inch = 25.4 mm.

**FIGURE R602.10.5
EXTERIOR CORNER FRAMING**

❖ This figure illustrates corner framing details specifically required by Sections R602.10.5 and R602.10.11 (Exception).

the wood structural panel sheathing shall be provided on both faces, sheathing edge nailing spacing shall not exceed four inches on center, at least three anchor bolts shall be placed at one-fifth points, and tie-down device uplift capacity shall not be less than 3,000 pounds (1360.8 kg).

❖ This section allows the panel width to be reduced to 32 inches (813 mm) provided the panel is based with a minimum $3/_8$-inch (9.5 mm) wood structural panel and hold-down fasteners are provided at each end of the panel. This method of construction provides for a very rigid and strong shear wall that is positively anchored to the foundation to resist the forces caused by wind or earthquakes. These are best suited for the limited width typically available adjacent to garage doors.

TABLE R602.10.5
LENGTH REQUIREMENTS FOR BRACED WALL PANELS IN A CONTINUOUSLY SHEATHED WALL[a,b]

MINIMUM LENGTH OF BRACED WALL PANEL (inches)			MAXIMUM OPENING HEIGHT NEXT TO THE BRACED WALL PANEL (% of wall height)
8-foot wall	9-foot wall	10-foot wall	
48	54	60	100%
32	36	40	85%
24	27	30	65%

For SI: 1 inch = 25.4 mm, 1 foot = 305 mm, 1 pound per square foot = 0.0479 kN/m^2.

a. Linear interpolation shall be permitted.

b. Full-height sheathed wall segments to either side of garage openings that support light frame roofs with roof covering dead loads of 3 psf or less shall be permitted to have a 4:1 aspect ratio.

❖ See the commentary for Section R610.10.5. Note "b" provides an exception to the braced wall panel lengths located on either side of a garage door opening. It allows an aspect ratio of 4:1. While "aspect ratio" is not defined in the IRC, for a shear wall it represents the height-width ratio; see Section 2305.3.3 of the *International Building Code*. The height is the clear height from the top of the foundation to the bottom of the diaphragm framing above. Typically the latter would also represent the top plate of the wall. The shear wall (segment) width is the horizontal dimension of full-height sheathing adjacent to unrestrained openings.

R602.10.7 Panel joints. All vertical joints of panel sheathing shall occur over studs. Horizontal joints in braced wall panels shall occur over blocking of a minimum of $1^1/_2$ inch (38 mm) thickness.

Exception: Blocking is not required behind horizontal joints in Seismic Design Categories A and B and detached dwellings in Seismic Design Category C when constructed in accordance with R602.10.3, Braced-wall-panel construction method 3 and Table R602.10.1, method 3, or where permitted by the manufacturer's installation requirements for the specific sheathing material.

❖ To provide a surface for panel nailing, all vertical joints of panel sheathing should occur at studs, and all horizontal joints should occur over blocking. The exception recognizes that in lesser seismic areas the blocking behind horizontal joints in Method 3 is unnecessary to ensure stability of the structure.

R602.10.8 Connections. Braced wall panel sole plates shall be fastened to the floor framing and top plates shall be connected to the framing above in accordance with Table R602.3(1). Sills shall be fastened to the foundation or slab in accordance with Sections R403.1.6 and R602.11. Where joists are perpendicular to the braced wall lines above, blocking shall be provided under and in line with the braced wall panels.

❖ The cross references provided in this section refer the reader to the various sections specifying connections that provide a continuous load path for lateral loads as described in Section 301.

R602.10.9 Interior braced wall support. In one-story buildings located in Seismic Design Category D_2, interior braced wall lines shall be supported on continuous foundations at intervals not exceeding 50 feet (15 240 mm). In two story buildings located in Seismic Design Category D_2, all interior braced wall panels shall be supported on continuous foundations.

Exception: Two-story buildings shall be permitted to have interior braced wall lines on continuous foundations at intervals not exceeding 50 feet (15 240 mm) provided that:

1. The height of cripple walls does not exceed 4 feet (1219 mm).
2. First-floor braced wall panels are supported on doubled floor joists, continuous blocking or floor beams.
3. The distance between bracing lines does not exceed twice the building width measured parallel to the braced wall line.

❖ When an interior braced wall line (as required by Section R602.10.11) is not supported by a continuous foundation, it is necessary for the supporting floor (diaphragm) to distribute the lateral force to a resisting element (foundation). If a floor diaphragm is overloaded, it could lead to a failure, which in turn results in an incomplete load path. This section limits this condition for buildings classified as Seismic Design Category D_2. Commentary Figure R602.10.9(1) shows the limitations for one- and two-story buildings. Commentary Figure R602.10.9(2) shows the exception to the requirement for two-story buildings.

R602.10.10 Design of structural elements. Where a building, or portion thereof, does not comply with one or more of the bracing requirements in this section, those portions shall be designed and constructed in accordance with accepted engineering practice.

❖ This section essentially reiterates Section R301.1.2 in requiring design of elements not conforming to conventional light-frame construction requirements.

R602.10.11 Bracing in Seismic Design Categories D_1 and D_2. Structures located in Seismic Design Categories D_1 and D_2 shall be provided with exterior and interior braced wall lines. Spacing between braced wall lines in each story shall not exceed 25 feet (7620 mm) on center in both the longitudinal and transverse directions.

Figure R602.10.9(1)
SEISMIC DESIGN CATEGORY D₂ INTERIOR BRACED WALL SUPPORT

Exception: In one- and two-story buildings, spacing between braced wall lines shall not exceed 35 feet (10 363 mm) on center in order to accommodate one single room not exceeding 900 square feet (83.61 m^2) in each dwelling unit. The length of wall bracing in braced wall lines spaced greater or less other than 25 feet (7620 mm) apart shall be the length required by Table R602.10.1 multiplied by the appropriate adjustment factor from Table R602.10.11.

Exterior braced wall lines shall have a braced wall panel located at each end of the braced wall line.

Exception: For braced wall panel construction Method 3 of Section R602.10.3, the braced wall panel shall be permitted to begin no more than 8 feet (2438 mm) from each end of the braced wall line provided one of the following is satisfied:

1. A minimum 24-inch-wide (610 mm) panel is applied to each side of the building corner and the two 24-inch-wide (610 mm) panels at the corner shall be attached to framing in accordance with Figure R602.10.5 or,

2. The end of each braced wall panel closest to the corner shall have a tie-down device fastened to the stud at the edge of the braced wall panel closest to the corner and to the foundation or framing below. The tie-down device shall be capable of providing an uplift allowable design value of at least 1,800 pounds (816.5 kg). The tie-down device shall be installed in accordance with the manufacturer's recommendations.

A designed collector shall be provided if the bracing is not located at each end of a braced wall line as indicated above or more than 8 feet (2438 mm) from each end of a braced wall line as indicated in the exception.

❖ The conventional light-frame provisions assume that every exterior wall serves as a braced wall line; this is considered sufficient for Seismic Design Categories A, B and C. By establishing a maximum spacing for braced wall lines in Seismic Design Categories D₁ and D₂, the code is essentially limiting the span of floor or

WALL CONSTRUCTION
FIGURE R602.10.9(2) - TABLE R602.10.11

Figure R602.10.9(2)
SEISMIC DESIGN CATEGORY D$_2$ INTERIOR BRACED WALL SUPPORT

roof diaphragms as well as the loading on the braced wall panels. The exception permits this limit to be increased under certain conditions.

For Seismic Design Categories D$_1$ and D$_2$, this section modifies the basic requirements of Section R602.10.1 by requiring installation of a braced wall panel at each end of an exterior braced wall line unless a designed collector is provided. The exception permits braced wall panels of wood structural panel sheathing (see Section R602.10, Method 3) to begin up to 8 feet (2438 mm) from the end of a braced wall line as long as either of the two conditions is met. Additionally, a designed collector must be provided if this braced wall panel begins more than 8 feet (2438 mm) from the end of the exterior braced wall line.

TABLE R602.10.11
ADJUSTMENT OF BRACING AMOUNTS FOR INTERIOR BRACED WALL LINES ACCORDING TO BRACED WALL LINE SPACING[a,b]

BRACED WALL LINE SPACING (feet)	MULTIPLY BRACING AMOUNT IN TABLE R602.10.1 BY:
15 or less	0.6
20	0.8
25	1.0
30	1.2
35	1.4

For SI: 1 foot = 304.8 mm.

a. Linear interpolation is permissible.

b. The adjustment is limited to the larger spacing between braced wall lines to either side of an interior braced wall line.

2003 INTERNATIONAL RESIDENTIAL CODE® COMMENTARY—VOLUME 1

❖ This table is used in the exception to Section R602.10.11. The exception provides some flexibility in arranging braced wall lines. If spacing greater than or less than 25 feet (7620 mm) is provided, the appropriate adjustment may be made to the amount of bracing required by Table R602.10.1.

R602.10.11.1 Cripple wall bracing. In addition to the requirements of Section R602.10.2, where interior braced wall lines occur without a continuous foundation below, the length of parallel exterior cripple wall bracing shall be one and one-half times the length required by Table R602.10.1. Where cripple walls braced using Method 3 of Section R602.10.3 cannot provide this additional length, the capacity of the sheathing shall be increased by reducing the spacing of fasteners along the perimeter of each piece of sheathing to 4 inches (102 mm) on center.

❖ Section R602.10.2 contains the basic requirements for bracing cripple walls for all seismic design categories. This section requires additional cripple wall bracing in Seismic Design Categories D_1 and D_2 where an interior braced wall line is permitted to be installed without being supported on a continuous foundation. Commentary Figure R602.10.9(1) shows an example of this condition that would require the added cripple wall bracing.

R602.10.11.2 Sheathing attachment. Adhesive attachment of wall sheathing shall not be permitted in Seismic Design Categories C, D_1 and D_2.

❖ Generally sheathing must be attached to the wall framing using fasteners specified in Table 602.3(1). Attachment with adhesives, while not covered under the prescriptive fastening requirements, is certainly a possibility as an alternative method under Section R104.11. However, attachment of sheathing using adhesives results in a considerable increase in the strength and stiffness of the wall assembly. While there may be advantages to this approach, the added stiffness may generate higher seismic loads. See the discussion of inertial force due to earthquakes in Section R301.2.2. The higher seismic forces could, in turn, overload other elements in the lateral-force-resisting system, resulting in premature failure of those elements. Therefore, in buildings classified as Seismic Design Category C, D_1 or D_2, this method of wall sheathing attachment is not permitted.

R602.11 Framing and connections for Seismic Design Categories D_1 and D_2. The framing and connection details of buildings located in Seismic Design Categories D_1 and D_2 shall be in accordance with Sections R602.11.1 through R602.11.3.

❖ This section provides the connection details necessary to transfer lateral forces in a building classified as Seismic Design Category D_1 or D_2.

R602.11.1 Wall anchorage. Braced wall line sills shall be anchored to concrete or masonry foundations in accordance with Sections R403.1.6 and R602.11. For all buildings in Seismic Design Categories D_1 and D_2 and townhouses in Seismic Design Category C, plate washers, a minimum of $1/4$ inch by 3 inches by 3 inches (6.4 mm by 76 mm by 76 mm) in size, shall be provided between the foundation sill plate and the nut.

❖ This section reiterates the requirements of Section R403.1.6.1.

R602.11.2 Interior braced wall panel connections. Interior braced wall lines shall be fastened to floor and roof framing in accordance with Table R602.3(1), to required foundations in accordance with Section R602.11.1, and in accordance with the following requirements:

1. Floor joists parallel to the top plate shall be toe-nailed to the top plate with at least 8d nails spaced a maximum of 6 inches (150 mm) on center.
2. Top plate laps shall be face-nailed with at least eight 16d nails on each side of the splice.

❖ To provide the required lateral load path, fastening of braced wall panels to floor and roof framing must be according to the fastener schedule in Table R602.3(1). Anchorage to the foundation is also specified.

R602.11.3 Stepped foundations. Where stepped foundations occur, the following requirements apply:

1. Where the height of a required braced wall panel that extends from foundation to floor above varies more than 4 feet (1220 mm), the braced wall panel shall be constructed in accordance with Figure R602.11.3.
2. Where the lowest floor framing rests directly on a sill bolted to a foundation not less than 8 feet (2440 mm) in length along a line of bracing, the line shall be considered as braced. The double plate of the cripple stud wall beyond the segment of footing that extends to the lowest framed floor shall be spliced by extending the upper top plate a minimum of 4 feet (1219 mm) along the foundation. Anchor bolts shall be located a maximum of 1 foot and 3 feet (305 and 914 mm) from the step in the foundation.
3. Where cripple walls occur between the top of the foundation and the lowest floor framing, the bracing requirements for a story shall apply.
4. Where only the bottom of the foundation is stepped and the lowest floor framing rests directly on a sill bolted to the foundations, the requirements of Section R602.11.1 shall apply.

❖ Where foundations are stepped (see Section R403.1.5), this section provides additional criteria for Seismic Design Category D_1 or D_2. See Figure R602.11.3.

This section is applicable only where the elevation of the top of the foundation varies by more than 4 feet (1220 mm).

Also, if the floor framing is supported directly on a sill plate that is anchored to a foundation 8 feet (2338 mm) or more in length, then the braced wall line is adequately braced. If not, Item 3 applies, and the cripple wall must be considered a story.

WALL CONSTRUCTION

For SI: 1 inch = 25.4 mm, 1 foot = 304.8 mm.

Note: Where footing Section "A" is less than 8 feet long in a 25 feet total length wall, provide bracing at cripple stud wall.

FIGURE R602.11.3
STEPPED FOUNDATION CONSTRUCTION

❖ See the commentary for Section R602.11.3.

SECTION R603
STEEL WALL FRAMING

R603.1 General. Elements shall be straight and free of any defects that would significantly affect structural performance. Cold-formed steel wall framing members shall comply with the requirements of this section.

❖ The provisions of this section apply to the construction of load-bearing wall systems using cold-formed steel framing, which is that type of construction made up in part or entirely of steel structural members cold-formed to shape from sheet or strip steel. The general shape of the framing members is in the form of the letter "C" (C-shape) and is described in further detail in Section R603.2. The use of cold-formed steel-framed designs not consistent with these provisions is beyond the scope of this code, and an analysis by a design professional is recommended.

R603.1.1 Applicability limits. The provisions of this section shall control the construction of exterior steel wall framing and interior load-bearing steel wall framing for buildings not greater than 60 feet (18 288 mm) in length perpendicular to the joist or truss span, not greater than 36 feet (10 973 mm) in width parallel to the joist span or truss, and not greater than two stories in height with each story not greater than 10 feet (3048 mm) high. All exterior walls installed in accordance with the provisions of this section shall be considered as load-bearing walls. Steel walls constructed in accordance with the provisions of this section shall be limited to sites subjected to a maximum design wind speed of 110 miles per hour Exposure A, B or C and a maximum ground snow load of 70 pounds per foot (3.35 kN/m^2).

❖ See the commentary for Section R505.1.1.

R603.1.2 In-line framing. Load-bearing steel studs constructed in accordance with Section R603 shall be located directly in-line with joists, trusses and rafters with a maximum tolerance of $^3/_4$ inch (19.1 mm) between their center lines. Interior load-bearing steel stud walls shall be supported on foundations or shall be located directly above load-bearing walls with a maximum tolerance of $^3/_4$ inch (19.1 mm) between the centerline of the studs.

❖ In-line framing is the preferred framing method. The advantage of in-line framing is that it provides a direct load path for the transfer of forces from joists to studs. In-line framing maximizes framing alignment in order to minimize secondary moments on the framing members, taking into account that the track cannot function as a load transfer member. The $^3/_4$-inch (19 mm) alignment value is accepted industry practice and addresses those conditions where two framing members (roof rafter and ceiling joist, for example) are to be aligned over a wall stud. In the absence of in-line framing, a load distribution member, such as a structural track, may be required for this force transfer.

R603.2 Structural framing. Load-bearing steel wall framing members shall comply with Figure R603.2(1) and the dimensional and minimum thickness requirements specified in Tables R603.2(1) and R603.2(2). Tracks shall comply with Figure R603.2(2) and shall have a minimum flange width of 1 1/4 inches (32 mm). The maximum inside bend radius for load-bearing members shall be the greater of 3/32 inch (2.4 mm) or twice the uncoated steel thickness. Holes in wall studs and other structural members shall not exceed 1.5 inches (38 mm) in width or 4 inches (102 mm) in length as shown in Figure R603.2(3). Holes shall be permitted only along the centerline of the web of the framing member. Holes shall not be less than 24 inches (610 mm) center to center and shall not be located less than 10 inches (254 mm) from edge of hole to end of member unless patched in accordance with Section R603.3.5.

❖ Stud framing must comply with the dimensional and minimum thickness requirements specified in Tables R603.2(1) and R603.2(2) respectively. Note "a" to Table R603.2(1) explains the meaning of the alphanumeric member designation used for all steel framing members in the code. This system replaces the varied designation approaches that were produced by each individual manufacturer. In addition, the designation is used to identify not only a specific steel framing member but also the section properties of that same member through the use of the product technical information document.

Steel thickness is expressed in mils, which pertains to the base metal thickness measured prior to painting or the application of metallic (corrosion resistant) coatings. The base metal thickness is typically stamped or embossed on the member by the manufacturer. Table R603.2(2) also references the equivalent thickness in inches and provides a further reference to gage number. The use of gage numbers to define the thickness (a specific mean decimal thickness for which rolling tolerances were established) of steel sheet is a practice that has been discontinued by the steel industry.

To allow for routing utilities, web holes (also referred to as "penetrations," "utility holes" and "punchouts") are permitted in the webs of studs. To avoid adversely affecting the strength of wall framing, the web holes are limited to the locations along the centerline of the web of the framing member as illustrated in Figure R603.2(3). Where these holes occur within 10 inches (254 mm) of the edge of a member, they must be patched as illustrated in Figure R603.3.5. Patching of web holes must extend at least 1 inch (25.4 mm) beyond all edges of the hole, and it must be fastened to the web with No. 8 screws having a minimum edge distance of 1/2 inch (12.7 mm).

Steel track sections are to serve in nonload-bearing applications only and are shown in Figure R603.2(2). They must have a minimum flange width of 1 1/4 inches (32 mm).

R603.2.1 Material. Load-bearing steel framing members shall be cold-formed to shape from structural quality sheet steel complying with the requirements of one of the following:

1. ASTM A 653: Grades 33, 37, 40 and 50 (Classes 1 and 3).
2. ASTM A 792: Grades 33, 37, 40 and 50A.
3. ASTM A 875: Grades 33, 37, 40 and 50 (Classes 1 and 3).
4. Steels that comply with ASTM A 653, except for tensile and elongation, shall be permitted, provided the ratio of tensile strength to yield point is at least 1.08 and the total elongation is at least 10 percent for a 2-inch (51 mm) gage length or 7 percent for an 8-inch (203 mm) gage length.

❖ Load-bearing steel framing members must be cold-formed to shape from structural quality sheet steel complying with appropriate material standards. The steel grades specified have the ductility and strength to meet the intent of these provisions.

R603.2.2 Identification. Load-bearing steel framing members shall have a legible label, stencil, stamp or embossment with the following information as a minimum:

1. Manufacturer's identification.
2. Minimum uncoated steel thickness in inches (mm).
3. Minimum coating designation.
4. Minimum yield strength, in kips per square inch (ksi) (kPa).

❖ Load-bearing steel framing members must have a legible label, stencil, stamp or embossment. This identification allows for verification that materials installed are consistent with the design and meet the intent of the code.

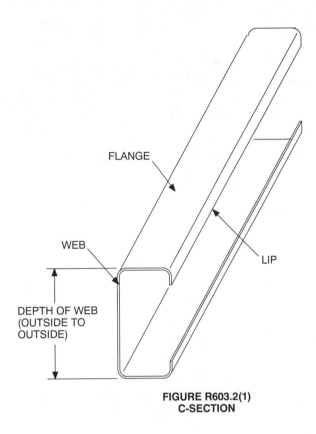

**FIGURE R603.2(1)
C-SECTION**

❖ See the commentary for Section R603.2.

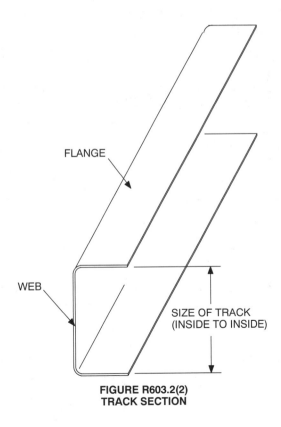

**FIGURE R603.2(2)
TRACK SECTION**

❖ See the commentary for Section R603.2.

FIGURE R603.2(3) - TABLE 603.2(2)   WALL CONSTRUCTION

For SI: 1 inch = 25.4 mm.

**FIGURE R603.2(3)
WEB HOLES**

❖ Holes are permitted in the web of steel wall studs as shown in order to allow for the installation of plumbing, wiring, etc. in a wall. The limits on size, location and spacing of these holes accommodate these utilities without adversely affecting the stud's load-carrying capacity.

**TABLE R603.2(1)
LOAD-BEARING COLD-FORMED STEEL STUD SIZES**

MEMBER DESIGNATION[a]	WEB DEPTH (inches)	MINIMUM FLANGE WIDTH (inches)	MAXIMUM FLANGE WIDTH (inches)	MINIMUM LIP SIZE (inches)
350S162-t	3.5	1.625	2	0.5
550S162-t	5.5	1.625	2	0.5

For SI: 1 inch = 25.4 mm.

a. The member designation is defined by the first number representing the member depth in $1/100$ inches, the letter "S" representing a stud or joist member, the second number representing the flange width in $1/100$ inches, and the letter "t" shall be a number representing the minimum base metal thickness in mils [See Table R603.2(2)].

❖ See the commentary for Section R603.2.

**TABLE R603.2(2)
MINIMUM THICKNESS OF COLD-FORMED STEEL STUDS**

DESIGNATION (mils)	MINIMUM UNCOATED THICKNESS (inches)	REFERENCE GAGE NUMBER
33	0.033	20
43	0.043	18
54	0.054	16
68	0.068	14

For SI: 1 inch = 25.4 mm, 1 mil = 0.0254 mm.

❖ See the commentary for Section R603.2.

R603.2.3 Corrosion protection. Load-bearing steel framing shall have a metallic coating complying with one of the following:

1. A minimum of G 60 in accordance with ASTM A 653.
2. A minimum of AZ 50 in accordance with ASTM A 792.
3. A minimum of GF 60 in accordance with ASTM A 875.

❖ The metallic coatings specified correspond to requirements in the referenced material standards. The minimum coating designations assume normal exposure conditions that are best defined as having the framing members enclosed within a building envelope or wall assembly within a controlled environment. When more severe exposure conditions are probable, such as industrial or marine atmospheres, consideration should be given to specifying a heavier coating.

R603.2.4 Fastening requirements. Screws for steel- to-steel connections shall be installed with a minimum edge distance and center-to-center spacing of $^1/_2$ inch (12.7 mm), shall be self-drilling tapping and shall conform to SAE J 78. Structural sheathing shall be attached to steel studs with minimum No. 8 self-drilling tapping screws that conform to SAE J78. Screws for attaching structural sheathing to steel wall framing shall have a minimum head diameter of 0.292 inch (7.4 mm) with countersunk heads and shall be installed with a minimum edge distance of $^3/_8$ inch (9.5 mm). Gypsum board shall be attached to steel wall framing with minimum No. 6 screws conforming to ASTM C 954 and shall be installed in accordance with Section R702. For all connections, screws shall extend through the steel a minimum of three exposed threads. All self-drilling tapping screws conforming to SAE J 78 shall have a Type II coating in accordance with ASTM B 633.

Where No. 8 screws are specified in a steel to steel connection the required number of screws in the connection is permitted to be reduced in accordance with the reduction factors in Table R505.2.4, when larger screws are used or when one of the sheets of steel being connected is thicker than 33 mils (0.84 mm). When applying the reduction factor the resulting number of screws shall be rounded up.

❖ Fasteners meeting the referenced standards and installed in accordance with this section will provide a load capacity consistent with these prescriptive provisions.

TABLE R603.2.4
SCREW SUBSTITUTION FACTOR

SCREW SIZE	THINNEST CONNECTED STEEL SHEET (mils)	
	33	43
#8	1.0	0.67
#10	0.93	0.62
#12	0.86	0.56

For SI: 1 mil = 0.0254 mm.

❖ In providing prescriptive connection requirements, the code typically specifies the quantity of No. 8 screws necessary to resist the required design forces transferred through a given connection. If the quantity of screws required for a particular connection is large, it may be desirable to reduce the quantity of screws by substituting a larger screw size. Where No. 8 screws are specified for a steel-to-steel connection, the required quantity of screws may be reduced if a larger screw size is substituted in accordance with the reduction factors of Table R603.2.4. Likewise, if the thinnest of the connected sheets is at least 43 mils (1.1 mm), the required quantity of screws may be reduced.

R603.3 Wall construction. All exterior steel framed walls and interior load-bearing steel framed walls shall be constructed in accordance with the provisions of this section and Figure R603.3.

❖ The balance of this section is comprised of the prescriptive design provisions for steel-framed walls.

R603.3.1 Wall to foundation or floor connections. Steel framed walls shall be anchored to foundations or floors in accordance with Table R603.3.1 and Figure R603.3.1(1) or R603.3.1(2).

❖ Steel framed walls anchored directly to foundations must be in accordance with Figure R603.3.1(1) with $^1/_2$-inch (12.7 mm) diameter anchor bolts spaced as required by Table R603.3.1 based on either the seismic design category or the wind speed/exposure. Wall tracks connected to wood sills must be in accordance with Figure R603.3.1(2) and Table R603.3.1. This connection is the same as required for floors in Figure R505.3.1(2).

R603.3.2 Load-bearing walls. Steel studs shall comply with Tables R603.3.2(2) through R603.3.2(7) for steels with minimum yield strength of 33 ksi (227.7 MPa) and Tables R603.3.2(8) through R603.3.2(13) for steels with minimum yield strength of 50 ksi (345 MPa). Fastening requirements shall be in accordance with Section R603.2.4 and Table R603.3.2(1). Tracks shall have the same minimum steel thickness as the wall studs. Exterior walls with a minimum of $^1/_2$-inch (12.7 mm) gypsum board installed in accordance with Section R702 on the interior surface and wood structural panels of minimum $^7/_{16}$-inch-thick (11.1 mm) oriented strand board or $^{15}/_{32}$-inch-thick (11.9 mm) plywood installed in accordance with Table R603.3.2(1) on the outside surface shall be permitted to use the next thinner stud, from Tables R603.3.2(2) through R603.3.2(13) but not less than 33 mils (0.84 mm). Interior load-bearing walls with a minimum of $^1/_2$-inch (12.7 mm) gypsum board installed in accordance with Section R702 on both sides of the wall shall be permitted to use the next thinner stud, from Tables R603.3.2(2) through R603.3.2(13) but not less than 33 mils (0.84 mm).

❖ Steel studs must comply with Tables R603.3.2(2) through R603.3.2(7) for steels with a minimum yield strength of 33 ksi (227,502 kPa) and Tables R603.3.2(8) through R603.3.2(13) for steels with a minimum yield strength of 50 ksi (344,700 kPa). These tables specify the minimum steel stud thickness based on wall height, wind speed/exposure, ground snow load, building width and stud depth and spacing.

In recognition of the stiffening effects of sheathing, the code allows the required stud thickness to be reduced to the next thinner size for exterior walls with minimum $1/2$-inch (12.7 mm) gypsum board installed on the interior surface and wood structural panels [minimum 7/16-inch (11.1 mm) thick oriented strand board or 15/32-inch (11.9 mm) thick plywood] installed on the exterior. Stud thickness must not be less than 33 mils (0.84 mm). Similarly, interior load-bearing walls with minimum $1/2$-inch (12.7 mm) gypsum board installed on both sides of the wall may use the next thinner stud, but not less than 33 mils (0.84 mm).

Fastening of studs to tracks and structural sheathing to studs must be as specified in Section R603.2.4 and Table R603.3.2(1). Track thickness must be no less than that of the wall studs.

For SI: 1 inch = 25.4 mm.

**FIGURE R603.3
STEEL WALL CONSTRUCTION**

❖ This figure illustrates typical steel-framed wall construction and many of the steel-framed wall elements regulated by this section.

WALL CONSTRUCTION FIGURE R603.3.1(1) - TABLE R603.3.1

For SI: 1 inch = 25.4 mm.

**FIGURE R603.3.1(1)
WALL TO FOUNDATION CONNECTION**

❖ This connection detail depicts the item on the third line of Table R603.3.1.

**TABLE R603.3.1
WALL TO FOUNDATION OR FLOOR CONNECTION REQUIREMENTS[a,b,c]**

	BASIC WIND SPEED (mph) AND EXPOSURE		
FRAMING CONDITION	85 A/B or Seismic Design Categories A, B and C	85 C or less than 110 A/B	Less than 110 C
Wall bottom track to floor joist or track	1-No. 8 screw at 12″ o.c.	1-No. 8 screw at 12″ o.c.	2-No. 8 screw at 12″ o.c.
Wall bottom track to wood sill per Figure R603.3.1(2)	Steel plate spaced at 4′ o.c., with 4-No. 8 screws and 4-10d or 6-8d common nails	Steel plate spaced at 3′ o.c., with 4-No. 8 screws and 4-10d or 6-8d common nails	Steel plate spaced at 2′ o.c., with 4-No. 8 screws and 4-10d or 6-8d common nails
Wall bottom track to foundation per Figure R603.3.1(1)	$1/2$″ minimum diameter anchor bolt at 6′ o.c.	$1/2$″ minimum diameter anchor bolt at 6′ o.c.	$1/2$″ minimum diameter anchor bolt at 4′ o.c.
Wind uplift connector capacity for 16-inch stud spacing[c]	N/R	N/R	65 lbs.
Wind uplift connector capacity for 24-inch stud spacing[c]	N/R	N/R	100 lbs.

For SI: 1 inch = 25.4 mm, 1 foot = 304.8 mm, 1 mile per hour = 1.609 km/hr, 1 pound = 4.4 N.

a. Anchor bolts shall be located not more than 12 inches from corners or the termination of bottom tracks (e.g. at door openings or corners). Bolts shall extend a minimum of 7 inches into concrete or masonry.
b. All screw sizes shown are minimum.
c. N/R = uplift connector not required. Uplift connectors are in addition to other connection requirements.

❖ This table specifies connections for wall tracks, which will provide appropriate load capacity based on the seismic design category and the basic wind speed.

FIGURE R603.3.1(2) - TABLE R603.3.2(1)   WALL CONSTRUCTION

For SI: 1 inch = 25.4 mm, 1 mil = 0.0254 mm.

FIGURE R603.3.1(2)
WALL TO WOOD SILL CONNECTION

❖ This connection detail illustrates the item on the second line of Table R603.3.1 and is identical to that for floors in Figure R505.3.1(2).

TABLE R603.3.2(1)
WALL FASTENING SCHEDULE[a]

DESCRIPTION OF BUILDING ELEMENT	NUMBER AND SIZE OF FASTENERS[a]	SPACING OF FASTENERS
Floor joist to track of load-bearing wall	2-No. 8 screws	Each joist
Wall stud to top or bottom track	2-No. 8 screws	Each end of stud, one per flange
Structural sheathing to wall studs	No. 8 screws	6" o.c. on edges and 12" o.c. at intermediate supports
Roof framing to wall	Approved design or tie down in accordance with Section R802.11	

For SI: 1 inch = 25.4 mm.

a. All screw sizes shown are minimum.

❖ This table provides fastener size and spacing for various portions of steel stud walls.

WALL CONSTRUCTION

TABLE R603.3.2(2)
COLD-FORMED STEEL STUD THICKNESS FOR 8-FOOT WALLS
Studs supporting roof and ceiling only (one-story building or second floor of a two-story building) 33 ksi steel

WIND SPEED		MEMBER SIZE[c]	MEMBER SPACING (inches)	STUD THICKNESS (mils)[a,b] Building width (feet)[d]															
Exp. A/B	Exp. C			24 Ground snow load (psf)				28 Ground snow load (psf)				32 Ground snow load (psf)				36 Ground snow load (psf)			
				20	30	50	70	20	30	50	70	20	30	50	70	20	30	50	70
85 mph	—	350S162	16	33	33	33	33	33	33	33	33	33	33	33	33	33	33	33	33
			24	33	33	33	33	33	33	33	33	33	33	33	33	33	33	33	43
		550S162	16	33	33	33	33	33	33	33	33	33	33	33	33	33	33	33	33
			24	33	33	33	33	33	33	33	33	33	33	33	33	33	33	33	33
100 mph	85 mph	350S162	16	33	33	33	33	33	33	33	33	33	33	33	33	33	33	33	33
			24	33	33	33	33	33	33	33	33	33	33	33	43	33	33	43	43
		550S162	16	33	33	33	33	33	33	33	33	33	33	33	33	33	33	33	33
			24	33	33	33	33	33	33	33	33	33	33	33	33	33	33	33	33
110 mph	100 mph	350S162	16	33	33	33	33	33	33	33	33	33	33	33	33	33	33	33	33
			24	33	33	43	43	33	33	43	43	33	43	43	43	43	43	43	43
		550S162	16	33	33	33	33	33	33	33	33	33	33	33	33	33	33	33	33
			24	33	33	33	33	33	33	33	33	33	33	33	33	33	33	33	33
120 mph	110 mph	350S162	16	33	33	33	33	33	33	33	33	33	33	33	33	33	33	33	33
			24	43	43	43	43	43	43	43	54	43	43	54	54	43	43	54	54
		550S162	16	33	33	33	33	33	33	33	33	33	33	33	33	33	33	33	33
			24	33	33	33	33	33	33	33	33	33	33	33	33	33	33	33	33
130 mph	120 mph	350S162	16	33	33	43	43	33	43	43	43	43	43	43	43	43	43	43	43
			24	54	54	54	54	54	54	54	68	54	54	68	68	54	54	68	68
		550S162	16	33	33	33	33	33	33	33	33	33	33	33	33	33	33	33	33
			24	33	33	33	33	33	33	33	43	33	33	33	43	33	33	43	43
—	130 mph	350S162	16	43	43	43	43	43	43	43	43	43	43	43	54	43	43	43	54
			24	68	68	68	68	68	68	68	68	68	68	68	68	68	68	68	(d)
		550S162	16	33	33	33	33	33	33	33	33	33	33	33	33	33	33	33	33
			24	33	43	43	43	43	43	43	43	43	43	43	43	43	43	43	43

For SI: 1 inch = 25.4 mm, 1 foot = 304.8 mm, 1 mil = 0.0254 mm, 1 mile per hour = 1.609 km/h, 1 pound per square foot = 0.0479 kN/m^2, 1 kilogram per square inch = 6.895 MPa.

a. Deflection criteria: $L/240$.
b. Building width is in the direction of horizontal framing members supported by the wall studs.
c. Design load assumptions:
   Roof dead load is 12 psf.
   Attic live load is 10 psf.
d. 68-mil-thick stud is allowed if wall is fully sheathed per Section R603.3.2.

❖ See the commentary for Section R603.3.2.

TABLE R603.3.2(3)
COLD-FORMED STEEL STUD THICKNESS FOR 8-FOOT WALLS
Studs supporting one floor, roof and ceiling (first story of a two-story building) 33 ksi steel

WIND SPEED		MEMBER SIZE[c]	MEMBER SPACING (inches)	STUD THICKNESS (mils)[a,b] Building width (feet)[d]															
				24				28				32				36			
Exp. A/B	Exp. C			Ground snow load (psf)				Ground snow load (psf)				Ground snow load (psf)				Ground snow load (psf)			
				20	30	50	70	20	30	50	70	20	30	50	70	20	30	50	70
85 mph	—	350S162	16	33	33	33	33	33	33	33	33	33	33	33	43	33	33	33	43
			24	43	43	43	43	43	43	43	43	43	43	43	54	43	43	54	54
		550S162	16	33	33	33	33	33	33	33	33	33	33	33	33	33	33	33	33
			24	33	33	33	33	33	33	33	43	33	33	43	43	33	43	43	54
100 mph	85 mph	350S162	16	33	33	33	33	33	33	33	33	33	33	33	43	33	33	43	43
			24	43	43	43	54	43	43	54	54	54	54	54	54	54	54	54	54
		550S162	16	33	33	33	33	33	33	33	33	33	33	33	33	33	33	33	43
			24	33	33	33	33	33	33	33	43	33	33	43	43	43	43	43	54
110 mph	100 mph	350S162	16	33	33	33	33	33	33	43	43	43	43	43	43	43	43	43	43
			24	54	54	54	54	54	54	54	54	54	54	54	68	54	54	68	68
		550S162	16	33	33	33	33	33	33	33	33	33	33	33	33	33	33	33	33
			24	33	33	33	33	33	33	43	43	43	43	43	43	43	43	43	54
120 mph	110 mph	350S162	16	43	43	43	43	43	43	43	43	43	43	43	43	43	43	43	54
			24	54	54	54	68	54	68	68	68	68	68	68	68	68	68	68	68
		550S162	16	33	33	33	33	33	33	33	33	33	33	33	33	33	33	33	33
			24	33	43	43	43	43	43	43	43	43	43	43	43	43	43	43	54
130 mph	120 mph	350S162	16	43	43	43	54	43	54	54	54	54	54	54	54	54	54	54	54
			24	68	68	68	68	68	68	(d)	(d)	(d)	(d)	(d)	(d)	(d)	(d)	(d)	(d)
		550S162	16	33	33	33	33	33	33	33	33	33	33	33	33	33	33	33	43
			24	43	43	43	43	43	43	43	54	43	43	54	54	54	54	54	54
—	130 mph	350S162	16	54	54	54	54	54	54	54	54	54	54	54	68	54	54	68	68
			24	(d)	(d)	(d)	(d)	(d)	(d)	(d)	(d)	(d)	(d)	(d)	(d)	(d)	(d)	(d)	(d)
		550S162	16	33	33	33	33	33	33	33	43	33	33	43	43	43	43	43	43
			24	43	43	54	54	54	54	54	54	54	54	54	54	54	54	54	54

For SI: 1 inch = 25.4 mm, 1 foot = 304.8 mm, 1 mil = 0.0254 mm, 1 mile per hour = 1.609 km/h, 1 pound per square foot = 0.0479 kN/m^2, 1 kilogram per square inch = 6.895 MPa.

a. Deflection criteria: $L/240$.
b. Building width is in the direction of horizontal framing members supported by the wall studs.
c. Design load assumptions:
    Roof dead load is 12 psf.
    Attic live load is 10 psf.
d. 68-mil-thick stud is allowed used if wall is fully sheathed per Section R603.3.2.

❖ See the commentary for Section R603.3.2.

TABLE R603.3.2(4)
COLD-FORMED STEEL STUD THICKNESS FOR 9-FOOT WALLS
Studs supporting roof and ceiling only (one-story building or second floor of a two-story building) 33 ksi steel

WIND SPEED		MEMBER SIZE[c]	MEMBER SPACING (inches)	STUD THICKNESS (mils)[a,b] Building width (feet)[d]															
Exp. A/B	Exp. C			24				28				32				36			
				Ground snow load (psf)				Ground snow load (psf)				Ground snow load (psf)				Ground snow load (psf)			
				20	30	50	70	20	30	50	70	20	30	50	70	20	30	50	70
85 mph	—	350S162	16	33	33	33	33	33	33	33	33	33	33	33	33	33	33	33	33
			24	33	33	33	33	33	33	33	33	33	33	33	33	33	33	33	43
		550S162	16	33	33	33	33	33	33	33	33	33	33	33	33	33	33	33	33
			24	33	33	33	33	33	33	33	33	33	33	33	33	33	33	33	33
100 mph	85 mph	350S162	16	33	33	33	33	33	33	33	33	33	33	33	33	33	33	33	33
			24	33	33	43	43	33	33	43	43	33	43	43	43	43	43	43	43
		550S162	16	33	33	33	33	33	33	33	33	33	33	33	33	33	33	33	33
			24	33	33	33	33	33	33	33	33	33	33	33	33	33	33	33	33
110 mph	100 mph	350S162	16	33	33	33	33	33	33	33	33	33	33	33	33	33	33	33	33
			24	43	43	43	43	43	43	43	43	43	43	43	54	43	43	43	54
		550S162	16	33	33	33	33	33	33	33	33	33	33	33	33	33	33	33	33
			24	33	33	33	33	33	33	33	33	33	33	33	33	33	33	33	33
120 mph	110 mph	350S162	16	33	33	33	33	33	33	33	43	33	33	43	43	33	33	43	43
			24	54	54	54	54	54	54	54	54	54	54	54	54	54	54	54	68
		550S162	16	33	33	33	33	33	33	33	33	33	33	33	33	33	33	33	33
			24	33	33	33	33	33	33	33	33	33	33	33	33	33	33	33	43
130 mph	120 mph	350S162	16	43	43	43	43	43	43	43	43	43	43	43	54	43	43	43	54
			24	68	68	68	68	68	68	68	68	68	68	68	68	68	68	68	(d)
		550S162	16	33	33	33	33	33	33	33	33	33	33	33	33	33	33	33	33
			24	33	43	43	43	43	43	43	43	43	43	43	43	43	43	43	43
—	130 mph	350S162	16	54	54	54	54	54	54	54	54	54	54	54	54	54	54	54	54
			24	(d)	(d)	(d)	(d)	(d)	(d)	(d)	(d)	(d)	(d)	(d)	(d)	(d)	(d)	(d)	(d)
		550S162	16	33	33	33	33	33	33	33	33	33	33	33	33	33	33	33	33
			24	43	43	43	43	43	43	43	43	43	43	43	54	43	43	43	54

For SI: 1 inch = 25.4 mm, 1 foot = 304.8 mm, 1 mil = 0.0254 mm, 1 mile per hour = 1.609 km/h, 1 pound per square foot = 0.0479 kN/m^2, 1 kilogram per square inch = 6.895 MPa.

a. Deflection criteria: $L/240$.
b. Building width is in the direction of horizontal framing members supported by the wall studs.
c. Design load assumptions:
   Roof dead load is 12 psf.
   Attic live load is 10 psf.
d. 68-mil-thick stud is allowed if wall is fully sheathed per Section R603.3.2.

❖ See the commentary for Section R603.3.2.

TABLE R603.3.2(5)
COLD-FORMED STEEL STUD THICKNESS FOR 9-FOOT WALLS
Studs supporting one floor, roof and ceiling (first story of a two-story building) 33 ksi steel

WIND SPEED		MEMBER SIZE[c]	MEMBER SPACING (inches)	STUD THICKNESS (mils)[a,b] Building width (feet)[d]															
				24				28				32				36			
				Ground snow load (psf)				Ground snow load (psf)				Ground snow load (psf)				Ground snow load (psf)			
Exp. A/B	Exp. C			20	30	50	70	20	30	50	70	20	30	50	70	20	30	50	70
85 mph	—	350S162	16	33	33	33	33	33	33	33	33	33	33	33	33	33	33	33	43
			24	43	43	43	43	43	43	43	54	43	43	54	54	54	54	54	54
		550S162	16	33	33	33	33	33	33	33	33	33	33	33	33	33	33	33	33
			24	33	33	33	33	33	33	33	43	33	33	43	43	33	33	43	43
100 mph	85 mph	350S162	16	33	33	33	33	33	33	33	43	33	43	43	43	43	43	43	43
			24	43	54	54	54	54	54	54	54	54	54	54	54	54	54	68	68
		550S162	16	33	33	33	33	33	33	33	33	33	33	33	33	33	33	33	33
			24	33	33	33	33	33	33	33	43	33	33	43	43	43	43	43	43
110 mph	100 mph	350S162	16	43	43	43	43	43	43	43	43	43	43	43	43	43	43	43	43
			24	54	54	54	54	54	54	54	54	54	54	54	54	54	54	68	68
		550S162	16	33	33	33	33	33	33	33	33	33	33	33	33	33	33	33	33
			24	33	33	43	43	43	43	43	43	43	43	43	43	43	43	43	54
120 mph	110 mph	350S162	16	43	43	43	43	43	43	43	54	43	43	54	54	54	54	54	54
			24	68	68	68	68	68	68	68	68	68	68	68	(d)	68	68	(d)	(d)
		550S162	16	33	33	33	33	33	33	33	33	33	33	33	33	33	33	33	33
			24	43	43	43	43	43	43	43	43	43	43	43	54	43	43	54	54
130 mph	120 mph	350S162	16	54	54	54	54	54	54	54	54	54	54	54	68	54	54	68	68
			24	(d)	(d)	(d)	(d)	(d)	(d)	(d)	(d)	(d)	(d)	(d)	(d)	(d)	(d)	(d)	(d)
		550S162	16	33	33	33	33	33	33	33	43	33	33	43	43	33	43	43	43
			24	43	43	54	54	54	54	54	54	54	54	54	54	54	54	54	54
—	130 mph	350S162	16	68	68	68	68	68	68	68	68	68	68	68	68	68	68	68	68
			24	(d)	(d)	(d)	(d)	(d)	(d)	(d)	—	(d)	(d)	—	—	—	—	—	—
		550S162	16	33	33	43	43	43	43	43	43	43	43	43	43	43	43	43	43
			24	54	54	54	54	54	54	54	54	54	54	54	68	54	54	68	68

For SI: 1 inch = 25.4 mm, 1 foot = 304.8 mm, 1 mil = 0.0254 mm, 1 mile per hour = 1.609 km/h, 1 pound per square foot = 0.0479 kN/m^2, 1 kilogram per square inch = 6.895 MPa.

a. Deflection criteria: $L/240$.
b. Building width is in the direction of horizontal framing members supported by the wall studs.
c. Design load assumptions:
   Roof dead load is 12 psf.
   Attic live load is 10 psf.
d. 68-mil-thick stud is allowed if wall is fully sheathed per Section R603.3.2.

❖ See the commentary for Section R603.3.2.

WALL CONSTRUCTION

TABLE R603.3.2(6)
COLD-FORMED STEEL STUD THICKNESS FOR 10-FOOT WALLS
Studs supporting roof and ceiling only (one-story building or second floor of a two-story building) 33 ksi steel

WIND SPEED		MEMBER SIZE[c]	MEMBER SPACING (inches)	STUD THICKNESS (mils)[a,b] Building width (feet)[d]															
				24				28				32				36			
				Ground snow load (psf)				Ground snow load (psf)				Ground snow load (psf)				Ground snow load (psf)			
Exp. A/B	Exp. C			20	30	50	70	20	30	50	70	20	30	50	70	20	30	50	70
85 mph	—	350S162	16	33	33	33	33	33	33	33	33	33	33	33	33	33	33	33	33
			24	33	33	33	43	33	33	43	43	33	33	43	43	33	43	43	43
		550S162	16	33	33	33	33	33	33	33	33	33	33	33	33	33	33	33	33
			24	33	33	33	33	33	33	33	33	33	33	33	33	33	33	33	33
100 mph	85 mph	350S162	16	33	33	33	33	33	33	33	33	33	33	33	33	33	33	33	33
			24	43	43	43	43	43	43	43	54	43	43	43	54	43	43	54	54
		550S162	16	33	33	33	33	33	33	33	33	33	33	33	33	33	33	33	33
			24	33	33	33	33	33	33	33	33	33	33	33	33	33	33	33	33
110 mph	100 mph	350S162	16	33	33	33	43	33	33	43	43	33	33	43	43	33	33	43	43
			24	54	54	54	54	54	54	54	54	54	54	54	54	54	54	54	68
		550S162	16	33	33	33	33	33	33	33	33	33	33	33	33	33	33	33	33
			24	33	33	33	33	33	33	33	33	33	33	33	33	33	33	33	43
120 mph	110 mph	350S162	16	43	43	43	43	43	43	43	43	43	43	43	43	43	43	43	54
			24	68	68	68	68	68	68	68	68	68	68	68	68	68	68	68	68
		550S162	16	33	33	33	33	33	33	33	33	33	33	33	33	33	33	33	33
			24	33	33	43	43	33	33	43	43	33	43	43	43	43	43	43	43
130 mph	120 mph	350S162	16	54	54	54	54	54	54	54	54	54	54	54	54	54	54	54	68
			24	(d)	(d)	(d)	(d)	(d)	(d)	(d)	(d)	(d)	(d)	(d)	(d)	(d)	(d)	(d)	(d)
		550S162	16	33	33	33	33	33	33	33	33	33	33	33	33	33	33	33	33
			24	43	43	43	43	43	43	43	54	43	43	54	54	43	43	54	54
—	130 mph	350S162	16	68	68	68	68	68	68	68	68	68	68	68	68	68	68	68	68
			24	(d)	(d)	(d)	—	(d)	(d)	—	—	(d)	(d)	—	—	(d)	—	—	—
		550S162	16	33	33	33	43	33	33	43	43	33	33	43	43	33	43	43	43
			24	54	54	54	54	54	54	54	54	54	54	54	54	54	54	54	68

For SI: 1 inch = 25.4 mm, 1 foot = 304.8 mm, 1 mil = 0.0254 mm, 1 mile per hour = 1.609 km/h, 1 pound per square foot = 0.0479 kN/m^2, 1 kilogram per square inch = 6.895 MPa.

a. Deflection criteria: $L/240$.
b. Building width is in the direction of horizontal framing members supported by the wall studs.
c. Design load assumptions:
   Roof dead load is 12 psf.
   Attic live load is 10 psf.
d. 68-mil-thick stud is allowed if wall is fully sheathed per Section R603.3.2.

❖ See the commentary for Section R603.3.2.

TABLE R603.3.2(7)
COLD-FORMED STEEL STUD THICKNESS FOR 10-FOOT WALLS
Studs supporting one floor, roof and ceiling (first story of a two-story building) 33 ksi steel

WIND SPEED		MEMBER SIZE[c]	MEMBER SPACING (inches)	STUD THICKNESS (mils)[a,b] Building width (feet)[d]															
				24				28				32				36			
Exp. A/B	Exp. C			Ground snow load (psf)				Ground snow load (psf)				Ground snow load (psf)				Ground snow load (psf)			
				20	30	50	70	20	30	50	70	20	30	50	70	20	30	50	70
85 mph	—	350S162	16	33	33	33	43	33	33	43	43	43	43	43	43	43	43	43	43
			24	54	54	54	54	54	54	54	54	54	54	54	68	54	54	68	68
		550S162	16	33	33	33	33	33	33	33	33	33	33	33	33	33	33	33	33
			24	33	33	33	33	33	33	33	43	33	33	43	43	43	43	43	54
100 mph	85 mph	350S162	16	43	43	43	43	43	43	43	43	43	43	43	43	43	43	43	54
			24	54	54	68	68	68	68	68	68	68	68	68	68	68	68	68	(d)
		550S162	16	33	33	33	33	33	33	33	33	33	33	33	33	33	33	33	33
			24	33	33	43	43	43	43	43	43	43	43	43	43	43	43	43	54
110 mph	100 mph	350S162	16	43	43	43	43	43	43	54	54	43	54	54	54	54	54	54	54
			24	68	68	68	68	68	68	68	(d)	68	68	(d)	(d)	(d)	(d)	(d)	(d)
		550S162	16	33	33	33	33	33	33	33	33	33	33	33	33	33	33	33	33
			24	43	43	43	43	43	43	43	43	43	43	43	54	43	43	54	54
120 mph	110 mph	350S162	16	54	54	54	54	54	54	54	54	54	54	54	54	54	54	54	54
			24	(d)	(d)	(d)	(d)	(d)	(d)	(d)	(d)	(d)	(d)	(d)	(d)	(d)	(d)	(d)	(d)
		550S162	16	33	33	33	33	33	33	33	33	33	33	33	43	33	33	43	43
			24	43	43	43	54	43	54	54	54	54	54	54	54	54	54	54	54
130 mph	120 mph	350S162	16	68	68	68	68	68	68	68	68	68	68	68	68	68	68	68	(4)
			24	(d)	(d)	—	—	—	—	—	—	—	—	—	—	—	—	—	—
		550S162	16	43	43	43	43	43	43	43	43	43	43	43	43	43	43	43	43
			24	54	54	54	54	54	54	54	68	54	68	68	68	68	68	68	68
—	130 mph	350S162	16	68	(d)	(d)	(d)	(d)	(d)	(d)	(d)	(d)	(d)	(d)	(d)	(d)	(d)	(d)	(d)
			24	—	—	—	—	—	—	—	—	—	—	—	—	—	—	—	—
		550S162	16	43	43	43	43	43	43	43	43	43	43	43	54	43	43	54	54
			24	68	68	68	68	68	68	68	68	68	68	68	68	68	68	68	(d)

For SI: 1 inch = 25.4 mm, 1 foot = 304.8 mm, 1 mil = 0.0254 mm, 1 mile per hour = 1.609 km/h, 1 pound per square foot = 0.0479 kN/m^2, 1 kilogram per square inch = 6.895 MPa.

a. Deflection criteria: $L/240$.
b. Building width is in the direction of horizontal framing members supported by the wall studs.
c. Design load assumptions:
   Roof dead load is 12 psf.
   Attic live load is 10 psf.
d. 68-mil-thick stud is allowed if wall is fully sheathed per Section R603.3.2.

❖ See the commentary for Section R603.3.2.

WALL CONSTRUCTION

TABLE R603.3.2(8)
COLD-FORMED STEEL STUD THICKNESS FOR 8-FOOT WALLS
Studs supporting roof and ceiling only (one-story building or second floor of a two-story building) 50 ksi steel

WIND SPEED		MEMBER SIZE[c]	MEMBER SPACING (inches)	STUD THICKNESS (mils)[a,b] Building width (feet)[d]															
				24				28				32				36			
				Ground snow load (psf)				Ground snow load (psf)				Ground snow load (psf)				Ground snow load (psf)			
Exp. A/B	Exp. C			20	30	50	70	20	30	50	70	20	30	50	70	20	30	50	70
85 mph	—	350S162	16	33	33	33	33	33	33	33	33	33	33	33	33	33	33	33	33
			24	33	33	33	33	33	33	33	33	33	33	33	33	33	33	33	33
		550S162	16	33	33	33	33	33	33	33	33	33	33	33	33	33	33	33	33
			24	33	33	33	33	33	33	33	33	33	33	33	33	33	33	33	33
100 mph	85 mph	350S162	16	33	33	33	33	33	33	33	33	33	33	33	33	33	33	33	33
			24	33	33	33	33	33	33	33	33	33	33	33	33	33	33	33	33
		550S162	16	33	33	33	33	33	33	33	33	33	33	33	33	33	33	33	33
			24	33	33	33	33	33	33	33	33	33	33	33	33	33	33	33	33
110 mph	100 mph	350S162	16	33	33	33	33	33	33	33	33	33	33	33	33	33	33	33	33
			24	33	33	33	33	33	33	33	33	33	33	33	33	33	33	33	43
		550S162	16	33	33	33	33	33	33	33	33	33	33	33	33	33	33	33	33
			24	33	33	33	33	33	33	33	33	33	33	33	33	33	33	33	33
120 mph	110 mph	350S162	16	33	33	33	33	33	33	33	33	33	33	33	33	33	33	33	33
			24	33	33	33	43	33	33	43	43	33	33	43	43	33	43	43	43
		550S162	16	33	33	33	33	33	33	33	33	33	33	33	33	33	33	33	33
			24	33	33	33	33	33	33	33	33	33	33	33	33	33	33	33	33
130 mph	120 mph	350S162	16	33	33	33	33	33	33	33	33	33	33	33	33	33	33	33	33
			24	43	43	43	43	43	43	43	43	43	43	43	54	43	43	43	54
		550S162	16	33	33	33	33	33	33	33	33	33	33	33	33	33	33	33	33
			24	33	33	33	33	33	33	33	33	33	33	33	33	33	33	33	33
—	130 mph	350S162	16	33	33	33	33	33	33	33	43	33	33	43	43	33	33	43	43
			24	43	54	54	54	54	54	54	54	54	54	54	54	54	54	54	54
		550S162	16	33	33	33	33	33	33	33	33	33	33	33	33	33	33	33	33
			24	33	33	33	33	33	33	33	33	33	33	33	33	33	33	33	33

For SI: 1 inch = 25.4 mm, 1 foot = 304.8 mm, 1 mil = 0.0254 mm, 1 mile per hour = 1.609 km/h, 1 pound per square foot = 0.0479 kN/m^2, 1 kilogram per square inch = 6.895 MPa.

a. Deflection criteria: $L/240$.
b. Building width is in the direction of horizontal framing members supported by the wall studs.
c. Design load assumptions:
   Roof dead load is 12 psf.
   Attic live load is 10 psf.
d. 68-mil-thick stud is allowed if wall is fully sheathed per Section R603.3.2.

❖ See the commentary for Section R603.3.2.

TABLE R603.3.2(9)
COLD-FORMED STEEL STUD THICKNESS FOR 8-FOOT WALLS
Studs supporting one floor, roof and ceiling (first story of a two-story building) 50 ksi steel

WIND SPEED		MEMBER SIZE[c]	MEMBER SPACING (inches)	STUD THICKNESS (mils)[a,b] Building width (feet)[d]															
				24				28				32				36			
				Ground snow load (psf)				Ground snow load (psf)				Ground snow load (psf)				Ground snow load (psf)			
Exp. A/B	Exp. C			20	30	50	70	20	30	50	70	20	30	50	70	20	30	50	70
85 mph	—	350S162	16	33	33	33	33	33	33	33	33	33	33	33	33	33	33	33	43
			24	33	33	33	43	33	33	43	43	43	43	43	43	43	43	43	54
		550S162	16	33	33	33	33	33	33	33	33	33	33	33	33	33	33	33	33
			24	33	33	33	33	33	33	33	33	33	33	33	43	33	33	43	43
100 mph	85 mph	350S162	16	33	33	33	33	33	33	33	33	33	33	33	33	33	33	33	43
			24	33	43	43	43	43	43	43	43	43	43	43	43	43	33	33	54
		550S162	16	33	33	33	33	33	33	33	33	33	33	33	33	33	33	33	43
			24	33	33	33	33	33	33	33	33	33	33	33	43	33	33	43	43
110 mph	100 mph	350S162	16	33	33	33	33	33	33	33	33	33	33	33	33	33	33	33	43
			24	43	43	43	43	43	43	43	43	43	43	54	54	43	54	54	54
		550S162	16	33	33	33	33	33	33	33	33	33	33	33	33	33	33	33	33
			24	33	33	33	33	33	33	33	33	33	33	33	43	33	33	43	43
120 mph	110 mph	350S162	16	33	33	33	33	33	33	33	43	33	33	43	43	33	43	43	43
			24	43	43	43	54	54	54	54	54	54	54	54	54	54	54	54	54
		550S162	16	33	33	33	33	33	33	33	33	33	33	33	33	33	33	33	33
			24	33	33	33	33	33	33	33	33	33	33	33	43	33	33	43	43
130 mph	120 mph	350S162	16	43	43	43	43	43	43	43	43	43	43	43	43	43	43	43	43
			24	54	54	54	54	54	54	54	68	54	54	68	68	68	68	68	68
		550S162	16	33	33	33	33	33	33	33	33	33	33	33	33	33	33	33	43
			24	33	33	33	43	33	43	43	43	43	43	43	43	43	43	43	43
—	130 mph	350S162	16	43	43	43	43	43	43	43	43	43	43	43	54	43	43	54	54
			24	54	68	68	68	68	68	68	68	68	68	68	68	68	68	68	68
		550S162	16	33	33	33	33	33	33	33	33	33	33	33	33	33	33	33	33
			24	43	43	43	43	43	43	43	43	43	43	43	43	43	43	43	43

For SI: 1 inch = 25.4 mm, 1 foot = 304.8 mm, 1 mil = 0.0254 mm, 1 mile per hour = 1.609 km/h, 1 pound per square foot = 0.0479 kN/m^2, 1 kilogram per square inch = 6.895 MPa.

a. Deflection criteria: $L/240$.
b. Building width is in the direction of horizontal framing members supported by the wall studs.
c. Design load assumptions:
   Roof dead load is 12 psf.
   Attic live load is 10 psf.
d. 68-mil-thick stud is allowed if wall is fully sheathed per Section R603.3.2.

❖ See the commentary for Section R603.3.2.

TABLE R603.3.2(10)
COLD-FORMED STEEL STUD THICKNESS FOR 9-FOOT WALLS
Studs supporting roof and ceiling only (one-story building or second floor of a two-story building) 50 ksi steel

WIND SPEED		MEMBER SIZE[c]	MEMBER SPACING (inches)	STUD THICKNESS (mils)[a,b] Building width (feet)[d]															
Exp. A/B	Exp. C			24 Ground snow load (psf)				28 Ground snow load (psf)				32 Ground snow load (psf)				36 Ground snow load (psf)			
				20	30	50	70	20	30	50	70	20	30	50	70	20	30	50	70
85 mph	—	350S162	16	33	33	33	33	33	33	33	33	33	33	33	33	33	33	33	33
			24	33	33	33	33	33	33	33	33	33	33	33	33	33	33	33	33
		550S162	16	33	33	33	33	33	33	33	33	33	33	33	33	33	33	33	33
			24	33	33	33	33	33	33	33	33	33	33	33	33	33	33	33	33
100 mph	85 mph	350S162	16	33	33	33	33	33	33	33	33	33	33	33	33	33	33	33	33
			24	33	33	33	33	33	33	33	33	33	33	33	33	33	33	33	33
		550S162	16	33	33	33	33	33	33	33	33	33	33	33	33	33	33	33	33
			24	33	33	33	33	33	33	33	33	33	33	33	33	33	33	33	33
110 mph	100 mph	350S162	16	33	33	33	33	33	33	33	33	33	33	33	33	33	33	33	33
			24	33	33	33	33	33	33	33	43	33	33	33	43	33	33	43	43
		550S162	16	33	33	33	33	33	33	33	33	33	33	33	33	33	33	33	33
			24	33	33	33	33	33	33	33	33	33	33	33	33	33	33	33	33
120 mph	110 mph	350S162	16	33	33	33	33	33	33	33	33	33	33	33	33	33	33	33	33
			24	43	43	43	43	43	43	43	43	43	43	43	43	43	43	43	43
		550S162	16	33	33	33	33	33	33	33	33	33	33	33	33	33	33	33	33
			24	33	33	33	33	33	33	33	33	33	33	33	33	33	33	33	33
130 mph	120 mph	350S162	16	33	33	33	33	33	33	33	33	33	33	33	43	33	33	43	43
			24	43	43	54	54	43	54	54	54	54	54	54	54	54	54	54	54
		550S162	16	33	33	33	33	33	33	33	33	33	33	33	33	33	33	33	33
			24	33	33	33	33	33	33	33	33	33	33	33	33	33	33	33	33
—	130 mph	350S162	16	43	43	43	43	43	43	43	43	43	43	43	43	43	43	43	43
			24	54	54	54	54	54	54	54	54	54	54	54	68	54	54	54	68
		550S162	16	33	33	33	33	33	33	33	33	33	33	33	33	33	33	33	33
			24	33	33	33	33	33	33	33	43	33	33	43	43	33	33	43	43

For SI: 1 inch = 25.4 mm, 1 foot = 304.8 mm, 1 mil = 0.0254 mm, 1 mile per hour = 1.609 km/h, 1 pound per square foot = 0.0479 kN/m^2, 1 kilogram per square inch = 6.895 MPa.

a. Deflection criteria: $L/240$.
b. Building width is in the direction of horizontal framing members supported by the wall studs.
c. Design load assumptions:
   Roof dead load is 12 psf.
   Attic live load is 10 psf.
d. 68-mil-thick stud is allowed if wall is fully sheathed per Section R603.3.2.

❖ See the commentary for Section R603.3.2.

TABLE R603.3.2(11)
COLD-FORMED STEEL STUD THICKNESS FOR 9-FOOT WALLS
Studs supporting one floor, roof and ceiling (first story of a two-story building) 50 ksi steel

WIND SPEED		MEMBER SIZE[c]	MEMBER SPACING (inches)	STUD THICKNESS (mils)[a,b] Building width (feet)[d]															
				24				28				32				36			
				Ground snow load (psf)				Ground snow load (psf)				Ground snow load (psf)				Ground snow load (psf)			
Exp. A/B	Exp. C			20	30	50	70	20	30	50	70	20	30	50	70	20	30	50	70
85 mph	—	350S162	16	33	33	33	33	33	33	33	33	33	33	33	33	33	33	33	33
			24	33	33	33	43	33	43	43	43	43	43	43	43	43	43	43	43
		550S162	16	33	33	33	33	33	33	33	33	33	33	33	33	33	33	33	33
			24	33	33	33	33	33	33	33	33	33	33	33	33	33	33	33	43
100 mph	85 mph	350S162	16	33	33	33	33	33	33	33	33	33	33	33	33	33	33	33	33
			24	43	43	43	43	43	43	43	43	43	43	43	43	43	43	54	54
		550S162	16	33	33	33	33	33	33	33	33	33	33	33	33	33	33	33	33
			24	33	33	33	33	33	33	33	33	33	33	33	33	33	33	33	43
110 mph	100 mph	350S162	16	33	33	33	33	33	33	33	33	33	33	43	33	33	33	43	43
			24	43	43	43	43	43	43	43	54	43	54	54	54	54	54	54	54
		550S162	16	33	33	33	33	33	33	33	33	33	33	33	33	33	33	33	33
			24	33	33	33	33	33	33	33	33	33	33	33	33	33	33	33	43
120 mph	110 mph	350S162	16	33	33	33	43	33	33	43	43	43	43	43	43	43	43	43	43
			24	54	54	54	54	54	54	54	54	54	54	54	54	54	54	54	68
		550S162	16	33	33	33	33	33	33	33	33	33	33	33	33	33	33	33	33
			24	33	33	33	33	33	33	33	43	33	33	43	43	43	43	43	43
130 mph	120 mph	350S162	16	43	43	43	43	43	43	43	43	43	43	43	43	43	43	43	54
			24	54	54	68	68	68	68	68	68	68	68	68	68	68	68	68	68
		550S162	16	33	33	33	33	33	33	33	33	33	33	33	33	33	33	33	33
			24	43	43	43	43	43	43	43	43	43	43	43	43	43	43	43	43
—	130 mph	350S162	16	43	43	43	54	43	43	54	54	54	54	54	54	54	54	54	54
			24	68	68	68	68	68	68	68	68	68	68	(d)	(d)	68	(d)	(d)	(d)
		550S162	16	33	33	33	33	33	33	33	33	33	33	33	33	33	33	33	33
			24	43	43	43	43	43	43	43	43	43	43	43	43	43	43	43	54

For SI: 1 inch = 25.4 mm, 1 foot = 304.8 mm, 1 mil = 0.0254 mm, 1 mile per hour = 1.609 km/h, 1 pound per square foot = 0.0479 kN/m^2, 1 kilogram per square inch = 6.895 MPa.

a. Deflection criteria: $L/240$.
b. Building width is in the direction of horizontal framing members supported by the wall studs.
c. Design load assumptions:
   Roof dead load is 12 psf.
   Attic live load is 10 psf.
d. 68-mil-thick stud is allowed if wall is fully sheathed per Section R603.3.2.

❖ See the commentary for Section R603.3.2.

WALL CONSTRUCTION

TABLE R603.3.2(12)

TABLE R603.3.2(12)
COLD-FORMED STEEL STUD THICKNESS FOR 10-FOOT WALLS
Studs supporting roof and ceiling only (one-story building or second floor of a two-story building) 50 ksi steel

WIND SPEED		MEMBER SIZE[c]	MEMBER SPACING (inches)	STUD THICKNESS (mils)[a,b] Building width (feet)[d]															
				24				28				32				36			
				Ground snow load (psf)				Ground snow load (psf)				Ground snow load (psf)				Ground snow load (psf)			
Exp. A/B	Exp. C			20	30	50	70	20	30	50	70	20	30	50	70	20	30	50	70
85 mph	—	350S162	16	33	33	33	33	33	33	33	33	33	33	33	33	33	33	33	33
			24	33	33	33	33	33	33	33	33	33	33	33	33	33	33	33	43
		550S162	16	33	33	33	33	33	33	33	33	33	33	33	33	33	33	33	33
			24	33	33	33	33	33	33	33	33	33	33	33	33	33	33	33	33
100 mph	85 mph	350S162	16	33	33	33	33	33	33	33	33	33	33	33	33	33	33	33	33
			24	33	33	33	43	33	33	33	43	33	33	43	43	33	33	43	43
		550S162	16	33	33	33	33	33	33	33	33	33	33	33	33	33	33	33	33
			24	33	33	33	33	33	33	33	33	33	33	33	33	33	33	33	33
110 mph	100 mph	350S162	16	33	33	33	33	33	33	33	33	33	33	33	33	33	33	33	33
			24	43	43	43	43	43	43	43	43	43	43	43	43	43	43	43	54
		550S162	16	33	33	33	33	33	33	33	33	33	33	33	33	33	33	33	33
			24	33	33	33	33	33	33	33	33	33	33	33	33	33	33	33	33
120 mph	110 mph	350S162	16	33	33	33	33	33	33	33	33	33	33	33	43	33	33	33	43
			24	43	43	54	54	43	43	54	54	43	54	54	54	54	54	54	54
		550S162	16	33	33	33	33	33	33	33	33	33	33	33	33	33	33	33	33
			24	33	33	33	33	33	33	33	33	33	33	33	33	33	33	33	33
130 mph	120 mph	350S162	16	43	43	43	43	43	43	43	43	43	43	43	43	43	43	43	43
			24	54	54	54	68	54	54	68	68	54	54	68	68	54	68	68	68
		550S162	16	33	33	33	33	33	33	33	33	33	33	33	33	33	33	33	33
			24	33	33	33	43	33	33	43	43	33	33	43	43	33	43	43	43
—	130 mph	350S162	16	43	43	54	54	43	43	54	54	43	54	54	54	54	54	54	54
			24	68	68	68	68	68	68	68	(d)	68	68	68	(d)	68	68	(d)	(d)
		550S162	16	33	33	33	33	33	33	33	33	33	33	33	33	33	33	33	33
			24	43	43	43	43	43	43	43	43	43	43	43	43	43	43	43	43

For SI: 1 inch = 25.4 mm, 1 foot = 304.8 mm, 1 mil = 0.0254 mm, 1 mile per hour = 1.609 km/h, 1 pound per square foot = 0.0479 kN/m^2, 1 kilogram per square inch = 6.895 MPa.

a. Deflection criteria: $L/240$.
b. Building width is in the direction of horizontal framing members supported by the wall studs.
c. Design load assumptions:
   Roof dead load is 12 psf.
   Attic live load is 10 psf.
d. 68-mil-thick stud is allowed if wall is fully sheathed per Section R603.3.2.

❖ See the commentary for Section R603.3.2.

TABLE R603.3.2(13)
COLD-FORMED STEEL STUD THICKNESS FOR 10-FOOT WALLS
Studs supporting one floor, roof and ceiling (first story of a two-story building) 50 ksi steel

WIND SPEED		MEMBER SIZE[c]	MEMBER SPACING (inches)	STUD THICKNESS (mils)[a,b] Building width (feet)[d]															
				24				28				32				36			
Exp. A/B	Exp. C			Ground snow load (psf)				Ground snow load (psf)				Ground snow load (psf)				Ground snow load (psf)			
				20	30	50	70	20	30	50	70	20	30	50	70	20	30	50	70
85 mph	—	350S162	16	33	33	33	33	33	33	33	33	33	33	33	33	33	33	33	43
			24	43	43	43	43	43	43	43	43	43	43	54	54	43	54	54	54
		550S162	16	33	33	33	33	33	33	33	33	33	33	33	33	33	33	33	33
			24	33	33	33	33	33	33	33	33	33	33	33	43	33	33	33	43
100 mph	85 mph	350S162	16	33	33	33	33	33	33	33	43	33	33	43	43	43	43	43	43
			24	43	43	54	54	54	54	54	54	54	54	54	54	54	54	54	54
		550S162	16	33	33	33	33	33	33	33	33	33	33	33	33	33	33	33	33
			24	33	33	33	33	33	33	33	33	33	33	33	43	33	33	43	43
110 mph	100 mph	350S162	16	33	33	43	43	43	43	43	43	43	43	43	43	43	43	43	43
			24	54	54	54	54	54	54	54	54	54	54	68	68	54	68	68	68
		550S162	16	33	33	33	33	33	33	33	33	33	33	33	33	33	33	33	33
			24	33	33	33	33	33	33	33	43	33	43	43	43	43	43	43	43
120 mph	110 mph	350S162	16	43	43	43	43	43	43	43	43	43	43	43	54	43	43	54	54
			24	54	54	68	68	68	68	68	68	68	68	68	68	68	68	68	(d)
		550S162	16	33	33	33	33	33	33	33	33	33	33	33	33	33	33	33	33
			24	43	43	43	43	43	43	43	43	43	43	43	43	43	43	43	43
130 mph	120 mph	350S162	16	54	54	54	54	54	54	54	54	54	54	54	54	54	54	54	54
			24	68	68	(d)	(d)	(d)	(d)	(d)	(d)	(d)	(d)	(d)	(d)	(d)	(d)	(d)	(d)
		550S162	16	33	33	33	33	33	33	33	33	33	33	33	33	33	33	33	43
			24	43	43	43	43	43	43	43	43	43	43	43	54	43	43	54	54
—	130 mph	350S162	16	54	54	54	54	54	54	54	54	54	54	68	68	54	68	68	68
			24	(d)	(d)	(d)	(d)	(d)	(d)	(d)	(d)	(d)	(d)	(d)	(d)	(d)	(d)	(d)	(d)
		550S162	16	33	33	33	33	33	33	43	43	33	43	43	43	43	43	43	43
			24	43	43	54	54	54	54	54	54	54	54	54	54	54	54	54	54

For SI: 1 inch = 25.4 mm, 1 foot = 304.8 mm, 1 mil = 0.0254 mm, 1 mile per hour = 1.609 km/h, 1 pound per square foot = 0.0479 kN/m², 1 kilogram per square inch = 6.895 MPa.

a. Deflection criteria: $L/240$.
b. Building width is in the direction of horizontal framing members supported by the wall studs.
c. Design load assumptions:
   Roof dead load is 12 psf.
   Attic live load is 10 psf.
d. 68-mil-thick stud is allowed if wall is fully sheathed per Section R603.3.2.

❖ See the commentary for Section R603.3.2.

R603.3.3 Stud bracing. The flanges of steel studs shall be laterally braced in accordance with one of the following:

1. Gypsum board installed with minimum No. 6 screws in accordance with Section R702 or structural sheathing installed in accordance with Table R603.3.2.

2. Horizontal steel strapping installed in accordance with Figure R603.3 at mid-height for 8-foot (2438 mm) walls, and one-third points for 9-foot and 10-foot (2743 mm and 3048 mm) walls. Steel straps shall be at least 1.5 inches in width and 33 mils in thickness (38 mm by 0.836 mm). Straps shall be attached to the flanges of studs with at least one No. 8 screw. In-line blocking shall be installed between studs at the termination of all straps. Straps shall be fastened to the blocking with at least two No. 8 screws.

❖ To provide resistance against buckling, lateral bracing of steel-stud flanges must be provided by gypsum board installed according to Section R702, structural sheathing installed in accordance with Table R603.3.2 or horizontal steel strapping as indicated in Figure R603.3. Steel strap bracing must be installed at mid-height for 8-foot (2338 mm) walls as shown in Figure R603.3 and at one-third points for 9-foot (2743 mm) and 10-foot (3048 mm) walls (two rows of bracing). These straps must be attached to stud flanges with at least one No. 8 screw. In-line blocking must be installed between studs at the termination of all straps. Straps must be fastened to this blocking with at least two No. 8 screws.

R603.3.4 Cutting and notching. Flanges and lips of steel studs and headers shall not be cut or notched.

❖ Flanges and lips of load-bearing wall framing must not be cut or notched; thus, they will retain structural integrity. The only holes permitted in steel wall framing are those conforming to Section R603.2.

R603.3.5 Hole patching. Stud web holes with dimensions conforming to Section R603.2 that are closer than 10 inches (254 mm) from the edge of the hole to the end of the member shall be patched with a solid steel plate, C-section or track section in accordance with Figure R603.3.5. The patch shall be of a minimum thickness as the stud member and shall extend at least 1 inch (25.4 mm) beyond all edges of the hole. The patch shall be fastened to the web with No. 8 screws (minimum) spaced not greater than 1 inch (25.4 mm) center to center along the edges of the patch, with a minimum edge distance of $^1/_2$ inch (12.7 mm).

❖ See the commentary for Section R603.2.

R603.3.6 Splicing. Steel studs and other structural members shall not be spliced. Tracks shall be spliced in accordance with Figure R603.3.6.

❖ Splicing of structural members, such as studs, is not permitted. Splicing of tracks is allowed since they do not serve as gravity load-carrying members. The splice shown in Figure R603.3.6 provides nominal load-carrying capability; however, it is necessary to resist the lateral loads produced by moderate wind and earthquake loads. In these cases, the floor system must act as a diaphragm with the tracks serving as the diaphragm chords. See Section R603.8.1.4 for additional track splicing requirements in areas of high seismicity or high-wind regions.

R603.4 Corner framing. Corner studs and the top tracks shall be installed in accordance with Figure R603.4.

❖ Figure R603.4 illustrates the required configuration of corner studs and overlapping of tracks at corners.

R603.5 Exterior wall covering. The method of attachment of exterior wall covering materials to cold-formed steel stud wall framing shall conform to the manufacturer's installation instructions.

❖ Attachment of exterior wall coverings must be according to the manufacturer's instructions.

For SI: 1 inch = 25.4 mm.

**FIGURE R603.3.5
HOLE PATCH**

❖ See the commentary for Section R603.3.5.

For SI: 1 inch = 25.4 mm.

**FIGURE R603.3.6
TRACK SPLICE**

❖ See the commentary for Section R603.3.6.

R603.6 Headers. Headers shall be installed above wall openings in all exterior walls and interior load-bearing walls in accordance with Figure R603.6 and Table R603.6(1), Table R603.6(2), and Table R603.6(3), or shall be designed and installed in accordance with the AISI Standard for Cold-formed Steel Framing-Header Design (COFS/Header Design).

❖ Headers must be located immediately below the ceiling or roof framing (at the top track, for example) above wall openings in all exterior walls and interior load-bearing walls in accordance with Figure R603.6. This figure illustrates the two types of design options—either back-to-back or as a box beam. Tables R603.6(1), R603.6(2) and R603.6(3) specify allowable header spans for various combinations of loading and header sizes. These tables give a very limited range of design options and are very conservative in many design scenarios. Therefore, the code provides the option to use the referenced AISI standard, which will provide an opportunity for a more economical design in many circumstances.

R603.6.1 Jack and king studs. The number of jack and king studs shall comply with Table R603.6(4). King and jack studs shall be of the same dimension and thickness as the adjacent wall studs. Headers constructed of C-shape framing members shall be connected to king studs in accordance with Table R603.6(5). One-half of the total number of screws shall be applied to the header and one-half to the king stud by use of a minimum 2-inch by 2-inch (51 mm by 51 mm) clip angle or 4-inch (102 mm) wide steel plate. The clip angle or plate shall extend the depth of the header minus $^1/_2$ inch (12.7 mm) and shall have a minimum thickness of the header members or the wall studs, whichever is thicker.

❖ Table R603.6(4) specifies the number of jack and king studs based on the header span. King and jack studs must be the same dimension as adjacent wall studs, with a steel thickness not less than that of the adjacent wall studs. The size and number of screws required for the header-to-king stud connection are listed in Table R603.6(5). One-half of the screws are to be applied to the header and one-half to the king stud with a clip angle as shown in Figure R603.6. The clip angle must be a minimum 2-inch by 2-inch (51 mm x 51 mm) angle with a minimum steel thickness not less than the header members and the wall studs.

R603.7 Structural sheathing. In areas where the basic wind speed is less than 110 miles per hour (177 km/h), wood structural sheathing panels shall be installed on all exterior walls of buildings in accordance with this section. Wood structural sheathing panels shall consist of minimum $^7/_{16}$-inch (11.1 mm) thick oriented strand board or $^{15}/_{32}$-inch (11.9 mm) thick plywood and shall be installed on all exterior wall surfaces in accordance with Section R603.7.1 and Figure R603.3. The minimum length of full height sheathing on exterior walls shall be determined in accordance with Table R603.7, but shall not be less than 20 percent of the braced wall length in any case. The minimum percentage of full height sheathing in Table R603.7 shall include only those sheathed wall sections, uninterrupted by openings, which are a minimum of 48 inches (1120 mm) wide. The minimum percentage of full-height structural sheathing shall be multiplied by 1.10 for 9-foot-high (2743 mm) walls and multiplied by 1.20 for 10-foot-high (3048 mm) walls. In addition, structural sheathing shall:

1. Be installed with the long dimension parallel to the stud framing and shall cover the full vertical height of studs, from the bottom of the bottom track to the top of the top track of each story.

2. Be applied to each end (corners) of each of the exterior walls with a minimum 48-inch-wide (1220 mm) panel.

❖ Wood structural sheathing panels must be installed on all exterior wall surfaces as shown in Figure R603.3. Where the basic wind speed is less than 110 miles per

WALL CONSTRUCTION

FIGURE R603.4

hour (177 km/h), and the building is classified as Seismic Design Category A, B or C, wood structural panel sheathing must be installed in accordance with this section.

Wood structural-panel sheathing must be a minimum 7/16-inch (11.1 mm) thick oriented strand board or 15/32-inch (11.9 mm) thick plywood fastened to studs and tracks in accordance with Table R603.3.2.

The minimum length of full-height sheathing required on exterior walls is determined from Table R603.7. The minimum tabulated lengths are expressed as a percentage of the wall length. Because the table is based on an 8-foot (2338 mm) wall height, the lengths must be increased by the appropriate multiplier for 9-foot (2743 mm) and 10-foot (3048 mm) high walls as well as the building aspect ratio found in Note "c" of the table. Additional adjustments are permitted for buildings with hipped roofs and for hold-downs in accordance with Section R603.7.2. After all adjustments are applied to these tabulated values, the resulting minimum percentage of full-height sheathing must not be less than 20 percent of the wall length.

For SI: 1 inch = 25.4 mm.

**FIGURE R603.4
CORNER FRAMING**

❖ See the commentary for Section R603.4.

Only sheathed wall sections that are a minimum of 48 inches (1219 mm) wide and uninterrupted by openings may be considered in meeting the minimum length requirement. This structural sheathing must be installed with the long dimension parallel to the stud framing, and it must cover the full vertical height of studs, from the bottom of the bottom track to the top of the top track of each story. Additionally, structural sheathing must be applied to each end (corners) of all exterior walls with a minimum 48-inch (1219 mm) wide panel.

TABLE R603.6(1)
ALLOWABLE HEADER SPANS[a,b]
Headers supporting roof and ceiling only (33 ksi steel)

NOMINAL MEMBER	GROUND SNOW LOAD (20 psf) Building width[c] (feet)				GROUND SNOW LOAD (30 psf) Building width[c] (feet)				GROUND SNOW LOAD (50 psf) Building width[c] (feet)				GROUND SNOW LOAD (70 psf) Building width[c] (feet)			
	24	28	32	36	24	28	32	36	24	28	32	36	24	28	32	36
2-350S162-33	3'-11"	3'-8"	3'-5"	3'-3"	3'-8"	3'-5"	3'-2"	2'-10"	3'-0"	2'-7"	2'-4"	2'-1"	2'-4"	2'-1"	—	—
2-350S162-43	4'-9"	4'-5"	4'-2"	4'-0"	4'-5"	4'-2"	3'-11"	3'-9"	3'-10"	3'-7"	3'-4"	3'-2"	3'-5"	3'-2"	3'-0"	2'-9"
2-350S162-54	5'-4"	5'-0"	4'-9"	4'-6"	5'-0"	4'-8"	4'-5"	4'-2"	4'-3"	4'-0"	3'-9"	3'-7"	3'-10"	3'-7"	3'-4"	3'-2"
2-350S162-68	6'-0"	5'-7"	5'-3"	5'-0"	5'-7"	5'-3"	4'-11"	4'-8"	4'-10"	4'-6"	4'-3"	4'-0"	4'-3"	4'-0"	3'-9"	3'-7"
2-550S162-33	3'-11"	3'-5"	3'-0"	2'-9"	3'-5"	3'-0"	2'-8"	2'-5"	2'-6"	2'-2"	—	—	—	—	—	—
2-550S162-43	6'-5"	6'-0"	5'-8"	5'-5"	6'-0"	5'-8"	5'-4"	5'-0"	5'-2"	4'-10"	4'-4"	3'-11"	4'-5"	3'-10"	3'-5"	3'-1"
2-550S162-54	7'-3"	6'-10"	6'-5"	6'-1"	6'-9"	6'-4"	6'-0"	5'-8"	5'-10"	5'-5"	5'-1"	4'-10"	5'-2"	4'-10"	4'-7"	4'-4"
2-550S162-68	8'-2"	7'-8"	7'-2"	6'-11"	7'-7"	7'-2"	6'-9"	6'-4"	6'-6"	6'-1"	5'-9"	5'-6"	5'-10"	5'-5"	5'-1"	4'-10"
2-800S162-33	3'-0"	2'-8"	2'-4"	2'-1"	2'-7"	2'-3"	—	—	—	—	—	—	—	—	—	—
2-800S162-43	6'-8"	5'-10"	5'-2"	4'-8"	5'-10"	5'-1"	4'-6"	4'-1"	4'-3"	3'-9"	3'-4"	3'-0"	3'-4"	2'-11"	2'-7"	2'-4"
2-800S162-54	9'-6"	8'-10"	8'-4"	7'-11"	8'-10"	8'-3"	7'-9"	7'-5"	7'-7"	7'-1"	6'-7"	5'-11"	6'-9"	5'-10"	5'-3"	4'-8"
2-800S162-68	10'-8"	10'-0"	9'-5"	8'-11"	9'-11"	9'-4"	8'-9"	8'-4"	8'-6"	8'-0"	7'-6"	7'-2"	7'-7"	7'-1"	6'-8"	6'-4"
2-1000S162-43	5'-7"	4'-10"	4'-4"	3'-11"	4'-10"	4'-3"	3'-9"	3'-5"	3'-7"	3'-1"	2'-9"	2'-6"	2'-10"	2'-6"	2'-2"	—
2-1000S162-54	10'-6"	9'-8"	8'-7"	7'-9"	9'-8"	8'-5"	7'-6"	6'-9"	7'-1"	6'-2"	5'-6"	4'-11"	5'-7"	4'-11"	4'-4"	3'-11"
2-1000S162-68	12'-7"	11'-9"	11'-1"	10'-6"	11'-9"	10'-12"	10'-4"	9'-10"	10'-1"	9'-5"	8'-10"	8'-5"	8'-11"	8'-4"	7'-11"	7'-6"
2-1200S162-43	—	—	—	—	—	—	—	—	—	—	—	—	—	—	—	—
2-1200S162-54	9'-6"	8'-3"	7'-4"	6'-7"	8'-3"	7'-3"	6'-5"	5'-9"	6'-1"	5'-4"	4'-9"	4'-3"	4'-10"	4'-2"	3'-9"	3'-4"
2-1200S162-68	13'-5"	12'-7"	11'-10"	11'-3"	12'-6"	11'-9"	11'-1"	10'-6"	10'-9"	10'-1"	9'-6"	8'-6"	9'-7"	8'-5"	7'-6"	6'-9"

For SI: 1 inch = 25.4 mm, 1 foot = 304.8 mm, 1 pound per square foot = 0.0479 kN/m^2, 1 kilogram per square inch = 6.895 MPa.

a. Deflection criteria: $L/360$ for live loads, $L/240$ for total loads.
b. Design load assumptions:
   Roof dead load is 7 psf.
   Ceiling dead load is 5 psf.
   Attic live load is 10 psf.
c. Building width is in the direction of horizontal framing members supported by the header.

❖ See the commentary for Section R603.6.

WALL CONSTRUCTION

TABLE R603.6(2)
ALLOWABLE HEADER SPANS[a,b]
Headers supporting one floor, roof and ceiling (33 ksi steel)

NOMINAL MEMBER	GROUND SNOW LOAD (20 psf) Building width[c] (feet)				GROUND SNOW LOAD (30 psf) Building width[c] (feet)				GROUND SNOW LOAD (50 psf) Building width[c] (feet)				GROUND SNOW LOAD (70 psf) Building width[c] (feet)			
	24	28	32	36	24	28	32	36	24	28	32	36	24	28	32	36
2-350S162-42	3'-3"	3'-1"	2'-11"	2'-8"	3'-2"	3'-0"	2'-9"	2'-6"	2'-11"	2'-8"	2'-4"	2'-2"	2'-10"	2'-6"	2'-3"	2'-1"
2-350S162-54	3'-8"	3'-6"	3'-3"	3'-2"	3'-6"	3'-4"	3'-2"	3'-0"	3'-4"	3'-1"	2'-11"	2'-9"	3'-3"	3'-0"	2'-10"	2'-9"
2-350S162-68	4'-2"	3'-11"	3'-8"	3'-6"	4'-0"	3'-9"	3'-6"	3'-4"	3'-8"	3'-6"	3'-3"	3'-1"	3'-7"	3'-5"	3'-2"	3'-1"
2-550S162-43	4'-1"	3'-8"	3'-3"	3'-0"	3'-10"	3'-4"	3'-0"	2'-9"	3'-3"	2'-11"	2'-7"	2'-4"	3'-2"	2'-9"	2'-6"	2'-3"
2-550S162-54	5'-0"	4'-9"	4'-6"	4'-3"	4'-10"	4'-6"	4'-3"	4'-1"	4'-6"	4'-2"	4'-0"	3'-9"	4'-4"	4'-1"	3'-11"	3'-8"
2-550S162-68	5'-7"	5'-4"	5'-0"	4'-9"	5'-5"	5'-1"	4'-10"	4'-7"	5'-0"	4'-9"	4'-6"	4'-3"	4'-11"	4'-7"	4'-4"	4'-2"
2-800S162-43	3'-2"	2'-10"	2'-6"	2'-3"	2'-11"	2'-7"	2'-4"	—	2'-6"	2'-3"	—	—	2'-5"	2'-2"	—	—
2-800S162-54	6'-3"	5'-7"	5'-0"	4'-6"	5'-10"	5'-2"	4'-7"	4'-2"	5'-0"	4'-5"	4'-0"	3'-7"	4'-10"	4'-3"	3'-10"	3'-5"
2-800S162-68	7'-4"	6'-11"	6'-7"	6'-3"	7'-1"	6'-8"	6'-3"	6'-0"	6'-7"	6'-2"	5'-10"	5'-7"	6'-5"	6'-0"	5'-8"	5'-5"
2-1000S162-43	2'-8"	2'-4"	—	—	2'-5"	—	—	—	—	—	—	—	—	—	—	—
2-1000S162-54	5'-3"	4'-8"	4'-2"	3'-9"	4'-10"	4'-3"	3'-10"	3'-6"	4'-2"	3'-8"	3'-4"	3'-0"	4'-0"	3'-6"	3'-2"	2'-10"
2-1000S162-68	8'-8"	8'-2"	7'-9"	7'-4"	8'-4"	7'-10"	7'-5"	7'-0"	7'-9"	7'-3"	6'-8"	6'-0"	7'-7"	7'-1"	6'-5"	5'-9"
2-1000S162-54	4'-6"	4'-0"	3'-7"	3'-3"	4'-2"	3'-8"	3'-3"	3'-0"	3'-7"	3'-2"	2'-10"	2'-7"	3'-5"	3'-0"	2'-9"	2'-5"
2-1000S162-68	9'-1"	8'-0"	7'-2"	6'-6"	8'-4"	7'-5"	6'-8"	6'-0"	7'-3"	6'-5"	5'-9"	5'-2"	6'-11"	6'-1"	5'-6"	4'-11"

For SI: 1 inch = 25.4 mm, 1 foot = 304.8 mm, 1 pound per square foot = 0.0479 kN/m², 1 kilogram per square inch = 6.895 MPa.

a. Deflection criteria: $L/360$ for live loads, $L/240$ for total loads.
b. Design load assumptions:
  Roof dead load is 7 psf.
  Ceiling dead load is 5 psf.
  Attic live load is 10 psf.
  Second floor live load is 30 psf.
  Second floor dead load is 10 psf.
  Second floor wall dead load is 10 psf.
c. Building width is in the direction of horizontal framing members supported by the header.

❖ See the commentary for Section R603.6.

TABLE R603.6(3)
ALLOWABLE HEADER SPANS[a,b,c]
Headers supporting one floor, roof and ceiling first story of a two-story building with center load bearing beam (33 ksi steel)

NOMINAL MEMBER	GROUND SNOW LOAD (20 psf) Building width[c] (feet)				GROUND SNOW LOAD (30 psf) Building width[c] (feet)				GROUND SNOW LOAD (50 psf) Building width[c] (feet)				GROUND SNOW LOAD (70 psf) Building width[c] (feet)			
	24	28	32	36	24	28	32	36	24	28	32	36	24	28	32	36
2-350S162-33	2'-10"	2'-6"	2'-3"	—	2'-7"	2'-3"	—	—	2'-2"	—	—	—	—	—	—	—
2-350S162-43	3'-9"	3'-6"	3'-4"	3'-2"	3'-7"	3'-4"	3'-2"	3'-0"	3'-3"	3'-1"	2'-11"	2'-8"	3'-0"	2'-9"	2'-6"	2'-3"
2-350S162-54	4'-2"	4'-0"	3'-9"	3'-7"	4'-0"	3'-9"	3'-7"	3'-5"	3'-8"	3'-5"	3'-3"	3'-1"	3'-5"	3'-2"	3'-0"	2'-10"
2-350S162-68	4'-8"	4'-5"	4'-2"	4'-0"	4'-6"	4'-2"	4'-0"	3'-10"	4'-1"	3'-10"	3'-8"	3'-6"	3'-9"	3'-7"	3'-4"	3'-2"
2-550S162-33	2'-5"	2'-2"	—	—	2'-2"	—	—	—	—	—	—	—	—	—	—	—
2-550S162-43	5'-1"	4'-9"	4'-3"	3'-10"	4'-10"	4'-3"	3'-10"	3'-6"	4'-0"	3'-7"	3'-2"	2'-11"	3'-5"	3'-1"	2'-9"	2'-6"
2-550S162-54	5'-8"	5'-4"	5'-1"	4'-10"	5'-5"	5'-1"	4'-10"	4'-7"	4'-11"	4'-8"	4'-5"	4'-2"	4'-7"	4'-4"	4'-1"	3'-11"
2-550S162-68	6'-5"	6'-0"	5'-9"	5'-6"	6'-1"	5'-9"	5'-5"	5'-2"	5'-7"	5'-3"	4'-11"	4'-9"	5'-2"	4'-10"	4'-7"	4'-4"
2-800S162-33	4'-2"	3'-8"	3'-3"	3'-0"	3'-8"	3'-3"	2'-11"	2'-8"	3'-1"	2'-9"	2'-5"	2'-3"	2'-8"	2'-4"	—	—
2-800S162-43	7'-5"	7'-0"	6'-6"	6'-0"	7'-1"	6'-6"	5'-10"	5'-4"	6'-2"	5'-5"	4'-11"	4'-5"	5'-3"	4'-8"	4'-2"	3'-9"
2-800S162-54	8'-4"	7'-11"	7'-6"	7'-1"	7'-11"	7'-6"	7'-1"	6'-9"	7'-3"	6'-10"	6'-6"	6'-2"	6'-9"	6'-4"	6'-0"	5'-8"
2-800S162-68	10'-0"	9'-9"	9'-0"	8'-6"	9'-6"	9'-0"	8'-6"	8'-1"	8'-8"	8'-2"	7'-9"	7'-5"	8'-1"	7'-7"	7'-2"	6'-10"
2-1000S162-43	3'-5"	3'-0"	2'-9"	2'-6"	3'-1"	2'-9"	2'-6"	2'-3"	2'-7"	2'-3"	2'-1"	—	2'-3"	—	—	—
2-1000S162-54	6'-10"	6'-0"	5'-6"	5'-0"	6'-2"	5'-5"	4'-11"	4'-5"	5'-1"	4'-6"	4'-1"	3'-8"	4'-5"	3'-11"	3'-6"	3'-2"
2-1000S162-68	9'-10"	9'-4"	8'-10"	8'-5"	9'-4"	8'-10"	8'-4"	8'-0"	8'-7"	8'-1"	7'-8"	7'-3"	7'-	7'-6"	7'-0"	6'-4"
2-1200S162-43	2'-11"	2'-7"	2'-4"	2'-1"	2'-8"	2'-4"	—	—	2'-3"	—	—	—	—	—	—	—
2-1200S162-54	5'-10"	5'-2"	4'-8"	4'-3"	5'-3"	4'-8"	4'-2"	3'-10"	4'-5"	3'-11"	3'-6"	3'-2"	3'-9"	3'-4"	3'-0"	2'-8"
2-1200S162-68	10'-6"	10'-0"	9'-5"	8'-6"	9'-12"	9'-5"	8'-5"	7'-8"	8'-10"	7'-10"	7'-0"	6'-4"	7'-7"	6'-9"	6'-0"	5'-5"

For SI: 1 inch = 25.4 mm, 1 foot = 304.8 mm, 1 pound per square foot = 0.0479 kN/m², 1 kilogram per square inch = 6.895 MPa.

a. Deflection criteria: $L/360$ for live loads, $L/240$ for total loads.
b. Design load assumptions:
  Roof dead load is 7 psf.
  Ceiling dead load is 5 psf.
  Attic live load is 10 psf.
  Second floor live load is 30 psf.

(continued)

FOOTNOTES TO TABLE R603.6(3)—continued

    Second floor dead load is 10 psf.
    Second floor wall dead load is 10 psf.
c. Building width is in the direction of horizontal framing members supported by the header.

❖ See the commentary for Section R603.6.

TABLE R603.6(4)
TOTAL NUMBER OF JACK AND KING STUDS REQUIRED AT EACH END OF AN OPENING

SIZE OF OPENING (feet-inches)	24" O.C. STUD SPACING		16" O.C. STUD SPACING	
	No. of jack studs	No. of king studs	No. of jack studs	No. of king studs
Up to 3'-6"	1	1	1	1
>3'-6" to 5'-0"	1	2	1	2
>5'-0" to 5'-6"	1	2	2	2
>5'-6" to 8'-0"	1	2	2	2
>8'-0" to 10'-6"	2	2	2	3
>10'-6" to 12'-0"	2	2	3	3
>12'-0" to 13'-0"	2	3	3	3
>13'-0" to 14'-0"	2	3	3	4

For SI: 1 inch = 25.4 mm, 1 foot = 304.8 mm.

❖ See the commentary for Section R603.6.

TABLE R603.6(5)
HEADER TO KING STUD CONNECTION REQUIREMENTS[a,b,c,d]

HEADER SPAN (feet)	BASIC WIND SPEED (mph), EXPOSURE		
	85 A/B or Seismic Design Categories A, B, C and D	85 C or less than 110 A/B	Less than 110 C
≤ 4'	4-No. 8 screws	4-No. 8 screws	6-No. 8 screws
> 4' to 8'	4-No. 8 screws	4-No. 8 screws	8-No. 8 screws
> 8' to 12'	4-No. 8 screws	6-No. 8 screws	10-No. 8 screws
> 12' to 16'	4-No. 8 screws	8-No. 8 screws	12-No. 8 screws

For SI: 1 inch = 25.4 mm, 1 foot = 304.8 mm, 1 mile per hour = 1.609 km/h, 1 pound = 0.454 kg.

a. All screw sizes shown are minimum.
b. For headers located on the first floor of a two-story building, the total number of screws may be reduced by two screws, but the total number of screws shall be no less than four.
c. For roof slopes of 6:12 or greater, the required number of screws may be reduced by half, but the total number of screws shall be no less than four.
d. Screws can be replaced by a up-lift connector which has a capacity of the number of screws multiplied by 164 pounds (e.g., 12-No. 8 screws can be replaced by an uplift connector whose capacity exceeds 12 by 164 pounds = 1,968 pounds).

❖ See the commentary for Section R603.6.

FIGURE R603.6
HEADER DETAIL

For SI: 1 inch = 25.4 mm.

❖ See the commentary for Section R603.6.

TABLE R603.7
MINIMUM PERCENTAGE OF FULL HEIGHT STRUCTURAL SHEATHING ON EXTERIOR WALLS[a,b,c,d,e]

WALL SUPPORTING	ROOF SLOPE	WIND SPEED (mph) AND EXPOSURE				
		85 A/B	100 A/B	110 A/B or 85 C	100 C	110 C
Roof and ceiling only	3:12	8	9	12	16	20
	6:12	12	15	20	26	35
	9:12	21	25	30	50	58
	12:12	30	35	40	66	75
One story, roof and ceiling	3:12	24	30	35	50	66
	6:12	25	30	40	58	74
	9:12	35	40	55	74	91
	12:12	40	50	65	100	115

For SI: 1 mile per hour = 1.609 km/h.

a. Linear interpolation shall be permitted.

b. Bracing amount shall not be less than 20 percent of the wall length after all applicable adjustments are made.

c. Minimum percentages are based on a building aspect ratio of 1:1. Minimum percentages for the shorter walls of a building shall be multiplied by a factor of 1.5 and 2.0 for building aspect ratios of 1.5:1 and 2:1 respectively.

d. For hip roofed homes with continuous structural sheathing, the amount of bracing shall be permitted to be multiplied by a factor of 0.95 for roof slopes not exceeding 7:12 and a factor of 0.9 for roof slopes greater than 7:12.

e. Sheathing percentage are permitted to be reduced in accordance with Section R603.7.2.

❖ See the commentary for Section R603.7.

R603.7.1 Structural sheathing fastening. All edges and interior areas of wood, structural sheathing panels shall be fastened to a framing member and tracks in accordance with Table R603.3.2(1).

❖ The wall-fastening schedule specifies the required size and spacing of screws.

R603.7.2 Hold-down requirements. The percent of structural sheathing, required in Table R603.7, is permitted to be multiplied by 0.6 where a hold down anchor with a capacity of 4,300 pounds (1952.2 kg) is provided at each end of exterior walls. A single hold-down anchor is permitted to be installed at wall corners.

❖ This section affords the option of using shorter braced-wall lengths and providing hold-down anchorage at the ends of the braced walls to resist uplift. The length of wall may be reduced by up to 40 percent where the specified hold-down capacities are installed. In accordance with Note "e" of Table R603.7, a reduction to the required full height sheathing is permitted if hold-downs are installed.

SECTION R604
WOOD STRUCTURAL PANELS

R604.1 Identification and grade. Wood structural panels shall conform to DOC PS 1 or DOC PS 2. All panels shall be identified by a grade mark or certificate of inspection issued by an approved agency.

❖ To verify the acceptability of wood structural panel products, this section requires labeling in accordance with the referenced standards. The purpose is the same as the purpose of Section R602.1 for dimension lumber. This label provides the inspector with the necessary information to determine the acceptability of the wood structural panel. See the commentary for Sections R503.2 and R503.2.1.

R604.2 Allowable spans. The maximum allowable spans for wood structural panel wall sheathing shall not exceed the values set forth in Table R602.3(3).

❖ The allowable span for wood structural panel sheathing is a function of its thickness and grade. Table R602.3(3) of the code provides allowable spans based on the thickness and the panel index.

R604.3 Installation. Wood structural panel wall sheathing shall be attached to framing in accordance with Table R602.3(1). Wood structural panels marked Exposure 1 or Exterior are considered water-repellent sheathing under the code.

❖ Table R602.3(1) of the code provides the necessary fastening requirements for proper wood structural panel installation.

SECTION R605
PARTICLEBOARD

R605.1 Identification and grade. Particleboard shall conform to ANSI A208.1 and shall be so identified by a grade mark or certificate of inspection issued by an approved agency. Particleboard shall comply with the grades specified in Table R602.3(4).

❖ Similar to wood structural panels, particleboard must be labeled in accordance with the referenced standard. Table R602.3.(4) provides prescriptive requirements relative to allowable spans. As with wood structural panel spans, these spans are a function of the thickness and grade. See the commentary for Sections R503.3 and R603.3.1.

SECTION R606
GENERAL MASONRY CONSTRUCTION

R606.1 General. Masonry construction shall be designed and constructed in accordance with the provisions of this section or in accordance with the provisions of ACI 530/ASCE 5/TMS 402.

❖ Masonry wall construction under the code must comply with this section or ACI 530/ASCE 5/TMS 402: *Building Code Requirements for Masonry Structures.*

R606.1.1 Professional registration not required. When the empirical design provisions of ACI 530/ASCE 5/TMS 402 Chapter 5 or the provisions of this section are used to design masonry, project drawings, typical details and specifications are not required to bear the seal of the architect or engineer responsible for design, unless otherwise required by the state law of the jurisdiction having authority.

❖ These empirical provisions are meant for use without the services of a professional engineer or architect. State law, however, may dictate otherwise and would take precedence.

R606.2 Thickness of masonry. The nominal thickness of masonry walls shall conform to the requirements of Sections R606.2.1 through R606.2.4.

❖ This section establishes minimum thicknesses for various masonry wall constructions.

R606.2.1 Minimum thickness. The minimum thickness of masonry bearing walls more than one story high shall be 8 inches (203 mm). Solid masonry walls of one-story dwellings and garages shall not be less than 6 inches (152 mm) in thickness when not greater than 9 feet (2743 mm) in height, provided that when gable construction is used, an additional 6 feet (1829 mm) is permitted to the peak of the gable. Masonry walls shall be laterally supported in either the horizontal or vertical direction at intervals as required by Section R606.8.

❖ The minimum thickness of bearing walls and exterior nonbearing walls of masonry construction more than one story high is limited to a nominal 8 inches (203 mm), which means a net thickness of not less than $7^1/_2$ inches (191 mm). A lesser thickness of 6-inch (152 mm) nominal or $5^1/_2$-inch (140 mm) net of solid masonry is permitted if the wall does not exceed 9 feet (2743 mm) in height other than at the peak of a gable, where such height may extend to 15 feet (4572 mm) as shown in Commentary Figure R606.2.1.

R606.2.2 Rubble stone masonry wall. The minimum thickness of rough, random or coursed rubble stone masonry walls shall be 16 inches (406 mm).

❖ Rubble stone walls must be at least 16 inches (406 mm) in thickness. See the commentary for Section R404.1.8.

R606.2.3 Change in thickness. Where walls of masonry of hollow units or masonry bonded hollow walls are decreased in thickness, a course of solid masonry shall be constructed between the wall below and the thinner wall above, or special units or construction shall be used to transmit the loads from face shells or wythes above to those below.

❖ Where the thickness of hollow masonry wall construction is decreased, a solid masonry course provides bearing for the face shells of the thinner hollow unit above.

For SI: 1 inch = 25.4 mm, 1 foot = 304.8 mm.

Figure R606.2.1
MINIMUM THICKNESS—MASONRY BEARING AND EXTERIOR WALLS FOR A ONE-STORY DWELLING

R606.2.4 Parapet walls. Unreinforced solid masonry parapet walls shall not be less than 8 inches (203 mm) in thickness and their height shall not exceed four time their thickness. Unreinforced hollow unit masonry parapet walls shall be not less that 8 inches (203 mm) in thickness, and their height shall not exceed three times their thickness. Masonry parapets in areas subject to wind loads of 30 pounds per square foot (1.44 kN/m^2) located in Seismic Design Category D_1, or D_2, or on townhouses in Seismic Design Category C shall be reinforced in accordance with Section R606.11.

❖ A parapet is an extension of a wall that extends above the roof line. Generally, parapet walls are only required to comply with fire-resistance requirements of the code. Parapet walls are required to have a certain minimum thickness with height limitations based on the thickness and reinforcement in the wall.

R606.3 Corbeled masonry. Solid masonry units shall be used for corbeling. The maximum corbeled projection beyond the face of the wall shall not be more than one-half of the wall thickness or one-half the wythe thickness for hollow walls; the maximum projection of one unit shall not exceed one-half the height of the unit or one-third the thickness at right angles to the wall. The top course of corbels shall be a header course when the corbeled masonry is used to support floor or roof-framing members.

❖ Corbeling of masonry walls is permitted within certain limitations. The maximum projection for each unit is limited to one-third the unit bed depth (*D*) or one-half the unit height (*H*), whichever is less, as shown in Commentary Figure R606.3(1). The total horizontal projection of the corbeled courses is limited to one-half the thickness of a solid wall or one-half the thickness of a wythe of a cavity wall as shown in Commentary Figure R606.3(2).

R606.3.1 Support conditions. Cavity wall or masonry veneer construction may be supported on an 8-inch (203 mm) foundation wall, provided the 8-inch (203 mm) wall is corbeled with solid masonry to the width of the wall system above. The total horizontal projection of the corbel shall not exceed 2 inches (51 mm) with individual corbels projecting not more than one-third the thickness of the unit or one-half the height of the unit. The top course of all corbels shall be a header course.

❖ Corbeling limitations of this section as shown in Commentary Figure R606.3.1 provide for the proper vertical load transfer from cavity wall construction to the supporting foundation wall.

R606.4 Allowable stresses. Allowable compressive stresses in masonry shall not exceed the values prescribed in Table R606.4. In determining the stresses in masonry, the effects of all loads and conditions of loading and the influence of all forces affecting the design and strength of the several parts shall be taken into account.

❖ Masonry compressive stresses are determined from the type and compressive strength of the unit and the type of mortar used in construction. Table R606.4 lists the maximum permitted stresses. Stresses must be computed based on the actual dimensions of the masonry.

Example:
How does a Type S versus Type N mortar affect the allowable compressive stress in unit masonry for a wall composed of solid concrete brick having a unit compressive stress of 2,500 psi (17,235 kPa)?

Solution:
From Table R606.4, the allowable compressive strength is a function of the mortar and would be:
2,500 (17,235 kPa) psi solid units (17,235 kPa): using Type S mortar = 160 psi (1103 kPa).
2,500 (17,235 kPa) psi solid units (17,235 kPa): using Type N mortar = 140 psi (965 kPa).

Figure R606.3(1)
MASONRY CORBELING LIMITATIONS PER COURSE

WALL CONSTRUCTION

Figure R606.3(2)
MASONRY CORBELS—TOTAL HORIZONTAL PROJECTION LIMITATIONS FOR WALLS

Figure R606.3.1
MASONRY CORBEL AT FOUNDATION WALL

R606.4.1 Combined units. In walls or other structural members composed of different kinds or grades of units, materials or mortars, the maximum stress shall not exceed the allowable stress for the weakest of the combination of units, materials and mortars of which the member is composed. The net thickness of any facing unit that is used to resist stress shall not be less than 1.5 inches (38 mm).

❖ Walls comprised of different grades of units, materials, or mortars must be limited to the maximum compressive stresses for the weakest combination of units and mortar.
  Example:
    What is the allowable compressive stress in unit masonry for a wall constructed of 8,000 psi

(55,152 kPa) solid brick units, 2,000 psi (13,788 kPa) hollow masonry units, and Type S mortar?
Solution:
From Table R606.4, the allowable compressive stress would be limited to the smaller of:
8,000 psi (55,152 kPa) solid brick units using Type S mortar = 350 psi (2,413 kPa);

2,000 psi (13,788 kPa) hollow masonry units using Type S mortar = 140 psi (965 kPa).

The allowable unit masonry stress used in design would be the weakest of the two materials constructed with Type S mortar. Therefore, the allowable stress would be 140 psi (965 kPa).

TABLE R606.4
ALLOWABLE COMPRESSIVE STRESSES FOR EMPIRICAL DESIGN OF MASONRY

CONSTRUCTION; COMPRESSIVE STRENGTH OF UNIT, GROSS AREA	ALLOWABLE COMPRESSIVE STRESSES[a] GROSS CROSS-SECTIONAL AREA[b]	
	Type M or S mortar	Type N mortar
Solid masonry of brick and other solid units of clay or shale; sand-lime or concrete brick:		
8,000 + psi	350	300
4,500 psi	225	200
2,500 psi	160	140
1,500 psi	115	100
Grouted[c] masonry, of clay or shale; sand-lime or concrete:		
4,500+ psi	225	200
2,500 psi	160	140
1,500 psi	115	100
Solid masonry of solid concrete masonry units:		
3,000+ psi	225	200
2,000 psi	160	140
1,200 psi	115	100
Masonry of hollow load-bearing units:		
2,000+ psi	140	120
1,500 psi	115	100
1,000 psi	75	70
700 psi	60	55
Hollow walls (cavity or masonry bonded[d]) solid units:		
2,500+ psi	160	140
1,500 psi	115	100
Hollow units	75	70
Stone ashlar masonry:		
Granite	720	640
Limestone or marble	450	400
Sandstone or cast stone	360	320
Rubble stone masonry:		
Coarse, rough or random	120	100

For SI: 1 pound per square inch = 6.895 kPa.

a. Linear interpolation shall be used for determining allowable stresses for masonry units having compressive strengths that are intermediate between those given in the table.
b. Gross cross-sectional area shall be calculated on the actual rather than nominal dimensions.
c. See Section R607.
d. Where floor and roof loads are carried upon one wythe, the gross cross-sectional area is that of the wythe under load; if both wythes are loaded, the gross cross-sectional area is that of the wall minus the area of the cavity between the wythes. Walls bonded with metal ties shall be considered as cavity walls unless the collar joints are filled with mortar or grout.

❖ See the commentary for Section R606.4.

R606.5 Piers. The unsupported height of masonry piers shall not exceed ten times their least dimension. When structural clay tile or hollow concrete masonry units are used for isolated piers to support beams and girders, the cellular spaces shall be filled solidly with concrete or Type M or S mortar, except that unfilled hollow piers may be used if their unsupported height is not more than four times their least dimension. Where hollow masonry units are solidly filled with concrete or Type M, S or N mortar, the allowable compressive stress shall be permitted to be increased as provided in Table R606.4.

❖ Isolated masonry piers conforming to this section may be used to support beams and girders. The minimum pier area should be based on the design load supported and the allowable stress of the construction per Table R606.4.

R606.5.1 Pier cap. Hollow piers shall be capped with 4 inches (102 mm) of solid masonry or concrete or shall have cavities of the top course filled with concrete or grout or other approved methods.

WALL CONSTRUCTION

❖ See the commentary for Section R606.5.

R606.6 Chases. Chases and recesses in masonry walls shall not be deeper than one-third the wall thickness, and the maximum length of a horizontal chase or horizontal projection shall not exceed 4 feet (1219 mm), and shall have at least 8 inches (203 mm) of masonry in back of the chases and recesses and between adjacent chases or recesses and the jambs of openings. Chases and recesses in masonry walls shall be designed and constructed so as not to reduce the required strength or required fire resistance of the wall and in no case shall a chase or recess be permitted within the required area of a pier. Masonry directly above chases or recesses wider than 12 inches (305 mm) shall be supported on noncombustible lintels.

❖ A chase is a continuous recess in a masonry wall that receives a pipe, conduit, etc. Generally, most chases are vertical. It is best to construct chases as the masonry wall is built so the strength of the wall is not reduced.

R606.7 Stack bond. In unreinforced masonry where masonry units are laid in stack bond, longitudinal reinforcement consisting of not less than two continuous wires each with a minimum aggregate cross-sectional area of 0.017 square inch (11 mm^2) shall be provided in horizontal bed joints spaced not more than 16 inches (406 mm) on center vertically.

❖ To control cracking where masonry units are placed in a stack bond and where the wall is of unreinforced masonry construction, longitudinal reinforcement must be used in the horizontal bed joints as shown in Commentary Figure R606.7(1). Premanufactured ladder or truss-type reinforcement as shown in Commentary Figure R606.7(2) is generally used to meet the requirements for longitudinal steel.

R606.8 Lateral support. Masonry walls shall be laterally supported in either the horizontal or the vertical direction. The maximum spacing between lateral supports shall not exceed the distances in Table R606.8. Lateral support shall be provided by cross walls, pilasters, buttresses or structural frame mem-

For SI: 1 square inch = 645.2 mm^2.

Figure R606.7(1)
MINIMUM HORIZONTAL JOINT REINFORCEMENT FOR STACK-BOND MASONRY

Figure R606.7(2)
HORIZONTAL JOINT REINFORCEMENT – LADDER AND TRUSS TYPES

bers when the limiting distance is taken horizontally, or by floors or roofs when the limiting distance is taken vertically.

❖ The limitations on the maximum unsupported height or length of masonry walls specified in Table R606.8 provides reasonable performance. For purposes of applying the unsupported height limitations, Figures R606.10(1), R606.10(2) and R606.10(3) provide details recognized as appropriate methods of anchorage. At the base of the wall, footings are a lateral support point. Thus, the unsupported height from the footing to the anchorage point at the floor or roof is the unsupported height, which must be limited to the values in Table R606.8.

In lieu of unsupported dimensional limitations being measured vertically from footing to supporting floor or roof, the span limitations in Table R604.8 of the code may be met with the use of pilasters, columns, piers, cross walls, or similar elements whose relative stiffness is greater than the wall. These elements are anchored to the roof or floor structural elements in such a manner as to transmit imposed lateral forces.

Commentary Figure R606.8 illustrates the lateral support limitations specified in the table.

Example:
The height or length limitation between lateral supports (see Figure R606.8) for a solid-grouted masonry bearing wall of 8-inch (203 mm) nominal units would be 8 x 20 = 160 inches (4064 mm). For a nonbearing interior wall using two wythes of 4-inch (102 mm) brick, the allowable limitation would be (4 + 4) x 36 = 288 inches (7315 mm).

TABLE R606.8
SPACING OF LATERAL SUPPORT FOR MASONRY WALLS

CONSTRUCTION	MAXIMUM WALL LENGTH TO THICKNESS OR WALL HEIGHT TO THICKNESS[a,b]
Bearing walls: Solid or solid grouted All other	 20 18
Nonbearing walls: Exterior Interior	 18 36

For SI: 1 foot = 304.8 mm.

a. Except for cavity walls and cantilevered walls, the thickness of a wall shall be its nominal thickness measured perpendicular to the face of the wall. For cavity walls, the thickness shall be determined as the sum of the nominal thicknesses of the individual wythes. For cantilever walls, except for parapets, the ratio of height to nominal thickness shall not exceed 6 for solid masonry, or 4 for hollow masonry. For parapets, see Section R606.2.4.

b. An additional unsupported height of 6 feet is permitted for gable end walls.

❖ Limitations on ratios of distance between lateral supports to wall thickness listed in this table are primarily traditional ratios based on successful performance. The thickness in this table is based on the nominal thickness of the wall for other than cavity wall construction and the sum of the nominal thickness of the wythes not including the cavity for cavity walls; see Note "a." To allow for additional height at gable-end walls, Note "b" permits an additional 6 feet (1829 mm). This is similar to the allowance discussed in the commentary to Figure R606.2.1.

WALL CONSTRUCTION

FIGURE R606.8 - R606.8.2.1

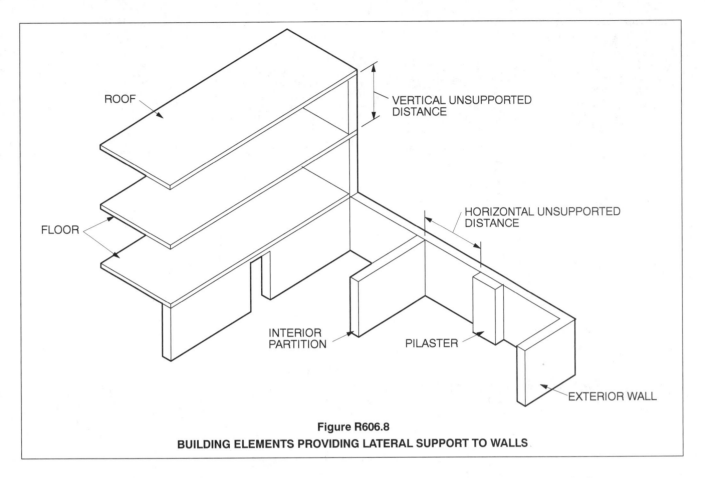

Figure R606.8
BUILDING ELEMENTS PROVIDING LATERAL SUPPORT TO WALLS

R606.8.1 Horizontal lateral support. Lateral support in the horizontal direction provided by intersecting masonry walls shall be provided by one of the methods in Section R606.8.1.1 or Section R606.8.1.2.

❖ When masonry walls span horizontally between intersecting walls, anchorage is achieved by either of the following sections.

R606.8.1.1 Bonding pattern. Fifty percent of the units at the intersection shall be laid in an overlapping masonry bonding pattern, with alternate units having a bearing of not less than 3 inches (76 mm) on the unit below.

❖ This method of anchorage is accomplished by offsetting (overlapping) alternating courses of masonry.

R606.8.1.2 Metal reinforcement. Interior nonload-bearing walls shall be anchored at their intersections, at vertical intervals of not more than 16 inches (406 mm) with joint reinforcement of at least 9 gage, or $1/4$ inch (6.4 mm) galvanized mesh hardware cloth. Intersecting masonry walls, other than interior nonloadbearing walls, shall be anchored at vertical intervals of not more than 8 inches (203 mm) with joint reinforcement of at least 9 gage and shall extend at least 30 inches (762 mm) in each direction at the intersection. Other metal ties, joint reinforcement or anchors, if used, shall be spaced to provide equivalent area of anchorage to that required by this section.

❖ See Commentary Figure R606.8.1.2.

R606.8.2 Vertical lateral support. Vertical lateral support of masonry walls in Seismic Design Category A, B or C shall be provided in accordance with one of the methods in Section R606.8.2.1 or Section R606.8.2.2.

❖ Where walls span vertically in buildings classified as Seismic Design Category A, B or C, masonry walls must be anchored to the floor and roof diaphragms to transmit the anticipated lateral forces. The unsupported height from the footing to the floor or roof must be within the limits of Table 606.8.

R606.8.2.1 Roof structures. Masonry walls shall be anchored to roof structures with metal strap anchors spaced in accordance with the manufacturer's instructions, $1/2$-inch (12.7 mm) bolts spaced not more than 6 feet (1829 mm) on center, or other approved anchors. Anchors shall be embedded at least 16 inches (406 mm) into the masonry, or be hooked or welded to bond beam reinforcement placed not less than 6 inches (152 mm) from the top of the wall.

❖ If $1/2$-inch (12.7 mm) diameter anchor bolts are used, they must be spaced no farther apart than 6 feet (1829 mm) on center. Where approved metal strap anchors are used, the spacing must be in accordance with the manufacturer's recommendations. Anchors are embedded in the masonry wall and connected to the wooden ledger as illustrated in Commentary Figures R606.8.2.1(1) and R606.8.2.1(2).

Figure R606.8.1.2
INTERSECTING MASONRY WALLS

Figure R606.8.2.1(1)
ANCHORAGE OF ROOF STRUCTURES

Figure R606.8.2.1(2)
ANCHORAGE OF ROOF STRUCTURES

R606.8.2.2 Floor diaphragms. Masonry walls shall be anchored to floor diaphragm framing by metal strap anchors spaced in accordance with the manufacturer's instructions, $^1/_2$-inch-diameter (12.7 mm) bolts spaced at intervals not to exceed 6 feet (1829 mm) and installed as shown in Figure R606.10(1), or by other approved methods.

❖ Masonry walls must be anchored to the floor diaphragm with bolts spaced no farther apart than 6 feet (1829 mm) on center. Fasteners may be either metal strap anchors installed per the manufacturer's recommendations or bolts that are embedded in the masonry walls and connected to the wooden ledger. Ledge fasteners transfer shear forces when loading is in the plane of the wall; when forces are out of plane, ledge fasteners prevent the wall and flooring system from separating.

R606.9 Lintels. Masonry over openings shall be supported by steel lintels, reinforced concrete or masonry lintels or masonry arches, designed to support load imposed.

❖ Masonry wall openings require a structural member designed to support the masonry above. Note that wood is not permitted for a lintel supporting a masonry wall.

R606.10 Anchorage. Masonry walls shall be anchored to floor and roof systems in accordance with the details shown in Figure R606.10(1), R606.10(2) or R606.10(3). Footings may be considered as points of lateral support.

❖ Masonry walls depend on floors and roofs for out-of-plane lateral support. Inadequate anchorage of masonry walls in areas of high, and even moderate, seismicity can be problematic. The referenced figures show anchorage requirements that vary based on seismic design category. They illustrate details that provide adequate load transfer under lateral loads.

R606.11 Seismic requirements. The seismic requirements of this section shall apply to the design of masonry and the construction of masonry building elements located in Seismic Design Category D_1 or D_2. Townhouses in Seismic Design Category C shall comply with the requirements in Section R606.11.2. These requirements shall not apply to glass unit masonry conforming to Section R610 or masonry veneer conforming to Section R703.7.

❖ Although the code generally permits unreinforced walls, this section requires reinforcement in walls of buildings classified as Seismic Design Category C, D_1 or D_2.

FIGURE R606.10(1)

WALL CONSTRUCTION

LEDGER BOLT SIZE AND SPACING

JOIST SPAN	BOLT SIZE AND SPACING	
	ROOF	FLOOR
10 FT.	1/2 AT 2 FT. 6 IN. 7/8 AT 3 FT. 6 IN.	1/2 AT 2 FT. 0 IN. 7/8 AT 2 FT. 9 IN.
10-15 FT.	1/2 AT 1 FT. 9 IN. 7/8 AT 2 FT. 6 IN.	1/2 AT 1 FT. 4 IN. 7/8 AT 2 FT. 0 IN.
15-20 FT.	1/2 AT 1 FT. 3 IN. 7/8 AT 2 FT. 0 IN.	1/2 AT 1 FT. 0 IN. 7/8 AT 1 FT. 6 IN.

Note: Where bolts are located in hollow masonry, the cells in the courses receiving the bolt shall be grouted solid.

For SI: 1 inch = 25.4 mm, 1 foot = 304.8 mm, 1 pound per square foot = 0.0479 kN/m².

FIGURE R606.10(1)
ANCHORAGE REQUIREMENTS FOR MASONRY WALLS LOCATED IN SEISMIC DESIGN CATEGORY A, B OR C AND WHERE WIND LOADS ARE LESS THAN 30 PSF

❖ See the commentary for Sections R606.10 and R606.11.1.

WALL CONSTRUCTION

FIGURE R606.10(2)

For SI: 1 inch = 25.4 mm, 1 foot = 304.8 mm.

FIGURE R606.10(2)
REQUIREMENTS FOR REINFORCED GROUTED MASONRY CONSTRUCTION IN SEISMIC DESIGN CATEGORY C

❖ See the commentary for Sections R606.10 and R606.11.1.

FIGURE R606.10(3)

WALL CONSTRUCTION

Note: A full bed joint must be provided. All cells containing vertical bars are to be filled to the top of wall and provide inspection opening as shown on detail "A." Horizontal bars are to be laid as shown on detail "B." Lintel bars are to be laid as shown on Section C.

For SI: 1 inch = 25.4 mm, 1 foot = 304.8 mm.

FIGURE R606.10(3)
REQUIREMENTS FOR REINFORCED MASONRY CONSTRUCTION IN SEISMIC DESIGN CATEGORY D_1 OR D_2

❖ See the commentary for Sections R606.10 and R606.11.1.

WALL CONSTRUCTION

R606.11.1 General. Masonry structures and masonry elements shall comply with the requirements of Sections R606.11.2 through R606.11.4 based on the seismic design category established in Table R301.2(1). Masonry structures and masonry elements shall comply with the requirements of Section R606.11 and Figures R606.10(1), R606.10(2) and R606.10(3) or shall be designed in accordance with ACI 530/ASCE 5/TMS 402.

❖ Masonry walls must comply with this section unless a design is provided. The three cited figures show the reinforcing requirements as a function of the seismic design category and wind load.

R606.11.1.1 Floor and roof diaphragm construction. Floor and roof diaphragms shall be constructed of structural wood sheathing panels, attached to wood framing in accordance with Table R602.3(1) or to cold-formed steel floor framing in accordance with Table R505.3.1(2) or to cold-formed steel roof framing in accordance with Table R804.3. Additionally, sheathing panel edges perpendicular to framing members shall be backed by blocking, and sheathing shall be connected to the blocking with fasteners at the edge spacing. For Seismic Design Categories C, D_1 and D_2, where the width-to-thickness dimension of the diaphragm exceeds 2-to-1, edge spacing of fasteners shall be 4 inches (102 mm) on center.

❖ This section reiterates the requirements of Chapters 5 and 8 for fastening sheathing to floor framing and roof framing. Additional blocking is specified for the sheathing edges perpendicular to the floor or roof framing, thus providing a blocked diaphragm. Edge fasteners as specified for panel edges must be provided at this blocking. Furthermore, if a diaphragm's depth-to-span ratio is greater than 2 in a building classified as Seismic Design Category C, D_1 or D_2, the edge nail spacing must be reduced to 4 inches (102 mm) on center. These requirements provide added strength and stiffness for diaphragms that support masonry walls.

R606.11.2 Seismic Design Category C. Townhouses located in Seismic Design Category C shall comply with the requirements of this section.

❖ Buildings classified as Seismic Design Category C must conform to the requirements of this section as well as Figures R606.10(1) and R606.10(2). The requirements are based on whether or not an element is part of the lateral force-resisting system. They are aimed at providing ductility by specifying minimum reinforcement. This section is similar to Section 2106.4 of the *International Building Code*.

R606.11.2.1 Design of elements not part of the lateral force-resisting system.

R606.11.2.1.1 Load-bearing frames or columns. Elements not part of the lateral-force-resisting system shall be analyzed as to their effect on the response of the system. Such frames or columns shall be adequate for vertical load carrying capacity and induced moment due to the design story drift.

❖ Load-bearing elements not part of a lateral force-resisting system must be analyzed for their effect of the response of the system and must be capable of supporting all loads in combination with drift-induced bending stresses. These empirical masonry provisions do not cover the design of elements referred to in this section. They would require an engineering design using ACI 530/ASCE 5/TMS 402. See the commentary for Section R606.11.2.

R606.11.2.1.2 Masonry partition walls. Masonry partition walls, masonry screen walls and other masonry elements that are not designed to resist vertical or lateral loads, other than those induced by their own weight, shall be isolated from the structure so that vertical and lateral forces are not imparted to these elements. Isolation joints and connectors between these elements and the structure shall be designed to accommodate the design story drift.

❖ Nonload-bearing elements that are not designed to resist vertical or lateral loads from earthquake effects on the building lateral force system must be isolated from the structure with isolation joints and connectors that can accommodate the design drift. The "design story drift" should be taken as one percent of the story height as allowed in Section 1617.5.3 of the *International Building Code* unless an analysis is performed.

R606.11.2.1.3 Reinforcement requirements for masonry elements. Masonry elements listed in Section R606.11.2.1.2 shall be reinforced in either the horizontal or vertical direction as shown in Figure R606.10(2) and in accordance with the following:

1. Horizontal reinforcement. Horizontal joint reinforcement shall consist of at least two longitudinal W1.7 wires spaced not more than 16 inches (406 mm) for walls greater than 4 inches (102 mm) in width and at least one longitudinal W1.7 wire spaced not more than 16 inches (406 mm) for walls not exceeding 4 inches (102 mm) in width; or at least one No. 4 bar spaced not more than 48 inches (1219 mm). Where two longitudinal wires of joint reinforcement are used, the space between these wires shall be the widest that the mortar joint will accommodate. Horizontal reinforcement shall be provided within 16 inches (406 mm) of the top and bottom of these masonry elements.

2. Vertical reinforcement. Vertical reinforcement shall consist of at least one No. 4 bar spaced not more than 48 inches (1219 mm). Vertical reinforcement shall be located within 16 inches (406 mm) of the ends of masonry walls.

❖ This section specifies reinforcement for masonry walls that are not part of the building lateral force system, as described in the previous section. See the commentary for Section R606.11.2.

R606.11.2.2 Design of elements part of the lateral-force-resisting system.

R606.11.2.2.1 Connections to masonry shear walls. Connectors shall be provided to transfer forces between masonry walls and horizontal elements in accordance with the requirements of

Section 2.1.8 of ACI 530/ASCE 5/TMS 402. Connectors shall be designed to transfer horizontal design forces acting either perpendicular or parallel to the wall, but not less than 200 pounds per linear foot (2919 N/m) of wall. The maximum spacing between connectors shall be 4 feet (1219 mm). Such anchorage mechanisms shall not induce tension stresses perpendicular to grain in ledgers or nailers.

❖ The connection of horizontal elements such as diaphragms to masonry walls or columns must be designed in accordance with ACI 530/ASCE 5/TMS 402. These connections should also be in accordance with Figure R606.10(2) and meet the minimum requirements specified in this section. Because these connections have been a source of earthquake-related failures, the code does not allow diaphragm anchorage mechanisms that cause cross-grain bending. This addresses the problem of premature failure of wood ledgers. See the commentary for Section R606.11.2.

R606.11.2.2.2 Connections to masonry columns. Connectors shall be provided to transfer forces between masonry columns and horizontal elements in accordance with the requirements of Section 2.1.8 of ACI 530/ASCE 5/TMS 402. Where anchor bolts are used to connect horizontal elements to the tops of columns, anchor bolts shall be placed within lateral ties. Lateral ties shall enclose both the vertical bars in the column and the anchor bolts. There shall be a minimum of two No. 4 lateral ties provided in the top 5 inches (127 mm) of the column.

❖ These empirical masonry provisions do not cover the design of masonry columns. A design per ACI 530/ASCE 5/TMS 402 must be provided. In that event, this section provides a reference to the appropriate section of the standard for the column-to-diaphragm connection design. See the commentary for Section R606.11.2.

R606.11.2.2.3 Minimum reinforcement requirements for masonry shear walls. Vertical reinforcement of at least one No. 4 bar shall be provided at corners, within 16 inches (406 mm) of each side of openings, within 8 inches (203 mm) of each side of movement joints, within 8 inches (203 mm) of the ends of walls, and at a maximum spacing of 10 feet (3048 mm).

Horizontal joint reinforcement shall consist of at least two wires of W1.7 spaced not more than 16 inches (406 mm); or bond beam reinforcement shall be provided of at least one No. 4 bar spaced not more than 10 feet (3048 mm). Horizontal reinforcement shall also be provided at the bottom and top of wall openings and shall extend not less than 24 inches (610 mm) nor less than 40 bar diameters past the opening; continuously at structurally connected roof and floor levels; and within 16 inches (406 mm) of the top of walls.

❖ Masonry shear walls resist the lateral loads from wind or earthquakes. This section specifies reinforcing requirements, which are illustrated in Figure R606.10(2). Refer to the commentary for Section R606.11.2.

R606.11.3 Seismic Design Category D_1. All structures in Seismic Design Category D_1 shall comply with the requirements of Seismic Design Category C and to the additional requirements of this section.

❖ Buildings classified as Seismic Design Category D_1 must conform to the requirements of this section as well as the requirements for Seismic Design Category C (see Section R606.11.2) and Figure R606.10(2). This section contains provisions that are similar to those in Section 2106.5 of the *International Building Code*.

R606.11.3.1 Design requirements. Masonry elements other than those covered by Section R606.11.2.1.2 shall be designed in accordance with the requirements of Chapter 1 and Sections 2.1 and 2.3 of ACI 530/ASCE 5/TMS 402 and shall meet the minimum reinforcement requirements contained in Sections R606.11.3.2 and R606.11.3.2.1.

Exception: Masonry walls limited to one story in height and 9 feet (2743 mm) between lateral supports need not be designed provided they comply with the minimum reinforcement requirements of Sections R606.11.3.2 and R606.11.3.2.1.

❖ This section specifies the design criteria for masonry elements in Seismic Design Category D_1.
Except for nonloadbearing elements that are not part of the lateral force-resisting system, masonry elements must be designed using ACI 530/ASCE 5/TMS 402. The exception allows certain one story masonry walls not to be designed to the engineering standards but satisfy the specified prescriptive code requirements.

R606.11.3.2 Minimum reinforcement requirements for masonry walls. Masonry walls other than those covered by Section R606.11.2.1.3 shall be reinforced in both the vertical and horizontal direction. The sum of the cross-sectional area of horizontal and vertical reinforcement shall be at least 0.002 times the gross cross-sectional area of the wall, and the minimum cross-sectional area in each direction shall be not less than 0.0007 times the gross cross-sectional area of the wall. Reinforcement shall be uniformly distributed. Table R606.11.3.2 provides the minimum reinforcing bar sizes required for varying thicknesses of masonry walls. The maximum spacing of reinforcement shall be 48 inches (1219 mm) provided that the walls are solid grouted and constructed of hollow open-end units, hollow units laid with full head joints or two wythe of solid units. The maximum spacing of reinforcement shall be 24 inches (610 mm) for all other masonry.

❖ The sum of horizontal and vertical reinforcement must be at least 0.2 percent of the cross-sectional area of the wall, and the reinforcing should be distributed as shown in Table R606.11.3.2. Refer to the commentary for Figure R606.10(3) and Section R606.11.3.

TABLE R606.11.3.2
MINIMUM DISTRIBUTED WALL REINFORCEMENT FOR BUILDINGS ASSIGNED TO SEISMIC DESIGN CATEGORY D_1

NOMINAL WALL THICKNESS (inches o.c.)	MINIMUM SUM OF THE VERTICAL AND HORIZONTAL REINFORCEMENT AREAS[a] (inches per square foot)	MINIMUM REINFORCEMENT AS DISTRIBUTED IN BOTH HORIZONTAL AND VERTICAL DIRECTIONS[b] (inches per square foot)	MINIMUM BAR SIZE FOR REINFORCEMENT SPACED AT 48 INCHES
6	0.135	0.047	#4
8	0.183	0.064	#5
10	0.231	0.081	#6
12	0.279	0.098	#6

For SI: 1 inch = 25.4 mm, 1 foot = 304.8 mm, 1 square inch per foot = 2.12 mm²/mm.
a. Based on the minimum reinforcing ratio of 0.002 times the gross cross-sectional area of the wall.
b. Based on the minimum reinforcing ratio each direction of 0.0007 times the gross cross-sectional area of the wall.

❖ This table lists the minimum reinforcement areas specified in Section R606.11.3.2. The minimum bar size shown in the far right column satisfies the requirement for 0.0007 times the gross cross-sectional area of the wall. This must be provided as a minimum in both the vertical and horizontal directions. Additional reinforcement is necessary in either direction in order to meet the requirement for a total of 0.002 times the gross cross-sectional area.

R606.11.3.2.1 Shear wall reinforcement requirements. The maximum spacing of vertical and horizontal reinforcement shall be the smaller of one-third the length of the shear wall, one-third the height of the shear wall, or 48 inches (1219 mm). The minimum cross-sectional area of vertical reinforcement shall be one-third of the required shear reinforcement. Shear reinforcement shall be anchored around vertical reinforcing bars with a standard hook.

❖ See the commentary for Section R606.11.3.

R606.11.3.3 Minimum reinforcement for masonry columns. Lateral ties in masonry columns shall be spaced not more than 8 inches (203 mm) on center and shall be at least $^3/_8$ inch (9.5 mm) diameter. Lateral ties shall be embedded in grout.

❖ See Figure R606.10(3). Also see the commentary for Sections R606.11.2.2.2 and R606.11.3.

R606.11.3.4 Material restrictions. Type N mortar or masonry cement shall not be used as part of the lateral-force-resisting system.

❖ This restates the requirement of Section R607.1.3.

R606.11.3.5 Lateral tie anchorage. Standard hooks for lateral tie anchorage shall be either a 135-degree (2.4 rad) standard hook or a 180-degree (3.2 rad) standard hook.

❖ Lateral ties (required by Section R606.11.3.3) must be anchored as specified. Also see the commentary for Section R606.11.3.

R606.11.4 Seismic Design Category D_2. All structures in Seismic Design Category D_2 shall comply with the requirements of Seismic Design Category D_1 and to the additional requirements of this section.

❖ Buildings classified as Seismic Design Category D_2 must conform to the requirements of this section as well as all the requirements for Seismic Design Category D_1 and Figure R606.10(3).

R606.11.4.1 Design of elements not part of the lateral-force-resisting system. Stack bond masonry that is not part of the lateral-force-resisting system shall have a horizontal cross-sectional area of reinforcement of at least 0.0015 times the gross cross-sectional area of masonry. Table R606.11.4.1 provides minimum reinforcing bar sizes for masonry walls. The maximum spacing of horizontal reinforcement shall be 24 inches (610 mm). These elements shall be solidly grouted and shall be constructed of hollow open-end units or two wythes of solid units.

❖ For elements that are not part of the lateral force-resisting system, horizontal reinforcement must be at least 0.0015 times the cross-sectional area of the wall as shown in Table R606.11.4.1.

TABLE R606.11.4.1
MINIMUM REINFORCING FOR STACKED BONDED MASONRY WALLS IN SEISMIC DESIGN CATEGORY D_2

NOMINAL WALL THICKNESS (inches)	MINIMUM BAR SIZE SPACED AT 24 INCHES
6	#4
8	#5
10	#5
12	#6

For SI: 1 inch = 25.4 mm.

❖ See the commentary for Section R606.11.4.1.

R606.11.4.2 Design of elements part of the lateral-force-resisting system. Stack bond masonry that is part of the lateral-force-resisting system shall have a horizontal cross-sectional area of reinforcement of at least 0.0025 times the gross cross-sectional area of masonry. Table R606.11.4.2 provides minimum reinforcing bar sizes for masonry walls. The maximum spacing of horizontal reinforcement shall be 16 inches (406 mm). These elements shall be solidly grouted and shall be constructed of hollow open-end units or two wythes of solid units.

❖ Elements that are part of the lateral force-resisting system must have horizontal reinforcement of at least 0.0025 times the cross-sectional area of the wall as shown in Table R606.11.4.2. These elements may be constructed of solid grouted hollow open-end units or of two wythes of solid units solidly grouted.

TABLE R606.11.4.2
MINIMUM REINFORCING FOR STACKED BONDED MASONRY WALLS IN SEISMIC DESIGN CATEGORY D_2

NOMINAL WALL THICKNESS (inches)	MINIMUM BAR SIZE SPACED AT 16 INCHES
6	#4
8	#5
10	#5
12	#6

For SI: 1 inch = 25.4 mm.

❖ See the commentary for Section R606.11.4.2.

R606.12 Protection for reinforcement. All bars shall be completely embedded in mortar or grout. Joint reinforcement embedded in horizontal mortar joints shall not have less than $^5/_8$-inch (15.9 mm) mortar coverage from the exposed face. All other reinforcement shall have a minimum coverage of one bar diameter over all bars, but not less than $^3/_4$ inch (19.1 mm), except where exposed to weather or soil, in which case the minimum coverage shall be 2 inches (51 mm).

❖ Commentary Figure R606.12 shows placement restrictions for reinforcement relative to exposed faces and coverage. It also illustrates the distance from the face of the masonry unit exposed to weather or soil.

R606.13 Beam supports. Beams, girders or other concentrated loads supported by a wall or column shall have a bearing of at least 3 inches (76 mm) in length measured parallel to the beam upon solid masonry not less than 4 inches (102 mm) in thickness, or upon a metal bearing plate of adequate design and dimensions to distribute the load safely, or upon a continuous reinforced masonry member projecting not less than 4 inches (102 mm) from the face of the wall.

❖ To provide for the transfer of vertical loads from beams, girders, or other elements to the masonry wall or column, a minimum bearing as shown in Commentary Figure R606.13 is required. An alternative to the minimum bearing is the use of metal bearing plates that are based on structural design.

R606.13.1 Joist bearing. Joists shall have a bearing of not less than $1^1/_2$ inches (38 mm), except as provided in Section

Figure R606.12
REINFORCING STEEL CLEARANCES

WALL CONSTRUCTION

Figure R606.13
BEAM OR GIRDER BEARING ON MASONRY

R606.13, and shall be supported in accordance with Figure R606.10(1).

❖ Where joists bear on masonry wall or column elements, the minimum bearings as shown in Figure R606.10(1) and Commentary Figure R606.13.1 are considered to be adequate. The code does not provide guidance distinguishing whether a member is a joist or a beam; however, joists are generally considered to be members that are a nominal 2 inches (51 mm) maximum in thickness and placed not more than 24 inches (610 mm) on center.

R606.14 Metal accessories. Joint reinforcement, anchors, ties and wire fabric shall conform to the following: ASTM A 82 for wire anchors and ties; ASTM A 36 for plate, headed and bent-bar anchors; ASTM A 510 for corrugated sheet metal anchors and ties; ASTM A 951 for joint reinforcement; ASTM B 227 for copper-clad steel wire ties; or ASTM A 167 for stainless steel hardware.

Figure R606.13.1
JOIST-BEARING MINIMUMS

❖ This section provides reference standards for metal accessories used in masonry wall construction.

R606.14.1 Corrosion protection. Minimum corrosion protection of joint reinforcement, anchor ties and wire fabric for use in masonry wall construction shall conform to Table R606.14.1.

❖ Table R606.14.1 provides reference standards for protecting metal accessories used in masonry wall construction.

The corrosion protection requirements are dependent on type of steel and exposure. Because of its high resistance to corrosion, stainless steel is not required to be coated. Joint reinforcement, anchors, ties and accessories of other than stainless steel must be protected by zinc coatings (galvanizing). The protective value of the zinc coating increases with increasing coating thickness; therefore, the amount of galvanizing required increases with the potential severity of exposure.

TABLE R606.14.1
MINIMUM CORROSION PROTECTION

MASONRY METAL ACCESSORY	STANDARD
Joint reinforcement, interior walls	ASTM A 641, Class 1
Wire ties or anchors in exterior walls completely embedded in mortar or grout	ASTM A 641, Class 3
Wire ties or anchors in exterior walls not completely embedded in mortar or grout	ASTM A 153, Class B-2
Joint reinforcement in exterior walls or interior walls exposed to moist environment	ASTM A 153, Class B-2
Sheet metal ties or anchors exposed to weather	ASTM A 153, Class B-2
Sheet metal ties or anchors completely embedded in mortar or grout	ASTM A 525, Class G-60
Stainless steel hardware for any exposure	ASTM A 167, Type 304

❖ See the commentary for Section R606.14.1.

SECTION R607
UNIT MASONRY

R607.1 Mortar. Mortar for use in masonry construction shall comply with ASTM C 270. The type of mortar shall be in accordance with Sections R607.1.1, R607.1.2 and R607.1.3 and shall meet the proportion specifications of Table R607.1 or the property specifications of ASTM C 270.

❖ This section specifies the requirements for mortar used in masonry construction. Mortar is the bonding agent that separates masonry units while bonding them together. It is an integral part of any masonry wall and must be strong, durable and capable of keeping the wall intact. In addition, it should help to create a moisture-resistant barrier.

Mortar makes up approximately 25 percent of a standard modular brick wall and about 10 percent of a block wall. It is therefore a significant part of the structure. ASTM C 270 discusses the materials and methods for mixing mortar. For purposes of economy, a good rule of thumb is to specify the lowest strength mortar that will satisfy the structural requirements of the project. Mortar consists of cementitious materials and well-graded sand with sufficient fines. Mortar is used for the following purposes:
- It is a bedding or seating materials for the masonry unit.
- It allows the unit to be leveled and properly placed.
- It bonds the units together.
- It provides compressive strength.
- It provides shear strength, particularly parallel to the wall.
- It allows some movement and elasticity between units.
- It seals irregularities of the masonry unit and provides a weather-tight wall, preventing penetration of wind and water into and through the wall.
- It can provide color to the wall when a mineral color additive is used.
- It can provide an architectural appearance by using various types of joints.

R607.1.1 Foundation walls. Masonry foundation walls constructed as set forth in Tables R404.1.1(1) through R404.1.1(4) and mortar shall be Type M or S.

❖ This section specifies the applicable tables in Chapter 4 that apply to masonry foundation walls.

R607.1.2 Masonry in Seismic Design Categories A, B and C. Mortar for masonry serving as the lateral-force-resisting system in Seismic Design Categories A, B and C shall be Type M, S or N mortar.

❖ Masonry construction in buildings classified as Seismic Design Category A, B or C must use Type M, S or N mortar.

R607.1.3 Masonry in Seismic Design Categories D_1 and D_2. Mortar for masonry serving as the lateral-force-resisting system in Seismic Design Categories D_1 and D_2 shall be Type M or S portland cement-lime or mortar cement mortar.

❖ In buildings classified as Seismic Design category D_1 or D_2 or in foundation walls in any building, only Type M or S mortar is acceptable.

WALL CONSTRUCTION

TABLE R607.1
MORTAR PROPORTIONS[a,b]

Mortar	Type	Portland cement or blended cement	Mortar cement M	Mortar cement S	Mortar cement N	Masonry cement M	Masonry cement S	Masonry cement N	Hydrated lime[c] or lime putty	Aggregate ratio (measured in damp, loose conditions)
Cement-lime	M	1	—	—	—	—	—	—	$1/4$	
	S	1	—	—	—	—	—	—	over $1/4$ to $1/2$	
	N	1	—	—	—	—	—	—	over $1/2$ to $1\,1/4$	
	O	1	—	—	—	—	—	—	over $1\,1/4$ to $2\,1/2$	
Mortar cement	M	1	—	—	1	—	—	—	—	Not less than $2\,1/4$ and not more than 3 times the sum of separate volumes of lime, if used, and cement
	M	—	1	—	—	—	—	—	—	
	S	$1/2$	—	—	1	—	—	—	—	
	S	—	—	1	—	—	—	—	—	
	N	—	—	—	1	—	—	—	—	
	O	—	—	—	1	—	—	—	—	
Masonry cement	M	1	—	—	—	—	—	1	—	
	M	—	—	—	—	1	—	—	—	
	S	$1/2$	—	—	—	—	—	1	—	
	S	—	—	—	—	—	1	—	—	
	N	—	—	—	—	—	—	1	—	
	O	—	—	—	—	—	—	1	—	

For SI: 1 cubic foot = 0.0283 m^3, 1 pound = 0.454 kg.

a. For the purpose of these specifications, the weight of 1 cubic foot of the respective materials shall be considered to be as follows:

Portland Cement	94 pounds	Masonry Cement	Weight printed on bag
Mortar Cement	Weight printed on bag	Hydrated Lime	40 pounds
Lime Putty (Quicklime)	80 pounds	Sand, damp and loose	80 pounds of dry sand

b. Two air-entraining materials shall not be combined in mortar.
c. Hydrated lime conforming to the requirements of ASTM C 207.

❖ The mortar required by Sections R607.1.1 through R607.1.3 must be proportioned as shown in this table or ASTM C 270. Type M mortar is suited for structures below or against grade such as retaining walls. It is also suited for masonry construction subject to high compressive loads, severe frost action or high lateral loads from earth pressures, hurricane winds or earthquakes. Type S mortar is appropriate for use in structures requiring high flexural bond strength that are subject to compressive and lateral loads. Type N mortar is meant for general use in above-grade masonry, residential basement construction, interior walls and partitions, masonry veneer and nonstructural masonry partitions.

R607.2 Placing mortar and masonry units.

R607.2.1 Bed and head joints. Unless otherwise required or indicated on the project drawings, head and bed joints shall be $3/8$ inch (9.5 mm) thick, except that the thickness of the bed joint of the starting course placed over foundations shall not be less than $1/4$ inch (6.4 mm) and not more than $3/4$ inch (19.1 mm).

❖ The provisions of Section R607 set forth specific requirements that are unique to construction utilizing hollow-unit masonry. All hollow masonry units are to be laid and set in mortar. The properties of masonry walls, including strength and appearance, vary significantly depending on the thickness of the mortar joints. The initial bed joint, which is the bed joint between the first course of masonry and the foundation wall, is limited to a thickness of not less than $1/4$ inch (6.4 mm) and a thickness not greater than $3/4$ inch (19 mm). This larger variation is to allow for the inherent unevenness of concrete or concrete masonry foundation walls. The remainders of bed joints, which can be controlled by the mason, are limited to $3/8$ inch (9.5 mm), unless specified otherwise on the project drawings.

R607.2.1.1 Mortar joint thickness tolerance. Mortar joint thickness shall be within the following tolerances from the specified dimensions:

1. Bed joint: + $1/8$ inch (3.2 mm).
2. Head joint: $1/4$ inch (6.4 mm), + $3/8$ inch (9.5 mm).
3. Collar joints: $1/4$ inch (6.4 mm), + $3/8$ inch (9.5 mm).

Exception: Nonload-bearing masonry elements and masonry veneers designed and constructed in accordance with Section R703.7 are not required to meet these tolerances.

❖ Joint-thickness tolerances limit the eccentricity of applied loads and reflect typical industry practice. They also provide a more uniform appearance for exposed masonry. The bed joint is a horizontal joint; the head joint is a vertical joint.

R607.2.2 Masonry unit placement. The mortar shall be sufficiently plastic and units shall be placed with sufficient pressure to extrude mortar from the joint and produce a tight joint. Deep furrowing of bed joints that produces voids shall not be permitted. Any units disturbed to the extent that initial bond is broken after initial placement shall be removed and relaid in fresh mortar. Surfaces to be in contact with mortar shall be clean and free of deleterious materials.

❖ This section gives placement requirements meant to provide adequate bonding between mortar and masonry units. Mortar is the bonding agent that integrates masonry units into a wall; it is used to bind masonry units into a single element by developing a complete, strong, and durable bond. Mortar is usually placed between absorbent masonry units and loses water upon contact with the units. Mortars have a high water-cement ratio when mixed, but the ratio decreases when the mortar comes into contact with the absorbent unit.

Once the mortar has begun to set or harden, tapping or attempting to otherwise move masonry units can be detrimental to the bond. Movement at this time will break the bond between the masonry unit and mortar. The partially dried mortar will not have sufficient plasticity to re-adhere sufficiently to the masonry units. After the initial bond has been broken, the masonry units should be removed and replaced in fresh mortar.

R607.2.2.1 Solid masonry. All solid masonry units shall be laid with full head and bed joints and all interior vertical joints that are designed to receive mortar shall be filled.

❖ This section applies to solid masonry construction (see the definition in Section 202).

R607.2.2.2 Hollow masonry. For hollow masonry units, all head and bed joints shall be filled solidly with mortar for a distance in from the face of the unit not less than the thickness of the face shell.

❖ See the definition for "Masonry Unit, Hollow" in Section 202. Normally, cross webs are not mortared. When individual cells are to be grouted, mortar should be placed on the webs on both sides of the cell to be grouted in order to prevent leakage of the grout.

R607.3 Installation of wall ties. The installation of wall ties shall be as follows:

1. The ends of wall ties shall be embedded in mortar joints. Wall tie ends shall engage outer face shells of hollow units by at least $^1/_2$ inch (12.7 mm). Wire wall ties shall be embedded at least $1^1/_2$ inches (38 mm) into the mortar bed of solid masonry units or solid grouted hollow units.

2. Wall ties shall not be bent after being embedded in grout or mortar.

❖ Proper installation of masonry wall ties is necessary so that bonding of multiple wythes is achieved (also see Section R608.1.2.1). In addition, wall ties should not be bent after placement in the mortar joints. This can break the initial bond between the masonry unit and mortar and possibly lead to an increased amount of moisture infiltration into the wall assembly. Bending the wall ties after initial placement can also be detrimental to the strength of the wall system in regard to the ties' ability to transfer lateral loads to the backing wall assembly.

SECTION R608
MULTIPLE WYTHE MASONRY

R608.1 General. The facing and backing of multiple wythe masonry walls shall be bonded in accordance with Section R608.1.1, R608.1.2 or R608.1.3. In cavity walls, neither the facing nor the backing shall be less than 3 inches (76 mm) nominal in thickness and the cavity shall not be more than 4 inches (102 mm) nominal in width. The backing shall be at least as thick as the facing.

Exception: Cavities shall be permitted to exceed the 4-inch (102 mm) nominal dimension provided tie size and tie spacing have been established by calculation.

❖ The provisions of Section R608 apply to masonry walls more than one masonry unit in thickness. In cases where a wall thickness is made up of multiple wythes of masonry units, bonding is required so that the wythes are tied together to act as a unit. Section R608 contains the specifics as to what constitutes an acceptable bonding mechanism. Two basic types of bonding are permissible: the first involves various overlapping arrangements; the second is achieved by installation of corrosion-resistant metal ties as stipulated for cavity-wall masonry construction.

R608.1.1 Bonding with masonry headers. Bonding with solid or hollow masonry headers shall comply with Sections R608.1.1.1 and R608.1.1.2.

❖ There are two methods of bonding multiwythe walls through the use of headers; they can be solid or hollow unit, depending on the wall system in question.

R608.1.1.1 Solid units. Where the facing and backing (adjacent wythes) of solid masonry construction are bonded by means of masonry headers, no less than 4 percent of the wall surface of each face shall be composed of headers extending not less than 3 inches (76 mm) into the backing. The distance between adjacent full-length headers shall not exceed 24 inches (610 mm) either vertically or horizontally. In walls in which a single header does not extend through the wall, headers from the opposite sides shall overlap at least 3 inches (76 mm), or headers from opposite sides shall be covered with another header course overlapping the header below at least 3 inches (76 mm).

❖ See the commentary for Section R608.1.1 and Commentary Figure R608.1.1.1.

R608.1.1.2 Hollow units. Where two or more hollow units are used to make up the thickness of a wall, the stretcher courses shall be bonded at vertical intervals not exceeding 34 inches (864 mm) by lapping at least 3 inches (76 mm) over the unit below, or by lapping at vertical intervals not exceeding 17 inches (432 mm) with units that are at least 50 percent greater in thickness than the units below.

❖ See the commentary for Section R608.1.1 and Commentary Figure R608.1.1.2.

Figure R608.1.1.1
BONDING WITH SOLID UNIT MASONRY

Figure R608.1.1.2
BONDING WITH HOLLOW UNIT MASONRY

R608.1.2 Bonding with wall ties or joint reinforcement. Bonding with wall ties or joint reinforcement shall comply with Sections R608.1.2.1 through R608.1.2.3.

❖ See the commentary for Section R608.1.

R608.1.2.1 Bonding with wall ties. Bonding with wall ties, except as required by Section R610, where the facing and backing (adjacent wythes) of masonry walls are bonded with $^3/_{16}$-inch-diameter (4.8 mm) wall ties embedded in the horizontal mortar joints, there shall be at least one metal tie for each 4.5 square feet (0.418 m^2) of wall area. Ties in alternate courses shall be staggered. The maximum vertical distance between ties shall not exceed 24 inches (610 mm), and the maximum horizontal distance shall not exceed 36 inches (914 mm). Rods or ties bent to rectangular shape shall be used with hollow masonry units laid with the cells vertically. In other walls, the ends of ties shall be bent to 90-degree (0.79 rad) angles to provide hooks no less than 2 inches (51 mm) long. Additional bonding ties shall be provided at all openings, spaced not more than 3 feet (914 mm) apart around the perimeter and within 12 inches (305 mm) of the opening.

❖ Multiwythe walls can be bonded with the use of wire-type ties. Generally, this type of tie consists of metal Z ties or rectangular ties. The code requires certain maximum spacing for wire tires similar to that for masonry headers. This section provides minimum wire diameters and requires spacing based on a square footage area of the wall in question. Additional bonding ties are required around wall openings, such as windows or doors. These wire ties are used to anchor wythes of masonry together to resist loads or to transfer loads across air spaces to the backing wall material. See Commentary Figures R608.1.2.1(1) and R608.1.2.1(2).

R608.1.2.2 Bonding with adjustable wall ties. Where the facing and backing (adjacent wythes) of masonry are bonded with adjustable wall ties, there shall be at least one tie for each 2.67 square feet (0.248 m^2) of wall area. Neither the vertical nor horizontal spacing of the adjustable wall ties shall exceed 24 inches (610 mm). The maximum vertical offset of bed joints from one wythe to the other shall be 1.25 inches (32 mm). The maximum clearance between connecting parts of the ties shall be $^1/_{16}$ inch (1.6 mm). When pintle legs are used, ties shall have at least two $^3/_{16}$-inch-diameter (4.8 mm) legs.

❖ This section pertains to the use of adjustable ties for anchoring the facing to the backing. Reduced tie spacing is required when these types of ties are specified. Maximum vertical offsets are prescribed so that the adjustable tie will not disengage through use. There are many forms of adjustable wall ties available for masonry construction. See Commentary Figure R608.1.2.2.

R608.1.2.3 Bonding with prefabricated joint reinforcement. Where the facing and backing (adjacent wythes) of masonry are bonded with prefabricated joint reinforcement, there shall be at least one cross wire serving as a tie for each 2.67 square feet (0.248 m^2) of wall area. The vertical spacing of the joint reinforcement shall not exceed 16 inches (406 mm). Cross wires on prefabricated joint reinforcement shall not be smaller than No. 9 gage. The longitudinal wires shall be embedded in the mortar.

❖ Bonding of adjacent wythes of masonry could also be accomplished with prefabricated joint reinforcements. They can be used in these wall types when air spaces are provided between two wythes of masonry, or they can be used in double-wythe grouted-wall construction. Cross wires must be of minimum size to provide proper load transfer. This reinforcement is placed in the bed joints. Commentary Figure R608.1.2.3 illustrates two of the more common types available.

Figure R608.1.2.1(1)
MASONRY BONDING-METAL TIES

R608.1.3 Bonding with natural or cast stone. Bonding with natural and cast stone shall conform to Sections R608.1.3.1 and R608.1.3.2.

❖ Ashlar masonry and rubble stone masonry require bonding through masonry headers. Although quite different than the usual masonry headers for clay or concrete masonry walls, bonder stones are required based on the square footage of wall area. Bonder stones must be embedded in mortar and have the proper extension into the wall surfaces on both sides. Each masonry type has separate bonding stone spacing requirements based on the construction. The bonding requirements in this section are primarily drawn from ACI 530/ASCE 5/TMS402.

R608.1.3.1 Ashlar masonry. In ashlar masonry, bonder units, uniformly distributed, shall be provided to the extent of not less than 10 percent of the wall area. Such bonder units shall extend not less than 4 inches (102 mm) into the backing wall.

❖ For examples of ashlar masonry, see the listings in Table R606.4.

R608.1.3.2 Rubble stone masonry. Rubble stone masonry 24 inches (610 mm) or less in thickness shall have bonder units with a maximum spacing of 3 feet (914 mm) vertically and 3 feet (914 mm) horizontally, and if the masonry is of greater thickness than 24 inches (610 mm), shall have one bonder unit for each 6 square feet (0.557 m^2) of wall surface on both sides.

❖ See the commentary for Section R608.1.3.

R608.2 Masonry bonding pattern. Masonry laid in running and stack bond shall conform to Sections R608.2.1 and R608.2.2.

❖ Masonry must be laid in running bond (see the definition in Chapter 2) or provided with horizontal reinforcement as specified in this section.

R608.2.1 Masonry laid in running bond. In each wythe of masonry laid in running bond, head joints in successive courses shall be offset by not less than one-fourth the unit length, or the masonry walls shall be reinforced longitudinally as required in Section R608.2.2.

❖ When masonry block is placed with an offset, it is called running bond. See Commentary Figure R608.1.2.1(1). Running bond is the most commonly used bonding pattern for masonry wall construction.

R608.2.2 Masonry laid in stack bond. Where unit masonry is laid with less head joint offset than in Section R607.2.1, the minimum area of horizontal reinforcement placed in mortar bed joints or in bond beams spaced not more than 48 inches (1219 mm) apart, shall be 0.0007 times the vertical cross-sectional area of the wall.

❖ Blocks stacked vertically are in the stack bond pattern. Refer to the definition in Section 202 of the code and Commentary Figure R606.7(1). The amount of steel in this section is an arbitrary amount to provide continuity across the head joints.

For SI: 1 inch = 25.4 mm, 1 foot = 304.8 mm.

**Figure R608.1.2.1(2)
BONDING AROUND OPENING IN MASONRY WALLS**

SECTION R609
GROUTED MASONRY

R609.1 General. Grouted multiple-wythe masonry is a form of construction in which the space between the wythes is solidly filled with grout. It is not necessary for the cores of masonry units to be filled with grout. Grouted hollow unit masonry is a form of construction in which certain cells of hollow units are continuously filled with grout.

❖ The provisions of Section 609 pertain to grouted masonry. The two types of grouted masonry, grouted multiple-wythe and grouted hollow unit masonry, are defined in the wording of this section and illustrated in Commentary Figure 609.1. In order to provide for an adequate bond between the masonry units and the grout, the units must be free of excessive dust or dirt.

R609.1.1 Grout. Grout shall consist of cementitious material and aggregate in accordance with ASTM C 476 and the proportion specifications of Table R609.1.1. Type M or Type S mortar to which sufficient water has been added to produce pouring consistency can be used as grout.

❖ ASTM C 476 contains the requirements for grout. The code identifies two types of grout for masonry construction: fine grout and coarse grout. They differ primarily in the maximum allowable size of aggregates. The proportions in Table R609.1.1 provide a minimum compressive strength of 2,000 psi (13,780 kPa). Type M or Type S mortar (see Section R607.1) may also be used as grout provided that sufficient water is added to produce a pouring consistency.

R609.1.2 Grouting requirements. Maximum pour heights and the minimum dimensions of spaces provided for grout

FIGURE R608.1.2.2 - FIGURE R608.1.2.3 WALL CONSTRUCTION

placement shall conform to Table R609.1.2. If the work is stopped for one hour or longer, the horizontal construction joints shall be formed by stopping all tiers at the same elevation and with the grout 1 inch (25.4 mm) below the top.

❖ The selection of fine or coarse grout is made based on the size of grout space and the height of grout pour. Table R609.1.2 covers requirements for selection of grout type. Fine grout is required where grout space is

Figure R608.1.2.2
BONDING WITH ADJUSTABLE WALL TIES

For SI: 1 inch = 25.4 mm.

Figure R608.1.2.3
BONDING WITH PREFABRICATED JOINT REINFORCEMENT

WALL CONSTRUCTION

Figure R609.1
GROUTED MULTIPLE-WYTHE MASONRY

small, narrow, or too congested with reinforcing steel. Fine grout also can be used whre coarse grout is permitted. It is preferable, however, to use coarse grout, because it reduces shrinkage and is less expensive. See Commentary Figure R609.1.2 for an illustration of grout placement.

R609.1.3 Grout space (cleaning). Provision shall be made for cleaning grout space. Mortar projections that project more than 0.5 inch (12.7 mm) into grout space and any other foreign matter shall be removed from grout space prior to inspection and grouting.

❖ See Commentary Figure R609.1.2.

R609.1.4 Grout placement. Grout shall be a plastic mix suitable for pumping without segregation of the constituents and shall be mixed thoroughly. Grout shall be placed by pumping or by an approved alternate method and shall be placed before any initial set occurs and in no case more than $1^1/_2$ hours after water has been added. Grouting shall be done in a continuous pour, in lifts not exceeding 5 feet (1524 mm). It shall be consolidated by puddling or mechanical vibrating during placing and reconsolidated after excess moisture has been absorbed but before plasticity is lost.

❖ See Commentary Figure R609.1.2.

R609.1.4.1 Grout pumped through aluminum pipes. Grout shall not be pumped through aluminum pipes.

❖ Aluminum can adversely affect grout and unprotected reinforcing steel through galvanic action. This restriction protects against possible contamination during grout placement.

Figure R609.1.2
GROUTING PROCEDURE FOR MASONRY

R609.1.5 Cleanouts. Where required by the building official, cleanouts shall be provided as specified in this section. The cleanouts shall be sealed before grouting and after inspection.

❖ Cleanouts should be furnished as described in this section. These aid the grout-spacing cleaning required by Section R609.1.2.

R609.1.5.1 Grouted multiple-wythe masonry. Cleanouts shall be provided at the bottom course of the exterior wythe at each pour of grout where such pour exceeds 5 feet (1524 mm) in height.

❖ See the commentary for Section R609.1.5.

R609.1.5.2 Grouted hollow unit masonry. Cleanouts shall be provided at the bottom course of each cell to be grouted at each pour of grout, where such pour exceeds 4 feet (1219 mm) in height.

❖ See the commentary for Section R609.1.5.

R609.2 Grouted multiple-wythe masonry. Grouted multiple-wythe masonry shall conform to all the requirements specified in Section R609.1 and the requirements of this section.

❖ All grouted multiple-wythe masonry walls must comply with this section.

R609.2.1 Bonding of backup wythe. Where all interior vertical spaces are filled with grout in multiple-wythe construction, masonry headers shall not be permitted. Metal wall ties shall be used in accordance with Section R608.1.2 to prevent spreading of the wythes and to maintain the vertical alignment of the wall. Wall ties shall be installed in accordance with Section R608.1.2 when the backup wythe in multiple-wythe construction is fully grouted.

❖ Masonry headers are not permitted in double-wythe grouted masonry construction because they span across the wythes and would interfere with the placement of the grout. Metal ties in accordance with Section R608.1.2 of the code prevent the spreading of wythes and maintain alignment during the grouting process.

WALL CONSTRUCTION

TABLE R609.1.1
GROUT PROPORTIONS BY VOLUME FOR MASONRY CONSTRUCTION

TYPE	PORTLAND CEMENT OR BLENDED CEMENT SLAG CEMENT	HYDRATED LIME OR LIME PUTTY	AGGREGATE MEASURED IN A DAMP, LOOSE CONDITION	
			Fine	Coarse
Fine	1	0 to 1/10	$2^1/_4$ to 3 times the sum of the volume of the cementitious materials	—
Coarse	1	0 to 1/10	$2^1/_4$ to 3 times the sum of the volume of the cementitious materials	1 to 2 times the sum of the volumes of the cementitious materials

TABLE R609.1.2
GROUT SPACE DIMENSIONS AND POUR HEIGHTS

GROUT TYPE	GROUT POUR MAXIMUM HEIGHT (feet)	MINIMUM WIDTH OF GROUT SPACES^{a,b} (inches)	MINIMUM GROUT^{b,c} SPACE DIMENSIONS FOR GROUTING CELLS OF HOLLOW UNITS (inches × inches)
Fine	1	0.75	1.5 × 2
	5	2	2 × 3
	12	2.5	2.5 × 3
	24	3	3 × 3
Coarse	1	1.5	1.5 × 3
	5	2	2.5 × 3
	12	2.5	3 × 3
	24	3	3 × 4

For SI: 1 inch = 25.4 mm, 1 foot = 304.8 mm.

a. For grouting between masonry wythes.

b. Grout space dimension is the clear dimension between any masonry protrusion and shall be increased by the horizontal projection of the diameters of the horizontal bars within the cross section of the grout space.

c. Area of vertical reinforcement shall not exceed 6 percent of the area of the grout space.

❖ See the commentary for Section R609.1.2.

R609.2.2 Grout spaces. Fine grout shall be used when interior vertical space to receive grout does not exceed 2 inches (51 mm) in thickness. Interior vertical spaces exceeding 2 inches (51 mm) in thickness shall use coarse or fine grout.

❖ Fine grout must be used for grout spaces not more than 2 inches wide (51 mm). This clearance should account for any horizontal wall reinforcing. ASTM C476 differentiates between fine and course grout.

R609.2.3 Grout barriers. Vertical grout barriers or dams shall be built of solid masonry across the grout space the entire height of the wall to control the flow of the grout horizontally. Grout barriers shall not be more than 25 feet (7620 mm) apart. The grouting of any section of a wall between control barriers shall be completed in one day with no interruptions greater than one hour.

❖ Grout barriers allow large walls to be properly grouted in sections.

R609.3 Reinforced grouted multiple-wythe masonry. Reinforced grouted multiple-wythe masonry shall conform to all the requirements specified in Sections R609.1 and R609.2 and the requirements of this section.

❖ Reinforced grouted multiple-wythe masonry walls must comply with this section in addition to Section R609.2.

R609.3.1 Construction. The thickness of grout or mortar between masonry units and reinforcement shall not be less than $^1/_4$ inch (6.4 mm), except that $^1/_4$-inch (6.4 mm) bars may be laid in horizontal mortar joints at least $^1/_2$ inch (12.7 mm) thick, and steel wire reinforcement may be laid in horizontal mortar joints at least twice the thickness of the wire diameter.

❖ When reinforcing is used in grouted masonry, the dimensional constraints given in this section provide clearances necessary for the adequate placement of the grout within the cavity as well as the proper bond between the reinforcement and the grout. These minimums are illustrated in Commentary Figure R609.3.1.

R609.4 Reinforced hollow unit masonry. Reinforced hollow unit masonry shall conform to all the requirements of Section R609.1 and the requirements of this section.

❖ Reinforced hollow-unit masonry employs hollow masonry units with reinforcement embedded in mortar or grout within certain cells throughout the wall length.

For SI: 1 inch = 25.4 mm.

Figure R609.3.1
REINFORCEMENT PLACEMENT LIMITATIONS

Figure R609.4.1(1)
REINFORCED HOLLOW UNIT MASONRY PLACEMENT

WALL CONSTRUCTION

Figure R609.4.1(2)
REINFORCED HOLLOW UNIT MASONRY VERTICAL CELL ALIGNMENT

R609.4.1 Construction. Requirements for construction shall be as follows:

1. All reinforced hollow-unit masonry shall be built to preserve the unobstructed vertical continuity of the cells to be filled. Walls and cross webs forming such cells to be filled shall be full-bedded in mortar to prevent leakage of grout. All head and end joints shall be solidly filled with mortar for a distance in from the face of the wall or unit not less than the thickness of the longitudinal face shells. Bond shall be provided by lapping units in successive vertical courses.
2. Cells to be filled shall have vertical alignment sufficient to maintain a clear, unobstructed continuous vertical cell of dimensions prescribed in Table R609.1.2.
3. Vertical reinforcement shall be held in position at top and bottom and at intervals not exceeding 200 diameters of the reinforcement.
4. All cells containing reinforcement shall be filled solidly with grout. Grout shall be poured in lifts of 8-foot (2438 mm) maximum height. When total grout pour exceeds 8 feet (2438 mm) in height, the grout shall be placed in lifts not exceeding 5 feet (1524 mm) and special inspection during grouting shall be required.
5. Horizontal steel shall be fully embedded by grout in an uninterrupted pour.

❖ The general purpose of the five requirements contained in the section is to provide for the proper placement of reinforcing steel and the grouting of cells so that the construction will act structurally as a unit.

Item 2 specifies minimum vertical cell-dimension limitations to provide for sufficient clear space in which to place the grout as illustrated in Commentary Figure R609.4.1(2).

Following the provision in Item 3 is necessary in order to maintain the vertical alignment of reinforcing steel during the grout pour. Also, the reinforcement will be maintained in a proper position relative to the block thickness as required by the design if the vertical reinforcement is secured at the top, bottom, and intervening points so that the distance between such anchorage does not exceed 200 bar diameters. Commentary Figure R609.4.1(3) illustrates methods of securing the reinforcement.

Items 4 and 5 provide limitations on the method of grouting the cells containing reinforcement so that the cells to be grouted will be filled solidly and the grout will be reasonably consolidated around the reinforcing steel. Refer to Commentary Figures R609.4.1(4) and R609.4.1(5) for illustrations of the provisions.

SECTION R610
GLASS UNIT MASONRY

R610.1 General. Panels of glass unit masonry located in load-bearing and nonload-bearing exterior and interior walls shall be constructed in accordance with this section.

❖ Code provisions for glass unit masonry are empirical. By definition (see the definition for "Masonry Unit, Glass" in Section 202) these units are non-loadbearing. Therefore, when they are used in a wall opening, a properly designed header or lintel must be provided. Furthermore, panels of glass unit masonry must be

Figure R609.4.1(3)
MASONRY REINFORCING STEEL ALIGNMENT

isolated from the structure so that in-plane loads are not transferred to the glass.

R610.2 Materials. Hollow glass units shall be partially evacuated and have a minimum average glass face thickness of $3/16$ inch (4.8 mm). The surface of units in contact with mortar shall be treated with a polyvinyl butyral coating or latex-based paint. The use of reclaimed units is prohibited.

❖ Glass masonry units may be hollow or solid. Hollow units must have the required minimum face thickness. The materials specified for treating edges in contact

WALL CONSTRUCTION

Figure R609.4.1(4)
REINFORCED HOLLOW UNIT MASONRY CLEANOUTS

with mortar improve the bonding between glass blocks and mortar.

R610.3 Units. Hollow or solid glass block units shall be standard or thin units.

❖ Hollow or solid-glass units are classified as either standard units or thin units.

R610.3.1 Standard units. The specified thickness of standard units shall be at least $3^{7}/_{8}$ inches (98 mm).

❖ See the commentary for Section R610.3.

R610.3.2 Thin units. The specified thickness of thin units shall be at least $3^{1}/_{8}$ inches (79 mm) for hollow units and at least 3 inches (76 mm) for solid units.

❖ See the commentary for Section R610.3.

R610.4 Isolated panels. Isolated panels of glass unit masonry shall conform to the requirements of this section.

❖ The code limits hollow glass unit masonry panel sizes based on structural and performance considerations. Height limits are more restrictive than length limits, primarily due to historical requirements rather than actual field experience or engineering principles. Exterior panels constructed of standard units may be up to 144 square feet (13.4 m^2). Exterior panels constructed of thin units are limited to 85 square feet (7.9 m^2). Interior panels are limited to 250 square feet (23.2 m^2) for standard units and 150 square feet (13.9 m^2) for thin units. These limitations are derived from ACI 530/ASCE5/TMS402: *Building Code Requirements for Masonry Structures*.

R610.4.1 Exterior standard-unit panels. The maximum area of each individual standard-unit panel shall be 144 square feet (13.4 m^2) when the design wind pressure is 20 psf (958 Pa). The maximum area of such panels subjected to design wind pressures other than 20 psf (958 Pa) shall be in accordance with Figure R610.4.1. The maximum panel dimension between structural supports shall be 25 feet (7620 mm) in width or 20 feet (6096 mm) in height.

❖ The area limit specified in this section corresponds to a wind pressure of 20 psf (958 Pa). The area should be

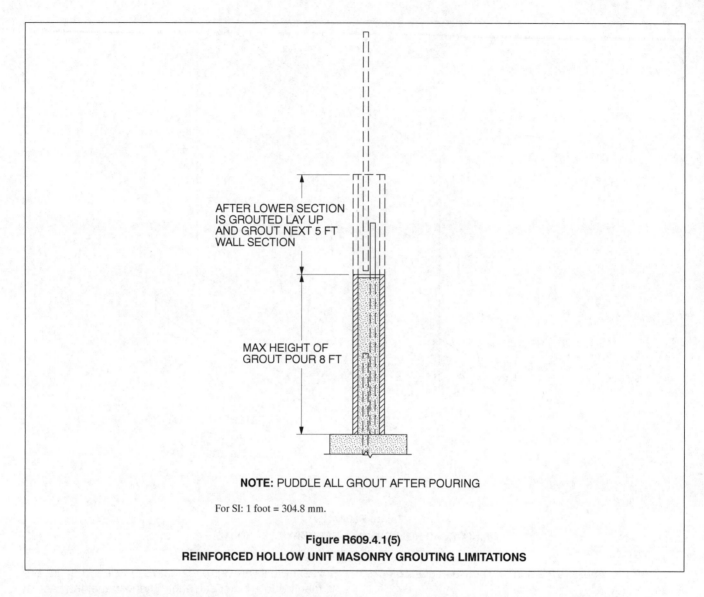

Figure R609.4.1(5)
REINFORCED HOLLOW UNIT MASONRY GROUTING LIMITATIONS

adjusted using Figure R610.4.1 for other values of wind pressure. See the commentary for Section R610.4.

R610.4.2 Exterior thin-unit panels. The maximum area of each individual thin-unit panel shall be 85 square feet (7.9 m^2). The maximum dimension between structural supports shall be 15 feet (4572 mm) in width or 10 feet (3048 mm) in height. Thin units shall not be used in applications where the design wind pressure as stated in Table R301.2(1) exceeds 20 psf (958 Pa).

❖ See the commentary for Section R610.4.

R610.4.3 Interior panels. The maximum area of each individual standard-unit panel shall be 250 square feet (23.2 m^2). The maximum area of each thin-unit panel shall be 150 square feet (13.9 m^2). The maximum dimension between structural supports shall be 25 feet (7620 mm) in width or 20 feet (6096 mm) in height.

❖ See the commentary for Section R610.4.

R610.4.4 Curved panels. The width of curved panels shall conform to the requirements of Sections R610.4.1, R610.4.2 and R610.4.3, except additional structural supports shall be provided at locations where a curved section joins a straight section, and at inflection points in multicurved walls.

❖ Support is required for standard and thin units, exterior and interior panels, at locations where a curved panel joins a straight section, and at inflection points. See the commentary for Section R610.4.

R610.5 Panel support. Glass unit masonry panels shall conform to the support requirements of this section.

❖ Structural members supporting glass-unit masonry panels must meet the deflection limit of this section in order to minimize the potential for cracking in the glass panels. Lateral support is required for the top and sides of panels and may be accomplished by anchors or channel-type restraints.

For SI: 1 square foot = 0.0929 m², 1 pound per square foot = 0.0479 kN/m².

**FIGURE R610.4.1
GLASS UNIT MASONRY DESIGN WIND LOAD RESISTANCE**

R610.5.1 Deflection. The maximum total deflection of structural members that support glass unit masonry shall not exceed $1/600$.

❖ See the commentary for Section R610.5.

R610.5.2 Lateral support. Glass unit masonry panels shall be laterally supported along the top and sides of the panel. Lateral supports for glass unit masonry panels shall be designed to resist a minimum of 200 pounds per lineal feet (2918 N/m) of panel, or the actual applied loads, whichever is greater. Except for single unit panels, lateral support shall be provided by panel anchors along the top and sides spaced a maximum of 16 inches (406 mm) on center or by channel-type restraints. Single unit panels shall be supported by channel-type restraints.

Exceptions:
1. Lateral support is not required at the top of panels that are one unit wide.
2. Lateral support is not required at the sides of panels that are one unit high.

❖ See the commentary for Section R610.5.

R610.5.2.1 Panel anchor restraints. Panel anchors shall be spaced a maximum of 16 inches (406 mm) on center in both jambs and across the head. Panel anchors shall be embedded a minimum of 12 inches (305 mm) and shall be provided with two fasteners so as to resist the loads specified in Section R610.5.2.

❖ See Commentary Figure R610.5.2.1.

R610.5.2.2 Channel-type restraints. Glass unit masonry panels shall be recessed at least 1 inch (25.4 mm) within channels and chases. Channel-type restraints shall be oversized to accommodate expansion material in the opening, packing and sealant between the framing restraints, and the glass unit masonry perimeter units.

❖ This type of lateral support is an option to the anchors described in the previous section. No specific material is specified for the channel, which means that any channel-type restraint providing the lateral support specified by Section R610.5.2 is acceptable.

R610.6 Sills. Before bedding of glass units, the sill area shall be covered with a water base asphaltic emulsion coating. The coating shall shall be a minimum of $1/8$ inch (3.2 mm) thick.

❖ Application of asphaltic emulsion is recommended by glass block manufacturers. See Commentary Figure R610.5.2.1.

R610.7 Expansion joints. Glass unit masonry panels shall be provided with expansion joints along the top and sides at all

Figure R610.5.2.1
GLASS UNIT MASONRY

structural supports. Expansion joints shall be a minimum of $3/8$ inch (9.5 mm) in thickness and shall have sufficient thickness to accommodate displacements of the supporting structure. Expansion joints shall be entirely free of mortar and other debris and shall be filled with resilient material.

❖ See Commentary Figure R610.5.2.1.

R610.8 Mortar. Glass unit masonry shall be laid with Type S or N mortar. Mortar shall not be retempered after initial set. Mortar unused within $1^{1}/_{2}$ hours after initial mixing shall be discarded.

❖ See the commentary for Section R607.1.

R610.9 Reinforcement. Glass unit masonry panels shall have horizontal joint reinforcement spaced a maximum of 16 inches (406 mm) on center located in the mortar bed joint. Horizontal joint reinforcement shall extend the entire length of the panel but shall not extend across expansion joints. Longitudinal wires shall be lapped a minimum of 6 inches (152 mm) at splices. Joint reinforcement shall be placed in the bed joint immediately below and above openings in the panel. The reinforcement shall have not less than two parallel longitudinal wires of size W1.7 or greater, and have welded cross wires of size W1.7 or greater.

❖ See Commentary Figure R610.5.2.1.

R610.10 Placement. Glass units shall be placed so head and bed joints are filled solidly. Mortar shall not be furrowed. Head and bed joints of glass unit masonry shall be $1/4$ inch (6.4 mm) thick, except that vertical joint thickness of radial panels shall not be less than $1/8$ inch (3.2 mm) or greater than $5/8$ inch (15.9 mm). The bed joint thickness tolerance shall be minus $1/16$ inch (1.6 mm) and plus $1/8$ inch (3.2 mm). The head joint thickness tolerance shall be plus or minus $1/8$ inch (3.2 mm).

❖ This section specifies the joint thicknesses and tolerances consistent with ACI 530/ASCE 5/TMS 402.

SECTION R611
INSULATING CONCRETE FORM
WALL CONSTRUCTION

R611.1 General. Insulating Concrete Form walls shall be designed and constructed in accordance with the provisions of this section or in accordance with the provisions of ACI 318. When ACI 318 or the provisions of this section are used to design insulating concrete form walls, project drawings, typical details and specifications are not required to bear the seal of the architect or engineer responsible for design, unless otherwise required by the state law of the jurisdiction having authority.

❖ Insulating Concrete Forms refer to hollow blocks, planks, or panels of rigid foam plastic insulation that are used as stay-in-place forms for concrete walls.

WALL CONSTRUCTION

These provisions cover three types of ICF walls; flat, waffle-grid and screen-grid. Wall type descriptions are based on the shape of the concrete inside the form, since it is the concrete that provides the load-carrying capability. This section provides prescriptive requirements for design and construction of ICF walls. Designs per ACI 318: *Building Code Requirements for Structural Concrete* are acceptable alternatives.

R611.2 Applicability limits. The provisions of this section shall apply to the construction of insulating concrete form walls for buildings not greater than 60 feet (18 288 mm) in plan dimensions, and floors not greater than 32 feet (9754 mm) or roofs not greater than 40 feet (12 192 mm) in clear span. Buildings shall not exceed two stories in height above-grade. Insulating concrete form walls shall comply with the requirements in Table R611.2. Walls constructed in accordance with the provisions of this section shall be limited to buildings subjected to a maximum design wind speed of 150 miles per hour (241 km/h), a maximum ground snow load of 70 psf (3.35 kN/m^2), and Seismic Design Categories A, B, C, D_1, and D_2. The provisions of this section shall not apply to the construction of insulating concrete form walls for buildings or portions of buildings considered irregular as defined in Section R301.2.2.2.2.

For townhouses in Seismic Design Category C and all buildings in Seismic Design Category D, the provisions of this section shall only apply to buildings meeting the following requirements.

1. Rectangular buildings with a maximum building aspect ratio of 2:1. The building aspect ratio shall be determined by dividing the longest dimension of the building by the shortest dimension of the building.
2. Walls are aligned vertically with the walls below.
3. Cantilever and setback construction shall not be permitted.
4. The weight of interior and exterior finishes applied to ICF walls shall not exceed 8 psf (0.38 kN/m^2).
5. The gable portion of ICF walls shall be constructed of light-frame construction.

❖ To keep the prescriptive criteria from becoming overly complex, this section establishes practical limits, which, in general, are related to the complexity of the structure in dealing with seismic design or the overall size of the structure. An engineered design in accordance with ACI 318 is required for buildings that fall outside these limits. The prescriptive details given in Section R611 do not include irregular buildings because such buildings require a higher level of design complexity in seismic design than can be adequately dealt with in the prescriptive requirements of this section.

The code contains even more restrictions for townhouses in Seismic Design Category C and all buildings in Seismic Design Categories D_1 and D_2. These structures require a higher level of design complexity in higher seismic categories. Therefore, these limitations are required in order to limit the number of permutations possible in the design as practical limits for the amount of prescriptive tables and details that can be reasonably installed in this code. Again, buildings outside of these limitations would be possible, but an engineered design in accordance with ACI 318 would be required.

R611.3 Flat insulating concrete form wall systems. Flat ICF wall systems shall comply with Figure R611.3 and shall have reinforcement in accordance with Tables R611.3(1) and R611.3(2) and Section R611.7.

❖ Figure R611.3 illustrates flat ICF wall construction. Vertical wall reinforcing should be provided where required by Table 611.3(2). Table 611.3(1) provides conversion of wind speed to design wind pressure for different wind exposure categories. These pressures are used in the design wind pressure column of Table 611.3(2). Reinforcing requirements vary based on basic wind speed, wall height, elements being supported, and wall thickness.

TABLE R611.2
REQUIREMENTS FOR ICF WALLS[a,b,c]

WALL TYPE AND NOMINAL SIZE	MAXIMUM WALL WEIGHT (psf)	MINIMUM WIDTH OF VERTICAL CORE (inches)	MINIMUM THICKNESS OF VERTICAL CORE (inches)	MAXIMUM SPACING OF VERTICAL CORES (inches)	MAXIMUM SPACING OF HORIZONTAL CORES (inches)	MINIMUM WEB THICKNESS (inches)
3.5" Flat	44	N/A	N/A	N/A	N/A	N/A
5.5" Flat	69	N/A	N/A	N/A	N/A	N/A
7.5" Flat	94	N/A	N/A	N/A	N/A	N/A
9.5" Flat	119	N/A	N/A	N/A	N/A	N/A
6" Waffle-Grid	56	6.25	5	12	16	2
8" Waffle-Grid	76	7	7	12	16	2
6" Screen-Grid	53	5.5	5.5	12	12	N/A

For SI: 1 inch = 25.4 mm

a. For width "W", thickness "T", spacing, and web thickness, refer to Figures R611.4 and R611.5.
b. N/A indicates not applicable.
c. Wall weight is based on a unit weight of concrete of 150 pcf (23.6kN/m3). The tabulated values do not include any allowance for interior and exterior finishes.

❖ See the commentary for Section R611.2.

FIGURE R611.2 - TABLE R611.3(1) WALL CONSTRUCTION

**Figure R611.2
ICF APPLICABILITY LIMITS**

**TABLE R611.3(1)
DESIGN WIND PRESSURE FOR USE WITH TABLES R611.3(2), R611.4(1), AND R611.5 FOR ABOVE GRADE WALLS[a]**

WIND SPEED (mph)[e]	DESIGN WIND PRESSURE (psf)					
	Enclosed[b]			Partially Enclosed[b]		
	Exposure[c]			Exposure[c]		
	B	C	D	B	C	D
85	18	24	29	23	31	37
90	20	27	32	25	35	41
100	24	34	39	31	43	51
110	29	41	48	38	52	61
120	35	48	57	45	62	73
130	41	56	66	53	73	85[d]
140	47	65	77	61	84[d]	99[d]
150	54	75	88[d]	70	96[d]	114[d]

For SI: 1 psf = 0.0479 kN/m^2; 1 mph = 1.6093 km/hr

a. This table is based on ASCE 7-98 components and cladding wind pressures using a mean roof height of 35 ft (10.7 m) and a tributary area of 10 ft^2 (0.9 m^2).
b. Buildings in wind borne debris regions as defined in Section R202 shall be considered as "Partially Enclosed" unless glazed openings are protected in accordance with Section R301.2.2.2 whereby the building shall be considered as "Enclosed." All other buildings shall be classified as "Enclosed."
c. Exposure Categories shall be determined in accordance with Section R301.2.1.4.
d. For wind pressures greater than 80 psf (3.8 kN/m^2), design is required in accordance with ACI 318 and approved manufacturer guidelines.
e. Interpolation is permitted between wind speeds.

❖ See the commentary for Sections R611.3 through R611.5.

WALL CONSTRUCTION

TABLE R611.3(2)
MINIMUM VERTICAL WALL REINFORCEMENT FOR FLAT ICF ABOVE-GRADE WALLS[a,b,c,d]

Design Wind Pressure [Table R611.3(1)] (psf)	Maximum Unsupported Wall Height (feet)	Minimum Vertical Reinforcement[d,e,f]					
		Non-Load Bearing Wall or Supporting Roof		Supporting Light-Frame Second Story and Roof		Supporting ICF Second Story and Roof	
		Minimum Wall Thickness (inches)					
		3.5[g]	5.5	3.5[g]	5.5	3.5[g]	5.5
20	8	#4@48	#4@48	#4@48	#4@48	#4@48	#4@48
	9	#4@48	#4@48	#4@48	#4@48	#4@48	#4@48
	10	#4@38	#4@48	#4@40	#4@48	#4@42	#4@48
30	8	#4@42	#4@48	#4@46	#4@48	#4@48	#4@48
	9	#4@32; #5@48	#4@48	#4@34; #5@48	#4@48	#4@34; #5@48	#4@48
	10	Design Required	#4@48	Design Required	#4@48	Design Required	#4@48
40	8	#4@30; #5@48	#4@48	#4@30; #5@48	#4@48	#4@32; #5@48	#4@48
	9	Design Required	#4@42	Design Required	#4@46	Design Required	#4@48
	10	Design Required	#4@32; #5@48	Design Required	#4@34; #5@48	Design Required	#4@38
50	8	#4@20; #5@30	#4@42	#4@22; #5@34	#4@46	#4@24; #5@36	#4@48
	9	Design Required	#4@34; #5@48	Design Required	#4@34; #5@48	Design Required	#4@38
	10	Design Required	#4@26; #5@38	Design Required	#4@26; #5@38	Design Required	#4@28; #5@46
60	8	Design Required	#4@34; #5@48	Design Required	#4@36	Design Required	#4@40
	9	Design Required	#4@26; #5@38	Design Required	#4@28; #5@46	Design Required	#4@34; #5@48
	10	Design Required	#4@22; #5@34	Design Required	#4@22; #5@34	Design Required	#4@26; #5@38
70	8	Design Required	#4@28; #5@46	Design Required	#4@30; #5@48	Design Required	#4@34; #5@48
	9	Design Required	#4@22; #5@34	Design Required	#4@22; #5@34	Design Required	#4@24; #5@36
	10	Design Required	#4@16; #5@26	Design Required	#4@18; #5@28	Design Required	#4@20; #5@30
80	8	Design Required	#4@26; #5@38	Design Required	#4@26; #5@38	Design Required	#4@28; #5@46
	9	Design Required	#4@20; #5@30	Design Required	#4@20; #5@30	Design Required	#4@21; #5@34
	10	Design Required	#4@14; #5@24	Design Required	#4@14; #5@24	Design Required	#4@16; #5@26

For SI: 1 foot = 0.3048 m; 1 inch = 25.4 mm; 1 mph = 1.6093 km/hr.

a. This table is based on reinforcing bars with a minimum yield strength of 40,000 psi (276 MPa) and concrete with a minimum specified compressive strength of 2,500 psi (17.2 MPa). For Seismic Design Category D, reinforcing bars shall have a minimum yield strength of 60,000 psi (414 MPa). See Section R611.6.2.
b. Deflection criterion is $L/240$, where L is the height of the wall story in inches.
c. Interpolation shall not be permitted.
d. Reinforcement spacing for 3.5 inch (88.9 mm) walls shall be permitted to be multiplied by 1.6 when reinforcing steel with a minimum yield strength of 60,000 psi (414 MPa) is used. Reinforcement shall not be less than one #4 bar at 48 inches (1.2 m) on center.
e. Reinforcement spacing for 5.5 inch (139.7 mm) walls shall be permitted to be multiplied by 1.5 when reinforcing steel with a minimum yield strength of 60,000 psi (414 MPa) is used. Reinforcement shall not be less than one #4 bar at 48 inches (1.2 m) on center.
f. See Section R611.7.1.2 for limitations on maximum spacing of vertical reinforcement in Seismic Design Categories C, D_1 and D_2.
g. A 3.5-inch wall shall not be permitted if wood ledgers are used to support floor or roof loads. See Section R611.8.

❖ See the commentary for Section R611.3.

**FIGURE R611.3
FLAT ICF WALL SYSTEM**

❖ See the commentary for Section R611.3.

R611.4 Waffle-grid insulating concrete form wall systems. Waffle-grid wall systems shall comply with Figure R611.4 and shall have reinforcement in accordance with Tables R611.3(1) and R611.4(1) and Section R611.7. The minimum core dimensions shall comply with Table R611.4(2).

❖ Figure R611.4 illustrates waffle-grid ICF construction, which must conform to the dimensions in Table R611.4(2). Vertical wall reinforcing should be provided where required by Table R611.4(1). Table 611.3(1) provides conversion of wind speed to design wind pressure for different wind exposure catagories. These pressures are used in the design wind pressure column of Table R611.4(1). Reinforcing requirements vary based on basic wind speed, wall height, elements being supported, and wall thickness.

R611.5 Screen-grid insulating concrete form wall systems. Screen-grid ICF wall systems shall comply with Figure R611.5 and shall have reinforcement in accordance with Tables R611.3(1) and R611.5 and Section R611.7. The minimum core dimensions shall comply with Table R611.4(2).

❖ Figure R611.5 illustrates screen-grid ICF construction, which must conform to the dimensions in Table R611.4(2). Vertical wall reinforcing should be provided where required by Table R611.5. Table 611.3(1) provides conversion of wind speed to design wind pressure for different wind exposure catagories. These pressures are used in the design wind pressure column of Table R611.5. Reinforcing requirements vary based on basic wind speed, wall height, elements being supported, and wall thickness.

R611.6 Material. Insulating concrete form wall materials shall comply with this section.

❖ ICFs are composed of three basic materials: concrete, reinforcing steel for concrete and foam plastic insulation. This section provides the scoping text for the three subsections that follow.

R611.6.1 Concrete material. Ready-mixed concrete for insulating concrete form walls shall be in accordance with Section R402.2. Maximum slump shall not be greater than 6 inches (152 mm) as determined in accordance with ASTM C 143. Maximum aggregate size shall not be larger than 3/4 inch (19.1 mm).

Exception: Concrete mixes conforming to the ICF manufacturer's recommendations.

In Seismic Design Categories D_1 and D_2 the minimum concrete compressive strength shall be 3,000 psi (20.5 MPa).

❖ Concrete strengths are determined using Section R402.2 and the requirements of this section. The limits on concrete slump and aggregate size are consistent with current practice in ICF construction. These limits consider ease of concrete placement and the ability to completely fill forms while limiting the fluid pressure on these forms from the freshly poured concrete. Minimum concrete strengths for Seismic Design Categories D_1 and D_2 are increased because of the greater loads imposed on these structures in these seismic design categories.

R611.6.2 Reinforcing steel. Reinforcing steel shall meet the requirements of ASTM A 615, A 706 or A 996. Except in Seismic Design Categories D_1 and D_2, the minimum yield strength of reinforcing steel shall be 40,000 psi (Grade 40) (276 MPa). In Seismic Design Categories D_1 and D_2, reinforcing steel shall meet the requirements of ASTM A706 for low-alloy steel with a minimum yield strength of 60,000 psi (Grade 60) (414 MPa).

❖ The prescriptive reinforcement requirements for ICF walls of buildings in Seismic Design Categories A, B and C were developed using Grade 40 reinforcing steel. Since this grade is not readily available, it is likely that Grade 60 steel will be furnished. In this case, the reinforcement required by the tables may be conser-

WALL CONSTRUCTION

vative. Because of the higher loads imposed on buildings in Seismic Design Categories D_1 and D_2, Grade 60 reinforcing steel is required. In addition, the material specification must be ASTM A 706, for low-alloy steel. This is consistent with ACI 318 requirements for design in high seismic areas. The A 706 material is more consistent in the actual yield stress of the material produced versus the required one. This is important to avoid using a material with a tensile strength that is too high, which could lead to undesirable brittle failure in concrete.

R611.6.3 Insulation materials. Insulating concrete forms material shall meet the surface burning characteristics of Section R314.1.1. A thermal barrier shall be provided on the building interior in accordance with Section R314.1.2 or Section R702.3.4.

❖ Insulating form material must meet the surface-burning characteristics specified in Section R314.1.1, and a thermal barrier is required in accordance with Section R314.1.2.

TABLE R611.4(1)
MINIMUM VERTICAL WALL REINFORCEMENT FOR WAFFLE-GRID ICF ABOVE-GRADE WALLS[a, b, c]

Design Wind Pressure [Table R611.3(1)] (psf)	Maximum Unsupported Wall Height (feet)	Minimum Vertical Reinforcement[d, e]					
		Non-Load Bearing Wall or Supporting Roof		Supporting Light-Frame Second Story and Roof		Supporting ICF Second Story and Roof	
		Minimum Wall Thickness (inches)					
		6	8	6	8	6	8
20	8	#4@48	#4@48	#4@48	#4@48	#4@48	#4@48
	9	#4@48	#4@48	#4@48	#4@48	#4@48	#4@48
	10	#4@48	#4@48	#4@48	#4@48	#4@48	#4@48
30	8	#4@48	#4@48	#4@48	#4@48	#4@48	#4@48
	9	#4@48	#4@48	#4@48	#4@48	#4@48	#4@48
	10	#4@36; #5@48	#4@48	#4@36; #5@48	#4@48	#4@36; #5@48	#4@48
40	8	#4@36; #5@48	#4@48	#4@48	#4@48	#4@48	#4@48
	9	#4@36; #5@48	#4@48	#4@36; #5@48	#4@48	#4@36; #5@48	#4@48
	10	#4@24; #5@36	#4@36; #5@48	#4@24; #5@36	#4@48	#4@24; #5@36	#4@48
50	8	#4@36; #5@48	#4@48	#4@36; #5@48	#4@48	#4@36; #5@48	#4@48
	9	#4@24; #5@36	#4@36; #5@48	#4@24; #5@36	#4@48	#4@24; #5@48	#4@48
	10	Design Required	#4@36; #5@48	Design Required	#4@36; #5@48	Design Required	#4@36; #5@48
60	8	#4@24; #5@36	#4@48	#4@24; #5@36	#4@48	#4@24; #5@48	#4@48
	9	Design Required	#4@36; #5@48	Design Required	#4@36; #5@48	Design Required	#4@36; #5@48
	10	Design Required	#4@24; #5@36	Design Required	#4@24; #5@36	Design Required	#4@24; #5@48
70	8	#4@24; #5@36	#4@36; #5@48	#4@24; #5@36	#4@36; #5@48	#4@24; #5@36	#4@48
	9	Design Required	#4@24; #5@36	Design Required	#4@24; #5@48	Design Required	#4@24; #5@48
	10	Design Required	#4@12; #5@36	Design Required	#4@24; #5@36	Design Required	#4@24; #5@36
80	8	#4@12; #5@24	#4@24; #5@48	#4@12; #5@24	#4@24; #5@48	#4@12; #5@24	#4@36; #5@48
	9	Design Required	#4@24; #5@36	Design Required	#4@24; #5@36	Design Required	#4@24; #5@36
	10	Design Required	#4@12; #5@24	Design Required	#4@12; #5@24	Design Required	#4@12; #5@24

For SI: 1 foot = 0.3048 m; 1 inch = 25.4 mm; 1 mph = 1.6093 km/hr.

a. This table is based on reinforcing bars with a minimum yield strength of 40,000 psi (276 MPa) and concrete with a minimum specified compressive strength of 2,500 psi (17.2 MPa). For Seismic Design Category D, reinforcing bars shall have a minimum yield strength of 60,000 psi (414 MPa). See Section R611.6.2.
b. Deflection criterion is $L/240$, where L is the height of the wall story in inches.
c. Interpolation shall not be permitted.
d. Reinforcement spacing shall be permitted to be increased by 12 inches (305 mm) when reinforcing steel with a minimum yield strength of 60,000 psi (414 MPa) is used or No. 4 reinforcing bars shall be permitted to be substituted for #5 bars when reinforcing steel with a minimum yield strength of 60,000 psi (414 MPa) is used at the same spacing required for #5 bars. Reinforcement shall not be less than one #4 bar at 48 inches (1.2 m) on center.
e. See Section R611.7.1.2 for limitations on maximum spacing of vertical reinforcement in Seismic Design Categories C, D_1 and D_2.

❖ See the commentary for Section R611.4.

TABLE R611.4(2)
DIMENSIONAL REQUIREMENTS FOR CORES AND WEBS IN WAFFLE-GRID AND SCREEN-GRID ICF WALLS[a,b]

WALL TYPE AND NOMINAL SIZE (inches)	MINIMUM WIDTH OF CORE (inches)	MINIMUM THICKNESS OF VERTICAL CORE (inches)	MAXIMUM SPACING OF VERTICAL CORES (inches)	MAXIMUM SPACING OF HORIZONTAL CORES (inches)	MINIMUM WEB THICKNESS (inches)
6″ Waffle-Grid	6.25	5	12	16	2
8″ Waffle-Grid	7	7	12	16	2
6″ Screen-Grid	5.5	5.5	12	12	N/A

For SI: 1 inch = 25.4 mm.
a. For width "W," thickness "T"; spacing, and web thickness, refer to Figures R611.4 and R611.5.
b. N/A indicates not applicable.

❖ See the commentary for Section R611.4.

For SI: 1 inch = 25.4 mm.

**FIGURE R611.4
WAFFLE-GRID ICF WALL SYSTEM**

❖ See the commentary for Section R611.4.

TABLE R611.5
MINIMUM VERTICAL WALL REINFORCEMENT FOR SCREEN-GRID ICF ABOVE-GRADE WALLS[a,b,c]

Design Wind Pressure Table R611.3(1) (psf)	Maximum Unsupported Wall Height (feet)	Minimum Vertical Reinforcement[d,e]		
		Non-Load Bearing Wall or Supporting Roof	Supporting Light-Frame Second Story and Roof	Supporting ICF Second Story and Roof
20	8	#4@48	#4@48	#4@48
	9	#4@48	#4@48	#4@48
	10	#4@48	#4@48	#4@48
30	8	#4@48	#4@48	#4@48
	9	#4@48	#4@48	#4@48
	10	#4@36; #5@48	#4@48	#4@48
40	8	#4@48	#4@48	#4@48
	9	#4@36; #5@48	#4@36; #5@48	#4@48
	10	#4@24; #5@48	#4@24; #5@48	#4@24; #5@48
50	8	#4@36; #5@48	#4@36; #5@48	#4@48
	9	#4@24; #5@48	#4@24; #5@48	#4@24; #5@48
	10	Design Required	Design Required	Design Required
60	8	#4@24; #5@48	#4@24; #5@48	#4@36; #5@48
	9	#4@24; #5@36	#4@24; #5@36	#4@24; #5@36
	10	Design Required	Design Required	Design Required
70	8	#4@24; #5@36	#4@24; #5@36	#4@24; #5@36
	9	Design Required	Design Required	Design Required
	10	Design Required	Design Required	Design Required
80	8	#4@12; #5@36	#4@24; #5@36	#4@24; #5@36
	9	Design Required	Design Required	Design Required
	10	Design Required	Design Required	Design Required

For SI: 1 inch = 25.4 mm, 1 foot = 304.8 mm, 1 mile per hour = 1.6093 km/h.

a. This table is based on reinforcing bars with a minimum yield strength of 40,000 psi (276 MPa) and concrete with a minimum specified compressive strength of 2,500 psi (17.2 MPa). For Seismic Design Category D, reinforcing bars shall have a minimum yield strength of 60,000 psi (414 MPa). See Section R611.6.2.
b. Deflection criterion is $L/240$, where L is the height of the wall story in inches.
c. Interpolation shall not be permitted.
d. Reinforcement spacing shall be permitted to be increased by 12 inches (305 mm) when reinforcing steel with a minimum yield strength of 60,000 psi (414 MPa) is used. Reinforcement shall not be less than one #4 bar at 48 inches (1.2 m) on center.
e. See Section R611.7.1.2 for limitations on maximum spacing of vertical reinforcement in Seismic Design Categories C, D_1 and D_2.

❖ See the commentary for Section R611.5.

For SI: 1 inch = 25.4 mm.

**FIGURE R611.5
SCREEN-GRID IFC WALL SYSTEM**

❖ See the commentary for Section R611.5.

R611.7 Wall construction. Insulating concrete form walls shall be constructed in accordance with the provisions of this section and Figure R611.7(1).

❖ Figure R611.7(1) illustrates correct ICF wall construction and the story limits of Section R611.2. The section provides references to sections that specify details for anchoring roof and floor assemblies.

R611.7.1 Reinforcement.

R611.7.1.1 Location. Vertical and horizontal wall reinforcement shall be placed within the middle third of the wall. Steel reinforcement shall have a minimum concrete cover in accordance with ACI 318.

> **Exception:** Where insulated concrete forms are used and the form remains in place as cover for the concrete, the minimum concrete cover for the reinforcing steel is permitted to be reduced to $3/4$ inch (19.1 mm).

❖ The limits placed on the location of wall reinforcement are consistent with current practice. All reinforcing must have the minimum cover required in ACI 318. The cover may be reduced to $3/4$ inch (19.1 mm) where insulating concrete forms are used and are left in place.

R611.7.1.2 Vertical steel. Above-grade concrete walls shall have reinforcement in accordance with Sections R611.3, R611.4, or R611.5 and R611.7.2. All vertical reinforcement in the top-most ICF story shall terminate with a bend or standard hook and be provided with a minimum lap splice of 24 inches (610 mm) with the top horizontal reinforcement.

For townhouses in Seismic Design Category C, the minimum vertical reinforcement shall be one No. 5 at 24 inches (610 m) on center or one No. 4 at 16 inches (407 mm) on center. For all buildings in Seismic Design Categories D_1 and D_2, the minimum vertical reinforcement shall be one No. 5 at 18 inches (457 mm) on center or one No. 4 at 12 inches (305 mm) on center.

Above-grade ICF walls shall be supported on concrete foundations reinforced as required for the above-grade wall immediately above, or in accordance with Tables R404.4(1) through R404.4(5), whichever requires the greater amount of reinforcement.

Vertical reinforcement shall be continuous from the bottom of the foundation wall to the roof. Lap splices, if required, shall comply with R611.7.1.5. Where vertical reinforcement in the above-grade wall is not continuous with the foundation wall reinforcement, dowel bars with a size and spacing to match the vertical ICF wall reinforcement shall be embedded $40d_b$ into the foundation wall and shall be lapspliced with the above-grade wall reinforcement. Alternatively, for No. 6 and larger bars, the portion of the bar embedded in the foundation wall shall be embedded 24 inches in the foundation wall and shall have a standard hook.

❖ Vertical reinforcing must be provided according to the tables and the type of ICF wall construction. In addition, wall openings must be provided with reinforcing as shown in Figure R611.7(2).

For higher seismic design categories and higher seismic performance demands, the amount of vertical reinforcement needs to be greater in order to deal with the larger demands on the performance of the wall.

R611.7.1.3 Horizontal steel. Concrete walls with a minimum thickness of 4 inches (102 mm) shall have a minimum of one continuous No. 4 horizontal reinforcing bar placed at 32 inches (812 mm) on center with one bar within 12 inches (305 mm) of the top of the wall story. Concrete walls 5.5 inches (140 mm) thick or greater shall have a minimum of one continuous No.4 horizontal reinforcing bar placed at 48 inches (1219 mm) on center with one bar located with in 12 inches (305 mm) of the top of the wall story.

For townhouses in Seismic Design Category C, the minimum horizontal reinforcement shall be one No. 5 at 24 inches (610 m) on center or one No. 4 at 16 inches (407 mm) on center. For all buildings in Seismic Design Categories D_1 and D_2, the minimum horizontal reinforcement shall be one No. 5 at 18 inches (457 mm) on center or one No. 4 at 12 inches (305 mm) on center.

Horizontal reinforcement shall be continuous around building corners using corner bars or by bending the bars. In either case, the minimum lap splice shall be 24 inches (610 mm). For townhouses in Seismic Design Category C and for all buildings in Seismic Design Categories D_1 and D_2, each end of all horizontal reinforcement shall terminate with a standard hook or lap splice.

❖ These minimums are provided for temperature and shrinkage. In more demanding seismic design categories the additional horizontal steel is needed to improve the performance of walls in resisting horizontal movement.

**TABLE R611.7(1)
MINIMUM WALL OPENING REINFORCEMENT REQUIREMENTS IN ICF WALLS**

WALL TYPE AND OPENING WIDTH, L FEET (m)	MINIMUM HORIZONTAL OPENING REINFORCEMENT	MINIMUM VERTICAL OPENING REINFORCEMENT
Flat, Waffle-, and Screen-Grid: L < 2 (0.61)	None Required	None Required
Flat, Waffle-, and Screen-Grid: L ≥ 2 (0.61)	Provide lintels in accordance with Section R611.7.3. Provide one No. 4 bar within 12 inches (305 mm) from the bottom of the opening. Top and bottom lintel reinforcement shall extend a minimum of 24 inches (610 mm) beyond the limits of the opening.	In locations with wind speeds less than or equal to 110 mph (177 km/hr) or in Seismic Design Categories A and B, provide one No. 4 bar for the full height of the wall story within 12 inches (305 mm) of each side of the opening. In locations with wind speeds greater than 110 mph (177 km/hr), townhouses in Seismic Design Category C, or all buildings in Seismic Design Categories D_1 and D_2, provide two No. 4 bars or one No. 5 bar for the full height of the wall story within 12 inches (305 mm) of each side of the opening.

For SI: 1 inch = 25.4 mm, 1 foot = 304.8 mm, 1 mile per hour = 1.609 km/h.

NOTE: This table is based on concrete with a minimum specified compressive strength of 2,500 psi, reinforcing steel with a minimum yield strength of 40,000 psi and an assumed equivalent rectangular cross section. This table is not intended to prohibit the use of ICF manufacturer's tables based on engineering analysis in accordance with ACI 318.

❖ See the commentary for Section R611.7.2.

TABLE R611.7(2)
MAXIMUM ALLOWABLE CLEAR SPANS FOR ICF LINTELS FOR FLAT LOAD-BEARING WALLS[a,b,c,d,f]
NO. 4 BOTTOM BAR SIZE

MINIMUM LINTEL THICKNESS, T (inches)	LINTEL DEPTH, D (inches)	MAXIMUM CLEAR SPAN, (feet-inches) (Number is Middle of Span, A)[e]					
		Supporting Roof Only		Supporting Light Framed 2nd Story and Roof		Supporting ICF Second Story and Roof	
		Ground Snow Load					
		30 psf	70 psf	30 psf	70 psf	30 psf	70 psf
3.5	8	4-9 (1-2)	4-2 (0-9)	3-10 (0-8)	3-4 (0-6)	3-5 (0-6)	3-1 (0-5)
	12	6-8 (1-11)	5-5 (1-3)	5-0 (1-1)	4-5 (0-10)	4-6 (0-10)	4-0 (0-8)
	16	7-11 (2-9)	6-5 (1-9)	6-0 (1-6)	5-3 (1-2)	5-4 (1-2)	4-10 (1-0)
	20	8-11 (3-5)	7-4 (2-3)	6-9 (1-11)	6-0 (1-6)	6-1 (1-7)	5-6 (1-3)
	24	9-10 (4-1)	8-1 (2-9)	7-6 (2-4)	6-7 (1-10)	6-9 (1-11)	6-1 (1-6)
5.5	8	5-2 (1-10)	4-2 (1-2)	3-10 (1-0)	3-5 (0-9)	3-5 (0-10)	3-1 (0-8)
	12	6-8 (3-0)	5-5 (2-0)	5-0 (1-9)	4-5 (1-4)	4-6 (1-4)	4-1 (1-1)
	16	7-10 (4-1)	6-5 (2-9)	6-0 (2-5)	5-3 (1-10)	5-4 (1-11)	4-10 (1-7)
	20	8-10 (5-3)	7-3 (3-6)	6-9 (3-1)	6-0 (2-4)	6-1 (2-5)	5-6 (2-0)
	24	9-8 (6-3)	8-0 (4-3)	7-5 (3-8)	6-7 (2-11)	6-8 (3-0)	6-0 (2-5)
7.5	8	5-2 (2-6)	4-2 (1-8)	3-11 (1-5)	3-5 (1-1)	3-6 (1-1)	3-2 (0-11)
	12	6-7 (4-0)	5-5 (2-8)	5-0 (2-4)	4-5 (1-10)	4-6 (1-10)	4-1 (1-6)
	16	7-9 (5-5)	6-5 (3-8)	5-11 (3-3)	5-3 (2-6)	5-4 (2-7)	4-10 (2-2)
	20	8-8 (6-10)	7-2 (4-8)	6-8 (4-2)	5-11 (3-3)	6-0 (3-4)	5-5 (2-9)
	24	9-6 (8-2)	7-11 (5-8)	7-4 (5-1)	6-6 (3-11)	6-7 (4-1)	6-0 (3-4)
9.5	8	5-2 (3-1)	4-2 (2-1)	3-11 (1-9)	3-5 (1-5)	3-6 (1-5)	3-2 (1-2)
	12	6-7 (5-0)	5-5 (3-4)	5-0 (3-0)	4-5 (2-4)	4-6 (2-5)	4-1 (1-11)
	16	7-8 (6-9)	6-4 (4-7)	5-11 (4-2)	5-3 (3-3)	5-4 (3-4)	4-10 (2-8)

For SI: 1 inch = 25.4 mm, 1 foot = 304.8 mm, 1 pound per square inch = 6.895 kN/m², 1 pound per square foot = 0.0479 kN/m².

a. This table is based on concrete with a minimum specified compressive strength of 2,500 psi, reinforcing steel with a minimum yield strength of 40,000 psi and an assumed equivalent rectangular cross section. When reinforcement with a minimum yield strength of 60,000 psi is used, the span lengths in the shaded cells shall be increased by 1.2 times the table values.
b. This table is not intended to prohibit the use of ICF manufacturer's tables based on engineering analysis in accordance with ACI 318.
c. Deflection criteria: $L/240$.
d. Design load assumptions:
   Floor dead load is 10 psf   Attic live load is 20 psf
   Floor live load is 30 psf   Roof dead load is 15 psf
   Building width is 32 feet   ICF wall dead load is 69 psf
   Light-framed wall dead load is 10 psf
e. No. 3 stirrups are required at $d/2$ spacing except no stirrups are required for the distance, (A), shown in the middle portion of the span in accordance with Figure R611.7(2) and Section R611.7.3.2.
f. Interpolation is permitted between ground snow loads and between lintel depths.

❖ See the commentary for Section R611.7.3.1.

WALL CONSTRUCTION

TABLE R611.7(3)
MAXIMUM ALLOWABLE CLEAR SPANS FOR ICF LINTELS FOR FLAT LOAD-BEARING WALLS[a,b,c,d,f]
NO. 5 BOTTOM BAR SIZE

MINIMUM LINTEL THICKNESS, T (inches)	LINTEL DEPTH, D (inches)	MAXIMUM CLEAR SPAN, (feet-inches) (Number is Middle of Span, A)[e]					
		Supporting Roof		Supporting Light Framed 2nd Story and Roof		Supporting ICF Second Story and Roof	
		Ground Snow Load					
		30 psf	70 psf	30 psf	70 psf	30 psf	70 psf
3.5	8	4-9 (1-2)	4-2 (0-9)	3-11 (0-8)	3-7 (0-6)	3-7 (0-6)	3-5 (0-5)
	12	7-2 (1-11)	6-3 (1-3)	5-11 (1-1)	5-5 (0-10)	5-5 (0-10)	5-0 (0-8)
	16	9-6 (2-9)	8-0 (1-9)	7-4 (1-6)	6-6 (1-2)	6-7 (1-2)	5-11 (1-0)
	20	11-1 (3-5)	9-1 (2-3)	8-4 (1-11)	7-5 (1-6)	7-6 (1-7)	6-9 (1-3)
	24	12-2 (4-1)	10-0 (2-9)	9-3 (2-4)	8-2 (1-10)	8-4 (1-11)	7-6 (1-6)
5.5	8	5-6 (1-10)	4-10 (1-2)	4-7 (1-0)	4-2 (0-9)	4-2 (0-10)	3-10 (0-8)
	12	8-3 (3-0)	6-9 (2-0)	6-3 (1-9)	5-6 (1-4)	5-7 (1-4)	5-0 (1-1)
	16	9-9 (4-1)	8-0 (2-9)	7-5 (2-5)	6-6 (1-10)	6-7 (1-11)	6-0 (1-7)
	20	10-11 (5-3)	9-0 (3-6)	8-4 (3-1)	7-5 (2-4)	7-6 (2-5)	6-9 (2-0)
	24	12-0 (6-3)	9-11 (4-3)	9-3 (3-8)	8-2 (2-11)	8-3 (3-0)	7-6 (2-5)
7.5	8	6-1 (2-6)	5-2 (1-8)	4-9 (1-5)	4-3 (1-1)	4-3 (1-1)	3-10 (0-11)
	12	8-2 (4-0)	6-9 (2-8)	6-3 (2-4)	5-6 (1-10)	5-7 (1-10)	5-0 (1-6)
	16	9-7 (5-5)	7-11 (3-8)	7-4 (3-3)	6-6 (2-6)	6-7 (2-7)	6-0 (2-2)
	20	10-10 (6-10)	8-11 (4-8)	8-4 (4-2)	7-4 (3-3)	7-6 (3-4)	6-9 (2-9)
	24	11-10 (8-2)	9-10 (5-8)	9-2 (5-1)	8-1 (3-11)	8-3 (4-1)	7-5 (3-4)
9.5	8	6-4 (3-1)	5-2 (2-1)	4-10 (1-9)	4-3 (1-5)	4-4 (1-5)	3-11 (1-2)
	12	8-2 (5-0)	6-8 (3-4)	6-2 (3-0)	5-6 (2-4)	5-7 (2-5)	5-0 (1-11)
	16	9-6 (6-9)	7-11 (4-7)	7-4 (4-2)	6-6 (3-3)	6-7 (3-4)	5-11 (2-8)
	20	10-8 (8-4)	8-10 (5-10)	8-3 (5-4)	7-4 (4-2)	7-5 (4-3)	6-9 (3-6)
	24	11-7 (10-0)	9-9 (6-11)	9-0 (6-5)	8-1 (5-0)	8-2 (5-2)	7-5 (4-3)

For SI: 1 inch = 25.4 mm, 1 foot = 304.8 mm, 1 pound per square inch = 6.895 kN/m^2, 1 pound per square foot = 0.0479 kN/m^2.

a. This table is based on concrete with a minimum specified compressive strength of 2,500 psi, reinforcing steel with a minimum yield strength of 40,000 psi and an assumed equivalent rectangular cross section. When reinforcement with a minimum yield strength of 60,000 psi is used the span lengths in the shaded cells shall be increased by 1.2 times the table values.
b. This table is not intended to prohibit the use of ICF manufacturer's tables based on engineering analysis in accordance with ACI 318.
c. Deflection criterion: $L/240$.
d. Design load assumptions:
   Floor dead load is 10 psf   Attic live load is 20 psf
   Floor live load is 30 psf   Roof dead load is 15 psf
   Building width is 32 feet   ICF wall dead load is 69 psf
   Light-framed wall dead load is 10 psf
e. No. 3 stirrups are required at $d/2$ spacing except no stirrups are required for the distance, (A), shown in the middle portion of the span in accordance with Figure R611.7(2) and Section R611.7.3.2.
f. Interpolation is permitted between ground snow loads and between lintel depths.

❖ See the commentary for Section R611.7.3.1.

TABLE R611.7(4)
MAXIMUM ALLOWABLE CLEAR SPANS FOR WAFFLE-GRID ICF WALL LINTELS[a,b,c,d,f]
NO. 4 BOTTOM BAR SIZE

NOMINAL LINTEL THICKNESS T[g,h] (inches)	LINTEL DEPTH D (inches)	MAXIMUM CLEAR SPAN (feet-inches) (Number is Middle of Span, A)[e]					
		Supporting Roof		Supporting Light Framed 2nd Story and Roof		Supporting ICF Second Story and Roof	
		Ground Snow Load					
		30 psf	70 psf	30 psf	70 psf	30 psf	70 psf
6	8	5-2 (0-10)	4-2 (0-7)	3-10 (0-6)	3-5 (0-4)	3-6 (0-5)	3-2 (0-4)
	12	6-8 (1-5)	5-5 (0-11)	5-0 (0-9)	4-5 (0-7)	4-7 (0-8)	4-2 (0-6)
	16	7-11 (1-11)	6-6 (1-4)	6-0 (1-1)	5-3 (0-10)	5-6 (0-11)	4-11 (0-9)
	20	8-11 (2-6)	7-4 (1-8)	6-9 (1-5)	6-0 (1-1)	6-3 (1-2)	5-7 (0-11)
	24	9-10 (3-0)	8-1 (2-0)	7-6 (1-9)	6-7 (1-4)	6-10 (1-5)	6-2 (1-2)
8	8	5-2 (0-10)	4-3 (0-7)	3-11 (0-6)	3-5 (0-4)	3-7 (0-5)	3-2 (0-4)
	12	6-8 (1-5)	5-5 (0-11)	5-1 (0-9)	4-5 (0-7)	4-8 (0-8)	4-2 (0-6)
	16	7-10 (1-11)	6-5 (1-4)	6-0 (1-1)	5-3 (0-10)	5-6 (0-11)	4-11 (0-9)
	20	8-10 (2-6)	7-3 (1-8)	6-9 (1-5)	6-0 (1-1)	6-2 (1-2)	5-7 (0-11)
	24	9-8 (3-0)	8-0 (2-0)	7-5 (1-9)	6-7 (1-4)	6-10 (1-5)	6-2 (1-2)

For SI: 1 inch = 25.4 mm, 1 foot = 0.3048 m, 1 psi = 6.895 kN/m^2, 1 psf = 0.0479 kN/m^2.

a. This table is based on concrete with a minimum specified compressive strength of 2,500 psi, reinforcing steel with a minimum yield strength of 40,000 psi and an assumed equivalent rectangular cross section. When reinforcement with a minimum yield strength of 60,000 psi is used the span lengths in the shaded cells shall be increased by 1.2 times the table values.

b. This table is not intended to prohibit the use of ICF manufacturer's tables based on engineering analysis in accordance with ACI 318.

c. Deflection criterion: $L/240$.

d. Design load assumptions:
   Floor dead load is 10 psf   Attic live load is 20 psf
   Floor live load is 30 psf   Roof dead load is 15 psf
   Building width is 32 feet   ICF wall dead load is 55 psf
   Light-framed wall dead load is 10 psf

e. No. 3 stirrups are required at $d/2$ spacing except no stirrups are required for the distance, (A), shown in the middle portion of the span in accordance with Figure R611.7(2) and Section R611.7.3.2.

f. Interpolation is permitted between ground snow loads and between lintel depths.

g. For actual wall lintel width, refer to Table R611.4(2).

h. Lintel width corresponds to the nominal waffle-grid ICF wall thickness with a minimum thickness of 2 inches.

❖ See the commentary for Section R611.7.3.1.

WALL CONSTRUCTION

TABLE R611.7(5)
MAXIMUM ALLOWABLE CLEAR SPANS FOR WAFFLE-GRID ICF WALL LINTELS[a,b,c,d,f]
NO. 5 BOTTOM BAR SIZE

NOMINAL LINTEL THICKNESS, T[g, h] (inches)	LINTEL DEPTH D (inches)	MAXIMUM CLEAR SPAN (feet-inches) (Number is Middle of Span, A)[e]					
		Supporting Roof		Supporting Light Framed 2nd Story and Roof		Supporting ICF Second Story and Roof	
		Ground Snow Load					
		30 psf	70 psf	30 psf	70 psf	30 psf	70 psf
6	8	5-4 (0-10)	4-8 (0-7)	4-5 (0-6)	4-1 (0-4)	4-5 (0-5)	3-10 (0-4)
6	12	8-0 (1-5)	6-9 (0-11)	6-3 (0-9)	5-6 (0-7)	6-3 (0-8)	5-1 (0-6)
6	16	9-9 (1-11)	8-0 (1-4)	7-5 (1-1)	6-6 (0-10)	7-5 (0-11)	6-1 (0-9)
6	20	11-0 (2-6)	9-1 (1-8)	8-5 (1-5)	7-5 (1-1)	8-5 (1-2)	6-11 (0-11)
6	24	12-2 (3-0)	10-0 (2-0)	9-3 (1-9)	8-2 (1-4)	9-3 (1-5)	7-8 (1-2)
8	8	6-0 (0-10)	5-2 (0-7)	4-9 (0-6)	4-3 (0-4)	4-9 (0-5)	3-11 (0-4)
8	12	8-3 (1-5)	6-9 (0-11)	6-3 (0-9)	5-6 (0-7)	6-3 (0-8)	5-2 (0-6)
8	16	9-9 (1-11)	8-0 (1-4)	7-5 (1-1)	6-6 (0-10)	7-5 (0-11)	6-1 (0-9)
8	20	10-11 (2-6)	9-0 (1-8)	8-4 (1-5)	7-5 (1-1)	8-4 (1-2)	6-11 (0-11)
8	24	12-0 (3-0)	9-11 (2-0)	9-2 (1-9)	8-2 (1-4)	9-2 (1-5)	7-8 (1-2)

For SI: 1 inch = 25.4 mm, 1 foot = 0.3048 m, 1 psi = 6.895 kN/m^2, 1 psf = 0.0479 kN/m^2.

a. This table is based on concrete with a minimum specified compressive strength of 2,500 psi, reinforcing steel with a minimum yield strength of 40,000 psi and an assumed equivalent rectangular cross section. When reinforcement with a minimum yield strength of 60,000 psi is used the span lengths in the shaded cells shall be increased by 1.2 times the table values.
b. This table is not intended to prohibit the use of ICF manufacturer's tables based on engineering analysis in accordance with ACI 318.
c. Deflection criterion: $L/240$.
d. Design load assumptions:
   Floor dead load is 10 psf   Attic live load is 20 psf
   Floor live load is 30 psf   Roof dead load is 15 psf
   Building width is 32 feet   ICF wall dead load is 53 psf
   Light-framed wall dead load is 10 psf
e. No. 3 stirrups are required at $d/2$ spacing except no stirrups are required for the distance, (A), shown in the middle portion of the span in accordance with Figure R611.7(2) and Section R611.7.3.2.
f. Interpolation is permitted between ground snow loads and between lintel depths.
g. For actual wall lintel width, refer to Table R611.4(2)
h. Lintel width corresponds to the nominal waffle-grid ICF wall thickness with a minimum thickness of 2 inches.

❖ See the commentary for Section R611.7.3.1.

TABLE R611.7(6)
MAXIMUM ALLOWABLE CLEAR SPANS FOR SCREEN-GRID ICF LINTELS IN LOAD-BEARING WALLS[a,b,c,d,e,f,g]
NO. 4 BOTTOM BAR SIZE

MINIMUM LINTEL THICKNESS, T (inches)[h,i]	MINIMUM LINTEL DEPTH, D (inches)	MAXIMUM CLEAR SPAN (feet-inches)					
		Supporting Roof		Supporting Light-Frame Second Story and Roof		Supporting ICF Second Story and Roof	
		Maximum Ground Snow Load (psf)					
		30	70	30	70	30	70
6	12	3-7	2-10	2-5	2-0	2-0	NA
	24	9-10	8-1	7-6	6-7	6-11	6-2

For SI: 1 inch = 25.4 mm, 1 foot = 0.3048 m, 1 psi = 6.895 kN/m^2, 1 psf = 0.0479 kN/m^2.

a. This table is based on concrete with a minimum specified compressive strength of 2,500 psi, reinforcing steel with a minimum yield strength of 40,000 psi and an assumed equivalent rectangular cross section. When reinforcement with a minimum yield strength of 60,000 psi is used the span lengths in the shaded cells shall be increased by 1.2 times the table values.

b. This table is not intended to prohibit the use of ICF manufacturer's tables based on engineering analysis in accordance with ACI 318.

c. Deflection criterion: $L/240$

d. Design load assumptions:
   Floor dead load is 10 psf Attic live load is 20 psf
   Floor live load is 30 psf Roof dead load is 15 psf
   Maximum floor clear span is 32 ft ICF wall dead load is 53 psf
   Light-frame wall dead load is 10 psf

e. Stirrup Requirements:
   Stirrups are not required for lintels 12 inches deep.
   One No. 3 stirrup is required in each vertical core for lintels 24 inches deep.

f. Interpolation is permitted between ground snow loads.

g. Flat ICF lintels may be used in lieu of screen-grid lintels.

h. For actual wall lintel width, refer to Table R611.4(2).

i. Lintel width corresponds to the nominal screen-grid ICF wall thickness.

❖ See the commentary for Sections R611.7.3.1 and R611.7.3.4.

TABLE R611.7(7)
MAXIMUM ALLOWABLE CLEAR SPANS FOR SCREEN-GRID ICF LINTELS IN LOAD-BEARING WALLS[a,b,c,d,e,f,g]
NO. 5 BOTTOM BAR SIZE

MINIMUM LINTEL THICKNESS, T (inches)[h,i]	MINIMUM LINTEL DEPTH, D (inches)	MAXIMUM CLEAR SPAN (feet-inches)					
		Supporting Roof		Supporting Light-Frame Second Story and Roof		Supporting ICF Second Story and Roof	
		Maximum Ground Snow Load (psf)					
		30	70	30	70	30	70
6	12	3-7	2-10	2-5	2-0	2-0	NA
	24	12-3	10-0	9-3	8-3	8-7	7-8

For SI: 1 inch = 25.4 mm, 1 foot = 304.8 mm, 1 pound per square inch = 6.895 kN/m^2, 1 pound per square foot = 0.0479 kN/m^2.

a. This table is based on concrete with a minimum specified compressive strength of 2,500 psi, reinforcing steel with a minimum yield strength of 40,000 psi and an assumed equivalent rectangular cross section. When reinforcement with a minimum yield strength of 60,000 psi is used the span lengths in the shaded cells shall be increased by 1.2 times the table values.

b. This table is not intended to prohibit the use of ICF manufacturer's tables based on engineering analysis in accordance with ACI 318.

c. Deflection criterion: $L/240$

d. Design load assumptions:
   Floor dead load is 10 psf Attic live load is 20 psf
   Floor live load is 30 psf Roof dead load is 15 psf
   Maximum floor clear span is 32 ft ICF wall dead load is 53 psf
   Light-frame wall dead load is 10 psf

e. Stirrup Requirements:
   Stirrups are not required for lintels 12 inches deep.
   One No. 3 stirrup is required in each vertical core for lintels 24 inches deep.

f. Interpolation is permitted between ground snow loads.

g. Flat ICF lintels may be used in lieu of screen-grid lintels.

h. For actual wall lintel width, refer to Table R611.4(2).

i. Lintel width corresponds to the nominal screen-grid ICF wall thickness.

❖ See the commentary for Sections R611.7.3.1 and R611.7.3.4.

WALL CONSTRUCTION   TABLE R611.7(8)

TABLE R611.7(8)
MAXIMUM ALLOWABLE CLEAR SPANS FOR ICF LINTELS WITHOUT STIRRUPS IN LOAD-BEARING WALLS[a,b,c,d,e,f,g,h]
(NO. 4 OR NO. 5) BOTTOM BAR SIZE

MINIMUM LINTEL THICKNESS, T (inches)	MINIMUM LINTEL DEPTH, D (inches)	MAXIMUM CLEAR SPAN (feet-inches)					
		Supporting Roof Only		Supporting Light-Frame Second Story and Roof		Supporting ICF Second Story and Roof	
		MAXIMUM GROUND SNOW LOAD (psf)					
		30	70	30	70	30	70
Flat ICF Lintel							
3.5	8	2-6	2-6	2-6	2-4	2-5	2-2
	12	4-2	4-2	4-1	3-10	3-10	3-7
	16	4-11	4-8	4-6	4-2	4-2	3-11
	20	6-3	5-3	4-11	4-6	4-6	4-3
	24	7-7	6-4	6-0	5-6	5-6	5-2
5.5	8	2-10	2-6	2-6	2-5	2-6	2-2
	12	4-8	4-4	4-3	3-11	3-10	3-7
	16	6-5	5-1	4-8	4-2	4-3	3-11
	20	8-2	6-6	6-0	5-4	5-5	5-0
	24	9-8	7-11	7-4	6-6	6-7	6-1
7.5	8	3-6	2-8	2-7	2-5	2-5	2-2
	12	5-9	4-5	4-4	4-0	3-10	3-7
	16	7-9	6-1	5-7	4-10	4-11	4-5
	20	8-8	7-2	6-8	5-11	6-0	5-5
	24	9-6	7-11	7-4	6-6	6-7	6-0
9.5	8	4-2	3-1	2-9	2-5	2-5	2-2
	12	6-7	5-1	4-7	3-11	4-0	3-7
	16	7-10	6-4	5-11	5-3	5-4	4-10
	20	8-7	7-2	6-8	5-11	6-0	5-5
	24	9-4	7-10	7-3	6-6	6-7	6-0
Waffle-Grid ICF Lintel							
6 or 8	8	2-6	2-6	2-6	2-4	2-4	2-2
	12	4-2	4-2	4-1	3-8	3-9	3-7
	16	5-9	5-8	5-7	5-1	5-2	4-8
	20	7-6	7-4	6-9	6-0	6-3	5-7
	24	9-2	8-1	7-6	6-7	6-10	6-2

For SI: 1 inch = 25.4 mm; 1 psf = 0.0479 kN/m^2; 1 ft = 0.3 m.

a. Table values are based on tensile reinforcement with a minimum yield strength of 40,000 psi (276 MPa), concrete with a minimum specified compressive strength of 2,500 psi (17.2 MPa), and a building width (clear span) of 32 feet (9.8m).
b. Spans located in shaded cells shall be permitted to be multiplied by 1.05 when concrete with a minimum compressive strength of 3,000 psi (20.7 MPa) is used or by 1.1 when concrete with a minimum compressive strength of 4,000 psi (27.6 MPa) is used.
c. Deflection criterion is *L/240*, where *L* is the clear span of the lintel in inches.
d. Linear interpolation shall be permitted between ground snow loads and between lintel depths.
e. Lintel depth, D, shall be permitted to include the available height of ICF wall located directly above the lintel, provided that the increased lintel depth spans the entire length of the opening.
f. Spans shall be permitted to be multiplied by 1.05 for a building width (clear span) of 28 feet (8.5 m).
g. Spans shall be permitted to be multiplied by 1.1 for a building width (clear span) of 24 feet (7.3 m) or less.
h. ICF wall dead load is 69 psf (3.3 kPa).

❖ See the commentary for Sections R611.7.3.1 and R611.7.3.4.

TABLE R611.7(9)
MINIMUM BOTTOM BAR ICF LINTEL REINFORCEMENT FOR LARGE CLEAR SPANS IN LOAD-BEARING WALLS[a,b,c,d,e,f,h]

MINIMUM LINTEL THICKNESS, T[e,g] (inches)	MINIMUM LINTEL DEPTH, D (inches)	Supporting Light-Frame Roof Only		Supporting Light-Frame Second Story and Roof		Supporting ICF Second Story and Light-Frame Roof	
		Maximum Ground Snow Load (psf)					
		30	70	30	70	30	70
Flat ICF Lintel, 12 Feet—3 inches Maximum Clear Span							
3.5	24	1 #5	1 #7	D/R	D/R	D/R	D/R
5.5	20	1 #6	1 #7	D/R	D/R	D/R	D/R
	24	1 #5	1 #7	1 #7	1 #8	1 #8	D/R
7.5	16	1 #7; 2 #5	D/R	D/R	D/R	D/R	D/R
	20	1 #6; 2 #4	1 #7; 2 #5	1 #8; 2 #6	D/R	D/R	D/R
	24	1 #6; 2 #4	1 #7; 2 #5	1 #7; 2 #5	1 #8; 2 #6	1 #8; 2 #6	1 #8; 2 #6
9.5	16	1 #7; 2 #5	D/R	D/R	D/R	D/R	D/R
	20	1 #6; 2 #4	1 #7; 2 #5	1 #8; 2 #6	1 #8; 2 #6	1 #8; 2 #6	1 #9; 2 #6
	24	1 #6; 2 #4	1 #7; 2 #5	1 #7; 2 #5	1 #7; 2 #6	1 #8; 2 #6	1 #9; 2 #6
Flat ICF Lintel, 16 Feet—3 inches Maximum Clear Span							
5.5	24	1 #7	D/R	D/R	D/R	D/R	D/R
7.5	24	1 #7; 2 #5	D/R	D/R	D/R	D/R	D/R
9.5	24	1 #7; 2 #5	1 #9; 2 #6	1 #9; 2 #6	D/R	D/R	D/R
Waffle-Grid ICF Lintel, 12 Feet—3 inches Maximum Clear Span							
6	20	1 #6	D/R	D/R	D/R	D/R	D/R
	24	1 #5	1 #7; 2 #5	1 #7; 2 #5	1 #8; 2 #6	1 #8; 2 #6	D/R
8	16	1 #7; 2 #5	D/R	D/R	D/R	D/R	D/R
	20	1 #6; 2 #4	1 #7; 2 #5	1 #8; 2 #6	D/R	D/R	D/R
	24	1 #5	1 #7; 2 #5	1 #7; 2 #5	1 #8; 2 #6	1 #8; 2 #6	1 #8; 2 #6
Screen-Grid ICF Lintel, 12 Feet—3 Inches Maximum Clear Span							
6	24	1 #5	1 #7	D/R	DR	D/R	D/R

For SI: 1 inch = 25.4 mm, 1 foot = 0.3048 m, 1 psi = 6.895 kN/m^2, 1 psf = 0.0479 kN/m^2.

a. This table is based on concrete with a minimum specified compressive strength of 2,500 psi, reinforcing steel with a minimum yield strength of 40,000 psi and an assumed equivalent rectangular cross section. When reinforcement with a minimum yield strength of 60,000 psi is used the span lengths in the shaded cells shall be increased by 1.2 times the table values.
b. This table is not intended to prohibit the use of ICF manufacturers tables based on engineering analysis in accordance with ACI 318.
c. D/R indicates design is required.
d. Deflection criterion: $L/240$.
e. Interpolation is permitted between ground snow loads and between lintel depths.
f. No. 3 stirrups are required a maximum d/2 spacing for spans greater than 4 feet.
g. Actual thickness is shown for flat lintels while nominal thickness is given for waffle-grid and screen-grid lintels. Lintel thickness corresponds to the nominal waffle- grid and screen-grid ICF wall thickness. Refer to Section R611.4(2) for actual wall thickness.
h. ICF wall dead load varies based on wall thickness using 150pcf concrete density.

❖ See the commentary for Sections R611.7.3.1 and R611.7.3.4.

WALL CONSTRUCTION   TABLE R611.7(9A)

TABLE R611.7(9A)
MINIMUM SOLID END WALL LENGTH REQUIREMENTS FOR FLAT ICF WALLS (WIND PERPENDICULAR TO RIDGE)[a,b,c]

WALL CATEGORY	BUILDING SIDE WALL LENGTH, L (feet)	Roof Slope	WIND VELOCITY PRESSURE FROM TABLE R611.7.4 (psf)							
			20	25	30	35	40	45	50	60
			Minimum Solid Wall Length on Building End Wall (feet)							
One-Story or Top Story of Two-Story	16	≤ 1:12	4.00	4.00	4.00	4.00	4.00	4.00	4.00	4.00
		5:12	4.00	4.00	4.00	4.00	4.00	4.00	4.25	4.50
		7:12[d]	4.00	4.25	4.25	4.50	4.75	4.75	5.00	5.50
		12:12[d]	4.25	4.50	4.75	5.00	5.25	5.50	5.75	6.25
	24	≤ 1:12	4.00	4.00	4.00	4.00	4.00	4.00	4.25	4.50
		5:12	4.00	4.00	4.00	4.25	4.25	4.50	4.50	4.75
		7:12[d]	4.25	4.50	4.75	5.00	5.25	5.50	5.75	6.25
		12:12[d]	4.75	5.00	5.25	5.75	6.00	6.50	6.75	7.50
	32	≤ 1:12	4.00	4.00	4.00	4.00	4.25	4.25	4.50	4.75
		5:12	4.00	4.00	4.25	4.50	4.50	4.75	5.00	5.25
		7:12[d]	4.50	5.00	5.25	5.50	6.00	6.25	6.50	7.25
		12:12[d]	5.00	5.50	6.00	6.50	7.00	7.25	7.75	8.75
	40	≤ 1:12	4.00	4.00	4.25	4.25	4.50	4.50	4.75	5.00
		5:12	4.00	4.25	4.50	4.75	4.75	5.00	5.25	5.50
		7:12[d]	4.75	5.25	5.75	6.00	6.50	7.00	7.25	8.00
		12:12[d]	5.50	6.00	6.50	7.25	7.75	8.25	8.75	10.0
	50	≤ 1:12	4.00	4.25	4.25	4.50	4.75	4.75	5.00	5.50
		5:12	4.25	4.50	4.75	5.00	5.25	5.50	5.75	6.00
		7:12[d]	5.25	5.75	6.25	6.75	7.25	7.75	8.25	9.25
		12:12[d]	6.00	6.75	7.50	8.00	8.75	9.50	10.25	11.5
	60	≤ 1:12	4.00	4.25	4.50	4.75	5.00	5.25	5.25	5.75
		5:12	4.50	4.75	5.00	5.25	5.50	5.75	6.00	6.75
		7:12[d]	5.50	6.25	6.75	7.50	8.00	8.50	9.25	10.25
		12:12[d]	6.50	7.25	8.25	9.00	9.75	10.5	11.5	13.0
First Story of Two-Story	16	≤ 1:12	4.00	4.25	4.50	4.75	5.00	5.25	5.25	5.75
		5:12	4.50	4.75	5.00	5.25	5.50	5.75	6.00	6.75
		7:12[d]	4.50	5.00	5.25	5.75	6.00	6.25	6.75	7.25
		12:12[d]	5.00	5.25	5.75	6.25	6.50	7.00	7.25	8.25
	24	≤ 1:12	4.50	4.75	5.00	5.25	5.50	5.75	6.00	6.75
		5:12	4.75	5.25	5.50	6.00	6.25	6.75	7.00	7.75
		7:12[d]	5.25	5.75	6.25	6.75	7.00	7.50	8.00	9.00
		12:12[d]	5.50	6.25	6.75	7.25	8.00	8.50	9.00	10.25
	32	≤ 1:12	4.75	5.00	5.50	5.75	6.25	6.50	6.75	7.50
		5:12	5.25	5.75	6.25	6.75	7.25	7.50	8.00	9.00
		7:12[d]	5.75	6.50	7.00	7.75	8.25	9.00	9.50	10.75
		12:12[d]	6.25	7.00	7.75	8.50	9.25	10.0	10.75	12.25
	40	≤ 1:12	5.00	5.50	5.75	6.25	6.75	7.25	7.50	8.50
		5:12	5.50	6.25	6.75	7.25	8.00	8.50	9.00	10.25
		7:12[d]	6.25	7.00	7.75	8.75	9.50	10.25	11.0	12.5
		12:12[d]	7.00	8.00	8.75	9.75	10.75	11.5	12.5	14.25
	50	≤ 1:12	5.50	6.00	6.50	7.00	7.50	8.00	8.50	9.50
		5:12	6.00	6.75	7.50	8.25	9.00	9.75	10.5	11.75
		7:12[d]	7.00	8.00	9.00	10.0	10.75	11.75	12.75	14.5
		12:12[d]	7.75	9.00	10.0	11.25	12.25	13.50	14.75	17.0
	60	≤ 1:12	5.75	6.50	7.00	7.50	8.25	8.75	9.50	10.75
		5:12	6.75	7.50	8.25	9.25	10.0	10.75	11.75	13.25
		7:12[d]	7.75	9.00	10.0	11.0	12.25	13.25	14.5	16.75
		12:12[d]	8.75	10.0	11.5	12.75	14.0	15.5	16.75	19.5

(continued)

Footnotes to Table R611.7(9A)

For SI: 1 foot = 0.3048 m; 1 inch = 25.4 mm; 1 psf = 0.0479 kN/m^2.

a. Table values are based on a 3.5 in (88.9 mm) thick flat wall. For a 5.5 in (139.7 mm) thick flat wall, multiply the table values by 0.9. The adjusted values shall not result in solid wall lengths less than 4ft.
b. Table values are based on a maximum unsupported wall height of 10 ft (3.0 m).
c. Linear interpolation shall be permitted.
d. The minimum solid wall lengths shown in the table are based on a building with an end wall length "W" of 60 feet and a roof slope of less than 7:12. For roof slopes of 7:12 or greater and end wall length "W" greater than 30 feet, the minimum solid wall length determined from the table shall be multiplied by:
  $1 + 0.4[(W-30)/30]$.

❖ See the commentary for Section R611.7.4.

TABLE R611.7(9B)
MINIMUM SOLID SIDEWALL LENGTH REQUIREMENTS FOR FLAT ICF WALLS (WIND PARALLEL TO RIDGE) a, b, c, d

WALL CATEGORY	BUILDING END WALL WIDTH, W (feet)	WIND VELOCITY PRESSURE FROM TABLE R611.7.4 (psf)							
		20	25	30	35	40	45	50	60
		Minimum Solid Wall Length on Building Side Wall (feet)							
One-Story or Top Story of Two-Story	16	4.00	4.00	4.00	4.00	4.25	4.25	4.50	4.75
	24	4.00	4.25	4.50	4.75	4.75	5.00	5.25	5.50
	32	4.50	4.75	5.00	5.25	5.50	6.00	6.25	6.75
	40	5.00	5.50	5.75	6.25	6.75	7.00	7.50	8.25
	50	5.75	6.25	7.00	7.50	8.25	8.75	9.50	10.75
	60	6.50	7.50	8.25	9.25	10.0	10.75	11.75	13.25
First Story of Two-Story	16	4.25	4.50	4.75	5.00	5.25	5.50	5.75	6.50
	24	4.75	5.25	5.50	6.00	6.25	6.75	7.00	8.00
	32	5.50	6.00	6.50	7.00	7.50	8.00	8.75	9.75
	40	6.25	7.00	7.50	8.25	9.00	9.75	10.5	12.0
	50	7.25	8.25	9.25	10.25	11.25	12.25	13.25	15.25
	60	8.50	9.75	11.0	12.25	13.5	15.0	16.25	18.75

For SI: 1 foot = 0.3048 m; 1 inch = 25.4 mm; 1 psf = 0.0479 kN/m^2.

a. Table values are based on a 3.5 in (88.9 mm) thick flat wall. For a 5.5 in (139.7 mm) thick flat wall, multiply the table values by 0.9. The adjusted values shall not result in solid wall lengths less than 4ft.
b. Table values are based on a maximum unsupported wall height of 10 ft (3.0 m).
c. Table values are based on a maximum 12:12 roof pitch.
d. Linear interpolation shall be permitted.

❖ See the commentary for Section R611.7.4.

TABLE R611.7(10)
MAXIMUM ALLOWABLE CLEAR SPANS FOR ICF LINTELS IN NON-LOAD-BEARING WALLS WITHOUT STIRRUPS[a, b, c, d]
NO. 4 BOTTOM BAR

MINIMUM LINTEL THICKNESS, T (inches)	MINIMUM LINTEL DEPTH, D (inches)	MAXIMUM CLEAR SPAN Supporting Light-Frame Non-Bearing Wall (feet-inches)	MAXIMUM CLEAR SPAN Supporting ICF Second Story and Non-Bearing Wall (feet-inches)
Flat ICF Lintel			
3.5	8	11-1	3-1
3.5	12	15-11	5-1
3.5	16	16-3	6-11
3.5	20	16-3	8-8
3.5	24	16-3	10-5
5.5	8	16-3	4-4
5.5	12	16-3	7-0
5.5	16	16-3	9-7
5.5	20	16-3	12-0
5.5	24	16-3	14-3
7.5	8	16-3	5-6
7.5	12	16-3	8-11
7.5	16	16-3	12-2
7.5	20	16-3	15-3
7.5	24	16-3	16-3
9.5	8	16-3	6-9
9.5	12	16-3	10-11
9.5	16	16-3	14-10
9.5	20	16-3	16-3
9.5	24	16-3	16-3
Waffle-Grid ICF Lintel			
6 or 8	8	9-1	2-11
6 or 8	12	13-4	4-10
6 or 8	16	16-3	6-7
6 or 8	20	16-3	8-4
6 or 8	24	16-3	9-11
Screen-Grid Lintel			
6	12	5-8	4-1
6	24	16-3	9-1

For SI: 1 foot = 0.3048 m; 1 inch = 25.4 mm; 1 psf = 47.8804 Pa.

a. This table is based on concrete with a minimum specified compressive strength of 2,500 psi, reinforcing steel with a minimum yield strength of 40,000 psi and an assumed equivalent rectangular cross section.
b. This table is not intended to prohibit the use of ICF manufacturers tables based on engineering analysis in accordance with ACI 318.
c. Deflection criterion is *L/240*, where *L* is the clear span of the lintel in inches.
d. Linear interpolation is permitted between lintel depths.

❖ See the commentary for Sections R611.7.3.1 and R611.7.3.5.

TABLE R611.7(10A)
MINIMUM SOLID END WALL LENGTH REQUIREMENTS FOR WAFFLE AND SCREEN-GRID ICF WALLS (WIND PERPENDICULAR TO RIDGE)[a,b,c]

WALL CATEGORY	BUILDING SIDE WALL LENGTH, L (feet)	ROOF SLOPE	WIND VELOCITY PRESSURE FROM TABLE R611.7.4							
			20	25	30	35	40	45	50	60
			Minimum Solid Wall Length on Building End Wall (feet)							
One-Story or Top Story of Two-Story	16	≤1:12	4.00	4.00	4.00	4.00	4.00	4.00	4.00	4.25
		5:12	4.00	4.00	4.00	4.00	4.00	4.25	4.25	4.50
		7:12[d]	4.00	4.25	4.50	4.75	5.00	5.25	5.50	6.00
		12:12[d]	4.25	4.75	5.00	5.50	5.75	6.00	6.50	7.00
	24	≤1:12	4.00	4.00	4.00	4.00	4.00	4.25	4.25	4.50
		5:12	4.00	4.00	4.00	4.25	4.50	4.50	4.75	5.00
		7:12[d]	4.50	4.75	5.00	5.50	5.75	6.25	6.50	7.25
		12:12[d]	5.00	5.50	6.00	6.50	7.00	7.25	7.75	8.75
	32	≤1:12	4.00	4.00	4.00	4.25	4.25	4.50	4.75	5.00
		5:12	4.00	4.00	4.25	4.50	4.75	5.00	5.25	5.75
		7:12[d]	4.75	5.25	5.75	6.25	6.50	7.00	7.50	8.50
		12:12[d]	5.50	6.25	6.75	7.50	8.00	8.75	9.25	10.5
	40	≤1:12	4.00	4.00	4.25	4.50	4.50	4.75	5.00	5.50
		5:12	4.00	4.25	4.50	5.00	5.25	5.50	5.75	6.25
		7:12[d]	5.25	5.75	6.25	7.00	7.50	8.00	8.50	9.75
		12:12[d]	6.00	6.75	7.75	8.50	9.25	10.0	10.75	12.25
	50	≤1:12	4.00	4.25	4.50	4.75	5.00	5.25	5.50	6.00
		5:12	4.25	4.75	5.00	5.25	5.50	6.00	6.25	7.00
		7:12[d]	5.75	6.50	7.00	7.75	8.50	9.25	9.75	11.25
		12:12[d]	6.75	7.75	8.75	9.75	10.75	11.5	12.5	14.5
	60	≤1:12	4.25	4.50	4.75	5.00	5.25	5.50	5.75	6.50
		5:12	4.50	5.00	5.25	5.75	6.00	6.50	6.75	7.75
		7:12[d]	6.25	7.00	8.00	8.75	9.50	10.25	11.25	12.75
		12:12[d]	7.50	8.75	9.75	11.0	12.0	13.25	14.25	16.5
First Story of Two-Story	16	≤1:12	4.25	4.50	4.75	5.00	5.25	5.50	5.75	6.50
		5:12	4.50	5.00	5.25	5.75	6.00	6.50	6.75	7.75
		7:12[d]	4.75	5.25	5.75	6.25	6.75	7.25	7.75	8.75
		12:12[d]	5.25	5.75	6.50	7.00	7.50	8.00	8.75	9.75
	24	≤1:12	4.50	5.00	5.25	5.75	6.25	6.50	7.00	7.75
		5:12	5.00	5.75	6.25	6.75	7.25	7.75	8.25	9.25
		7:12[d]	5.75	6.25	7.00	7.75	8.25	9.00	9.75	11.0
		12:12[d]	6.25	7.00	7.75	8.50	9.50	10.25	11.0	12.75
	32	≤1:12	5.00	5.50	6.00	6.50	7.00	7.50	8.00	9.00
		5:12	5.75	6.25	7.00	7.75	8.25	9.00	9.75	11.0
		7:12[d]	6.50	7.25	8.25	9.00	10.0	10.75	11.75	13.5
		12:12[d]	7.25	8.25	9.25	10.25	11.25	12.5	13.5	15.5
	40	≤1:12	5.50	6.00	6.50	7.25	7.75	8.50	9.00	10.25
		5:12	6.25	7.00	7.75	8.75	9.50	10.25	11.0	12.75
		7:12[d]	7.25	8.25	9.25	10.5	11.5	12.5	13.75	15.75
		12:12[d]	8.00	9.50	10.75	12.0	13.25	14.5	15.75	18.25

(continued)

WALL CONSTRUCTION

TABLE R611.7(10A)—continued
MINIMUM SOLID END WALL LENGTH REQUIREMENTS FOR WAFFLE AND SCREEN-GRID ICF WALLS (WIND PERPENDICULAR TO RIDGE)[a,b,c]

WALL CATEGORY	BUILDING SIDE WALL LENGTH, L (feet)	ROOF SLOPE	WIND VELOCITY PRESSURE FROM TABLE R611.7.4							
			20	25	30	35	40	45	50	60
			Minimum Solid Wall Length on Building End Wall (feet)							
First Story of Two-Story (continued)	50	≤ 1:12	6.00	6.75	7.50	8.00	8.75	9.50	10.25	11.75
		5:12	7.00	8.00	9.00	10.0	11.0	12.0	13.0	14.75
		7:12[d]	8.25	9.50	10.75	12.25	13.5	14.75	16.0	18.75
		12:12[d]	9.25	11.0	12.5	14.0	15.5	17.25	18.75	22.0
	60	≤ 1:12	6.50	7.25	8.25	9.00	10.0	10.75	11.75	13.25
		5:12	7.75	8.75	10.0	11.25	12.25	13.5	14.75	17.0
		7:12[d]	9.25	10.75	12.25	14.0	15.5	17.0	18.5	21.75
		12:12[d]	10.5	12.25	14.25	16.25	18.0	20.0	21.75	25.5

For SI: 1 foot = 0.3048 m; 1 inch = 25.4 mm; 1 psf = 0.0479 kN/m^2.

a. Table values are based on a 6 in (152.4 mm) thick nominal waffle-grid wall. For a 8 in (203.2 mm) thick nominal waffle-grid wall, multiply the table values by 0.90.
b. Table values are based on a maximum unsupported wall height of 10 ft (3.0 m).
c. Linear interpolation is permitted.
d. The minimum solid wall lengths shown in the table are based on a building with an end wall length "W" of 60 feet and a roof slope of less than 7:12. For roof slopes of 7:12 or greater and end wall length "W" greater than 30 feet, the minimum solid wall length determined from the table shall be multiplied by: 1 + 0.4[(W-30)/30].

❖ See the commentary for Section R611.7.4.

TABLE R611.7(10B)
MINIMUM SOLID SIDE WALL LENGTH REQUIREMENTS for 6-INCH WAFFLE AND SCREEN-GRID ICF WALLS (WIND PARALLEL TO RIDGE)[a,b,c,d]

WALL CATEGORY	BUILDING END WALL WIDTH, W (feet)	WIND VELOCITY PRESSURE FROM TABLE R611.7.4 (psf)							
		20	25	30	35	40	45	50	60
		Minimum Solid Wall Length on Building Side Wall (feet)							
One-Story or Top Story of Two-Story	16	4.00	4.00	4.00	4.25	4.25	4.50	4.75	5.00
	24	4.00	4.25	4.50	5.00	5.25	5.50	5.75	6.25
	32	4.50	5.00	5.50	5.75	6.25	6.75	7.00	8.00
	40	5.25	6.00	6.50	7.00	7.75	8.25	8.75	10.0
	50	6.50	7.25	8.00	9.00	9.75	10.75	11.5	13.25
	60	7.75	8.75	10.0	11.25	12.25	13.5	14.5	17.0
First Story of Two-Story	16	4.50	4.75	5.25	5.50	5.75	6.25	6.50	7.25
	24	5.00	5.75	6.25	6.75	7.25	7.75	8.25	9.50
	32	6.00	6.75	7.50	8.25	9.00	9.75	10.5	12.0
	40	7.00	8.00	9.00	10.0	11.0	12.0	13.0	15.0
	50	8.50	9.75	11.25	12.5	14.0	15.25	16.75	19.5
	60	10.25	12.0	13.75	15.5	17.25	19.0	21.0	24.5

For SI: 1 foot = 0.3048 m; 1 inch = 25.4 mm; 1 psf = 0.0479 kN/m^2.

a. Table values are based on a 6 in (152.4 mm) thick nominal waffle-grid wall. For a 8 in (203.2 mm) thick nominal waffle-grid wall, multiply the table values by 0.90.
b. Table values are based on a maximum unsupported wall height of 10 ft (3.0 m).
c. Table values are based on a maximum 12:12 roof pitch.
d. Linear interpolation shall be permitted.

❖ See the commentary for Section R611.7.4.

TABLE R611.7(11)
MINIMUM PERCENTAGE OF SOLID WALL LENGTH ALONG EXTERIOR WALL LINES FOR TOWNHOUSES IN SEISMIC DESIGN CATEGORY C AND ALL BUILDINGS IN SEISMIC DESIGN CATEGORY D[a,b]

SEISMIC DESIGN CATEGORY (SDC)	MINIMUM SOLID WALL LENGTH (percent)		
	One-Story or Top Story of Two-Story	Wall Supporting Light Frame Second Story and Roof	Wall Supporting ICF Second Story and Roof
Townhouses in SDC C[c]	20 percent	25 percent	35 percent
D$_1$[d]	25 percent	30 percent	40 percent
D$_2$[d]	30 percent	35 percent	45 percent

For SI: 1 inch = 25.4 mm; 1 mph = 1.6093 km/hr.

a. Base percentages are applicable for maximum unsupported wall height of 10-feet (3.0-m), light-frame gable construction, and all ICF wall types. These percentages assume that the maximum weight of the interior and exterior wall finishes applied to ICF walls do not exceed 8 psf (0.38 kN/m^2).

b. For all walls, the minimum required length of solid walls shall be based on the table percent value multiplied by the minimum dimension of a rectangle inscribing the overall building plan.

c. Walls shall be reinforced with a minimum No. 5 bar (Grade 40 or 60) spaced a maximum of 24 inches (609.6 mm) on center each way or a No. 4 bar spaced a maximum of 16 inches on center each way. (Grade 40 or 60) spaced at a maximum of 16 inches (406.4 mm) on center each way.

d. Walls shall be constructed with a minimum concrete compressive strength of 3,000 psi (20.7 MPa) and reinforced with minimum #5 rebar (Grade 60 ASTM A706) spaced a maximum of 18 inches (457.2 mm) on center each way or No. 4 rebar (Grade 60 ASTM A706) spaced at a maximum of 12 inches (304.8 mm) on center each way. The minimum thickness of flat ICF walls shall be 5.5 inches.

❖ See the commentary for Section R611.7.4.

For SI: 1 foot = 304.8 mm.

Note: Section cut through flat wall or vertical core of waffle- or screen-grid wall.

FIGURE R611.7(1)
ICF WALL CONSTRUCTION

❖ See the commentary for Section R611.7.4.

WALL CONSTRUCTION — FIGURE R611.7(2) - FIGURE R611.7(3)

**FIGURE R611.7(2)
REINFORCEMENT OF OPENINGS**

❖ This figure illustrates the reinforcement of openings specified in Sections R611.7.2 and R611.7.3.

For SI: 1 inch = 25.4 mm.

Note: Section cut through flat wall.

**FIGURE R611.7(3)
ICF LINTELS FOR FLAT WALLS**

❖ See the commentary for Section R611.7.3.1.

For SI: 1 inch = 25.4 mm.

Note: Section cut through vertical core of a waffle-grid lintel.

**FIGURE R611.7(4)
SINGLE FORM HEIGHT WAFFLE-GRID LINTEL**

❖ See the commentary for Section R611.7.3.1.

For SI: 1 inch = 25.4 mm.

Note: Section cut through vertical core of a waffle-grid lintel.

**FIGURE R611.7(5)
DOUBLE FORM HEIGHT WAFFLE-GRID LINTEL**

❖ See the commentary for Section R611.7.3.1.

WALL CONSTRUCTION

FIGURE R611.7(6)
SINGLE FORM HEIGHT SCREEN-GRID LINTEL

❖ See the commentary for Section R611.7.3.1.

FIGURE R611.7(7)
DOUBLE FORM HEIGHT SCREEN-GRID LINTEL

❖ See the commentary for Section R611.7.3.1.

R611.7.1.4 Lap Splices. Where lap splicing of vertical or horizontal reinforcing steel is necessary, the lap splice shall be in accordance with Figure R611.7.1.4 and a minimum of $40d_b$, where d_b is the diameter of the smaller bar. The maximum distance between noncontact parallel bars at a lap splice shall not exceed $8d_b$.

❖ Where continuous bars are not provided, lap splices are necessary to allow forces to be transferred between sections of rebar. This section specifies the minimum lap splice length as 40 times the bar diameter. Noncontact lap splices are permitted provided the spacing of lapped bars does not exceed eight bar diameters. These requirements are based on ACI 318. Figure R611.7.1.4 illustrates the code text requirements.

R611.7.1.5 Standard hook. Where the free end of a reinforcing bar is required to have a standard hook, the hook shall be a 180-degree bend plus $4d_b$ extension but not less than $2^1/_2$ inches, or a 90-degree bend plus $12d_b$ extension.

❖ Requirements for hooks are consistent with the requirements of ACI 318.

R611.7.2 Wall openings. Wall openings shall have a minimum of 8 inches (203 mm) of depth of concrete for flat and waffle-

FIGURE R611.7.1.4
LAP SPLICES

❖ This figure illustrates the requirements of Section R 611.7.1.4.

grid ICF walls and 12 inches (305 mm) for screen-grid walls over the length of the opening. When the depth of concrete above the opening is less than 12 inches for flat or waffle-grid walls, lintels in accordance with Section R611.7.3 shall be provided. Reinforcement around openings shall be provided in accordance with Table R611.7(1) and Figure R611.7(2). All reinforcement placed horizontally above or below an opening shall extend a minimum of 24 inches (610 mm) beyond the limits of the opening. Wall opening reinforcement shall be provided in addition to the reinforcement required by Sections R611.3, R611.4, R611.5 and R611.7.1. The perimeter of all wall openings shall be framed with a minimum 2-inch by 4-inch plate, anchored to the wall with $1/2$-inch (12.7 mm) diameter anchor bolts spaced a maximum of 24 inches (610 mm) on center. The bolts shall be embedded into the concrete a minimum of 4 inches (102 mm) and have a minimum of $1^1/_2$ (38 mm) inches of concrete cover to the face of the wall.

Exception: The 2-inch by 4-inch plate is not required where the wall is formed to provide solid concrete around the perimeter of the opening with a minimum depth of 4 inches (102 mm) for the full thickness of the wall.

❖ See Table R611.7(1) for a listing of the minimum reinforcement required at wall openings. The requirements vary based on the length of the opening.

R611.7.3 Lintels.

R611.7.3.1 General requirements. Lintels shall be provided over all openings greater than or equal to 2 feet (610 mm) in width. Lintels for flat ICF walls shall be constructed in accordance with Figure R611.7(3) and Table R611.7(2) or R611.7(3). Lintels for waffle-grid ICF walls shall be constructed in accordance with Figure R611.7(4) or Figure R611.7(5) and Table R611.7(4) or R611.7(5). Lintels for screen-grid ICF walls shall be constructed in accordance with Figure R611.7(6) or Figure R611.7(7). Lintel construction in accordance with Figure R611.7(3) shall be permitted to be used with waffle-grid and screen-grid ICF wall construction. Lintel depths are permitted to be increased by the height of the ICF wall located directly above the opening, provided that the lintel depth spans the entire length of the opening.

❖ To support the loads from above, any wall opening of 2 feet (1219 mm) or more requires a lintel in accordance with Section R611.7.3. Figures R611.7(2) through R611.7(7) illustrate these requirements. By implication, therefore, openings smaller than 2 feet can be installed without the use of a lintel to transfer loads from above as a beam. The referenced figures indicate the need for horizontal reinforcement and stirrups for transverse reinforcement depending on the type of ICF wall being employed and the size of the opening. Commentary Table R611.7.3.1 provides a summary of the referenced figures and tables in Sections R611.7.3.1 through R611.7.3.5.

Table R611.7.3
APPLICABLE REQUIREMENTS FOR ICF LINTELS

Type of ICF Wall	Applicable Figures	Applicable Tables
Flat	R611.7(3)	R611.7(2) and R611.7(3) No stirrups: R611.7(8) Large spans: R611.7(9) Non-loadbearing: R611.7(10)
Waffle-Grid	R611.7(4) and R611.7(5)	R611.7(4) and R611.7(5) No stirrups: R611.7(8) Large spans: R611.7(9) Non-loadbearing: R611.7(10)
Screen-Grid	R611.7(6) and R611.7(7)	R611.7(6) and R611.7(7) Large spans: R611.7(9) Non-loadbearing: R611.7(10)

R611.7.3.2 Stirrups. Where required, No. 3 stirrups shall be installed in flat, waffle-grid and screen-grid wall lintels in accordance with the following:

1. For flat walls the stirrups shall be spaced at a maximum spacing of $d/2$ where d equals the depth of the lintel (D) minus the bottom cover of concrete as shown in Figure R611.7(3). Stirrups shall not be required in the middle portion of the span (A) per Figure R611.7(2), for flat walls for a length not to exceed the values shown in parenthesis in Tables R611.7(2) and R611.7(3) or for spans in accordance with Table R611.7(8).

2. For waffle-grid walls a minimum of two No. 3 stirrups shall be placed in each vertical core of waffle-grid lintels. Stirrups shall not be required in the middle portion of the span (A) per Figure R611.7(2), for waffle-grid walls for a length not to exceed the values shown in parenthesis in Tables R611.7(4) and R611.7(5) or for spans in accordance with Table R611.7(8).

3. For screen-grid walls one No. 3 stirrup shall be placed in each vertical core of screen-grid lintels.

Exception: Stirrups are not required in Screen-grid lintels meeting the following requirements:

1. Lintel Depth (D) = 12 inches (305 mm) – spans less than or equal 3 feet – 7 inches
2. Lintel Depth (D) = 24 inches (610 mm) – spans less than or equal 4 feet – 4 inches

❖ Stirrups are required for lintels in load-bearing walls for transverse reinforcement and shear reinforcement consistent with the requirements of ACI 318. Each type of ICF form has a different cross-sectional dimension that produces a different configuration of concrete member. Thus, the text and figures referenced provide varying sizes and locations of stirrups. See the commentary for Section R611.7.3.1.

R611.7.3.3 Horizontal reinforcement. One No. 4 horizontal bar shall be provided in the top of the lintel. Horizontal reinforcement placed within 12 inches (305 mm) of the top of the wall in accordance with Section R611.7.1.3 shall be permitted to serve as the top or bottom reinforcement in the lintel provided the reinforcement meets the location requirements in Figure R611.7(2), R611.7(3), R611.7(4), R611.7(5), R611.7(6), or R611.7(7), and the size requirements in Tables R611.7(2), R611.7(3), R611.7(4), R611.7(5), R611.7(6), R611.7(7), or R611.7(8).

❖ Horizontal reinforcement for lintels is necessary for the design of the lintel to transfer loads from above across a door or window opening. The text of Section R611.7.3.3 along with the tables and figures give prescriptive requirements based on the size of the building and the amount of design snow load. In other words, the size of the tributary load being supported by the exterior wall impacts the needed load carrying capacity of the lintels that carry this load over an opening. See the commentary for Section R611.7.3.1.

R611.7.3.4 Load-bearing walls. Lintels in flat ICF load-bearing walls shall comply with Table R611.7(2), Table RR611.7(3) or Table R611.7(8). Lintels in waffle-grid ICF load-bearing walls shall comply with Table R611.7(4), Table R611.7(5) or Table R611.7(8). Lintels in screen-grid ICF load-bearing walls shall comply with Table R611.7(6) or Table R611.7(7).

Where spans larger than those permitted in Table R611.7(2), Table R611.7(3), Table R611.7(4), Table R611.7(5), R611.7(6), R611.7(7) or R611.7(8) are required, the lintels shall comply with Table R611.7 (9).

❖ This section references several allowable span tables for lintels based on the type of ICF wall construction, lintel depth and thickness, reinforcing, number of stories supported, and snow load. Table R611.7(6) accommodates larger openings such as would be required for garage doors.

R611.7.3.5 Nonload-bearing walls. Lintels in nonload-bearing flat, waffle-grid and screen-grid ICF walls shall comply with Table R611.7 (10). Stirrups are not required.

❖ For non load-bearing walls, a single table provides allowable clear spans (opening length) based on lintel depth and support conditions.

R611.7.4 Minimum length of wall without openings. The wind velocity pressures of Table R611.7.4 shall be used to determine the minimum amount of solid wall length in accordance with Tables R611.7(9A) through R611.7(10B) and Figure R611.7.4. Table R611.7(11) shall be used to determine the minimum amount of solid wall length for townhouses in Seismic Design Category C, and all buildings in Seismic Design Categories D_1, and D_2 for all types of ICF walls. The greater amount of solid wall length required by wind loading or seismic loading shall apply. The minimum percentage of solid wall length shall include only those solid wall segments that are a minimum of 24 inches (610 mm) in length. The maximum distance between wall segments included in determining solid wall length shall not exceed 18 feet (5486 mm). A minimum length of 24 inches (610 mm) of solid wall segment, extending the full height of each wall story, shall occur at all interior and exterior corners of exterior walls.

For Seismic Design Categories D_1 and D_2, the amount of solid wall length shall include only those solid wall segments that are a minimum of 48 inches (1220 mm) in length. The minimum nominal wall thickness shall be 5.5 inches (140 mm) for all wall types.

❖ To provide lateral force resistance, the exterior walls must have the minimum lengths of solid wall specified in Tables R611.7(9A) through R611.7(10B) based on wind speed. Attached dwellings (townhouses) classified as Seismic Design Category C and all buildings in Seismic Design Categories D_1 and D_2 must also comply with the wall length required by Table R611.7(11). Figure 611.7.4 illustrates the requirements. The lengths in each of these tables are specified as a percentage of the total length of wall under consideration.

TABLE R611.7.4
WIND VELOCITY PRESSURE FOR DETERMINATION OF MINIMUM SOLID WALL LENGTH[a]

WIND SPEED (mph)[d]	VELOCITY PRESSURE (psf) Exposure[b]		
	B	C	D
85	14	19	23
90	16	21	25
100	19	26	31
110	23	32	37
120	27	38	44
130	32	44	52
140	37	51	60
150	43	59	69[c]

For SI: 1 psf = 0.0479 kN/m^2; 1 mph = 1.6093 km/hr

a. Table values are based on ASCE 7-98 Figure 6-4 using a mean roof height of 35 ft (10.7 m).
b. Exposure Categories shall be determined in accordance with Section R301.2.1.4.
c. Design is required in accordance with ACI 318 and approved manufacturer guidelines.
d. Interpolation is permitted between wind speeds.

❖ See the commentary for Section R611.7.4.

R611.8 ICF wall-to-floor connections.

R611.8.1 Top bearing. Floors bearing on the top of ICF foundation walls in accordance with Figure R611.8(1) shall have the wood sill plate anchored to the ICF wall with minimum $^1/_2$-inch (12.7 mm) diameter bolts embedded a minimum of 7 inches (178 mm) and placed at a maximum spacing of 6 feet (1829 mm) on center and not more than 12 inches (305 mm) from corners. Anchor bolts for waffle-grid and screen-grid walls shall be located in the cores. In conditions where wind speeds are in excess of 90 miles per hour (144 km/h), the $^1/_2$-inch (12.7 mm) diameter anchor bolts shall be placed at a maximum spacing of 4 feet (1219 mm) on center. Bolts shall extend a minimum of 7 inches (178 mm) into concrete. Sill plates shall be protected against decay where required by Section R319. Cold-formed steel framing systems shall be anchored to the concrete in accordance with Section R505.3.1 or Section R603.3.1.

❖ This section specifies requirements for bearings and connection of floors on top of ICF walls and illustrates details for a variety of ledger conditions.

TABLE R611.8(1)
FLOOR LEDGER-ICF WALL CONNECTION (SIDE-BEARING CONNECTION) REQUIREMENTS[a,b,c]

MAXIMUM FLOOR CLEAR SPAN[d] (feet)	MAXIMUM ANCHOR BOLT SPACING[e] (inches)			
	Staggered $^1/_2$-inch-diameter anchor bolts	Staggered $^5/_8$-inch-diameter anchor bolts	Two $^1/_2$-inch-diameter anchor bolts[f]	Two $^5/_8$-inch-diameter anchor bolts[f]
8	18	20	36	40
10	16	18	32	36
12	14	18	28	36
14	12	16	24	32
16	10	14	20	28
18	9	13	18	26
20	8	11	16	22
22	7	10	14	20
24	7	9	14	18
26	6	9	12	18
28	6	8	12	16
30	5	8	10	16
32	5	7	10	14

For SI: 1 inch = 25.4 mm, 1 foot = 304.8 mm.

a. Minimum ledger board nominal depth shall be 8 inches. The thickness of the ledger board shall be a minimum of 2 inches. Thickness of ledger board is in nominal lumber dimensions. Ledger board shall be minimum No. 2 Grade.
b. Minimum edge distance shall be 2 inches for $^1/_2$-inch-diameter anchor bolts and 2.5 inches for $^5/_8$-inch-diameter anchor bolts.
c. Interpolation is permitted between floor spans.
d. Floor span corresponds to the clear span of the floor structure (i.e., joists or trusses) spanning between load-bearing walls or beams.
e. Anchor bolts shall extend through the ledger to the center of the flat ICF wall thickness or the center of the horizontal or vertical core thickness of the waffle-grid or screen-grid ICF wall system.
f. Minimum vertical distance between bolts shall be 1.5 inches for $^1/_2$-inch-diameter anchor bolts and 2 inches for $^5/_8$-inch-diameter anchor bolts.

❖ See the commentary for Section R611.8.1.

WALL CONSTRUCTION

FIGURE R611.7.4

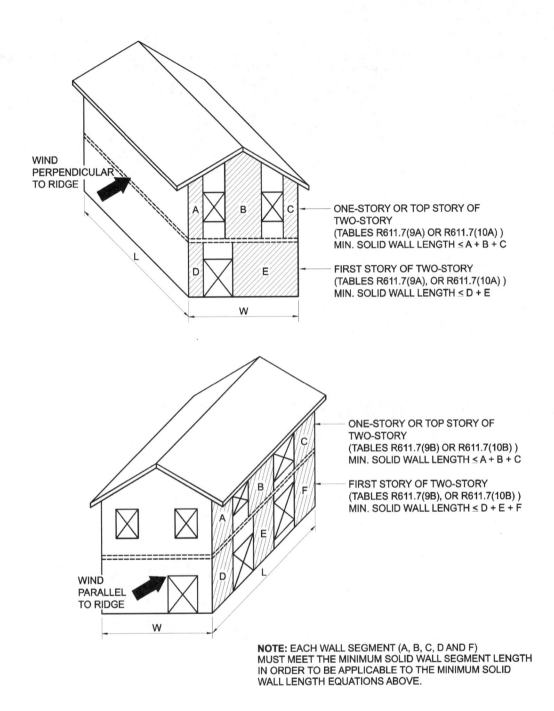

FIGURE R611.7.4
MINIMUM SOLID WALL LENGTH

❖ See the commentary for Section R611.7.4.

TABLE R611.8(2)
DESIGN VALUES (PLF) FOR FLOOR JOIST-TO-WALL ANCHORS REQUIRED FOR TOWNHOUSES IN SEISMIC DESIGN CATEGORY C AND ALL BUILDINGS IN SEISMIC DESIGN CATEGORY D[a,b]

WALL TYPE	SEISMIC DESIGN CATEGORY		
	C	D_1	D_2
Flat 3.5	193	NP	NP
Flat 5.5	303	502	708
Flat 7.5	413	685	965
Flat 9.5	523	867	1,223
Waffle 6	246	409	577
Waffle 8	334	555	782
Screen 6	233	387	546

For SI: 1plf = 14.59 N/m.

NP = Not Permitted

a. Table values are based on IBC Equation 16-64 using a tributary wall height of 11 feet (3353 mm). Table values shall be permitted to be reduced for tributary wall heights less than 11 feet (3353 mm) by multiplying the table values by X/11, where X is the tributary wall height.

b. Values may be reduced by 30 percent when used for ASD.

❖ See the commentary for Section R611.8.1.

FIGURE R611.8(1)
SECTION CUT THROUGH FLAT WALL OR VERTICAL CORE OF WAFFLE- OR SCREEN-GRID WALL

❖ See the commentary for Section R611.8.1.

WALL CONSTRUCTION

For SI: 1 inch = 25.4 mm.

Note: Section cut through flat wall or vertical core of a waffle- or screen-grid wall.

**FIGURE R611.8(2)
FLOOR LEDGER—ICF WALL CONNECTION (SIDE-BEARING CONNECTION)**

❖ See the commentary for Section R611.8.1.

For SI: 1 inch = 25.4 mm.

Note: Section cut through flat wall or vertical core of a waffle- or screen-grid wall.

**FIGURE R611.8(3)
FLOOR LEDGER—ICF WALL CONNECTION (LEDGE-BEARING CONNECTION)**

❖ See the commentary for Section R611.8.2.

FIGURE R611.8(4) - FIGURE R611.8(5)

WALL CONSTRUCTION

For SI: 1 inch = 25.4 mm.

Note: Section cut through flat wall.

**FIGURE R611.8(4)
WOOD FLOOR LEDGER—ICF WALL SYSTEM CONNECTION
(THROUGH-BOLT SIDE-BEARING CONNECTION)**

❖ See the commentary for Section R611.8.2.

For SI: 1 inch = 25.4 mm.

**FIGURE R611.8(5)
FLOOR LEDGER ICF WALL CONNECTION
(THROUGH-BOLT-LEDGE-BEARING CONNECTION)**

❖ See the commentary for Section R611.8.2.

WALL CONSTRUCTION   FIGURE R611.8(6) - FIGURE R611.8(7)

FIGURE R611.8(6)
ANCHORAGE REQUIREMENTS FOR TOP BEARING WALLS FOR TOWNHOUSES IN SEISMIC DESIGN CATEGORY C AND ALL BUILDINGS IN SEISMIC DESIGN CATEGORY D FOR FLOOR FRAMING PARALLEL TO WALL

❖ See the commentary for Section R611.8.1.1.

FIGURE R611.8(7)
ANCHORAGE REQUIREMENTS FOR LEDGER BEARING WALLS FOR TOWNHOUSES IN SEISMIC DESIGN CATEGORY C AND ALL BUILDINGS IN SEISMIC DESIGN CATEGORY D FOR FLOOR FRAMING PARALLEL TO WALL

❖ See the commentary for Section R611.8.2.1.

R611.8.1.1 Top bearing requirements for Seismic Design Categories C, D_1, and D_2. For townhouses in Seismic Design Category C, wood sill plates attached to ICF walls shall be anchored with Grade A 307, $^3/_8$-inch (9.5 mm) diameter headed anchor bolts embedded a minimum of 7 inches (178 mm) and placed at a maximum spacing of 36 inches (914 mm) on center. For all buildings in Seismic Design Category D_1, wood sill plates attached to ICF walls shall be anchored with ASTM A307, Grade A, $^3/_8$-inch (9.5 mm) diameter headed anchor bolts embedded a minimum of 7 inches (178 mm) and placed at a maximum spacing of 24 inches (610 mm) on center. For all buildings in Seismic Design Category D_2, wood sill plates attached to ICF walls shall be anchored with Grade A 307, $^3/_8$-inch (9.5 mm) diameter headed anchor bolts embedded a minimum of 7 inches (178 mm) and placed at a maximum spacing of 16 inches (406 mm) on center. Larger diameter bolts than specified herein shall not be used.

For townhouses in Seismic Design Category C, each floor joist perpendicular to an ICF wall shall be attached to the sill plate with an 18-gauge angle bracket using 3 - 8d common nails per leg in accordance with Figure R611.8(1). For all buildings in Seismic Design Category D_1, each floor joist perpendicular to an ICF wall shall be attached to the sill plate with an 18-gauge angle bracket using 4 - 8d common nails per leg in accordance with Figure R611.8(1). For all buildings in Seismic Design Category D_2, each floor joist perpendicular to an ICF wall shall be attached to the sill plate with an 18-gauge angle bracket using 6 - 8d common nails per leg in accordance with Figure R611.8(1).

For ICF walls parallel to floor framing in townhouses in Seismic Design Category C, full depth blocking shall be placed at 24 inches (610 mm) on center and shall be attached to the sill plate with an 18-gauge angle bracket using 5 - 8d common nails per leg in accordance with Figure R611.8(6). For ICF walls parallel to floor framing for all buildings in Seismic Design Category D_1, full depth blocking shall be placed at 24 inches (610 mm) on center and shall be attached to the sill plate with an 18-gauge angle bracket using 6 - 8d common nails per leg in accordance with Figure R611.8(6). For ICF walls parallel to floor framing for all buildings in Seismic Design Category D_2, full depth blocking shall be placed at 24 inches (610 mm) on center and shall be attached to the sill plate with an 18-gauge angle bracket using 9 - 8d common nails per leg in accordance with Fiqure R611.8(6).

❖ This section provides additional requirements for bearings and connections of floors on top of an ICF wall in a building used for a townhouse in Seismic Design Category C and in all buildings in Seismic Design Categories D_1 and D_2. As they do with all other structural provisions of this code, these buildings require a more rigorous set of detail requirements in order to perform acceptably in the higher seismic conditions.

R611.8.2 Ledger bearing. Wood ledger boards supporting bearing ends of joists or trusses shall be anchored to flat ICF walls with minimum thickness of 5.5 inches (140 mm) and to waffle- or screen-grid ICF walls with minimum nominal thickness of 6 inches (152 mm) in accordance with Figure R611.8(2), R611.8(3), R611.8(4) or R611.8(5) and Table R611.8(1). Wood ledger boards supporting bearing ends of joists or trusses shall be anchored to flat ICF walls with minimum thickness of 3.5 inches (140 mm) in accordance with Figure R611.8(4) or R611.8(5) and Table R611.8(1). The ledger shall be a minimum 2 by 8, No. 2 Southern Pine or No. 2 Douglas Fir. Ledgers anchored to nonload-bearing walls to support floor or roof sheathing shall be attached with $^1/_2$ inch (12.7 mm) diameter or headed anchor bolts spaced a maximum of 6 feet (1829 mm) on center. Anchor bolts shall be embedded a minimum of 4 inches (102 mm) into the concrete.

❖ This section specifies requirements for connections and bearing of wood ledger boards supporting joists or trusses on top of ICF walls and illustrates details for a variety of ledger conditions.

R611.8.2.1 Ledger bearing requirements for Seismic Design Categories C, D_1, and D_2. Additional anchorage mechanisms connecting the wall to floor system shall be installed at a maximum spacing of 6 feet (1829 mm) on center for townhouses in Seismic Design Category C and 4 feet (1220 mm) on center for all buildings Seismic Design Categories D_1 and D_2. The additional anchorage mechanisms shall be attached to the ICF wall reinforcement and joist, rafters, or blocking in accordance with Figures R611.8(1) through R611.8(7). The additional anchorage shall be installed through an oversized hole in the ledger board that is $^1/_2$ inch (12.7 mm) larger than the anchorage mechanism diameter to prevent combined tension and shear in the mechanism. The blocking shall be attached to floor or roof sheathing in accordance with edge fastener spacing. Such additional anchorage shall not be accomplished by the use of toe nails or nails subject to withdrawal nor shall such anchorage mechanisms induce tension stresses perpendicular to grain in ledgers or nailers. The capacity of such anchors shall result in connections capable of resisting the design values listed in Table R611.8(2). The diaphragm sheathing fasteners applied directly to a ledger shall not be considered effective in providing the additional anchorage required by this section.

Where the additional anchorage mechanisms consist of threaded rods with hex nuts or headed bolts complying with ASTM A307, Grade A or ASTM F1554, Grade 36, the design tensile strengths shown in Table R611.9 shall be equal to or greater than the product of the design values listed in Table R611.8(2) and the spacing of the bolts in feet. Anchor bolts shall be embedded as indicated in Table R611.9. Bolts with hooks shall not be used.

❖ This section provides additional requirements for bearings and connections of wood ledger boards supporting joists or trusses on top of ICF walls in a building used for a townhouse in Seismic Design Category C and in all buildings in Seismic Design Categories D_1 and D_2. As they do with all other structural provisions of this code, these buildings require a more rigorous set of detail requirements in order to perform acceptably in the higher seismic conditions.

R611.8.3 Floor and roof diaphragm construction. Floor and roof diaphragms shall be constructed of structural wood sheathing panels, attached to wood framing in accordance with Table R602.3(1) or Table R602.3(2) or to cold- formed steel floor framing in accordance with Table R505.3.1(2) or to cold-formed steel roof framing in accordance with Table R804.3.

❖ The requirements of this section are similar to those for masonry wall construction in Section R606.11.1.1. The addition of blocking for panel edges perpendicular to the framing provides added diaphragm capacity.

R611.8.3.1 Floor and roof diaphragm construction requirements in Seismic Design Categories D_1 and D_2. The requirements of this section shall apply in addition to those required by Section R611.8.3. Edge spacing of fasteners in floor and roof sheathing shall be 4 inches (102 mm) on center for Seismic Design Category D_1 and 3 inches (76 mm) on center for Seismic Design Category D_2. In Seismic Design Categories D_1 and D_2, all sheathing edges shall be attached to framing or blocking. Minimum sheathing fastener size shall be 0.113 inch (2.8 mm) diameter with a minimum penetration of $1^3/_8$ inches (35 mm) into framing members supporting the sheathing. Minimum wood structural panel thickness shall be $^7/_{16}$ inch (11 mm) for roof sheathing and $^{23}/_{32}$ inch (18 mm) for floor sheathing. Vertical offsets in floor framing shall not be permitted.

❖ This section provides additional requirements for design and detailing of floor and roof diaphragm construction in a building used for a townhouse in Seismic Design Category C and in all buildings in Seismic Design Categories D_1 and D_2. As they do with all other structural provisions of this code, these buildings require more rigorous requirements in order to perform acceptably in the higher seismic conditions.

R611.9 ICF wall to top sill plate (roof) connections. Wood sill plates attaching roof framing to ICF walls shall be anchored with minimum $^1/_2$ inch (12.7 mm) diameter anchor bolt embedded a minimum of 7 inches (178 mm) and placed at 6 feet (1829 mm) on center in accordance with Figure R611.9. Anchor bolts shall be located in the cores of waffle-grid and screen-grid ICF walls. Roof assemblies subject to wind uplift pressure of 20 pounds per square foot (1.44 kN/m^2) or greater as established in Table R301.2(2) shall have rafter or truss ties provided in accordance with Table R802.11.

❖ These provisions address the uplift forces on roofs caused by wind. See the commentary for Section R802.11.

R611.9.1 ICF wall to top sill plate (roof) connection for Seismic Design Categories C, D_1, and D_2. The requirements of this section shall apply in addition to those required by Section R611.9. The top of an ICF wall at a gable shall be attached to an attic floor in accordance with Section R611.8.1.1. For townhouses in Seismic Design Category C attic floor diaphragms shall be constructed of structural wood sheathing panels, attached to wood framing in accordance with Table R602.3(1) or Table R602.3(2). Edge spacing of fasteners in attic floor sheathing shall be 4 inches (102 mm) on center for Seismic Design Category D_1 and 3 inches (76 mm) on center for Seismic Design Category D_2. In Seismic Design Categories D_1, and D_2, all sheathing edges shall be attached to framing or blocking. Minimum sheathing fastener size shall be 0.113 inch (2.8 mm) diameter with a minimum penetration of $1^3/_8$ inches (35 mm) into framing members supporting the sheathing. Minimum wood structural panel thickness shall be $^7/_{16}$ inch (11 mm) for the attic floor sheathing. Where hipped roof construction is used, the use of a structural attic floor is not required.

For townhouses in Seismic Design Category C, wood sill plates attached to ICF walls shall be anchored with Grade A 307, $^3/_8$-inch (9.5 mm) diameter anchor bolts embedded a minimum of 7 inches (178 mm) and placed at a maximum spacing of 36 inches (914 mm) on center. For all buildings in Seismic Design Category D_1, wood sill plates attached to ICF walls shall be anchored with Grade A 307, $^3/_8$-inch (9.5 mm) diameter anchor bolts embedded a minimum of 7 inches (178 mm) and placed at a maximum spacing of 24 inches (610 mm) on center.

Note: Section cut through flat wall or vertical cove of waffle- or screen-grid wall.

**FIGURE R611.9
ROOF SILL PLATE—ICF WALL CONNECTION**

❖ See the commentary for Sections R611.9 and R802.11.

For all buildings in Seismic Design Category D_2 wood sill plates attached to ICF walls shall be anchored with Grade A 307, $^3/_8$-inch (9.5 mm) diameter anchor bolts embedded a minimum of 7 inches (178 mm) and placed at a maximum spacing of 16 inches (406 mm) on center.

For townhouses in Seismic Design Category C, each floor joist shall be attached to the sill plate with an 18-gauge angle bracket using 3 - 8d common nails per leg in accordance with Figure R611.8(1). For all buildings in Seismic Design Category D_1, each floor joist shall be attached to the sill plate with an 18-gauge angle bracket using 4 - 8d common nails per leg in accordance with Figure R611.8(1). For all buildings in Seismic Design Category D_2, each floor joist shall be attached to the sill plate with an 18-gauge angle bracket using 6 - 8d common nails per leg in accordance with Figure R611.8(1).

Where hipped roof construction is used without an attic floor, the following shall apply. For townhouses in Seismic Design Category C, each rafter shall be attached to the sill plate with an 18-gauge angle bracket using 3 - 8d common nails per leg in accordance with Figure R611.9. For all buildings in Seismic Design Category D_1, each rafter shall be attached to the sill plate with an 18-gauge angle bracket using 4 - 8d common nails per leg in accordance with Figure R611.9. For all buildings in Seismic Design Category D_2, each rafter shall be attached to the sill plate with an 18-gauge angle bracket using 6 - 8d common nails per leg in accordance with Figure R611.9.

❖ This section provides additional requirements for bearings and connections of roofs on top of an ICF wall in a building used for a townhouse in Seismic Design Category C and in all buildings in Seismic Design Categories D_1 and D_2. As they do with all other structural provisions of this code, these buildings require a more rigorous set of detail requirements in order to perform acceptably in the higher seismic conditions.

SECTION R612
CONVENTIONALLY FORMED CONCRETE WALL CONSTRUCTION

R612.1 General. Conventionally formed concrete walls with flat surfaces shall be designed and constructed in accordance with the provisions of Section R611 for Flat ICF walls or in accordance with the provisions of ACI 318.

❖ Conventionally formed concrete walls may be constructed under the provisions for ICF in Section R611 or in accordance with ACI 318.

SECTION R613
EXTERIOR WINDOWS AND GLASS DOORS

R613.1 General. This section prescribes performance and construction requirements for exterior window systems installed in wall systems. Waterproofing, sealing and flashing systems are not included in the scope of this section.

❖ Glass doors and windows are components of the exterior wall. Accordingly, this section specifies performance criteria for exterior windows and glass doors as well as their supporting elements in order to protect against glass breakage due to high wind pressure.

R613.2 Performance. Exterior windows and doors shall be designed to resist the design wind loads specified in Table R301.2(2) adjusted for height and exposure per Table R301.2(3).

❖ Exterior glazing is subject to wind loading as specified in Section 301.2.1. The code requires use of the component and cladding pressures of Table R301.2(2), illustrated in Figure R301.2(8).

TABLE R611.9
DESIGN TENSILE STRENGTH OF HEADED BOLTS CAST IN CONCRETE[a]

DIAMETER OF BOLT inches	MINIMUM EMBEDMENT DEPTH (in.)	DESIGN TENSILE STRENGTH[b] (pounds)
$^1/_4$	2	1040
$^3/_8$ with washer[c]	$2^3/_4$[d]	2540
$^1/_2$ with washer[c]	4[d]	4630

a. Applicable to concrete of all strengths. See Notes (c) and (d).
b. Values are based on ASTM F1554, Grade 36 bolts. Where ASTM A307, Grade A headed bolts are used, the strength shall be increased by 1.034.
c. A hardened washer shall be installed at the nut embedded in the concrete or head of the bolt to increase the bearing area. The washer is not required where the concrete strength is 4000 psi or greater.
d. Embedment depth shall be permitted to be reduced $^1/_4$-inch where 4000 psi concrete is used.

❖ See the commentary for Section R611.9.1.

WALL CONSTRUCTION

R613.3 Testing and labeling. Exterior windows and glass doors shall be tested by an approved independent laboratory, and bear a label identifying manufacturer, performance characteristics and approved inspection agency to indicate compliance with the requirements of one of the following specifications:

AAMA/NWWDA 101/I.S.2
AAMA/WDMA 101/I.S.2/NAFS

Exceptions:

1. Decorative glazed openings.
2. Exterior window and door assemblies not included within the scope of AAMA/NWWDA 101/I.S.2 or AAMA/WDMA 101/I.S.2/NAFS shall be tested in accordance with ASTM E 330. Assemblies covered by this exception containing glass shall comply with Section R308.5.
3. Structural wind load design pressures for window units smaller than the size tested in accordance with this section shall be permitted to be higher than the design value of the tested unit provided such higher pressures are determined by accepted engineering analysis. All components of the small unit shall be the same as the tested unit. Where such calculated design pressures are used they shall be validated by an additional test of the window unit having the highest allowable design pressure.

❖ AAMA/NWWDA 101/I.S.2, *Voluntary Specifications for Aluminum, Vinyl (PVC) and Wood Windows and Glass Doors* is the standard referenced by the code for evaluating exterior glazing components. Exception 2 is an exception to the original scope of section R613.3, giving requirements for all exterior window and door assemblies regardless of whether they are glass or non-glass. All of these doors and windows must be testing in accordance with ASTM E 330 because they are outside the scope of the referenced standard AAMA/NWWDA 101/I.S.2. ASTM E 330 will require testing to 1.5 times the design pressure.

R613.4 Windborne debris protection. Protection of exterior windows and glass doors in buildings located in hurricane-prone regions from windborne debris shall be in accordance with Section R301.2.1.2.

❖ See the commentary for Section R301.2.1.2.

R613.5 Anchorage methods. The methods cited in this section apply only to anchorage of window and glass door assemblies to the main force-resisting system.

❖ These requirements provide a mechanism for transferring the wind forces from the door or window to a lateral force-resisting element. Anchorage should be according to the manufacturer's recommendations. Variations should provide equal or greater anchorage performance.

R613.5.1 Anchoring requirements. Window and glass door assemblies shall be anchored in accordance with the published manufacturer's recommendations to achieve the design pressure specified. Substitute anchoring systems used for substrates not specified by the fenestration manufacturer shall provide equal or greater anchoring performance as demonstrated by accepted engineering practice.

❖ See the commentary for Section R613.5.

R613.5.2 Anchorage details. Products shall be anchored in accordance with the minimum requirements illustrated in Figures R613.5(1), R613.5(2), R613.5(3), R613.5(4), R613.5(5), R613.5(6), R613.5(7) and R613.5(8).

❖ The code incorporates prescriptive anchorage methods that provide for transfer of wind loads to the supporting construction. See the commentary for Section R613.5.

FIGURE R613.5(1)
THROUGH THE FRAME

APPLY FRAME CLIP TO WINDOW OR DOOR IN ACCORDANCE WITH PUBLISHED MANUFACTURER'S RECOMMENDATIONS.

FIGURE R613.5(2)
FRAME CLIP

FIGURE R613.5(3)
THROUGH THE FRAME

THROUGH THE FRAME ANCHORING METHOD. ANCHORS SHALL BE PROVIDED TO TRANSFER LOAD FROM THE WINDOW OR DOOR FRAME INTO THE ROUGH OPENING SUBSTRATE.

FIGURE R613.5(4)
FRAME CLIP

APPLY FRAME CLIP TO WINDOW OR DOOR FRAME IN ACCORDANCE WITH PUBLISHED MANUFACTURER'S RECOMMENDATIONS. ANCHORS SHALL BE PROVIDED TO TRANSFER LOAD FROM THE FRAME CLIP INTO THE ROUGH OPENING SUBSTRATE

FIGURE R613.5(5)
THROUGH THE FLANGE

APPLY ANCHORS THROUGH FLANGE IN ACCORDANCE WITH PUBLISHED MANUFACTURER'S RECOMMENDATIONS.

FIGURE R613.5(6)
THROUGH THE FRAME

FIGURE R613.5(7)
FRAME CLIP

FIGURE R613.5(8)
THROUGH THE FLANGE

❖ See the commentary for Sections R613.5 and R613.5.2.1.

R613.5.2.1 Masonry, concrete or other structural substrate. Where the wood shim or buck thickness is less than $1^1/_2$ inches (38 mm), window and glass door assemblies shall be anchored through the jamb, or by jamb clip and anchors shall be embedded directly into the masonry, concrete or other substantial substrate material. Anchors shall adequately transfer load from the window or door frame into the rough opening substrate [see Figures R613.5(1) and R613.5(2).]

Where the wood shim or buck thickness is 1.5 inches (38 mm) or greater, the buck is securely fastened to the masonry, concrete or other substantial substrate and the buck extends beyond the interior face of the window or door frame, window and glass door assemblies shall be anchored through the jamb, or by jamb clip, or through the flange to the secured wood buck. Anchors shall be embedded into the secured wood buck to adequately transfer load from the window or door frame assembly [Figures R613.5(3), R613.5(4) and R613.5(5)].

❖ When a wood member at least 1.5 inches (38 mm) thick is used as a shim, it must be anchored to the substrate. Otherwise window and door assemblies must be anchored directly to the substrate. See Figures R613.5(1) through R613.5(5).

R613.5.2.2 Wood or other approved framing material. Where the framing material is wood or other approved framing material, window and glass door assemblies shall be anchored through the frame, or by frame clip, or through the flange. Anchors shall be embedded into the frame construction to adequately transfer load [Figures R613.5(6), R613.5(7) and R613.5(8)].

❖ Door and window assemblies must be anchored directly to wood framing. See Figures R613.5(6) through R613.5(8).

R613.6 Mullions occurring between individual window and glass door assemblies.

R613.6.1 Mullions. Mullions shall be tested by an approved testing laboratory or be engineered in accordance with accepted engineering practice. Both methods shall use performance criteria cited in Sections R613.6.2, R613.6.3 and R613.6.4.

❖ Mullions provide support for adjacent window and/or door assemblies. Generally, they must be tested and perform as required for windows and glass doors. Deflection limits and a factor of safety maintain serviceability and provide minimum strength.

WALL CONSTRUCTION

R613.6.2 Load transfer. Mullions shall be designed to transfer the design pressure loads applied by the window and door assemblies to the rough opening substrate.

❖ See the commentary for Section R613.6.1.

R613.6.3 Deflection. Mullions shall be capable of resisting the design pressure loads applied by the window and door assemblies to be supported without deflecting more than $L/175$, where L = the span of mullion in inches.

❖ See the commentary for Section R613.6.1.

R613.6.4 Structural safety factor. Mullions shall be capable of resisting a load of 1.5 times the design pressure loads applied by the window and door assemblies to be supported without exceeding the appropriate material stress levels. If tested by an approved laboratory, the 1.5 times the design pressure load shall be sustained for 10 seconds, and the permanent deformation shall not exceed 0.4 percent of the mullion span after the 1.5 times design pressure load is removed.

❖ See the commentary for Section R613.6.1.

Bibliography

The following resource materials are referenced in this chapter or are relevant to the subject matter addressed in this chapter.

ACI 318-02, *Building Code Requirements for Structural Concrete*. Farmington Hills, MI: American Concrete Institute, 2002.

ACI 530-02, *Building Code Requirements for Masonry Structures*. Farmington Hills, MI: American Concrete Institute, 2002.

AFPA NDS-01, *National Design Specification (NDS) for Wood Construction—with 2001 Supplement*. Washington DC: American Forest and Paper Association, 2001.

AHA A 135.4-95, *Basic Hardware*. Palatine, IL: American Hardboard Association, 1995.

AHA A 194.1-85, *Cellulosic Fiber Board*. Palatine, IL: American Hardboard Association, 1985.

ANSI A208.1-99, *Particleboard*. New York: American National Standards Institute, 1999.

ANSI/AAMA/ADMA 101/I.S.2-97, *Voluntary Specifications for Aluminum Vinyl PVC and Wood Windows and Glass Doors*. Schaumburg, IL: American Architectural Manufacturers Association, 1997.

ASTM A 36/A 36M-01, *Specification for Carbon Structural Steel*. West Conshohocken, PA: ASTM International, 2001.

ASTM A 153-01a, *Specification for Zinc Coating (Hot Dip) on Iron and Steel Hardware*. West Conshohocken, PA: ASTM International, 2001.

ASTM A 167-99, *Specification for Stainless and Heat-Resisting Chromium-nickel Steel Plate, Sheet, and Strip*. West Conshohocken, PA: ASTM International, 1999.

ASTM A 510M-00, *Specification for General Requirements for Wire Rods and Coarse Round Wire, Carbon Steel Metric*. West Conshohocken, PA: ASTM International, 2000.

ASTM A 615/A 0615M-01b, *Specification for Deformed and Plain Billet-steel Bars for Concrete Reinforcement*. West Conshohocken, PA: ASTM International, 2001.

ASTM A 616M-96a, *Specifications for Rail Steel Deformed and Plain Bars for Concrete Reinforcement*. West Conshohocken, PA: ASTM International, 1996.

ASTM A 617/A 617M-96, *Specification for Axle Steel Deformed and Plain Bars for Concrete Reinforcement*. West Conshohocken, PA: ASTM International, 1996.

ASTM A 641/A 0641M-98, *Specification for Zinc-Coated (Galvanized) Carbon Steel Wire*. West Conshohocken, PA: ASTM International, 1998.

ASTM 653/A 0653M-00, *Specification for Steel Sheet, Zinc-Coated Galvanized) Zinc-Iron Alloy-Coated (Galvannealed) by the Hot-Dip Process*. West Conshohocken, PA: ASTM International, 2000.

ASTM A 706M-00, *Specification for Low-alloy Steel Deformed and Plain Bars for Concrete Reinforcement*. West Conshohocken, PA: ASTM International, 2000.

ASTM A 792M-99, *Specification for Steel Sheet, 55% Aluminum-zinc Alloy-coated by the Hot-dip Process*. West Conshohocken, PA: ASTM International, 1999.

ASTM A 875/A 0875M-01a, *Specification for Steel Sheet, Zinc-5% Aluminum Alloy-Coated by the Hot-Dip Process*. West Conshohocken, PA: ASTM International, 2001.

ASTM A 951-00, *Specification for Masonry Joint Reinforcement*. West Conshohocken, PA: ASTM International, 2000.

ASTM B 227-98, *Specification for Hard-drawn Copper-clad Steel Wire*. West Conshohocken, PA: ASTM International, 1998.

ASTM B 633-98e01, *Specification for Electrodeposited Coatings of Zinc on Iron and Steel*. West Conshohocken, PA: ASTM International, 2001.

ASTM C 79-01, *Specification for Treated Core and Nontreated Core Gypsum Sheathing Board*. West Conshohocken, PA: ASTM International, 2001.

ASTM C 143/C 0143M-00, *Test Method for Slump or Hydraulic Cement Concrete*. West Conshohocken, PA: ASTM International, 2000.

ASTM C 208-95(2001), *Specification for Cellulosic Fiber Insulating Board*. West Conshohocken, PA: ASTM International, 2001.

ASTM C 270-01a, *Specification for Mortar for Unit Masonry*. West Conshohocken, PA: ASTM International, 2001.

ASTM C 476-01, *Specification for Grout for Masonry*. West Conshohocken, PA: ASTM International, 2001.

ASTM C 954-00, *Specification for Steel Drill Screws for the Application of Gypsum Panel Products or Metal Plaster Bases to Steel Studs From 0.033 in. (0.84 mm) to 0.112 in. (2.84 mm) in Thickness*. West Conshohocken, PA: ASTM International, 2000.

ASTM E 330-97e[1], *Test Method for Structural Performance of Exterior Windows, Curtain Walls, and Doors by Uniform Static Air Pressure Difference*. West Conshohocken, PA: ASTM International, 1997.

DOC PS 1-95, *Construction and Industrial Plywood*. Gaithersburg, MD: United States Department of Commerce, National Institute of Standards and Technology, 1995.

DOC PS 2-92, *Performance Standard for Wood-based Structural-use Panels*. Gaithersburg, MD: United States Department of Commerce, National Institute of Standards and Technology, 1992.

DOC PS 20-99, *American Softwood Lumber Standard*. Gaithersburg, MD: United States Department of Commerce, National Institute of Standards and Technology, 1999.

FEMA 303, *Recommended Provisions for Seismic Regulations for New Buildings and Other Structures*. Washington, DC: Building Seismic Safety Council, 1997.

GA 253-99, *Recommended Standard Specification for the Application of Gypsum Sheathing*. Washington, DC: Gypsum Association, 1999.

Prescriptive Method for Insulating Concrete Forms in Residential Construction. Skokie, IL: Portland Cement Association, 1998.

SAE J 78-(1998): *Steel Self-drilling Tapping Screws*. Warrendale, PA: Society of Automotive Engineers, 1998.

Chapter 7:
Wall Covering

General Comments

Interior walls coverings are used for a variety of purposes. Often designed simply as an aesthetic element to finish the interior of the building, these wall coverings may also protect the structural elements from impact or moisture damage. Exterior wall coverings, in great part, provide the weather-resistant exterior envelope that protects the building's interior from the elements. As evidenced by the common use of exterior insulation finish systems in various regions of the country, new methods and materials are constantly introduced to provide for different appearances, improved insulating quality, sound transmission control and fire resistance. The code has developed prescriptive and performance regulations to control these aspects and the types and thickness of exterior wall coverings.

This chapter contains provisions for the design and construction of interior and exterior wall coverings. In addition to identifying the various types of wall covering regulated by the code, Chapter 7 references a number of material standards related to the specific types of wall covering materials. Of primary consideration are the application methods prescribed by the code for those wall-covering materials conforming to their applicable reference standard.

Section R701 establishes the scope of the chapter and provides a general requirement dealing with weather protection for wall covering materials sensitive to adverse weather conditions. Section R702 establishes the various types of materials, materials standards and methods of application permitted for use as interior coverings, including interior plaster, gypsum board, ceramic tile, wood veneer paneling, hardboard paneling, wood shakes and wood shingles. Section R703 addresses exterior wall coverings as well as the water-resistant sheathing paper often needed beneath the exterior materials. Exterior wall coverings regulated by this section include wood, hardboard and wood structural panel siding, wood shakes and shingles, exterior plaster, stone and masonry veneer and exterior insulation finish systems.

Purpose

The purpose of this chapter is to provide the minimum requirements applicable to wall covering materials used both in exterior and interior applications. It specifies the types of wall coverings addressed by the code, including such common materials as gypsum board and ceramic tile used on a building's interior, as well as horizontal siding and wood structural panel siding installed as the exterior membrane. This section also recognizes a relatively new exterior wall covering, the exterior insulation finish system (EIFS).

SECTION R701
GENERAL

R701.1 Application. The provisions of this chapter shall control the design and construction of the interior and exterior wall covering for all buildings.

❖ Chapter 7 deals with both interior wall coverings and exterior wall coverings. A variety of interior coverings are regulated, including interior plaster, gypsum board, ceramic tile, wood veneer paneling, hardboard paneling, wood shakes and wood shingles. Wood, siding, hardboard siding, wood structural panel siding, wood shakes, wood shingles, exterior plaster, stone veneer and masonry veneer are addressed where used as exterior wall coverings.

R701.2 Installation. Products sensitive to adverse weather shall not be installed until adequate weather protection for the installation is provided. Exterior sheathing shall be dry before applying exterior cover.

❖ Since many materials are subject to deterioration from the effects of moisture or other adverse weather conditions, the code limits the installation of such materials until proper weather protection is in place. Gypsum plaster systems, gypsum board and interior grade plywood are particularly sensitive to moisture. It is also critical that exterior sheathing be dry prior to installation of the surface cover materials. Entrapment of moisture within the exterior wall assembly can result in a reduction in the effectiveness of the wall over time.

To assist in the application of the provisions of this chapter, it is important to understand the terminology that is used. Commentary Figures R701.2(1) and R701.2(2) provide pictorial descriptions of several of these terms.

SECTION R702
INTERIOR COVERING

R702.1 General. Interior coverings or wall finishes shall be installed in accordance with this chapter and Tables R702.1(1), Table R702.1(2), Table R702.1(3) and Table R702.3.5. Interior masonry veneer shall comply with the requirements of Section R703.7.1 for support and Section R703.7.4 for anchorage, except an air space is not required.

Interior finishes and materials shall conform to the flame spread and smoke-density requirements of Section R315.

❖ This section contains the installation requirements for interior wall coverings. Interior wall coverings of plaster and gypsum board must be installed in accordance with the applicable tables. The installation requirements include material orientation, backing support, spacing and size as well as method of attachment to supports. These requirements can result in wall coverings that will perform as intended from durability, appearance, and structural points of view. A reference is made to Section R315, which addresses the flame-spread and smoke-density provisions applicable to interior wall finishes.

TABLE R702.1(1)
THICKNESS OF PLASTER

PLASTER BASE	FINISHED THICKNESS OF PLASTER FROM FACE OF LATH, MASONRY, CONCRETE (inches)	
	Gypsum plaster	Portland cement mortar
Expanded metal lath	⁵⁄₈, minimum[a]	⁵⁄₈, minimum[a]
Wire lath	⁵⁄₈, minimum[a]	³⁄₄, minimum (interior)[b] ⁷⁄₈, minimum (exterior)[b]
Gypsum lath[g]	¹⁄₂, minimum	³⁄₄, minimum (interior)[b]
Masonry walls[c]	¹⁄₂, minimum	¹⁄₂, minimum
Monolithic concrete walls[c,d]	⁵⁄₈, maximum	⁷⁄₈, maximum
Monolithic concrete ceilings[c,d]	³⁄₈, maximum[e]	¹⁄₂, maximum
Gypsum veneer base[f,g]	¹⁄₁₆, minimum	³⁄₄, minimum (interior)[b]
Gypsum sheathing[g]	—	³⁄₄, minimum (interior)[b] ⁷⁄₈, minimum (exterior)[b]

For SI: 1 inch = 25.4 mm.

a. When measured from back plane of expanded metal lath, exclusive of ribs, or self-furring lath, plaster thickness shall be ³⁄₄ inch minimum.
b. When measured from face of support or backing.
c. Because masonry and concrete surfaces may vary in plane, thickness of plaster need not be uniform.
d. When applied over a liquid bonding agent, finish coat may be applied directly to concrete surface.
e. Approved acoustical plaster may be applied directly to concrete or over base coat plaster, beyond the maximum plaster thickness shown.
f. Attachment shall be in accordance with Table R702.3.5.
g. Where gypsum board is used as a base for portland cement plaster, weather-resistant sheathing paper complying with Section R703.2 shall be provided.

❖ Based on several variables, this table sets forth the required plaster thickness measured from the face of the lath, masonry or concrete to the plaster surface. Both gypsum plaster and portland cement mortar are addressed, with the option of the plaster application occurring over a variety of plaster bases. In most cases, the minimum plaster thickness is mandated; however, the maximum thickness is regulated where plaster is applied over a monolithic concrete wall or ceiling.

TABLE R702.1(2)
GYPSUM PLASTER PROPORTIONS[a]

NUMBER	COAT	PLASTER BASE OR LATH	MAXIMUM VOLUME AGGREGATE PER 100 POUNDS NEAT PLASTER[b] (cubic feet)	
			Damp loose sand[a]	Perlite or vermiculite[c]
Two-coat work	Base coat	Gypsum lath	2.5	2
	Base coat	Masonry	3	3
Three-coat work	First coat	Lath	2[d]	2
	Second coat	Lath	3[d]	2[e]
	First and second coats	Masonry	3	3

For SI: 1 inch = 25.4 mm, 1 cubic foot = 0.0283 m³, 1 pound = 0.454 kg.

a. Wood-fibered gypsum plaster may be mixed in the proportions of 100 pounds of gypsum to not more than 1 cubic foot of sand where applied on masonry or concrete.
b. When determining the amount of aggregate in set plaster, a tolerance of 10 percent shall be allowed.
c. Combinations of sand and lightweight aggregate may be used, provided the volume and weight relationship of the combined aggregate to gypsum plaster is maintained.
d. If used for both first and second coats, the volume of aggregate may be 2.5 cubic feet.
e. Where plaster is 1 inch or more in total thickness, the proportions for the second coat may be increased to 3 cubic feet.

❖ This table covers the maximum permitted amount of aggregate for gypsum plaster, measured in cubic feet per 100 pounds. Applicable for both damp loose sand, perlite and vermiculite, the limits vary based upon the type of plaster base or lath. The maximum proportions for each coat of both two-coat and three-coat plaster systems are identified.

WALL COVERING

TABLE R702.1(3)
PORTLAND CEMENT PLASTER

Coat	MAXIMUM VOLUME AGGREGATE PER VOLUME CEMENTITIOUS MATERIAL[a]				MINIMUM PERIOD MOIST COATS	MINIMUM INTERVAL BETWEEN
	Portland cement plaster[b] maximum volume aggregate per volume cement	Portland cement-lime plaster[c]				
		Maximum volume lime per volume cement	Maximum volume sand per volume cement and lime	Approximate minimum thickness[d] curing (inches)		
First	4	3/4	4	3/8[e]	48[f] Hours	48[g] Hours
Second	5	3/4	5	First and second coats	48 Hours	7 Days[h]
Finished	3[i]	—	3[i]	1/8	—	Note[h]

For SI: 1 inch = 25.4 mm, 1 pound = 0.454 kg.

a. When determining the amount of aggregate in set plaster, a tolerance of 10 percent may be allowed.
b. From 10 to 20 pounds of dry hydrated lime (or an equivalent amount of lime putty) may be added as a plasticizing agent to each sack of Type I and Type II standard portland cement in base coat plaster.
c. No additions of plasticizing agents shall be made.
d. See Table R702.1(1).
e. Measured from face of support or backing to crest of scored plaster.
f. Twenty-four-hour minimum period for moist curing of interior portland cement plaster.
g. Twenty-four hour minimum interval between coats of interior portland cement plaster.
h. Finish coat plaster may be applied to interior portland cement base coats after a 48-hour period.
i. For finish coat, plaster up to an equal part of dry hydrated lime by weight (or an equivalent volume of lime putty) may be added to Type I, Type II and Type III standard portland cement.

❖ Where portland cement plaster is used rather than gypsum plaster, this table is applicable for determining the maximum permitted volume of aggregate. In addition, the minimum period for moist curing is established for interior portland cement plaster as is the minimum required interval between coats.

Figure R701.2(1)
PLASTER WALL COVERING—APPLICABLE TERMS

Figure R701.2(2)
GYPSUM APPLICATION—APPLICABLE TERMS

R702.2 Interior plaster. Gypsum plaster or portland cement plastering materials shall conform to ASTM C 5, C 28, C 35, C 37, C 59, C 61, C 587, C 588, C 631, C 847, C 897, C 933, C 1032 and C 1047, and shall be installed or applied in conformance with ASTM C 843, C 844 and C 1063. Plaster shall not be less than three coats when applied over metal lath and not less than two coats when applied over other bases permitted by this section, except that veneer plaster may be applied in one coat not to exceed $^3/_{16}$ inch (4.76 mm) thickness, provided the total thickness is as set forth in Table R702.1(1).

❖ Multicoat plastering has been commonplace for the past century. It is generally regarded by the industry that multicoat work is necessary for control of plaster thickness and density, particularly when applied by hand. Because of the uniformity created where plaster is applied in thin, successive layers, the code requires three-coat plastering over metal lath and two-coat work over other plaster bases approved for use by the code. Reducing the requirement for plaster bases other than metal lath depends on the rigidity of the plaster base itself. More-rigid plaster bases are not as susceptible to variations in thickness and flatness of the surface. In fact, the first coat applied in three-coat work on a flexible base, such as metal lath, is used to stiffen that base to provide the rigidity necessary to attain uniform thickness and surface flatness.

Both a gypsum plaster and a portland cement plaster interior wall covering system are shown in Commentary Figure R702.2(1). The finished thicknesses for both plastering methods determined by the type of base materials are set forth in Table R702.1(1). Commentary Figure R702.2(2) illustrates the veneer plaster interior wall covering system. This section also cites the materials standards to which gypsum plaster and portland cement plaster must conform.

R702.2.1 Support. Support spacing for gypsum or metal lath on walls or ceilings shall not exceed 16 inches (406 mm) for $^3/_8$ inch thick (9.5 mm) or 24 inches (610 mm) for $^1/_2$-inch-thick (12.7 mm) plain gypsum lath. Gypsum lath shall be installed at right angles to support framing with end joints in adjacent courses staggered by at least one framing space.

❖ To provide a relatively firm base for the first layer of plaster, the code mandates a maximum span between supporting elements of the lath. Applicable to both gypsum and metal lath for either wall or ceiling installations, the maximum spacing of supports is based on the lath thickness. Gypsum lath must be installed perpendicular to the direction of the support framing, with the end joints staggered in adjacent courses.

WALL COVERING

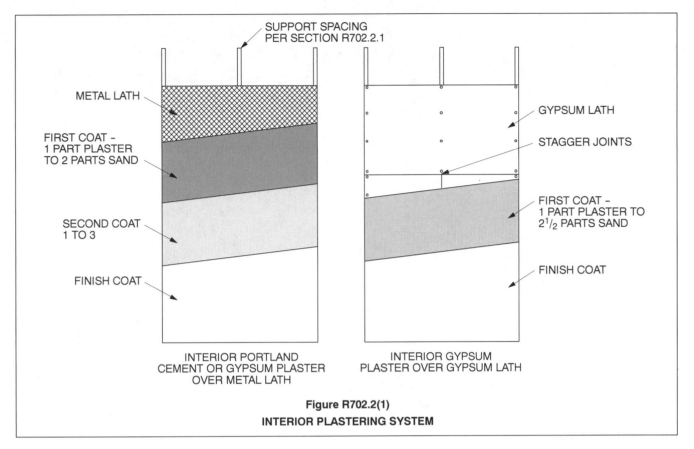

Figure R702.2(1)
INTERIOR PLASTERING SYSTEM

R702.3 Gypsum board.

❖ Gypsum wallboard is by far the most common interior covering in use today. The variety of types and sizes of the material allow its use in almost every construction application. This section sets forth the criteria for the application of gypsum board in wall and ceiling applications.

R702.3.1 Materials. All gypsum board materials and accessories shall conform to ASTM C 36, C 79, C 475, C 514, C 630, C 931, C 960, C 1002, C 1047, C 1177, C 1178, C 1278, C 1395 or C 1396 and shall be installed in accordance with the provisions of this section. Adhesives for the installation of gypsum board shall conform to ASTM C 557.

❖ This section sets forth the standards applicable to gypsum wallboard materials and accessories. Included in the list of standards is ASTM C 557, which deals with the use of adhesives to attach gypsum wallboard to wood framing members.

R702.3.2 Wood framing. Wood framing supporting gypsum board shall not be less than 2 inches (51 mm) nominal thickness in the least dimension except that wood furring strips not less than 1-inch-by-2 inch (25.4 mm by 51 mm) nominal dimension may be used over solid backing or framing spaced not more than 24 inches (610 mm) on center.

❖ For adequate structural backing, vertical support for gypsum wallboard used as an interior wall covering is required to have a least nominal dimension of 2 inches (51 mm). Dimensional lumber meeting this 2-inch (51 mm) nominal dimension would have an actual thickness of approximately $1^1/_2$ inches (38 mm).

Wood stripping used as furring over vertical supports on which the interior wall covering is to be applied is also required to meet the 2-inch (51 mm) nominal dimension in any direction unless the framing members are spaced at 24 inches (610 mm) on center or less. Where the spacing of the framing does not exceed 24 inches (610 mm) or where the furring strips are placed on solid backing, the minimum size requirements are reduced to 1 inch by 2 inches (25.4 mm by 51 mm).

Commentary Figures R702.3.2(1), R702.3.2(2) and R702.3.2(3) illustrate the dimensional constraints.

R702.3.3 Steel framing. Steel framing supporting gypsum board shall not be less than 1.25 inches (32 mm) wide in the least dimension. Light-gage nonload-bearing steel framing shall comply with ASTM C 645. Load-bearing steel framing and steel framing from 0.033 inch to 0.112 inch (0.838 mm to 2.84 mm) thick shall comply with ASTM C 955.

❖ In the same manner as wood framing, steel framing must be of a minimum cross-sectional dimension where used to support gypsum board. In the case of steel members, the least dimension must be at least 1.25 inches (32 mm).

This section also references two standards relating to steel framing members. ASTM C 645 addresses

Figure R702.2(2)
VENEER PLASTER SYSTEM

nonstructural steel framing members, while ASTM C 955 deals with the load-bearing steel studs, tracks and bracing involving gypsum panels.

R702.3.4 Insulating concrete form walls. Foam plastics for insulating concrete form walls constructed in accordance with Sections R404.4 and R611 on the interior of habitable spaces shall be covered in accordance with Section R314.1.2. Adhesives are permitted to be used in conjunction with mechanical fasteners. Adhesives used for interior and exterior finishes shall be compatible with the insulating form materials.

❖ Reinforcing the provisions of Section R318.1.2 regarding the thermal barrier requirement for foam plastic insulation, this section references the requirements in Chapters 4 and 6 regulating insulating concrete form foundation walls and insulating concrete form wall systems. Foam plastic insulation used in these wall applications must be separated from the interior of the building by a minimum $1/2$-inch (12.7 mm) gypsum board or equivalent material.

Adhesives may be used to fasten the protective membrane (typically gypsum board) over the insulating concrete form wall. In all cases, however, mechanical fasteners must supplement the adhesive in accordance with Section R318.1.2. A total reliance on the adhesive to ensure the continued protection afforded by the gypsum board during a fire is misguided. Where used in conjunction with mechanical fasteners, adhesives must also be of a type that is compatible with the insulating form materials.

R702.3.5 Application. Maximum spacing of supports and the size and spacing of fasteners used to attach gypsum board shall comply with Table R702.3.5. Gypsum sheathing shall be attached to exterior walls in accordance with Table R602.3(1). Gypsum board shall be applied at right angles or parallel to framing members. All edges and ends of gypsum board shall occur on the framing members, except those edges and ends that are perpendicular to the framing members. Interior gypsum board shall not be installed where it is directly exposed to the weather or to water.

❖ To provide adequate backing for nailing the gypsum wallboard to the support framing, all edges and ends except those that are perpendicular to framing support members must occur on framing members. Gypsum wallboard may be installed in either a perpendicular or parallel manner. This is illustrated in Commentary Figure R702.3.5(1). Already addressed in a general sense in Section R701.2, the installation of interior gypsum wallboard must not be done while the interior of the building is exposed to weather. Direct exposure of interior gypsum board to weather or water is not permitted.

For SI: 1 inch = 25.4 mm.

Figure R702.3.2(1)
INTERIOR WALL COVERING—DIRECT APPLICATION

For SI: 1 inch = 25.4 mm.

Figure R702.3.2(2)
INTERIOR WALL COVERING—FURRED APPLICATION OVER MASONRY OR CONCRETE

Figure R702.3.2(3)
INTERIOR WALL COVERING—FURRED APPLICATION OVER SOLID BACKING OR OVER STUDS

R702.3.6 Fastening. Screws for attaching gypsum board to wood framing shall be Type W or Type S in accordance with ASTM C 1002 and shall penetrate the wood not less than $^5/_8$ inch (15.9 mm). Screws for attaching gypsum board to light-gage steel framing shall be Type S in accordance with ASTM C 1002 and shall penetrate the steel not less than $^3/_8$ inch (9.5 mm). Screws for attaching gypsum board to steel framing 0.033 inch to 0.112 inch (0.838 mm to 2.84 mm) thick shall comply with ASTM C 954.

❖ Screws may be used in place of nails in the application of gypsum wallboard. This section restates the provisions of Table R702.3.5 regarding the use of screws. Gypsum wallboard screws, either Type S or Type W, are to be spaced in accordance with the table and must penetrate wood framing members a minimum of $^5/_8$ inch (15.9 mm). Where gypsum wallboard is attached to light-gage steel framing members, the screws must penetrate a minimum of $^3/_8$ inch (9.5 mm).

R702.4 Ceramic tile.

❖ Based on the code's requirement for nonabsorbent surfaces in bathtub and shower areas, ceramic tile is one of several materials used in such areas subject to water splash. Ceramic tile is regulated wherever installed, particularly when gypsum board is used as a base or backer board. This section regulates the installation of ceramic tile in residential construction.

R702.4.1 General. Ceramic tile surfaces shall be installed in accordance with ANSI A108.1, A108.4, A108.5, A108.6, A108.11, A118.1, A118.3, A136.1 and A137.1.

❖ This section references a number of ANSI standards that apply to the installation of ceramic tile. Commentary Figure R702.4.1 illustrates two possible methods of ceramic tile applications.

R702.4.2 Gypsum backer. Gypsum board utilized as the base or backer for adhesive application of ceramic tile or other nonabsorbent finish material shall conform with ASTM C630 or C1178. Water-resistant gypsum backing board shall be permitted to be used on ceilings where framing spacing does not exceed 12 inches (305 mm) on center for $^1/_2$-inch-thick (12.7 mm) or 16 inches (406 mm) for $^5/_8$ inch-thick (15.9 mm) gypsum board. Water-resistant gypsum board shall not be installed over a vapor retarder in a shower or tub compartment. All cut or exposed edges, including those at wall intersections, shall be sealed as recommended by the manufacturer.

❖ Water-resistant gypsum backing board must be used as a base for ceramic tile or other nonabsorbent finish materials where attached with an adhesive. It is required to be used in conjunction with the tile or wall panels because of its moisture-resistant qualities.

The use of water-resistant backing board is limited in a ceiling application, as $^1/_2$-inch-thick (12.7 mm) board is prohibited on ceilings where the framing members are spaced in excess of 12 inches (305 mm) on center. In addition, $^5/_8$-inch-thick water-resistant gypsum backing board is permitted only where ceiling members are spaced at 16 inches (406 mm) on center.

WALL COVERING

TABLE R702.3.5
MINIMUM THICKNESS AND APPLICATION OF GYPSUM BOARD

THICKNESS OF GYPSUM BOARD (inches)	APPLICATION	ORIENTATION OF GYPSUM BOARD TO FRAMING	MAXIMUM SPACING OF FRAMING MEMBERS (inches o.c.)	MAXIMUM SPACING OF FASTENERS (inches) Nails[a]	MAXIMUM SPACING OF FASTENERS (inches) Screws[b]	SIZE OF NAILS FOR APPLICATION TO WOOD FRAMING[c]
Application without adhesive						
$3/8$	Ceiling[d]	Perpendicular	16	7	12	13 gage, $1^{1}/_{4}''$ long, $^{19}/_{64}''$ head; 0.098 diameter, $1^{1}/_{4}''$ long, annular-ringed; or 4d cooler nail, 0.080'' diameter, $1^{3}/_{8}''$ long, $^{7}/_{32}''$ head.
$3/8$	Wall	Either direction	16	8	16	
$1/2$	Ceiling	Either direction	16	7	12	13 gage, $1^{3}/_{8}''$ long, $^{19}/_{64}''$ head; 0.098 diameter, $1^{1}/_{4}''$ long, annular-ringed; 5d cooler nail, 0.086 diameter, $1^{5}/_{8}''$ long, $^{15}/_{64}''$ head; or gypsum board nail, 0.086 diameter, $1^{5}/_{8}''$ long, $^{9}/_{32}''$ head.
$1/2$	Ceiling[d]	Perpendicular	24	7	12	
$1/2$	Wall	Either direction	24	8	12	
$1/2$	Wall	Either direction	16	8	16	
$5/8$	Ceiling	Either direction	16	7	12	13 gage, $1^{5}/_{8}''$ long, $^{19}/_{64}''$ head; 0.098 diameter, $1^{3}/_{8}''$ long, annular-ringed; 6d cooler nail, 0.092 diameter, $1^{7}/_{8}''$ long, $1/4''$ head; or gypsum board nail, 0.0915 diameter, $1^{7}/_{8}''$ long, $^{19}/_{64}''$ head.
$5/8$	Ceiling[e]	Perpendicular	24	7	12	
$5/8$	Wall	Either direction	24	8	12	
$5/8$	Wall	Either direction	16	8	16	
Application with adhesive						
$3/8$	Ceiling[d]	Perpendicular	16	16	16	Same as above for $3/8''$ gypsum board
$3/8$	Wall	Either direction	16	16	24	
$1/2$ or $5/8$	Ceiling	Either direction	16	16	16	Same as above for $1/2''$ and $5/8''$ gypsum board, respectively
$1/2$ or $5/8$	Ceiling[d]	Perpendicular	24	12	16	
$1/2$ or $5/8$	Wall	Either direction	24	16	24	
two $3/8$ layers	Ceiling	Perpendicular	16	16	16	Base ply nailed as above for $1/2''$ gypsum board; face ply installed with adhesive
two $3/8$ layers	Wall	Either direction	24	24	24	

For SI: 1 inch = 25.4 mm.

a. For application without adhesive, a pair of nails spaced not less than 2 inches apart or more than $2^{1}/_{2}$ inches apart may be used with the pair of nails spaced 12 inches on center.

b. Screws shall be Type S or W per ASTM C 1002 and shall be sufficiently long to penetrate wood framing not less than $5/8$ inch and metal framing not less than $3/8$ inch.

c. Where metal framing is used with a clinching design to receive nails by two edges of metal, the nails shall be not less than $5/8$ inch longer than the gypsum board thickness and shall have ringed shanks. Where the metal framing has a nailing groove formed to receive the nails, the nails shall have barbed shanks or be 5d, $13^{1}/_{2}$ gage, $1^{5}/_{8}$ inches long, $^{15}/_{64}$-inch head for $1/2$-inch gypsum board; and 6d, 13 gage, $1^{7}/_{8}$ inches long, $^{15}/_{64}$-inch head for $5/8$-inch gypsum board.

d. Three-eighths-inch-thick single-ply gypsum board shall not be used on a ceiling where a water-based textured finish is to be applied, or where it will be required to support insulation above a ceiling. On ceiling applications to receive a water-based texture material, either hand or spray applied, the gypsum board shall be applied perpendicular to framing. When applying a water-based texture material, the minimum gypsum board thickness shall be increased from $3/8$ inch to $1/2$ inch for 16-inch on center framing, and from $1/2$ inch to $5/8$ inch for 24-inch on center framing or $1/2$-inch sag-resistant gypsum ceiling board shall be used.

e. Type X gypsum board for garage ceilings beneath habitable rooms shall be installed perpendicular to the ceiling framing and shall be fastened at maximum 6 inches o.c. by minimum $1^{7}/_{8}$ inches 6d coated nails or equivalent drywall screws.

❖ Table R702.3.5 is a comprehensive table identifying the minimum thickness and fastening requirements for gypsum wallboard. The wallboard thickness [$3/8$-inch, $1/2$-inch or $5/8$-inch (10 mm, 12.7 mm or 16 mm)], location of the wallboard (wall or ceiling), orientation of the wallboard to the framing members (parallel or perpendicular), spacing of framing members [16 inches or 24 inches (406 mm or 610 mm)] on center and type of fasteners (nails or screws) are all set forth in the table. Where gypsum sheathing is to be attached to exterior walls, the provisions of Table R602.3(1) apply. Commentary Figure R702.3.5(2) shows a single-nailing method of attachment; however, Note "a" of Table R702.3.5 permits a double-nailing method as illustrated in Commentary Figure 702.3.5(3).

Figure R702.3.5(1)
GYPSUM BOARD APPLICATION

For SI: 1 inch = 25.4 mm.

Figure R702.3.5(2)
GYPSUM WALLBOARD—SINGLE-NAILING APPLICATION

WALL COVERING

Figure R702.3.5(3)
GYPSUM WALLBOARD—DOUBLE-NAILING APPLICATION

Figure R702.4.1
INTERIOR WALL TILE INSTALLATION

R702.4.3 Limitations. Water-resistant gypsum backing board shall not be used in the following locations:

1. Over a vapor retarder in a shower or bathtub compartment.
2. Where there will be direct exposure to water, or in areas subject to continuous high humidity.

❖ Although there are many gypsum board sheet products that are manufactured and approved for use in wet areas or areas exposed to moisture or humidity, there are still some extreme conditions where even water-resistant gypsum board will not provide the level of moisture protection necessary.

  1. Gypsum board installed on the walls and ceilings at shower and bathtub areas must be finished to prevent moisture from penetrating the wall or ceiling finish and contacting the gypsum board. The finish applied to the exposed face of the gypsum board must create a water-resistant barrier that not only stops water from getting to the gypsum board but also prevents the release of moisture from within the wall or the gypsum board itself. For this reason, gypsum board must not be installed over the outboard side of any vapor barrier or retarder. This will create a waterproof membrane on both faces of the gypsum board, causing moisture to be trapped in the gypsum board that will ultimately cause it to decompose and fail.

2. Water-resistant gypsum board is not to be used in areas that will be subject to a continuous exposure to moisture or humidity at locations such as saunas, steam rooms, gang showers or indoor pools. Gypsum board products, including the water-resistant type, are not intended for these extreme conditions and will provide unsatisfactory performance. Nongypsum wall and ceiling materials such as concrete masonry, ceramic tile on cement backer board, cement plaster (stucco) or other materials designed and recommended for high moisture exposure must be used in these locations.

R702.5 Other finishes. Wood veneer paneling and hardboard paneling shall be placed on wood or cold-formed steel framing spaced not more than 16 inches (406 mm) on center. Wood veneer and hard board paneling less than $1/4$ inch (6.4 mm) nominal thickness shall not have less than a $3/8$-inch (9.5 mm) gypsum board backer. Wood veneer paneling not less than $1/4$-inch (6.4 mm) nominal thickness shall conform to ANSI/HPVA HP-1. Hardboard paneling shall conform to ANSI/AHA A135.5.

❖ This section provides requirements for wood veneer and hardboard paneling that are not addressed in previous sections. Modular home construction will generally use hardboard paneling applied directly to the studs. For this reason, there are specific provisions for wood veneer and hardboard siding. Wall finishes with adequate thickness will resist excessive bending and possible failure.

R702.6 Wood shakes and shingles. Wood shakes and shingles shall conform to CSSB *Grading Rules for Wood Shakes and Shingles* and shall be permitted to be installed directly to the studs with maximum 24 inches (610 mm) on center spacing.

❖ Complying wood shakes and wood shingles may be attached directly to the studs in an interior application provided the studs are spaced at no more than 24 inches (610 mm) on center. The shakes and shingles must conform to the appropriate grading rules set forth by the Cedar Shake and Shingle Bureau.

R702.6.1 Attachment. Nails, staples or glue are permitted for use in attaching shakes or shingles to the wall, and the shakes or shingles shall be permitted to be attached directly to the surface provided the fasteners are appropriate for the type of wall surface material. When nails or staples are used, two fasteners shall be provided and shall be placed so that they are covered by the course above.

❖ Since there is no concern for weather protection, the shakes or shingles may be attached directly to the surface with nails, staples or glue. At least two fasteners are required per shake or shingle where nails or staples are used, located in a manner so that the course above will cover the fasteners.

R702.6.2 Furring strips. Where furring strips are used, they shall be 1 inch by 2 inches or 1 inch by 3 inches (25.4 mm by 51 mm or 25.4 mm by 76 mm), spaced a distance on center equal to the desired exposure, and shall be attached to the wall by nailing through other wall material into the studs.

❖ In addition to direct attachment of wood shakes and shingles to studs or a wall surface, the use of furring strips is permitted under specific conditions. Furring strips should be either 1 inch by 2 inches (25.4 mm by 51 mm) or 1 inch by 3 inches (25.4 mm by 76 mm). The spacing of the strips is determined by the desired exposure of the shake or shingle, attached by nailing through other wall surface materials into the studs.

SECTION R703
EXTERIOR COVERING

R703.1 General. Exterior walls shall provide the building with a weather-resistant exterior wall envelope. The exterior wall envelope shall include flashing as described in Section R703.8. The exterior wall envelope shall be designed and constructed in such a manner as to prevent the accumulation of water within the wall assembly by providing a water-resistive barrier behind the exterior veneer as required by Section R703.2.

❖ One of the functions of an exterior wall covering is to create an enclosed space free from the effects of the environment. The covering system, including flashing where necessary, must be designed not only to resist the penetration of moisture into the building's interior spaces but also to prevent the accumulation of water within the wall assembly. Such moisture build-up can lead to the rapid deterioration of the wall assembly and possibly the structural members. Exterior coverings and veneers applied to walls do not carry a load other than their own weight. Veneer and coverings add no strength to the wall.

R703.2 Weather-resistant sheathing paper. Asphalt-saturated felt free from holes and breaks, weighing not less than 14 pounds per 100 square feet (0.683 kg/m^2) and complying with ASTM D 226 or other approved weather-resistant material shall be applied over studs or sheathing of all exterior walls as required by Table R703.4. Such felt or material shall be applied horizontally, with the upper layer lapped over the lower layer not less than 2 inches (51 mm). Where joints occur, felt shall be lapped not less than 6 inches (152 mm).

Exception: Such felt or material is permitted to be omitted in the following situations:

1. In detached accessory buildings.
2. Under panel siding with shiplap joints or battens.
3. Under exterior wall finish materials as permitted in Table R703.4.
4. Under paperbacked stucco lath.

❖ Asphalt-saturated felt or any other approved weather-resistant material is required behind certain types of siding material because of the possibility of moisture penetrating behind the siding. This felt or other material protects the wall construction from potential rotting.

The weather-resistant membrane may be omitted where there is a low possibility of moisture penetration or the potential for moisture penetration is not a great concern. Another application where the material may be omitted is the exterior joint treatment of panel siding with shiplap joints or battens as illustrated in Commentary Figures R703.3.1(1) and R703.3.1(2). The weather-resistant membrane may be eliminated when approved paperbacked stucco lath is used; the paper backing functions as the membrane if the backing paper is not punctured.

R703.3 Wood, hardboard and wood structural panel siding.

❖ This section addresses the use of wood siding products as exterior wall coverings. Both horizontal siding and panel siding are regulated.

R703.3.1 Panel siding. Joints in wood, hardboard or wood structural panel siding shall be made as follows unless otherwise approved. Vertical joints in panel siding shall occur over framing members, unless wood or wood structural panel sheathing is used, and shall be shiplapped or covered with a batten. Horizontal joints in panel siding shall be lapped a minimum of 1 inch (25.4 mm) or shall be shiplapped or shall be flashed with Z-flashing and occur over solid blocking, wood or wood structural panel sheathing.

❖ Plywood is acceptable as an exterior wall covering provided that the plywood is approved for use in an exterior location and the joints are made waterproof. Typical methods for accomplishing the joint treatment include: 1) lapping joints horizontally, and 2) having the joints occur over framing and protecting such joints with a continuous wood batten using approved caulking, flashing or vertical or horizontal shiplaps. See Commentary Figures R703.3.1(1) and R703.3.1(2) for illustrations of joint treatment.

R703.3.2 Horizontal siding. Horizontal lap siding shall be lapped a minimum of 1 inch (25.4 mm), or 0.5 inch (12.7 mm) if rabbeted, and shall have the ends caulked, covered with a batten, or sealed and installed over a strip of flashing.

❖ In order to maintain the necessary weather protection, horizontal siding must have a minimum lap of at least 1 inch (25.4 mm), unless the siding is rabbeted, in which case the joint need only be $^1/_2$ inch (12.7 mm). Various methods can be used to protect the ends of horizontal siding, including the use of flashing, caulking, or battens.

R703.4 Attachments. Unless specified otherwise, all wall coverings shall be securely fastened in accordance with Table R703.4 or with other approved aluminum, stainless steel, zinc-coated or other approved corrosion-resistive fasteners.

❖ The presence of frequent moisture in the atmosphere necessitates that wall covering be fastened with approved corrosion-resistant fasteners such as aluminum, stainless steel, zinc or zinc-coated fasteners. Table R703.4 sets forth the size of fasteners and other attachment requirements.

Figure R703.3.1(1)
VERTICAL JOINT TREATMENT OF PANEL SIDING

R703.5 Wood shakes and shingles. Wood shakes and shingles shall conform to CSSB *Grading Rules for Wood Shakes and Shingles*.

❖ Grading rules established by the Cedar Shake and Shingle Bureau are applicable to wood shakes and shingles used as exterior wall covering materials. The provisions of this section deal with the specific application and attachment of wood shakes and shingles to the exterior of a building.

R703.5.1 Application. Wood shakes or shingles shall be applied either single-course or double-course over nominal $^1/_2$-inch (12.7 mm) wood-based sheathing or to furring strips over $^1/_2$-inch (12.7 mm) nominal nonwood sheathing. A weather-resistant permeable membrane shall be provided over all sheathing, with horizontal overlaps in the membrane of not less than 2 inches (51 mm) and vertical overlaps of not less than 6 inches (152 mm). Where furring strips are used, they shall be 1 inch by 3 inches or 1 inch by 4 inches (25.4 mm by 76 mm or 25.4 mm by 102 mm) and shall be fastened horizontally to the studs with 7d or 8d box nails and shall be spaced a distance on center equal to the actual weather exposure of the shakes or shingles, not to exceed the maximum exposure specified in Table R703.5.2. The spacing between adjacent shingles to allow for expansion shall not exceed $^1/_4$ inch (6.4 mm), and between adjacent shakes, it shall not exceed $^1/_2$ inch (12.7 mm). The offset spacing between joints in adjacent courses shall be a minimum of $1^1/_2$ inches (38 mm).

❖ Where wood shakes and shingles are applied as an exterior wall covering, the sheathing must be covered with a weather-resistant permeable membrane prior to shingle or shake installation. The application of wood shakes and shingles may be over either solid sheathing or furring strips applied directly to the studs. The minimum sheathing thickness and furring strip size and fastening is prescribed by the code. To allow for expansion, a small space should be established between each shake or shingle. The spacing is limited to $^1/_4$ inch (6.4 mm) between adjacent shingles and $^1/_2$ inch (12.7 mm) between adjacent shakes. In order to avoid channels between the shakes or shingles where water penetration might occur, joints must be offset at least $1^1/_2$ inches (38 mm) in adjacent courses.

R703.5.2 Weather exposure. The maximum weather exposure for shakes and shingles shall not exceed that specified in Table R703.5.2.

❖ Table R703.5.2 specifies the maximum weather exposure permitted for wood shakes and wood shingles installed on exterior walls. Where a single course of wood shingles is used, the maximum exposure varies from $7^1/_2$ inches to $11^1/_2$ inches (190.5 mm to 292 mm), depending on the length of the shingles. Maximum wood shake exposures are either $8^1/_2$ inches or $11^1/_2$ inches (216 mm to 292 mm) for a single course application. In double course applications, the maximum exposures are increased for both shingles and shakes.

R703.5.3 Attachment. Each shake or shingle shall be held in place by two hot-dipped zinc-coated, stainless steel, or aluminum nails or staples. The fasteners shall be long enough to penetrate the sheathing or furring strips by a minimum of $^1/_2$ inch (12.7 mm) and shall not be overdriven.

❖ Hot-dipped zinc-coated, stainless steel, or aluminum fasteners are to be used to attach wood shakes and shingles. Either nails or staples are permitted, with two fasteners used for each shake or shingle. The length of the nails or staples must be adequate to where the sheathing or furring strips are penetrated a minimum of $^1/_2$ inch (12.7 mm). As the nails or staples are driven, it is important that they do not excessively penetrate the shake or shingle, which could result in reduced holding force.

Figure R703.3.1(2)
HORIZONTAL JOINT TREATMENT OF PANEL SIDING

WALL COVERING

TABLE R703.4
WEATHER-RESISTANT SIDING ATTACHMENT AND MINIMUM THICKNESS

SIDING MATERIAL		NOMINAL THICKNESS[a] (inches)	JOINT TREATMENT	SHEATHING PAPER REQUIRED	Wood or wood structural panel sheathing	Fiberboard sheathing into stud	Gypsum sheathing into stud	Foam plastic sheathing into stud	Direct to studs	Number or spacing of fasteners
Horizontal aluminum[e]	Without insulation	0.019[f]	Lap	No	0.120 nail $1^{1}/_{2}$″ long	0.120 nail 2″ long	0.120 nail 2″ long	0.120 nail[z]	Not allowed	Same as stud spacing
		0.024	Lap	No	0.120 nail $1^{1}/_{2}$″ long	0.120 nail 2″ long	0.120 nail 2″ long	0.120 nail[z]	Not allowed	
	With insulation	0.019	Lap	No	0.120 nail $1^{1}/_{2}$″ long	0.120 nail $2^{1}/_{2}$″ long	0.120 nail $2^{1}/_{2}$″ long	0.120 nail[z]	0.120 nail $1^{1}/_{2}$″ long	
Brick veneer Concrete masonry veneer		2 2	Section R703	Yes (Note m)	See Section R703 and Figure R703.7[h]					
Hardboard[l] Panel siding-vertical		$7/_{16}$	Note g	See R703.2	Note o	Note o	Note o	Note o	Note o	6″ panel edges 12″ inter. sup.[p]
Hardboard[l] Lap-siding-horizontal		$7/_{16}$	Note r	Yes	Note q	Note q	Note q	Note q	Note q	Same as stud spacing 2 per bearing
Steel[i]		29 ga.	Lap	No	0.113 nail $1^{3}/_{4}$″ Staple-$1^{3}/_{4}$″	0.113 nail $2^{3}/_{4}$″ Staple-$2^{1}/_{2}$″	0.113 nail $2^{1}/_{2}$″ Staple-$2^{1}/_{4}$″	0.113 nail[z] Staple[z]	Not allowed	Same as stud spacing
Stone veneer		2	Section R703	Yes (Note m)	See Section R703 and Figure R703.7[h]					
Particleboard panels		$3/_{8}$ - $1/_{2}$	Note g	Note g	6d box nail	6d box nail	6d box nail	box nail[z]	6d box nail, $3/_{8}$ not allowed	6″ panel edge 12″ inter. sup.
		$5/_{8}$	Note g	Note g	6d box nail	8d box nail	8d box nail	box nail[z]	6d box nail	
Plywood panel[j] (exterior grade)		$3/_{8}$	Note g	Note g	0.099 nail-2″	0.113 nail-$2^{1}/_{2}$″	0.099 nail-2″	0.113 nail[z]	0.099 nail-2″	6″ on edges
Vinyl siding[n]		0.035	Lap	No	0.120 nail $1^{1}/_{2}$″ Staple-$1^{3}/_{4}$″	0.120 nail 2″ Staple-$2^{1}/_{2}$″	0.120 nail 2″ Staple-$2^{1}/_{2}$″	0.120 nail[z] Staple[z]	Not allowed	Same as stud spacing
Wood[k] Rustic, drop		$3/_{8}$ Min	Lap	No	Fastener penetration into stud-1″				0.113 nail-$2^{1}/_{2}$″ Staple-2″	Face nailing up to 6″ widths, 1 nail per bearing; 8″ widths and over, 2 nails per bearing
Shiplap		$19/_{32}$ Average	Lap	No						
Bevel		$7/_{16}$								
Butt tip		$3/_{16}$	Lap	No						
Fiber cement panel siding[s]		$5/_{16}$	Note t	Yes Note y	6d corrosion resistant nail[u]	6d corrosion resistant nail[u]	6d corrosion resistant nail[u]	—	4d corrosion resistant nail[v]	6″ oc on edges, 12″ oc on intermed. studs
Fiber cement lap siding[s]		$5/_{16}$	Note w	Yes Note y	6d corrosion resistant nail[u]	6d corrosion resistant nail[u]	6d corrosion resistant nail[u]	—	6d corrosion resistant nail[x]	Note x

For SI: 1 inch = 25.4 mm.

a. Based on stud spacing of 16 inches on center. Where studs are spaced 24 inches, siding shall be applied to sheathing approved for that spacing.
b. Nail is a general description and shall be T-head, modified round head, or round head with smooth or deformed shanks.
c. Staples shall have a minimum crown width of $7/_{16}$-inch outside diameter and be manufactured of minimum No. 16 gage wire.
d. Nails or staples shall be aluminum, galvanized, or rust-preventive coated and shall be driven into the studs for fiberboard or gypsum backing.
e. Aluminum nails shall be used to attach aluminum siding.
f. Aluminum (0.019 inch) shall be unbacked only when the maximum panel width is 10 inches and the maximum flat area is 8 inches. The tolerance for aluminum siding shall be +0.002 inch of the nominal dimension.
g. If boards or panels are applied over sheathing or a weather-resistant membrane, joints need not be treated. Otherwise, vertical joints shall occur at studs and be covered with battens or be lapped.
h. All attachments shall be coated with a corrosion-resistive coating.
i. Shall be of approved type.
j. Three-eighths-inch plywood shall not be applied directly to studs spaced greater than 16 inches on center when long dimension is parallel to studs. One-half-inch plywood shall not may be applied directly to studs spaced greater than 24 inches on center. The stud spacing shall not exceed the panel span rating provided by the manufacturer unless the panels are installed with the face grain perpendicular to studs or over sheathing approved for that stud spacing.

(continued)

Footnotes to Table R703.4—continued

k. Woodboard sidings applied vertically shall be nailed to horizontal nailing strips or blocking set 24 inches on center. Nails shall penetrate 1.5 inches into studs, studs and wood sheathing combined, or blocking. A weather-resistant membrane shall be installed weatherboard fashion under the vertical siding unless the siding boards are lapped or battens are used.
l. Hardboard siding shall comply with AHA A135.6.
m. For masonry veneer, a weather-resistant membrane or building paper is not required over water-repellent sheathing materials when a 1-inch air space is provided between the veneer and the sheathing. When the 1-inch space is filled with mortar, a weather-resistant membrane or building paper is required over studs or sheathing.
n. Vinyl siding shall comply with ASTM D 3679.
o. Minimum shank diameter of 0.092 inch, minimum head diameter of 0.225 inch, and nail length must accommodate sheathing and penetrate framing 1.5 inches.
p. When used to resist shear forces, the spacing must be 4 inches at panel edges and 8 inches on interior supports.
q. Minimum shank diameter of 0.099 inch, minimum head diameter of 0.240 inch, and nail length must accommodate sheathing and penetrate framing 1.5 inches.
r. Vertical end joints shall occur at studs and shall be covered with a joint cover or shall be caulked.
s. Fiber cement siding shall comply with the requirements of ASTM C 1186.
t. See Section R703.10.1.
u. Minimum 0.102 inch smooth shank, 0.255 inch round head.
v. Minimum 0.099 inch smooth shank, 0.250 inch round head.
w See Section R703.10.2.
x. Face nailing: 2 nails at each stud. Concealed nailing: one 11 gage $1^1/_2$ galv. roofing nail (0.371 inch head diameter, 0.120 inch shank) or 6d galv. box nail at each stud.
y. See Section R703.2, Exceptions.
z. Minimum nail length must accommodate sheathing and penetrate framing 1.5 inches.

❖ This table should be used in addition to all other applicable requirements of the code for the specific material under consideration. Testing and experience have determined that the minimum thicknesses tabulated will be durable and protect the building against the elements for relatively long periods of time when the siding is attached and maintained as indicated.

TABLE R703.5.2
MAXIMUM WEATHER EXPOSURE FOR WOOD SHAKES AND SHINGLES ON EXTERIOR WALLS[a,b,c]
(Dimensions are in inches)

LENGTH	EXPOSURE FOR SINGLE COURSE	EXPOSURE FOR DOUBLE COURSE
Shingles[a]		
16	$7^1/_2$	12[b]
18	$8^1/_2$	14[c]
24	$11^1/_2$	16
Shakes[a]		
18	$8^1/_2$	14
24	$11^1/_2$	18

For SI: 1 inch = 25.4 mm.
a. Dimensions given are for No. 1 grade.
b. A maximum 10-inch exposure is permitted for No. 2 grade.
c. A maximum 11-inch exposure is permitted for No. 2 grade.

❖ Depending on the length of the shingle and the number of courses, the table specifies the maximum length of exposure for weathering purposes. Shingles are available in three different lengths, while shakes are regulated for two lengths. Both a single-course application and double-course application are addressed.

R703.5.3.1 Staple attachment. Staples shall not be less than 16 gage and shall have a crown width of not less than $^7/_{16}$ inch (11.1 mm), and the crown of the staples shall be parallel with the butt of the shake or shingle. In single-course application, the fasteners shall be concealed by the course above and shall be driven approximately 1 inch (25.4 mm) above the butt line of the succeeding course and $^3/_4$ inch (19.1 mm) from the edge. In double-course applications, the exposed shake or shingle shall be face-nailed with two casing nails, driven approximately 2 inches (51 mm) above the butt line and $^3/_4$ inch (19.1 mm) from each edge. In all applications, staples shall be concealed by the course above. With shingles wider than 8 inches (203 mm) two additional nails shall be required and shall be nailed approximately 1 inch (25.4 mm) apart near the center of the shingle.

❖ This section describes in detail the proper attachment of either staples or nails where used as fasteners for wood shakes and shingles applied to exterior walls. Both single-course and double-course application methods are described. The general requirement of Section R703.5.3 for two fasteners per shingle is modified by this section, as two additional fasteners are needed for shingles having a width of more than 8 inches (203 mm).

WALL COVERING

R703.5.4 Bottom courses. The bottom courses shall be doubled.

❖ In all cases, the bottom course of shakes and shingles in an exterior wall application is to be doubled.

R703.6 Exterior plaster.

❖ Portland cement plaster is the only material approved by the code for exterior plaster. Gypsum plaster deteriorates under conditions of weather and moisture, which are prevalent on the exterior surfaces of buildings. Commentary Figures R703.6(1) and R703.6(2) illustrate exterior plastering systems.

R703.6.1 Lath. All lath and lath attachments shall be of corrosion-resistant materials. Expanded metal or woven wire lath shall be attached with $1^1/_2$-inch-long (38 mm), 11 gage nails having a $^7/_{16}$-inch (11.1 mm) head, or $^7/_8$-inch-long (22.2 mm), 16 gage staples, spaced at no more than 6 inches (152 mm), or as otherwise approved.

❖ As expected, lath and its attachments must be of corrosion-resistant materials. Unless another fastening method is approved, lath must be attached with the nails or staples specified in this section and spaced at a maximum of 6 inches (152 mm) on center.

R703.6.2 Plaster. Plastering with portland cement plaster shall be not less than three coats when applied over metal lath or wire lath and shall be not less than two coats when applied over masonry, concrete or gypsum backing. If the plaster surface is completely covered by veneer or other facing material or is completely concealed, plaster application need be only two coats, provided the total thickness is as set forth in Table R702.1(1).

On wood-frame construction with an on-grade floor slab system, exterior plaster shall be applied in such a manner as to cover, but not extend below, lath, paper and screed.

The proportion of aggregate to cementitious materials shall be as set forth in Table R702.1(3).

❖ The code requires that exterior portland cement plaster be applied in not less than three coats when applied over metal or wire-fabric lath, for the same reasons as discussed for interior plaster. Where the portland cement plaster is applied over other approved plaster bases, the code requires only two-coat work. The code permits plaster work that is completely concealed to be of only two coats provided the total thickness is that required by Table R702.1(1), since the finish coat of

For SI: 1 inch = 25.4 mm.

Figure R703.6(1)
EXTERIOR PLASTERING SYSTEM

Figure R703.6(2)
EXTERIOR PLASTER—WOOD FRAME ON CONCRETE FLOOR SLAB

plaster is to provide a surface for exterior finishes and to provide an aesthetic appearance. Thus, where the plaster surface is to be completely concealed, it is not necessary to provide a finish coat.

The code requires that the exterior plaster be installed to completely cover, but not extend below, the lath and paper-on-wood exterior wall construction supported by an on-grade concrete slab. This requirement, combined with the presence of a weep screed, prevents the entrapment of free moisture and the subsequent channeling of the moisture to the interior of the building.

R703.6.2.1 Weep screeds. A minimum 0.019-inch (0.48 mm) (No. 26 galvanized sheet gage), corrosion-resistant weep screed or plastic weep screed, with a minimum vertical attachment flange of $3^1/_2$ inches (89 mm) shall be provided at or below the foundation plate line on exterior stud walls in accordance with ASTM C 926. The weep screed shall be placed a minimum of 4 inches (102 mm) above the earth or 2 inches (51 mm) above paved areas and shall be of a type that will allow trapped water to drain to the exterior of the building. The weather-resistant barrier shall lap the attachment flange. The exterior lath shall cover and terminate on the attachment flange of the weep screed.

❖ Water and moisture can penetrate an exterior plaster wall for a variety of reasons and in a number of ways. It is expected that some moisture will penetrate the plaster in an exterior wall; therefore, the design of the wall should include a weep screed, which will provide a way to release the moisture; see Commentary Figure R703.6(3). Once water or moisture penetrates the plaster, it will migrate down the exterior face of the weather-resistive barrier until it reaches the sill plate or mud sill. At this point, the water will seek a way out of the wall. If the exterior plaster system is not detailed and constructed with provisions to allow the moisture to escape to the exterior, it will find its own way out. This exit will almost certainly be through the interior of the wall and cause leaking and therefore damage to the interior of the building. For this reason, the code requires a continuous weep screed at the bottom of exterior walls to permit the moisture to escape to the exterior of the building.

R703.7 Stone and masonry veneer, general. All stone and masonry veneer shall be installed in accordance with this chapter, Table R703.4 and Figure R703.7. Such veneers installed over a backing of wood or cold-formed steel shall be

WALL COVERING

Figure R703.6(3)
WEEP SCREED FOR EXTERIOR PLASTER ON STUD WALL AT SLAB ON GRADE

limited to the first story above grade and shall not exceed 5 inches (127 mm) in thickness.

Exceptions:

1. In Seismic Design Categories A and B, exterior masonry veneer with a backing of wood or cold-formed steel framing shall not exceed 30 feet (9144 mm) in height above the noncombustible foundation, with an additional 8 feet (2348 mm) permitted for ends.

2. In Seismic Design Category C, exterior masonry veneer with a backing of wood or cold-formed steel framing shall not exceed 30 feet (9144 mm) in height above the noncombustible foundation, with an additional 8 feet (2348 mm) permitted for gabled ends. In other than the topmost story, the length of bracing shall be 1.5 times the length otherwise required in Chapter 6.

3. For detached one- or two- family dwellings with a maximum nominal thickness of 4 inches (102 mm) of exterior masonry veneer with a backing of wood frame located in Seismic Design Category D_1, the masonry veneer shall not exceed 20 feet (6096 mm) in height above a noncombustible foundation, with an additional 8 feet (2438 mm) permitted for gabled ends, or 30 feet (9144 mm) in height with an additional 8 feet (2438 mm) permitted for gabled ends where the lower 10 feet (3048 mm) has a backing of concrete or masonry wall, provided the following criteria are met:

   3.1. Braced wall panels shall be constructed with a minimum of $7/16$ inch (11.1 mm) thick sheathing fastened with 8d common nails at 4 inches (102 mm) on center on panel edges and at 12 inches (305 mm) on center on intermediate supports.

   3.2. The bracing of the top story shall be located at each end and at least every 25 feet (7620 mm) on center but not less than 45% of the braced wall line. The bracing of the first story shall be as provided in Table R602.10.1.

   3.3. Hold down connectors shall be provided at the ends of braced walls for the second floor to first floor wall assembly with an allowable design of 2100 lbs. (952.5 kg) Hold down connectors shall be provided at the ends of each wall segment of the braced walls for the first floor to foundation assembly with an allowable design of 3700 lbs. (1678 kg). In all cases, the hold down connector force shall be transferred to the foundation.

   3.4. Cripple walls shall not be permitted.

4. For detached one- and two-family dwellings with a maximum actual thickness of 3 inches (76 mm) of exterior masonry veneer with a backing of wood frame located in Seismic Design Category D_2, the masonry veneer shall not exceed 20 feet (6096 mm) in height

above a noncombustible foundation, with an additional 8 feet (2438 mm) permitted for gabled ends, or 30 feet (9144 mm) in height with an additional 8 feet (2438 mm) permitted for gabled ends where the lower 10 feet (3048 mm) has a backing of concrete or masonry wall, provided the following criteria are met:

4.1. Braced wall panels shall be constructed with a minimum of $^7/_{16}$ inch (11.1 mm) thick sheathing fastened with 8d common nails at 4 inches (102 mm) on center on panel edges and at 12 inches (305 mm) on center on intermediate supports.

4.2. The bracing of the top story shall be located at each end and at least every 25 feet (7620 mm) on center but not less than 55% of the braced wall line. The bracing of the first story shall be as provided in Table R602.10.1.

4.3. Hold down connectors shall be provided at the ends of braced walls for the second floor to first floor wall assembly with an allowable design of 2300 lbs.(1043 kg). Hold down connectors shall be provided at the ends of each wall segment of the braced walls for the first floor to foundation assembly with an allowable design of 3900 lbs. (1769 kg). In all cases, the hold down connector force shall be transferred to the foundation.

4.4. Cripple walls shall not be permitted.

❖ For the purposes of this section, stone and masonry veneer are nonstructural facing materials providing ornamentation, protection or insulation. To be considered a veneer, the material cannot act structurally with the backing insofar as the structural strength of the assembly is concerned. Stone and masonry veneer can be used as either an interior or exterior wall finish.

In addition to the provisions of this section, Table R703.4 provides details for the installation of stone and masonry veneer. The height of such veneer is limited, primarily based upon the seismic design category of the building. For the lesser seismic design categories, a maximum height of 30 feet (9144 mm) above a noncombustible foundation is permitted for exterior masonry veneer. In the higher seismic design categories, stone and masonry veneer is limited to the first story above grade and a maximum thickness of 5 inches (127 mm).

Masonry veneer two stories (20 feet) (6096 mm) in height is allowed for detached one- and two-family dwellings of light wood frame construction in Seismic Design Categories D_1 and D_2. Three stories (30 feet) (9144 mm) is allowed if the first story (lower 10 feet) (3048 mm) is concrete or masonry construction. In both cases, additional wall bracing and hold-down anchorage are required.

R703.7.1 Interior veneer support. Veneers used as interior wall finishes shall be permitted to be supported on wood or cold-formed steel floors that are designed to support the loads imposed.

❖ Where designed to support the weight of the veneer, a wood-framed or cold-formed steel floor system may be used to support stone or masonry veneer used in an interior application.

R703.7.2 Exterior veneer support. Except in Seismic Design Categories D_1 and D_2, exterior masonry veneers having an installed weight of 40 pounds per square foot (195 kg/m^2) or less shall be permitted to be supported on wood or cold-formed steel construction. When masonry veneer supported by wood or cold-formed steel construction adjoins masonry veneer supported by the foundation, there shall be a movement joint between the veneer supported by the wood or cold-formed steel construction and the veneer supported by the foundation. The wood or cold-formed steel construction supporting the masonry veneer shall be designed to limit the deflection to $^1/_{600}$ of the span for the supporting members. The design of the wood or cold-formed steel construction shall consider the weight of the veneer and any other loads.

❖ Two criteria must be met in order to support exterior masonry veneer directly on wood or cold-formed steel construction, rather than on a concrete foundation wall. First, the construction must be in Seismic Design Category A, B or C. Second, the total installed weight of the veneer must not exceed 40 pounds per square foot (195 kg/m^2). The prescriptive provisions of the code are not designed to address conditions where the potential for extensive lateral loading from earthquakes is relatively high or where the weight of the masonry veneer is excessive. Even in those seismic design categories permitted by this section, lateral movement must be addressed. Therefore, a movement joint must be provided between the exterior veneer supported by wood or steel and the adjacent exterior veneer supported by the foundation. The deflection of the wood or cold-formed steel members supporting the masonry veneer is also limited to a maximum of 1/600 of the span of the supporting members.

WALL COVERING

FIGURE R703.7

For SI: 1 inch = 25.4 mm.

FIGURE R703.7
MASONRY VENEER WALL DETAILS

(continued)

FIGURE R703.7 — WALL COVERING

For SI: 1 inch = 25.4 mm.

a. See Sections R703.7.5, R703.7.6 and R703.8.
b. See Sections R703.2 and R703.7.4.
c. See Sections R703.7.4.2 and R703.7.4.3.
d. See Section R703.7.2.1.

**FIGURE R703.7—continued
MASONRY VENEER WALL DETAILS**

❖ This figure outlines a number of methods for attaching masonry veneer to framed walls. The fundamental application and attachment methods are listed, including the use of building paper or water-repellent sheathing, the provision for weepholes and the need for an airspace unless mortared. References to the appropriate code requirements are listed in the notes.

R703.7.2.1 Support by steel angle. A minimum 6 inches by 4 inches by $^5/_{16}$ inch (152 mm by 102 mm by 8 mm) steel angle, with the long leg placed vertically, shall be anchored to double 2 inches by 4 inches (51 mm by 102 mm) wood studs at a maximum on center spacing of 16 inches (406 mm). Anchorage of the steel angle at every double stud spacing shall be a minimum of two $^7/_{16}$ inch (11.1 mm) diameter by 4 inches (102 mm) lag screws. The steel angle shall have a minimum clearance to underlying construction of $^1/_{16}$ inch (1.6 mm). A minimum of two-thirds the width of the masonry veneer thickness shall bear on the steel angle. Flashing and weep holes shall be located in the masonry veneer wythe in accordance with Figure R703.7.1. The maximum height of masonry veneer above the steel angle support shall be 12 feet, 8 inches (3861 mm). The air space separating the masonry veneer from the wood backing shall be in accordance with R703.7.4 and R703.7.4.2. The method of support for the masonry veneer on wood construction shall be constructed in accordance with Figure R703.7.2.1.

The maximum slope of the roof construction without stops shall be 7:12. Roof construction with slopes greater than 7:12 but not more than 12:12 shall have stops of a minimum 3 inches × 3 inches × $^1/_4$ inch (76 mm × 76 mm × 6 mm) steel plate welded to the angle at 24 inches (610 mm) on center along the angle or as approved by the building official.

❖ As illustrated in Figure R703.7.2.1, a steel angle may be used to support exterior masonry veneer in wood-frame construction. The minimum size of the angle is specified, as are the details of anchorage and the minimum bearing length. Two limits are placed on this method of veneer support: the maximum veneer height bearing on the steel angle must be 12 feet, 8 inches (3861 mm), and the roof slope is limited to 7:12. A roof slope greater than 7:12 is permitted if stops are welded to the steel angle to prevent the veneer from sliding down the angle.

R703.7.2.2 Support by roof construction. A steel angle shall be placed directly on top of the roof construction. The roof supporting construction for the steel angle shall consist of a minimum of three 2-inch by 6-inch (51 mm by 152 mm) wood members. The wood member abutting the vertical wall stud construction shall be anchored with a minimum of three $^5/_8$-inch (15.9 mm) diameter by 5-inch (127 mm) lag screws to every wood stud spacing. Each additional roof member shall be anchored by the use of two 10d nails at every wood stud spacing. A minimum of two-thirds the width of the masonry veneer thickness shall bear on the steel angle. Flashing and weep holes shall be located in the masonry veneer wythe in accordance with Figure R703.7.1. The maximum height of the masonry veneer above the steel angle support shall be 12 feet, 8 inches (3861 mm). The air space separating the masonry veneer from the wood backing shall be in accordance with R703.7.4 and R703.7.4.2. The method of support for the ma-

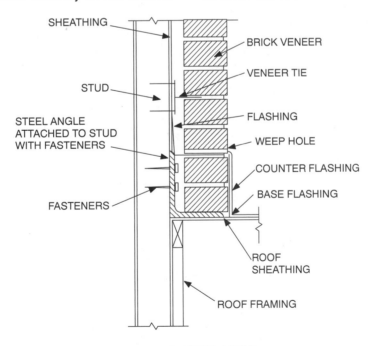

SUPPORT BY STEEL ANGLE

FIGURE R703.7.2.1
EXTERIOR MASONRY VENEER SUPPORT BY STEEL ANGLES

❖ Under most conditions, exterior masonry veneer will be supported by the foundation of the building. Where support of the veneer is by wood construction, the code describes special methods of veneer support. Figures R703.7.2.1 and R703.7.2.2 illustrate support by either steel angle or roof members. The figures show the fundamental elements of attachment, while Sections R703.7.2.1 and R703.7.2.2 contain the details of the installation materials and methods.

sonry veneer on wood construction shall be constructed in accordance with Figure R703.7.2.2.

The maximum slope of the roof construction without stops shall be 7:12. Roof construction with slopes greater than 7:12 but not more than 12:12 shall have stops of a minimum 3 inches × 3 inches × $^1/_4$ inch (76 mm × 76 mm × 6 mm) steel plate welded to the angle at 24 inches (610 mm) on center along the angle or as approved by the building official.

❖ In addition to the option of supporting exterior masonry veneer by a steel angle, the code also permits the wood roof construction itself to support the exterior veneer. Detailed in Figure R703.7.2.2, the steel angle supporting the masonry veneer must be fully supported by at least three 2-inch by 6-inch (51 mm by 152 mm) wood members. This section also addresses the attachment details and bearing conditions. As described for masonry veneer supported by steel angles, the maximum veneer height and maximum roof slope are also regulated. The roof slope is limited to 7:12; however, a roof slope greater than 7:12 is permitted if stops are welded to the steel angle to prevent the veneer from sliding down the angle.

R703.7.3 Lintels. Masonry veneer shall not support any vertical load other than the dead load of the veneer above. Veneer above openings shall be supported on lintels of non-combustible materials and the allowable span shall not exceed the values set forth in Table R703.7.3. The lintels shall have a length of bearing of not less than 4 inches (102 mm).

❖ Masonry veneer, like all veneers, is a non load-bearing wall covering. It is not intended to support any loads, except for the dead load of the masonry veneer above. Where masonry veneer occurs above an opening created for a doorway, window, or similar feature in a building, the veneer must be supported on a noncombustible lintel with a minimum length-of-bearing of 4 inches (102 mm) on each end. Table R703.7.1 identifies the maximum spans permitted for four different sizes of steel lintels where supporting masonry veneer. The allowable spans vary based upon the number of stories of veneer supported by the lintel.

R703.7.4 Anchorage. Masonry veneer shall be anchored to the supporting wall with corrosion-resistant metal ties. Where veneer is anchored to wood backings through the use of corrugated sheet metal ties, the distance separating the veneer from the sheathing material shall be a maximum of 1 inch (25.4 mm). Where the veneer is anchored to wood backings through the use of metal strand wire ties, the distance separating the veneer from the sheathing material shall be a maximum of $4^1/_2$ inches (114 mm). Where the veneer is anchored to cold-formed steel backings, adjustable metal strand wire ties shall be used. Where veneer is anchored to cold-formed steel backings, the distance separating the veneer from the sheathing material shall be a maximum of 4.5 inches (114 mm).

❖ In order to maintain the attachment of the masonry veneer to the wall construction, proper anchorage must be provided. In all cases, corrosion-resistant metal ties must be used. Two types of ties are addressed for attachment to wood construction: corrugated sheet

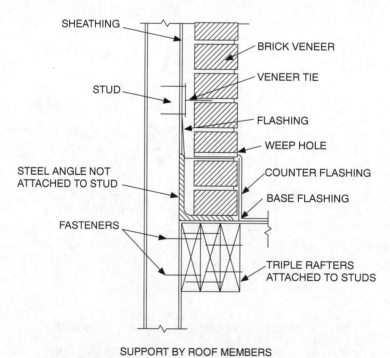

**FIGURE R703.7.2.2
EXTERIOR MASONRY VENEER SUPPORT BY ROOF MEMBERS**

❖ See the commentary for Figure R703.7.2.1.

metal ties and metal strand wire ties. The distance between the veneer and the sheathing is limited and varies based upon the type of tie used. Only adjustable metal strand wire ties may be used where the backing is of cold-formed steel. The distance between the steel backing and the masonry veneer is also limited.

R703.7.4.1 Size and spacing. Veneer ties, if strand wire, shall not be less in thickness than No. 9 U.S. gage wire and shall have a hood embedded in the mortar joint, or if sheet metal, shall be not less than No. 22 U.S. gage by $7/8$ inch (22.3 mm) corrugated. Each tie shall be spaced not more than 24 inches (610 mm) on center horizontally and vertically and shall support not more than 2.67 square feet (0.248 m^2) of wall area.

Exception: In Seismic Design Category D_1 or D_2 and townhouses in Seismic Design Category C and in wind areas of more than 30 pounds per square foot pressure (1.44 kN/m^2), each tie shall support not more than 2 square feet (0.186 m^2) of wall area.

❖ This section sets forth the minimum size of metal veneer ties for use with masonry veneer applications. The spacing of the ties is also limited, with a maximum horizontal distance between ties of 24 inches (610 mm). In addition, no single tie can support more than 2.67 square feet (0.248 m^2) of wall area. An increased number of ties are required in those areas having a high seismic design category or a high wind load.

R703.7.4.1.1 Veneer ties around wall openings. Veneer ties around wall openings. Additional metal ties shall be provided around all wall openings greater than 16 inches (406 mm) in either dimension. Metal ties around the perimeter of openings shall be spaced not more than 3 feet (9144 mm) on center and placed within 12 inches (305 mm) of the wall opening.

❖ Due to the discontinuity of the veneer adjacent to wall openings such as windows and doors, it is often necessary to increase the number of metal ties around wall openings. The location of ties around openings is specified, including the maximum spacing and the maximum distance from the opening. Small openings less than 16 inches in either dimension are exempt from these provisions.

R703.7.4.2 Air space. The veneer shall be separated from the sheathing by an air space of a minimum of 1 inch (25.4 mm) but not more than 4.5 inches (114 mm). The weather-resistant membrane or asphalt-saturated felt required by Section R703.2 is not required over water-repellent sheathing materials.

❖ There must be an air space of at least 1 inch (25.4 mm) between the veneer and the sheathing unless the space is filled with mortar or grout per Section R703.7.4.3. The maximum air-space width of $4^1/_2$ inches (114 mm) is based on the maximum permitted separation between the sheathing material and metal strand wire ties. Where the air space is provided, it is not necessary to provide weather-resistant sheathing paper over any water-repellant sheathing materials.

TABLE R703.7.3
ALLOWABLE SPANS FOR LINTELS SUPPORTING MASONRY VENEER[a,b,c]

SIZE OF STEEL ANGLE[a,c] (inches)	NO STORY ABOVE	ONE STORY ABOVE	TWO STORIES ABOVE	NO. OF $^1/_2$″ OR EQUIVALENT REINFORCING BARS[c]
$3 \times 3 \times 1/_4$	6′-0″	4′-6″	3′-0″	1
$4 \times 3 \times 1/_4$	8′-0″	6′-0″	4′-6″	1
$5 \times 3^1/_2 \times ^5/_{16}$	10′-0″	8′-0″	6′-0″	2
$6 \times 3^1/_2 \times ^5/_{16}$	14′-0″	9′-6″	7′-0″	2
$2-6 \times 3^1/_2 \times ^5/_{16}$	20′-0″	12′-0″	9′-6″	4

For SI: 1 inch = 25.4 mm, 1 foot = 304.8 mm.

a. Long leg of the angle shall be placed in a vertical position.

b. Depth of reinforced lintels shall not be less than 8 inches and all cells of hollow masonry lintels shall be grouted solid. Reinforcing bars shall extend not less than 8 inches into the support.

c. Steel members indicated are adequate typical examples; other steel members meeting structural design requirements may be used.

❖ Where masonry veneer extends above an opening in a wall, it must be supported by a noncombustible lintel. The maximum span for the lintel is based on the number of stories located above the opening. This measurement provides a reasonable estimate of the amount of veneer to be carried. The allowable span is increased based on the size of the steel angle used as the lintel. The table addresses conditions for one-, two- and three-story masonry veneer applications and describes the required reinforcing necessary based on the size of the steel angle.

R703.7.4.3 Mortar or grout fill. As an alternate to the air space required by Section R703.7.4.2, mortar or grout shall be permitted to fill the air space. When the 1-inch (25.4 mm) space is filled with mortar, a weather-resistant membrane or building paper is required over studs or sheathing. When filling the air space, it is permitted to replace the sheathing and weather-resistant membrane or asphalt-saturated felt paper with a wire mesh and approved paper or an approved paper-backed reinforcement attached directly to the studs.

❖ Where an air space complying with Section R703.7.4.2 is not provided between the masonry veneer and the supporting sheathing, mortar or grout may be used to fill the void. Under these conditions, it is important to cover the studs or sheathing with a weather-resistant membrane, a system of wire mesh and approved paper, or an approved paper-backed reinforcement attached directly to the studs.

R703.7.5 Flashing. Flashing shall be located beneath the first course of masonry above finished ground level above the foundation wall or slab and at other points of support, including structural floors, shelf angles and lintels when masonry veneers are designed in accordance with Section R703.7. See Section R703.8 for additional requirements.

❖ Flashing is necessary to close off the points of water entry at the first course of masonry above the finished ground level, as well as at other points of support such as at shelf angles and lintels. As always, flashing must be of an approved corrosion-resistant material.

R703.7.6 Weepholes. Weepholes shall be provided in the outside wythe of masonry walls at a maximum spacing of 33 inches (838 mm) on center. Weepholes shall not be less than $^3/_{16}$ inch (4.8 mm) in diameter. Weepholes shall be located immediately above the flashing.

❖ At a point directly above the flashing mandated by Section R703.7.5, weepholes must be provided to allow for the escape of any moisture that may have penetrated the masonry veneer. As moisture will adversely affect the integrity of the wall if not removed from the wall assembly, weepholes must be installed within the maximum spacing set forth in this section. The minimum diameter of the weepholes is also regulated.

R703.8 Flashing. Approved corrosion-resistive flashing shall be provided in the exterior wall envelope in such a manner as to prevent entry of water into the wall cavity or penetration of water to the building structural framing components. The flashing shall extend to the surface of the exterior wall finish and shall be installed to prevent water from reentering the exterior wall envelope. Approved corrosion-resistant flashings shall be installed at all of the following locations:

1. At top of all exterior window and door openings in such a manner as to be leakproof, except that self-flashing windows having a continuous lap of not less than $1^1/_8$ inches (28 mm) over the sheathing material around the perimeter of the opening, including corners, do not require additional flashing; jamb flashing may also be omitted when specifically approved by the building official.
2. At the intersection of chimneys or other masonry construction with frame or stucco walls, with projecting lips on both sides under stucco copings.
3. Under and at the ends of masonry, wood or metal copings and sills.
4. Continuously above all projecting wood trim.
5. Where exterior porches, decks or stairs attach to a wall or floor assembly of wood-frame construction.
6. At wall and roof intersections.
7. At built-in gutters.

❖ The code requires that all points subject to the entry of moisture be appropriately flashed. Roof and wall intersections and parapets create significant challenges, as do exterior wall openings exposed to the weather. Where wind-driven rain is expected, the concerns are even greater. While the code identifies a number of locations where flashing is specifically required, the entire exterior envelope must be weather-tight to protect the interior from weather. Therefore, any location on the exterior envelope that provides a route for the admission of water or moisture into the building must be properly protected. Commentary Figure R703.8 illustrates examples of flashing.

R703.9 Exterior insulation finish systems, general. All Exterior Insulation Finish Systems (EIFS) shall be installed in accordance with the manufacturer's installation instructions and the requirements of this section. Decorative trim shall not be face nailed through the EIFS. The EIFS shall terminate not less than 6 inches (152 mm) above the finished ground level.

❖ The installation requirements for exterior insulation finish systems, commonly referred to as EIFS, are for the most part provided by the manufacturer of the specific system under consideration. Manufacturer's installation instructions must be followed for a durable, weather-tight exterior envelope. The code sets forth additional general requirements that must be followed.

As a basic requirement, EIFS must not be located within 6 inches of the finished ground level. In addition, any decorative trim applied over the EIFS must not be nailed through the EIFS. Both provisions prevent the penetration of moisture into the exterior wall system.

R703.9.1 Weather-resistive barrier. All EIFS shall have a weather-resistive barrier applied between the underlying water-sensitive building components and the exterior insulation, and a means of draining water to the exterior of the veneer. A weather-resistive barrier shall be compliant with ASTM D 226 Type I asphalt saturated felt or equivalent, shall be applied horizontally with the upper layer lapped over the lower layer not less than 2 inches (51 mm), and shall have all vertical joints lapped not less than 6 inches (152 mm).

❖ The use of a weather-resistant barrier with EIFS is necessary between the structural components and the

WALL COVERING

Figure R703.8
FLASHING EXAMPLES

wall covering, as it is with most other wall-covering systems. In addition, there must be a means of removal for any moisture that penetrates into the wall system. The type and installation of the required weather-resistant barrier for EIFS is consistent with that required for other types of exterior wall systems. The code allows drainable EIFS.

R703.9.2 Flashing, general. Flashing of EIFS shall be provided in accordance with the requirements of Section R703.8.

❖ The general provisions for flashing at the exterior wall envelope are also applicable to exterior insulation finish systems.

R703.10 Fiber cement siding.

❖ Fiber cement siding is a recent addition to the list of material used for weather-resistant siding. It is manufactured from fiber-reinforced cement, and this code permits its use as either panel siding or horizontal lap siding.

R703.10.1 Panel siding. Panels shall be installed with the long dimension parallel to framing. Vertical joints shall occur over framing members and shall be sealed with caulking or covered with battens. Horizontal joints shall be flashed with Z-flashing and blocked with solid wood framing.

❖ Panel siding of fiber cement is acceptable exterior wall covering provided the joints are made waterproof. Typical methods for accomplishing the joint treatment include (1) having the joints occur over framing and protecting such joints with caulking or a continuous batten and (2) flashing and blocking horizontal joints. See the commentary for Figures R703.3.1(1) and R703.3.1(2) for examples of these joint treatments.

R703.10.2 Horizontal lap siding. Lap siding shall be lapped a minimum of $1^1/_4$ inches and shall have the ends sealed with caulking, covered with an H-section joint cover, or located over a strip of flashing. Lap siding courses may be installed with the fastener heads exposed or concealed, according to approved manufacturers' installation instructions.

❖ In order to maintain the necessary weather protection, horizontal siding must have a minimum lap of at least $1^1/_4$ inches (31.8 mm). The end joints must be treated with one of the following methods: (1) sealed with caulk, (2) covered with an H-section joint cover or (3) located over a strip of flashing. Fastener heads may be exposed or concealed in accordance with approved manufacturers' installation instructions.

Bibliography

The following resource materials are referenced in this chapter or are relevant to the subject matter addressed in this chapter.

ANSI/AHA A135.5-95, *Prefinished Hardboard Paneling*. Palatine, IL: American Hardboard Association, 1995.

ANSI/AHA A136.6-98, *Hardboard Siding*. Palatine, IL: American Hardboard Association, 1998.

ANSI A108.1A-99, *Installation of Ceramic Tile in the Wet-set Method, with Portland Cement Mortar*. New York: American National Standards Institute, 1999.

ANSI A108.1B-99, *Installation of Ceramic Tile, Quarry Tile on a Cured Portland Cement Mortar Setting Bed with Dry-set or Latex-Portland Mortar*. New York: American National Standards Institute, 1999.

ANSI A108.4-99, *Installation of Ceramic Tile with Organic Adhesives or Water Cleanable Tile-setting Epoxy Adhesive*. New York: American National Standards Institute, 1999.

ANSI A108.5-99, *Installation of Ceramic Tile with Dry-set Portland Cement Mortar or Latex-Portland Cement Mortar*. New York: American National Standards Institute, 1999.

ANSI A108.6-99, Installation of Ceramic Tile with Chemical Resistant, Water Cleanable Tile-setting and -grouting Epoxy. New York: American National Standards Institute, 1999.

ANSI A108.11-99, *Interior Installation of Cementitious Backer Units*. New York: American National Standards Institute, 1999.

ANSI A118.1-99, *American National Standard Specifications for Dry-set Portland Cement Mortar*. New York: American National Standards Institute, 1999.

ANSI A118.3-99, *American National Standard Specification for Chemical Resistant, Water Cleanable Tile-setting and Grouting Epoxy and Water Cleanable Tile-setting Epoxy Adhesive*. New York: American National Standards Institute, 1999.

ANSI A136.1-99, *American National Standard Specifications for Organic Adhesives for Installation of Ceramic Tile*. New York: American National Standards Institute, 1999.

ANSI A137.1-88, *American National Standard Specifications for Ceramic Tile*. New York: American National Standards Institute, 1988.

ANSI/HPVA HP-1-2000, *American National Standard for Hardwood and Decorative Plywood*. Reston, VA: Hardwood Plywood & Veneer Association, 2000.

ASTM C 5-98, *Standard Specification for Quicklime for Structural Purposes*. West Conshohocken, PA: ASTM International, 1998.

ASTM C 28/C28M-00, *Standard Specification for Gypsum Plasters*. West Conshohocken, PA: ASTM International, 2000.

ASTM C 35-95(2001), *Standard Specification for Inorganic Aggregates for Use in Gypsum Plaster*. West Conshohocken, PA: ASTM International, 2001.

ASTM C 36/C0036M-01, *Specification for Gypsum Wallboard*. West Conshohocken, PA: ASTM International, 2001.

ASTM C 37/C0037M-01, *Specification for Gypsum Lath*. West Conshohocken, PA: ASTM International, 2001.

ASTM C 59/C0059M-00, *Specification for Gypsum Casting Plaster and Gypsum Molding Plaster*. West Conshohocken, PA: ASTM International, 2000.

ASTM C 61/C0061M-00, *Specification for Gypsum Keene's Cement*. West Conshohocken, PA: ASTM International, 2000.

ASTM C 79-01, *Specification for Treated Core and Nontreated Core Gypsum Sheathing Board*. West Conshohocken, PA: ASTM International, 2001.

ASTM C 475-01, *Specification for Joint Compound and Joint Tape for Finishing Gypsum Board*. West Conshohocken, PA: ASTM International, 2001.

ASTM C 514-01, *Specification for Nails for the Application of Gypsum Wallboard*. West Conshohocken, PA: ASTM International, 2001.

ASTM C 557-99, *Standard Specification for Adhesives for Fastening Gypsum Wallboard to Wood Framing*. West Conshohocken, PA: ASTM International, 1999.

ASTM C 587-97, *Standard Specification for Gypsum Veneer Plaster*. West Conshohocken, PA: ASTM International, 1997.

ASTM C 588/C0588-99, *Specification for Gypsum Base for Plaster Veneers*. West Conshohocken, PA: ASTM International, 1999.

ASTM C 630/0630M-01, *Specification for Water-resistant Gypsum Backing Board*. West Conshohocken, PA: ASTM International, 2001.

ASTM C 631-(2000), *Specification for Bonding Compounds for Interior Gypsum Plastering*. West Conshohocken, PA: ASTM International, 2000.

ASTM C 645-00, *Specification for Nonstructural Steel Framing Members*. West Conshohocken, PA: ASTM International, 2000.

ASTM C 843-99, *Specification for Application of Gypsum Veneer Plaster*. West Conshohocken, PA: ASTM International, 1999.

ASTM C 844-99, *Specification for Application of Gypsum Base to Receive Gypsum Veneer Plaster*. West Conshohocken, PA: ASTM International, 1999.

WALL COVERING

ASTM C 847-(2000), *Specification for Metal Lath*. West Conshohocken, PA: ASTM International, 2000.

ASTM C 897-00, *Specification for Aggregate for Job-Mixed Portland Cement-Based Plasters*. West Conshohocken, PA: ASTM International, 2000.

ASTM C 933-96a(2001), *Specification for Welded Wire Lath*. West Conshohocken, PA: ASTM International, 2001.

ASTM C 954-00, *Specification for Steel Drill Screws for the Application of Gypsum Panel Products or Metal Plaster Bases to Steel Studs From 0.033 in. (0.84 mm) to 0.112 in. (2.84 mm) in Thickness*. West Conshohocken, PA: ASTM International, 2000.

ASTM C 955-01, *Specification for Load-bearing Transverse and Axial Steel Studs, Runner Tracks, and Bracing or Bridging for Screw Application of Gypsum Panel Products and Metal Plaster Bases*. West Conshohocken, PA: ASTM International, 2001.

ASTM C 960-01, *Specification for Predecorated Gypsum Board*. West Conshohocken, PA: ASTM International, 2001.

ASTM C 1002-01, *Specification for Steel Drill Screws for the Application of Gypsum Panel Products or Metal Plaster Bases*. West Conshohocken, PA: ASTM International, 2001.

ASTM C 1032-96, *Specification for Woven Wire Plaster Base*. West Conshohocken, PA: ASTM International, 1996.

ASTM C 1047-99, *Specification for Accessories for Gypsum Wallboard and Gypsum Veneer Base*. West Conshohocken, PA: ASTM International, 1999.

ASTM C 1063-99, *Specification for Installation of Lathing and Furring to Receive Interior and Exterior Portland Cement-Based Plaster*. West Conshohocken, PA: ASTM International, 1999.

ASTM C 1177/C 1177M-01, *Specification for Glass Mat Gypsum Substrate for Use as Sheathing*. West Conshohocken, PA: ASTM International, 2001.

ASTM C 1178/C 1178M-01, *Specification for Glass Mat Water-resistant Gypsum Backing Panel*. West Conshohocken, PA: ASTM International, 2001.

ASTM C 1278/C 1278M-01, *Specification for Fiber-reinforced Gypsum Panels*. West Conshohocken, PA: ASTM International, 2001.

ASTM C 1395/C 1395M-01, *Specification for Gypsum Ceiling Board*. West Conshohocken, PA: ASTM International, 2001.

ASTM D 226-97a, *Specification for Asphalt-saturated Organic Felt Used in Roofing and Waterproofing*. West Conshohocken, PA: ASTM International, 1997.

CSSB-97, *Grading and Packing Rules for Western Red Cedar Shakes and Western Red Shingles of the Cedar Shake and Shingle Bureau*. Bellevue, WA: Cedar Shake and Shingle Bureau, 1997.

Chapter 8:
Roof-ceiling Construction

General Comments

Section R801 establishes the scope of the chapter as well as performance requirements for roof and ceiling construction. There are two roof-ceiling framing systems: wood framing and steel framing. Section R802 addresses wood roof and ceiling framing, while Section R804 deals with steel roof and ceiling framing. Section R803 specifies requirements for roof sheathing to be used with either of these framing systems. Other topics covered in this chapter are the application of ceiling finishes in Section R805, the proper ventilation of concealed spaces in roofs (e.g. enclosed attics and rafter spaces) in Section R806, the "access into attic" criteria in Section R807 and the proper clearance of combustible insulation from heat-producing devices in Section R808.

Purpose

Chapter 8 regulates the design and construction of roof-ceiling systems. Proper roof-member design provides for the support of the required design live and snow loads as well as providing for the transfer of these loads to supporting walls. Attics must be identified as "no storage" or "limited storage" zones so that ceiling joists can be appropriately sized. The allowable span tables are formatted to simplify the selection of rafter and ceiling joist size.

Roof systems must resist wind uplift, and tiedowns must be provided when necessary. Also, the roof system comprised of sheathing fastened to either wood or steel framing serves as a diaphragm, which is a key lateral load path element in resisting the forces of wind and earthquakes.

This chapter also includes roof ventilation requirements, aimed at minimizing the adverse effects of moisture. The requirements for insulation clearance protect against combustible insulation coming into contact with a heat source.

SECTION R801
GENERAL

R801.1 Application. The provisions of this chapter shall control the design and construction of the roof-ceiling system for all buildings.

❖ The provisions of Chapter 8 address conventionally framed roof-ceiling construction.

R801.2 Requirements. Roof and ceiling construction shall be capable of accommodating all loads imposed according to Section R301 and of transmitting the resulting loads to the supporting structural elements.

❖ This section states the performance expectations for roof and ceiling construction.

R801.3 Roof drainage. In areas where expansive or collapsible soils are known to exist, all dwellings shall have a controlled method of water disposal from roofs that will collect and discharge all roof drainage to the ground surface at least 5 feet (1524 mm) from foundation walls or to an approved drainage system.

❖ Saturated expansive or collapsible soils can lead to foundation failures because their additional loads are imposed on the foundation wall. To minimize the potential for the soil adjacent to the foundation wall to become saturated due to roof drainage, the code requires that the roofs drain 5 feet (1524 mm) from the foundation. This requirement applies to foundation walls only and does not apply to slab-on-grade foundations.

SECTION R802
WOOD ROOF FRAMING

R802.1 Identification. Load-bearing dimension lumber for rafters, trusses and ceiling joists shall be identified by a grade mark of a lumber grading or inspection agency that has been approved by an accreditation body that complies with DOC PS 20. In lieu of a grade mark, a certificate of inspection issued by a lumber grading or inspection agency meeting the requirements of this section shall be accepted.

❖ See the commentary for Section R502.1 and Figure R502.1.

R802.1.1 Blocking. Blocking shall be a minimum of utility grade lumber.

❖ Since blocking is a less demanding task than load carrying, the code permits the use of utility grade lumber.

R802.1.2 End-jointed lumber. Approved end-jointed lumber identified by a grade mark conforming to Section R802.1 may be used interchangeably with solid-sawn members of the same species and grade.

❖ See the commentary for Section R502.1.3.

R802.1.3 Fire-retardant-treated wood. Fire-retardant-treated wood is any wood product which, when impregnated with chemicals by a pressure process or other means during manufacture, shall have, when tested in accordance with ASTM E84, a listed flame spread index of 25 or less and show no evidence of significant progressive combustion when the test is continued for an additional 20-minute period. In addition, the flame front shall not progress more than 10.5 feet (3200 mm) beyond the center line of the burners at any time during the test.

❖ Fire-retardant-treated wood (FRTW) is plywood and lumber that has been pressure impregnated with chemicals to improve its flame spread characteristics beyond those of untreated wood. The effectiveness of the pressure-impregnated fire retardant treatment is determined by subjecting the material to tests conducted in accordance with ASTM E 84, with the modification that the test is extended to 30 minutes rather than 15 minutes. Using this procedure, a flame spread index is established during the standard 10-minute test period. The test is continued for an additional 20 minutes. During this added time period, there must not be any significant flame spread. At no time must the flame spread more than $10^1/_2$ feet (3200 mm) past the centerline of the burners.

The result of impregnating wood with fire retardant chemicals is a chemical reaction at certain temperature ranges. This reaction reduces the release of certain intermediate products that contribute to the flaming of wood, and the reaction also results in the formation of a greater percentage of charcoal and water. Some chemicals are effective in reducing the oxidation rate for charcoal residue. Fire retardant chemicals also reduce the heat release rate of FRTW when it is burning over a wide range of temperatures. This section gives provisions for the treatment and use of fire-retardant-treated wood.

R802.1.3.1 Labeling. Fire-retardant-treated lumber and wood structural panels shall be labeled. The label shall contain:

1. The identification mark of an approved agency in accordance with Section 1703.5 of the *International Building Code.*
2. Identification of the treating manufacturer.
3. The name of the fire-retardant treatment.
4. The species of wood treated.
5. Flame spread and smoke developed rating.
6. Method drying after treatment.
7. Conformance with appropriate standards in accordance with Sections R802.1.3.2 through R802.1.3.5.
8. For FRTW exposed to weather, damp or wet location, the words "No increase in the listed classification when subjected to the Standard Rain Test" (ASTM D2898).

❖ For continued quality, each piece of fire-retardant-treated wood must be identified by an approved agency having a reinspection service. The identification must show the performance rating of the material, including the 30-minute ASTM E 84 test results determined in Section R802.1.3 and the design adjustment values determined in Section R802.1.3.2. The third-party agency that provides the FRTW label is also required to state on the label that the FRTW complies with the requirements of Section R802.1.3 and that design adjustment values have been determined for the FRTW in compliance with the provisions of Section R802.1.3.2.

The FRTW label must be distinct from the grading label to avoid confusion between the two. The grading label provides information about the properties of wood before it is fire-retardant treated; the FRTW label provides properties of the wood after FRTW treatment. It is imperative that the FRTW label be presented in such a manner that it complements the grading label and does not create confusion over which label takes precedence.

R802.1.3.2 Strength adjustments. Design values for untreated lumber and wood structural panels as specified in Section R802.1, shall be adjusted for fire retardant- treated wood. Adjustments to design values shall be based upon an approved method of investigation which takes into consideration the effects of the anticipated temperature and humidity to which the fire-retardant-treated wood will be subjected, the type of treatment and redrying procedures.

❖ Experience has shown that certain factors can affect the physical properties of FRTW. Among these factors are the pressure treatment and redrying processes used and the extremes of temperature and humidity that the FRTW will be subjected to once installed. The design values for all FRTW must be adjusted for the effects of the treatment and environmental conditions, such as high temperature and humidity in attic installations. This section requires the determination of these design adjustment values, based on an investigation procedure that includes subjecting the FRTW to similar temperatures and humidities and that has been approved by the code official. The FRTW tested must be identical to that which is produced. Items to be considered by the code official reviewing the test procedure include species and grade of the untreated wood and conditioning of the wood, such as drying before the fire retardant treatment process. A fire-retardant wood treater may choose to have its treatment process evaluated by model code evaluation services.

The FRTW is required by Section R802.1.3.1 to be labeled with the design adjustment values. These can take the form of factors that are multiplied by the original design values of the untreated wood to determine its allowable stresses or new allowable stresses that have already been factored down in consideration of the FRTW treatment.

R802.1.3.2.1 Wood structural panels. The effect of treatment and the method of redrying after treatment, and exposure to high temperatures and high humidities on the flexure properties of fire-retardant-treated softwood plywood shall be determined in accordance with ASTM D 5516. The test data developed by ASTM D 5516 shall be used to develop adjustment factors, maximum loads and spans, or both for untreated plywood design values in accordance with ASTM D 6305. Each manufacturer shall publish the allowable maximum loads and spans for service as floor and roof sheathing for their treatment.

❖ This section references the test standard developed to evaluate the flexural properties of fire-retardant-treated plywood that is exposed to high temperatures. Note that while the section title refers to wood structural panels, the standard is specifically for softwood plywood. Therefore judgment is required in determining the effects of elevated temperature and humidity on other types of wood structural panels.

R802.1.3.2.2 Lumber. For each species of wood treated the effect of the treatment and the method of redrying after treatment and exposure to high temperatures and high humidities on the allowable design properties of fire-retardant-treated lumber shall be determined in accordance with ASTM D 5664. The test data developed by ASTM D 5664 shall be used to develop modification factors for use at or near room temperature and at elevated temperatures and humidity in accordance with an approved method of investigation. Each manufacturer shall publish the modification factors for service at temperatures of not less than 80°F (26.7°C) and for roof framing. The roof framing modification factors shall take into consideration the climatological location.

❖ This section references the test standard developed to determine the necessary adjustments to design values for lumber that has been fire-retardant treated and includes the effects of elevated temperatures.

R802.1.3.3 Exposure to weather. Where fire-retardant-treated wood is exposed to weather, or damp or wet locations, it shall be identified as "Exterior" to indicate there is no increase in the listed flamespread index as defined in Section R802.1.3 when subjected to ASTM D 2898.

❖ Some fire-retardant treatments are soluble when exposed to the weather or used under high-humidity conditions. The humidity threshold established for interior applications in Section R802.1.3.4 is 92 percent. Therefore fire-retardant-treated wood used in an interior location that will exceed this threshold must comply with this section. When a fire-retardant-treated wood product is to be exposed to any of the conditions noted in this section, it must be further tested in accordance with ASTM D 2898. Testing requires the material to meet the performance criteria listed in Section R802.1.3. The material is then subjected to the ASTM weathering test and retested after drying. There must not be any significant differences in the performance recorded before and after the weathering test.

R802.1.3.4 Interior applications. Interior fire-retardant-treated wood shall have a moisture content of not over 28 percent when tested in accordance with ASTM D 3201 procedures at 92 percent relative humidity. Interior fire-retardant-treated wood shall be tested in accordance with Section R802.1.3.2.1 or R802.1.3.2.2. Interior fire-retardant-treated wood designated as Type A shall be tested in accordance with the provisions of this section.

❖ The environment in which the fire-retardant-treated wood is used can affect its performance. In order to make sure that the performance will be adequate in a humid interior condition, the testing in accordance with ASTM D 3201 as well as testing specified in Sections R802.1.3.2.1 or R802.1.3.2.2 is required for all interior wood. Requiring all interior wood to be tested for the effects of high temperature and humidity reduces the chance of a premature failure due to improper use of fire-retardant-treated wood.

R802.1.3.5 Moisture content. Fire-retardant-treated wood shall be dried to a moisture content of 19 percent or less for lumber and 15 percent or less for wood structural panels before use. For wood kiln dried after treatment (KDAT) the kiln temperatures shall not exceed those used in kiln drying the lumber and plywood submitted for the tests described in Section R802.1.3.2.1 for plywood and R802.1.3.2.2 for lumber.

❖ These moisture content thresholds are necessary to prevent leaching of the fire retardant from the wood. Section R802.1.3.2 requires that the strength adjustments consider the redrying procedure, and this section clarifies that the drying temperatures for wood kiln-dried after treatment must be consistent with those on which the adjustment factors were based.

R802.1.4 Structural glued laminated timbers. Glued laminated timbers shall be manufactured and identified as required in AITC A190.1 and ASTM D3737.

❖ Glulam timbers are required by this section to be manufactured following ANSI/AITC 90.1 and ASTM D 3737. Knowing the standard these products must meet makes it easier to determine that the product found in the field will meet the design requirements.

R802.2 Design and construction. Roof-ceilings shall be designed and constructed in accordance with the provisions of this chapter and Figures R606.10(1), R606.10(2) and R606.10(3) or in accordance with AFPA/NDS. Components of roof-ceilings shall be fastened in accordance with Table R602.3(1).

❖ Wood roof-ceiling construction must comply with Section R802 or AF&PA/NDS, *National Design Specification for Wood Construction*. The references to Chapter 6 figures are meant for buildings using masonry wall construction.

R802.3 Framing details. Rafters shall be framed to ridge board or to each other with a gusset plate as a tie. Ridge board shall be at least 1-inch (25.4 mm) nominal thickness and not less in depth than the cut end of the rafter. At all valleys and hips there shall be a valley or hip rafter not less than 2-inch (51 mm) nominal thickness and not less in depth than the cut end of the rafter. Hip and valley rafters shall be supported at the ridge by a brace to a bearing partition or be designed to carry and distribute the specific load at that point. Where the roof pitch is less than three units vertical in 12 units horizontal (25-percent slope), structural members that support rafters and ceiling joists, such as ridge beams, hips and valleys, shall be designed as beams.

❖ Traditional practice is to provide a ridgeboard between opposite rafters as a nailing base and to provide a full bearing for the rafter. Rafters must be placed directly opposite each other, and the ridgeboard must have a depth equal to the end of the rafter as illustrated in Commentary Figure R802.3(1). Commentary Figure R802.3(2) shows an option for framing opposing rafters using a gusset plate.

R802.3.1 Ceiling joist and rafter connections. Ceiling joists and rafters shall be nailed to each other in accordance with Tables R602.3(1) and R802.5.1(9), and the assembly shall be nailed to the top wall plate in accordance with Table R602.3(1). Ceiling joists shall be continuous or securely joined where they meet over interior partitions and nailed to adjacent rafters to provide a continuous tie across the building when such joists are parallel to the rafters.

Where ceiling joists are not parallel to rafters, subflooring or metal straps attached to the ends of the rafters shall be installed in a manner to provide a continuous tie across the building, or rafters shall be tied to 1-inch by 4-inch (25.4 mm by 102 mm) (nominal) minimum-size crossties. The connections shall be in accordance with Table R602.3(1) or connections of equivalent capacities shall be provided. Where ceiling joists or rafter ties are not provided at the top plate, the ridge formed by these rafters shall also be supported by a girder designed in accordance with accepted engineering practice.

Rafter ties shall be spaced not more than 4 feet (1219 mm) on center.

For **SI:** 1 inch = 25.4 mm.

Figure R802.3(1)
RIDGE BOARD

Figure R802.3(2)
RIDGE ALTERNATIVE—GUSSET PLATE TIE

ROOF-CEILING CONSTRUCTION

❖ To transfer roof loads supported by hip and valley rafters to the supporting structure below, the rafters must frame into a ridge supported by a structural member, such as a brace, to a supporting element below.

So that joists do not become accidentally displaced and so that they transfer thrust from the rafters to the ceiling joists, a mechanical connection to supporting members as shown in Commentary Figure 802.3.1(1) must be provided. To resist the horizontal thrust generated at the exterior walls by the loading of rafters, a continuous tie between the exterior walls is required. Commentary Figures R802.3.1(2) and R802.3.1(3) illustrate methods of accomplishing this. In Commentary Figure R802.3.1(3), the ceiling joists running parallel with the roof rafter framing provide the continuous tie. Where ceiling joists are not parallel with the roof rafters, separate cross ties (rafter ties) are to be provided, as shown in Commentary Figure R802.3.1(2).

Cathedral ceilings (where ceiling joists or rafter ties are not provided) must be designed so the wall or ridge beam carries the full load of the roof. This would require the walls to be supported by a continuous foundation and/or beams and girders. The ridge beam must also be capable of supporting the full load exerted by the tributary area of the rafters. The ridge beam must be stiff enough to minimize the deflection from the rafter loads so that rafter thrust does not displace the walls.

The rafter-to-ceiling joist connection requirements of Table R802.5.1(9) apply when all of the following conditions exist:
- Snow load is at least 30 pounds per square foot (psf) (1.44 Kn/m^2)
- Roof slope is at least 3:12
- Roof ridge is not supported by a load-bearing member.

Otherwise, the fastening schedule of Table R602.3(1) applies.

R802.3.2 Ceiling joists lapped. Ends of ceiling joists shall be lapped a minimum of 3 inches (76 mm) or butted over bearing partitions or beams and toenailed to the bearing member. When ceiling joists are used to provide resistance to rafter thrust, lapped joists shall be nailed together in accordance with Table R602.3(1) and butted joists shall be tied together in a manner to resist such thrust.

❖ Where ceiling joists function as a structural element, providing resistance to rafter thrust, proper connection through a lapped joint or butted joint connection to form a continuous tie is required. See Commentary Figures R802.3.2(1) and R802.3.2(2).

R802.4 Allowable ceiling joist spans. Spans for ceiling joists shall be in accordance with Tables R802.4(1) and R802.4(2). For other grades and species and for other loading conditions, refer to the AF&PA Span Tables for Joists and Rafters.

Figure R802.3.1(1)
CEILING JOIST AT SUPPORTS

❖ Tables R802.4(1) and R802.4(2) list allowable ceiling joist spans for common lumber sizes, species, and grades based on spacing and design loads. These tables are similar to the rafter tables explained in the commentary for Section R802.5.

R802.5 Allowable rafter spans. Spans for rafters shall be in accordance with Tables R802.5.1(1) through R802.5.1(8). For other grades and species and for other loading conditions, refer to the AF&PA Span Tables for Joists and Rafters. The span of each rafter shall be measured along the horizontal projection of the rafter.

❖ Tables R802.5.1(1) through R802.5.1(8) list allowable rafter spans for common lumber sizes, species, and grades based on spacing and design loads. Tables R802.5.1(2) through R802.5.1(8) are based on ground snow loads of 30 psf, 50 psf and 70 psf. The snow loads are based on ASCE 7-95 criteria as follows:

for balanced condition $p_s = C_s * p_f = 0.7 * C_e * C_t * C_s * p_g = 0.77 * p_g$ where $C_e = 1.9$, $C_t = 1.1$, and $C_s = 1.0$.

for unbalanced condition $p_s = 1.3\ p_f/C_e = 0.7*1.3*C_t*C_s*p_g = 1.00*p_g$ where $C_t = 1.1$ and $C_s = 1.0$.

The unbalanced condition controls. By inspection, the unbalanced condition is equal to the ground snow load.

Example 1 in the commentary for Table R802.5.1(1) explains the allowable spans for clear-span roof rafters using the code tables. As illustrated in Figure R802.5.1, the horizontal projection of the rafter may be used for the span length rather than using the true length of the rafter measured along the slope of the roof.

Figure R802.3.1(2)
ROOF FRAMING WITH CEILING JOISTS NOT PARALLEL TO RAFTERS

R802.5.1 Purlins. Purlins are permitted to be installed to reduce the span of rafters as shown in Figure R802.5.1. Purlins shall be sized no less than the required size of the rafters that they support. Purlins shall be continuous and shall be supported by 2-inch by 4-inch (51 mm by 102 mm) braces installed to bearing walls at a slope not less than 45 degrees from the horizontal. The braces shall be spaced not more than 4 feet (1219 mm) on center and the unbraced length of braces shall not exceed 8 feet (2438 mm).

❖ This section provides specific instructions for the installation of purlins that will provide intermediate support for the rafters and thus reduce the rafter span. Figure R802.5.1 illustrates purlin installation. Example 2 (below) shows the effect on rafter design and roof purlin requirements when struts to interior walls support the roof rafters.

R802.6 Bearing. The ends of each rafter or ceiling joist shall have not less than $1^1/_2$ inches (38 mm) of bearing on wood or metal and not less than 3 inches (76 mm) on masonry or concrete.

❖ The requirements for roof rafters and ceiling joists bearing on wood, metal, and masonry are similar to the requirements for floor joists. See the commentary for Section R502.6 relative to the bearing requirements for wood, metal, and masonry.

R802.6.1 Finished ceiling material. If the finished ceiling material is installed on the ceiling prior to the attachment of the ceiling to the walls, such as in construction at a factory, a compression strip of the same thickness as the finish ceiling material shall be installed directly above the top plate of bearing walls if the compressive strength of the finish ceiling material is less than the loads it will be required to withstand. The compression strip shall cover the entire length of such top plate and shall be at least one-half the width of the top plate. It shall be of material capable of transmitting the loads transferred through it.

❖ In prefabricated (panelized) construction where the finished ceiling material is attached to the roof assembly before it is attached to the walls, care must be taken to provide a compression strip to line up with the top plate of the supporting walls in order to provide an adequate bearing surface.

ROOF-CEILING CONSTRUCTION

Figure R802.3.1(3)
ROOF FRAMING WITH CEILING JOISTS PARALLEL TO RAFTERS

Figure R802.3.2(1)
LAPPED CEILING JOIST CONNECTION PROVIDING RESISTANCE TO RAFTER THRUST

Figure R802.3.2(2)
BUTTED CEILING JOIST CONNECTION PROVIDING RESISTANCE TO RAFTER THRUST

TABLE R802.4(1)
CEILING JOIST SPANS FOR COMMON LUMBER SPECIES
(Uninhabitable attics without storage, live load = 10 psf, L/Δ = 240)

CEILING JOIST SPACING (inches)	SPECIES AND GRADE		DEAD LOAD = 5 psf			
			2x4	2x6	2x8	2x10
			Maximum ceiling joist spans			
			(feet - inches)	(feet - inches)	(feet - inches)	(feet - inches)
12	Douglas fir-larch	SS	13-2	20-8	Note a	Note a
	Douglas fir-larch	#1	12-8	19-11	Note a	Note a
	Douglas fir-larch	#2	12-5	19-6	25-8	Note a
	Douglas fir-larch	#3	10-10	15-10	20-1	24-6
	Hem-fir	SS	12-5	19-6	25-8	Note a
	Hem-fir	#1	12-2	19-1	25-2	Note a
	Hem-fir	#2	11-7	18-2	24-0	Note a
	Hem-fir	#3	10-10	15-10	20-1	24-6
	Southern pine	SS	12-11	20-3	Note a	Note a
	Southern pine	#1	12-8	19-11	Note a	Note a
	Southern pine	#2	12-5	19-6	25-8	Note a
	Southern pine	#3	11-6	17-0	21-8	25-7
	Spruce-pine-fir	SS	12-2	19-1	25-2	Note a
	Spruce-pine-fir	#1	11-10	18-8	24-7	Note a
	Spruce-pine-fir	#2	11-10	18-8	24-7	Note a
	Spruce-pine-fir	#3	10-10	15-10	20-1	24-6

(continued)

ROOF-CEILING CONSTRUCTION

TABLE R802.4(1)—continued
CEILING JOIST SPANS FOR COMMON LUMBER SPECIES
(Uninhabitable attics without storage, live load = 10 psf, L/Δ = 240)

CEILING JOIST SPACING (inches)	SPECIES AND GRADE		DEAD LOAD = 5 psf			
			2x4	2x6	2x8	2x10
			Maximum ceiling joist spans			
			(feet - inches)	(feet - inches)	(feet - inches)	(feet - inches)
16	Douglas fir-larch	SS	11-11	18-9	24-8	Note a
	Douglas fir-larch	#1	11-6	18-1	23-10	Note a
	Douglas fir-larch	#2	11-3	17-8	23-0	Note a
	Douglas fir-larch	#3	9-5	13-9	17-5	21-3
	Hem-fir	SS	11-3	17-8	23-4	Note a
	Hem-fir	#1	11-0	17-4	22-10	Note a
	Hem-fir	#2	10-6	16-6	21-9	Note a
	Hem-fir	#3	9-5	13-9	17-5	21-3
	Southern pine	SS	11-9	18-5	24-3	Note a
	Southern pine	#1	11-6	18-1	23-1	Note a
	Southern pine	#2	11-3	17-8	23-4	Note a
	Southern pine	#3	10-0	14-9	18-9	22-2
	Spruce-pine-fir	SS	11-0	17-4	22-10	Note a
	Spruce-pine-fir	#1	10-9	16-11	22-4	Note a
	Spruce-pine-fir	#2	10-9	16-11	22-4	Note a
	Spruce-pine-fir	#3	9-5	13-9	17-5	21-3
19.2	Douglas fir-larch	SS	11-3	17-8	23-3	Note a
	Douglas fir-larch	#1	10-10	17-0	22-5	Note a
	Douglas fir-larch	#2	10-7	16-7	21-0	25-8
	Douglas fir-larch	#3	8-7	12-6	15-10	19-5
	Hem-fir	SS	10-7	16-8	21-11	Note a
	Hem-fir	#1	10-4	16-4	21-6	Note a
	Hem-fir	#2	9-11	15-7	20-6	25-3
	Hem-fir	#3	8-7	12-6	15-10	19-5
	Southern -pine	SS	11-0	17-4	22-10	Note a
	Southern pine	#1	10-10	17-0	22-5	Note a
	Southern pine	#2	10-7	16-8	21-11	Note a
	Southern pine	#3	9-1	13-6	17-2	20-3
	Spruce-pine-fir	SS	10-4	16-4	21-6	Note a
	Spruce-pine-fir	#1	10-2	15-11	21-0	25-8
	Spruce-pine-fir	#2	10-2	15-11	21-0	25-8
	Spruce-pine-fir	#3	8-7	12-6	15-10	19-5
24	Douglas fir-larch	SS	10-5	16-4	21-7	Note a
	Douglas fir-larch	#1	10-0	15-9	20-1	24-6
	Douglas fir-larch	#2	9-10	14-10	18-9	22-11
	Douglas fir-larch	#3	7-8	11-2	14-2	17-4
	Hem-fir	SS	9-10	15-6	20-5	Note a
	Hem-fir	#1	9-8	15-2	19-7	23-11
	Hem-fir	#2	9-2	14-5	18-6	22-7
	Hem-fir	#3	7-8	11-2	14-2	17-4
	Southern pine	SS	10-3	16-1	21-2	Note a
	Southern pine	#1	10-0	15-9	20-10	Note a
	Southern pine	#2	9-10	15-6	20-1	23-11
	Southern pine	#3	8-2	12-0	15-4	18-1
	Spruce-pine-fir	SS	9-8	15-2	19-11	25-5
	Spruce-pine-fir	#1	9-5	14-9	18-9	22-11
	Spruce-pine-fir	#2	9-5	14-9	18-9	22-11
	Spruce-pine-fir	#3	7-8	11-2	14-2	17-4

Check sources for availability of lumber in lengths greater than 20 feet.

For SI: 1 inch = 25.4 mm, 1 foot = 304.8 mm, 1 pound per square foot = 0.0479 kN/m^2.

a. Span exceeds 26 feet in length.

❖ See the commentary for Section R802.4.

TABLE R802.4(2)
CEILING JOIST SPANS FOR COMMON LUMBER SPECIES
(Uninhabitable attics with limited storage, live load = 20 psf, L/Δ = 240)

CEILING JOIST SPACING (inches)	SPECIES AND GRADE		DEAD LOAD = 10 psf			
			2x4	2x6	2x8	2x10
			Maximum ceiling joist spans			
			(feet - inches)	(feet - inches)	(feet - inches)	(feet - inches)
12	Douglas fir-larch	SS	10-5	16-4	21-7	Note a
	Douglas fir-larch	#1	10-0	15-9	20-1	24-6
	Douglas fir-larch	#2	9-10	14-10	18-9	22-11
	Douglas fir-larch	#3	7-8	11-2	14-2	17-4
	Hem-fir	SS	9-10	15-6	20-5	Note a
	Hem-fir	#1	9-8	15-2	19-7	23-11
	Hem-fir	#2	9-2	14-5	18-6	22-7
	Hem-fir	#3	7-8	11-2	14-2	17-4
	Southern pine	SS	10-3	16-1	21-2	Note a
	Southern pine	#1	10-0	15-9	20-10	Note a
	Southern pine	#2	9-10	15-6	20-1	23-11
	Southern pine	#3	8-2	12-0	15-4	18-1
	Spruce-pine-fir	SS	9-8	15-2	19-11	25-5
	Spruce-pine-fir	#1	9-5	14-9	18-9	22-11
	Spruce-pine-fir	#2	9-5	14-9	18-9	22-11
	Spruce-pine-fir	#3	7-8	11-2	14-2	17-4
16	Douglas fir-larch	SS	9-6	14-11	19-7	25-0
	Douglas fir-larch	#1	9-1	13-9	17-5	21-3
	Douglas fir-larch	#2	8-9	12-10	16-3	19-10
	Douglas fir-larch	#3	6-8	9-8	12-4	15-0
	Hem-fir	SS	8-11	14-1	18-6	23-8
	Hem-fir	#1	8-9	13-5	16-10	20-8
	Hem-fir	#2	8-4	12-8	16-0	19-7
	Hem-fir	#3	6-8	9-8	12-4	15-0
	Southern pine	SS	9-4	14-7	19-3	24-7
	Southern pine	#1	9-1	14-4	18-11	23-1
	Southern pine	#2	8-11	13-6	17-5	20-9
	Southern pine	#3	7-1	10-5	13-3	15-8
	Spruce-pine-fir	SS	8-9	13-9	18-1	23-1
	Spruce-pine-fir	#1	8-7	12-10	16-3	19-10
	Spruce-pine-fir	#2	8-7	12-10	16-3	19-10
	Spruce-pine-fir	#3	6-8	9-8	12-4	15-0
19.2	Douglas fir-larch	SS	8-11	14-0	18-5	23-4
	Douglas fir-larch	#1	8-7	12-6	15-10	19-5
	Douglas fir-larch	#2	8-0	11-9	14-10	18-2
	Douglas fir-larch	#3	6-1	8-10	11-3	13-8
	Hem-fir	SS	8-5	13-3	17-5	22-3
	Hem-fir	#1	8-3	12-3	15-6	18-11
	Hem-fir	#2	7-10	11-7	14-8	17-10
	Hem-fir	#3	6-1	8-10	11-3	13-8
	Southern pine	SS	8-9	13-9	18-1	23-1
	Southern pine	#1	8-7	13-6	17-9	21-1
	Southern pine	#2	8-5	12-3	15-10	18-11
	Southern pine	#3	6-5	9-6	12-1	14-4
	Spruce-pine-fir	SS	8-3	12-11	17-1	21-8
	Spruce-pine-fir	#1	8-0	11-9	14-10	18-2
	Spruce-pine-fir	#2	8-0	11-9	14-10	18-2
	Spruce-pine-fir	#3	6-1	8-10	11-3	13-8

(continued)

ROOF-CEILING CONSTRUCTION

TABLE R802.4(2)—continued
CEILING JOIST SPANS FOR COMMON LUMBER SPECIES
(Uninhabitable attics with limited storage, live load = 20 psf, L/Δ = 240)

CEILING JOIST SPACING (inches)	SPECIES AND GRADE		DEAD LOAD = 10 psf			
			2x4	2x6	2x8	2x10
			Maximum Ceiling Joist Spans			
			(feet - inches)	(feet - inches)	(feet - inches)	(feet - inches)
24	Douglas fir-larch	SS	8-3	13-0	17-1	20-11
	Douglas fir-larch	#1	7-8	11-2	14-2	17-4
	Douglas fir-larch	#2	7-2	10-6	13-3	16-3
	Douglas fir-larch	#3	5-5	7-11	10-0	12-3
	Hem-fir	SS	7-10	12-3	16-2	20-6
	Hem-fir	#1	7-6	10-11	13-10	16-11
	Hem-fir	#2	7-1	10-4	13-1	16-0
	Hem-fir	#3	5-5	7-11	10-0	12-3
	Southern pine	SS	8-1	12-9	16-10	21-6
	Southern pine	#1	8-0	12-6	15-10	18-10
	Southern pine	#2	7-8	11-0	14-2	16-11
	Southern pine	#3	5-9	8-6	10-10	12-10
	Spruce-pine-fir	SS	7-8	12-0	15-10	19-5
	Spruce-pine-fir	#1	7-2	10-6	13-3	16-3
	Spruce-pine-fir	#2	7-2	10-6	13-3	16-3
	Spruce-pine-fir	#3	5-5	7-11	10-0	12-3

Check sources for availability of lumber in lengths greater than 20 feet.

For SI: 1 inch = 25.4 mm, 1 foot = 304.8 mm, 1 pound per square foot = 0.0479 kN/m^2.

a. Span exceeds 26 feet in length

❖ See the commentary for Section R802.4.

TABLE R802.5.1(1)
RAFTER SPANS FOR COMMON LUMBER SPECIES
(Roof live load=20 psf, ceiling not attached to rafters, L/Δ=180)

RAFTER SPACING (inches)	SPECIES AND GRADE		DEAD LOAD = 10 psf					DEAD LOAD = 20 psf				
			2x4	2x6	2x8	2x10	2x12	2x4	2x6	2x8	2x10	2x12
			(feet-inches)	(feet-inches)	(feet-inches)	(feet-inches)	(feet-inches)	(feet-inches)	(feet-inches)	(feet-inches)	(feet-inches)	(feet-inches)
12	Douglas fir-larch	SS	11-6	18-0	23-9	Note b	Note b	11-6	18-0	23-5	Note b	Note b
	Douglas fir-larch	#1	11-1	17-4	22-5	Note b	Note b	10-6	15-4	19-5	23-9	Note b
	Douglas fir-larch	#2	10-10	16-7	21-0	25-8	Note b	9-10	14-4	18-2	22-3	25-9
	Douglas fir-larch	#3	8-7	12-6	15-10	19-5	22-6	7-5	10-10	13-9	16-9	19-6
	Hem-fir	SS	10-10	17-0	22-5	Note b	Note b	10-10	17-0	22-5	Note b	Note b
	Hem-fir	#1	10-7	16-8	21-10	Note b	Note b	10-3	14-11	18-11	23-2	Note b
	Hem-fir	#2	10-1	15-11	20-8	25-3	Note b	9-8	14-2	17-11	21-11	25-5
	Hem-fir	#3	8-7	12-6	15-10	19-5	22-6	7-5	10-10	13-9	16-9	19-6
	Southern pine	SS	11-3	17-8	23-4	Note b	Note b	11-3	17-8	23-4	Note b	Note b
	Southern pine	#1	11-1	17-4	22-11	Note b	Note b	11-1	17-3	21-9	25-10	Note b
	Southern pine	#2	10-10	17-0	22-5	Note b	Note b	10-6	15-1	19-5	23-2	Note b
	Southern pine	#3	9-1	13-6	17-2	20-3	24-1	7-11	11-8	14-10	17-6	20-11
	Spruce-pine-fir	SS	10-7	16-8	21-11	Note b	Note b	10-7	16-8	21-9	Note b	Note b
	Spruce-pine-fir	#1	10-4	16-3	21-0	25-8	Note b	9-10	14-4	18-2	22-3	25-9
	Spruce-pine-fir	#2	10-4	16-3	21-0	25-8	Note b	9-10	14-4	18-2	22-3	25-9
	Spruce-pine-fir	#3	8-7	12-6	15-10	19-5	22-6	7-5	10-10	13-9	16-9	19-6
16	Douglas fir-larch	SS	10-5	16-4	21-7	Note b	Note b	10-5	16-0	20-3	24-9	Note b
	Douglas fir-larch	#1	10-0	15-4	19-5	23-9	Note b	9-1	13-3	16-10	20-7	23-10
	Douglas fir-larch	#2	9-10	14-4	18-2	22-3	25-9	8-6	12-5	15-9	19-3	22-4
	Douglas fir-larch	#3	7-5	10-10	13-9	16-9	19-6	6-5	9-5	11-11	14-6	16-10
	Hem-fir	SS	9-10	15-6	20-5	Note b	Note b	9-10	15-6	19-11	24-4	Note b
	Hem-fir	#1	9-8	14-11	18-11	23-2	Note b	8-10	12-11	16-5	20-0	23-3
	Hem-fir	#2	9-2	14-2	17-11	21-11	25-5	8-5	12-3	15-6	18-11	22-0
	Hem-fir	#3	7-5	10-10	13-9	16-9	19-6	6-5	9-5	11-11	14-6	16-10
	Southern pine	SS	10-3	16-1	21-2	Note b	Note b	10-3	16-1	21-2	Note b	Note b
	Southern pine	#1	10-0	15-9	20-10	25-10	Note b	10-0	15-0	18-10	22-4	Note b
	Southern pine	#2	9-10	15-1	19-5	23-2	Note b	9-1	13-0	16-10	20-1	23-7
	Southern pine	#3	7-11	11-8	14-10	17-6	20-11	6-10	10-1	12-10	15-2	18-1
	Spruce-pine-fir	SS	9-8	15-2	19-11	25-5	Note b	9-8	14-10	18-10	23-0	Note b
	Spruce-pine-fir	#1	9-5	14-4	18-2	22-3	25-9	8-6	12-5	15-9	19-3	22-4
	Spruce-pine-fir	#2	9-5	14-4	18-2	22-3	25-9	8-6	12-5	15-9	19-3	22-4
	Spruce-pine-fir	#3	7-5	10-10	13-9	16-9	19-6	6-5	9-5	11-11	14-6	16-10
19.2	Douglas fir-larch	SS	9-10	15-5	20-4	25-11	Note b	9-10	14-7	18-6	22-7	Note b
	Douglas fir-larch	#1	9-5	14-0	17-9	21-8	25-2	8-4	12-2	15-4	18-9	21-9
	Douglas fir-larch	#2	8-11	13-1	16-7	20-3	23-6	7-9	11-4	14-4	17-7	20-4
	Douglas fir-larch	#3	6-9	9-11	12-7	15-4	17-9	5-10	8-7	10-10	13-3	15-5
	Hem-fir	SS	9-3	14-7	19-2	24-6	Note b	9-3	14-4	18-2	22-3	25-9
	Hem-fir	#1	9-1	13-8	17-4	21-1	24-6	8-1	11-10	15-0	18-4	21-3
	Hem-fir	#2	8-8	12-11	16-4	20-0	23-2	7-8	11-2	14-2	17-4	20-1
	Hem-fir	#3	6-9	9-11	12-7	15-4	17-9	5-10	8-7	10-10	13-3	15-5
	Southern pine	SS	9-8	15-2	19-11	25-5	Note b	9-8	15-2	19-11	25-5	Note b
	Southern pine	#1	9-5	14-10	19-7	23-7	Note b	9-3	13-8	17-2	20-5	24-4
	Southern pine	#2	9-3	13-9	17-9	21-2	24-10	8-4	11-11	15-4	18-4	21-6
	Southern pine	#3	7-3	10-8	13-7	16-0	19-1	6-3	9-3	11-9	13-10	16-6
	Spruce-pine-fir	SS	9-1	14-3	18-9	23-11	Note b	9-1	13-7	17-2	21-0	24-4
	Spruce-pine-fir	#1	8-10	13-1	16-7	20-3	23-6	7-9	11-4	14-4	17-7	20-4
	Spruce-pine-fir	#2	8-10	13-1	16-7	20-3	23-6	7-9	11-4	14-4	17-7	20-4
	Spruce-pine-fir	#3	6-9	9-11	12-7	15-4	17-9	5-10	8-7	10-10	13-3	15-5

(continued)

ROOF-CEILING CONSTRUCTION

TABLE R802.5.1(1)—continued
RAFTER SPANS FOR COMMON LUMBER SPECIES
(Roof live load=20 psf, ceiling not attached to rafters, $L/\Delta=180$)

RAFTER SPACING (inches)	SPECIES AND GRADE		DEAD LOAD = 10 psf					DEAD LOAD = 20 psf				
			2x4	2x6	2x8	2x10	2x12	2x4	2x6	2x8	2x10	2x12
			Maximum rafter spans[a]									
			(feet - inches)	(feet - inches)	(feet - inches)	(feet - inches)	(feet - inches)	(feet - inches)	(feet - inches)	(feet - inches)	(feet - inches)	(feet - inches)
24	Douglas fir-larch	SS	9-1	14-4	18-10	23-4	Note b	8-11	13-1	16-7	20-3	23-5
	Douglas fir-larch	#1	8-7	12-6	15-10	19-5	22-6	7-5	10-10	13-9	16-9	19-6
	Douglas fir-larch	#2	8-0	11-9	14-10	18-2	21-0	6-11	10-2	12-10	15-8	18-3
	Douglas fir-larch	#3	6-1	8-10	11-3	13-8	15-11	5-3	7-8	9-9	11-10	13-9
	Hem-fir	SS	8-7	13-6	17-10	22-9	Note b	8-7	12-10	16-3	19-10	23-0
	Hem-fir	#1	8-4	12-3	15-6	18-11	21-11	7-3	10-7	13-5	16-4	19-0
	Hem-fir	#2	7-11	11-7	14-8	17-10	20-9	6-10	10-0	12-8	15-6	17-11
	Hem-fir	#3	6-1	8-10	11-3	13-8	15-11	5-3	7-8	9-9	11-10	13-9
	Southern pine	SS	8-11	14-1	18-6	23-8	Note b	8-11	14-1	18-6	22-11	Note b
	Southern pine	#1	8-9	13-9	17-9	21-1	25-2	8-3	12-3	15-4	18-3	21-9
	Southern pine	#2	8-7	12-3	15-10	18-11	22-2	7-5	10-8	13-9	16-5	19-3
	Southern pine	#3	6-5	9-6	12-1	14-4	17-1	5-7	8-3	10-6	12-5	14-9
	Spruce-pine-fir	SS	8-5	13-3	17-5	21-8	25-2	8-4	12-2	15-4	18-9	21-9
	Spruce-pine-fir	#1	8-0	11-9	14-10	18-2	21-0	6-11	10-2	12-10	15-8	18-3
	Spruce-pine-fir	#2	8-0	11-9	14-10	18-2	21-0	6-11	10-2	12-10	15-8	18-3
	Spruce-pine-fir	#3	6-1	8-10	11-3	13-8	15-11	5-3	7-8	9-9	11-10	13-9

Check sources for availability of lumber in lengths greater than 20 feet.

For SI: 1 inch = 25.4 mm, 1 foot = 304.8 mm, 1 pound per square foot = 0.0479 kN/m².

a. The tabulated rafter spans assume that ceiling joists are located at the bottom of the attic space or that some other method of resisting the outward push of the rafters on the bearing walls, such as rafter ties, is provided at that location. When ceiling joists or rafter ties are located higher in the attic space, the rafter spans shall be multiplied by the factors given below:

H_C/H_R	Rafter Span Adjustment Factor
2/3 or greater	0.50
1/2	0.58
1/3	0.67
1/4	0.76
1/5	0.83
1/6	0.90
1/7.5 and less	1.00

where: H_C = Height of ceiling joists or rafter ties measured vertically above the top of the rafter support walls.

H_R = Height of roof ridge measured vertically above the top of the rafter support walls.

b. Span exceeds 26 feet in length.

❖ Example 1:

You wish to design a rafter system using Douglas-Fir-Larch #2 lumber for a 20 psf (0.96 kN/m²) live load, with rafters spaced 24 inches (610 mm) on center and having a horizontal projection of the rafter span of 14 feet, 6 inches (4420 mm). The rafters do not support a ceiling, and the roof slope is $2^1/_2$ units vertical in 12 units horizontal. Assume the roof dead load is 10 psf (0.48 kN/m²).

Solution:

The object is to locate a tabulated span value of 14 feet, 6 inches (4420 mm) or more. Going to Table R802.5.1(1) of the code under the portion headed "DEAD LOAD = 10 psf," note that 2-inch by 8-inch (51 mm x 208 mm) rafters at 24 inches (610 mm) on center will span 14 feet, 10 inches (4521 mm).

Example 2:

To find a substitute for the 2-inch by 8-inch (51 mm x 208 mm) rafters of Example 1, determine the required rafter size for a condition that uses a purlin and 2-inch by 4-inch (51 mm x 102 mm) braces (located at mid span of the rafters) at 4 feet (1219 mm) on center as depicted in Figure R802.5.1. Assume the same loads, rafter spacing, and 14-feet, 6-inch (4420 mm) horizontal rafter projection. Between the exterior wall and ridge, use Hem Fir #3 lumber.

Solution:

The rafter span, identified in Figure R802.5.1, is reduced to 7 feet, 3 inches (2210 mm) by adding the purlin at midspan. In Table R802.5.1(1) of the code, 2-inch by 6-inch (51 mm x 152 mm) rafters at 24 inches (610 mm) on center will span 8 feet, 10 inches (2692 mm).

TABLE R802.5.1(2)
RAFTER SPANS FOR COMMON LUMBER SPECIES
(Roof live load=20 psf, ceiling attached to rafters, L/Δ=240)

RAFTER SPACING (inches)	SPECIES AND GRADE		DEAD LOAD = 10 psf					DEAD LOAD = 20 psf				
			2x4	2x6	2x8	2x10	2x12	2x4	2x6	2x8	2x10	2x12
			Maximum rafter spans[a]									
			(feet - inches)	(feet - inches)	(feet - inches)	(feet - inches)	(feet - inches)	(feet - inches)	(feet - inches)	(feet - inches)	(feet - inches)	(feet - inches)
12	Douglas fir-larch	SS	10-5	16-4	21-7	Note b	Note b	10-5	16-4	21-7	Note b	Note b
	Douglas fir-larch	#1	10-0	15-9	20-10	Note b	Note b	10-0	15-4	19-5	23-9	Note b
	Douglas fir-larch	#2	9-10	15-6	20-5	25-8	Note b	9-10	14-4	18-2	22-3	25-9
	Douglas fir-larch	#3	8-7	12-6	15-10	19-5	22-6	7-5	10-10	13-9	16-9	19-6
	Hem-fir	SS	9-10	15-6	20-5	Note b	Note b	9-10	15-6	20-5	Note b	Note b
	Hem-fir	#1	9-8	15-2	19-11	25-5	Note b	9-8	14-11	18-11	23-2	Note b
	Hem-fir	#2	9-2	14-5	19-0	24-3	Note b	9-2	14-2	17-11	21-11	25-5
	Hem-fir	#3	8-7	12-6	15-10	19-5	22-6	7-5	10-10	13-9	16-9	19-6
	Southern pine	SS	10-3	16-1	21-2	Note b	Note b	10-3	16-1	21-2	Note b	Note b
	Southern pine	#1	10-0	15-9	20-10	Note b	Note b	10-0	15-9	20-10	25-10	Note b
	Southern pine	#2	9-10	15-6	20-5	Note b	Note b	9-10	15-1	19-5	23-2	Note b
	Southern pine	#3	9-1	13-6	17-2	20-3	24-1	7-11	11-8	14-10	17-6	20-11
	Spruce-pine-fir	SS	9-8	15-2	19-11	25-5	Note b	9-8	15-2	19-11	25-5	Note b
	Spruce-pine-fir	#1	9-5	14-9	19-6	24-10	Note b	9-5	14-4	18-2	22-3	25-9
	Spruce-pine-fir	#2	9-5	14-9	19-6	24-10	Note b	9-5	14-4	18-2	22-3	25-9
	Spruce-pine-fir	#3	8-7	12-6	15-10	19-5	22-6	7-5	10-10	13-9	16-9	19-6
16	Douglas fir-larch	SS	9-6	14-11	19-7	25-0	Note b	9-6	14-11	19-7	24-9	Note b
	Douglas fir-larch	#1	9-1	14-4	18-11	23-9	Note b	9-1	13-3	16-10	20-7	23-10
	Douglas fir-larch	#2	8-11	14-1	18-2	22-3	25-9	8-6	12-5	15-9	19-3	22-4
	Douglas fir-larch	#3	7-5	10-10	13-9	16-9	19-6	6-5	9-5	11-11	14-6	16-10
	Hem-fir	SS	8-11	14-1	18-6	23-8	Note b	8-11	14-1	18-6	23-8	Note b
	Hem-fir	#1	8-9	13-9	18-1	23-1	Note b	8-9	12-11	16-5	20-0	23-3
	Hem-fir	#2	8-4	13-1	17-3	21-11	25-5	8-4	12-3	15-6	18-11	22-0
	Hem-fir	#3	7-5	10-10	13-9	16-9	19-6	6-5	9-5	11-11	14-6	16-10
	Southern pine	SS	9-4	14-7	19-3	24-7	Note b	9-4	14-7	19-3	24-7	Note b
	Southern pine	#1	9-1	14-4	18-11	24-1	Note b	9-1	14-4	18-10	22-4	Note b
	Southern pine	#2	8-11	14-1	18-6	23-2	Note b	8-11	13-0	16-10	20-1	23-7
	Southern pine	#3	7-11	11-8	14-10	17-6	20-11	6-10	10-1	12-10	15-2	18-1
	Spruce-pine-fir	SS	8-9	13-9	18-1	23-1	Note b	8-9	13-9	18-1	23-0	Note b
	Spruce-pine-fir	#1	8-7	13-5	17-9	22-3	25-9	8-6	12-5	15-9	19-3	22-4
	Spruce-pine-fir	#2	8-7	13-5	17-9	22-3	25-9	8-6	12-5	15-9	19-3	22-4
	Spruce-pine-fir	#3	7-5	10-10	13-9	16-9	19-6	6-5	9-5	11-11	14-6	16-10
19.2	Douglas fir-larch	SS	8-11	14-0	18-5	23-7	Note b	8-11	14-0	18-5	22-7	Note b
	Douglas fir-larch	#1	8-7	13-6	17-9	21-8	25-2	8-4	12-2	15-4	18-9	21-9
	Douglas fir-larch	#2	8-5	13-1	16-7	20-3	23-6	7-9	11-4	14-4	17-7	20-4
	Douglas fir-larch	#3	6-9	9-11	12-7	15-4	17-9	5-10	8-7	10-10	13-3	15-5
	Hem-fir	SS	8-5	13-3	17-5	22-3	Note b	8-5	13-3	17-5	22-3	25-9
	Hem-fir	#1	8-3	12-11	17-1	21-1	24-6	8-1	11-10	15-0	18-4	21-3
	Hem-fir	#2	7-10	12-4	16-3	20-0	23-2	7-8	11-2	14-2	17-4	20-1
	Hem-fir	#3	6-9	9-11	12-7	15-4	17-9	5-10	8-7	10-10	13-3	15-5
	Southern pine	SS	8-9	13-9	18-1	23-1	Note b	8-9	13-9	18-1	23-1	Note b
	Southern pine	#1	8-7	13-6	17-9	22-8	Note b	8-7	13-6	17-2	20-5	24-4
	Southern pine	#2	8-5	13-3	17-5	21-2	24-10	8-4	11-11	15-4	18-4	21-6
	Southern pine	#3	7-3	10-8	13-7	16-0	19-1	6-3	9-3	11-9	13-10	16-6
	Spruce-pine-fir	SS	8-3	12-11	17-1	21-9	Note b	8-3	12-11	17-1	21-0	24-4
	Spruce-pine-fir	#1	8-1	12-8	16-7	20-3	23-6	7-9	11-4	14-4	17-7	20-4
	Spruce-pine-fir	#2	8-1	12-8	16-7	20-3	23-6	7-9	11-4	14-4	17-7	20-4
	Spruce-pine-fir	#3	6-9	9-11	12-7	15-4	17-9	5-10	8-7	10-10	13-3	15-5

(continued)

ROOF-CEILING CONSTRUCTION TABLE R802.5.1(2)

TABLE R802.5.1(2)—continued
RAFTER SPANS FOR COMMON LUMBER SPECIES
(Roof live load=20 psf, ceiling attached to rafters, L/Δ=240)

RAFTER SPACING (inches)	SPECIES AND GRADE		DEAD LOAD = 10 psf					DEAD LOAD = 20 psf				
			2x4	2x6	2x8	2x10	2x12	2x4	2x6	2x8	2x10	2x12
			Maximum rafter spans[a]									
			(feet-inches)	(feet-inches)	(feet-inches)	(feet-inches)	(feet-inches)	(feet-inches)	(feet-inches)	(feet-inches)	(feet-inches)	(feet-inches)
24	Douglas fir-larch	SS	8-3	13-0	17-2	21-10	Note b	8-3	13-0	16-7	20-3	23-5
	Douglas fir-larch	#1	8-0	12-6	15-10	19-5	22-6	7-5	10-10	13-9	16-9	19-6
	Douglas fir-larch	#2	7-10	11-9	14-10	18-2	21-0	6-11	10-2	12-10	15-8	18-3
	Douglas fir-larch	#3	6-1	8-10	11-3	13-8	15-11	5-3	7-8	9-9	11-10	13-9
	Hem-fir	SS	7-10	12-3	16-2	20-8	25-1	7-10	12-3	16-2	19-10	23-0
	Hem-fir	#1	7-8	12-0	15-6	18-11	21-11	7-3	10-7	13-5	16-4	19-0
	Hem-fir	#2	7-3	11-5	14-8	17-10	20-9	6-10	10-0	12-8	15-6	17-11
	Hem-fir	#3	6-1	8-10	11-3	13-8	15-11	5-3	7-8	9-9	11-10	13-9
	Southern pine	SS	8-1	12-9	16-10	21-6	Note b	8-1	12-9	16-10	21-6	Note b
	Southern pine	#1	8-0	12-6	16-6	21-1	25-2	8-0	12-3	15-4	18-3	21-9
	Southern pine	#2	7-10	12-3	15-10	18-11	22-2	7-5	10-8	13-9	16-5	19-3
	Southern pine	#3	6-5	9-6	12-1	14-4	17-1	5-7	8-3	10-6	12-5	14-9
	Spruce-pine-fir	SS	7-8	12-0	15-10	20-2	24-7	7-8	12-0	15-4	18-9	21-9
	Spruce-pine-fir	#1	7-6	11-9	14-10	18-2	21-0	6-11	10-2	12-10	15-8	18-3
	Spruce-pine-fir	#2	7-6	11-9	14-10	18-2	21-0	6-11	10-2	12-10	15-8	18-3
	Spruce-pine-fir	#3	6-1	8-10	11-3	13-8	15-11	5-3	7-8	9-9	11-10	13-9

Check sources for availability of lumber in lengths greater than 20 feet.

For SI: 1 inch = 25.4 mm, 1 foot = 304.8 mm, 1 pound per square foot = 0.0479 kN/m^2.

a. The tabulated rafter spans assume that ceiling joists are located at the bottom of the attic space or that some other method of resisting the outward push of the rafters on the bearing walls, such as rafter ties, is provided at that location. When ceiling joists or rafter ties are located higher in the attic space, the rafter spans shall be multiplied by the factors given below:

H_C/H_R	Rafter Span Adjustment Factor
2/3 or greater	0.50
1/2	0.58
1/3	0.67
1/4	0.76
1/5	0.83
1/6	0.90
1/7.5 and less	1.00

where: H_C = Height of ceiling joists or rafter ties measured vertically above the top of the rafter support walls.

H_R = Height of roof ridge measured vertically above the top of the rafter support walls.

b. Span exceeds 26 feet in length.

❖ See the commentary for Sections R802.5 and R802.5.1 and Table R802.5.1(1).

TABLE R802.5.1(3)
RAFTER SPANS FOR COMMON LUMBER SPECIES
(Ground snow load=30 psf, ceiling not attached to rafters, L/Δ=180)

RAFTER SPACING (inches)	SPECIES AND GRADE		DEAD LOAD = 10 psf					DEAD LOAD = 20 psf				
			2x4	2x6	2x8	2x10	2x12	2x4	2x6	2x8	2x10	2x12
			\multicolumn{10}{c}{Maximum rafter spans[a]}
			(feet - inches)	(feet - inches)	(feet - inches)	(feet - inches)	(feet - inches)	(feet - inches)	(feet - inches)	(feet - inches)	(feet - inches)	(feet - inches)
12	Douglas fir-larch	SS	10-0	15-9	20-9	Note b	Note b	10-0	15-9	20-1	24-6	Note b
	Douglas fir-larch	#1	9-8	14-9	18-8	22-9	Note b	9-0	13-2	16-8	20-4	23-7
	Douglas fir-larch	#2	9-5	13-9	17-5	21-4	24-8	8-5	12-4	15-7	19-1	22-1
	Douglas fir-larch	#3	7-1	10-5	13-2	16-1	18-8	6-4	9-4	11-9	14-5	16-8
	Hem-fir	SS	9-6	14-10	19-7	25-0	Note b	9-6	14-10	19-7	24-1	Note b
	Hem-fir	#1	9-3	14-4	18-2	22-2	25-9	8-9	12-10	16-3	19-10	23-0
	Hem-fir	#2	8-10	13-7	17-2	21-0	24-4	8-4	12-2	15-4	18-9	21-9
	Hem-fir	#3	7-1	10-5	13-2	16-1	18-8	6-4	9-4	11-9	14-5	16-8
	Southern pine	SS	9-10	15-6	20-5	Note b	Note b	9-10	15-6	20-5	Note b	Note b
	Southern pine	#1	9-8	15-2	20-0	24-9	Note b	9-8	14-10	18-8	22-2	Note b
	Southern pine	#2	9-6	14-5	18-8	22-3	Note b	9-0	12-11	16-8	19-11	23-4
	Southern pine	#3	7-7	11-2	14-3	16-10	20-0	6-9	10-0	12-9	15-1	17-11
	Spruce-pine-fir	SS	9-3	14-7	19-2	24-6	Note b	9-3	14-7	18-8	22-9	Note b
	Spruce-pine-fir	#1	9-1	13-9	17-5	21-4	24-8	8-5	12-4	15-7	19-1	22-1
	Spruce-pine-fir	#2	9-1	13-9	17-5	21-4	24-8	8-5	12-4	15-7	19-1	22-1
	Spruce-pine-fir	#3	7-1	10-5	13-2	16-1	18-8	6-4	9-4	11-9	14-5	16-8
16	Douglas fir-larch	SS	9-1	14-4	18-10	23-9	Note b	9-1	13-9	17-5	21-3	24-8
	Douglas fir-larch	#1	8-9	12-9	16-2	19-9	22-10	7-10	11-5	14-5	17-8	20-5
	Douglas fir-larch	#2	8-2	11-11	15-1	18-5	21-5	7-3	10-8	13-6	16-6	19-2
	Douglas fir-larch	#3	6-2	9-0	11-5	13-11	16-2	5-6	8-1	10-3	12-6	14-6
	Hem-fir	SS	8-7	13-6	17-10	22-9	Note b	8-7	13-6	17-1	20-10	24-2
	Hem-fir	#1	8-5	12-5	15-9	19-3	22-3	7-7	11-1	14-1	17-2	19-11
	Hem-fir	#2	8-0	11-9	14-11	18-2	21-1	7-2	10-6	13-4	16-3	18-10
	Hem-fir	#3	6-2	9-0	11-5	13-11	16-2	5-6	8-1	10-3	12-6	14-6
	Southern pine	SS	8-11	14-1	18-6	23-8	Note b	8-11	14-1	18-6	23-8	Note b
	Southern pine	#1	8-9	13-9	18-1	21-5	25-7	8-8	12-10	16-2	19-2	22-10
	Southern pine	#2	8-7	12-6	16-2	19-3	22-7	7-10	11-2	14-5	17-3	20-2
	Southern pine	#3	6-7	9-8	12-4	14-7	17-4	5-10	8-8	11-0	13-0	15-6
	Spruce-pine-fir	SS	8-5	13-3	17-5	22-1	25-7	8-5	12-9	16-2	19-9	22-10
	Spruce-pine-fir	#1	8-2	11-11	15-1	18-5	21-5	7-3	10-8	13-6	16-6	19-2
	Spruce-pine-fir	#2	8-2	11-11	15-1	18-5	21-5	7-3	10-8	13-6	16-6	19-2
	Spruce-pine-fir	#3	6-2	9-0	11-5	13-11	16-2	5-6	8-1	10-3	12-6	14-6
19.2	Douglas fir-larch	SS	8-7	13-6	17-9	21-8	25-2	8-7	12-6	15-10	19-5	22-6
	Douglas fir-larch	#1	7-11	11-8	14-9	18-0	20-11	7-1	10-5	13-2	16-1	18-8
	Douglas fir-larch	#2	7-5	10-11	13-9	16-10	19-6	6-8	9-9	12-4	15-1	17-6
	Douglas fir-larch	#3	5-7	8-3	10-5	12-9	14-9	5-0	7-4	9-4	11-5	13-2
	Hem-fir	SS	8-1	12-9	16-9	21-4	24-8	8-1	12-4	15-7	19-1	22-1
	Hem-fir	#1	7-9	11-4	14-4	17-7	20-4	6-11	10-2	12-10	15-8	18-2
	Hem-fir	#2	7-4	10-9	13-7	16-7	19-3	6-7	9-7	12-2	14-10	17-3
	Hem-fir	#3	5-7	8-3	10-5	12-9	14-9	5-0	7-4	9-4	11-5	13-2
	Southern pine	SS	8-5	13-3	17-5	22-3	Note b	8-5	13-3	17-5	22-0	25-9
	Southern pine	#1	8-3	13-0	16-6	19-7	23-4	7-11	11-9	14-9	17-6	20-11
	Southern pine	#2	7-11	11-5	14-9	17-7	20-7	7-1	10-2	13-2	15-9	18-5
	Southern pine	#3	6-0	8-10	11-3	13-4	15-10	5-4	7-11	10-1	11-11	14-2
	Spruce-pine-fir	SS	7-11	12-5	16-5	20-2	23-4	7-11	11-8	14-9	18-0	20-11
	Spruce-pine-fir	#1	7-5	10-11	13-9	16-10	19-6	6-8	9-9	12-4	15-1	17-6
	Spruce-pine-fir	#2	7-5	10-11	13-9	16-10	19-6	6-8	9-9	12-4	15-1	17-6
	Spruce-pine-fir	#3	5-7	8-3	10-5	12-9	14-9	5-0	7-4	9-4	11-5	13-2

(continued)

ROOF-CEILING CONSTRUCTION

TABLE R802.5.1(3)—continued
RAFTER SPANS FOR COMMON LUMBER SPECIES
(Ground snow load=30 psf, ceiling not attached to rafters, $L/\Delta=180$)

RAFTER SPACING (inches)	SPECIES AND GRADE		DEAD LOAD = 10 psf					DEAD LOAD = 20 psf				
			2x4	2x6	2x8	2x10	2x12	2x4	2x6	2x8	2x10	2x12
			\multicolumn{10}{c}{Maximum rafter spans[a]}									
			(feet-inches)	(feet-inches)	(feet-inches)	(feet-inches)	(feet-inches)	(feet-inches)	(feet-inches)	(feet-inches)	(feet-inches)	(feet-inches)
24	Douglas fir-larch	SS	7-11	12-6	15-10	19-5	22-6	7-8	11-3	14-2	17-4	20-1
	Douglas fir-larch	#1	7-1	10-5	13-2	16-1	18-8	6-4	9-4	11-9	14-5	16-8
	Douglas fir-larch	#2	6-8	9-9	12-4	15-1	17-6	5-11	8-8	11-0	13-6	15-7
	Douglas fir-larch	#3	5-0	7-4	9-4	11-5	13-2	4-6	6-7	8-4	10-2	11-10
	Hem-fir	SS	7-6	11-10	15-7	19-1	22-1	7-6	11-0	13-11	17-0	19-9
	Hem-fir	#1	6-11	10-2	12-10	15-8	18-2	6-2	9-1	11-6	14-0	16-3
	Hem-fir	#2	6-7	9-7	12-2	14-10	17-3	5-10	8-7	10-10	13-3	15-5
	Hem-fir	#3	5-0	7-4	9-4	11-5	13-2	4-6	6-7	8-4	10-2	11-10
	Southern pine	SS	7-10	12-3	16-2	20-8	25-1	7-10	12-3	16-2	19-8	23-0
	Southern pine	#1	7-8	11-9	14-9	17-6	20-11	7-1	10-6	13-2	15-8	18-8
	Southern pine	#2	7-1	10-2	13-2	15-9	18-5	6-4	9-2	11-9	14-1	16-6
	Southern pine	#3	5-4	7-11	10-1	11-11	14-2	4-9	7-1	9-0	10-8	12-8
	Spruce-pine-fir	SS	7-4	11-7	14-9	18-0	20-11	7-1	10-5	13-2	16-1	18-8
	Spruce-pine-fir	#1	6-8	9-9	12-4	15-1	17-6	5-11	8-8	11-0	13-6	15-7
	Spruce-pine-fir	#2	6-8	9-9	12-4	15-1	17-6	5-11	8-8	11-0	13-6	15-7
	Spruce-pine-fir	#3	5-0	7-4	9-4	11-5	13-2	4-6	6-7	8-4	10-2	11-10

Check sources for availability of lumber in lengths greater than 20 feet.

For SI: 1 inch = 25.4 mm, 1 foot = 304.8 mm, 1 pound per square foot = 0.0479 kN/m^2.

a. The tabulated rafter spans assume that ceiling joists are located at the bottom of the attic space or that some other method of resisting the outward push of the rafters on the bearing walls, such as rafter ties, is provided at that location. When ceiling joists or rafter ties are located higher in the attic space, the rafter spans shall be multiplied by the factors given below:

H_C/H_R	Rafter Span Adjustment Factor
2/3 or greater	0.50
1/2	0.58
1/3	0.67
1/4	0.76
1/5	0.83
1/6	0.90
1/7.5 and less	1.00

where: H_C = Height of ceiling joists or rafter ties measured vertically above the top of the rafter support walls.

H_R = Height of roof ridge measured vertically above the top of the rafter support walls.

b. Span exceeds 26 feet in length.

❖ See the commentary for Sections R802.5 and R802.5.1 and Table R802.5.1(1).

TABLE R802.5.1(4)
RAFTER SPANS FOR COMMON LUMBER SPECIES
(Ground snow load=50 psf, ceiling not attached to rafters, L/Δ=180)

RAFTER SPACING (inches)	SPECIES AND GRADE		DEAD LOAD = 10 psf					DEAD LOAD = 20 psf				
			2x4	2x6	2x8	2x10	2x12	2x4	2x6	2x8	2x10	2x12
			Maximum rafter spans[a]									
			(feet-inches)	(feet-inches)	(feet-inches)	(feet-inches)	(feet-inches)	(feet-inches)	(feet-inches)	(feet-inches)	(feet-inches)	(feet-inches)
12	Douglas fir-larch	SS	8-5	13-3	17-6	22-4	26-0	8-5	13-3	17-0	20-9	24-0
	Douglas fir-larch	#1	8-2	12-0	15-3	18-7	21-7	7-7	11-2	14-1	17-3	20-0
	Douglas fir-larch	#2	7-8	11-3	14-3	17-5	20-2	7-1	10-5	13-2	16-1	18-8
	Douglas fir-larch	#3	5-10	8-6	10-9	13-2	15-3	5-5	7-10	10-0	12-2	14-1
	Hem-fir	SS	8-0	12-6	16-6	21-1	25-6	8-0	12-6	16-6	20-4	23-7
	Hem-fir	#1	7-10	11-9	14-10	18-1	21-0	7-5	10-10	13-9	16-9	19-5
	Hem-fir	#2	7-5	11-1	14-0	17-2	19-11	7-0	10-3	13-0	15-10	18-5
	Hem-fir	#3	5-10	8-6	10-9	13-2	15-3	5-5	7-10	10-0	12-2	14-1
	Southern pine	SS	8-4	13-0	17-2	21-11	Note b	8-4	13-0	17-2	21-11	Note b
	Southern pine	#1	8-2	12-10	16-10	20-3	24-1	8-2	12-6	15-9	18-9	22-4
	Southern pine	#2	8-0	11-9	15-3	18-2	21-3	7-7	10-11	14-1	16-10	19-9
	Southern pine	#3	6-2	9-2	11-8	13-9	16-4	5-9	8-5	10-9	12-9	15-2
	Spruce-pine-fir	SS	7-10	12-3	16-2	20-8	24-1	7-10	12-3	15-9	19-3	22-4
	Spruce-pine-fir	#1	7-8	11-3	14-3	17-5	20-2	7-1	10-5	13-2	16-1	18-8
	Spruce-pine-fir	#2	7-8	11-3	14-3	17-5	20-2	7-1	10-5	13-2	16-1	18-8
	Spruce-pine-fir	#3	5-10	8-6	10-9	13-2	15-3	5-5	7-10	10-0	12-2	14-1
16	Douglas fir-larch	SS	7-8	12-1	15-10	19-5	22-6	7-8	11-7	14-8	17-11	20-10
	Douglas fir-larch	#1	7-1	10-5	13-2	16-1	18-8	6-7	9-8	12-2	14-11	17-3
	Douglas fir-larch	#2	6-8	9-9	12-4	15-1	17-6	6-2	9-0	11-5	13-11	16-2
	Douglas fir-larch	#3	5-0	7-4	9-4	11-5	13-2	4-8	6-10	8-8	10-6	12-3
	Hem-fir	SS	7-3	11-5	15-0	19-1	22-1	7-3	11-5	14-5	17-8	20-5
	Hem-fir	#1	6-11	10-2	12-10	15-8	18-2	6-5	9-5	11-11	14-6	16-10
	Hem-fir	#2	6-7	9-7	12-2	14-10	17-3	6-1	8-11	11-3	13-9	15-11
	Hem-fir	#3	5-0	7-4	9-4	11-5	13-2	4-8	6-10	8-8	10-6	12-3
	Southern pine	SS	7-6	11-10	15-7	19-11	24-3	7-6	11-10	15-7	19-11	23-10
	Southern pine	#1	7-5	11-7	14-9	17-6	20-11	7-4	10-10	13-8	16-2	19-4
	Southern pine	#2	7-1	10-2	13-2	15-9	18-5	6-7	9-5	12-2	14-7	17-1
	Southern pine	#3	5-4	7-11	10-1	11-11	14-2	4-11	7-4	9-4	11-0	13-1
	Spruce-pine-fir	SS	7-1	11-2	14-8	18-0	20-11	7-1	10-9	13-8	15-11	19-4
	Spruce-pine-fir	#1	6-8	9-9	12-4	15-1	17-6	6-2	9-0	11-5	13-11	16-2
	Spruce-pine-fir	#2	6-8	9-9	12-4	15-1	17-6	6-2	9-0	11-5	13-11	16-2
	Spruce-pine-fir	#3	5-0	7-4	9-4	11-5	13-2	4-8	6-10	8-8	10-6	12-3
19.2	Douglas fir-larch	SS	7-3	11-4	14-6	17-8	20-6	7-3	10-7	13-5	16-5	19-0
	Douglas fir-larch	#1	6-6	9-6	12-0	14-8	17-1	6-0	8-10	11-2	13-7	15-9
	Douglas fir-larch	#2	6-1	8-11	11-3	13-9	15-11	5-7	8-3	10-5	12-9	14-9
	Douglas fir-larch	#3	4-7	6-9	8-6	10-5	12-1	4-3	6-3	7-11	9-7	11-2
	Hem-fir	SS	6-10	10-9	14-2	17-5	20-2	6-10	10-5	13-2	16-1	18-8
	Hem-fir	#1	6-4	9-3	11-9	14-4	16-7	5-10	8-7	10-10	13-3	15-5
	Hem-fir	#2	6-0	8-9	11-1	13-7	15-9	5-7	8-1	10-3	12-7	14-7
	Hem-fir	#3	4-7	6-9	8-6	10-5	12-1	4-3	6-3	7-11	9-7	11-2
	Southern pine	SS	7-1	11-2	14-8	18-9	22-10	7-1	11-2	14-8	18-7	21-9
	Southern pine	#1	7-0	10-8	13-5	16-0	19-1	6-8	9-11	12-5	14-10	17-8
	Southern pine	#2	6-6	9-4	12-0	14-4	16-10	6-0	8-8	11-2	13-4	15-7
	Southern pine	#3	4-11	7-3	9-2	10-10	12-11	4-6	6-8	8-6	10-1	12-0
	Spruce-pine-fir	SS	6-8	10-6	13-5	16-5	19-1	6-8	9-10	12-5	15-3	17-8
	Spruce-pine-fir	#1	6-1	8-11	11-3	13-9	15-11	5-7	8-3	10-5	12-9	14-9
	Spruce-pine-fir	#2	6-1	8-11	11-3	13-9	15-11	5-7	8-3	10-5	12-9	14-9
	Spruce-pine-fir	#3	4-7	6-9	8-6	10-5	12-1	4-3	6-3	7-11	9-7	11-2

(continued)

ROOF-CEILING CONSTRUCTION

TABLE R802.5.1(4)—continued
RAFTER SPANS FOR COMMON LUMBER SPECIES
(Ground snow load=50 psf, ceiling not attached to rafters, $L/\Delta=180$)

RAFTER SPACING (inches)	SPECIES AND GRADE		DEAD LOAD = 10 psf					DEAD LOAD = 20 psf				
			2x4	2x6	2x8	2x10	2x12	2x4	2x6	2x8	2x10	2x12
			Maximum rafter spans[a]									
			(feet-inches)	(feet-inches)	(feet-inches)	(feet-inches)	(feet-inches)	(feet-inches)	(feet-inches)	(feet-inches)	(feet-inches)	(feet-inches)
24	Douglas fir-larch	SS	6-8	10-3	13-0	15-10	18-4	6-6	9-6	12-0	14-8	17-0
	Douglas fir-larch	#1	5-10	8-6	10-9	13-2	15-3	5-5	7-10	10-0	12-2	14-1
	Douglas fir-larch	#2	5-5	7-11	10-1	12-4	14-3	5-0	7-4	9-4	11-5	13-2
	Douglas fir-larch	#3	4-1	6-0	7-7	9-4	10-9	3-10	5-7	7-1	8-7	10-0
	Hem-fir	SS	6-4	9-11	12-9	15-7	18-0	6-4	9-4	11-9	14-5	16-8
	Hem-fir	#1	5-8	8-3	10-6	12-10	14-10	5-3	7-8	9-9	11-10	13-9
	Hem-fir	#2	5-4	7-10	9-11	12-1	14-1	4-11	7-3	9-2	11-3	13-0
	Hem-fir	#3	4-1	6-0	7-7	9-4	10-9	3-10	5-7	7-1	8-7	10-0
	Southern pine	SS	6-7	10-4	13-8	17-5	21-0	6-7	10-4	13-8	16-7	19-5
	Southern pine	#1	6-5	9-7	12-0	14-4	17-1	6-0	8-10	11-2	13-3	15-9
	Southern pine	#2	5-10	8-4	10-9	12-10	15-1	5-5	7-9	10-0	11-11	13-11
	Southern pine	#3	4-4	6-5	8-3	9-9	11-7	4-1	6-0	7-7	9-0	10-8
	Spruce-pine-fir	SS	6-2	9-6	12-0	14-8	17-1	6-0	8-10	11-2	13-7	15-9
	Spruce-pine-fir	#1	5-5	7-11	10-1	12-4	14-3	5-0	7-4	9-4	11-5	13-2
	Spruce-pine-fir	#2	5-5	7-11	10-1	12-4	14-3	5-0	7-4	9-4	11-5	13-2
	Spruce-pine-fir	#3	4-1	6-0	7-7	9-4	10-9	3-10	5-7	7-1	8-7	10-0

Check sources for availability of lumber in lengths greater than 20 feet.

For SI: 1 inch = 25.4 mm, 1 foot = 304.8 mm, 1 pound per square foot = 0.0479 kN/m^2.

a. The tabulated rafter spans assume that ceiling joists are located at the bottom of the attic space or that some other method of resisting the outward push of the rafters on the bearing walls, such as rafter ties, is provided at that location. When ceiling joists or rafter ties are located higher in the attic space, the rafter spans shall be multiplied by the factors given below:

H_C/H_R	Rafter Span Adjustment Factor
2/3 or greater	0.50
1/2	0.58
1/3	0.67
1/4	0.76
1/5	0.83
1/6	0.90
1/7.5 and less	1.00

where: H_C = Height of ceiling joists or rafter ties measured vertically above the top of the rafter support walls.

H_R = Height of roof ridge measured vertically above the top of the rafter support walls.

b. Span exceeds 26 feet in length.

❖ See the commentary for Sections R802.5 and R802.5.1 and Table R802.5.1(1).

TABLE R802.5.1(5)
RAFTER SPANS FOR COMMON LUMBER SPECIES
(Ground snow load=30 psf, ceiling attached to rafters, $L/\Delta=240$)

RAFTER SPACING (inches)	SPECIES AND GRADE		DEAD LOAD = 10 psf					DEAD LOAD = 20 psf				
			2x4	2x6	2x8	2x10	2x12	2x4	2x6	2x8	2x10	2x12
			\multicolumn{10}{c}{Maximum rafter spans[a]}									
			(feet - inches)	(feet - inches)	(feet - inches)	(feet - inches)	(feet - inches)	(feet - inches)	(feet - inches)	(feet - inches)	(feet - inches)	(feet - inches)
12	Douglas fir-larch	SS	9- 1	14- 4	18-10	24- 1	Note b	9- 1	14- 4	18-10	24- 1	Note b
	Douglas fir-larch	#1	8- 9	13- 9	18- 2	22- 9	Note b	8- 9	13- 2	16- 8	20- 4	23- 7
	Douglas fir-larch	#2	8- 7	13- 6	17- 5	21- 4	24- 8	8- 5	12- 4	15- 7	19- 1	22- 1
	Douglas fir-larch	#3	7- 1	10- 5	13- 2	16- 1	18- 8	6- 4	9- 4	11- 9	14- 5	16- 8
	Hem-fir	SS	8- 7	13- 6	17-10	22- 9	Note b	8- 7	13- 6	17-10	22- 9	Note b
	Hem-fir	#1	8- 5	13- 3	17- 5	22- 2	25- 9	8- 5	12-10	16- 3	19-10	23- 0
	Hem-fir	#2	8- 0	12- 7	16- 7	21- 0	24- 4	8- 0	12- 2	15- 4	18- 9	21- 9
	Hem-fir	#3	7- 1	10- 5	13- 2	16- 1	18- 8	6- 4	9- 4	11- 9	14- 5	16- 8
	Southern pine	SS	8-11	14- 1	18- 6	23- 8	Note b	8-11	14- 1	18- 6	23- 8	Note b
	Southern pine	#1	8- 9	13- 9	18- 2	23- 2	Note b	8- 9	13- 9	18- 2	22- 2	Note b
	Southern pine	#2	8- 7	13- 6	17-10	22- 3	Note b	8- 7	12-11	16- 8	19-11	23- 4
	Southern pine	#3	7- 7	11- 2	14- 3	16-10	20- 0	6- 9	10- 0	12- 9	15- 1	17-11
	Spruce-pine-fir	SS	8- 5	13- 3	17- 5	22- 3	Note b	8- 5	13- 3	17- 5	22- 3	Note b
	Spruce-pine-fir	#1	8- 3	12-11	17- 0	21- 4	24- 8	8- 3	12- 4	15- 7	19- 1	22- 1
	Spruce-pine-fir	#2	8- 3	12-11	17- 0	21- 4	24- 8	8- 3	12- 4	15- 7	19- 1	22- 1
	Spruce-pine-fir	#3	7- 1	10- 5	13- 2	16- 1	18- 8	6- 4	9- 4	11- 9	14- 5	16- 8
16	Douglas fir-larch	SS	8- 3	13- 0	17- 2	21-10	Note b	8- 3	13- 0	17- 2	21- 3	24- 8
	Douglas fir-larch	#1	8- 0	12- 6	16- 2	19- 9	22-10	7-10	11- 5	14- 5	17- 8	20- 5
	Douglas fir-larch	#2	7-10	11-11	15- 1	18- 5	21- 5	7- 3	10- 8	13- 6	16- 6	19- 2
	Douglas fir-larch	#3	6- 2	9- 0	11- 5	13-11	16- 2	5- 6	8- 1	10- 3	12- 6	14- 6
	Hem-fir	SS	7-10	12- 3	16- 2	20- 8	25- 1	7-10	12- 3	16- 2	20- 8	24- 2
	Hem-fir	#1	7- 8	12- 0	15- 9	19- 3	22- 3	7- 7	11- 1	14- 1	17- 2	19-11
	Hem-fir	#2	7- 3	11- 5	14-11	18- 2	21- 1	7- 2	10- 6	13- 4	16- 3	18-10
	Hem-fir	#3	6- 2	9- 0	11- 5	13-11	16- 2	5- 6	8- 1	10- 3	12- 6	14- 6
	Southern pine	SS	8- 1	12- 9	16-10	21- 6	Note b	8- 1	12- 9	16-10	21- 6	Note b
	Southern pine	#1	8- 0	12- 6	16- 6	21- 1	25- 7	8- 0	12- 6	16- 2	19- 2	22-10
	Southern pine	#2	7-10	12- 3	16- 2	19- 3	22- 7	7-10	11- 2	14- 5	17- 3	20- 2
	Southern pine	#3	6- 7	9- 8	12- 4	14- 7	17- 4	5-10	8- 8	11- 0	13- 0	15- 6
	Spruce-pine-fir	SS	7- 8	12- 0	15-10	20- 2	24- 7	7- 8	12- 0	15-10	19- 9	22-10
	Spruce-pine-fir	#1	7- 6	11- 9	15- 1	18- 5	21- 5	7- 3	10- 8	13- 6	16- 6	19- 2
	Spruce-pine-fir	#2	7- 6	11- 9	15- 1	18- 5	21- 5	7- 3	10- 8	13- 6	16- 6	19- 2
	Spruce-pine-fir	#3	6- 2	9- 0	11- 5	13-11	16- 2	5- 6	8- 1	10- 3	12- 6	14- 6
19.2	Douglas fir-larch	SS	7- 9	12- 3	16- 1	20- 7	25- 0	7- 9	12- 3	15-10	19- 5	22- 6
	Douglas fir-larch	#1	7- 6	11- 8	14- 9	18- 0	20-11	7- 1	10- 5	13- 2	16- 1	18- 8
	Douglas fir-larch	#2	7- 4	10-11	13- 9	16-10	19- 6	6- 8	9- 9	12- 4	15- 1	17- 6
	Douglas fir-larch	#3	5- 7	8- 3	10- 5	12- 9	14- 9	5- 0	7- 4	9- 4	11- 5	13- 2
	Hem-fir	SS	7- 4	11- 7	15- 3	19- 5	23- 7	7- 4	11- 7	15- 3	19- 1	22- 1
	Hem-fir	#1	7- 2	11- 4	14- 4	17- 7	20- 4	6-11	10- 2	12-10	15- 8	18- 2
	Hem-fir	#2	6-10	10- 9	13- 7	16- 7	19- 3	6- 7	9- 7	12- 2	14-10	17- 3
	Hem-fir	#3	5- 7	8- 3	10- 5	12- 9	14- 9	5- 0	7- 4	9- 4	11- 5	13- 2
	Southern pine	SS	7- 8	12- 0	15-10	20- 2	24- 7	7- 8	12- 0	15-10	20- 2	24- 7
	Southern pine	#1	7- 6	11- 9	15- 6	19- 7	23- 4	7- 6	11- 9	14- 9	17- 6	20-11
	Southern pine	#2	7- 4	11- 5	14- 9	17- 7	20- 7	7- 1	10- 2	13- 2	15- 9	18- 5
	Southern pine	#3	6- 0	8-10	11- 3	13- 4	15-10	5- 4	7-11	10- 1	11-11	14- 2
	Spruce-pine-fir	SS	7- 2	11- 4	14-11	19- 0	23- 1	7- 2	11- 4	14- 9	18- 0	20-11
	Spruce-pine-fir	#1	7- 0	10-11	13- 9	16-10	19- 6	6- 8	9- 9	12- 4	15- 1	17- 6
	Spruce-pine-fir	#2	7- 0	10-11	13- 9	16-10	19- 6	6- 8	9- 9	12- 4	15- 1	17- 6
	Spruce-pine-fir	#3	5- 7	8- 3	10- 5	12- 9	14- 9	5- 0	7- 4	9- 4	11- 5	13- 2

(continued)

ROOF-CEILING CONSTRUCTION

TABLE R802.5.1(5)—continued
RAFTER SPANS FOR COMMON LUMBER SPECIES
(Ground snow load=30 psf, ceiling attached to rafters, L/Δ=240)

RAFTER SPACING (inches)	SPECIES AND GRADE		DEAD LOAD = 10 psf					DEAD LOAD = 20 psf				
			2x4	2x6	2x8	2x10	2x12	2x4	2x6	2x8	2x10	2x12
			Maximum rafter spans[a]									
			(feet-inches)	(feet-inches)	(feet-inches)	(feet-inches)	(feet-inches)	(feet-inches)	(feet-inches)	(feet-inches)	(feet-inches)	(feet-inches)
24	Douglas fir-larch	SS	7-3	11-4	15-0	19-1	22-6	7-3	11-3	14-2	17-4	20-1
	Douglas fir-larch	#1	7-0	10-5	13-2	16-1	18-8	6-4	9-4	11-9	14-5	16-8
	Douglas fir-larch	#2	6-8	9-9	12-4	15-1	17-6	5-11	8-8	11-0	13-6	15-7
	Douglas fir-larch	#3	5-0	7-4	9-4	11-5	13-2	4-6	6-7	8-4	10-2	11-10
	Hem-fir	SS	6-10	10-9	14-2	18-0	21-11	6-10	10-9	13-11	17-0	19-9
	Hem-fir	#1	6-8	10-2	12-10	15-8	18-2	6-2	9-1	11-6	14-0	16-3
	Hem-fir	#2	6-4	9-7	12-2	14-10	17-3	5-10	8-7	10-10	13-3	15-5
	Hem-fir	#3	5-0	7-4	9-4	11-5	13-2	4-6	6-7	8-4	10-2	11-10
	Southern pine	SS	7-1	11-2	14-8	18-9	22-10	7-1	11-2	14-8	18-9	22-10
	Southern pine	#1	7-0	10-11	14-5	17-6	20-11	7-0	10-6	13-2	15-8	18-8
	Southern pine	#2	6-10	10-2	13-2	15-9	18-5	6-4	9-2	11-9	14-1	16-6
	Southern pine	#3	5-4	7-11	10-1	11-11	14-2	4-9	7-1	9-0	10-8	12-8
	Spruce-pine-fir	SS	6-8	10-6	13-10	17-8	20-11	6-8	10-5	13-2	16-1	18-8
	Spruce-pine-fir	#1	6-6	9-9	12-4	15-1	17-6	5-11	8-8	11-0	13-6	15-7
	Spruce-pine-fir	#2	6-6	9-9	12-4	15-1	17-6	5-11	8-8	11-0	13-6	15-7
	Spruce-pine-fir	#3	5-0	7-4	9-4	11-5	13-2	4-6	6-7	8-4	10-2	11-10

Check sources for availability of lumber in lengths greater than 20 feet.

For SI: 1 inch = 25.4 mm, 1 foot = 304.8 mm, 1 pound per square foot = 0.0479 kN/m^2.

a. The tabulated rafter spans assume that ceiling joists are located at the bottom of the attic space or that some other method of resisting the outward push of the rafters on the bearing walls, such as rafter ties, is provided at that location. When ceiling joists or rafter ties are located higher in the attic space, the rafter spans shall be multiplied by the factors given below:

H_C/H_R	Rafter Span Adjustment Factor
2/3 or greater	0.50
1/2	0.58
1/3	0.67
1/4	0.76
1/5	0.83
1/6	0.90
1/7.5 and less	1.00

where: H_C = Height of ceiling joists or rafter ties measured vertically above the top of the rafter support walls.

H_R = Height of roof ridge measured vertically above the top of the rafter support walls.

b. Span exceeds 26 feet in length.

❖ See the commentary for Sections R802.5 and R802.5.1 and Table R802.5.1(1).

TABLE R802.5.1(6)
RAFTER SPANS FOR COMMON LUMBER SPECIES
(Ground snow load=50 psf, ceiling attached to rafters, L/Δ=240)

RAFTER SPACING (inches)	SPECIES AND GRADE		DEAD LOAD = 10 psf					DEAD LOAD = 20 psf				
			2x4	2x6	2x8	2x10	2x12	2x4	2x6	2x8	2x10	2x12
			Maximum rafter spans[a]									
			(feet-inches)	(feet-inches)	(feet-inches)	(feet-inches)	(feet-inches)	(feet-inches)	(feet-inches)	(feet-inches)	(feet-inches)	(feet-inches)
12	Douglas fir-larch	SS	7-8	12-1	15-11	20-3	24-8	7-8	12-1	15-11	20-3	24-0
	Douglas fir-larch	#1	7-5	11-7	15-3	18-7	21-7	7-5	11-2	14-1	17-3	20-0
	Douglas fir-larch	#2	7-3	11-3	14-3	17-5	20-2	7-1	10-5	13-2	16-1	18-8
	Douglas fir-larch	#3	5-10	8-6	10-9	13-2	15-3	5-5	7-10	10-0	12-2	14-1
	Hem-fir	SS	7-3	11-5	15-0	19-2	23-4	7-3	11-5	15-0	19-2	23-4
	Hem-fir	#1	7-1	11-2	14-8	18-1	21-0	7-1	10-10	13-9	16-9	19-5
	Hem-fir	#2	6-9	10-8	14-0	17-2	19-11	6-9	10-3	13-0	15-10	18-5
	Hem-fir	#3	5-10	8-6	10-9	13-2	15-3	5-5	7-10	10-0	12-2	14-1
	Southern pine	SS	7-6	11-10	15-7	19-11	24-3	7-6	11-10	15-7	19-11	24-3
	Southern pine	#1	7-5	11-7	15-4	19-7	23-9	7-5	11-7	15-4	18-9	22-4
	Southern pine	#2	7-3	11-5	15-0	18-2	21-3	7-3	10-11	14-1	16-10	19-9
	Southern pine	#3	6-2	9-2	11-8	13-9	16-4	5-9	8-5	10-9	12-9	15-2
	Spruce-pine-fir	SS	7-1	11-2	14-8	18-9	22-10	7-1	11-2	14-8	18-9	22-4
	Spruce-pine-fir	#1	6-11	10-11	14-3	17-5	20-2	6-11	10-5	13-2	16-1	18-8
	Spruce-pine-fir	#2	6-11	10-11	14-3	17-5	20-2	6-11	10-5	13-2	16-1	18-8
	Spruce-pine-fir	#3	5-10	8-6	10-9	13-2	15-3	5-5	7-10	10-0	12-2	14-1
16	Douglas fir-larch	SS	7-0	11-0	14-5	18-5	22-5	7-0	11-0	14-5	17-11	20-10
	Douglas fir-larch	#1	6-9	10-5	13-2	16-1	18-8	6-7	9-8	12-2	14-11	17-3
	Douglas fir-larch	#2	6-7	9-9	12-4	15-1	17-6	6-2	9-0	11-5	13-11	16-2
	Douglas fir-larch	#3	5-0	7-4	9-4	11-5	13-2	4-8	6-10	8-8	10-6	12-3
	Hem-fir	SS	6-7	10-4	13-8	17-5	21-2	6-7	10-4	13-8	17-5	20-5
	Hem-fir	#1	6-5	10-2	12-10	15-8	18-2	6-5	9-5	11-11	14-6	16-10
	Hem-fir	#2	6-2	9-7	12-2	14-10	17-3	6-1	8-11	11-3	13-9	15-11
	Hem-fir	#3	5-0	7-4	9-4	11-5	13-2	4-8	6-10	8-8	10-6	12-3
	Southern pine	SS	6-10	10-9	14-2	18-1	22-0	6-10	10-9	14-2	18-1	22-0
	Southern pine	#1	6-9	10-7	13-11	17-6	20-11	6-9	10-7	13-8	16-2	19-4
	Southern pine	#2	6-7	10-2	13-2	15-9	18-5	6-7	9-5	12-2	14-7	17-1
	Southern pine	#3	5-4	7-11	10-1	11-11	14-2	4-11	7-4	9-4	11-0	13-1
	Spruce-pine-fir	SS	6-5	10-2	13-4	17-0	20-9	6-5	10-2	13-4	16-8	19-4
	Spruce-pine-fir	#1	6-4	9-9	12-4	15-1	17-6	6-2	9-0	11-5	13-11	16-2
	Spruce-pine-fir	#2	6-4	9-9	12-4	15-1	17-6	6-2	9-0	11-5	13-11	16-2
	Spruce-pine-fir	#3	5-0	7-4	9-4	11-5	13-2	4-8	6-10	8-8	10-6	12-3
19.2	Douglas fir-larch	SS	6-7	10-4	13-7	17-4	20-6	6-7	10-4	13-5	16-5	19-0
	Douglas fir-larch	#1	6-4	9-6	12-0	14-8	17-1	6-0	8-10	11-2	13-7	15-9
	Douglas fir-larch	#2	6-1	8-11	11-3	13-9	15-11	5-7	8-3	10-5	12-9	14-9
	Douglas fir-larch	#3	4-7	6-9	8-6	10-5	12-1	4-3	6-3	7-11	9-7	11-2
	Hem-fir	SS	6-2	9-9	12-10	16-5	19-11	6-2	9-9	12-10	16-1	18-8
	Hem-fir	#1	6-1	9-3	11-9	14-4	16-7	5-10	8-7	10-10	13-3	15-5
	Hem-fir	#2	5-9	8-9	11-1	13-7	15-9	5-7	8-1	10-3	12-7	14-7
	Hem-fir	#3	4-7	6-9	8-6	10-5	12-1	4-3	6-3	7-11	9-7	11-2
	Southern pine	SS	6-5	10-2	13-4	17-0	20-9	6-5	10-2	13-4	17-0	20-9
	Southern pine	#1	6-4	9-11	13-1	16-0	19-1	6-4	9-11	12-5	14-10	17-8
	Southern pine	#2	6-2	9-4	12-0	14-4	16-10	6-0	8-8	11-2	13-4	15-7
	Southern pine	#3	4-11	7-3	9-2	10-10	12-11	4-6	6-8	8-6	10-1	12-0
	Spruce-pine-fir	SS	6-1	9-6	12-7	16-0	19-1	6-1	9-6	12-5	15-3	17-8
	Spruce-pine-fir	#1	5-11	8-11	11-3	13-9	15-11	5-7	8-3	10-5	12-9	14-9
	Spruce-pine-fir	#2	5-11	8-11	11-3	13-9	15-11	5-7	8-3	10-5	12-9	14-9
	Spruce-pine-fir	#3	4-7	6-9	8-6	10-5	12-1	4-3	6-3	7-11	9-7	11-2

(continued)

ROOF-CEILING CONSTRUCTION

TABLE R802.5.1(6)

TABLE R802.5.1(6)—continued
RAFTER SPANS FOR COMMON LUMBER SPECIES
(Ground snow load=50 psf, ceiling attached to rafters, L/Δ=240)

RAFTER SPACING (inches)	SPECIE AND GRADE		DEAD LOAD = 10 psf					DEAD LOAD = 20 psf				
			2x4	2x6	2x8	2x10	2x12	2x4	2x6	2x8	2x10	2x12
			Maximum rafter spans[a]									
			(feet - inches)	(feet - inches)	(feet - inches)	(feet - inches)	(feet - inches)	(feet - inches)	(feet - inches)	(feet - inches)	(feet - inches)	(feet - inches)
24	Douglas fir-larch	SS	6-1	9-7	12-7	15-10	18-4	6-1	9-6	12-0	14-8	17-0
	Douglas fir-larch	#1	5-10	8-6	10-9	13-2	15-3	5-5	7-10	10-0	12-2	14-1
	Douglas fir-larch	#2	5-5	7-11	10-1	12-4	14-3	5-0	7-4	9-4	11-5	13-2
	Douglas fir-larch	#3	4-1	6-0	7-7	9-4	10-9	3-10	5-7	7-1	8-7	10-0
	Hem-fir	SS	5-9	9-1	11-11	15-2	18-0	5-9	9-1	11-9	14-5	15-11
	Hem-fir	#1	5-8	8-3	10-6	12-10	14-10	5-3	7-8	9-9	11-10	13-9
	Hem-fir	#2	5-4	7-10	9-11	12-1	14-1	4-11	7-3	9-2	11-3	13-0
	Hem-fir	#3	4-1	6-0	7-7	9-4	10-9	3-10	5-7	7-1	8-7	10-0
	Southern pine	SS	6-0	9-5	12-5	15-10	19-3	6-0	9-5	12-5	15-10	19-3
	Southern pine	#1	5-10	9-3	12-0	14-4	17-1	5-10	8-10	11-2	13-3	15-9
	Southern pine	#2	5-9	8-4	10-9	12-10	15-1	5-5	7-9	10-0	11-11	13-11
	Southern pine	#3	4-4	6-5	8-3	9-9	11-7	4-1	6-0	7-7	9-0	10-8
	Spruce-pine-fir	SS	5-8	8-10	11-8	14-8	17-1	5-8	8-10	11-2	13-7	15-9
	Spruce-pine-fir	#1	5-5	7-11	10-1	12-4	14-3	5-0	7-4	9-4	11-5	13-2
	Spruce-pine-fir	#2	5-5	7-11	10-1	12-4	14-3	5-0	7-4	9-4	11-5	13-2
	Spruce-pine-fir	#3	4-1	6-0	7-7	9-4	10-9	3-10	5-7	7-1	8-7	10-0

Check sources for availability of lumber in lengths greater than 20 feet.

For SI: 1 inch = 25.4 mm, 1 foot = 304.8 mm, 1 pound per square foot = 0.0479 kN/m^2.

a. The tabulated rafter spans assume that ceiling joists are located at the bottom of the attic space or that some other method of resisting the outward push of the rafters on the bearing walls, such as rafter ties, is provided at that location. When ceiling joists or rafter ties are located higher in the attic space, the rafter spans shall be multiplied by the factors given below:

H_C/H_R	Rafter Span Adjustment Factor
2/3 or greater	0.50
1/2	0.58
1/3	0.67
1/4	0.76
1/5	0.83
1/6	0.90
1/7.5 and less	1.00

where: H_C = Height of ceiling joists or rafter ties measured vertically above the top of the rafter support walls.

H_R = Height of roof ridge measured vertically above the top of the rafter support walls.

b. Span exceeds 26 feet in length.

❖ See the commentary for Sections R802.5 and R802.5.1 and Table R802.5.1(1).

TABLE R802.5.1(7)
TABLE R802.5.1(7)
RAFTER SPANS FOR 70 PSF GROUND SNOW LOAD
(Ceiling not attached to rafters, L/Δ=180)

RAFTER SPACING (inches)	SPECIES AND GRADE		DEAD LOAD = 10 psf					DEAD LOAD = 20 psf				
			2x4	2x6	2x8	2x10	2x12	2x4	2x6	2x8	2x10	2x12
			Maximum Rafter Spans[a]									
			(feet-inches)	(feet-inches)	(feet-inches)	(feet-inches)	(feet-inches)	(feet-inches)	(feet-inches)	(feet-inches)	(feet-inches)	(feet-inches)
12	Douglas fir-larch	SS	7-7	11-10	15-8	19-5	22-6	7-7	11-10	15-0	18-3	21-2
	Douglas fir-larch	#1	7-1	10-5	13-2	16-1	18-8	6-8	9-10	12-5	15-2	17-7
	Douglas fir-larch	#2	6-8	9-9	12-4	15-1	17-6	6-3	9-2	11-8	14-2	16-6
	Douglas fir-larch	#3	5-0	7-4	9-4	11-5	13-2	4-9	6-11	8-9	10-9	12-5
	Hem-fir	SS	7-2	11-3	14-9	18-10	22-1	7-2	11-3	14-8	18-0	20-10
	Hem-fir	#1	6-11	10-2	12-10	15-8	18-2	6-6	9-7	12-1	14-10	17-2
	Hem-fir	#2	6-7	9-7	12-2	14-10	17-3	6-2	9-1	11-5	14-0	16-3
	Hem-fir	#3	5-0	7-4	9-4	11-5	13-2	4-9	6-11	8-9	10-9	12-5
	Southern pine	SS	7-5	11-8	15-4	19-7	23-10	7-5	11-8	15-4	19-7	23-10
	Southern pine	#1	7-3	11-5	14-9	17-6	20-11	7-3	11-1	13-11	16-6	19-8
	Southern pine	#2	7-1	10-2	13-2	15-9	18-5	6-8	9-7	12-5	14-10	17-5
	Southern pine	#3	5-4	7-11	10-1	11-11	14-2	5-1	7-5	9-6	11-3	13-4
	Spruce-pine-fir	SS	7-0	11-0	14-6	18-0	20-11	7-0	11-0	13-11	17-0	19-8
	Spruce-pine-fir	#1	6-8	9-9	12-4	15-1	17-6	6-3	9-2	11-8	14-2	16-6
	Spruce-pine-fir	#2	6-8	9-9	12-4	15-1	17-6	6-3	9-2	11-8	14-2	16-6
	Spruce-pine-fir	#3	5-0	7-4	9-4	11-5	13-2	4-9	6-11	8-9	10-9	12-5
16	Douglas fir-larch	SS	6-10	10-9	13-9	16-10	19-6	6-10	10-3	13-0	15-10	18-4
	Douglas fir-larch	#1	6-2	9-0	11-5	13-11	16-2	5-10	8-6	10-9	13-2	15-3
	Douglas fir-larch	#2	5-9	8-5	10-8	13-1	15-2	5-5	7-11	10-1	12-4	14-3
	Douglas fir-larch	#3	4-4	6-4	8-1	9-10	11-5	4-1	6-0	7-7	9-4	10-9
	Hem-fir	SS	6-6	10-2	13-5	16-6	19-2	6-6	10-1	12-9	15-7	18-0
	Hem-fir	#1	6-0	8-9	11-2	13-7	15-9	5-8	8-3	10-6	12-10	14-10
	Hem-fir	#2	5-8	8-4	10-6	12-10	14-11	5-4	7-10	9-11	12-1	14-1
	Hem-fir	#3	4-4	6-4	8-1	9-10	11-5	4-1	6-0	7-7	9-4	10-9
	Southern pine	SS	6-9	10-7	14-0	17-10	21-8	6-9	10-7	14-0	17-10	21-0
	Southern pine	#1	6-7	10-2	12-9	15-2	18-1	6-5	9-7	12-0	14-4	17-1
	Southern pine	#2	6-2	8-10	11-5	13-7	16-0	5-10	8-4	10-9	12-10	15-1
	Southern pine	#3	4-8	6-10	8-9	10-4	12-3	4-4	6-5	8-3	9-9	11-7
	Spruce-pine-fir	SS	6-4	10-0	12-9	15-7	18-1	6-4	9-6	12-0	14-8	17-1
	Spruce-pine-fir	#1	5-9	8-5	10-8	13-1	15-2	5-5	7-11	10-1	12-4	14-3
	Spruce-pine-fir	#2	5-9	8-5	10-8	13-1	15-2	5-5	7-11	10-1	12-4	14-3
	Spruce-pine-fir	#3	4-4	6-4	8-1	9-10	11-5	4-1	6-0	7-7	9-4	10-9
19.2	Douglas fir-larch	SS	6-5	9-11	12-7	15-4	17-9	6-5	9-4	11-10	14-5	16-9
	Douglas fir-larch	#1	5-7	8-3	10-5	12-9	14-9	5-4	7-9	9-10	12-0	13-11
	Douglas fir-larch	#2	5-3	7-8	9-9	11-11	13-10	5-0	7-3	9-2	11-3	13-0
	Douglas fir-larch	#3	4-0	5-10	7-4	9-0	10-5	3-9	5-6	6-11	8-6	9-10
	Hem-fir	SS	6-1	9-7	12-4	15-1	17-4	6-1	9-2	11-8	14-2	15-5
	Hem-fir	#1	5-6	8-0	10-2	12-5	14-5	5-2	7-7	9-7	11-8	13-7
	Hem-fir	#2	5-2	7-7	9-7	11-9	13-7	4-11	7-2	9-1	11-1	12-10
	Hem-fir	#3	4-0	5-10	7-4	9-0	10-5	3-9	5-6	6-11	8-6	9-10
	Southern pine	SS	6-4	10-0	13-2	16-9	20-4	6-4	10-0	13-2	16-5	19-2
	Southern pine	#1	6-3	9-3	11-8	13-10	16-6	5-11	8-9	11-0	13-1	15-7
	Southern pine	#2	5-7	8-1	10-5	12-5	14-7	5-4	7-7	9-10	11-9	13-9
	Southern pine	#3	4-3	6-3	8-0	9-5	11-2	4-0	5-11	7-6	8-10	10-7
	Spruce-pine-fir	SS	6-0	9-2	11-8	14-3	16-6	5-11	8-8	11-0	13-5	15-7
	Spruce-pine-fir	#1	5-3	7-8	9-9	11-11	13-10	5-0	7-3	9-2	11-3	13-0
	Spruce-pine-fir	#2	5-3	7-8	9-9	11-11	13-10	5-0	7-3	9-2	11-3	13-0
	Spruce-pine-fir	#3	4-0	5-10	7-4	9-0	10-5	3-9	5-6	6-11	8-6	9-10

(continued)

ROOF-CEILING CONSTRUCTION

TABLE R802.5.1(7)

TABLE R802.5.1(7)—continued
RAFTER SPANS FOR 70 PSF GROUND SNOW LOAD
(Ceiling not attached to rafters, L/Δ=180)

RAFTER SPACING (inches)	SPECIES AND GRADE		DEAD LOAD = 10 psf					DEAD LOAD = 20 psf				
			2x4	2x6	2x8	2x10	2x12	2x4	2x6	2x8	2x10	2x12
			\multicolumn{10}{c}{Maximum rafter spans[a]}
			(feet-inches)	(feet-inches)	(feet-inches)	(feet-inches)	(feet-inches)	(feet-inches)	(feet-inches)	(feet-inches)	(feet-inches)	(feet-inches)
24	Douglas fir-larch	SS	6-0	8-10	11-3	13-9	15-11	5-9	8-4	10-7	12-11	15-0
	Douglas fir-larch	#1	5-0	7-4	9-4	11-5	13-2	4-9	6-11	8-9	10-9	12-5
	Douglas fir-larch	#2	4-8	6-11	8-9	10-8	12-4	4-5	6-6	8-3	10-0	11-8
	Douglas fir-larch	#3	3-7	5-2	6-7	8-1	9-4	3-4	4-11	6-3	7-7	8-10
	Hem-fir	SS	5-8	8-8	11-0	13-6	13-11	5-7	8-3	10-5	12-4	12-4
	Hem-fir	#1	4-11	7-2	9-1	11-1	12-10	4-7	6-9	8-7	10-6	12-2
	Hem-fir	#2	4-8	6-9	8-7	10-6	12-2	4-4	6-5	8-1	9-11	11-6
	Hem-fir	#3	3-7	5-2	6-7	8-1	9-4	3-4	4-11	6-3	7-7	8-10
	Southern pine	SS	5-11	9-3	12-2	15-7	18-2	5-11	9-3	12-2	14-8	17-2
	Southern pine	#1	5-7	8-3	10-5	12-5	14-9	5-3	7-10	9-10	11-8	13-11
	Southern pine	#2	5-0	7-3	9-4	11-1	13-0	4-9	6-10	8-9	10-6	12-4
	Southern pine	#3	3-9	5-7	7-1	8-5	10-0	3-7	5-3	6-9	7-11	9-5
	Spruce-pine-fir	SS	5-6	8-3	10-5	12-9	14-9	5-4	7-9	9-10	12-0	12-11
	Spruce-pine-fir	#1	4-8	6-11	8-9	10-8	12-4	4-5	6-6	8-3	10-0	11-8
	Spruce-pine-fir	#2	4-8	6-11	8-9	10-8	12-4	4-5	6-6	8-3	10-0	11-8
	Spruce-pine-fir	#3	3-7	5-2	6-7	8-1	9-4	3-4	4-11	6-3	7-7	8-10

Check sources for availability of lumber in lengths greater than 20 feet.

For SI: 1 inch = 25.4 mm, 1 foot = 304.8 mm, 1 pound per square foot = 0.0479 kN/m^2.

a. The tabulated rafter spans assume that ceiling joists are located at the bottom of the attic space or that some other method of resisting the outward push of the rafters on the bearing walls, such as rafter ties, is provided at that location. When ceiling joists or rafter ties are located higher in the attic space, the rafter spans shall be multiplied by the factors given below:

H_C/H_R	Rafter Span Adjustment Factor
2/3 or greater	0.50
1/2	0.58
1/3	0.67
1/4	0.76
1/5	0.83
1/6	0.90
1/7.5 and less	1.00

where: H_C = Height of ceiling joists or rafter ties measured vertically above the top of the rafter support walls.

H_R = Height of roof ridge measured vertically above the top of the rafter support walls.

❖ See the commentary for Sections R802.5 and R802.5.1 and Table R802.5.1(1).

TABLE R802.5.1(8)
RAFTER SPANS FOR 70 PSF GROUND SNOW LOAD
(Ceiling attached to rafters, L/Δ=240)

RAFTER SPACING (inches)	SPECIES AND GRADE		DEAD LOAD = 10 psf					DEAD LOAD = 20 psf				
			2x4	2x6	2x8	2x10	2x12	2x4	2x6	2x8	2x10	2x12
			(feet-inches)	(feet-inches)	(feet-inches)	(feet-inches)	(feet-inches)	(feet-inches)	(feet-inches)	(feet-inches)	(feet-inches)	(feet-inches)
12	Douglas fir-larch	SS	6-10	10-9	14-3	18-2	22-1	6-10	10-9	14-3	18-2	21-2
	Douglas fir-larch	#1	6-7	10-5	13-2	16-1	18-8	6-7	9-10	12-5	15-2	17-7
	Douglas fir-larch	#2	6-6	9-9	12-4	15-1	17-6	6-3	9-2	11-8	14-2	16-6
	Douglas fir-larch	#3	5-0	7-4	9-4	11-5	13-2	4-9	6-11	8-9	10-9	12-5
	Hem-fir	SS	6-6	10-2	13-5	17-2	20-10	6-6	10-2	13-5	17-2	20-10
	Hem-fir	#1	6-4	10-0	12-10	15-8	18-2	6-4	9-7	12-1	14-10	17-2
	Hem-fir	#2	6-1	9-6	12-2	14-10	17-3	6-1	9-1	11-5	14-0	16-3
	Hem-fir	#3	5-0	7-4	9-4	11-5	13-2	4-9	6-11	8-9	10-9	12-5
	Southern pine	SS	6-9	10-7	14-0	17-10	21-8	6-9	10-7	14-0	17-10	21-8
	Southern pine	#1	6-7	10-5	13-8	17-6	20-11	6-7	10-5	13-8	16-6	19-8
	Southern pine	#2	6-6	10-2	13-2	15-9	18-5	6-6	9-7	12-5	14-10	17-5
	Southern pine	#3	5-4	7-11	10-1	11-11	14-2	5-1	7-5	9-6	11-3	13-4
	Spruce-pine-fir	SS	6-4	10-0	13-2	16-9	20-5	6-4	10-0	13-2	16-9	19-8
	Spruce-pine-fir	#1	6-2	9-9	12-4	15-1	17-6	6-2	9-2	11-8	14-2	16-6
	Spruce-pine-fir	#2	6-2	9-9	12-4	15-1	17-6	6-2	9-2	11-8	14-2	16-6
	Spruce-pine-fir	#3	5-0	7-4	9-4	11-5	13-2	4-9	6-11	8-9	10-9	12-5
16	Douglas fir-larch	SS	6-3	9-10	12-11	16-6	19-6	6-3	9-10	12-11	15-10	18-4
	Douglas fir-larch	#1	6-0	9-0	11-5	13-11	16-2	5-10	8-6	10-9	13-2	15-3
	Douglas fir-larch	#2	5-9	8-5	10-8	13-1	15-2	5-5	7-11	10-1	12-4	14-3
	Douglas fir-larch	#3	4-4	6-4	8-1	9-10	11-5	4-1	6-0	7-7	9-4	10-9
	Hem-fir	SS	5-11	9-3	12-2	15-7	18-11	5-11	9-3	12-2	15-7	18-0
	Hem-fir	#1	5-9	8-9	11-2	13-7	15-9	5-8	8-3	10-6	12-10	14-10
	Hem-fir	#2	5-6	8-4	10-6	12-10	14-11	5-4	7-10	9-11	12-1	14-1
	Hem-fir	#3	4-4	6-4	8-1	9-10	11-5	4-1	6-0	7-7	9-4	10-9
	Southern pine	SS	6-1	9-7	12-8	16-2	19-8	6-1	9-7	12-8	16-2	19-8
	Southern pine	#1	6-0	9-5	12-5	15-2	18-1	6-0	9-5	12-0	14-4	17-1
	Southern pine	#2	5-11	8-10	11-5	13-7	16-0	5-10	8-4	10-9	12-10	15-1
	Southern pine	#3	4-8	6-10	8-9	10-4	12-3	4-4	6-5	8-3	9-9	11-7
	Spruce-pine-fir	SS	5-9	9-1	11-11	15-3	18-1	5-9	9-1	11-11	14-8	17-1
	Spruce-pine-fir	#1	5-8	8-5	10-8	13-1	15-2	5-5	7-11	10-1	12-4	14-3
	Spruce-pine-fir	#2	5-8	8-5	10-8	13-1	15-2	5-5	7-11	10-1	12-4	14-3
	Spruce-pine-fir	#3	4-4	6-4	8-1	9-10	11-5	4-1	6-0	7-7	9-4	10-9
19.2	Douglas fir-larch	SS	5-10	9-3	12-2	15-4	17-9	5-10	9-3	11-10	14-5	16-9
	Douglas fir-larch	#1	5-7	8-3	10-5	12-9	14-9	5-4	7-9	9-10	12-0	13-11
	Douglas fir-larch	#2	5-3	7-8	9-9	11-11	13-10	5-0	7-3	9-2	11-3	13-0
	Douglas fir-larch	#3	4-0	5-10	7-4	9-0	10-5	3-9	5-6	6-11	8-6	9-10
	Hem-fir	SS	5-6	8-8	11-6	14-8	17-4	5-6	8-8	11-6	14-2	15-5
	Hem-fir	#1	5-5	8-0	10-2	12-5	14-5	5-2	7-7	9-7	11-8	13-7
	Hem-fir	#2	5-2	7-7	9-7	11-9	13-7	4-11	7-2	9-1	11-1	12-10
	Hem-fir	#3	4-0	5-10	7-4	9-0	10-5	3-9	5-6	6-11	8-6	9-10
	Southern pine	SS	5-9	9-1	11-11	15-3	18-6	5-9	9-1	11-11	15-3	18-6
	Southern pine	#1	5-8	8-11	11-8	13-10	16-6	5-8	8-9	11-0	13-1	15-7
	Southern pine	#2	5-6	8-1	10-5	12-5	14-7	5-4	7-7	9-10	11-9	13-9
	Southern pine	#3	4-3	6-3	8-0	9-5	11-2	4-0	5-11	7-6	8-10	10-7
	Spruce-pine-fir	SS	5-5	8-6	11-3	14-3	16-6	5-5	8-6	11-0	13-5	15-7
	Spruce-pine-fir	#1	5-3	7-8	9-9	11-11	13-10	5-0	7-3	9-2	11-3	13-0
	Spruce-pine-fir	#2	5-3	7-8	9-9	11-11	13-10	5-0	7-3	9-2	11-3	13-0
	Spruce-pine-fir	#3	4-0	5-10	7-4	9-0	10-5	3-9	5-6	6-11	8-6	9-10

Maximum rafter spans[a]

(continued)

ROOF-CEILING CONSTRUCTION

TABLE R802.5.1(8)—continued
RAFTER SPANS FOR 70 PSF GROUND SNOW LOAD[a]
(Ceiling attached to rafters, L/Δ=240)

RAFTER SPACING (inches)	SPECIES AND GRADE		DEAD LOAD = 10 psf					DEAD LOAD = 20 psf				
			2x4	2x6	2x8	2x10	2x12	2x4	2x6	2x8	2x10	2x12
			Maximum rafter spans[a]									
			(feet-inches)	(feet-inches)	(feet-inches)	(feet-inches)	(feet-inches)	(feet-inches)	(feet-inches)	(feet-inches)	(feet-inches)	(feet-inches)
24	Douglas fir-larch	SS	5-5	8-7	11-3	13-9	15-11	5-5	8-4	10-7	12-11	15-0
	Douglas fir-larch	#1	5-0	7-4	9-4	11-5	13-2	4-9	6-11	8-9	10-9	12-5
	Douglas fir-larch	#2	4-8	6-11	8-9	10-8	12-4	4-5	6-6	8-3	10-0	11-8
	Douglas fir-larch	#3	3-7	5-2	6-7	8-1	9-4	3-4	4-11	6-3	7-7	8-10
	Hem-fir	SS	5-2	8-1	10-8	13-6	13-11	5-2	8-1	10-5	12-4	12-4
	Hem-fir	#1	4-11	7-2	9-1	11-1	12-10	4-7	6-9	8-7	10-6	12-2
	Hem-fir	#2	4-8	6-9	8-7	10-6	12-2	4-4	6-5	8-1	9-11	11-6
	Hem-fir	#3	3-7	5-2	6-7	8-1	9-4	3-4	4-11	6-3	7-7	8-10
	Southern pine	SS	5-4	8-5	11-1	14-2	17-2	5-4	8-5	11-1	14-2	17-2
	Southern pine	#1	5-3	8-3	10-5	12-5	14-9	5-3	7-10	9-10	11-8	13-11
	Southern pine	#2	5-0	7-3	9-4	11-1	13-0	4-9	6-10	8-9	10-6	12-4
	Southern pine	#3	3-9	5-7	7-1	8-5	10-0	3-7	5-3	6-9	7-11	9-5
	Spruce-pine-fir	SS	5-0	7-11	10-5	12-9	14-9	5-0	7-9	9-10	12-0	12-11
	Spruce-pine-fir	#1	4-8	6-11	8-9	10-8	12-4	4-5	6-6	8-3	10-0	11-8
	Spruce-pine-fir	#2	4-8	6-11	8-9	10-8	12-4	4-5	6-6	8-3	10-0	11-8
	Spruce-pine-fir	#3	3-7	5-2	6-7	8-1	9-4	3-4	4-11	6-3	7-7	8-10

Check sources for availability of lumber in lengths greater than 20 feet.

For SI: 1 inch = 25.4 mm, 1 foot = 304.8 mm, 1 pound per square foot = 0.0479 kN/m^2.

a. The tabulated rafter spans assume that ceiling joists are located at the bottom of the attic space or that some other method of resisting the outward push of the rafters on the bearing walls, such as rafter ties, is provided at that location. When ceiling joists or rafter ties are located higher in the attic space, the rafter spans shall be multiplied by the factors given below:

H_C/H_R	Rafter Span Adjustment Factor
2/3 or greater	0.50
1/2	0.58
1/3	0.67
1/4	0.76
1/5	0.83
1/6	0.90
1/7.5 and less	1.00

where: H_C = Height of ceiling joists or rafter ties measured vertically above the top of the rafter support walls.

H_R = Height of roof ridge measured vertically above the top of the rafter support walls.

❖ See the commentary for Sections R802.5 and R802.5.1 and Table R802.5.1(1).

TABLE R802.5.1(9)
RAFTER/CEILING JOIST HEEL JOINT CONNECTIONS[a,b,c,d,e,f]

RAFTER SLOPE	RAFTER SPACING (inches)	GROUND SNOW LOAD (psf)											
		30 psf				50 psf				70 psf			
		Roof span (feet)											
		12	20	28	36	12	20	28	36	12	20	28	36
		Required number of 16d common nails[a,b] per heel joint splices[c,d,e,f]											
3:12	12	4	6	8	11	5	8	12	15	6	11	15	20
	16	5	8	11	14	6	11	15	20	8	14	20	26
	24	7	11	16	21	9	16	23	30	12	21	30	39
4:12	12	3	5	6	8	4	6	9	11	5	8	12	15
	16	4	6	8	11	5	8	12	15	6	11	15	20
	24	5	9	12	16	7	12	17	22	9	16	23	29
5:12	12	3	4	5	7	3	5	7	9	4	7	9	12
	16	3	5	7	9	4	7	9	12	5	9	12	16
	24	4	7	10	13	6	10	14	18	7	13	18	23
7:12	12	3	3	4	5	3	4	5	7	3	5	7	9
	16	3	4	5	6	3	5	7	9	4	6	9	11
	24	3	5	7	9	4	7	10	13	5	9	13	17
9:12	12	3	3	3	4	3	3	4	5	3	4	5	7
	16	3	3	4	5	3	4	5	7	3	5	7	9
	24	3	4	6	7	3	6	8	10	4	7	10	13
12:12	12	3	3	3	3	3	3	3	4	3	3	4	5
	16	3	3	3	4	3	3	4	5	3	4	5	7
	24	3	3	4	6	3	4	6	8	3	6	8	10

For SI: 1 inch = 25.4 mm, 1 foot = 304.8 mm, 1 pound per square foot = 0.0479 kN/m².

a. 40d box nails shall be permitted to be substituted for 16d common nails.
b. Nailing requirements shall be permitted to be reduced 25 percent if nails are clinched.
c. Heel joint connections are not required when the ridge is supported by a load-bearing wall, header or ridge beam.
d. When intermediate support of the rafter is provided by vertical struts or purlins to a loadbearing wall, the tabulated heel joint connection requirements shall be permitted to be reduced proportionally to the reduction in span.
e. Equivalent nailing patterns are required for ceiling joist to ceiling joist lap splices.
f. When rafter ties are substituted for ceiling joists, the heel joint connection requirement shall be taken as the tabulated heel joint connection requirement for two-thirds of the actual rafter-slope.

❖ This table provides requirements for the connection between roof rafters and ceiling joists to be used in areas with ground snow loads of 30 psf (144 kN/m²) up to 70 psf (3.35 kN/m²). The connection is essential to resist the thrust in the roof rafter where the ridge board does not provide vertical support of the roof rafter Section R802.3 requires sa designed ridge beam if the roof slope is less than 3:12.

R802.7 Cutting and notching. Structural roof members shall not be cut, bored or notched in excess of the limitations specified in this section.

❖ The limitations on cutting and notching roof rafters and ceiling joists are similar to those for cutting and notching of floor joists. See the commentary for Section R502.8.

R802.7.1 Sawn lumber. Notches in solid lumber joists, rafters and beams shall not exceed one-sixth of the depth of the member, shall not be longer than one-third of the depth of the member and shall not be located in the middle one-third of the span. Notches at the ends of the member shall not exceed one-fourth the depth of the member. The tension side of members 4 inches (102 mm) or greater in nominal thickness shall not be notched except at the ends of the members. The diameter of the holes bored or cut into members shall not exceed one-third the depth of the member. Holes shall not be closer than 2 inches (51 mm) to the top or bottom of the member, or to any other hole located in the member. Where the member is also notched, the hole shall not be closer than 2 inches (51 mm) to the notch.

Exception: Notches on cantilevered portions of rafters are permitted provided the dimension of the remaining portion of the rafter is not less than 4-inch nominal (102 mm) and the length of the cantilever does not exceed 24 inches (610 mm).

❖ Figure R502.8 illustrates cutting and notching limitations. The tension side of a member 4 inches (102 mm) or more in thickness canonly be notched at the ends. Also see the commentary for Section R502.8.

R802.7.2 Engineered wood products. Cuts, notches and holes bored in laminated veneer lumber, glue-laminated members or I-joists are not permitted unless the effect of such penetrations are specifically considered in the design of the member.

ROOF-CEILING CONSTRUCTION

For SI: 1 inch = 25.4 mm, 1 foot = 304.8 mm, 1 degree = 0.018 rad.

Note: Where ceiling joists run perpendicular to the rafters, rafter ties shall be nailed to the rafter near the plate line and spaced not more than 4 feet on center.

**FIGURE R802.5.1
BRACED RAFTER CONSTRUCTION**

❖ This figure illustrates the continuous use of purlins to reduce rafter spans as described in Section R802.5.1.

❖ The cutting and notching limitations for sawn lumber do not apply to engineered wood products. The code does not allow drilling and notching of engineered wood products without documentation that the design has taken these modifications into account.

R802.8 Lateral support. Rafters and ceiling joists having a depth-to-thickness ratio exceeding 5 to 1 based on nominal dimensions shall be provided with lateral support at points of bearing to prevent rotation.

❖ Rafters and ceiling joists must be laterally supported to prevent twisting at the supports.

R802.8.1 Bridging. Rafters and ceiling joists having a depth-to-thickness ratio exceeding 6 to 1 based on nominal dimensions shall be supported laterally by solid blocking, diagonal bridging (wood or metal) or a continuous 1-inch by 3-inch (25.4 mm by 76 mm) wood strip nailed across the rafters or ceiling joists at intervals not exceeding 8 feet (2438 mm).

❖ Bridging must be provided for rafters and ceiling joists. This is similar to the requirements for floor joists in Section R502.7.1.

R802.9 Framing of openings. Openings in roof and ceiling framing shall be framed with header and trimmer joists. When the header joist span does not exceed 4 feet (1219 mm), the header joist may be a single member the same size as the ceiling joist or rafter. Single trimmer joists may be used to carry a single header joist that is located within 3 feet (914 mm) of the trimmer joist bearing. When the header joist span exceeds 4 feet (1219 mm), the trimmer joists and the header joist shall be doubled and of sufficient cross section to support the ceiling joists or rafter framing into the header. Approved hangers shall be used for the header joist to trimmer joist connections when the header joist span exceeds 6 feet (1829 mm). Tail joists over 12 feet (3658 mm) long shall be supported at the header by framing anchors or on ledger strips not less than 2 inches by 2 inches (51 mm by 51 mm).

❖ Requirements for framing at roof openings are similar to those of Section R502.10 for floor openings.

R802.10 Wood trusses.

R802.10.1 Truss design drawings. Truss design drawings, prepared in conformance with Section R802.10.1, shall be provided to the building official and approved prior to installation. Truss design drawings shall include, at a minimum, the information specified below. Truss design drawing shall be provided with the shipment of trusses delivered to the jobsite.

1. Slope or depth, span and spacing.
2. Location of all joints.
3. Required bearing widths.
4. Design loads as applicable.
   - 4.1. Top chord live load (including snow loads).
   - 4.2. Top chord dead load.
   - 4.3. Bottom chord live load.
   - 4.4. Bottom chord dead load.
   - 4.5. Concentrated loads and their points of application.
   - 4.6. Controlling wind and earthquake loads.
5. Adjustments to lumber and joint connector design values for conditions of use.
6. Each reaction force and direction.
7. Joint connector type and description (e.g., size, thickness or gauge) and the dimensioned location of each joint connector except where symmetrically located relative to the joint interface.
8. Lumber size, species and grade for each member.
9. Connection requirements for:
   - 9.1. Truss to truss girder.
   - 9.2. Truss ply to ply.
   - 9.3. Field splices.
10. Calculated deflection ratio and/or maximum description for live and total load.
11. Maximum axial compression forces in the truss members to enable the building designer to design the size, connections and anchorage of the permanent continuous lateral bracing. Forces shall be shown on the truss design drawing or on supplemental documents.
12. Required permanent truss member bracing location.

❖ See the definition in Chapter 2 of "Truss Design Drawing." Also see the commentary for Section R802.10.2.

R802.10.2 Design. Wood trusses shall be designed in accordance with accepted engineering practice. The design and manufacture of metal plate connected wood trusses shall comply with ANSI/TPI 1. The truss design drawings shall be prepared by a registered professional where required by the statutes of the jurisdiction in which the project is to be constructed in accordance with Section R106.1.

❖ The code contains no prescriptive provisions for the design and installation of wood trusses. A design is required in accordance with accepted engineering practice. For snow load the truss design must be checked for both the balanced and unbalanced condition. While the unbalanced condition controls for rafter design, that is not necessarily the case for truss design. The balanced condition could control. See the commentary for Section R802.5 for the snow load criteria. In recognition of the extensive use of trusses in residential construction, the code references ANSI/TPI 1, *National Design Standard for Metal-Plate-Connected Wood Truss Construction*. This standard provides regulations for the design and installation of metal-plate-connected wood trusses, including the procedures for full-scale tests and testing methods for evaluating metal-plate connectors. In addition to adequate design, the handling and erection of the trusses must be performed properly so the performance capabilities of the trusses are not compromised. A truss member should never be cut without approval from the design engineer.

Usually trusses are delivered in bundles, which reduces the potential for damage. Sufficient slings and spreader bars should be used to reduce stresses caused by sway and bending.

Shop drawings showing the lumber schedule, design loads, and panel point details (size, location and attachment of plates) should be filled with the building permit application and should be available at the time of inspection.

R802.10.3 Bracing. Trusses shall be braced to prevent rotation and provide lateral stability in accordance with the requirements specified in the construction documents for the building and on the individual truss design drawings. In the absence of specific bracing requirements, trusses shall be braced in accordance with TPI/HIB.

❖ To prevent their collapse during construction, and until permanent bracing is installed, trusses should be adequately braced temporarily. When braced for use, trusses should be positioned as close to vertical as possible; tilted trusses will not perform as required.

The construction documents and the individual truss design drawings should specify the bracing. If the bracing is not specified, then trusses are to be braced in accordance with the Truss Plate Institute's *Commentary and Recommendations for Handling and Bracing Metal Plate Connected Wood Trusses, TPI/HIB-91*.

R802.10.4 Alterations to trusses. Truss members shall not be cut, notched, drilled, spliced or otherwise altered in any way without the approval of a registered design professional. Alterations resulting in the addition of load (e.g., HVAC equipment, water heater) that exceeds the design load for the truss shall not be permitted without verification that the truss is capable of supporting such additional loading.

❖ The addition of loads in excess of the design load is allowed only if it can be verified that the truss has additional capacity. See the commentary for Section R502.8.2.

R802.10.5 Truss to wall connection. Trusses shall be connected to wall plates by the use of approved connectors having a resistance to uplift of not less than 175 pounds (79.45 kg.) and shall be installed in accordance with the manufacturer's specifications. For roof assemblies subject to wind uplift pressures of 20 pounds per square foot (0.958 kN/m2) or greater, as established in Table R301.2(2), adjusted for height and exposure per Table R301.2(3), see section R802.11.

❖ In conventional rafter-ceiling joist construction, the ceiling joist is toenailed to the top plate, and the rafter is toenailed to the top plate with the ceiling joist face nailed to the rafter. Toenailing of trusses to the top plate does not provide an equivalent connection. Toenailing may splinter the truss chord and compromise the truss stability. This section requires that the proper framing anchor be used for the truss-to-wall connection in order to maintain the same structural integrity as expected in a conventionally framed roof.

R802.11 Roof tie-down.

❖ Since roof uplift caused by wind can be a significant factor in roof-system damage, the code requires roof-to-wall connections capable of resisting this force.

R802.11.1 Uplift resistance. Roof assemblies which are subject to wind uplift pressures of 20 pounds per square foot (0.958 kN/m^2) or greater shall have roof rafters or trusses attached to their supporting wall assemblies by connections capable of providing the resistance required in Table R802.11. Wind uplift pressures shall be determined using an effective wind area of 100 square feet (9.3 m^2) and Zone 1 in Table R301.2(2), as adjusted for height and exposure per Table R301.2(3).

A continuous load path shall be provided to transmit the uplift forces from the rafter or truss ties to the foundation.

❖ This section requires compliance with Table R602.3(1) for rafters and Section R802.10.5 for trusses in addition to Table R802.11, which specifies the minimum uplift resistance to be provided when the roof assemblies are subjected to wind uplift pressures of 20 pounds per square foot or more. The toenail connections required by Table R602.3(1) could conceivably provide enough uplift resistance to satisfy Table R802.11 for short roof spans in the lower wind speed areas, but in almost all instances, additional uplift resistance is required in the form of approved connectors.

Basic wind speed for any site is determined from Figure R301.2(4) and recorded in Table R301.2(1). Using this wind speed, the designer selects the appropriate uplift pressure from Table R301.2(2) for Zone 1 [see Figure R301.2(8)] using an effective wind area of 100 square feet (9.3 m^2). This pressure must then be adjusted for height and exposure (see Section R301.2) using Table 301.2(3). If the resulting adjusted wind-uplift pressure is 20 psf (0.96 kN/m^2) or more, then uplift resistance must be provided in accordance with Table R802.11. Determination of the need for uplift resistance is illustrated in the following example. The connection of the rafter or truss to the wall must have the minimum uplift strength specified in Table R802.11.

Example:

A roof has an 18-degree slope and a mean height of 26 feet (7925 mm). The basic wind speed is 100 mph (145 km/h), and the site is Exposure B. Determine the adjusted wind-uplift pressure. Is uplift resistance required?

Solution:

Using Table R301.2(2), find the section covering a roof slope greater than 10 through 30 degrees. For Zone 1, with an effective wind area of 100 square feet (9.2 m^2), the component pressures for 100 mph (145 km/h) are +10 psf and -14.9 psf. The negative values in this table represent uplift, and only these values should be used in complying with this section. The adjustment factor for Exposure B with a 26-foot (7925 mm) mean roof height is 1.00 from Table R301.2(3). Therefore, the adjusted uplift pressure = (14.9) (1.00) = 14.9 psf.

Since the adjusted uplift pressure is less than 20 psf, no uplift resistance is required.

When uplift connectors are required by the table, one is required on every rafter or truss to the stud below, assuming the roof framing is spaced 24 inches (610 mm) on center, in accordance with Note b.

Additionally, the other notes to Table R802.11 may modify the tabular requirements. Also note that the text of this section requires that in addition to having the roof framing tied to the wall below, a load path must then be established to the foundation. Note f modifies the load requirements for the connections that establish the load path.

TABLE R802.11
REQUIRED STRENGTH OF TRUSS OR RAFTER CONNECTIONS TO RESIST WIND UPLIFT FORCES[a,b,c,e,f]
(Pounds per connection)

BASIC WIND SPEED (3-second gust)	ROOF SPAN (feet)							OVERHANGS[d] (pounds/feet)
	12	20	24	28	32	36	40	
85	-72	-120	-145	-169	-193	-217	-241	-38.55
90	-91	-151	-181	-212	-242	-272	-302	-43.22
100	-131	-218	-262	-305	-349	-393	-436	-53.36
110	-175	-292	-351	-409	-467	-526	-584	-64.56

For SI: 1 inch = 25.4 mm, 1 foot = 305 mm, 1 mph = 1.61 km/hr, 1 pound/foot = 14.5939 N/m, 1 pound = 0.454 kg.

a. The uplift connection requirements are based on a 30 foot mean roof height located in Exposure B. For Exposures C and D and for other mean roof heights, multiply the above loads by the Adjustment Coefficients in Table R-301.2(3).

b. The uplift connection requirements are based on the framing being spaced 24 inches on center. Multiply by 0.67 for framing spaced 16 inches on center and multiply by 0.5 for framing spaced 12 inches on center.

c. The uplift connection requirements include an allowance for 10 pounds of dead load.

d. The uplift connection requirements do not account for the effects of overhangs. The magnitude of the above loads shall be increased by adding the overhang loads found in the table. The overhang loads are also based on framing spaced 24 inches on center. The overhang loads given shall be multiplied by the overhang projection and added to the roof uplift value in the table.

e. The uplift connection requirements are based upon wind loading on end zones as defined in Section M1609.6 of the *International Building Code*. Connection loads for connections located a distance of 20% of the least horizontal dimension of the building from the corner of the building are permitted to be reduced by multiplying the table connection value by 0.7 and multiplying the overhang load by 0.8.

f. For wall-to-wall and wall-to-foundation connections, the capacity of the uplift connector is permitted to be reduced by 100 pounds for each full wall above. (For example, if a 600-pound rated connector is used on the roof framing, a 500-pound rated connector is permitted at the next floor level down.)

❖ See the commentary for Section R802.11.1.

SECTION R803 ROOF SHEATHING

R803.1 Lumber sheathing. Allowable spans for lumber used as roof sheathing shall conform to Table R803.1. Spaced lumber sheathing for wood shingle and shake roofing shall conform to the requirements of Sections R905.7 and R905.8. Spaced lumber sheathing is not allowed in Seismic Design Category D_2.

❖ Table R803.1 specifies the minimum thickness of lumber roof sheathing based on rafter spacing. Spaced lumber sheathing used in conjunction with wood shingle or shake roofs is permitted when installed in accordance with Section R905. Since spaced lumber sheathing does not provide a roof diaphragm capable of transferring high in-plane loads to the supporting shear walls or braced wall lines, it is prohibited in structures classified as Seismic Design Category D_2.

TABLE R803.1
MINIMUM THICKNESS OF LUMBER ROOF SHEATHING

RAFTER OR BEAM SPACING (inches)	MINIMUM NET THICKNESS (inches)
24	$^5/_8$
48[a]	$1^1/_2$ T & G
60[b]	
72[c]	

For SI: 1 inch = 25.4 mm.

a. Minimum 270 F_b, 340,000 E.
b. Minimum 420 F_b, 660,000 E.
c. Minimum 600 F_b, 1,150,000 E.

❖ See the commentary for Section R803.1.

R803.2 Wood structural panel sheathing.

R803.2.1 Identification and grade. Wood structural panels shall conform to DOC PS 1, DOC PS 2 or, when manufactured in Canada, CSA 0437, and shall be identified by grade mark or certificate of inspection issued by an approved agency. Wood structural panels shall comply with the grades specified in Table R503.2.1.1(1).

❖ Span rating and installation requirements for wood structural panels used as roof sheathing are similar to those of Section R503.2 for floor sheathing.

R803.2.1.1 Exposure durability. All wood structural panels, when designed to be permanently exposed in outdoor applications, shall be of an exterior exposure durability. Wood structural panel roof sheathing exposed to the underside may be of interior type bonded with exterior glue, identified as Exposure 1.

❖ Exterior-type wood structural panels are required when the wood structural panels are exposed to the environment. Exterior wood structural panels use a resistant-to-moisture glue that may penetrate the outer veneer lamination. Additionally, exterior wood structural panels typically use a higher grade of veneer for the inner plies, which are compatible with the exterior glue-line performance. Wood structural panel roof sheathing exposed only on the underside, such as eave overhangs, is permitted to be of the interior type with exterior glue because of the decrease in exposure hazard relative to moisture.

R803.2.1.2 Fire-retardant-treated plywood. The allowable unit stresses for fire-retardant-treated plywood, including fas-

ROOF-CEILING CONSTRUCTION

tener values, shall be developed from an approved method of investigation that considers the effects of anticipated temperature and humidity to which the fire-retardant-treated plywood will be subjected, the type of treatment and redrying process. The fire-retardant-treated plywood shall be graded by an approved agency.

❖ See the commentary for Section R802.1.3.

R803.2.2 Allowable spans. The maximum allowable spans for wood structural panel roof sheathing shall not exceed the values set forth in Table R503.2.1.1(1).

❖ The lower portion of Table 503.2.1.1(1) lists allowable spans and loads for wood structural panel roof sheathing. Also see the commentary for Section R503.2.2.

R803.2.3 Installation. Wood structural panel used as roof sheathing shall be installed with joints staggered or nonstaggered in accordance with Table R602.3(1), or APA E30 for wood roof framing or with Table R804.3 for steel roof framing.

❖ This section provides a reference to the fastening schedule in Chapter 6 for attachment to wood roof framing. Notes f and g to Table R602.3(1) establish zones requiring increased nailing based on criteria such as basic wind speed and/or mean roof height. APA-E30, *APA Design and Construction Guide* is referenced as an alternative to the Chapter 6 fastening schedule.

This section also gives a cross reference to the Roof Framing Fastening Schedule, Table R804.3, for attachments to cold-formed steel roof framing.

SECTION R804
STEEL ROOF FRAMING

R804.1 General. Elements shall be straight and free of any defects that would significantly affect their structural performance. Cold-formed steel roof framing members shall comply with the requirements of this section.

❖ The provisions of this section govern cold-formed steel roof and ceiling framing.

R804.1.1 Applicability limits. The provisions of this section shall control the construction of steel roof framing for buildings not greater than 60 feet (18 288 mm) in length perpendicular to the joist, rafter or truss span, not greater than 36 feet (10 973 mm) in width parallel to the joist span or truss, not greater than two stories in height with each story not greater than 10 feet (3048 mm) high, and roof slopes not smaller than 3:12 (25-percent slope) or greater than 12:12 (100-percent slope). Steel roof framing constructed in accordance with the provisions of this section shall be limited to sites subjected to a maximum design wind speed of 110 miles per hour (209 km/h) Exposure A, B or C and a maximum ground snow load of 70 psf (3.35 kN/m^2).

❖ Other than providing roof slope limits, this section is identical to Section R505.1.1 and the limitations illustrated in Figure R505.1.1.

R804.1.2 In-line framing. Steel roof framing constructed in accordance with Section R804 shall be located directly in-line with load-bearing studs below with a maximum tolerance of $^3/_4$ inch (19.1 mm) between the centerline of the stud and roof joist/rafter.

❖ The code does not anticipate loads on the top plate of a wall that would cause bending of the top plate. Therefore, roof-framing members must be located directly in line with load-bearing studs of the supporting steel-framed walls as indicated in Figure R603.3. The permitted offset provides some tolerance for rafter location while not adversely affecting the top plate.

R804.1.3 Roof trusses. The design, quality assurance, installation and testing of cold-formed steel trusses shall be in accordance with the AISI Standard for Cold-formed Steel Framing-Truss Design (COFS/Truss).

❖ The standard provides design and installation provisions for trusses constructed of cold-formed steel framing and establishes provisions that do not currently exist in the *International Residential Code*.

R804.2 Structural framing. Load-bearing steel roof framing members shall comply with Figure R804.2(1) and the dimensional and minimum thickness requirements specified in Tables R804.2(1) and R804.2(2). Tracks shall comply with Figure R804.2(2) and shall have a minimum flange width of $1^1/_4$ inches (32 mm). The maximum inside bend radius for load-bearing members shall be the greater of $^3/_{32}$ inch (2.4 mm) or twice the uncoated steel thickness. Holes in roof framing members shall not exceed 1.5 inches (38 mm) in width or 4 inches (102 mm) in length as shown in Figure R804.2(3). Holes shall be permitted only along the centerline of the web of the framing member. Holes shall not be less than 24 inches (610 mm) center-to-center and shall not be located less than 10 inches (254 mm) from the edge of the hole to the edge of the bearing surface or support unless patched in accordance with Section R804.3.6.

❖ Figure R804.2(1) shows a typical load-bearing steel roof or ceiling member, which must comply with the dimensional and minimum thickness requirements specified in Tables R804.2(1) and R804.2(2). These tables are similar to those in Section R603 for walls and those in Section R505 for floors. Note "a" to Table R804.2(1) explains the member designation used for all steel framing members in the code. Steel thickness is expressed in mils, and Table R804.2(2) provides the equivalent thickness in inches as well as gage number. The referenced thickness pertains to the base metal thickness measured prior to painting or the application of galvanized coatings. The base metal thickness is typically stamped or embossed on the member by the manufacturer.

Holes are permitted only along the centerline of the web of roof-framing members. Their location and size are limited as shown in Figure R804.2(3), except holes may be located within 10 inches (254 mm) of the bearing surface or support, provided they are patched in accordance with Section R804.3.6. These patching requirements are similar to those of Section R505.3.6. Also see the commentary for Section R505.2.

R804.2.1 Material. Load-bearing steel framing members shall be cold-formed to shape from structural quality sheet steel complying with the requirements of one of the following:

1. ASTM A 653; Grades 33, 37, 40 and 50 (Classes 1 and 3).
2. ASTM A 792; Grades 33, 37, 40 and 50A.
3. ASTM A 875; Grades 33, 37, 40 and 50 (Classes 1 and 3).
4. Steels that comply with ASTM A 653, except for tensile and elongation, shall be permitted provided the ratio of tensile strength to yield point is at least 1.08 and the total elongation is at least 10 percent for a 2-inch (51 mm) gage length or 7 percent for an 8-inch (203 mm) gage length.

❖ Load-bearing steel framing members must be cold-formed to shape from structural quality sheet steel complying with the appropriate material standards. The steel grades specified have the ductility and strength to meet the intent of these provisions.

R804.2.2 Identification. Load-bearing steel framing members shall have a legible label, stencil, stamp or embossment with the following information as a minimum:

1. Manufacturer's identification.
2. Minimum uncoated steel thickness in inches (mm).
3. Minimum coating designation.
4. Minimum yield strength, in kips per square inch (ksi).

❖ Load-bearing steel framing members must have a legible label, stencil, stamp, or embossment. This identification allows for verification that materials that are installed are consistent with the design and meet the intent of the code.

FIGURE R804.2(1)
C-SECTION

FIGURE R804.2(2)
TRACK SECTION

For SI: 1 inch = 25.4 mm.

FIGURE R804.2(3)
WEB HOLES

❖ See the commentary for Section R804.2 for Figures R804.2(1) through R804.2(3).

ROOF-CEILING CONSTRUCTION

TABLE R804.2(1)
LOAD-BEARING COLD-FORMED STEEL MEMBER SIZES

NOMINAL MEMBER SIZE MEMBER DESIGNATION[a]	WEB DEPTH (inches)	MINIMUM FLANGE WIDTH (inches)	MAXIMUM FLANGE WIDTH (inches)	MINIMUM LIP SIZE (inches)
350S162-t	3.5	1.625	2	0.5
550S162-t	5.5	1.625	2	0.5
800S162-t	8	1.625	2	0.5
1000S162-t	10	1.625	2	0.5
1200S162-t	12	1.625	2	0.5

For SI: 1 inch = 25.4 mm.

a. The member designation is defined by the first number representing the member depth in $1/100$ inch, the letter "s" representing a stud or joist member, the second number representing the flange width in $1/100$ inch, and the letter "t" shall be a number representing the minimum base metal thickness in mils [see Table R804.2(2)].

❖ See the commentary for Section R804.2.

TABLE R804.2(2)
MINIMUM THICKNESS OF COLD-FORMED STEEL ROOF FRAMING MEMBERS

DESIGNATION (mils)	MINIMUM UNCOATED THICKNESS (inches)	REFERENCED GAGE NUMBER
33	0.033	20
43	0.043	18
54	0.054	16
68	0.068	14

For SI: 1 inch = 25.4 mm, 1 mil = 0.0254 mm.

❖ See the commentary for Section R804.2.

R804.2.3 Corrosion protection. Load-bearing steel framing shall have a metallic coating complying with one of the following:

1. A minimum of G 60 in accordance with ASTM A 653.
2. A minimum of AZ 50 in accordance with ASTM A 792.
3. A minimum of GF 60 in accordance with ASTM A 875.

❖ The metallic coatings specified correspond to requirements in the referenced material standards in Section R804.2.1.

R804.2.4 Fastening requirements. Screws for steel-to-steel connections shall be installed with a minimum edge distance and center-to-center spacing of $1/2$ inch (12.7 mm), shall be self-drilling tapping, and shall conform to SAE J78. Structural sheathing shall be attached to roof rafters with minimum No. 8 self-drilling tapping screws that conform to SAE J78. Screws for attaching structural sheathing to steel roof framing shall have a minimum head diameter of 0.292 inch (7.4 mm) with countersunk heads and shall be installed with a minimum edge distance of $3/8$ inch (9.5 mm). Gypsum board ceilings shall be attached to steel joists with minimum No. 6 screws conforming to ASTM C 954 and shall be installed in accordance with Section R805. For all connections, screws shall extend through the steel a minimum of three exposed threads. All self-drilling tapping screws conforming to SAE J78 shall have a minimum Type II coating in accordance with ASTM B 633.

Where No. 8 screws are specified in a steel-to-steel connection, the required number of screws in the connection is permitted to be reduced in accordance with the reduction factors in Table R804.2.4 when larger screws are used or when one of the sheets of steel being connected is thicker that 33 mils (0.84 mm). When applying the reduction factor, the resulting number of screws shall be rounded up.

❖ Fasteners meeting the referenced standards and installed in accordance with this section will provide a load capacity consistent with these prescriptive provisions.

TABLE R804.2.4
SCREW SUBSTITUTION FACTOR

SCREW SIZE	THINNEST CONNECTED STEEL SHEET (mils)	
	33	43
#8	1.0	0.67
#10	0.93	0.62
#12	0.86	0.56

For SI: 1 mil = 0.0254 mm.

❖ See the commentary for Table R505.2.4.

R804.3 Roof construction. Steel roof systems constructed in accordance with the provisions of this section shall consist of both ceiling joists and rafters in accordance with Figure R804.3 and fastened in accordance with Table R804.3.

❖ Figure R804.3 illustrates steel roof system components. The roof-framing fastening schedule in Table R804.3 specifies the minimum connection requirements for components of the roof-framing system.

FIGURE R804.3 – TABLE R804.3   ROOF-CEILING CONSTRUCTION

For SI: 1 inch = 25.4 mm, 1 foot = 304.8 mm, 1 mil = 0.0254 mm.

**FIGURE R804.3
STEEL ROOF CONSTRUCTION**

❖ This figure is a schematic of typical steel-framed roof construction, providing section references to the code requirements for many of the roof components.

**TABLE R804.3
ROOF FRAMING FASTENING SCHEDULE[a,b]**

DESCRIPTION OF BUILDING ELEMENTS	NUMBER AND SIZE OF FASTENERS	SPACING OF FASTENERS
Ceiling joist to top track of load-bearing wall	2 No. 10 screws	Each joist
Roof sheathing (oriented strand board or plywood) to rafters	No. 8 screws	6" o.c. on edges and 12" o.c. at interior supports. 6" o.c. at gable end truss
Truss to bearing wall[a]	2 No. 10 screws	Each truss
Gable end truss to endwall top track	No. 10 screws	12" o.c.
Rafter to ceiling joist	Minimum No. 10 screws, per Table R804.3.1(3)	Evenly spaced, less than $1/2$" from all edges.

For SI: 1 inch = 25.4 mm, 1 foot = 304.8 mm, 1 pound per square foot = 0.0479 kN/m^2, 1 mil = 0.0254 mm.

a. Screws shall be applied through the flanges of the truss or ceiling joist or a 54 mil clip angle shall be used with two No. 10 screws in each leg. See Section R804.4 for additional requirements to resist uplift forces.
b. Spacing of fasteners on roof sheathing panel edges applies to panel edges supported by framing members and at all roof plane perimeters. Blocking of roof sheathing panel edges perpendicular to the framing members shall not be required except at the intersection of adjacent roof planes. Roof perimeter shall be supported by framing members or cold-formed blocking of the same depth and gauge as the floor members.

❖ See the commentary for Section R804.3.

R804.3.1 Allowable ceiling joist spans. The clear span of cold-formed steel ceiling joists shall not exceed the limits set forth in Table R804.3.1(1) or R804.3.1(2). Ceiling joists shall have a minimum bearing length of 1.5 inches (38 mm) and shall be connected to rafters (heel joint) in accordance with Figure R804.3.1(1) and Table R804.3.1(3). When continuous joists are framed across interior bearing supports, the interior bearing supports shall be located within 24 inches (610 mm) of mid-span of the ceiling joist, and the individual spans shall not exceed the applicable spans in Table R804.3.1(1) or R804.3.1(2). Where required in Table R804.3.1(1) or R804.3.1(2), bearing stiffeners shall be installed at each bearing location in accordance with Section R804.3.8 and Figure R804.3.8. When the attic is to be used as an occupied space, the ceiling joists shall be designed in accordance with Section R505.

❖ Allowable spans for ceiling joists are given in Tables R804.3.1(1) and R804.3.1(2) for live loads of 10 psf (0.48 kN m^2) and 20 psf (0.96 kN/m^2) respectively. An attic with occupied space requires the use of higher design live loads, and the ceiling joists are required to be designed in accordance with Section R505.3.2. Ceiling joists must be connected to rafters in accordance with Figure R804.3.1(1) and Table R804.3.1(3). Use of the allowable span tables is illustrated below.

Example:
Steel ceiling joists at 16 inches on center support a ceiling with limited attic storage and are braced at mid-span. Their span is 17 feet, 6 inches (5334 mm) in a 35-foot wide (10 668 mm) building with a roof slope of 8:12 and 30 psf (1436 Pa) snow load. Determine the required joist size and the required connection to rafter.

Solution:
The live load is 20 psf (0.96 kN/m^2) for limited attic storage. In Table R804.3.1(2), under the column headed mid-span bracing, look under 16-inch joist spacing. Note that 1000S162-68 has an allowable span of 18 feet, 5 inches (5613 mm), which is greater than the required span.

The joist-rafter connection in Table R804.3.1(3) requires 4 No. 10 screws fastened per figure R804.3.1(1).

When continuous joists are framed across interior bearing supports, the interior bearing supports must be located within 24 inches (610 mm) of midspan of the ceiling joist, and the individual spans must not exceed the applicable spans in Table R804.3.1(1) or Table R804.3.1(2).

Bearing stiffeners are required at each bearing point as well as at concentrated loads. See Note c to Tables R804.3.1(1) and R804.3.1(2). Bearing stiffeners must be installed in accordance with Section R804.3.8.

TABLE R804.3.1(1)
ALLOWABLE SPANS FOR COLD-FORMED STEEL CEILING JOISTS[a,b,c] 10 psf live load (no attic storage)

NOMINAL JOIST SIZE	LATERAL SUPPORT OF TOP (COMPRESSION) FLANGE					
	Unbraced		Mid-span bracing		Third-point bracing	
	Spacing (inches)		Spacing (inches)		Spacing (inches)	
	16	24	16	24	16	24
350S162-33	9'-2"	8'-3"	11'-9"	10'-1"	11'-9"	10'-4"
350S162-43	9'-11"	8'-10"	12'-10"	11'-2"	12'-10"	11'-2"
350S162-54	10'-8"	9'-6"	13'-9"	12'-0"	13'-9"	12'-0"
350S162-68	11'-7"	10'-4"	14'-8"	12'-10"	14'-8"	12'-10"
550S162-33	10'-5"	9'-5"	14'-5"	12'-8"[c]	16'-4"	13'-10"[c]
550S162-43	11'-2"	10'-1"	15'-7"	13'-10"	18'-0"	15'-5"
550S162-54	12'-0"	10'-9"	16'-7"	14'-9"	19'-5"	16'-8"
550S162-68	12'-11"	11'-7"	17'-8"	15'-10"	20'-11"	18'-1"
800S162-33	11'-8"[c]	10'-6"[c]	16'-5"[c]	14'-9"[c]	19'-5"[c]	16'-7"[c]
800S162-43	12'-6"	11'-3"	17'-6"	15'-10"	21'-2"	18'-7"
800S162-54	13'-4"	11'-11"	18'-7"	16'-9"	22'-7"	20'-0"
800S162-68	14'-3"	12'-9"	19'-8"	17'-8"	23'-11"	21'-4"
1000S162-43	13'-4"[c]	12'-1"[c]	18'-9"[c]	16'-11"[c]	22'-11"[c]	20'-6"[c]
1000S162-54	14'-2"	12'-9"	19'-10"	17'-10"	24'-2"	21'-9"
1000S162-68	15'-2"	13'-7"	21'-0"	18'-11"	25'-6"	23'-0"
1200S162-43	14'-1"[c]	12'-8"[c]	19'-10"[c]	17'-11"[c]	24'-3"[c]	21'-6"[c]
1200S162-54	15'-0"[c]	13'-5"[c]	20'-11"[c]	18'-11"[c]	25'-7"[c]	23'-1"[c]
1200S162-68	15'-11"	14'-4"	22'-2"	19'-11"	27'-0"	24'-4"

For SI: 1 inch = 25.4 mm, 1 foot = 304.8 mm, 1 pound per square foot = 0.0479 kN/m^2.
a. Deflection criteria: $L/240$ for total loads.
b. Ceiling dead load = 5 psf.
c. Bearing stiffeners are required at all bearing points and concentrated load locations.

❖ See the commentary for Section R804.3.1.

TABLE R804.3.1(2)
ALLOWABLE SPANS FOR COLD-FORMED STEEL CEILING JOISTS[a,b,c]
20 psf live load (Limited attic storage where development of future rooms is not possible)

NOMINAL JOIST SIZE	LATERAL SUPPORT OF TOP FLANGE					
	Unbraced		Mid-span bracing		Third-point bracing	
	Spacing (inches)		Spacing (inches)		Spacing (inches)	
	16	24	16	24	16	24
350S162-33	8'-0"	6'-0"	9'-8"	6'-0"	9'-0"	6'-0"
350S162-43	8'-8"	7'-8"	10'-9"	9'-1"	10'-10"	9'-5"
350S162-54	9'-3"	8'-3"	11'-7"	9'-11"	11'-7"	10'-1"
350S162-68	10'-0"	8'-11"	12'-5"	10'-10"	12'-5"	10'-10"
550S162-33	9'-2"	6'-0"	12'-2"[c]	10'-5"[c]	13'-3"	11'-0"[c]
550S162-43	9'-10"	8'-10"	13'-4"	11'-6"	14'-9"	12'-5"
550S162-54	10'-5"	9'-5"	14'-4"	12'-6"	16'-1"	13'-7"
550S162-68	11'-3"	10'-0"	15'-4"	13'-5"	17'-5"	14'-10"
800S162-33	10'-3"[c]	9'-3"[c]	14'-4"[c]	12'-5"[c]	15'-11"[c]	13'-4"[c]
800S162-43	10'-11"	9'-10"	15'-5"	13'-8"[c]	17'-11"[c]	15'-5"[c]
800S162-54	11'-8"	10'-6"	16'-3"	14'-7"	19'-3"	16'-8"
800S162-68	12'-5"	11'-2"	17'-3"	15'-6"	20'-7"	18'-0"
1000S162-43	11'-9"	10'-7"[c]	16'-6"[c]	14'-10"[c]	19'-10"[c]	17'-1"[c]
1000S162-54	12'-5"	11'-2"	17'-5"	15'-8"	21'-1"	18'-7"
1000S162-68	13'-3"	11'-10"	18'-5"	16'-7"	22'-4"	19'-11"
1200S162-43	12'-5"[c]	11'-2"[c]	17'-5"[c]	15'-8"[c]	20'-9"[c]	18'-0"[c]
1200S162-54	13'-1"[c]	11'-9"[c]	18'-5"[c]	16'-7"[c]	22'-5"[c]	20'-1"[c]
1200S162-68	13'-11"	12'-6"	19'-5"	17'-6"	23'-8"	21'-3"

For SI: 1 inch = 25.4 mm, 1 foot = 304.8 mm, 1 pound per square foot = 0.0479 kN/m^2.
a. Deflection criteria: L/240 for total loads.
b. Ceiling dead load = 5 psf.
c. Bearing stiffeners are required at all bearing points and concentrated load locations.

❖ See the commentary for Section R804.3.1.

TABLE R804.3.1(3)
NUMBER OF SCREWS REQUIRED FOR CEILING JOIST TO RAFTER CONNECTION[a]

ROOF SLOPE	BUILDING WIDTH (feet)															
	24				28				32				36			
	Ground snow load (psf)				Ground snow load (psf)				Ground snow load (psf)				Ground snow load (psf)			
	20	30	50	70	20	30	50	70	20	30	50	70	20	30	50	70
3/12	5	6	9	12	6	7	10	13	7	8	12	15	8	9	13	17
4/12	4	5	7	9	5	6	8	10	6	6	9	12	6	7	10	13
5/12	4	4	6	7	4	5	7	9	5	5	8	10	5	6	9	11
6/12	3	4	5	7	4	4	6	8	4	5	7	9	4	5	7	10
7/12	3	3	5	6	3	4	5	7	4	4	6	8	4	5	7	9
8/12	3	3	4	5	3	3	5	6	3	4	5	7	4	4	6	8
9/12	2	3	4	5	3	3	4	6	3	4	5	6	3	4	6	7
10/12	2	3	4	5	3	3	4	5	3	3	5	6	3	4	5	7
11/12	2	3	4	4	3	3	4	5	3	3	5	6	3	4	5	6
12/12	2	3	3	4	2	3	4	5	3	3	4	6	3	4	5	6

For SI: 1 foot = 304.8 mm, 1 pound per square foot = 0.0479 kN/m^2.
a. Screws shall be No. 10 minimum.

❖ This table lists the connections required between the roof rafter and the ceiling joists. The connection is essential to resist the thrust in the roof rafter where the ridge board does not provide vertical support the roof rafter.

ROOF-CEILING CONSTRUCTION

For SI: 1 mil = 0.0254 mm.

**FIGURE R804.3.1(1)
JOIST TO RAFTER CONNECTION**

❖ The fastening to the wall track is illustrated. See the commentary for Table R804.3.1(3) for a discussion of the joist-to-rafter connection.

R804.3.2 Ceiling joist bracing. The bottom flanges of steel ceiling joists shall be laterally braced in accordance with Section R702. The top flanges of steel ceiling joists shall be laterally braced with a minimum of 33 mil (0.84 mm) C-section, 33 mil (0.84 mm) track section, or $1^1/_2$ inch by 33 mil (38 mm by 0.84 mm) continuous steel strapping as required in Table R804.3.1(1) or R804.3.1(2). Lateral bracing shall be installed in accordance with Figure R804.3. C-section, tracks or straps shall be fastened to the top flange at each joist with at least one No. 8 screw and shall be fastened to blocking with at least two No. 8 screws. Blocking or bridging (X-bracing) shall be installed between joists in-line with strap bracing at a maximum spacing of 12 feet (3658 mm) measured perpendicular to the joists, and at the termination of all straps. The third point bracing span values from Table R804.3.1(1) or R804.3.1(2) shall be used for straps installed at closer spacings than third point bracing, or when sheathing is applied to the top of the ceiling joists.

❖ The bottom flanges of steel ceiling joists must be laterally braced by a finished ceiling in accordance with Section R702. Tables R804.3.1(1) and R804.3.1(2) provide allowable spans for ceiling joists that are unbraced, braced at the midpoint (see Figure R804.3), or braced at the third-points. Top flanges of steel ceiling joists must be laterally braced according to the bracing option selected from these tables and as shown in Figure R804.3. The allowable spans for third-point bracing should be used when top flange bracing is installed at closer intervals or when sheathing fastened to the top flange provides bracing. Strap bracing requires the installation of blocking or bridging similar to that of floor joists described in the commentary for Section R505.3.

R804.3.3 Allowable rafter spans. The horizontal projection of the rafter span, as shown in Figure R804.3, shall not exceed the limits set forth in Table R804.3.3(1). Wind speeds shall be converted to equivalent ground snow loads in accordance with Table R804.3.3(2). Rafter spans shall be selected based on the higher of the ground snow load or the equivalent snow load converted from the wind speed. When required, a rafter support brace shall be a minimum of 350S162-33 C-section with maximum length of 8 feet (2438 mm) and shall be connected to a ceiling joist and rafter with four No. 10 screws at each end.

❖ Table R804.3.3(1) lists allowable rafter spans based on ground-snow loads. In order to account for wind loading on rafters, Table R804.3.3(2) must be used to convert wind speed to an equivalent ground-snow load. Then the allowable rafter span is determined from Table R804.3.3(1) based on the larger of either the ground-snow load or the equivalent snow load converted from the wind speed. As in Section 505.3.2 (for floor joists), a table must be used to determine allowable rafter spans.

When installed as illustrated in Figure R804.3, a rafter brace may be used to reduce the rafter span.

R804.3.3.1 Rafter framing. Rafters shall be connected to a parallel ceiling joist to form a continuous tie between exterior walls in accordance with Figures R804.3 and R804.3.1(1) and Table R804.3.1(3). Rafters shall be connected to a ridge member with a minimum 2-inch by 2-inch (51 mm by 51 mm) clip angle fastened with minimum No. 10 screws to the ridge member in accordance with Figure R804.3.3.1 and Table R804.3.3.1. The clip angle shall have a minimum steel thickness as the rafter member and shall extend the full depth of the rafter member. The ridge member shall be fabricated from a C-section and a track section, which shall be of a minimum size and steel thickness as the adjacent rafters and shall be installed in accordance with Figure R804.3.3.1.

❖ Ceiling joists connected to rafters as required in Section R804.3.1(1) provide the required continuous tie between exterior walls; see Figures R804.3 and R804.3.1(1). Rafters must be connected to a ridge member as illustrated in Figure R804.3.3.1.

R804.3.3.2 Roof cantilevers. Roof cantilevers shall not exceed 24 inches (610 mm) in accordance with Figure R804.3. Roof cantilevers shall be supported by a header in accordance with Section R603.6 or shall be supported by the floor framing in accordance with Section R505.3.7.

❖ Figure R804.3 shows the permitted roof cantilevers. In addition to limiting the gravity load, this provision places a cap on the amount of wind uplift on the cantilevered member, its hold-down and the supporting construction.

TABLE R804.3.3(1)
ALLOWABLE HORIZONTAL RAFTER SPANS[a,b]

NOMINAL RAFTER SIZE	GROUND SNOW LOAD							
	20 psf		30 psf		50 psf		70 psf	
	Spacing (inches-feet)		Spacing (inches-feet)		Spacing (inches-feet)		Spacing (inches-feet)	
	16	24	16	24	16	24	16	24
550S162-33	12'-8"	10'-4"	11'-9"	9'-7"	9'-11"	8'-1"	8'-10"	7'-2"
550S162-43	15'-5"	12'-7"	14'-3"	11'-8"	12'-1"	9'-10"	10'-8"	8'-9"
550S162-54	13'-0"	14'-2"	16'-1"	13'-1"	13'-8"	11'-2"	12'-1"	9'-10"
550S162-68	18'-1"	15'-10"	17'-3"	14'-9"	15'-4"	12'-6"	13'-6"	11'-1"
800S162-33	15'-5"	11'-5"	14'-4"	9'-10"	10'-7"	7'-1"	8'-3"	5'-6"
800S162-43	19'-1"	15'-7"	17'-9"	14'-6"	15'-1"	12'-3"	13'-3"	10'-9"
800S162-54	22'-7"	18'-5"	21'-0"	17'-1"	17'-9"	14'-6"	15'-9"	12'-10"
800S162-68	24'-7"	20'-9"	23'-4"	19'-3"	20'-0"	16'-4"	17'-8"	14'-5"
1000S162-43	21'-2"	17'-3"	19'-8"	16'-0"	16'-8"	13'-1"	14'-9"	10'-3"
1000S162-54	25'-1"	20'-6"	23'-3"	19'-0"	19'-9"	16'-1"	17'-5"	14'-3"
1000S162-68	29'-6"	24'-6"	27'-9"	22'-9"	23'-8"	19'-3"	21'-0"	17'-1"
1200S162-43	23'-0"	18'-2"	21'-4"	15'-7"	16'-9"	11'-3"	13'-2"	8'-9"
1200S162-54	27'-3"	22'-3"	25'-3"	20'-7"	21'-5"	17'-6"	18'-11"	15'-5"
1200S162-68	32'-1"	26'-2"	29'-9"	24'-3"	25'-3"	20'-7"	22'-4"	18'-2"

For SI: 1 inch = 25.4 mm, 1 foot = 304.8 mm, 1 pound per square foot = 0.0479 kN/m^2.

a. Deflection criteria: $L/240$ for live loads and $L/180$ for total loads.

b. Roof dead load = 12 pounds per square foot.

❖ See the commentary for Section R804.3.3.

TABLE R804.3.3(2)
BASIC WIND SPEED TO EQUIVALENT SNOW LOAD CONVERSION

BASIC WIND SPEED AND EXPOSURE		EQUIVALENT GROUND SNOW LOAD (psf)									
		Roof slope									
Exp. A/B	Exp. C	3:12	4:12	5:12	6:12	7:12	8:12	9:12	10:12	11:12	12:12
85 mph	—	20	20	20	20	20	20	30	30	30	30
100 mph	85 mph	20	20	20	20	30	30	30	30	50	50
110 mph	100 mph	20	20	20	20	30	50	50	50	50	50
—	110 mph	30	30	30	50	50	50	70	70	70	—

For SI: 1 mile per hour = 0.447 m/s, 1 pound per square foot = 0.0479 kN/m^2.

❖ See the commentary for Section R804.3.3.

ROOF-CEILING CONSTRUCTION

TABLE R804.3.3.1
NUMBER OF SCREWS REQUIRED AT EACH LEG OF CLIP ANGLE FOR RAFTER TO RIDGE MEMBER CONNECTION[a]

BUILDING WIDTH (feet)	GROUND SNOW LOAD (psf)			
	0 to 20	21 to 30	31 to 50	51 to 70
24	2	3	4	4
28	2	3	4	5
32	3	3	4	5
36	3	4	5	6

For SI: 1 foot = 304.8 mm, 1 pound per square foot = 0.0479 kN/m^2.
a. Screws shall be No. 10 minimum.

❖ See the commentary for Section R804.3.3.1.

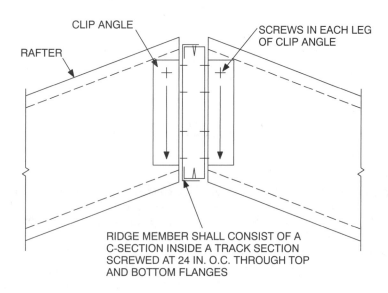

For SI: 1 inch = 25.4 mm.

FIGURE R804.3.3.1
RIDGE BOARD CONNECTION

❖ See the commentary for Section R804.3.3.1.

R804.3.4 Rafter bottom flange bracing. The bottom flanges of steel rafters shall be continuously braced with a minimum 33-mil (0.84 mm) C-section, 33-mil (0.84 mm) track section, or a 1^1/$_2$-inch by 33-mil (38 mm by 0.84 mm) steel strapping at a maximum spacing of 8 feet (2438 mm) as measured parallel to the rafters. Bracing shall be installed in accordance with Figure R804.3. The C-section, track section, or straps shall be fastened to blocking with at least two No. 8 screws. Blocking or bridging (X-bracing) shall be installed between rafters in-line with the continuous bracing at a maximum spacing of 12 feet (3658 mm) measured perpendicular to the rafters and at the termination of all straps. The ends of continuous bracing shall be fastened to blocking with at least two No. 8 screws.

❖ The bottom flanges of steel rafters requires continuous bracing as illustrated in Figure R804.3. Installation of bracing is necessary to provide resistance to twisting and/or lateral displacement of the rafters.

R804.3.5 Cutting and notching. Flanges and lips of load-bearing steel roof framing members shall not be cut or notched. Holes in webs shall be in accordance with Section R804.2.

❖ Flanges and lips of load-bearing steel roof framing members must not be cut or notched since this would affect their structural integrity.

R804.3.6 Hole patching. Holes in ceiling joist and rafters with dimensions conforming to Section R804.2 that are closer than 10 inches (254 mm) from the edge of the hole to the edge of bearing surface shall be patched with a solid steel plate, C-section or track section in accordance with Figure R804.3.6. The steel patch shall be of a minimum thickness as the receiving member and shall extend at least 1 inch (25.4 mm) beyond all edges of the hole. The steel patch shall be fastened to the web with No. 8 screws (minimum) spaced no greater than 1 inch (25.4 mm) center-to-center along the edges of the patch, with a minimum edge distance of ^{1}/$_2$ inch (12.7 mm).

❖ The commentary for Section R804.2 discusses the requirements for patching holes in the webs of roof framing members.

**FIGURE R804.3.6
HOLE PATCHING**

❖ See the commentary for Section R804.2.

R804.3.7 Splicing. Rafters and other structural members, except ceiling joists, shall not be spliced. Splices in ceiling joists shall only be permitted at interior bearing points and shall be constructed in accordance with Figure R804.3.7(1). Spliced ceiling joists shall be connected with the same number and size of screws on connection. Splicing of tracks shall conform with Figure R804.3.7(2).

❖ Splicing of load-carrying steel framing members is not permitted except when lapped ceiling joists occur at interior bearing points in accordance with Figure R804.3.7(1). Since Section R804.3.3.1 requires that ceiling joists provide a continuous tie between the exterior walls, the quantity of screws required for this splice is the same as required by Table R804.3.1(3) for the heel joint connection shown in Figure R804.3.1(1).
Splicing of tracks is allowed since, in accordance with Section R804.2, they are not permitted to act as load-carrying members; see Figure R804.3.7(2).

R804.3.8 Bearing stiffener. A bearing stiffener shall be fabricated from a minimum 33-mil (0.84 mm) C-section or track section. Each stiffener shall be fastened to the web of the ceiling joist with a minimum of four No. 8 screws equally spaced as shown in Figure R804.3.8. Stiffeners shall extend across the full depth of the web and shall be installed on either side of the web.

❖ A bearing stiffener (also referred to as a "transverse stiffener" or "web stiffener") must be installed as shown in Figure R804.3.8. The stiffener at the joist/rafter connection is similar to the requirements in Section R505.3.4, except a 33 mil (0.84 mm) track may be used rather than the 43 mil (1.09 mm) required in Section R505.3.

R804.3.9 Headers. Roof-ceiling framing above wall openings shall be supported on headers. The allowable spans for headers in bearing walls shall not exceed the values set forth in Table R603.6(1).

❖ Headers must be located immediately below the ceiling or roof framing (for example, at the top of the track), above wall openings, in all exterior walls and interior load-bearing walls in accordance with Figure R603.6. Table R603.6(1) gives the allowable spans for headers in bearing walls.

R804.3.10 Framing of opening. Openings in roof and ceiling framing shall be framed with headers and trimmers between ceiling joists or rafters. Header joist spans shall not exceed 4 feet (1219 mm). Header and trimmer joists shall be fabricated from joist and track sections, which shall be of a minimum size and thickness in accordance with Figures R804.3.10(1) and R804.3.10(2). Each header joist shall be connected to trimmer joist with a minimum of four 2-inch by 2-inch (51 by 51 mm) clip angles. Each clip angle shall be fastened to both the header and trimmer joists with four No. 8 screws, evenly spaced, through each leg of the clip angle. The clip angles shall have a steel thickness not less than that of the floor joist.

❖ Roof openings must be framed as illustrated in Figures R804.3.10(1) and R804.3.10(2). Each header joist must be connected to the trimmer joists with a minimum of four 2-inch by 2-inch (51 mm by 51 mm) clip angles, which must be fastened to both the header and trimmer joists with four No. 8 screws, evenly spaced, through each leg. The clip angle's thickness must not be less than that of the floor joist.

ROOF-CEILING CONSTRUCTION

For SI: 1 inch = 25.4 mm.

**FIGURE R804.3.7(1)
SPLICED CEILING JOISTS**

❖ See the commentary for Section R804.3.7.

For SI: 1 inch = 25.4 mm.

**FIGURE R804.3.7(2)
TRACK SPLICE**

❖ See the commentary for Section R804.3.7.

**FIGURE R804.3.8
BEARING STIFFENER**

❖ See the commentary for Section R804.3.8.

For SI: 1 foot = 304.8 mm.

**FIGURE R804.3.10(1)
ROOF OPENING**

❖ See the commentary for Section R804.3.10.

ROOF-CEILING CONSTRUCTION

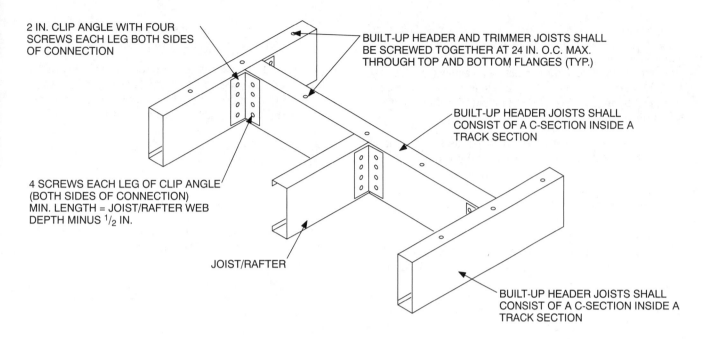

For SI: 1 inch = 25.4 mm.

**FIGURE R804.3.10(2)
HEADER TO TRIMMER CONNECTION**

❖ See the commentary for Section R804.3.10.

R804.4 Roof tie-down. Roof assemblies subject to wind uplift pressures of 20 pounds per square foot (0.96 kN/m^2) or greater, as established in Table R301.2(2), shall have rafter-to-bearing wall ties provided in accordance with Table R802.11.

❖ Steel-framed roof assemblies require rafter-to-bearing wall ties with a capacity determined in the manner described in Section R802.11.

SECTION R805
CEILING FINISHES

R805.1 Ceiling installation. Ceilings shall be installed in accordance with the requirements for interior wall finishes as provided in Section R702.

❖ This section provides a reference to the ceiling installation requirements found in Section R702.

SECTION R806
ROOF VENTILATION

R806.1 Ventilation required. Enclosed attics and enclosed rafter spaces formed where ceilings are applied directly to the underside of roof rafters shall have cross ventilation for each separate space by ventilating openings protected against the entrance of rain or snow. Ventilating openings shall be provided with corrosion-resistant wire mesh, with $^1/_8$ inch (3.2 mm) minimum to $^1/_4$ inch (6.4 mm) maximum openings.

❖ Large amounts of water vapor migrate by air movement or diffusion through the building envelope materials because of a vapor pressure difference. The sources of water vapor include cooking, laundering, bathing, and human breathing and perspiration. These can account for an average daily production of 25 pounds of water vapor in a typical family-of-four dwelling. The average can be much higher where appliances such as humidifiers, washers, and dryers are used.

As the vapor moves into the attic, it may reach its dew point, condensing on wood roof components. This wetting and drying action will cause rotting and decay. To avoid this, the attic must be ventilated to prevent the accumulation of water on building components. The installation of a vapor retarder acts to prevent the passage of moisture to the attic, and an effective vapor retarder allows a decrease in ventilation. Vapor retarders are ineffective when openings in the barrier allow moisture to be carried by air into the attic. This is also the reason exhaust fans must terminate outdoors and not in the attic. Care should be exercised to assure that attic vent openings remain unobstructed.

To minimize condensation problems within attic and enclosed rafter spaces, free-flow ventilation of such spaces is required. Ventilation openings must be screened to prevent the entry of animals.

R806.2 Minimum area. The total net free ventilating area shall not be less than 1 to 150 of the area of the space ventilated except that the total area is permitted to be reduced to 1 to 300, provided at least 50 percent and not more than 80 percent of the required ventilating area is provided by ventilators located in the upper portion of the space to be ventilated at least 3 feet (914 mm) above eave or cornice vents with the balance of the required ventilation provided by eave or cornice vents. As an alternative, the net free cross-ventilation area may be reduced to 1 to 300 when a vapor barrier having a transmission rate not exceeding 1 perm (57.4 mg/s · m^2 · Pa) is installed on the warm side of the ceiling.

❖ The attic vent size required by the code should not be overlooked. The net-free area can be as much as 50 percent less than the gross opening area. For example, one manufacturer's 24-inch square (610 mm) gable vent (gross area equals 576 square inches [0.37 m^2]) is listed in their catalog as having a net free area of 308 square inches (0.20 m^2), which is about 53 percent of the gross area. The manufacturer's literature should be consulted to obtain free-area information.

R806.3 Vent clearance. Where eave or cornice vents are installed, insulation shall not block the free flow of air. A minimum of a 1-inch (25.4 mm) space shall be provided between the insulation and the roof sheathing at the location of the vent.

❖ Vent openings must be maintained clear, and they must not block the free flow of air; therefore, the code requires that insulation be held back from the vent opening a minimum of 1 inch (25.4 mm).

SECTION R807
ATTIC ACCESS

R807.1 Attic access. In buildings with combustible ceiling or roof construction, an attic access opening shall be provided to attic areas that exceed 30 square feet (2.8 m^2) and have a vertical height of 30 inches (762 mm) or greater.

The rough-framed opening shall not be less than 22 inches by 30 inches (559 mm by 762 mm) and shall be located in a hallway or other readily accessible location. A 30-inch (762 mm) minimum unobstructed headroom in the attic space shall be provided at some point above the access opening. See Section M1305.1.3 for access requirements where mechanical equipment is located in attics.

❖ The requirement for an attic access is predicated on the likelihood that during the life of the structure, access to an attic space for repair of piping, electrical, and mechanical systems will be required.

SECTION R808
INSULATION CLEARANCE

R808.1 Combustible insulation. Combustible insulation shall be separated a minimum of 3 inches (76 mm) from recessed lighting fixtures, fan motors and other heat-producing devices.

Exception: When heat-producing devices are listed for lesser clearances, combustible insulation complying with the listing requirements shall be separated in accordance with the conditions stipulated in the listing.

Recessed lighting fixtures installed in the building thermal envelope shall meet the requirements of Section N1101.3.

❖ The minimum clearance required between combustible insulation and light fixtures, fan motors, etc., minimizes the chance of accidental ignition of the insulation. If a heat-producing device is listed for less than the minimum clearance specified in this section and all conditions of the listing are satisfied, then the reduced clearance would apply.

Bibliography

The following resource materials are referenced in this chapter or are relevant to the subject matter addressed in this chapter.

AFPA NDS-01, *National Design Specification (NDS) for Wood Construction—with 2001 Supplement*. Washington, DC: American Forest and Paper Association, 2001.

AFPA 93, *Span Tables for Joists and Rafters*. Washington DC: American Forest and Paper Association, 1993.

APA E30-01, *Engineered Wood Construction Guide*. Tacoma, WA: APA—The Engineered Wood Association, 2001.

ASTM A 653/653M-00, *Specification for Steel Sheet, Zinc-coated (Galvanized) or Zinc-iron Alloy-coated (Galvanized) by the Hot-dip Process*. West Conshohocken, PA: ASTM International, 2000.

ASTM A 792M-99, *Specification for Steel Sheet, 55% Aluminum-Zinc Alloy-Coated by the Hot-Dip Process*. West Conshohocken, PA: ASTM International, 1999.

ASTM A 875/0875M—01a, *Specification for Steel Sheet, Zinc-5% Aluminum Alloy-Coated by the Hot-Dip Process*. West Conshohocken, PA: ASTM International, 2001.

ASTM C954-00, *Specification for Steel Drill Screws for the Application of Gypsum Panel Products or Metal Plaster Bases to Steel Studs from 0.033 in. (0.84 mm) to 0.112 in. (2.84 mm) in Thickness*. West Conshohocken, PA: ASTM International, 2000.

AWPA C20-99, *Structural Lumber—Fire-Retardant Treatment by Pressure Processes*. Granbury, TX: American Wood-Preservers' Association, 1999.

AWPA C27-88, *Plywood—Fire-Retardant Treatment by Pressure Processes*. Granbury, TX: American Wood-Preservers' Association, 1988

CSA 0437-Series 93, *Standards on OSB and Waferboard (Reaffirmed 2001)*. Toronto, Ontario, Canada: Canadian Standards Association, 1993.

DOC PS 1-95, *Construction and Industrial Plywood*. Gaithersburg, MD: United States Department of Commerce, National Institute of Standards and Technology, 1995.

ROOF-CEILING CONSTRUCTION

DOC PS 2-92, *Performance Standard for Wood-Based Structural-Use Panels*. Gaithersburg, MD: United States Department of Commerce, National Institute of Standards and Technology, 1992.

DOC PS 20-99, *American Softwood Lumber Standard*. Gaithersburg, MD: United States Department of Commerce, National Institute of Standards and Technology, 1999.

SAE J78-(1998), *Steel Self-Drilling Tapping Screws*. Warrendale, PA: Society of Automotive Engineers, 1998.

TPI 1-2002, *National Design Standard for Metal-Plate-Connected Wood Truss Construction.* Madison, WI: Truss Plate Institute, 2002.

TPI HIB—91, *Commentary and Recommendations for Handling, Installing and Bracing Metal Plate Connected Wood Trusses*. Madison, WI: Truss Plate Institute, 1991.

Chapter 9:
Roof Assemblies

General Comments

While a small portion of this chapter regulates the fire classification of roof covering materials, the major emphasis is on the materials and installation methods that will provide a weathertight exterior envelope. The *International Residential Code®* limits the requirements for fire-retardant roof coverings to two scenarios: 1) where the local or state ordinances in effect mandate a classified roof covering and 2) where the edge of the roof is in close proximity to a property line. Otherwise, the provisions of the chapter focus on the weather-protection features of a roof assembly.

The chapter contains a number of specific code requirements for various types of roof covering materials. These requirements may place limitations on roof slope, identify the proper attachment methods, mandate the use of underlayment and address flashing at appropriate points. In addition, the code often references a specification or installation standard for a particular type of roof covering. In all cases, it is important that the manufacturer's installation instructions be followed.

This chapter contains provisions for the design and construction of roof assemblies, primarily focusing on the materials and installation of roof coverings. The general requirements for roof assemblies found in the IRC deal with the role of the roof as a portion of the exterior envelope that provides weather protection for the building. Fire classifications, roof drainage, roof insulation and reroofing operations are also addressed.

Section R901 states the scope of the chapter, including both the materials and quality of roof construction. Section R902 deals with fire-classification requirements for roof covering materials where the roof extends very close to an adjoining property line. Section R903 establishes the general requirements for weather protection, including provisions for flashing and roof drainage.

Section R904 establishes general requirements for roofing materials. Section R905 discusses the many different types of roof covering materials regulated by the IRC, including asphalt shingles, clay and concrete tile, metal roof shingles, mineral-surfaced roll roofing, wood shakes and shingles, slate and slate-type shingles, built-up roofs, metal roof panels, modified bitumen roofing, thermoset and thermoplastic single-ply roofing, sprayed polyurethane foam roofing and liquid-applied coatings. Section R906 allows the installation of above-deck thermal installation where appropriately listed, and Section R907 establishes the methods and materials acceptable for recovering or replacing an existing roof.

Purpose

Roof assemblies provide a variety of functions; however, this chapter deals primarily with roof coverings for weather protection. The requirements address the design and construction of roof assemblies, with a focus on the materials and quality of installation. While a number of requirements are specified directly from the code, the majority of the regulations applicable to roof coverings come from the material standards. In addition, the manufacturer's installation instructions should always be followed for the specific roof covering material under consideration. Flashing and roof drainage criteria are also provided as important aspects of the weather-protection features. Many of the common roof coverings are addressed by detailed provisions regulating the roof deck, roof slope, underlayment, application and attachment. Where a new roof is required for an existing building, the code sets forth specific guidelines for either recovering or replacing the existing roof covering.

SECTION R901
GENERAL

R901.1 Scope. The provisions of this chapter shall govern the design, materials, construction and quality of roof assemblies.

❖ The focus of this chapter is on roof coverings. A variety of roof covering materials are included, including asphalt shingles, clay and concrete tile, metal roof shingles, wood shingles and shakes and metal roof panels. Roof classification, drainage, insulation and flashing are also regulated, as is the reroofing of an existing roof.

SECTION R902
ROOF CLASSIFICATION

R902.1 Roofing covering materials. Roofs shall be covered with materials as set forth in Sections R904 and R905. Class A, B or C roofing shall be installed in areas designated by law as requiring their use or when the edge of the roof is less than 3 feet (914 mm) from a property line. Classes A, B and C roofing required to be listed by this section shall be tested in accordance with UL 790 or ASTM E 108. Roof assemblies with coverings of brick, masonry, slate, clay or concrete roof tile, exposed concrete roof deck, ferrous or copper shingles or sheets, and metal sheets and shingles, shall be considered Class A roof coverings.

❖ The IRC identifies two conditions under which roof assembly classifications A, B, or C are required. First, where local, state or other governing ordinance requires the use of Class A, B or C roofing, it must be installed as designated by such law. Second, on roofs that project within 3 feet (914 mm) of the property line, a Class A, B or C roof must be provided. Testing for such roofing must be in accordance with either ASTM E 108 or UL 790.

Roof coverings classified as Class A, B or C have been shown through testing to provide protection of the roof against severe, moderate and light fire exposures, respectively. These exposures are external and are generally created by fires in adjoining structures, wild fires and fires from the subject building that extend up the exterior and onto the top surface of the roof. Wild fires and some structure fires create flying and flaming brands that can ignite nonclassified roof coverings.

By their nature, roof coverings composed of brick, masonry, slate, clay tile, concrete tile, ferrous or copper shingles or sheets and metal sheets or shingles are considered excellent materials for resisting fire spread and are classified as Class A. An exposed concrete roof deck also qualifies for this classification.

R902.2 Fire-retardant-treated shingles and shakes. Fire-retardant-treated wood shakes and shingles shall be treated by impregnation with chemicals by the full-cell vacuum-pressure process, in accordance with AWPA C1. Each bundle shall be marked to identify the manufactured unit and the manufacturer, and shall also be labeled to identify the classification of the material in accordance with the testing required in Section R902.1, the treating company and the quality control agency.

❖ This section recognizes the process of impregnating wood shakes and shingles with chemicals in order to make them fire retardant. AWPA C1, regulating the preservative treatment of timber products by pressure processes, is the referenced standard. During the impregnation process the cells of wood shakes and shingles are exposed to a vacuum that draws all of the air and moisture from the material. Fire-retardant chemicals are then impregnated into the shakes and shingles by pressurization. After proper curing, the shakes and shingles are then classified. The fire classification must be marked on each bundle of materials, along with other manufacturing information. Identification will also include the name of the company performing the pressure-treatment process, as well as the quality control agency reviewing for compliance with the test standard.

SECTION R903
WEATHER PROTECTION

R903.1 General. Roof decks shall be covered with approved roof coverings secured to the building or structure in accordance with the provisions of this chapter. Roof assemblies shall be designed and installed in accordance with this code and the approved manufacturer's installation instructions such that the roof assembly shall serve to protect the building or structure.

❖ In all cases, a roof must be designed to provide protection from the elements. For the roof to adequately perform this function, it must be designed in accordance with this chapter. This section requires flashing where the roof intersects vertical elements such as walls, chimneys, dormers, plumbing stacks, plumbing vents and other penetrations of the weather-protective barrier. Roof drainage to remove water from the roof to an approved location is also regulated.

Materials and installation methods for roof decks and supporting elements are regulated by Chapter 8 of the IRC. A complying roof deck must then be covered in an approved manner to provide the necessary weather protection mandated by the code. This chapter includes the necessary provisions to provide such a weatherproof barrier. In addition, and just as important, the approved manufacturer's installation instructions must be adhered to.

R903.2 Flashing. Flashings shall be installed in such a manner so as to prevent moisture entering the wall and roof through joints in copings, through moisture permeable materials, and at intersections with parapet walls and other penetrations through the roof plane.

❖ As with flashing required for other locations at the building's exterior, it is critical that those locations where moisture may penetrate the exterior membrane at the roof be adequately protected. General locations identified include joints at the coping, intersections with parapet walls and other penetrations through the roof plane.

R903.2.1 Locations. Flashings shall be installed at wall and roof intersections; wherever there is a change in roof slope or direction; and around roof openings. Where flashing is of metal, the metal shall be corrosion resistant with a thickness of not less than 0.019 inch (No. 26 galvanized sheet).

❖ This section identifies specific roof locations for the installation of flashings. In those cases where metal is used as the flashing material, it must be corrosion-resistant and a minimum No. 26 galvanized sheet.

R903.3 Coping. Parapet walls shall be properly coped with noncombustible, weatherproof materials of a width no less than the thickness of the parapet wall.

❖ The top of an exterior wall, typically a masonry parapet wall, must have a coping. It is typically sloped to prevent standing water, and it must be of noncombustible, weatherproofed materials. In order to maintain protection at the top of the wall, the coping must extend the full thickness of the parapet wall.

R903.4 Roof drainage. Unless roofs are sloped to drain over roof edges, roof drains shall be installed at each low point of the roof. Where required for roof drainage, scuppers shall be

ROOF ASSEMBLIES

placed level with the roof surface in a wall or parapet. The scupper shall be located as determined by the roof slope and contributing roof area.

❖ In most residential construction, drainage water from the roof simply flows down the roof's slope to and over the roof edge, often to a gutter and downspout system that carries the water away from the building. In situations where the roof design does not allow for water flow over the roof edges, roof drains must be provided in the roof surface. The drains will be at the low points of the roof, sized and located to remove all roof water. In a case where the low points occur at parapets of the exterior walls, scuppers through the parapets must be provided, constructed with the flow points level with the roof surface at the low points and located based on the roof slope and contributing roof area.

R903.4.1 Overflow drains and scuppers. Where roof drains are required, overflow drains having the same size as the roof drains shall be installed with the inlet flow line located 2 inches (51 mm) above the low point of the roof, or overflow scuppers having three times the size of the roof drains and having a minimum opening height of 4 inches (102 mm) shall be installed in the adjacent parapet walls with the inlet flow located 2 inches (51 mm) above the low point of the roof served. The installation and sizing of overflow drains, leaders and conductors shall comply with the *International Plumbing Code*.

Overflow drains shall discharge to an approved location and shall not be connected to roof drain lines.

❖ It is quite possible that roof drains may become clogged or blocked by debris, resulting in insufficient drainage. This could cause structural failure of portions of the roof system due to the increased weight on the roof caused by the impounded water. This section addresses such concerns by requiring overflow drains or overflow scuppers for those roofs that do not drain over the roof edges.

While the required roof drains must be located at the low points of the roof, the overflow drain inlets need to be located 2 inches (51 mm) vertically above the low points. The same criterion applies where overflow scuppers are provided. In addition, overflow scuppers must be sized with a minimum opening height of 4 inches (102 mm) and have an opening area of at least three times that of the roof drains. The *International Plumbing Code* sets forth the specific details for the installation and sizing of overflow drains, leaders, and conductors.

The overflow drains must not be connected to the roof drains. In the event a blockage occurs in the drainage line beyond the point of connection, both the roof drain and the overflow drain would be affected. The overflow drain should discharge to an approved location, typically a highly visible location that would indicate the overflow system is in use. This will provide an alert that there is some sort of blockage in the primary roof drain system.

SECTION R904
MATERIALS

R904.1 Scope. The requirements set forth in this section shall apply to the application of roof covering materials specified herein. Roof assemblies shall be applied in accordance with this chapter and the manufacturer's installation instructions. Installation of roof assemblies shall comply with the applicable provisions of Section R905.

❖ Roof assemblies and roof covering materials must conform to the provisions of this chapter, the applicable reference standards listed in Chapter 43, and the manufacturer's installation instructions. This section provides general conditions for the application of roof covering materials referenced elsewhere in Chapter 9. A cross reference to Section R905 is provided for requirements dealing with the installation of specific roof coverings.

R904.2 Compatibility of materials. Roof assemblies shall be of materials that are compatible with each other and with the building or structure to which the materials are applied.

❖ The materials that comprise a roof assembly must be compatible in order to avoid a decrease or failure in the performance of the roof. Not only must the roofing materials be compatible with each other, they must also be suitable for the building as a whole.

R904.3 Material specifications and physical characteristics. Roof covering materials shall conform to the applicable standards listed in this chapter. In the absence of applicable standards or where materials are of questionable suitability, testing by an approved testing agency shall be required by the building official to determine the character, quality and limitations of application of the materials.

❖ In most cases, materials used as roof coverings are regulated by an applicable standard. Identified in this chapter, they are also listed in Chapter 43. If a roof covering material is not regulated by a specific standard, the building official must have substantiating data to show that the materials are consistent with the intent of the code regarding quality and application limitations. Typically, testing the roof covering material under consideration is necessary to provide the information necessary to show code compliance.

R904.4 Product identification. Roof covering materials shall be delivered in packages bearing the manufacturer's identifying marks and approved testing agency labels when required. Bulk shipments of materials shall be accompanied with the same information issued in the form of a certificate or on a bill of lading by the manufacturer.

❖ Identification of roof covering materials provides much of the information necessary for acceptance of the materials by the building official. As it is often difficult, if not impossible, to ascertain complying roof covering materials by visual observation, other proper identification of the materials is mandatory. Packaging of roof covering materials must include the manufacturer's identification marks. If the code or applicable standard requires review by an approved testing agency, the identifying marks of the testing agency must also be provided. The code also addresses the identification requirements for bulk shipments of materials.

SECTION R905
REQUIREMENTS FOR ROOF COVERINGS

R905.1 Roof covering application. Roof coverings shall be applied in accordance with the applicable provisions of this section and the manufacturer's installation instructions.

❖ In addition to the minimum requirements specified for roof coverings in the code, manuals published by various associations provide detailed discussions of the proper methods of installing roof coverings. These methods have been established based on many years of experience with the materials and their performance. While the provisions in these documents are not specific code requirements, this section mandates the use of the manufacturer's installation instructions.

R905.2 Asphalt shingles. The installation of asphalt shingles shall comply with the provisions of this section.

❖ This section regulates asphalt shingles composed of organic felt or glass felt and coated with mineral granules. Provisions address requirements for sheathing, roof slope, underlayment, fasteners, and attachment.

R905.2.1 Sheathing requirements. Asphalt shingles shall be fastened to solidly sheathed decks.

❖ The code requires a solid roof surface for the installation of asphalt shingles. Section R803 regulates solid sheathing.

R905.2.2 Slope. Asphalt shingles shall only be used on roof slopes of two units vertical in 12 units horizontal (2:12) or greater. For roof slopes from two units vertical in 12 units horizontal (2:12) up to four units vertical in 12 units horizontal (4:12), double underlayment application is required in accordance with Section R905.2.7.

❖ The performance of all roof coverings is based in part on the slope of the roof surface. As the slope of the roof decreases, water drainage is slowed, and the potential for water intrusion increases due to the greater potential of water back-up under the roofing. Asphalt shingles, because of their configuration and installation methods, are restricted to use on roofs having a minimum slope of 2:12. Where the slope is no steeper than 4:12, the underlayment must be doubled to provide a greater barrier to leakage. Section R905.2.7 specifies the method of such double underlayment. Commentary Figure R905.2.2(1) shows an example of asphalt roof shingles installed on a high-slope roof (4:12 minimum), while Commentary Figure R905.2.2(2) depicts a low-slope roof installation (between 2:12 and 4:12).

R905.2.3 Underlayment. Unless otherwise noted, required underlayment shall conform with ASTM D 226, Type I, or ASTM D 4869, Type I.

Figure R905.2.2(1)
ASPHALT ROOFING SHINGLES APPLICATION HIGH SLOPE (4:12 MINIMUM)

Figure R905.2.2(2)
APPLICATION OF ASPHALT SHINGLE SLOPES BETWEEN 2:12 AND 4:12

Self-adhering polymer modified bitumen sheet shall comply with ASTM D 1970.

❖ Three types of underlayment are recognized for use with asphalt shingle roof coverings: asphalt-saturated organic felts as regulated by ASTM D 226, Type I; asphalt-saturated organic felt shingles per ASTM D 4869, Type I and self-adhering polymer modified bitumen sheet materials addressed in ASTM D 1970. Section R905.2.7 contains the methods prescribed for the installation of underlayment for asphalt shingles.

R905.2.4 Asphalt shingles. Asphalt shingles shall have self-seal strips or be interlocking, and comply with ASTM D 225 or D 3462.

❖ The IRC permits two types of asphalt shingles: interlocking and those with self-seal strips. Although the method of installation varies for these two types, the result is the same: a weather-resistant barrier between the elements and the interior of the building. Two test standards regulate asphalt shingles. ASTM D 225 addresses asphalt shingles made from organic felt, while ASTM D 3462 deals with shingles made from glass felt. Both shingle types are surfaced with mineral granules.

R905.2.5 Fasteners. Fasteners for asphalt shingles shall be galvanized steel, stainless steel, aluminum or copper roofing nails, minimum 12 gage [0.105 inch (2.67 mm)] shank with a minimum $^3/_8$-inch (9.5 mm) diameter head, ASTM F 1667, of a length to penetrate through the roofing materials and a minimum of $^3/_4$ inch (19.1 mm) into the roof sheathing. Where the roof sheathing is less than $^3/_4$ inch (19.1 mm) thick, the fasteners shall penetrate through the sheathing. Fasteners shall comply with ASTM F 1667.

❖ Roofing nails must be of a corrosion-resistant material, specified in this section as either galvanized steel, stainless steel, aluminum, or copper. A roofing nail must have a minimum 12 gage shank with a head at least $^3/_8$ inch (9.5 mm) in diameter. To provide the necessary holding power, roofing nails must penetrate the roof sheathing a minimum of $^3/_4$ inch (19.1 mm). For roof sheathing having a thickness less than $^3/_4$ inch (19.1 mm), the nails must penetrate completely through the sheathing. The material specification standard for roofing nails is ASTM F 1667.

R905.2.6 Attachment. Asphalt shingles shall have the minimum number of fasteners required by the manufacturer. For normal application, asphalt shingles shall be secured to the roof with not less than four fasteners per strip shingle or two fasteners per individual shingle. Where the roof slope exceeds 20 units vertical in 12 units horizontal (20:12), special methods of fastening are required. For roofs located where the basic wind speed per Figure R301.2(4) is 110 mph (177 km/h) or greater, special methods of fastening are required. Special fastening methods shall be tested in accordance with ASTM D 3161, modified to use a wind speed of 110 mph (177 km/h).

Shingles classified using ASTM D 3161 are acceptable for use in wind zones less than 110 mph. Shingles classified using ASTM D 3161 modified to use a wind speed of 110 mph are acceptable for use in all cases where special fastening is required.

❖ In all cases, the minimum number of fasteners for asphalt shingles must be the number required by the shingle manufacturer. It is typical for asphalt strip shingles to be attached to the roof sheathing by at least four fasteners for each strip shingle. If individual shingles are installed, at least two fasteners are required per shingle. For very steep roof slopes, the manufacturer's installation instructions should be reviewed for any special fastening methods.

Where the basic wind speeds in accordance with Figure R301.2(4) are greater than or equal to 110 mph (177.1 km/h), special testing is required to determine how such roofing should be fastened. The fastening methods must be specifically tested to ASTM D 3161, but the test must be modified to accommodate the wind speed of 110 mph (177.1 km/h).

ASTM D 3161 deals with several ranges of wind speeds. More basic wind speeds are considered up to 110 mph (177.1 km/h). Therefore, areas prone to higher winds must comply with the modified test.

R905.2.7 Underlayment application. For roof slopes from two units vertical in 12 units horizontal (17-percent slope), up to four units vertical in 12 units horizontal (33-percent slope), underlayment shall be two layers applied in the following manner. Apply a 19-inch (483 mm) strip of underlayment felt parallel with and starting at the eaves, fastened sufficiently to hold in place. Starting at the eave, apply 36-inch-wide (914 mm) sheets of underlayment, overlapping successive sheets 19 inches (483 mm), and fastened sufficiently to hold in place. For roof slopes of four units vertical in 12 units horizontal (33-percent slope) or greater, underlayment shall be one layer applied in the following manner. Underlayment shall be applied shingle fashion, parallel to and starting from the eave and lapped 2 inches (51 mm), fastened sufficiently to hold in place. End laps shall be offset by 6 feet (1829 mm).

❖ This section specifies the underlayment requirements for asphalt shingles installed on both low-slope and high-slope roofs. Low-slope roofs present a potential problem because water drains slowly, and this creates the opportunity for water back-up. Wind-driven rain can also pose a problem. Therefore, a special underlayment application method is used so that the roof remains weathertight. All portions of the roof will be protected by a minimum of two layers of underlayment if the installer follows the instructions found in this section. Commentary Figure R905.2.7 illustrates the required application of underlayment for a low-slope application. Only one layer of underlayment, applied in shingle fashion in accordance with this section, is mandated for high-slope roofs of asphalt shingles.

R905.2.7.1 Ice protection. In areas where the average daily temperature in January is 25°F (-4°C) or less or when Table R301.2(1) criteria so designates, an ice barrier that consists of a least two layers of underlayment cemented together or of a self-adhering polymer modified bitumen sheet, shall be used in lieu of normal underlayment and extend from the eave's edge to a point at least 24 inches (610 mm) inside the exterior wall line of the building.

For SI: 1 inch = 25.4 mm, 1 foot = 304.8 mm, °C = [(°F) - 32/1.8]

Figure R905.2.7
LOW-SLOPE DOUBLE-PLY UNDERLAYMENT APPLICATION

Exception: Detached accessory structures that contain no conditioned floor area.

❖ Where ice dams may be formed along the eave because snow continually freezes and thaws or frozen slush backs up in gutters, the underlayment application in the area of the eaves must be modified to prevent ice dams from forcing water under the roofing, which could damage ceilings, walls, and insulation; see Commentary Figure R905.2.7.1(1). Two layers of underlayment should be cemented together with asphalt cement from the edge of the roof and continue up the roof to a point that is at least 24 inches (610 mm) inside the interior wall line of the building as shown in Commentary Figure R905.2.7.1(2). The environment within the envelope of the building provides adequate warmth to prevent ice dams from forming above the heated space; therefore, the two layers of cemented underlayment are permitted to terminate 24 inches (610 mm) inside the interior wall line of the building. The local jurisdiction is responsible for determining if the eave flashing is required based on weather records, and it must so indicate in Table R301.2(1).

An exception to this section exempts accessory buildings from such restrictions as they are unheated structures where the need for protection against ice dams is unnecessary. The same exception is found in Sections R905.4.3, R905.5.3, R905.6.3, R905.7.3, and R905.8.3.

R905.2.7.2 Underlayment and high wind. Underlayment applied in areas subject to high winds [greater than 110 mph (177km/h) per Figure R301.2(4)] shall be applied with corrosion-resistant fasteners in accordance with manufacturer's installation instructions. Fasteners are to be applied along the overlap not farther apart than 36 inches (914 mm) on center.

❖ In high-wind areas, corrosion-resistant fasteners must be located in accordance with the manufacturer's installation instructions, but in no case can they be more than 36 inches (914 mm) on center. The general requirement for sufficient fastening to hold the underlayment in place is not adequate where increased wind loads are anticipated.

Figure R905.2.7.1(1)
POSSIBLE WATER DAMAGE FROM AN ICE DAM

Figure R905.2.7.1(2)
PROTECTIVE ICE SHIELD

R905.2.8 Flashing. Flashing for asphalt shingles shall comply with this section.

❖ This section sets forth the special conditions for flashing installed as a part of an asphalt shingle roof system. Specific flashing locations addressed include base and cap flashing, valleys, crickets and saddles and sidewall flashing.

R905.2.8.1 Base and cap flashing. Base and cap flashing shall be installed in accordance with manufacturer's installation instructions. Base flashing shall be of either corrosion-resistant metal of minimum nominal 0.019-inch (0.483 mm) thickness or mineral surface roll roofing weighing a minimum of 77 pounds per 100 square feet (3.76 kg/m^2). Cap flashing shall be corrosion-resistant metal of minimum nominal 0.019-inch (0.483 mm) thickness.

❖ If metal is used as cap or base flashing, it must be corrosion resistant and have a minimum nominal thickness of 0.019-inch (0.483 mm). Mineral surface roll roofing may also be used as base flashing, provided it has a minimum weight of 77 pounds per 100 square feet (3.76 kg/m^2).

R905.2.8.2 Valleys. Valley linings shall be installed in accordance with manufacturer's installation instructions before applying shingles. Valley linings of the following types shall be permitted:

1. For open valley (valley lining exposed) lined with metal, the valley lining shall be at least 24 inches (610 mm) wide and of any of the corrosion-resistant metals in Table R905.2.8.2.

2. For open valleys, valley lining of two plies of mineral surface roll roofing, complying with ASTM D 249, shall be permitted. The bottom layer shall be 18 inches (457 mm) and the top layer a minimum of 36 inches (914 mm) wide.

3. For closed valleys (valley covered with shingles), valley lining of one ply of smooth roll roofing complying with ASTM D 224 Type II or Type III and at least 36 inches (914 mm) wide or valley lining as described in Items 1 and 2 above shall be permitted. Specialty underlayment complying with ASTM D 1970 may be used in lieu of the lining material.

❖ Open valley linings may be of either metal or mineral surface roll roofing as set forth in this section. Closed valleys are also permitted with a number of lining alternatives available. Commentary Figures R905.2.8.2(1) and R905.2.8.2(2) illustrate typical valley flashings for open and closed (woven) valleys, respectively.

TABLE R905.2.8.2
VALLEY LINING MATERIAL

MATERIAL	MINIMUM THICKNESS (inches)	GAGE	WEIGHT (pounds)
Cold-rolled copper	0.0216 nominal	—	ASTM B370, 16 oz. per square foot
Lead-coated copper	0.0216 nominal	—	ASTM B101, 16 oz. per square foot
High-yield copper	0.0162 nominal	—	ASTM B370, 12 oz. per square foot
Lead-coated high-yield copper	0.0162 nominal	—	ASTM B101, 12 oz. per square foot
Aluminum	0.024	—	—
Stainless steel	—	28	—
Galvanized steel	0.0179	26 (zinc coated G90)	—
Zinc alloy	0.027	—	—
Lead	—	—	2^1/$_2$
Painted terne	—	—	20

For SI: 1 inch = 25.4 mm, 1 pound = 0.454 kg.

❖ If exposed corrosion-resistant metal materials are used as valley linings, they are regulated by this table.

ROOF ASSEMBLIES

The table identifies a variety of acceptable metal linings, including copper, aluminum, stainless steel, galvanized steel, zinc alloy, lead and painted terne. The materials are regulated by their thickness, gage or weight.

For SI: 1 inch = 25.4 mm.

**Figure R905.2.8.2(1)
FLASHING FOR OPEN ROOF VALLEY**

For SI: 1 inch = 25.4 mm.

**Figure R905.2.8.2(2)
FLASHING FOR WOVEN VALLEY**

R905.2.8.3 Crickets and saddles. A cricket or saddle shall be installed on the ridge side of any chimney greater than 30 inches (762 mm) wide. Cricket or saddle coverings shall be sheet metal or of the same material as the roof covering.

❖ Where a chimney is more than 30 inches (762 mm) wide, a cricket or saddle must be installed on the ridge side of the chimney to divert water away from the chimney penetration of the roof. A chimney of such width creates the potential for standing water that, despite appropriate flashing, can be detrimental to the roof's performance.

R905.2.8.4 Sidewall flashing. Flashing against a vertical sidewall shall be by the step-flashing method.

❖ Step flashing must be used to flash along the intersection of a roof and a vertical sidewall. Commentary Figure R905.2.8.4 shows an example of flashing at a side wall.

R905.2.8.5 Other flashing. Flashing against a vertical front wall, as well as soil stack, vent pipe and chimney flashing, shall be applied according to asphalt shingle manufacturer's printed instructions.

❖ The installation of flashing at vertical roof penetrations such as chimneys, soil stacks, and vent pipes must be in conformance with the instructions provided by the asphalt shingle manufacturer. Commentary Figure R905.2.8.5 shows an example of chimney flashing.

R905.3 Clay and concrete tile. The installation of clay and concrete shall comply with the provisions of this section. Clay roof tile shall comply with ASTM C1167.

❖ This section addresses both concrete tile and clay tile. The specifications for clay roof tiles must be in conformance with ASTM C 1167. Commentary Figure R905.3 depicts several examples of clay and concrete roof tile.

R905.3.1 Deck requirements. Concrete and clay tile shall be installed only over solid sheathing or spaced structural sheathing boards.

❖ Both solid sheathing and spaced structural sheathing boards are permitted as a base for the installation of clay or concrete roof tile.

R905.3.2 Deck slope. Clay and concrete roof tile shall be installed on roof slopes of two and one-half units vertical in 12 units horizontal ($2^1/_2$:12) or greater. For roof slopes from two and one-half units vertical in 12 units horizontal ($2^1/_2$:12) to four units vertical in 12 units horizontal (4:12), double underlayment application is required in accordance with Section R905.3.3.

❖ To keep the roof covering weathertight and promote proper drainage, clay and concrete tile cannot be installed on a roof having a slope of less than $2^1/_2$:12. The application of two layers of underlayment is mandated on roofs of 4:12 and less. The provisions of Section R905.3.3 address the full requirements for underlayment on concrete or clay tile roofs.

R905.3.3 Underlayment. Unless otherwise noted, required underlayment shall conform with ASTM D 226, Type II; ASTM D 2626, Type I; or ASTM D 249 mineral surfaced roll roofing.

❖ The code recognizes three types of underlayment for use with clay or concrete tile roof covering: asphalt-

Figure 905.2.8.4
FLASHING AT SIDE WALL

ROOF ASSEMBLIES

saturated organic felts as regulated by ASTM D 226, Type II; asphalt-saturated and coated organic felt base sheets per ASTM D 2626, Type I and asphalt organic felt roll roofing surfaced with mineral granules addressed in ASTM D 249. This section also contains the methods prescribed for the installation of underlayment for clay and concrete tile.

R905.3.3.1 Low slope roofs. For roof slopes from two and one-half units vertical in 12 units horizontal ($2^1/_2$:12), up to four units vertical in 12 units horizontal (4:12), underlayment shall be a minimum of two layers underlayment applies as follows:

1. Starting at the eave, a 19-inch (483 mm) strip of underlayment shall be applied parallel with the eave and fastened sufficiently in place.

Figure R905.2.8.5
CHIMNEY FLASHING DETAIL

Figure R905.3
ROLL AND FLAT TILE EXAMPLES

2. Starting at the eave, 36-inch-wide (914 mm) strips of underlayment felt shall be applied, overlapping successive sheets 19 inches (483 mm), and fastened sufficiently in place.

❖ Where the roof covering consists of clay tile or concrete tile, the roof is considered low-slope if the slope is at least 2^1/$_2$:12 but no steeper than 4:12. Low-slope roofs present a potential problem because water drains slowly; this creates an opportunity for water back-up. Wind-driven rain can also pose a problem. Therefore, a special underlayment application method is used so that the roof remains weathertight. All portions of the roof will be protected by a minimum of two layers of underlayment if the installer follows the instructions found in this section.

R905.3.3.2 High slope roofs. For roof slopes of four units vertical in 12 units horizontal (4:12) or greater, underlayment shall be a minimum of one layer of underlayment felt applied shingle fashion, parallel to and starting from the eaves and lapped 2 inches (51 mm), fastened sufficiently in place.

❖ Where the roof covering consists of clay tile or concrete tile, the roof is considered high-slope if the slope is at least 4:12. Only one layer of underlayment felt, applied in shingle fashion in accordance with this section, is mandated for high-slope roofs of concrete or clay tile.

R905.3.3.3 Underlayment and high wind. Underlayment applied in areas subject to high wind [greater than 110 miles per hour (177 km/h) per Figure R301.2(4)] shall be applied with corrosion-resistant fasteners in accordance with manufacturer's installation instructions. Fasteners are to be applied along the overlap not farther apart than 36 inches (914 mm) on center.

❖ In high-wind areas, corrosion-resistant fasteners must be located in accordance with the manufacturer's installation instructions, but in no case can they be more than 36 inches (914 mm) on center. The general requirement for sufficient fastening to hold the underlayment in place is not adequate where increased wind loads are anticipated.

R905.3.4 Tile. Clay roof tile shall comply with ASTM C 1167.

❖ ASTM C 1167 regulates clay roof tiles manufactured from clay, shale, or similar natural earthy substances. The standard addresses a variety of performance requirements, including durability, strength and permeability.

R905.3.5 Concrete tile. Concrete roof tile shall comply with ASTM C 1492.

❖ ASTM C 1492 regulates concrete roof tiles manufactured from portland cement, water and mineral aggregates with or without the inclusion of other materials. Lightweight or normal weight aggregates or both are used in the manufacture of the concrete tiles. The tiles are shaped during manufacturing by molding, pressing or extrusion. The tiles are usually planar or undulating rectangular shapes available in a variety of cross-sectional areas, profiles, shapes, sizes, surface textures and colors. The standard addresses a variety of performance tests including dimensional tolerances, freeze thaw, transverse strength, permeability and water absorption.

R905.3.6 Fasteners. Nails shall be corrosion-resistant and not less than 11 gage, ^{5}/$_{16}$-inch (10.6 mm) head, and of sufficient length to penetrate the deck a minimum of ^{3}/$_4$ inch (19.1 mm) or through the thickness of the deck, whichever is less. Attaching wire for clay or concrete tile shall not be smaller than 0.083 inch (2.1 mm). Perimeter fastening areas include three tile courses but not less than 36 inches (914 mm) from either side of hips or ridges and edges of eaves and gable rakes.

❖ Roofing nails for fastening clay or concrete tiles must be corrosion resistant. A roofing nail must have a minimum 11 gage shank with a head at least ^{5}/$_{16}$-inch (10.6 mm) in diameter. To provide the necessary holding power, roofing nails must penetrate the roof sheathing a minimum of ^{3}/$_4$ inch (19.1 mm). For roof sheathing having a thickness less than ^{3}/$_4$ inch (19.1 mm), the nails must penetrate completely through the sheathing. The portions of the roof considered as perimeter fastening areas include a minimum of three tile courses, but in no case are they less than 36 inches (914 mm) from the edges of eaves and gable rakes or from either side of hips and ridges.

R905.3.7 Application. Tile shall be applied in accordance with this chapter and the manufacturer's installation instructions, based on the following:

1. Climatic conditions.
2. Roof slope.
3. Underlayment system.
4. Type of tile being installed.

Clay and concrete roof tiles shall be fastened in accordance with this section and the manufacturer's installation instructions. Perimeter tiles shall be fastened with a minimum of one fastener per tile. Tiles with installed weight less than 9 pounds per square foot (0.43 kN/m^2) require a minimum of one fastener per tile regardless of roof slope. Clay and concrete roof tile attachment shall be in accordance with the manufacturer's installation instructions where applied in areas where the wind speed exceeds 100 miles per hour (161 km/h) and on buildings where the roof is located more than 40 feet (12 192 mm) above grade. In areas subject to snow, a minimum of two fasteners per tile is required. In all other areas, clay and concrete roof tiles shall be attached in accordance with Table R905.3.7.

❖ This section identifies four key issues that must be taken into consideration when using the code and the manufacturer's installation instructions in the proper application of clay and concrete roof tile. At least one fastener is required for each roof tile for those tiles at the perimeter of the roof, as well as any tiles weighing less than 9 pounds per square foot (0.43 kN/m^2). In areas where the roof is exposed to a roof snow load, at least two fasteners are required for each tile. Due to the special concerns in high-wind areas, or where the roof height is well above grade, the manufacturer's installation instructions must be referred to in the determination of the minimum number of fasteners.

If none of the above-mentioned conditions occur, Table R905.3.7 must be used to determine proper fastening. The type of sheathing (spaced or solid), the presence of battens and the slope of the roof all impact the minimum number of fasteners required. Commentary Figures R905.3.7(1) through (4) illustrate some of the common shapes of tile and installation techniques.

TABLE R905.3.7
CLAY AND CONCRETE TILE ATTACHMENT

SHEATHING	ROOF SLOPE	NUMBER OF FASTENERS
Solid without battens	All	One per tile
Spaced or solid with battens and slope < 5:12	Fasteners not required	—
Spaced sheathing without battens	5:12 ≤ slope < 12:12	One per tile/every other row
	12:12 ≤ slope < 24:12	One per tile

❖ This table sets forth the number of fasteners required to attach clay or concrete tile to various types of sheathing. The table is not applicable where the wind speed exceeds 100 miles per hour (161 km/h), where the building's roof is located more than 40 feet (12 192 mm) above grade or in locales subject to snow. Under all other conditions, the table indicates the number of fasteners required based on the type of sheathing and, in some cases, the slope of the roof.

R905.3.8 Flashing. At the juncture of roof vertical surfaces, flashing and counter flashing shall be provided in accordance with this chapter and the manufacturer's installation instructions and, where of metal, shall not be less than 0.019 inch (0.48 mm) (No. 26 galvanized sheet gage) corrosion-resistant metal. The valley flashing shall extend at least 11 inches (279 mm) from the centerline each way and have a splash diverter rib not less than 1 inch (25.4 mm) high at the flow line formed as part of the flashing. Sections of flashing shall have an end lap of not less than 4 inches (102 mm). For roof slopes of three units vertical in 12 units horizontal (25-percent slope) and greater, valley flashing shall have a 36-inch-wide (914 mm) underlayment of one layer of Type I underlayment running the full length of the valley, in addition to other required underlayment. In areas where the average daily temperature in January is 25°F (-4°C) or less, metal valley flashing underlayment shall be solid cemented to the roofing underlayment for slopes or be of self-adhering polymer modified bitumen sheet.

❖ Flashing for clay and concrete roof tile is important because the shape of the tile may terminate 2 or 3 inches (51 or 76 mm) above the roof sheathing. Flashing details, therefore, must be constructed in accordance with the code and the manufacturer's installation instructions for the tile shape being used. Some common methods of providing flashing at different locations are illustrated in Commentary Figures R905.3.8(1) through (4).

Figure R905.3.7(1)
APPLICATION OF ROLL TILE

Figure R905.3.7(2)
TILE ROOF GABLE RAKE SECTION

Figure R905.3.7(3)
TILE ROOF EAVE SECTION

Figure R905.3.7(4)
TILE ROOF RIDGE SECTION

ROOF ASSEMBLIES

Figure R905.3.8(1)
TILE ROOF FLASHING AT VALLEY

For SI: 1 degree = 0.01745 rad.

Figure R905.3.8(2)
TILE ROOF FLASHING AT ROOF ABUTTING CHIMNEY/WALL

Figure R905.3.8(3)
TILE ROOF CHIMNEY FLASHING DETAIL

Figure R905.3.8(4)
TILE ROOF FLASHING FOR A VENT PIPE

R905.4 Metal roof shingles. The installation of metal roof shingles shall comply with the provisions of this section.

❖ Metal roof shingles differ from the metal roof panels regulated by Section R905.10. By definition, a metal roof shingle must have an installed weather exposure of less than 3 square feet (0.28 m²) per shingle or sheet. While decking and slope limitations differ to some degree, the provisions for underlayment and flashing are very similar to those for other types of shingle applications.

R905.4.1 Deck requirements. Metal roof shingles shall be applied to a solid or closely fitted deck, except where the roof covering is specifically designed to be applied to spaced sheathing.

❖ Unless manufacturers of metal shingles indicate that their particular shingle may be placed over a roof deck having spaced sheathing, metal roof shingles must be installed on a solid or closely fitted deck.

R905.4.2 Deck slope. Metal roof shingles shall not be installed on roof slopes below three units vertical in 12 units horizontal (25-percent slope).

❖ Based on the characteristics of shape, size, and method of attachment, metal roof shingles must not be applied to roofs having a slope of less than 3:12. A lesser slope increases the potential for leakage to an unacceptable level.

R905.4.3 Underlayment. In areas where the average daily temperature in January is 25°F (-4°C) or less, or when Table R301.2(1) criteria so designate, an ice barrier that consists of at least two layers of underlayment cemented together or of a self-adhering polymer modified bitumen sheet, shall be used in lieu of normal underlayment and extend from the eave's edge to a point at least 24 inches (610 mm) inside the exterior wall line of the building. Underlayment shall comply with ASTM D 226, Type I.

Exception: Detached accessory structures that contain no conditioned floor area.

❖ Where ice dams may be formed along the eave because snow continually freezes and thaws or frozen slush backs up in gutters, the underlayment application in the area of the eaves must be modified to prevent ice dams from forcing water under the roofing, which could damage ceilings, walls and insulation. Two layers of underlayment should be cemented together with asphalt cement from the edge of the roof and continue up the roof to a point that is at least 24 inches (610 mm) inside the interior wall line of the building as shown in Commentary Figure R905.2.7.1(2). The environment within the envelope of the building provides adequate warmth to prevent ice dams from forming above the heated space; therefore, the two layers of cemented underlayment are permitted to terminate 24 inches (610 mm) inside the interior wall line of the building. The local jurisdiction is responsible for determining if the eave flashing is required based on weather records, and it must so indicate in Table R301.2(1).

The exception is used throughout Chapter 9 for unconditioned accessory buildings. See the commentary for Section R905.2.7.

R905.4.4 Material standards. Metal roof shingle roof coverings of galvanized steel shall be 0.013 inch (0.378 mm) minimum thickness. Metal roof shingle roof coverings of aluminum shall be of 0.024 inch (0.610 mm) minimum thickness.

❖ The code provides no specific standards for the regulation of metal roof shingles. The prescriptive provisions of this section mandate a minimum shingle thickness of 0.013 inch (0.378 mm) for galvanized steel

ROOF ASSEMBLIES

shingles and a minimum thickness of 0.024 inch (0610 mm) for aluminum roof shingles.

R905.4.5 Application. Metal roof shingles shall be secured to the roof in accordance with this chapter and the approved manufacturer's installation instructions.

❖ The proper application of metal roof shingles is typically based upon the installation instructions of the shingle manufacturer. Where the IRC sets forth specific requirements for the attachment of metal shingles, such requirements must be met.

R905.4.6 Flashing. Roof valley flashing shall be provided of corrosion-resistant metal of the same material as the roof covering or shall comply with the standards in Table R905.10.3. The valley flashing shall extend at least 8 inches (203 mm) from the center line each way and shall have a splash diverter rib not less than $^3/_4$ inch (19.1 mm) high at the flow line formed as part of the flashing. Sections of flashing shall have an end lap of not less than 4 inches (102 mm). The metal valley flashing shall have a 36-inch-wide (914 mm) underlayment directly under it consisting of one layer of underlayment running the full length of the valley, in addition to underlayment required for metal roof shingles. In areas where the average daily temperature in January is 25°F (-4°C) or less, the metal valley flashing underlayment shall be solid cemented to the roofing underlayment for roof slopes under seven units vertical in 12 units horizontal (58.3-percent slope) or self-adhering polymer modified bitumen sheet.

❖ This section specifies two approaches to roof valley flashing for metal-shingle roofs. The first is the use of material consistent with that of the shingles for the flashing. As an option, the flashing must be accomplished in compliance with the appropriate material standard listed in Table R905.10.3. This table identifies the various types of metal roof covering and the corresponding application standard. This section specifies installation details for the valley flashing as well as the requirement for underlayment running the full length of the valley. In cold-weather areas, the special underlayment techniques to eliminate ice dams apply to valley flashing underlayment.

R905.5 Mineral-surfaced roll roofing. The installation of mineral-surfaced rolling roofing shall comply with this section.

❖ This section addresses the decking requirements and slope, underlayment, materials standards, and application methods for mineral-surfaced roll roofing.

R905.5.1 Deck requirements. Mineral-surfaced roll roofing shall be fastened to solidly sheathed roofs.

❖ The deck must be solidly sheathed in order for mineral-surfaced roll roofing to be installed.

R905.5.2 Deck slope. Mineral-surfaced roll roofing shall not be applied on roof slopes below one unit vertical in 12 units horizontal (8-percent slope).

❖ While mineral-surfaced roll roofing provides a virtually weathertight roof surface, it is still necessary that water flow be positive toward the drainage points. Therefore, a minimum slope of 1:12 is mandated for roof decks that are covered with mineral-surfaced roll roofing.

R905.5.3 Underlayment. In areas where the average daily temperature in January is 25°F (-4°C) or less, or when Table R301.2(1) criteria so designates, an ice barrier that consists of at least two layers of underlayment cemented together or of a self-adhering polymer modified bitumen sheet, shall extend from the eave's edge to a point at least 24 inches (610 mm) inside the exterior wall line of the building. Underlayment shall conform with ASTM D 226, Type I.

Exception: Detached accessory structures that contain no conditioned floor area.

❖ Where ice dams may be formed along the eave because snow continually freezes and thaws or frozen slush backs up in gutters, the underlayment application in the area of the eaves is modified to prevent ice dams from forcing water under the roofing, which could damage ceilings, walls and insulation. Two layers of underlayment should be cemented together with asphalt cement from the edge of the roof and continue up the roof to a point that is at least 24 inches (610 mm) inside the interior wall line of the building as shown in Commentary Figure R905.2.7.1(2). The environment within the envelope of the building provides adequate warmth to prevent ice dams from forming above the heated space; therefore, the two layers of cemented underlayment are permitted to terminate 24 inches (610 mm) inside the interior wall line of the building. The local jurisdiction is responsible for determining if the eave flashing is required based on weather records, and it must so indicate in Table R301.2(1).

The exception is used throughout Chapter 9 for unconditioned accessory buildings. See the commentary for Section R905.2.7.1.

R905.5.4 Material standards. Mineral-surfaced roll roofing shall conform to ASTM D 224, D 249, D 371 or D 3909.

❖ The code recognizes several materials standards for the regulation of mineral-surfaced roll roofing, including ASTM D 224 for smooth-surfaced asphalt roll roofing of organic felt, ASTM D 249 for asphalt roll roofing of organic felt surfaced with mineral granules (ASTM D 371 for wide-selvage) and ASTM D3909 for asphalt roll roofing of glass felt surfaced with mineral granules.

R905.5.5 Application. Mineral-surfaced roll roofing shall be installed in accordance with this chapter and the manufacturer's installation instructions.

❖ The general provisions of this chapter apply to the installation of mineral-surfaced roll roofing, as are the installation instructions of the manufacturer of the specific roofing system.

R905.6 Slate and slate-type shingles. The installation of slate and slate-type shingles shall comply with the provisions of this section.

❖ This section establishes the materials standards and installation criteria for shingles of slate and slate-type materials. It includes specific provisions related to deck slope, underlayment, shingle application and flashing.

R905.6.1 Deck requirements. Slate shingles shall be fastened to solidly sheathed roofs.

❖ Only solidly-sheathed roofs can be used as the deck for the installation of slate and slate-type shingles.

R905.6.2 Deck slope. Slate shingles shall only be used on slopes of four units vertical in 12 units horizontal (33-percent slope) or greater.

❖ Shingles of slate or slate-type materials pose special problems due to the inconsistency of the shingles' shape and size. To keep the roof covering weathertight and promote proper drainage, slate and slate-type shingles cannot be installed on a roof having a slope of less than 4:12. Lower-sloped roofs present a potential problem because water drains slowly and, thus, creates the opportunity for water back-up. Wind-driven rain can also pose a problem.

R905.6.3 Underlayment. In areas where the average daily temperature in January is 25°F (-4°C) or less or when Table R301.2(1) criteria so designates, an ice barrier that consists of a least two layers of underlayment cemented together or of a self-adhering polymer modified bitumen sheet, shall extend from the eave's edge to a point at least 24 inches (610 mm) inside the exterior wall line of the building. Underlayment shall conform with ASTM D 226, Type I.

Exception: Detached accessory structures that contain no conditioned floor area.

❖ Where ice dams may be formed along the eave because snow continually freezes and thaws or frozen slush backs up in gutters, the underlayment application in the area of the eaves is modified to prevent ice dams from forcing water under the roofing, which could damage ceilings, walls and insulation. Two layers of underlayment should be cemented together with asphalt cement from the edge of the roof and continue up the roof to a point that is at least 24 inches (610 mm) inside the interior wall line of the building as shown in Commentary Figure R905.2.7.1(2). The required underlayment must comply with ASTM D 226, Type II. This is the rare condition under which Type II underlayment is mandated. The referenced underlayment for most other types of roof covering materials is Type I. The environment within the envelope of the building provides adequate warmth to prevent ice dams from forming above the heated space; therefore, the two layers of cemented underlayment are permitted to terminate 24 inches (610 mm) inside the interior wall line of the building. The local jurisdiction is responsible for determining if the eave flashing is required based on weather records, and it must so indicate in Table R301.2(1).

The exception is used throughout Chapter 9 for unconditioned accessory buildings. See the commentary for Section R905.2.7.1.

R905.6.4 Material standards. Slate shingles shall comply with ASTM C 406.

❖ The appropriate specification standard for roofing slate is ASTM C 406.

R905.6.5 Application. Minimum headlap for slate shingles shall be in accordance with Table R905.6.5. Slate shingles shall be secured to the roof with two fasteners per slate. Slate shingles shall be installed in accordance with this chapter and the manufacturer's installation instructions.

❖ The general provisions of this chapter apply to the installation of slate and slate-type shingles, as do the installation instructions of the manufacturer of the specific roofing materials. In addition, this section sets forth two requirements specific to slate roofing. First, two fasteners must be used to secure each slate shingle in place. Second, Table R905.6.5 is provided to mandate the minimum headlap at each shingle course. Taking into account the effects of both water traveling down the roof's surface and wind-driven rain, the minimum headlap is reduced as the slope of the roof increases.

**TABLE R905.6.5
SLATE SHINGLE HEADLAP**

SLOPE	HEADLAP (inches)
4:12 ≤ slope < 8:12	4
8:12 ≤ slope < 20:12	3
Slope ≥ 20:12	2

For SI: 1 inch = 25.4 mm.

❖ If slate shingles are installed, they must have a minimum lap at each course as required by this table. Due to the potential for water infiltration, as the slope of the roof decreases, the minimum headlap increases. The extended lap aids in resisting the effects of slow-traveling water and wind-driven rain.

R905.6.6 Flashing. Flashing and counterflashing shall be made with sheet metal. Valley flashing shall be a minimum of 15 inches (381 mm) wide. Valley and flashing metal shall be a minimum uncoated thickness of 0.0179-inch (0.455 mm) zinc coated G90. Chimneys, stucco or brick walls shall have a minimum of two plies of felt for a cap flashing consisting of a 4-inch-wide (102 mm) strip of felt set in plastic cement and extending 1 inch (25.4 mm) above the first felt and a top coating of plastic cement. The felt shall extend over the base flashing 2 inches (51 mm).

ROOF ASSEMBLIES

❖ Special requirements are necessary for the sheet metal flashing and counterflashing of roofs covered with slate or slate-type shingles. The provisions address flashing for roof valleys and for the roof intersections with chimneys and stucco or brick walls.

R905.7 Wood shingles. The installation of wood shingles shall comply with the provisions of this section.

❖ Wood shingles differ from wood shakes in that wood shingles are sawed materials and have a uniform butt thickness across the individual shingle length. Wood shakes are split from logs, shaped by the manufacturer, and have varying butt thicknesses. This section establishes the materials standards and installation criteria for wood shingles. Also included are specific provisions related to deck requirements and slope, underlayment, shingle application and valley flashing.

R905.7.1 Deck requirements. Wood shingles shall be installed on solid or spaced sheathing. Where spaced sheathing is used, sheathing boards shall not be less than 1-inch by 4-inch (25.4 mm by 102 mm) nominal dimensions and shall be spaced on centers equal to the weather exposure to coincide with the placement of fasteners.

❖ Either solid or spaced sheathing is permitted as a deck for the installation of wood shingles. If spaced sheathing is used, the minimum 1-inch by 4-inch (25.4 mm by 102 mm) sheathing boards are required. The placement of the spaced sheathing is important, with the center-to-center spacing of the sheathing boards equivalent to the weather exposure of the wood shingles. This method allows a consistent fastening pattern for applying the wood-shingle roof-covering materials. Table R905.7.5 contains the maximum permitted weather exposure for wood shingles.

R905.7.1.1 Solid sheathing required. In areas where the average daily temperature in January is 25°F (-4°C) or less, solid sheathing is required on that portion of the roof requiring the application of an ice shield.

❖ In cold-weather climates where an ice shield is required in accordance with Section R905.7.3, solid sheathing must be installed for decking. The solid sheathing need be located only where the ice shield is applied, typically a point from the eave's edge to at least 24 inches (610 mm) inside the exterior wall line of the building.

R905.7.2 Deck slope. Wood shingles shall be installed on slopes of three units vertical in 12 units horizontal (25-percent slope) or greater.

❖ Wood shingle roofs pose special problems for weather protection because of the texture of the materials. To keep the roof covering weathertight and to promote proper drainage, wood shingles cannot be installed on a roof having a slope of less than 3:12. Lower-sloped roofs present a potential problem because water drains slowly, creating an opportunity for water backup. Wind-driven rain can also pose a problem.

R905.7.3 Underlayment. In areas where the average daily temperature in January is 25°F (-4°C) or less, or when Table R301.2(1) criteria so designates, an ice barrier that consists of at least two layers of underlayment cemented together or of a self-adhering polymer modified bitumen sheet, shall extend from the eave's edge to a point at least 24 inches (610 mm) inside the exterior wall line of the building. Underlayment shall conform with ASTM D 226, Type I.

Exception: Detached accessory structures that contain no conditioned floor area.

❖ Where ice dams may be formed along the eave because snow continually freezes and thaws or frozen slush backs up in gutters, the underlayment application in the area of the eaves is modified to prevent ice dams from forcing water under the roofing, which could damage ceilings, walls and insulation. Two layers of underlayment should be cemented together with asphalt cement from the edge of the roof and continue up the roof to a point that is at least 24 inches (610 mm) inside the interior wall line of the building as shown in Commentary Figure R905.2.7.1(2). The environment within the envelope of the building provides adequate warmth to prevent ice dams from forming above the heated space; therefore, the two layers of cemented underlayment are permitted to terminate 24 inches (610 mm) inside the interior wall line of the building. The local jurisdiction is responsible for determining if the eave flashing is required based on weather records, and it must so indicate in Table R301.2(1).

The exception is used throughout Chapter 9 for unconditioned accessory buildings. See the commentary for Section R905.2.7.1.

R905.7.4 Material standards. Wood shingles shall be of naturally durable wood and comply with the requirements of Table R905.7.4.

❖ Wood shingles are manufactured in 24-, 18- and 16-inch (610-, 457- and 406-mm) lengths and are graded into three categories: 1, 2 and 3. The Cedar Shake and Shingle Bureau establishes the grading rules for wood shingles. Table R905.7.4 describes each grade, and Commentary Figure R905.7.4 shows an example label.

TABLE R905.7.4
WOOD SHINGLE MATERIAL REQUIREMENTS

MATERIAL	MINIMUM GRADES	APPLICABLE GRADING RULES
Wood shingles of naturally durable wood	1, 2 or 3	Cedar Shake and Shingle Bureau

❖ This table identifies the basic material requirements for wood roof shingles. It states that the shingles must be of naturally durable wood; graded as No. 1, No. 2 or No. 3 and subject to the applicable grading rules of the

Cedar Shake and Shingle Bureau. Each grade is described in Table R905.7.4. Commentary Figure R905.7.4 is an example of the required identification label of an approved grading or inspection bureau or agency.

R905.7.5 Application. Wood shingles shall be installed according to this chapter and the manufacturer's installation instructions. Wood shingles shall be laid with a side lap not less than $1^1/_2$ inches (38 mm) between joints in courses, and no two joints in any three adjacent courses shall be in direct alignment. Spacing between shingles shall not be less than $^1/_4$ inch to $^3/_8$ inch (6.4 mm to 9.5 mm). Weather exposure for wood shingles shall not exceed those set in Table R905.7.5. Fasteners for wood shingles shall be corrosion-resistant with a minimum penetration of $^1/_2$ inch (12.7 mm) into the sheathing. For sheathing less than $^1/_2$ inch (12.7 mm) in thickness, the fasteners shall extend through the sheathing. Wood shingles shall be attached to the roof with two fasteners per shingle, positioned no more than $^3/_4$ inch (19.1 mm) from each edge and no more than 1 inch (25.4 mm) above the exposure line.

❖ The general provisions of this chapter apply to the installation of wood shingles, as do the installation instructions of the manufacturer of the specific roofing materials. In addition, this section sets forth specific requirements regulating the installation method for wood shingles. Many of these requirements are shown in Commentary Figure R905.7.5(1).

Fasteners for applying wood shingles must be corrosion resistant. Only two nails or staples are to be used for each shingle, and they must be placed approximately ¾ inch (19 mm) from each side edge and 1 inch (25 mm) above the exposure line. The depth of fastener penetration must be a minimum of $^1/_2$ inch (12.7 mm) into the sheathing or completely through sheathing less than $^1/_2$ inch (12.7 mm) thick. Care in spacing and driving the fasteners is necessary to provide maximum service from the roof covering; see Commentary Figure R905.7.5(2).

TABLE R905.7.5
WOOD SHINGLE WEATHER EXPOSURE AND ROOF SLOPE

ROOFING MATERIAL	LENGTH (inches)	GRADE	EXPOSURE (inches) 3:12 pitch to < 4:12	EXPOSURE (inches) 4:12 pitch or steeper
Shingles of naturally durable wood	16	No. 1	$3^3/_4$	5
	16	No. 2	$3^1/_2$	4
	16	No. 3	3	$3^1/_2$
	18	No. 1	$4^1/_4$	$5^1/_2$
	18	No. 2	4	$4^1/_2$
	18	No. 3	$3^1/_2$	4
	24	No. 1	$5^3/_4$	$7^1/_2$
	24	No. 2	$5^1/_2$	$6^1/_2$
	24	No. 3	5	$5^1/_2$

For SI: 1 inch = 25.4 mm.

❖ Table R905.7.5 specifies wood shingle exposure. Depending on the grade of the material, the total shingle length and the slope of the roof deck, the table specifies the maximum length of exposure for weathering purposes. The grade of the shingle is listed in the label required for wood shingles. Each of the three different grades are available in three different lengths. For lower-sloped roofs (3:12 to 4:12) where wind uplift is more significant, the maximum weather exposure is always less than for the steeper sloped roofs (4:12 and greater).

R905.7.6 Valley flashing. Roof flashing shall be not less than No. 26 gage [0.019 inches (0.48 mm)] corrosion-resistant sheet metal and shall extend 10 inches (254 mm) from the centerline each way for roofs having slopes less than 12 units vertical in 12 units horizontal (100-percent slope), and 7 inches (178 mm) from the centerline each way for slopes of 12 units vertical in 12 units horizontal and greater. Sections of flashing shall have an end lap of not less than 4 inches (102 mm).

❖ Valleys formed by the intersection of two sloping roofs are vulnerable to leakage because of the high concentration of water; therefore, valley flashing must be

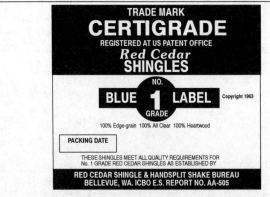

Figure R905.7.4
WOOD SHINGLE GRADE DESCRIPTION AND EXAMPLE LABEL

maintained to prevent blockage. This section requires the use of minimum No. 26 gage (0.019 inches/0.48 mm) corrosion-resistant sheet metal for this purpose. Commentary Figure R905.7.6 illustrates the code provisions for valley flashing.

R905.7.7 Label required. Each bundle of shingles shall be identified by a label of an approved grading or inspection bureau or agency.

❖ In order to verify compliance with the material standards set forth in Section R905.7.4, every bundle of wood shingles must be labeled. The label, a sample of which is shown in Commentary Figure R905.7.4, must indicate the name of the approved grading or inspection bureau or agency.

R905.8 Wood shakes. The installation of wood shakes shall comply with the provisions of this section.

❖ Wood shakes differ from wood shingles in that wood shakes are split from logs, shaped by the manufacturer and have varying butt thicknesses. Wood shingles are sawed materials and have a uniform butt thickness across the individual shingle length. This section establishes the materials standards and installation criteria for wood shakes. Also included are specific

Figure R905.7.5(1)
WOOD SHINGLE APPLICATION

Figure R905.7.5(2)
FASTENER PENETRATION

provisions related to deck requirements and slope, underlayment, interlayment, shingle application and valley flashing.

R905.8.1 Deck requirements. Wood shakes shall only be used on solid or spaced sheathing. Where spaced sheathing is used, sheathing boards shall not be less than 1-inch by 4-inch (25.4 mm by 102 mm) nominal dimensions and shall be spaced on centers equal to the weather exposure to coincide with the placement of fasteners. Where 1-inch by 4-inch (25.4 mm by 102 mm) spaced sheathing is installed at 10 inches (254 mm) on center, additional 1-inch by 4-inch (25.4 mm by 102 mm) boards shall be installed between the sheathing boards.

❖ Either solid or spaced sheathing is permitted as a deck for the installation of wood shakes. If spaced sheathing is used, the minimum 1-inch by 4-inch (25.4 mm by 102 mm) sheathing boards are required. The placement of the spaced sheathing is important, with the center-to-center spacing of the sheathing boards equivalent to the weather exposure of the wood shakes. This method allows for a consistent fastening pattern for applying wood shake roof covering materials. Table R905.8.6 contains the maximum permitted weather exposure for wood shakes. If the greatest allowable exposure of 10 inches is used per Table R905.8.6, an additional 1-inch by 4-inch (25.4 mm by 102 mm) board must be installed between the sheathing boards.

R905.8.1.1 Solid sheathing required. In areas where the average daily temperature in January is 25°F (-4°C) or less, solid sheathing is required on that portion of the roof requiring an ice barrier.

❖ In cold-weather climates where an ice shield is required in accordance with Section R905.7.3, solid sheathing must be installed for decking. The solid sheathing need be located only where the ice shield is applied, typically a point from the eave's edge to at least 24 inches (610 mm) inside the exterior wall line of the building.

R905.8.2 Deck slope. Wood shakes shall only be used on slopes of three units vertical in 12 units horizontal (25-percent slope) or greater.

❖ Wood shake roofs pose special problems for weather protection because of the texture of the materials. To keep the roof covering weathertight and to promote proper drainage, wood shakes cannot be installed on a roof having a slope of less than 3:12. Lower-sloped roofs present a potential problem because water drains slowly, creating an opportunity for water backup. Wind-driven rain can also pose a problem. The use of an interlayment as installed per Section R905.8.7 also plays an important part in maintaining the weathertightness of the roof.

R905.8.3 Underlayment. In areas where the average daily temperature in January is 25°F (-4°C) or less, or when Table R301.2(1) criteria so designates, an ice barrier that consists of

Figure R905.7.6
WOOD SHINGLE VALLEY FLASHING

at least two layers of underlayment cemented together or a self-adhering polymer modified bitumen sheet, shall extend from the edge of the eave to a point at least 24 inches (610 mm) inside the exterior wall line of the building. Underlayment shall conform with ASTM D 226, Type I.

Exception: Detached accessory structures that contain no conditioned floor area.

❖ Where ice dams may be formed along the eave because snow continually freezes and thaws or frozen slush backs up in gutters, the underlayment application in the area of the eaves is modified to prevent ice dams from forcing water under the roofing, which could damage ceilings, walls and insulation. Two layers of underlayment should be cemented together with asphalt cement from the edge of the roof and continue up the roof to a point that is at least 24 inches (610 mm) inside the interior wall line of the building as shown in Commentary Figure R905.2.7.1(2). The environment within the envelope of the building provides adequate warmth to prevent ice dams from forming above the heated space; therefore, the two layers of cemented underlayment are permitted to terminate 24 inches (610 mm) inside the interior wall line of the building. The local jurisdiction is responsible for determining if the eave flashing is required based on weather records, and it must so indicate in Table R301.2(1).

The exception is used throughout Chapter 9 for unconditioned accessory buildings. See commentary for Section R905.2.7.1.

R905.8.4 Interlayment. Interlayment shall comply with ASTM D 226, Type I.

❖ Wood shakes must be installed with an interlayment of minimum No. 30 felt shingled between each course in the manner described by Section R905.8.7. The interlayment must be in compliance with ASTM D 226, Type I for asphalt-saturated organic felt.

R905.8.5 Material standards. Wood shakes shall comply with the requirements of Table R905.8.5.

❖ Wood shakes are divided into five material classifications in the IRC, identified in Table R905.8.5. There are two grades of wood shakes, No. 1 and No. 2. The shakes are cut into 18-inch (457 mm) and 24-inch (610 mm) lengths.

R905.8.6 Application. Wood shakes shall be installed according to this chapter and the manufacturer's installation instructions. Wood shakes shall be laid with a side lap not less than $1^1/_2$ inches (38 mm) between joints in adjacent courses. Spacing between shakes in the same course shall be $^1/_8$ inch to $^5/_8$ inch (3.2 mm to 15.9 mm) for shakes and tapersawn shakes of naturally durable wood and shall be $^1/_4$ inch to $^3/_8$ inch (6.4 mm to 9.5 mm) for preservative tapersawn shakes. Weather exposure for wood shakes shall not exceed those set forth in Table R905.8.6. Fasteners for wood shakes shall be corrosion-resistant, with a minimum penetration of $^1/_2$ inch (12.7 mm) into the sheathing. For sheathing less than $^1/_2$ inch (12.7 mm) in thickness, the fasteners shall extend through the sheathing. Wood shakes shall be attached to the roof with two fasteners per shake, positioned no more than 1 inch (25.4 mm) from each edge and no more than 2 inches (51 mm) above the exposure line.

❖ The general provisions of this chapter apply to the installation of wood shakes, as do the installation instructions of the manufacturer of the specific roofing materials. In addition, this section sets forth specific requirements regulating the installation method for wood shakes. Commentary Figure R905.8.6 shows many of these requirements.

Fasteners for applying wood shakes must be corrosion-resistant. Only two nails or staples are to be used for each shake, and they must be placed approximately 1 inch (25.4 mm) from each side edge and no more than 2 inches (51 mm) above the exposure line. The depth of fastener penetration must be a minimum of $^1/_2$ inch (12.7 mm) into the sheathing, or completely through sheathing less than $^1/_2$ inch (12.7 mm) thick. Care in spacing and driving is necessary to provide maximum service from the roof covering; see Commentary Figure R905.7.5(2).

TABLE R905.8.5
WOOD SHAKE MATERIAL REQUIREMENTS

MATERIAL	MINIMUM GRADES	APPLICABLE GRADING RULES
Wood shakes of naturally durable wood	1	Cedar Shake and Shingle Bureau
Taper sawn shakes of naturally durable wood	1 or 2	Cedar Shake and Shingle Bureau
Preservative-treated shakes and shingles of naturally durable wood	1	Cedar Shake and Shingle Bureau
Fire-retardant-treated shakes and shingles of naturally durable wood	1	Cedar Shake and Shingle Bureau
Preservative-treated tapersawn shakes of southern yellow pine treated in accordance with AWPA Standard C2	1 or 2	Forest Products Laboratory of the Texas Forest Services

❖ Wood shakes are regulated based on the type of shake material, the treatment applied to the shake, and the cut of the shake. Most of the grading rules applicable to wood shakes come from the Cedar Shake and Shingle Bureau. If preservative-treated tapersawn southern yellow pine wood shakes are to be installed, they are governed under the grading rules of the Forest Products Laboratory of the Texas Forest Services. The minimum grade required by the table must be provided on the approved label required on each bundle of wood shakes.

Figure R905.8.6
WOOD SHAKE APPLICATION

TABLE R905.8.6
WOOD SHAKE WEATHER EXPOSURE AND ROOF SLOPE

ROOFING MATERIAL	LENGTH (inches)	GRADE	EXPOSURE (inches) 4:12 pitch or steeper
Shakes of naturally durable wood	18	No. 1	$7^1/_2$
	24	No. 1	10[a]
Preservative-treated taper sawn shakes of southern yellow pine	18	No. 1	$7^1/_2$
	24	No. 1	10
	18	No. 2	$5^1/_2$
	24	No. 2	$7^1/_2$
Taper-sawn shakes of naturally durable wood	18	No. 1	$7^1/_2$
	24	No. 1	10
	18	No. 2	$5^1/_2$
	24	No. 2	$7^1/_2$

For SI: 1 inch = 25.4 mm.

a. For 24-inch by $^3/_8$-inch handsplit shakes, the maximum exposure is $7^1/_2$ inches.

❖ Table R905.8.6 specifies wood shake exposure. Depending on the grade of the material and the total shingle length, the table specifies the maximum length of exposure for weathering purposes. The grade of the shingle will be listed in the label required for wood shingles.

R905.8.7 Shake placement. The starter course at the eaves shall be doubled and the bottom layer shall be either 15-inch (381 mm), 18-inch (457 mm) or 24-inch (610 mm) wood shakes or wood shingles. Fifteen-inch (381 mm) or 18-inch (457 mm) wood shakes may be used for the final course at the ridge. Shakes shall be interlaid with 18-inch-wide (457 mm) strips of not less than No. 30 felt shingled between each course in such a manner that no felt is exposed to the weather by positioning the lower edge of each felt strip above the butt end of the shake it covers a distance equal to twice the weather exposure.

❖ In order to provide as weathertight a roof as possible where using wood shakes as the roof covering material, a special installation method is described in this section. At the eaves, the starter course must be two layers of shakes. Going up the roof toward the ridge, the shakes must be supplemented with 18-inch (457 mm) interlayment strips of minimum No. 30 felt. The manner of interlayment must prevent any felt being exposed to the weather. This is illustrated in Commentary Figure R905.8.7.

R905.8.8 Valley flashing. Roof valley flashing shall not be less than No. 26 gage [0.019 inches (0.48 mm)] corrosion-resistant sheet metal and shall extend at least 11 inches (279 mm) from the centerline each way. Sections of flashing shall have an end lap of not less than 4 inches (102 mm).

❖ Valleys formed by the intersection of two sloping roofs are vulnerable to leakage because of the high concentration of water; therefore, valley flashing must be maintained to prevent blockage. This section requires the use of minimum No. 26 gage (0.019 in./0.48 mm) corrosion-resistant sheet metal for this purpose.

ROOF ASSEMBLIES

For SI: 1 inch = 25.4 mm.

Figure R905.8.7
WOOD SHAKE APPLICATION ON ROOFS WITH SLOPES NOT LESS THAN 3 IN 12

R905.8.9 Label required. Each bundle of shakes shall be identified by a label of an approved grading or inspection bureau or agency.

❖ In order to verify compliance with the material standards set forth in Section R905.7.4, every bundle of wood shakes must be labeled. The label, a sample of which is shown in Commentary Figure R905.8.9, must indicate the name of the approved grading or inspection bureau or agency.

R905.9 Built-up roofs. The installation of built-up roofs shall comply with the provisions of this section.

❖ A built-up roof is a roof covering system built at the job site. The provisions of this section deal with the fundamental requirements for such roofs. With all of the different products available, the possible combination of built-up roof systems is almost endless.

Asphalt and coal tar are the bitumens used for a built-up roof. They become fluid when heated and bond the felt layers together. To fuse the roof felts properly, the bitumens must be applied at the proper temperature. Too low a temperature will cause inadequate fusion, and overheating may change the properties of the bitumens or cause a fire hazard.

Figure R905.8.9
WOOD SHAKE GRADE DESCRIPTION AND EXAMPLE LABELS

R905.9.1 Slope. Built-up roofs shall have a design slope of a minimum of one-fourth unit vertical in 12 units horizontal (2-percent slope) for drainage, except for coal-tar built-up roofs, which shall have a design slope of a minimum one-eighth unit vertical in 12 units horizontal (1-percent slope).

❖ Built-up roofs are often chosen for very low-slope applications, as they may be installed on roof decks having a minimum slope of $1/4$ inch per foot ($1/4$:12), just enough to allow positive drainage. Only where coal-tar built-up roofs are installed must the roof deck slope to a greater degree. A minimum roof slope of 1:12 is required for coal-tar built-up roofs.

R905.9.2 Material standards. Built-up roof covering materials shall comply with the standards in Table R905.9.2.

❖ Table R905.9.2 of the IRC sets forth the appropriate material standards for the various types of built-up roofs. Eighteen different types of roofing systems are identified, and their corresponding specification standard is noted.

TABLE R905.9.2
BUILT-UP ROOFING MATERIAL STANDARDS

MATERIAL STANDARD	STANDARD
Acrylic coatings used in roofing	ASTM D 6083
Aggregate surfacing	ASTM D 1863
Asphalt adhesive used in roofing	ASTM D 3747
Asphalt cements used in roofing	ASTM D 3019; D 2822; D 4586
Asphalt-coated glass fiber base sheet	ASTM D 4601
Asphalt coatings used in roofing	ASTM D 1227; D 2823; D 2824; D 4479
Asphalt glass felt	ASTM D 2178
Asphalt primer used in roofing	ASTM D 41
Asphalt-saturated and asphalt-coated organic felt base sheet	ASTM D 2626
Asphalt-saturated organic felt (perforated)	ASTM D 226
Asphalt used in roofing	ASTM D 312
Coal tar cements used in roofing	ASTM D 4022; D 5643
Coal-tar saturated organic felt	ASTM D 227
Coal-tar primer used in roofing, dampproofing and waterproofing	ASTM D 43
Coal-tar used in roofing	ASTM D 450, Types I or II
Glass mat, coal tar	ASTM D 4990
Glass mat, venting type	ASTM D 4897
Mineral-surfaced inorganic cap sheet	ASTM D 3909
Thermoplastic fabrics used in roofing	ASTM D 5665; D 5726

❖ The types of materials used in built-up roofing must comply with the following standards:
- ASTM D 6083, which covers a liquid-applied water-dispersed 100-percent acrylic elastomeric latex coating used as a protective coating for roofs.
- ASTM D 1863, which covers the quality and grading of crushed stone, crushed slag, and water-worn gravel suitable for use as aggregate surfacing on built-up roofs.
- ASTM D 3747, which covers emulsified asphalt adhesive for use in adhering preformed roof insulation to steel roof decks with inclines up to 33 percent. When applied as a continuous film over an acceptable deck surface, the emulsion functions as both an adhesive and a vapor retarder.
- ASTM D 3019, D 2822 and D 4586, which cover asphalt cements used in roofing.
  - ASTM D 3019 covers lap cement consisting of asphalt dissolved in a volatile petroleum solvent with or without mineral or other stabilizers, or both, for use with roll roofing.
  - ASTM D 2822 covers asphalt roof cement used for trowel application to roofings and flashings.
  - ASTM D 4586 covers asbestos-free asphalt roof cement suitable for trowel application to roofings and flashings.
- ASTM D 4601, which covers asphalt-impregnated and coated glass fiber base sheet, with or without perforations, for use as the first ply of the built-up roofing. When not perforated, this sheet may be used as a vapor retarder under or between roof insulation with solid top coating of asphaltic material.
- ASTM D 1227, D 2823, D 2824 and D 4479, which cover asphalt coatings used in roofing.
  - ASTM D1227 covers emulsified asphalt suitable for use as a protective coating for built-up roofs and other exposed surfaces with inclines of not less than 4 percent.
  - ASTM D 2823 covers asphalt roof coatings of brushing or spraying consistency.

- ASTM D 2824 covers aluminum-pigmented asphalt roof coating, including non-fibered, asbestos-fibered, and fibered without asbestos.
- ASTM D 4479 covers asbestos-free asphalt roof coating of brushing or spraying consistency.
- ASTM D 2178, which covers glass felt impregnated to varying degrees with asphalt, which may be used with asphalts conforming to the requirements of ASTM D 312 in the construction of built-up roofs, and with asphalts conforming to the requirements of ASTM D 449 in the membrane system of waterproofing.
- ASTM D 41, which covers asphaltic primer suitable for use with asphalt in roofing, dampproofing, and waterproofing below or above ground level, for application to concrete, masonry, metal, and asphalt surfaces.
- ASTM D 2626, which covers base sheet with fine mineral surfacing on the top side, with or without perforations, for use as the first ply of a built-up roof. When not perforated, this sheet may be used as a vapor retarder under roof insulation.
- ASTM D 226, which covers asphalt-saturated organic felts, with perforations, that may be used with asphalts conforming to the requirements of ASTM D 312 in the construction of built-up roofs, and with asphalts conforming to the requirements of ASTM D 449 in the membrane system of waterproofing.
- ASTM D 312, which covers four types of asphalt intended for use in built-up roof construction. The specification is for general classification purposes only and does not imply restrictions on the slope at which an asphalt must be used. There are four classifications:
  - Type I includes asphalts that are relatively susceptible to flow at roof temperatures with good adhesive and self-sealing properties.
  - Type II includes asphalts that are moderately susceptible to flow at roof temperatures.
  - Type III includes asphalts that are relatively non-susceptible to flow at roof temperatures for use in built-up roof construction on slope inclines from 8.3 percent to 25 percent.
  - Type IV includes asphalts that are generally non-susceptible to flow at roof temperatures for use in built-up roof construction on slope inclines from approximately 16.7 percent to 50 percent.
- ASTM D 4022 and D 5643, which cover coal tar cements used in roofing.
  - ASTM D 4022 covers coal-tar roof cement suitable for trowel application in coal-tar roofing and flashing systems.
  - ASTM D 5643 covers asbestos-free coal-tar roof cement suitable for trowel application in coal-tar roofing and flashing systems.
- ASTM D 227, which covers coal-tar saturated organic felt that may be used with coal-tar pitches conforming to the appropriate requirements of ASTM D 450 in the construction of built-up roofs and in the membrane system of waterproofing.
- ASTM D 43 covers coal tar primer suitable for use with coal tar pitch in roofing, dampproofing, and waterproofing below or above ground level for application to concrete, masonry, and coal tar surfaces.
- ASTM D 450, Types I and II, which covers coal-tar pitch suitable for use in the construction of built-up roofing, dampproofing and membrane waterproofing systems.
  - Type I is suitable for use in built-up roofing systems with felts conforming to the requirements of ASTM D 227 or as specified by the manufacturer.
  - Type II is suitable for use in dampproofing and in membrane waterproofing systems.
- ASTM D 4990, which covers glass felt impregnated with coal tar intended to be used with coal-tar pitch conforming to the appropriate requirements of ASTM D 450 in construction of built-up roofs and in the construction waterproofing systems.
- ASTM D 4897, which covers asphalt impregnated and coated glass fiber base sheet with mineral surfacing on the top side and coarse mineral granules on the bottom side for use as the first ply of a roofing membrane. These base sheets provide for the lateral release of pressure in roofing systems because they are not solidly attached, and the coarse granular surface provides an open, porous channel in the horizontal plane beneath the membrane. The base sheets can be with or without perforations or embossings.
- ASTM D 3909, which covers asphalt-impregnated and coated glass felt roll roofing surfaced on the weather side with mineral granules, for use as a cap sheet in the construction of built-up roofs.
- ASTM D 5665 and D 5726, which cover thermoplastic fabrics used in roofing.
  - ASTM D 5665 covers thermoplastic fabrics such as polyester, polyester/polyamide bicomponent, or composites with fiberglass or polyester scrims that can be used during the construction of cold-applied roofing and waterproofing.
  - ASTM D 5726 covers thermoplastic fabrics such as polyester, polyester/polyamide bicomponent, or composites with fiberglass or polyester scrims that can be used during the construction of hot-applied roofing and waterproofing.

R905.9.3 Application. Built-up roofs shall be installed according to this chapter and the manufacturer's installation instructions.

❖ The general provisions of this chapter apply to the installation of built-up roofs, as do the installation instructions of the manufacturer of the specific roofing system. Commentary Figures R905.9.3(1) through (3) illustrate examples of the application of the base ply to nailable and non-nailable decks.

Figure R905.9.3(1)
BUILT-UP ROOF—NAILABLE DECK APPLICATION

Figure R905.9.3(2)
BUILT-UP ROOF—APPLICATION TO NONNAILABLE DECK BY SPOT MOPPING

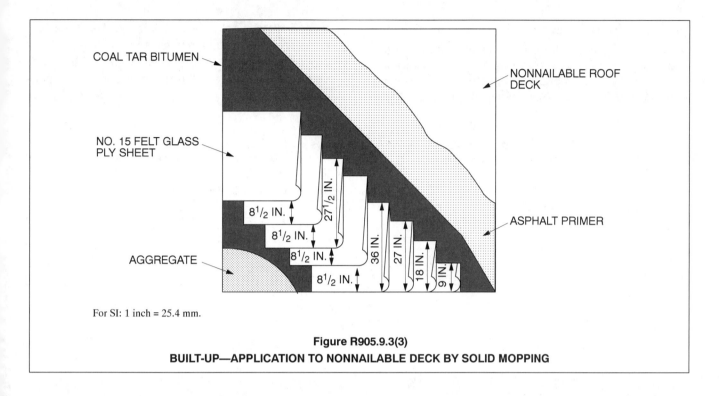

Figure R905.9.3(3)
BUILT-UP—APPLICATION TO NONNAILABLE DECK BY SOLID MOPPING

R905.10 Metal roof panels. The installation of metal roof panels shall comply with the provisions of this section.

❖ Metal roof panels differ from metal roof shingles regulated by Section R905.4. By definition, a metal roof panel must have an installed weather exposure of at least 3 square feet (0.28 m²) per shingle or sheet. This section addresses the deck requirements, deck slope and attachment requirements for metal roof panels.

R905.10.1 Deck requirements. Metal roof panel roof coverings shall be applied to a solid or spaced sheathing, except where the roof covering is specifically designed to be applied to spaced supports.

❖ Unless specifically limited, metal roof panels may be applied to either solidly sheathed decks or spaced sheathing. In those cases where the specific metal panels are designed to be installed only on spaced supports, the manufacturer's installation instructions govern.

R905.10.2 Slope. The minimum slope for lapped, nonsoldered seam metal roofs without applied lap sealant shall be three units vertical in 12 units horizontal (25-percent slope). The minimum slope for lapped, nonsoldered seam metal roofs with applied lap sealant shall be one-half vertical unit in 12 units horizontal (4-percent slope). The minimum slope for standing seam roof systems shall be one-fourth unit vertical in 12 units horizontal (2-percent slope).

❖ The minimum slope permitted for a roof deck supporting metal roof panels varies based upon the type of panels being installed. Three different types of metal roof panel systems are regulated: lapped, non-soldered seam metal roofs without applied lap sealant; lapped, nonsoldered seam metal roofs with applied lap sealant and standing seam metal roof systems. The slope limitations vary based on the ability of the metal panel roof system to shed water without water intrusion.

R905.10.3 Material standards. Metal-sheet roof covering systems that incorporate supporting structural members shall be designed in accordance with the *International Building Code*. Metal-sheet roof coverings installed over structural decking shall comply with Table R905.10.3.

❖ If the metal roof panel system uses the structural members of the roof construction for support, it is necessary to design the roof system is accordance with the *International Building Code*. Only where the metal roof coverings are installed and supported by structural decking may the provisions of the IRC be utilized. Under such conditions, IRC Table R905.10.3 must be consulted. The table references the appropriate material specification standard, prescriptive application or thickness requirement based on the specific type of metal roof covering.

R905.10.4 Attachment. Metal roofing shall be installed in accordance with this chapter and the manufacturer's installation instructions. Metal roofing fastened directly to steel framing shall be attached by approved fasteners. The following fasteners shall be used:

1. Galvanized fasteners shall be used for galvanized roofs.
2. Three hundred series stainless steel fasteners shall be used for copper roofs.

3. Stainless steel fasteners are acceptable for metal roofs.

❖ The general provisions of this chapter apply to the installation of metal panel roofing, as do the installation instructions of the manufacturer of the specific roofing system. In addition, if the metal roof panels are attached directly to steel roof framing members, the code requires attachment by specific types of approved fasteners that are designed to avoid any adverse reaction between the fasteners and the roof panels.

TABLE R905.10.3
METAL ROOF COVERINGS STANDARDS AND INSTALLATION

ROOF COVERING TYPE	STANDARD APPLICATION RATE/THICKNESS
Aluminum	ASTM B 209, 0.024 inch minimum thickness for rollformed panels and 0.019 inch minimum thickness for pressformed shingles.
Aluminum zinc alloy coated steel	ASTM A 792 AZ 50
Copper	ASTM B 370, 16 oz. per sq. ft. for metal sheet roof covering systems; 12 oz. per sq. ft. for preformed metal shingle systems; CDA 4115
Galvanized steel	ASTM A 653 G-90 zinc coated, 0.013 inch thick minimum
Hard lead	2 lbs. per sq. ft.
Lead-coated copper	ASTM B 101
Prepainted steel	ASTM A 755
Soft lead	3 lbs. per sq. ft.
Terne (tin) and terne coated stainless	Terne coating of 40 lb. per double base box, field painted where applicable in accordance with manufacturer's installation instructions.

For SI: 1 ounce per square foot = 0.0026 kg/m^2, 1 pound per square foot = 4.882 kg/m^2, 1 inch = 25.4 mm, 1 pound = 0.454 kg.

❖ This table sets forth the appropriate material specification standard, where applicable, or the standard application rate or thickness for a number of materials used as metal roof coverings. Metal roof coverings are regulated in the following manner:
- Aluminum roof coverings must comply with ASTM B 209, which covers aluminum and aluminum-alloy flat sheet, coil sheet, and plates. For rollformed panels, the minimum thickness must be 0.024 inch. For pressformed shingles the minimum thickness must be 0.019 inch.
- ASTM A 792 addresses aluminum zinc alloy coated steel roof coverings for coating designation AZ 50, which covers 55-percent aluminum-zinc alloy-coated steel sheet in coils and cut lengths.
- Copper roof coverings must comply with ASTM B 370, which covers copper sheet and strip used in building construction. For metal sheet roof covering systems, the minimum material weight must be 16 ounces per square foot. Where preformed metal shingle systems are installed, a minimum weight of 12 ounces per square foot is mandated. The provisions of CDA 4115-1929, *Copper in Architecture — Design Handbook* are also applicable.
- Galvanized steel roof coverings must comply with ASTM A 653 for coating designation G-90, which covers sheet steel that is zinc-coated (galvanized) by the hot-dip process in coils and cut lengths. The minimum thickness of the coating application is 0.013 inch.
- Hard lead roof covering must be installed so that the application weight is at least 2 pounds per square foot.
- Lead-coated copper roof coverings are regulated by ASTM B 101, which establishes the requirements for lead-coated copper sheet and strip in flat lengths or coils in ounce-weight thicknesses for roofing, flashing, gutters, downspouts, and for general sheet metalwork in building construction. The lead coating is applied by hot dipping.
- ASTM A 755 regulates pre-painted steel roof coverings. It covers sheet steel that is metallic coated by the hot-dip process and coil-coated with organic films for exterior exposed building products. Sheet of this designation is furnished in coils, cut lengths, and formed cut lengths. Building products include corrugated and various types of roll and brake-formed configurations.
- Soft lead roof coverings are to have a minimum application weight of 3 pounds per square foot.
- Terne (tin) and terne-coated stainless roof covering must have an installation weight of at least 40 pounds per double base box. Where applicable, the roof coating material is to be painted in the field in accordance with the installation instructions of the manufacturer.

R905.11 Modified bitumen roofing. The installation of modified bitumen roofing shall comply with the provisions of this section.

❖ This section establishes the materials standards and installation criteria for modified bitumen roofing systems.

R905.11.1 Slope. Modified bitumen membrane roofs shall have a design slope of a minimum of one-fourth unit vertical in 12 units horizontal (2-percent slope) for drainage.

❖ Where a modified bitumen roofing system is to be installed, the roof design slope must be a minimum of 1:48, more often recognized as $^1/_4$ inch per foot.

R905.11.2 Material standards. Modified bitumen roof coverings shall comply with the standards in Table R905.11.2.

❖ Table R905.11.2 must be referenced in order for the installer to determine the appropriate specification standard for various materials that may be used in the application of modified bitumen roofing. The table identifies reference standards for the membrane, primer, cement, adhesive, and coatings that may be a part of the roof installation.

TABLE R905.11.2
MODIFIED BITUMEN ROOFING MATERIAL STANDARDS

MATERIAL	STANDARD
Acrylic coating	ASTM D 6083
Asphalt adhesive	ASTM D 3747
Asphalt cement	ASTM D 3019
Asphalt coating	ASTM D 1227; D 2824
Asphalt primer	ASTM D 41
Modified bitumen roof membrane	ASTM D 6162; D 6163; D 6164; D 6222; D 6223; D 6298; CGSB 37-56M

❖ The materials used in modified bitumen roof coverings must comply with the following standards:
- ASTM D 6162, D 6163, D 6164, D 6222, D 6223, D6298 and CGSB 37-56M, which cover modified bitumen roof membranes.
  - ASTM D 6162 covers prefabricated modified bituminous sheet materials reinforced with a combination of polyester fabric and glass fiber, with or without granules, which use styrene butadiene styrene (SBS) thermoplastic elastomer as the primary modifier and are intended for use in the fabrication of multiple-ply roofing and waterproofing membranes.
  - ASTM D 6163 covers prefabricated modified bituminous sheet materials with glass fiber reinforcement, with or without granules, which use styrene butadiene styrene (SBS) thermoplastic elastomer as the primary modifier and are intended for use in the fabrication of multiple-ply roofing and waterproofing membranes.
  - ASTM D 6164 covers prefabricated modified bituminous sheet materials reinforced with polyester fabric, with or without granules, which use styrene butadiene styrene (SBS) thermoplastic elastomer as the primary modifier and are intended for use in the fabrication of multiple-ply roofing and waterproofing membranes.
  - ASTM D 6222 covers atactic polypropelene (APP) modified bituminous sheet materials using polyester reinforcement.
  - ASTM D 6223 covers atactic polypropelene (APP) modified bituminous sheet materials using a combination of polyester and glass fiber reinforcement.
  - ASTM D 6298 covers fiberglass reinforced modified bituminous sheet materials that use styrene-butadiene-styrene (SBS) thermoplastic elastomer as the primary modifier and are surfaced with a factory-applied continuous metal foil.
  - CGSB 37-56M covers prefabricated modified bituminous membrane reinforced for roofing applications.
- ASTM D 41, which covers asphaltic primer suitable for use with asphalt in roofing, dampproofing, and waterproofing below or above ground level, for application to concrete, masonry, metal, and asphalt surfaces.
- ASTM D 3019, which covers lap cement consisting of asphalt dissolved in a volatile petroleum solvent with or without mineral or other stabilizers, or both, for use with roll roofing.
- ASTM D 3747, which covers emulsified asphalt adhesive for use in adhering preformed roof insulation to steel roof decks with inclines up to 33 percent. When applied as a continuous film over an acceptable deck surface, the emulsion functions as both an adhesive and a vapor retarder.
- ASTM D 1227 and D 2824, which cover asphalt roof coatings.
  - ASTM D 1227 covers emulsified asphalt suitable for use as a protective coating for built-up roofs and other exposed surfaces with inclines of not less than 4 percent.
  - ASTM D 2824 covers aluminum-pigmented asphalt roof coatings, including non-fibered, asbestos-fibered, and asbestos-free fibered materials.
- ASTM D 6083, which covers a liquid-applied water-dispersed 100 percent acrylic elastomeric latex coating used as a protective coating for roofs.

R905.11.3 Application. Modified bitumen roofs shall be installed according to this chapter and the manufacturer's installation instructions.

❖ The general provisions of this chapter apply to the installation of modified bitumen roofing, as do the installation instructions of the manufacturer of the specific roofing materials or system.

R905.12 Thermoset single-ply roofing. The installation of thermoset single-ply roofing shall comply with the provisions of this section.

❖ This section establishes the materials standards and installation criteria for thermoset single-ply roofing systems.

R905.12.1 Slope. Thermoset single-ply membrane roofs shall have a design slope of a minimum of one-fourth unit vertical in 12 units horizontal (2-percent slope) for drainage.

❖ If a thermoset single-ply roof system is to be installed, the roof design slope must be a minimum of 1:48, more often recognized as $1/4$-inch per foot.

R905.12.2 Materials standards. Thermoset single-ply roof coverings shall comply with RMA RP-1, RP-2 or RP-3, or ASTM D 4637, ASTM D 5019 or CGSB 37-52M.

❖ The code recognizes several materials standards for the regulation of thermoset single-ply membrane roofs. The standards include RMA RP-1 for non-reinforced black ethylene-propylene-diene-terpolymer (EPDM) rubber sheets, RMA RP-2 for fabric-reinforced EPDM rubber sheets, RMA RP-3 for fabric-reinforced black polychloroprene rubber sheets, ASTM D 4637 for EPDM sheets, ASTM D 5019 for reinforced non-vulcanized polymeric sheet made from chlorosulfonated polyethylene (CSPE) and poly-isobutylene (PIB) and CGSB 37-52M for sheet-applied elastomeric roofing and waterproofing membrane.

R905.12.3 Application. Thermoset single-ply roofs shall be installed according to this chapter and the manufacturer's installation instructions.

❖ The general provisions of this chapter apply to the installation of thermoset single-ply roofing, as do the installation instructions of the manufacturer of the specific roofing system.

R905.13 Thermoplastic single-ply roofing. The installation of thermoplastic single-ply roofing shall comply with the provisions of this section.

❖ This section sets forth the materials standards and installation criteria for single-ply roof systems consisting of polyvinyl chloride material.

R905.13.1 Slope. Thermoplastic single-ply membrane roofs shall have a design slope of a minimum of one-fourth unit vertical in 12 units horizontal (2-percent slope).

❖ If a thermoplastic single-ply roof system is to be installed, the roof design slope must be a minimum of 1:48, more often recognized as $1/4$-inch per foot.

R905.13.2 Material standards. Thermoplastic single-ply roof coverings shall comply with ASTM D 4434 or CGSB 37-54M.

❖ Two materials standards, ASTM D 4434 and CGSB 37-54M, are recognized for the regulation of thermoplastic single-ply membrane roofs and are acceptable standards in determining the appropriateness of the roofing system.

R905.13.3 Application. Thermoplastic single-ply roofs shall be installed according to this chapter and the manufacturer's installation instructions.

❖ The general provisions of this chapter apply to the installation of thermoplastic single-ply roofing, as do the installation instructions of the manufacturer of the roofing system.

R905.14 Sprayed polyurethane foam roofing. The installation of sprayed polyurethane foam roofing shall comply with the provisions of this section.

❖ This section establishes the materials standards and installation criteria for sprayed polyurethane foam roofing systems.

R905.14.1 Slope. Sprayed polyurethane foam roofs shall have a design slope of a minimum of one-fourth unit vertical in 12 units horizontal (2-percent slope) for drainage.

❖ Where a sprayed polyurethane foam roof system is to be installed, the roof design slope must be a minimum of 1:48, more often recognized as $1/4$-inch per foot.

R905.14.2 Material standards. Spray-applied polyurethane-foam insulation shall comply with ASTM C 1029.

❖ ASTM C 1029 is the standard for the regulation of spray-applied rigid cellular polyurethane thermal insulation.

R905.14.3 Application. Foamed in place roof insulation shall be installed in accordance with this chapter and the manufacturer's installation instructions. A liquid-applied protective coating that complies with Section R905.15 shall be applied no less than 2 hours nor more than 72 hours following the application of the foam.

❖ The general provisions of this chapter address the installation of spray-applied rigid cellular polyurethane thermal membrane roofing, as do the installation instructions of the manufacturer of the roofing system. In addition, a liquid-applied protective coating must be provided to protect the exterior surface of the foam. The protective coating, regulated in greater detail by Section R905.15, must be applied between 2 and 72 hours following the installation of the foam insulation.

R905.14.4 Foam plastics. Foam plastic materials and installation shall comply with Section R314.

❖ The foam-plastics provisions of Section R314, particularly those of Section R314.2.2 specific to roof-covering assemblies, are applicable to roof assemblies containing sprayed polyurethane foam. The use of wood structural-panel roof sheathing is permitted in lieu of a thermal barrier for separating the interior of the building from the sprayed polyurethane foam-plastic roofing material. Sections R314.2.2 and R803 contain specific requirements for wood structural-panel sheathing.

R905.15 Liquid-applied coatings. The installation of liquid-applied coatings shall comply with the provisions of this section.

❖ This section establishes the materials standards and installation criteria for liquid-applied roof coatings.

R905.15.1 Slope. Liquid-applied roofs shall have a design slope of a minimum of one-fourth unit vertical in 12 units horizontal (2-percent slope).

❖ If a liquid-applied roof coating is to be installed, the roof design slope must be a minimum of 1:48, more often recognized as $^1/_4$-inch per foot.

R905.15.2 Material standards. Liquid-applied roof coatings shall comply with ASTM C 836, C 957, D 1227, D 3468 or D 6083.

❖ The code recognizes several materials standards for the regulation of liquid-applied roof coatings. They include ASTM C 836 for cold-liquid applied elastomeric waterproofing membranes that require the installation of separate wearing surfaces, ASTM C 957 for cold-liquid applied elastomeric waterproofing membranes that provide integral wearing surfaces, ASTM D 1227 for emulsified asphalt used as a protective coating for built-up roofing systems, ASTM D 3468 for liquid-applied neoprene and chlorosulfanated polyethylene materials used in roofing and waterproofing and ASTM D 6083 for liquid-applied acrylic roof coatings.

R905.15.3 Application. Liquid-applied roof coatings shall be installed according to this chapter and the manufacturer's installation instructions.

❖ The general provisions of this chapter apply to the installation of liquid-applied roof coatings, as do the installation instructions of the manufacturer of the roofing system.

SECTION R906
ROOF INSULATION

R906.1 General. The use of above deck thermal insulation shall be permitted provided such insulation is covered with an approved roof covering and passes FM 4450 or UL 1256.

❖ During these days of energy consciousness, roof insulation has become more and more prevalent. It has distinct benefits not only in energy conservation but also in building-occupant comfort. Insulation also provides a smooth, uniform substrate for application of the roofing materials. The code requires that above-deck thermal insulation be covered by an approved roof covering and be in compliance with FM 4450 or UL 1256. While UL 1256 specifies the testing methods for determining the fire resistance of roof covering materials, FM 4450 deals with Class I insulated steel deck roofs only.

SECTION R907
REROOFING

R907.1 General. Materials and methods of application used for recovering or replacing an existing roof covering shall comply with the requirements of Chapter 9.

Exception: Reroofing shall not be required to meet the minimum design slope requirement of one-quarter unit vertical in 12 units horizontal (2-percent slope) in Section 905 for roofs that provide positive roof drainage.

❖ This section addresses the concerns associated with unregulated reroofing operations. The provisions require that when an existing building is reroofed, the existing roof be structurally sound and in a proper condition to receive the new roofing. It is often necessary to remove the existing roof covering prior to installing the new roofing materials. This section identifies those situations where the reroofing cannot occur over an existing roof covering.

If new roof covering materials are to be installed as a replacement for an existing roof covering, or if the new roof covering is applied directly over an existing roof, the roof covering materials and the application process must conform with the provisions for a new roof covering installation.

There may be occasions where only a small portion of the existing roof covering is in need of replacement or a limited amount of the roof is in need of repair. In those situations, it is acceptable to use the provisions for existing buildings in Chapter 34 of the *International Building Code*, particularly Section 3402.3. This section of the IBC permits the alteration or repair of nonstructural elements to be made of the same materials that are already in place. In other words, the repair or replacement of existing roof covering may be made using the same materials and methods found on the existing roof. It is assumed that the repairs will not cause the building to be reduced in code compliance from the previous level.

In order to avoid a situation in which significant portions of a roof covering are repaired or replaced in small increments without being required to comply with the current provisions of the IRC, the code limits the amount of exempted work that can be done in any one-year period. Where more than 25 percent of a building's roof covering is replaced within a 12-month period, the roof covering must be in compliance with the provisions for new roofing materials and application methods.

R907.2 Structural and construction loads. The structural roof components shall be capable of supporting the roof covering system and the material and equipment loads that will be encountered during installation of the roof covering system.

❖ A fundamental requirement throughout the code is the recognition of the impact of any new repair or replacement work on the structural system. If a new roof covering system is installed, the structural members of the roof and any additional structural members that carry roof loads must be reviewed for their ability to support the loads that will be imposed during the installation process. This includes the weight of the new roof covering materials as well as any installation equipment that is placed on the roof.

R907.3 Recovering versus replacement. New roof coverings shall not be installed without first removing existing roof coverings where any of the following conditions occur:

1. Where the existing roof or roof covering is water-soaked or has deteriorated to the point that the existing roof or roof covering is not adequate as a base for additional roofing.
2. Where the existing roof covering is wood shake, slate, clay, cement or asbestos-cement tile.
3. Where the existing roof has two or more applications of any type of roof covering.

Exceptions:

1. Complete and separate roofing systems, such as standing-seam metal roof systems, that are designed to transmit the roof loads directly to the building's structural system and that do not rely on existing roofs and roof coverings for support shall not require the removal of existing roof coverings.
2. Metal panel, metal shingle, and concrete and clay tile roof coverings shall be permitted to be installed over existing wood shake roofs when applied in accordance with Section R907.4.

❖ The base for application of new roofing materials must provide a sound and consistent surface on which to install the new materials. The code will not permit the installation of new roof coverings over existing roof coverings where there is a potential for future problems with the roof's effectiveness. Therefore, the code lists four specific conditions where the existing roof covering materials must be removed prior to the installation of new roofing.

If the existing roof or roof covering is water-soaked, the concealment of the existing roof in such a condition will lead to problems of deterioration and failure in the future. Any situation where the existing roof construction does not provide for an acceptable base is reason to remove all existing roof covering materials prior to reroofing. In addition, roofing materials such as wood shakes, clay tiles and similar materials are not generally acceptable as a base for a new roof covering because of the variation in their surfaces. The installation of a new roof covering is permitted over wood shakes or shingle roofs only if the surface below the new roofing is properly protected per Section R907.4.

A maximum of two roof coverings are permitted by the code, based primarily on the dead load weight that is accumulated on the roof for every layer of roofing materials. Where two or more applications of any roof covering are present, the existing materials must be removed prior to installation of the new roof covering. Asphalt shingles present a special concern in those areas likely to have hail. Where severe hail damage can be expected, based upon the map shown in Figure R907.3, it is always necessary to remove an existing asphalt shingle roof covering prior to the application of any new roof covering. The areas designated as "subject to severe hail damage" in the table have a historical record of hail damage, particularly to asphalt-shingled roofs.

R907.4 Roof recovering. Where the application of a new roof covering over wood shingle or shake roofs creates a combustible concealed space, the entire existing surface shall be covered with gypsum board, mineral fiber, glass fiber or other approved materials securely fastened in place.

❖ A new roof covering can be applied over the top of an existing roof of wood shakes or shingles only if the existing roof surface of shakes or shingles is protected to address the concern of concealed combustible spaces. The application of a new roof over wood shakes or shingles creates an extensive amount of concealed areas, all with a high degree of combustible materials. By using gypsum board, mineral fiber, glass fiber or other similar materials in the reroofing installation, the combustible shakes or shingles will be protected by materials suitable for fireblocking purposes.

R907.5 Reinstallation of materials. Existing slate, clay or cement tile shall be permitted for reinstallation, except that damaged, cracked or broken slate or tile shall not be reinstalled. Existing vent flashing, metal edgings, drain outlets, collars and metal counterflashings shall not be reinstalled where rusted, damaged or deteriorated. Aggregate surfacing materials shall not be reinstalled.

❖ Because all roofing materials should make a building weathertight, their reuse is strictly limited. Unless damaged, existing tile of cement, slate, or clay may be reused. The reuse of all other roofing materials is prohibited. Roof accessories, such as metal edgings, flashing, drain outlets and collars, may be reinstalled only if they are in a suitable condition.

R907.6 Flashings. Flashings shall be reconstructed in accordance with approved manufacturer's installation instructions. Metal flashing to which bituminous materials are to be adhered shall be primed prior to installation.

❖ During reroofing operations, all flashings that are to remain must be reconstructed in a manner consistent with the manufacturer's instructions. In those cases where bituminous materials are applied to any existing metal flashing, a primer must be applied to the flashing to increase adhesion to the surface.

Bibliography

The following resource materials are referenced in this chapter or are relevant to the subject matter addressed in this chapter.

ASTM A 653M-00, *Specification for Steel Sheet, Zinc-Coated (Galvanized) or Zinc-Iron Alloy-Coated (Galvannealed) by the Hot-Dip Process*. West Conshohocken, PA: ASTM International, 2000.

ASTM A 755M-99, *Specification for Steel Sheet, Metallic-Coated by the Hot-Dip Process and Prepainted by the Coil Coating Process for Exterior Exposed Building Products*. West Conshohocken, PA: ASTM International, 1999.

ASTM A 792M-99, *Specification for Steel Sheet, 55% Aluminum-Zinc Alloy-Coated by the Hot-Dip Process*. West Conshohocken, PA: ASTM International, 1999.

ASTM B 101-96, *Specification for Copper Sheets and Strip for Building Construction*. West Conshohocken, PA: ASTM International, 1996.

ASTM B 209-96, *Specification for Aluminum and Aluminum-alloy Sheet and Plate*. West Conshohocken, PA: ASTM International, 1996.

ASTM B 370-98, *Specification for Lead-Coated Copper Sheet and Strip for Building Construction*. West Conshohocken, PA: ASTM International, 1998.

ASTM C 836-00, *Specification for High Solids Content, Cold Liquid-Applied Elastomeric Waterproofing Membrane for Use with Separate Wearing Course*. West Conshohocken, PA: ASTM International, 2000.

ASTM C 957-93(1998), *Specification for High Solids Content, Cold Liquid-Applied Elastomeric Waterproofing Membrane for Use with Integral Wearing Surface*. West Conshohocken, PA: ASTM International, 1993.

ASTM C 1029-96, *Specification for Spray-Applied Rigid Cellular Polyurethane Thermal Insulation*. West Conshohocken, PA: ASTM International, 1996.

ASTM C 1167-96, *Specification for Clay Roof Tiles*. West Conshohocken, PA: ASTM International, 1996.

ASTM D 41-94, *Specification for Asphalt Primer Used in Roofing, Dampproofing, and Waterproofing*. West Conshohocken, PA: ASTM International, 1994.

ASTM D 43-00, *Specification for Coal Tar Primer Used in Roofing, Dampproofing and Waterproofing*. West Conshohocken, PA: ASTM International, 2000.

ASTM D 225-01, *Specification for Asphalt Shingles (Organic Felt) Surfaced with Mineral Granules*. West Conshohocken, PA: ASTM International, 2001.

ASTM D 226-97a, *Specification for Asphalt-Saturated (Organic Felt) Used in Roofing and Waterproofing*. West Conshohocken, PA: ASTM International, 1997.

ASTM D 227-97a, *Specification for Coal-Tar Saturated (Organic Felt) Used in Roofing and Waterproofing*. West Conshohocken, PA: ASTM International, 1997.

ASTM D 249-89 (96), *Specification for Asphalt Roll Roofing (Organic Felt) Surfaced with Mineral Granules*. West Conshohocken, PA: ASTM International, 1989.

ASTM D 312-00, *Specification for Asphalt Used in Roofing*. West Conshohocken, PA: ASTM International, 2000.

ASTM D 371-89 (1996), *Specification for Asphalt Roll Roofing (Organic Felt) Surfaced with Mineral Granules: Wide-Selvage*. West Conshohocken, PA: ASTM International, 1989.

ASTM D 450-(00)e^1, *Specification for Coal-Tar Pitch Used in Roofing, Dampproofing and Waterproofing*. West Conshohocken, PA: ASTM International, 2000.

ASTM D 1227-00, *Specification for Emulsified Asphalt Used as a Protective Coating for Roofing*. West Conshohocken, PA: ASTM International, 2000.

ASTM D 1863-93 (2000), *Specification for Mineral Aggregate Used in Built-Up Roofs*. West Conshohocken, PA: ASTM International, 2000.

ASTM D 1970-01, *Specification for Self-Adhering Polymer Modified Bitumen Sheet Materials Used as Steep Roofing Underlayment for Ice-Dam Protection*. West Conshohocken, PA: ASTM International, 2001.

ASTM D 2178-97a, *Specification for Asphalt Glass Felt Used in Roofing and Waterproofing*. West Conshohocken, PA: ASTM International, 1997.

ASTM D 2626-97b, *Specification for Asphalt-Saturated and Coated Organic Felt Base Sheet Used in Roofing*. West Conshohocken, PA: ASTM International, 1997.

ASTM D 2822-91(1997)e^1, *Specification for Asphalt Roof Cement*. West Conshohocken, PA: ASTM International, 1991.

ASTM D 2823-90(1997)e^1, *Specification for Asphalt Roof Coatings*. West Conshohocken, PA: ASTM International, 1990.

ASTM D 2824-94, *Specification for Aluminum-Pigmented Asphalt Roof Coatings, Non-Fibered Asbestos Fibered, and Fibered without Asbestos*. West Conshohocken, PA: ASTM International, 1994.

ASTM D 3019-e01, *Specification for Lap Cement Used with Asphalt Roll Roofing, Nonfibered, Asbestos Fibered, and Nonasbestos Fibered*. West Conshohocken, PA: ASTM International, 2001.

ASTM D 3161-99a, *Test Method for Wind Resistance of Asphalt Shingles (Fan Induced Method)*. West Conshohocken, PA: ASTM International, 1999.

ASTM D 3462-01e01, *Specification for Asphalt Shingles Made From Glass Felt and Surfaced with Mineral*

Granules. West Conshohocken, PA: ASTM International, 2001.

ASTM D 3468-99, *Specification for Liquid-Applied Neoprene and Chlorosulfanated Polyethylene Used in Roofing and Waterproofings*. West Conshohocken, PA: ASTM International, 1999.

ASTM D 3747-e01, *Specification for Emulsified Asphalt Adhesive for Adhering Roof Insulation*. West Conshohocken, PA: ASTM International, 2001.

ASTM D 3909-97b, *Specification for Asphalt Roll Roofing (Glass Felt) Surfaced with Mineral Granules*. West Conshohocken, PA: ASTM International, 1997.

ASTM D 4022-94(00e1), *Specification for Coal Tar Roof Cement, Asbestos Containing*. West Conshohocken, PA: ASTM International, 1994.

ASTM D 4434-96, *Specification for Poly (Vinyl Chloride) Sheet Roofing*. West Conshohocken, PA: ASTM International, 1996.

ASTM D 4479-00, *Specification for Asphalt Roof Coatings, Asbestos-Free*. West Conshohocken, PA: ASTM International, 2000.

ASTM D 4586-00, *Specification for Asphalt Roof Cement, Asbestos-Free*. West Conshohocken, PA: ASTM International, 2000.

ASTM D 4601-98, *Specification for Asphalt-Coated Glass Fiber Base Sheet Used in Roofing*. West Conshohocken, PA: ASTM International, 1998.

ASTM D 4637-96, *Specification for EPDM Sheet Used in Single-Ply Roof Membrane*. West Conshohocken, PA: ASTM International, 1996.

ASTM D 4869-88(1993)e[1], *Specification for Asphalt-Saturated (Organic Felt) Shingle Underlayment Used in Roofing*. West Conshohocken, PA: ASTM International, 1988.

ASTM D 4897-01, *Specification for Asphalt Coated Glass-Fiber Venting Base Sheet Used in Roofing*. West Conshohocken, PA: ASTM International, 2001.

ASTM D 4990-97a, *Specification for Coal Tar Glass Felt Used in Roofing and Waterproofing*. West Conshohocken, PA: ASTM International, 1997.

ASTM D 5019-96, *Specification for Reinforced Non-vulcanized Polymeric Sheet Used in Roofing Membrane*. West Conshohocken, PA: ASTM International, 1996.

ASTM D 5643-94(00)e[1], *Specification for Coal Tar Cement Used in Roofing*. West Conshohocken, PA: ASTM International, 1994.

ASTM D 5665-99a, *Specification for Thermoplastic Fabrics Used in Cold-Applied Roofing and Waterproofing*. West Conshohocken, PA: ASTM International, 1999.

ASTM D 5726-98, *Specification for Thermoplastic Fabrics Used in Hot-Applied Roofing and Waterproofing*. West Conshohocken, PA: ASTM International, 1998.

ASTM D 6083-97a, *Specification for Liquid Applied Acrylic Coating Used in Roofing*. West Conshohocken, PA: ASTM International, 1997.

ASTM D 6162-00a, *Specification for Styrene Butadiene Styrene (SBS) Modified Bituminous Sheet Materials Using a Combination of Polyester and Glass Fiber Reinforcements*. West Conshohocken, PA: ASTM International, 2000.

ASTM D 6163-00, *Specification for Styrene Butadiene Styrene (SBS) Modified Bituminous Sheet Materials Using Glass Fiber Reinforcements*. West Conshohocken, PA: ASTM International, 2000.

ASTM D 6164-00, *Specification for Styrene Butadiene Styrene (SBS) Modified Bituminous Sheet Materials Using Polyester Reinforcements*. West Conshohocken, PA: ASTM International, 2000.

ASTM D 6298-98, *Specification for Fiberglass Reinforced Styrene-Butadiene-Styrene (SBS) Modified Bituminous Sheet with a Factory Applied Metal Surface*. West Conshohocken, PA: ASTM International, 1998.

ASTM E 108-99, *Test Method for Fire Tests of Roof Coverings*. West Conshohocken, PA: ASTM International, 1999.

ASTM F 1667-01a, *Specification for Driven Fasteners, Nails, Spikes, and Staples*. West Conshohocken, PA: ASTM International, 2001.

AWPA C1-00, *All Timber Products — Preservative Treatment by Pressure Processes*. Granbury, TX: American Wood-Preservers' Association, 2000.

AWPA C2-01, *Lumber, Timbers, Bridge Ties and Mine Ties — Preservative Treatment by Pressure Processes*. Granbury, TX: American Wood-Preservers' Association, 2001.

CDA 4115-1929, *Copper in Architecture - Design Handbook*. New York: The Copper Development Association, 1929.

CGSB 37-GP-52M-84, *Roofing and Waterproofing Membrane, Sheet Applied, Elastomeric*. Ottawa, Ontario: Canadian General Standards Board, 1984.

CGSB 37-54M-79, *Roofing and Waterproofing Membrane, Sheet Applied, Flexible, Polyvinyl Chloride*. Ottawa, Ontario: Canadian General Standards Board, 1979.

CGSB 37-GP-56M-80, *Membrane, Modified Bituminous, Prefabricated and Reinforced for Roofing* — with December 1985 Amendment. Ottawa, Ontario: Canadian General Standards Board, 1980.

CSSB-97, *Grading and Packing Rules for Western Red Cedar Shakes and Western Red Shingles of the Cedar Shake and Shingle Bureau*. Bellevue, WA: Cedar Shake & Shingle Bureau, 1997.

FM 4450-(1989), *Approved Standard for Class I Insulated Steel Deck Roofs - Supplements through 7/92*. Norwood, MA: Factory Mutual, 1989.

IPC 2003, *International Plumbing Code*. Falls Church, VA: International Code Council, 2003.

RMA RP1-90, *Minimum Requirements for Non-Reinforced Black EPDM Rubber Sheets*. Washington, DC: Rubber Manufacturer's Association, 1990.

RMA RP2-90, *Minimum Requirements for Fabric-Reinforced Black EPDM Rubber Sheets*. Washington, DC: Rubber Manufacturer's Association, 1990.

RMA RP3-85, *Minimum Requirements for Fabric-Reinforced Black Polychloroprene Rubber Sheets*. Washington, DC: Rubber Manufacturer's Association, 1985.

UL 790-98, *Tests for Fire Resistance of Roof Covering Materials*. Northbrook, IL: Underwriters Laboratories, 1998.

UL 1256-98, *Fire Test of Roof Deck Construction*. Northbrook, IL: Underwriters Laboratories, 1998.

Chapter 10:
Chimneys and Fireplaces

General Comments

Chapter 10 regulates two basic types of chimneys and fireplaces: factory-built and those constructed of masonry and other approved materials. Chimneys and fireplaces constructed of masonry rely on prescriptive requirements for the details of their construction, while the factory-built type rely on the listing and labeling method of approval. This chapter contains provisions for unvented gas log heaters as well as provisions for seismic issues related to fireplaces and chimneys.

Section R1001 regulates the construction of masonry chimneys. Section R1002 establishes the standards for the use and installation of factory-built chimneys. Section R1003 regulates the construction of masonry fireplaces. Section R1004 establishes the standards for the use and installation of factory-built fireplaces including the use of unvented gas log heaters. Section R1005 requires the installation of an exterior air supply for use with both factory-built and masonry fireplaces. Section R1006 establishes the standards for the installation of masonry heaters.

Purpose

The purpose of Chapter 10 is to provide requirements for the safe construction of masonry chimneys and fireplaces and to establish the standards for the use and installation of factory-built chimneys, fireplaces and masonry heaters.

SECTION R1001
MASONRY CHIMNEYS

R1001.1 General. Masonry chimneys shall be constructed, anchored, supported and reinforced as required in this chapter and the applicable provisions of Chapters 3, 4 and 6. In Seismic Design Categories D_1 and D_2, masonry and concrete chimneys shall be reinforced and anchored as detailed in Section R1003 for chimneys serving fireplaces. In Seismic Design Category A, B or C, reinforcement and seismic anchorage is not required. Chimneys shall be structurally sound, durable, smoke-tight and capable of conveying flue gases to the exterior safely.

❖ The provisions of this section apply to any masonry chimney constructed to exhaust or vent products of combustion for fireplaces and other residential appliances including gas, pellet fuel-burning and oil-fired. Specific attention must be paid to the construction of masonry and concrete chimneys in Seismic Design Categories D_1 and D_2. The seismic design category is to be determined from Section R301.2.2. These chimneys must be reinforced as detailed in Section R1003 for chimneys serving fireplaces.

Generally, fireplaces or other fuel burning appliances contain the combustion process, while the chimney or flue and necessary connectors transport the products of combustion from the interior to the exterior. Even though the combustion process may be safely contained within a conforming fireplace, the combustion products may accumulate inside the building and reach lethal proportions unless they are properly conveyed from the building. Thus, the chimney must be structurally sound, durable, smoke-tight and capable of conveying the flue gases to the exterior.

R1001.1.1 Support. Masonry chimneys shall be supported on foundations of solid masonry or concrete at least 12 inches (305 mm) thick and at least 6 inches (152 mm) beyond each side of the exterior dimensions of the chimney. Footings shall be founded on natural, undisturbed earth below the frostline. In areas not subject to freezing, footings shall be located a minimum of 12 inches (305 mm) below finished grade.

❖ Due to the large mass and thus great weight of the chimney, the chimney foundation will mostly likely be supporting a greater load than the surrounding foundation. The foundation must be of a size that will support this load. Consideration must also be given to the remainder of the foundation so that uneven settling will not occur between the chimney and the building.

The code prescribes minimum requirements for the foundation of masonry chimneys; see Commentary Figure R1001.1.1.

R1001.2 Corbeling. Masonry chimneys shall not be corbeled more than one-half of the chimney's wall thickness from a wall or foundation, nor shall a chimney be corbeled from a wall or foundation that is less than 12 inches (305 mm) in thickness unless it projects equally on each side of the wall, except that on the second story of a two-story dwelling, corbeling of chimneys on the exterior of the enclosing walls may equal the wall thickness. The projection of a single course shall not exceed one-half the unit height or one-third of the unit bed depth, whichever is less.

❖ Corbeling is the projection of masonry from the surface of the wall or fireplace in small increments for each course of masonry. The chimney should not be corbeled more than one-half of its wall thickness from the wall or foundation. A single course is not permitted to project more than one-half of the individual unit height or more than one-third of the individual unit depth. Ex-

Figure R1001.1.1
TYPICAL MASONRY CHIMNEY

ceeding these amounts could cause the unit to fail because it might not be able to carry the intended load. See Commentary Figure R1001.2(1).

Additionally, masonry walls are not to be corbeled more than one-half of the chimney's wall thickness from the wall or foundation unless it projects equally on each side of the wall. If the corbeling is not equal on each side, the chimney might overturn or crack; see Commentary Figure R1001.2(2).

R1001.3 Changes in dimension. The chimney wall or chimney flue lining shall not change in size or shape within 6 inches (152 mm) above or below where the chimney passes through floor components, ceiling components or roof components.

❖ Changes in the size or shape of a chimney change the stiffness of the chimney, and these areas are more likely to provide potential areas for leaks or cracks in the chimney. The code prohibits these changes within 6 inches (152 mm) of where the chimney passes through the floor, ceiling, or roof components.

R1001.4 Offsets. Where a masonry chimney is constructed with a fireclay flue liner surrounded by one wythe of masonry, the maximum offset shall be such that the centerline of the flue above the offset does not extend beyond the center of the chimney wall below the offset. Where the chimney offset is supported by masonry below the offset in an approved manner, the maximum offset limitations shall not apply. Each individual corbeled masonry course of the offset shall not exceed the projection limitations specified in Section R1001.2.

❖ Limitations on chimney offsets provide a gradual transition in order to prevent critical stress concentrations at the bottom of the offset. The maximum offset of the centerline of the flue to the centerline of the wall of the chimney below maintains structural stability for vertical loads. The slope of the offset is limited to that permitted for corbels in Section R1001.2. The offset limitation does not apply in cases where the offset is supported by masonry below the offset and in a manner approved by the building official.

R1001.5 Additional load. Chimneys shall not support loads other than their own weight unless they are designed and constructed to support the additional load. Masonry chimneys shall be permitted to be constructed as part of the masonry walls or reinforced concrete walls of the building.

❖ Chimneys are subject to considerable stresses due to thermal effects and therefore should not support any structural load other than their own weight, unless specifically designed as a supporting member for the additional load. Also, due to its heavy mass, a chimney will tend to settle more than the building structures; as a consequence, the chimney and any part of the building that it supports will settle at a greater rate and to a greater degree than the rest of the building, resulting in damage. If a chimney is subject to structural loads for which it is not designed, the additional stresses created could lead to cracks in the chimney. Thus, the chimney would become a hazard due to the potential of flames from the firebox penetrating the cracks and igniting combustible construction. Moreover, buildups of the products of combustion that leak through the cracks create hazardous conditions inside the building.

R1001.6 Termination. Chimneys shall extend at least 2 feet (610 mm) higher than any portion of a building within 10 feet (1048 mm), but shall not be less than 3 feet (914 mm) above the highest point where the chimney passes through the roof.

❖ The provisions for the termination height of chimneys above the roof provide for the necessary upward draft in the chimney. Experience has indicated that the required heights produce satisfactory operation of the chimney. See Commentary Figure R1001.1.1.

R1001.6.1 Spark arrestors. Where a spark arrestor is installed on a masonry chimney, the spark arrestor shall meet all of the following requirements:

1. The net free area of the arrestor shall not be less than four times the net free area of the outlet of the chimney flue it serves.
2. The arrestor screen shall have heat and corrosion resistance equivalent to 19-gage galvanized steel or 24-gage stainless steel.
3. Openings shall not permit the passage of spheres having a diameter greater than $1/2$ inch (12.7 mm) nor block the passage of spheres having a diameter less than $3/8$ inch (9.5 mm).
4. The spark arrestor shall be accessible for cleaning and the screen or chimney cap shall be removable to allow for cleaning of the chimney flue.

❖ This section provides specifications for spark arrestors, if they are provided. Their use is not mandated by the code, but owners and builders often install them.

R1001.7 Wall thickness. Masonry chimney walls shall be constructed of solid masonry units or hollow masonry units grouted solid with not less than a 4-inch (102 mm) nominal thickness.

❖ Masonry chimney walls are required to have a minimum nominal thickness of 4 inches (102 mm) of either solid masonry units or hollow masonry grouted solid for structural stability.

R1001.8 Flue lining (material). Masonry chimneys shall be lined. The lining material shall be appropriate for the type of

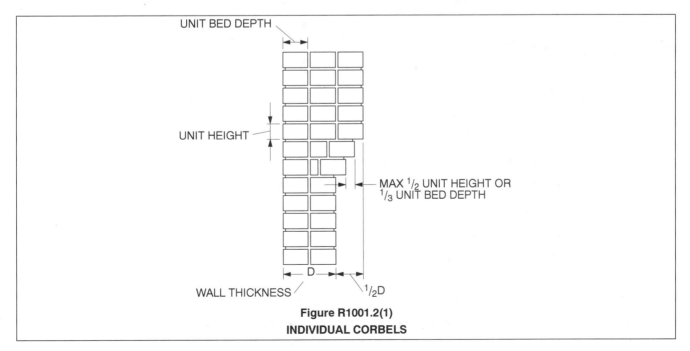

Figure R1001.2(1)
INDIVIDUAL CORBELS

appliance connected, according to the terms of the appliance listing and manufacturer's instructions.

❖ Masonry chimneys must be properly lined to allow for the smooth flow of products of combustion. This section regulates specific flue linings.

R1001.8.1 Residential-type appliances (general). Flue lining systems shall comply with one of the following:

1. Clay flue lining complying with the requirements of ASTM C 315 or equivalent.
2. Listed chimney lining systems complying with UL 1777.
3. Factory-built chimneys or chimney units listed for installation within masonry chimneys.
4. Other approved materials that will resist corrosion, erosion, softening, or cracking from flue gases and condensate at temperatures up to 1,800°F (982°C).

❖ Flue linings must comply with one of the standards listed in this section or be listed for installation within masonry chimneys. Other materials may be used, provided they will resist flue gases and condensate at temperatures up to 1,800°F (982°C) and will not crack, soften or corrode.

R1001.8.2 Flue linings for specific appliances. Flue linings other than these covered in Section R1001.8.1, intended for use with specific types of appliances, shall comply with Sections R1001.8.3 through R1001.8.6.

❖ Except as provided in Section R1001.8.1, flue linings for use with specific types of appliances must comply with Section R1001.8.3 through Section R1001.8.6.

R1001.8.3 Gas appliances. Flue lining systems for gas appliances shall be in accordance with Chapter 24.

❖ Chapter 24 covers flue linings for gas appliances.

R1001.8.4 Pellet fuel-burning appliances. Flue lining and vent systems for use in masonry chimneys with pellet fuel-burning appliances shall be limited to the following:

1. Flue lining systems complying with Section R1001.8.1.
2. Pellet vents listed for installation within masonry chimneys. (See Section R1001.8.6 for marking.)

❖ This section's provisions limit flue lining and vent systems.

R1001.8.5 Oil-fired appliances approved for use with Type L vent. Flue lining and vent systems for use in masonry chimneys with oil-fired appliances approved for use with Type L vent shall be limited to the following:

For SI: 1 inch = 25.4 mm.

Figure R1001.2(2)
CORBELING FROM FOUNDATION WALL

1. Flue lining systems complying with Section R1001.8.1.
2. Listed chimney liners complying with UL 641. (See Section R1001.8.6 for marking.)

❖ Flue lining and vent systems for use in masonry chimneys with oil-fired appliances approved for use with a Type L vent are limited to either flue lining systems complying with Section R1001.8.1 or listed chimney liners complying with UL 641.

R1001.8.6 Notice of usage. When a flue is relined with a material not complying with Section R1001.8.1, the chimney shall be plainly and permanently identified by a label attached to a wall, ceiling or other conspicuous location adjacent to where the connector enters the chimney. The label shall include the following message or equivalent language:

THIS CHIMNEY FLUE IS FOR USE ONLY WITH [TYPE OR CATEGORY OF APPLIANCE] APPLIANCES THAT BURN [TYPE OF FUEL]. DO NOT CONNECT OTHER TYPES OF APPLIANCES.

❖ Flue relined with materials not conforming to Section R1001.8.1 must be permanently identified in a conspicuous location adjacent to the connector's entrance into the building. Thus, current and future occupants of the building will know the type of appliance that may be attached to the flue.

R1001.9 Flue lining (installation). Flue liners shall be installed in accordance with ASTM C 1283 and extend from a point not less than 8 inches (203 mm) below the lowest inlet or, in the case of fireplaces, from the top of the smoke chamber to a point above the enclosing walls. The lining shall be carried up vertically, with a maximum slope no greater than 30 degrees from the vertical.

Fireclay flue liners shall be laid in medium-duty refractory mortar conforming to ASTM C 199 with tight mortar joints left smooth on the inside and installed to maintain an air space or insulation not to exceed the thickness of the flue liner separating the flue liners from the interior face of the chimney masonry walls. Flue lining shall be supported on all sides. Only enough mortar shall be placed to make the joint and hold the liners in position.

❖ Flue linings must be installed so there is no leakage through the liner to adjacent spaces, which would result in contaminated air. Flue linings must be installed in accordance with ASTM C 1283, beginning at least 8 inches (203 mm) below the lowest inlet and extending to a point above the top of the chimney walls. Masonry chimneys atop masonry fireplaces should have the lining start at the top of the smoke chamber of the fireplace. If the chimney is corbeled, the lining may not slope greater than 30 degrees from the vertical. A vent space is required around the lining and is not to be obstructed by the mortar used to set the liner. Only the amount of mortar needed to set the liners and hold them in place should be used. This air space will help limit the transfer of the heat build-up in the chimney to the masonry wall. This air space cannot be used for venting other appliances. See Commentary Figure R1001.10.

R1001.9.1 Listed materials. Listed materials used as flue linings shall be installed in accordance with the terms of their listings and manufacturer's instructions.

❖ Materials used as flue liners must be installed per the manufacturer's listing and instructions so the flue gases will be properly removed from the structure and not leak back in.

R1001.9.2 Space around lining. The space surrounding a chimney lining system or vent installed within a masonry chimney shall not be used to vent any other appliance.

Exception: This shall not prevent the installation of a separate flue lining in accordance with the manufacturer's installation instructions.

❖ The open space around a chimney lining system or vent that is installed within a masonry chimney is not to be used to vent any other appliance. This unused space is otherwise not designed or constructed to vent flue gases.

R1001.10 Multiple flues. When two or more flues are located in the same chimney, masonry wythes shall be built between adjacent flue linings. The masonry wythes shall be at least 4 inches (102 mm) thick and bonded into the walls of the chimney.

Exception: When venting only one appliance, two flues may adjoin each other in the same chimney with only the flue lining separation between them. The joints of the adjacent flue linings shall be staggered at least 4 inches (102 mm).

❖ Each separate fireplace should have its own flue. Even outdoor barbecues connected to a chimney should have their own separate flue. Flues in the same chimney are required to be separated by a 4-inch (102 mm) masonry wythe bonded into the walls of the chimney. Failure to properly separate flues within a common chimney could result in transfer of smoke or other products of combustion from one flue to the other due to the down-draft created by interior suction of the inactive flue. The draft could also be created by external wind effects forcing the smoke down the inactive flues as it exhausts the active flue. Two flues in the same chimney venting only one appliance need not be separated from each other; however, the flue joints must be staggered at least 4 inches (102 mm). See Commentary Figure 1001.10.

R1001.11 Flue area (appliance). Chimney flues shall not be smaller in area than that of the area of the connector from the appliance [see Tables R1001.11(1) and R1001.11(2)]. The sizing of a chimney flue to which multiple-appliance venting systems are connected shall be in accordance with Section M1805.3.

❖ Tables R1001.11(1) and R1001.11(2) determine the sizes of chimney flues for appliances. In no case may

FIGURE R1001.10 – TABLE R1001.11(2)   CHIMNEYS AND FIREPLACES

**Figure R1001.10
MULTIPLE FLUES**

the flue of the chimney have a smaller area than the area of the connector from the appliance. If the appliance is connected to a flue smaller than the appliance's flue, a back-draft can occur, which will result in the appliance not operating properly and the products of combustion not being vented to the outside.

**TABLE R1001.11(1)
NET CROSS-SECTIONAL AREA OF ROUND FLUE SIZES[a]**

FLUE SIZE, INSIDE DIAMETER (inches)	CROSS-SECTIONAL AREA (square inches)
6	28
7	38
8	50
10	78
$10^3/_4$	90
12	113
15	176
18	254

For SI: 1 inch = 25.4 mm, 1 square inch = 645.16 mm².

a. Flue sizes are based on ASTM C 315.

❖ Table R1001.11(1) lists the net cross-sectional area of round flues. The first column provides the interior flue diameter; the second column provides the cross-sectional area. For example, an 8-inch (203 mm) round flue requires a cross-sectional area of at least 50 square inches (32 258 mm²). Flue sizes are based on ASTM C 315.

**TABLE R1001.11(2)
NET CROSS-SECTIONAL AREA OF SQUARE AND RECTANGULAR FLUE SIZES[a]**

FLUE SIZE, OUTSIDE DIMENSIONS (inches)	CROSS-SECTIONAL AREA (square inches)
$4^1/_2 \times 13$	34
$7^1/_2 \times 7^1/_2$	37
$8^1/_2 \times 8^1/_2$	47
$7^1/_2 \times 11^1/_2$	58
$8^1/_2 \times 13$	74
$7^1/_2 \times 15^1/_2$	82
$11^1/_2 \times 11^1/_2$	91
$8^1/_2 \times 17^1/_2$	101
13×13	122
$11^1/_2 \times 15^1/_2$	124
$13 \times 17^1/_2$	165
$15^1/_2 \times 15^1/_2$	168
$15^1/_2 \times 19^1/_2$	214
$17^1/_2 \times 17^1/_2$	226
$19^1/_2 \times 19^1/_2$	269
20×20	286

For SI: 1 inch = 25.4 mm, 1 square inch = 645.16 mm².

a. Flue sizes are based on ASTM C 315.

❖ Table R1001.11(2) lists the net cross-sectional area of rectangular flues. The first column provides the outside dimensions of the flue. The second column provides the cross-sectional area for the flue. For example, a rectangular flue with an exterior dimension of $11^1/_2$ inches by $15^1/_2$ inches (292 mm by 394 mm) requires a net cross-sectional area of at least 124 square inches (80 000 mm²). Flue sizes are based on ASTM C 315.

R1001.12 Flue area (masonry fireplace). Flue sizing for chimneys serving fireplaces shall be in accordance with Section R1001.12.1 or Section R1001.12.2.

❖ There are two methods for determining the flue size for masonry fireplaces. Option 1 uses a fraction of the area of the fireplace opening. Option 2 allows the use of a chart in determining the flue size.

R1001.12.1 Option 1. Round chimney flues shall have a minimum net cross-sectional area of at least $1/12$ of the fireplace opening. Square chimney flues shall have a minimum net cross-sectional area of $1/10$ of the fireplace opening. Rectangular chimney flues with an aspect ratio less than 2 to 1 shall have a minimum net cross-sectional area of $1/10$ of the fireplace opening. Rectangular chimney flues with an aspect ratio of 2 to 1 or more shall have a minimum net cross-sectional area of $1/8$ of the fireplace opening.

❖ Option 1 indicates the use of different fractional amounts of the fireplace opening depending upon the shape of the flue. Round chimney flues must have a net cross-sectional area of 1/12 of the fireplace opening, and square flues must have a net cross-sectional area of 1/10 of the fireplace opening. Rectangular flues need either 1/8 or 1/10 of the net cross-sectional area of the fireplace opening depending upon the aspect ratio of the flue. These fractional amounts vary because the effective flue area (EFA) of a square, smooth flue is equal to that of a smooth, round flue where the diameter of the round flue is the same as one side of the square flue; see Commentary Figure 1001.12.1. The crosshatched areas within the flue of the square or rectangular vent provide no draft.

R1001.12.2 Option 2. The minimum net cross-sectional area of the chimney flue shall be determined in accordance with Figure R1001.12.2. A flue size providing at least the equivalent net cross-sectional area shall be used. Cross-sectional areas of clay flue linings are provided in Tables R1001.11(1) and R1001.11(2) or as provided by the manufacturer or as measured in the field. The height of the chimney shall be measured from the firebox floor to the top of the chimney flue.

❖ Option 2 entails using the chart in Figure R1001.12.2 to determine the required flue area. The flue must have the equivalent net cross-sectional area as determined by the chart. The cross-sectional area of clay flue linings are provided in Tables R1001.11(1) and R1001.11(2) or as provided by the manufacturer. Alternately, the cross-sectional area of the flue may be field measured.

R1001.13 Inlet. Inlets to masonry chimneys shall enter from the side. Inlets shall have a thimble of fireclay, rigid refractory material or metal that will prevent the connector from pulling out of the inlet or from extending beyond the wall of the liner.

❖ Chimney inlet thimbles must not project into the flue. Inlets must have a thimble of fireclay, rigid refractory material or other metal. Thimbles must be installed so they will not dislodge from the inlet opening. See Commentary Figure 1001.13.

R1001.14 Masonry chimney cleanout openings. Cleanout openings shall be provided within 6 inches (152 mm) of the base of each flue within every masonry chimney. The upper edge of the cleanout shall be located at least 6 inches (152 mm) below the lowest chimney inlet opening. The height of the opening shall be at least 6 inches (152 mm). The cleanout shall be provided with a noncombustible cover.

Exception: Chimney flues serving masonry fireplaces where cleaning is possible through the fireplace opening.

❖ The code requires that cleanout openings be provided within 6 inches (152 mm) of the base of the flue. Addi-

Figure R1001.12.1
EFFECTIVE FLUE AREAS

tionally, the upper edge of the cleanout opening must be at least 6 inches below the lowest chimney outlet. The cleanout openings should be at least 6 inches (152 mm) high and have a noncombustible cover. The interior of the chimney may become coated with carbonaceous deposits and acids, and as a result the brick in the mortar will be chemically attacked, causing a gradual deterioration. The deposits will fall to the bottom of the chimney, and a cleanout will be necessary to remove them. Additionally, chimney-cleaning operations, which should be done periodically, will result in the accumulation of deposits at the base of the chimney, necessitating a cleanout opening. In the case of a fireplace chimney, the fireplace opening itself provides the cleanout, and no further opening is required.

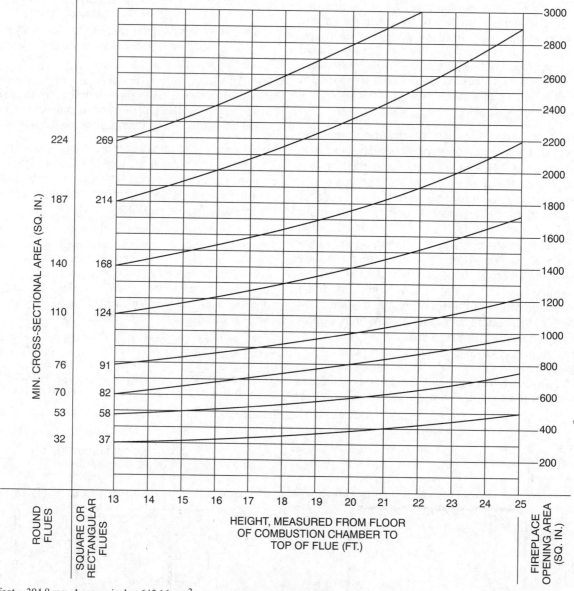

For SI: 1 foot = 304.8 mm, 1 square inch = 645.16 mm².

**FIGURE R1001.12.2
FLUE SIZES FOR MASONRY CHIMNEYS**

❖ Figure R1001.12.2 is used to determine the minimum flue size required for masonry chimneys. Two factors contribute to the determination of the flue size: 1) the height of the flue measured from the floor of the combustion chamber to the top of the flue and 2) the fireplace opening dimension. In using the chart, one would go to the intersection of the x-y axis of the flue height and the fireplace opening size and then to the curved line above. Follow the curved line to the left to determine the minimum flue size. For example, for a flue with a height of 20 feet (6096 mm) and a fireplace opening of 1200 square inches (.77 m²), the required net cross-sectional area would be 110 square inches (70 967 mm²) for a round flue or 124 square inches (80 000 mm²) for a rectangular flue.

Figure R1001.13
CHIMNEY WITH INLET CLEANOUT—MINIMUM CLEARANCES TO COMBUSTIBLE FIRESTOPPING

R1001.15 Chimney clearances. Any portion of a masonry chimney located in the interior of the building or within the exterior wall of the building shall have a minimum air space clearance to combustibles of 2 inches (51 mm). Chimneys located entirely outside the exterior walls of the building, including chimneys that pass through the soffit or cornice, shall have a minimum air space clearance of 1 inch (25.4 mm). The air space shall not be filled, except to provide fire blocking in accordance with Section R1001.16.

Exceptions:

1. Masonry chimneys equipped with a chimney lining system listed and labeled for use in chimneys in contact with combustibles in accordance with UL 1777 and installed in accordance with the manufacturer's installation instructions are permitted to have combustible material in contact with their exterior surfaces.

2. When masonry chimneys are constructed as part of masonry or concrete walls, combustible materials shall not be in contact with the masonry or concrete wall less than 12 inches (306 mm) from the inside surface of the nearest flue lining.

3. Exposed combustible trim and the edges of sheathing materials, such as wood siding and flooring, shall be permitted to abut the masonry chimney side walls, in accordance with Figure R1001.15 provided such combustible trim or sheathing is a minimum of 12 inches (306 mm) from the inside surface of the nearest flue lining. Combustible material and trim shall not overlap the corners of the chimney by more than 1 inch (25.4 mm).

❖ Because of the passage of hot gases and products of combustion, the chimney becomes heated. The code requires that combustibles not be placed directly against the chimney to avoid potential heat transfer and eventual combustion of the combustibles. Masonry chimneys located on the interior of a structure or within an exterior wall require a 2-inch (51 mm) airspace clearance to combustibles. Masonry chimneys located entirely outside of the exterior walls are permitted to have a 1-inch (25.4 mm) clearance to combustibles; see Figure R1001.15. Three exceptions permit reduced clearances.

R1001.16 Chimney fireblocking. All spaces between chimneys and floors and ceilings through which chimneys pass shall be fireblocked with noncombustible material securely fastened in place. The fireblocking of spaces between chimneys and wood joists, beams or headers shall be to a depth of 1 inch (25.4 mm) and shall only be placed on strips of metal or metal lath laid across the spaces between combustible material and the chimney.

❖ As in concealed spaces in wood frame construction, spaces such as floors and ceilings that chimneys pass through must be fireblocked. Section R1001.15 pro-

**FIGURE R1001.15
CLEARANCE FROM COMBUSTIBLES**

hibits chimneys from abutting combustible material. Therefore, the 1-inch (25.4 mm) to 2-inch (51 mm) air space required to separate the chimney from combustible material must be fireblocked. The fireblocking must be of noncombustible material to a depth of 1 inch (25.4 mm) and be supported by noncombustible material such as metal lath.

R1001.17 Chimney crickets. Chimneys shall be provided with crickets when the dimension parallel to the ridgeline is greater than 30 inches (762 mm) and does not intersect the ridgeline. The intersection of the cricket and the chimney shall be flashed and counterflashed in the same manner as normal roof-chimney intersections. Crickets shall be constructed in compliance with Figure R1001.17 and Table R1001.17.

❖ Chimney crickets are required when the chimney width parallel to the ridgeline is greater than 30 inches (762 mm), provided the chimney does not intersect the ridgeline. The cricket diverts the flow of water from the roof around the chimney. Flashing is required between the cricket and the chimney, as it is in other roof-chimney intersections.

**TABLE R1001.17
CRICKET DIMENSIONS**

ROOF SLOPE	H
12 - 12	$1/2$ of W
8 - 12	$1/3$ of W
6 - 12	$1/4$ of W
4 - 12	$1/6$ of W
3 - 12	$1/8$ of W

❖ Table R1001.17 establishes the height of chimney crickets as depicted in Figure R1001.17. For example, a cricket installed at the face of a chimney that is 4 feet (1219 mm) wide on a 6 - 12 roof slope, would need to be 1 foot (305 mm) high ($1/4$ of 4 feet).

For SI: 1 inch = 25.4 mm.

**FIGURE R1001.17
CHIMNEY CRICKET**

CHIMNEYS AND FIREPLACES

SECTION R1002
FACTORY-BUILT CHIMNEYS

R1002.1 Listing. Factory-built chimneys shall be listed and labeled and shall be installed and terminated in accordance with the manufacturer's installation instructions.

❖ Factory-built fireplaces can be used if they are listed and labeled and are installed following the manufacturer's installation instructions. "Listed" and "Labeled" are defined in Chapter 2.

R1002.2 Decorative shrouds. Decorative shrouds shall not be installed at the termination of factory-built chimneys except where such shrouds are listed and labeled for use with the specific factory-built chimney system and installed in accordance with the manufacturer's installation instructions.

❖ In general, decorative shrouds are not permitted on top of factory-built chimneys. Installing a shroud on top of a chimney that is not designed for such may result in improper venting of the appliance attached to the chimney. Decorative shrouds are permitted provided they are listed and labeled for use with the specific chimney and are installed in accordance with the manufacturers' installation instructions.

R1002.3 Solid fuel appliances. Factory-built chimneys for use with solid fuel-burning appliances shall comply with the Type HT requirements of UL 103.

Exception: Chimneys for use with fireplace stoves listed only to UL 737 shall comply with the requirements of UL 103.

❖ Factory-built chimneys for use with solid-fuel-burning appliances must comply with Type HT requirements of UL 103. The exception allows chimneys used in conjunction with UL 737- listed fireplace stoves to comply only with UL 103.

R1002.4 Factory-built fireplaces. Chimneys for use with factory-built fireplaces shall comply with the requirements of UL 127.

❖ Chimneys installed and used for factory-built fireplaces must comply with UL 127, *Factory-Built Fireplaces*.

R1002.5 Support. Where factory-built chimneys are supported by structural members, such as joists and rafters, such members shall be designed to support the additional load.

❖ Joists and rafters supporting factory-built chimneys must be designed to carry the load of the chimney in addition to other loads that the members carry.

R1002.6 Medium-heat appliances. Factory-built chimneys for medium-heat appliances producing flue gases having a temperature above 1,000°F (538°C), measured at the entrance to the chimney shall comply with UL 959.

❖ Factory-built chimneys for medium-heat appliances that produce flue gases with a temperature above 1,000°F (538°C) at the entry to the chimney must comply with UL 959, *Medium Heat Appliance Factory Built Chimneys*.

Factory-built fireplaces are permitted when they are listed and labeled and are installed in accordance with the manufacturers' installation instructions. The word "approved" is defined in Chapter 2. Both factory-built fireplaces and chimneys for solid-fuel-burning appliances must meet the appropriate UL standard as listed in the code. Agencies such as UL provide listing services for factory-built fireplaces and chimneys. The listing includes a description of the manufactured unit and requirements for clearance from combustible materials, hearth details if necessary and other limitations placed on the unit to provide a safe installation. Factory-built fireplaces are installed in conjunction with factory-built chimneys; thus, the complete installation of fireplace and chimney is factory built. See Commentary Figure R1002.6.

SECTION R1003
MASONRY FIREPLACES

R1003.1 General. Masonry fireplaces shall be constructed in accordance with this section and the applicable provisions of Chapters 3 and 4.

❖ This section covers details for masonry fireplaces. Table R1003.1 and Figure R1003.1 show details for masonry fireplaces and chimneys.

R1003.2 Footings and foundations. Footings for masonry fireplaces and their chimneys shall be constructed of concrete or solid masonry at least 12 inches (305 mm) thick and shall extend at least 6 inches (152 mm) beyond the face of the fireplace or foundation wall on all sides. Footings shall be founded on natural, undisturbed earth or engineered fill below frost depth. In areas not subjected to freezing, footings shall be at least 12 inches (305 mm) below finished grade.

❖ The dead-load bearing pressure for a building structure, particularly if it is a light wood-frame building, is usually quite low, while the bearing pressure on the foundation for a masonry or concrete fireplace and chimney can be several times higher. It is good practice to proportion the foundation of a masonry fireplace and chimney to have approximately the same bearing pressure as is present under the building structure itself, although the codes do not require such a design. Where the soil is compressible, differential settlements between the fireplace and chimney and its surrounding structure can cause cracking of the finish materials in the vicinity of the fireplace and chimney. Also, firestopping may be displaced so that a draft opening is created.

Figure R1002.6
EXAMPLE OF FACTORY-BUILT FIREPLACE AND CHIMNEY

CHIMNEYS AND FIREPLACES TABLE R1003.1

TABLE R1003.1
SUMMARY OF REQUIREMENTS FOR MASONRY FIREPLACES AND CHIMNEYS

ITEM	LETTER[a]	REQUIREMENTS
Hearth slab thickness	A	4″
Hearth extension (each side of opening)	B	8″ fireplace opening < 6 sq. ft. 12″ fireplace opening ≥ 6 sq. ft.
Hearth extension (front of opening)	C	16″ fireplace opening < 6 sq. ft. 20″ fireplace opening ≥ 6 sq. ft.
Hearth slab reinforcing	D	Reinforced to carry its own weight and all imposed loads.
Thickness of wall of firebox	E	10″ solid brick or 8″ where a firebrick lining is used. Joints in firebrick $1/4$″ max.
Distance from top of opening to throat	F	8″
Smoke chamber wall thickness Unlined walls	G	6″ 8″
Chimney Vertical reinforcing[b]	H	Four No. 4 full-length bars for chimney up to 40″ wide. Add two No. 4 bars for each additional 40″ or fraction of width or each additional flue.
Horizontal reinforcing	J	$1/4$″ ties at 18″ and two ties at each bend in vertical steel.
Bond beams	K	No specified requirement
Fireplace lintel	L	Noncombustible material.
Chimney walls with flue lining	M	Solid masonry units or hollow masonry units grouted solid with at least 4 inch nominal thickness.
Walls with unlined flue	N	8″ solid masonry.
Distances between adjacent flues	N	See Section R1001.10.
Effective flue area (based on area of fireplace opening)	P	See Section R1001.12.

Clearances:
  Combustible material
  Mantel and trim
  Above roof
	R	See Sections R1001.15 and R1003.12.
		See Section R1001.13.
		3′ at roofline and 2′ at 10′.

Anchorage[b]
  Strap
  Number
  Embedment into chimney
Fasten to
Bolts
	S	$3/16$″ × 1″
		Two
		12″ hooked around outer bar with 6″ extension
		4 joists
		Two $1/2$″ diameter.

Footing
  Thickness
  Width
	T	12″ min.
		6″ each side of fireplace wall.

For SI: 1 inch = 25.4 mm, 1 foot = 304.8 mm, 1 square foot = 0.0929 m².

NOTE: This table provides a summary of major requirements for the construction of masonry chimneys and fireplaces. Letter references are to Figure R1003.1, which shows examples of typical construction. This table does not cover all requirements, nor does it cover all aspects of the indicated requirements. For the actual mandatory requirements of the code, see the indicated section of text.

a. The letters refer to Figure R1003.1.
b. Not required in Seismic Design Category A, B or C.

❖ This table provides a summary of the requirements for masonry fireplaces and chimneys directly correlated to Figure R1002.1. The first column lists various parts of the fireplace. The second column provides a letter designation to the figure corresponding to the item in the figure with the same letter. The third column provides the requirements for that particular fireplace or chimney.

FIGURE R1003.1

For SI: 1 inch = 25.4 mm, 1 foot = 304.8 mm.

**FIGURE R1003.1
FIREPLACE AND CHIMNEY DETAILS**

R1003.2.1 Ash dump cleanout. Cleanout openings located within foundation walls below fireboxes, when provided, shall be equipped with ferrous metal or masonry doors and frames constructed to remain tightly closed except when in use. Cleanouts shall be accessible and located so that ash removal will not create a hazard to combustible materials.

❖ Noncombustible, tightly sealed cleanout doors are required in order to reduce the danger of fire spread through the cleanout openings. Cleanout openings must be easily accessible to allow ash to be readily removed. The ashes and other materials left in the fireplace may remain hot for some time. Cleanout doors must be tight closing to prevent air from entering the firebox and causing the unburned materials to re-ignite. The cleanout door must be located so that ashes and other unburned materials that may still be hot can be removed without creating a hazard to combustible materials.

R1003.3 Seismic reinforcing. Masonry or concrete chimneys in Seismic Design Categories D_1 and D_2 shall be reinforced. Reinforcing shall conform to the requirements set forth in Table R1003.1 and Section R609, Grouted Masonry.

❖ Masonry and concrete have inherently brittle natures, and they are therefore weak when subjected to tensile forces. Using steel, an inherently ductile material, provides ductility in masonry and concrete construction. Chimneys in general and unreinforced masonry and concrete chimneys in particular do not perform well in earthquakes due to their tall and slender geometry. Observation of buildings after earthquakes has shown that chimneys suffered severe damage and sometimes completely collapsed. To improve the performance of masonry and concrete chimneys under severe earthquake loading, steel reinforcing bars are required as shown in Table R1003.1. This reinforcement is required in Seismic Design Categories D_1 and D_2 only. Reinforcement is not required in Seismic Design Categories A, B and C, and buildings in seismic design categories higher than D must be designed in accordance with the *International Building Code*. See Section R301.2.2.

R1003.3.1 Vertical reinforcing. For chimneys up to 40 inches (1016 mm) wide, four No. 4 continuous vertical bars shall be placed between wythes of solid masonry or within the cells of hollow unit masonry and grouted in accordance with Section R609. Grout shall be prevented from bonding with the flue liner so that the flue liner is free to move with thermal expansion. For chimneys greater than 40 inches (1016 mm) wide, two additional No. 4 vertical bars shall be provided for each additional flue incorporated in the chimney or for each additional 40 inches (1016 mm) in width or fraction thereof.

❖ To keep the chimney structure together and prevent severe damage or fracture, four No. 4 bars, continuous from bottom to top, are required. The bars are typically placed at the four corners, similar to the placement of vertical bars in concrete columns. The hollow cells in hollow unit masonry are used for placement of vertical bars, and if solid masonry is used, the reinforcing steel must be placed between the wythes. For wider chimneys or chimneys with more than one flue, four No. 4 bars are not adequate, and reinforcing bars in increments of two are added for each additional flue or each additional 40 inches (1016 mm) of chimney width.

R1003.3.2 Horizontal reinforcing. Vertical reinforcement shall be placed within $^1/_4$-inch (6.4 mm) ties, or other reinforcing of equivalent net cross-sectional area, placed in the bed joints according to Section R607 at a minimum of every 18 inches (457 mm) of vertical height. Two such ties shall be provided at each bend in the vertical bars.

❖ Vertical reinforcement under high seismic loading tends to buckle, then burst out of the concrete or masonry. To prevent this action, the vertical reinforcing required under Section R1003.3.1 is required to be placed in $^1/_4$-inch (6.4 mm) ties that are placed in the bed joints in accordance with Section R 607. The ties are required at no more than 18-inch (457 mm) intervals vertically to form an effective cage for the vertical reinforcement. Where the vertical reinforcing bars have bends and abrupt changes in direction due to changes in sections of the chimney, two such horizontal ties are required at each bend to handle the additional effects of stress concentration.

R1003.4 Seismic anchorage. Masonry and concrete chimneys in Seismic Design Categories D_1 and D_2 shall be anchored at each floor, ceiling or roof line more than 6 feet (1829 mm) above grade, except where constructed completely within the exterior walls. Anchorage shall conform to the requirements of Section R1003.4.1.

❖ The sway of a chimney under earthquake lateral movement may cause it to impact the roof and floor framing and ultimately pull away from the framing. To control or prevent this action in Seismic Design Categories D_1 and D_2, masonry and concrete chimneys must be anchored at each floor, ceiling or roof line to provide lateral bracing. Such anchorage is not required in three situations: 1) in Seismic Design Categories A, B and C; 2) at floor, ceiling or roof lines 6 feet (1829 mm) or less above grade and 3) where the chimney is constructed completely within the exterior walls and is therefore braced by the building elements surrounding it.

In Seismic Design Categories E and F, the reinforcing and anchorage must be designed in accordance with the *International Building Code*.

R1003.4.1 Anchorage. Two $^3/_{16}$-inch by 1-inch (4.8 mm by 25.4 mm) straps shall be embedded a minimum of 12 inches (305 mm) into the chimney. Straps shall be hooked around the outer bars and extend 6 inches (152 mm) beyond the bend. Each strap shall be fastened to a minimum of four floor ceiling or floor joists or rafters with two $^1/_2$-inch (12.7 mm) bolts.

❖ The anchors required under Section R 1003.4 consist of a minimum of two straps each 3/16 inch by 1 inch (4.8 mm by 25.4 mm). The straps must be firmly connected to the building framing and chimney to control or prevent the chimney from pulling away or buckling. To accomplish this firm connection, the straps must be bolted to at least four joists or rafters using 2-inch (12.7 mm) bolts. At the other end they must be embedded at least 12 inches (305 mm) into the chimney and hooked around the outer vertical reinforcing bars. Short hooks might pull out in earthquakes, and for this reason the minimum length of the hook around the vertical bar is 6 inches (152 mm).

R1003.5 Firebox walls. Masonry fireboxes shall be constructed of solid masonry units, hollow masonry units grouted solid, stone or concrete. When a lining of firebrick at least 2 inches (51 mm) in thickness or other approved lining is provided, the minimum thickness of back and side walls shall each be 8 inches (203 mm) of solid masonry, including the lining. The width of joints between firebricks shall not be greater than $^1/_4$ inch (6.4 mm). When no lining is provided, the total minimum thickness of back and side walls shall be 10 inches (254 mm) of solid masonry. Firebrick shall conform to ASTM C 27 or C 1261 and shall be laid with medium-duty refractory mortar conforming to ASTM C 199.

❖ This section specifies the minimum thicknesses of refractory brick or solid masonry necessary to contain the generated heat.

Solid masonry walls forming the firebox are required to have a minimum total thickness of 8 inches (204 mm), including the refractory lining.

The refractory lining is to consist of a low-duty, fireclay refractory brick with a minimum thickness of 2 inches (52 mm), laid with medium-duty refractory mortar. Mortar joints are generally 1/16 to 3/16 inch (1.6 to 4.8 mm) thick but not thicker than 1/4 inch (6.4 mm) to reduce thermal movements and prevent joint deterioration.

Where a firebrick lining is not used in firebox construction, the wall thickness is not to be less than 10 inches (254 mm) of solid masonry. Firebrick is required to conform to ASTM C 27 or ASTM C 1261 and must be laid with medium-duty refractory mortar conforming to ASTM C 199.

R1003.5.1 Steel fireplace units. Steel fireplace units are permitted to be installed with solid masonry to form a masonry fireplace when installed either according to the requirements of their listing or according to the requirements of this section. Steel fireplace units incorporating a steel firebox lining, shall be constructed with steel not less than $^1/_4$ inch (6.4 mm) in thickness, and an air circulating chamber which is ducted to the interior of the building. The firebox lining shall be encased with solid masonry to provide a total thickness at the back and sides of not less than 8 inches (203 mm), of which not less than 4 inches (102 mm) shall be of solid masonry or concrete. Circulating airducts employed with steel fireplace units shall be constructed of metal or masonry.

❖ Steel fireplace units and an air chamber may be used as part of the required thickness of masonry fireplaces so that the combined thickness of the masonry and steel is 8 inches (203 mm). The steel may not be less than $^1/_4$ inch (6.4 mm) thick, and the masonry must be at least 4 inches thick. Where warm-air ducts are employed as part of a steel fireplace unit for circulating purposes, they must be constructed of metal or masonry.

R1003.6 Firebox dimensions. The firebox of a concrete or masonry fireplace shall have a minimum depth of 20 inches (508 mm). The throat shall not be less than 8 inches (203 mm) above the fireplace opening. The throat opening shall not be less than 4 inches (102 mm) in depth. The cross-sectional area of the passageway above the firebox, including the throat, damper and smoke chamber, shall not be less than the cross-sectional area of the flue.

Exception: Rumford fireplaces shall be permitted provided that the depth of the fireplace is at least 12 inches (305 mm) and at least one-third of the width of the fireplace opening, that the throat is at least 12 inches (305 mm) above the lintel and is at least $^1/_{20}$ the cross-sectional area of the fireplace opening.

❖ The proper functioning of the fireplace depends on the size of the face opening and the chimney dimensions, which in turn are related to the room size. This section specifies a minimum depth of 20 inches (508 mm) for the combustion chamber because that depth influences the draft requirement; see Commentary Figure R1003.6(1). The dimensions of the firebox (depth, opening size and shape) are usually based on two considerations: aesthetics and the need to prevent the room from overheating. Suggested dimensions for single-opening fireboxes are given in technical publications of the Brick Institute of America (BIA) and the National Concrete Masonry Association (NCMA).

This section also provides additional criteria for the throat's location and minimum cross-sectional area. Those criteria are based on many years of construction of successfully functioning fireplaces. The exception permits the use of Rumford fireplaces, which are tall, shallow fireplaces that can radiate a large amount of heat into a room.

The code reference to "depth of fireplace" is interpreted as the depth of the firebox. The throat must be at least 12 inches (305 mm) above the lintel and at least 5 percent (one-twentieth) of the cross-sectional area of the fireplace opening; see Commentary Figure R1003.6(2). Smoke chambers and flues for Rumford fireplaces should be sized and built like other masonry fireplaces. While those who build Rumford fireplaces do not totally agree about how they work, many books and guides address their construction.

CHIMNEYS AND FIREPLACES   FIGURE R1003.6(1) - R1003.8.1

Figure R1003.6(1)
SECTION THROUGH FIREPLACE

R1003.7 Lintel and throat. Masonry over a fireplace opening shall be supported by a lintel of noncombustible material. The minimum required bearing length on each end of the fireplace opening shall be 4 inches (102 mm). The fireplace throat or damper shall be located a minimum of 8 inches (203 mm) above the lintel.

❖ A noncombustible lintel with a minimum bearing length of 4 inches (102 mm) on each end must be provided to support masonry above the fireplace opening. Provisions should be made to allow for expansion of the steel lintels when they are heated.

R1003.7.1 Damper. Masonry fireplaces shall be equipped with a ferrous metal damper located at least 8 inches (203 mm) above the top of the fireplace opening. Dampers shall be installed in the fireplace or the chimney venting the fireplace, and shall be operable from the room containing the fireplace.

❖ A damper is used to close the chimney flue when the fireplace is not in use. This section provides guidance on its location and construction.

R1003.8 Smoke chamber. Smoke chamber walls shall be constructed of solid masonry units, hollow masonry units grouted solid, stone or concrete. Corbelling of masonry units shall not leave unit cores exposed to the inside of the smoke chamber. When a lining of firebrick at least 2 inches (51 mm) thick, or a lining of vitrified clay at least $^5/_8$ inch (15.9 mm) thick, is provided, the total minimum thickness of front, back and side walls shall be 6 inches (152 mm) of solid masonry, including the lining. Firebrick shall conform to ASTM C 27 or C 1261 and shall be laid with medium-duty refractory mortar conforming to ASTM C 199. Where no lining is provided, the total minimum thickness of front, back and side walls shall be 8 inches (203 mm) of solid masonry. When the inside surface of the smoke chamber is formed by corbeled masonry, the inside surface shall be parged smooth.

❖ The smoke chamber is located directly above the firebox and must be constructed of solid masonry. A smoke shelf located behind the damper deflects down drafts. As the downdraft hits the smoke shelf, it is turned upward by the damper assembly. Curved smoke shelves perform better; however, flat smoke shelves are acceptable.

R1003.8.1 Smoke chamber dimensions. The inside height of the smoke chamber from the fireplace throat to the beginning of the flue shall not be greater than the inside width of the fireplace opening. The inside surface of the smoke chamber shall not be inclined more than 45 degrees (0.39 rad) from vertical when prefabricated smoke chamber linings are used or when the smoke chamber walls are rolled or sloped rather than corbeled. When the inside surface of the smoke chamber is formed by corbeled masonry, the walls shall not be corbeled more than 30 degrees (0.26 rad) from vertical.

❖ This section provides specific dimensions for the smoke chamber of masonry fireplaces, which are necessary to provide proper flow of the smoke and products of combustion out of the fireplace and through the chimney. The height of the smoke chamber cannot exceed the inside width of the fireplace opening. The inside surface of the smoke chamber is permitted to be sloped 45 degrees (0.39 rad) from the vertical when constructed of a prefabricated smoke chamber lining or the smoke chamber walls are rolled or sloped. If the smoke chamber is sloped using corbeled masonry, the slope is limited to be no more than 30 degrees from the vertical. The slope is restricted more when using corbeled masonry because the surface is not smooth and will tend to create turbulence within the smoke chambers, thus affecting the flow of smoke to the flue.

R1003.9 Hearth and hearth extension. Masonry fireplace hearths and hearth extensions shall be constructed of concrete or masonry, supported by noncombustible materials, and reinforced to carry their own weight and all imposed loads. No combustible material shall remain against the underside of hearths and hearth extensions after construction.

❖ The hearth includes both the floor of the fire box and the projection in front of it. The hearth extension protects wood, carpet, and combustible materials from being ignited by sparks, hot embers, or ashes, which may fall from the fire box.

R1003.9.1 Hearth thickness. The minimum thickness of fireplace hearths shall be 4 inches (102 mm).

❖ The minimum thickness of fireplace hearths is 4 inches (102 mm).

R1003.9.2 Hearth extension thickness. The minimum thickness of hearth extensions shall be 2 inches (51 mm).

Exception: When the bottom of the firebox opening is raised at least 8 inches (203 mm) above the top of the hearth extension, a hearth extension of not less than $^3/_8$-inch-thick (9.5 mm) brick, concrete, stone, tile or other approved noncombustible material is permitted.

❖ The hearth extension must be 2 inches (51 mm) thick unless the bottom of the fireplace opening is raised at least 8 inches (203 mm) above the top of the hearth extension. Where the bottom of the fireplace opening is raised at least 8 inches above the hearth extension, the hearth extension is permitted to be reduced to a thickness of $^3/_8$-inch-thick (9.5 mm) brick, concrete, stone, tile or other approved noncombustible material.

Figure R1003.6(2)
RUMFORD FIREPLACE

R1003.10 Hearth extension dimensions. Hearth extensions shall extend at least 16 inches (406 mm) in front of and at least 8 inches (203 mm) beyond each side of the fireplace opening. Where the fireplace opening is 6 square feet (0.557 m^2) or larger, the hearth extension shall extend at least 20 inches (508 mm) in front of and at least 12 inches (305 mm) beyond each side of the fireplace opening.

❖ It is necessary for the hearth to extend in front of and to the side of the firebox opening a sufficient dimension to prevent burning embers from landing on combustible surfaces.

R1003.11 Fireplace clearance. All wood beams, joists, studs and other combustible material shall have a clearance of not less than 2 inches (51 mm) from the front faces and sides of masonry fireplaces and not less than 4 inches (102 mm) from the back faces of masonry fireplaces. The air space shall not be filled, except to provide fire blocking in accordance with Section R1003.13.

Exceptions:

1. Masonry fireplaces listed and labeled for use in contact with combustibles in accordance with UL 127 and installed in accordance with the manufacturer's installation instructions are permitted to have combustible material in contact with their exterior surfaces.

2. When masonry fireplaces are part of masonry or concrete walls, combustible materials shall not be in contact with the masonry or concrete walls less than 12 inches (305 mm) from the inside surface of the nearest firebox lining.

3. Exposed combustible trim and the edges of sheathing materials such as wood siding, flooring and drywall shall be permitted to abut the masonry fireplace side walls and hearth extension in accordance with Figure R1003.12, provided such combustible trim or sheathing is a minimum of 12 inches (305 mm) from the inside surface of the nearest firebox lining.

4. Exposed combustible mantels or trim may be placed directly on the masonry fireplace front surrounding the fireplace opening providing such combustible materials shall not be placed within 6 inches (152 mm) of a fireplace opening. Combustible material within 12 inches (306 mm) of the fireplace opening shall not project more than $^1/_8$ inch (3.2 mm) for each 1-inch (25.4 mm) distance from such an opening.

❖ A 2-inch (51 mm) clearance is required from the front and side surfaces of the fireplace. A 4-inch (102 mm) clearance is required from the back of the fireplace. This air space created around the fireplace cannot be filled, except for approved fire blocking.

There are four exceptions to this section that allow combustible materials to come in contact with the fireplace. Exception 1 allows factory-built fireplaces constructed in accordance with UL 127, *Factory-Built Fireplaces* to be in contact with combustibles. The installation must carefully follow the manufacturer's installation instructions.

Exception 2 applies to masonry fireplaces constructed in concrete or masonry walls. In this case combustible materials must be at least 12 inches (305 mm) from the inside surface of the firebox lining because of the conductive nature of the concrete or masonry walls.

Exception 3 permits combustible trim and edges of certain sheathing materials to abut the fireplace in accordance with Figure R1003.12. This exception permits the combustible trim to cover openings that would otherwise occur between the fireplace and other construction.

Exception 4 provides specific dimensions that protect combustible trim from radiant heat.

R1003.12 Mantel and trim. Woodwork or other combustible materials shall not be placed within 6 inches (152 mm) of a fireplace opening. Combustible material within 12 inches (305 mm) of the fireplace opening shall not project more than $^1/_8$

For SI: 1 inch = 25.4 mm.

FIGURE R1003.12
CLEARANCE FROM COMBUSTIBLES

❖ This figure depicts the provisions of Exception 3 to Section R1003.11 for combustible trim and sheathing adjacent to masonry fireplaces and hearth extensions.

inch (3.2 mm) for each 1-inch (25.4 mm) distance from such opening.

❖ Wood mantels and trim placed too close to the fireplace opening are susceptible to burning. To prevent this, the code requires that wood be kept specific distance away from the fireplace opening.

R1003.13 Fireplace fireblocking. Fireplace fireblocking shall comply with the provisions of Section R602.8.

❖ Refer to the commentary for Section R602.8.

SECTION R1004
FACTORY-BUILT FIREPLACES

R1004.1 General. Factory-built fireplaces shall be listed and labeled and shall be installed in accordance with the conditions of the listing. Factory-built fireplaces shall be tested in accordance with UL 127.

❖ The code requires that factory-built fireplaces be listed and labeled. In addition, factory-built fireplaces must be tested by an approved testing lab in accordance with UL 127, *Factory-Built Fireplaces*.

R1004.2 Hearth extensions. Hearth extensions of approved factory-built fireplaces shall be installed in accordance with the listing of the fireplace. The hearth extension shall be readily distinguishable from the surrounding floor area.

❖ The primary requirements here are that the hearth extension must be installed in accordance with the listing of the fireplace, and it must be easily distinguished from the surrounding floor areas.

R1004.3 Decorative shrouds. Decorative shrouds shall not be installed at the termination of chimneys for factory-built fireplaces except where such shrouds are listed and labeled for use with the specific factory-built fireplace system and installed in accordance with the manufacturer's installation instructions.

❖ These provisions are the same as found in Section R1002.2. See the commentary for Section R1002.2.

R1004.4 Unvented gas log heaters. An unvented gas log heater shall not be installed in a factory-built fireplace unless the fireplace system has been specifically tested, listed and labeled for such use in accordance with UL 127.

❖ Unvented gas log heaters must not be installed in factory-built fireplaces if the fireplace has not been specifically tested, listed, and labeled for such use. Without such testing, there is no way to know how the unvented gas log heater will function in the fireplace, and an untested heater may cause harm to both the appliance and the occupants of the building.

SECTION R1005
EXTERIOR AIR SUPPLY

R1005.1 Exterior air. Factory-built or masonry fireplaces covered in this chapter shall be equipped with an exterior air supply to assure proper fuel combustion unless the room is mechanically ventilated and controlled so that the indoor pressure is neutral or positive.

❖ An adequate supply of combustion air must be provided for the fireplace. The code requires that both masonry and factory-built fireplaces have an exterior air supply. The air ducts used in either type of fireplace must be listed.

R1005.1.1 Factory-built fireplaces. Exterior combustion air ducts for factory-built fireplaces shall be a listed component of the fireplace and shall be installed according to the fireplace manufacturer's instructions.

❖ An exterior combustion air supply for factory-built fireplaces must be listed components, and the ducts must be installed in accordance with the manufacturer's instructions.

R1005.1.2 Masonry fireplaces. Listed combustion air ducts for masonry fireplaces shall be installed according to the terms of their listing and manufacturer's instructions.

❖ If a masonry fireplace uses a listed combustion air duct, the air duct must be installed per its listing.

R1005.2 Exterior air intake. The exterior air intake shall be capable of providing all combustion air from the exterior of the dwelling or from spaces within the dwelling ventilated with outside air such as nonmechanically ventilated crawl or attic spaces. The exterior air intake shall not be located within the garage or basement of the dwelling nor shall the air intake be located at an elevation higher than the firebox. The exterior air intake shall be covered with a corrosion-resistant screen of $^1/_4$-inch (6.4 mm) mesh.

❖ The exterior air supply must be capable of providing all of the necessary combustion air. This air must be taken from the exterior of the building or from spaces such as crawl or attic spaces that are adequately ventilated. The crawl or attic spaces cannot be mechanically ventilated. Attic and/or crawl space mechanical ventilating systems are primarily used to remove air from those areas. They do this by exhausting the unwanted air or creating a negative pressure in those areas. If an air intake for a fireplace terminates in a crawl or attic space that has a mechanical ventilation system there is a potential for the air intake to perform the opposite of its designed intent. The exterior air intake cannot be in the garage or basement of the dwelling unit. The air intake must be lower than the fire box so that the firebox will properly draw in the combustion air. Where combustion air openings are located inside the firebox, the air intake opening on the outside of the dwelling cannot be located higher than the firebox. Such an installation could create a chimney effect, drawing the products of combustion up through the combustion air ducts. These ducts generally are not constructed of materials that can withstand the heat and sparks that could be drawn through them. Interior combustion air openings located outside the firebox are not subject to this requirement. The exterior air intake must be covered with a corrosion-resistant screen of $^1/_4$-inch (6.4 mm) mesh.

R1005.3 Clearance. Unlisted combustion air ducts shall be installed with a minimum 1-inch (25.4 mm) clearance to combustibles for all parts of the duct within 5 feet (1524 mm) of the duct outlet.

❖ Where masonry fireplaces use unlisted combustion air ducts, the ducts must be installed with a minimum clearance of 1 inch (25.4 mm) to combustibles for all parts of the ducts within 5 feet (1524 mm) of the outlets.

R1005.4 Passageway. The combustion air passageway shall be a minimum of 6 square inches (3870 mm²) and not more than 55 square inches (0.035 m²), except that combustion air systems for listed fireplaces shall be constructed according to the fireplace manufacturer's instructions.

❖ The combustion air passageway must be a minimum of 6 square inches (3870 mm²) and no more than 55 square inches (0.035 m²). Combustion air systems for listed places must have combustion air passageways in a size that is established by the manufacturer, which may be different from the dimensions indicated in this section. Combustion passageways allow combustion air to the firebox without relying on the air within the structure. Without this combustion air, the oxygen in the structure could be depleted to a level unacceptable for the occupants of the building.

R1005.5 Outlet. The exterior air outlet is permitted to be located in the back or sides of the firebox chamber or within 24 inches (610 mm) of the firebox opening on or near the floor. The outlet shall be closable and designed to prevent burning material from dropping into concealed combustible spaces.

❖ The combustion air opening may be located in the back or sides of the firebox chamber, or it may be located within 24 inches (610 mm) of the fire box opening on or near the floor. If the opening is on or near the floor, the outlet must be closeable and designed to prevent burning material from dropping into concealed combustible spaces. A noncombustible screen could be used for this purpose, provided the mesh is such that burning embers cannot penetrate the screen.

SECTION R1006
MASONRY HEATERS

R1006.1 Installation. A masonry heater shall be installed according to one of the following:

1. The terms of its listing; or

2. ASTM E 1602.

❖ ASTM E 1602 provides guidelines for installing masonry heaters.

R1006.2 Seismic reinforcing. Masonry heaters shall be anchored and reinforced as required in this chapter. All masonry heaters shall maintain a minimum clearance of 4 inches (100 mm) to adjacent framing from the body of the masonry heater. In Seismic Design Categories A, B and C, reinforcement and seismic anchorage shall not be required. In Seismic Design Categories D_1, and D_2, masonry heaters shall be anchored to the foundation. Where the masonry chimney shares a common wall with the facing of the masonry heater, the chimney portion of the structure shall be reinforced in accordance with Section R1003.3.

❖ Heat conducted through masonry heater walls can ignite combustible structural materials in contact with these walls. For this reason, a minimum required clearance to combustibles from masonry heaters has been established. Because masonry heaters typically generate more heat for a longer period of time than traditional fireplaces, greater clearances to combustible materials are needed to reduce the risk of fire.

Because of the large bulk and squat geometry of these heaters, seismic reinforcement is not typically required. Flexural tensile stresses, which typically cause damage to unreinforced masonry, rarely occur. Where chimneys extend above these heaters, however, seismic reinforcement is required by Section R1003.3. See the commentary to Sections R1003.3 and R1003.4 for seismic reinforcements and anchorage for chimneys.

BIBLIOGRAPHY

The following resource materials are referenced in this chapter or are relevant to the subject matter addressed in this chapter.

ASTM C 27-98, *Classification of Fireclay and High-Alumina Refractory Brick*. West Conshohocken, PA: ASTM International, 1998.

ASTM C 199-84 (2000), *Test for Pier Refractory Mortar*. West Conshohocken, PA: ASTM International, 1984.

ASTM C 315-00, *Specification for Clay Flue Linings*. West Conshohocken, PA: ASTM International, 2000.

ASTM C 1261-98, *Specification for Firebox Brick for Residential Fireplaces*. West Conshohocken, PA: ASTM International, 1998.

ASTM C 1283-00e01, *Practice for Installing Clay Flue Lining*. West Conshohocken, PA: ASTM International, 2000.

ASTM E 1602-94, *Guide for Construction of Solid Fuel Burning Masonry Heaters*. West Conshohocken, PA: ASTM International, 1994.

Amrhein, S. E. *Residential Masonry and Chimney Handbook*. Los Angeles, CA: Masonry Institute of America, 1989.

Handbook to the Uniform Building Code. Whittier, CA: International Conference of Building Officials, 1995.

Olin, Howard B. *Construction Principles, Materials and Methods*. Chicago: The Institute of Financial Education, 1983.

UL 103-98, *Factory-Built Chimneys for Residential Type and Building Heating Appliances*—with Revisions through March 1999. Northbrook, IL: Underwriters Laboratory, 1998.

UL 127-99, *Factory-Built Fireplaces*--with Revisions through November 1999. Northbrook, IL: Underwriters Laboratory, 1999.

UL 641-95, *Type L, Low-Temperature Venting Systems*-- with Revisions through April 1999, Type L. Northbrook, IL: Underwriters Laboratory, 1999.

UL 737-96, *Fireplace Stoves*—with Revisions through May 1998. Northbrook, IL: Underwriters Laboratory, 1996.

UL 959-01, *Medium-Heat Appliance Factory-Built Chimneys*. Northbrook, IL: Underwriters Laboratory, 2001.

UL 1777-98, *Chimney Liners*—with Revisions through July 1998. Northbrook, IL: Underwriters Laboratory, 1998.

Part IV — Energy Conservation

Chapter 11: Energy Efficiency

General Comments

Chapter 11 contains provisions that set forth the energy-efficiency related requirements for the design and construction of buildings regulated under this code. The applicable portions of the building must comply with the provisions within this chapter for energy efficiency.

Section N1101 contains the scope and application of the chapter and also regulates material identification and labeling. Section N1102 contains the insulation R-value requirements and the window U-factor requirements for the building envelope, which includes the roof ceiling assembly, wall assembly and floor assembly. Section 1103 contains the requirements for heating and cooling systems and includes requirements for equipment efficiency, duct installation and piping insulation. Section 1104 contains the requirements for service water heating performance.

Purpose

The purpose of this chapter is to define requirements for the portions of the building and building systems that impact energy use in new construction and to promote the effective use of energy. The provisions within the chapter promote energy efficiency in the building envelope, the heating and cooling system and the service water heating system of the building. Compliance with the chapter will provide a minimum level of energy efficiency for new construction. Greater levels of efficiency can be installed to decrease the energy use of new construction.

SECTION N1101
GENERAL

N1101.1 Scope. This chapter sets forth energy-efficiency-related requirements for the design and construction of buildings regulated by this code.

Exception: Provided that they are separated by building envelope assemblies from the remainder of the building, portions of the building that do not enclose conditioned space shall be exempt from the building envelope provisions, but shall comply with the provisions for building mechanical and service water heating systems.

❖ Chapter 11 applies to portions of the building thermal envelope that enclose conditioned space as shown in Commentary Figure N1101.1(1). Conditioned space is the area provided with heating and/or cooling either directly, through a positive heating/cooling supply such as registers located in the space, or indirectly, through an opening that allows heated or cooled air to communicate directly with the space. For example, a walk-in closet connected to a master bedroom may not contain a positive heating supply through a register, but it would be conditioned indirectly by the free passage of heated or cooled air into the space from the bedroom.

The purpose of the exception is twofold. First, it exempts areas that are separated from the conditioned spaces of the building by the building envelope from the normal building envelope requirements such as insulation and fenestration. A good example of this would be an unconditioned garage or an attic space. In the case of a garage, if the unconditioned garage area is separated from the conditioned areas of the residence by an assembly that meets the "building envelope" criteria (the wall between them is insulated), then the exterior walls of the garage would not be required to be insulated to separate the garage from the exterior climate. Since the garage is not conditioned space, and because it is separated from the conditioned space of the residence, then it does not need to be separated (insulated) from the exterior. The second portion of the exception points out that mechanical and service water heating equipment that may be located in these unconditioned areas still must meet the energy efficiency provisions. Again, an unconditioned garage that has a furnace that serves the residence located within it would be a good example. The equipment must comply with the minimum equipment efficiencies regardless of where it is installed. A furnace or water heater installed in an unheated space must still comply with the efficiency requirements.

The building thermal envelope consists of the wall, roof/ceiling, and floor assemblies that surround the conditioned space. Raised floors over a crawlspace, garage or directly exposed to outside air are part of the floor assembly. Walls surrounding a conditioned basement (in addition to surrounding conditioned spaces above grade) are part of the building envelope. The code defines above-grade walls surrounding conditioned space as exterior walls. This definition includes walls between the conditioned space and an unconditioned garage, roof and basement knee walls, dormer walls, gable end walls, walls enclosing a mansard roof and basement walls with an average below-grade area that is less than 50 percent of the total basement gross wall area. This definition would not include walls separating an unconditioned garage from the outdoors. The roof/ceiling assembly is the surface where insulation will be installed, typically on top of the gypsum board. See Commentary Figure N1101.1(2).

In addition to provisions covering the building envelope, the IRC has provisions covering heating and cooling systems and water heating. Federal law covers the minimum equipment efficiency requirements for most heating/cooling systems and water heating under the National Appliance Energy Conservation Act.

N1101.2 Compliance. Compliance with this chapter shall be demonstrated by meeting the requirements of the applicable sections and tables of this chapter. Where applicable, provisions are based on the climate zone where the building is located. The climate zone where the building is located shall be based on zone assignments in Table N1101.2 for the county and state in which the building is being constructed. Alternatively, the climate zone shall be permitted to be determined by the heating degree days assigned by the building official.

❖ There are two categories of requirements under this chapter. The first contains energy efficiency requirements that must be complied with regardless of the level of insulation installed in the building envelope and the quantity and type of glazing. Examples of these provisions include sealing the building envelope to reduce infiltration and meeting the minimum heating and cooling equipment efficiency requirements. Most of the requirements are not climate dependent, as they save energy in all climates, but some, the solar heat gain coefficient provision under Section N1102.2 for example, apply only to areas in the country that are cooling-load dominated. The second category of requirements is completely climate dependent and regulates minimum insulation levels (R-values) and maximum glazing efficiency (U-factors). These requirements are presented in Table N1102.1 of the code. The table can be used with either the climate zone designation or the Heating Degree Day (HDD) value for the jurisdiction. The HDD values are determined from NOAA, *Annual Degree Days to Selected Basis Derived from the 1960-1990 Normals*. Table N1101.2, at the end of the chapter, lists county names for all the counties in the applicable state, followed by the climate zone number for the county.

N1101.2.1 Detached one- and two-family dwellings. Compliance shall be demonstrated by either:

1. Meeting the requirements of this chapter for buildings with a glazing area that does not exceed 15 percent of the gross area of exterior walls; or
2. Meeting the requirements of the *International Energy Conservation Code* for detached one- and two-family dwellings.

❖ This section shows how compliance for detached one- and two-family dwellings may be demonstrated. In the 2000 edition of the code these one- and two-family dwelling units were designated as Type A-1. The new designation makes it much easier to understand the

Figure N1101.1(1)
CONDITIONED ENVELOPE

ENERGY EFFICIENCY FIGURE N1101.1(2) - N1101.3

**Figure N1101.1(2)
BUILDING ENVELOPE**

application of the provisions. The IRC places a limitation on glazing area for the building, setting the maximum amount at 15 percent of the gross area of exterior walls. Basement wall areas (defined by the code as walls 50 percent or more below grade) that surround conditioned space do not count in the gross exterior wall area total. The wall area between a conditioned space and an unconditioned garage is also an exterior wall. Walls separating two units with conditioned space on both sides are not exterior walls. The gross area includes glazing and door area in addition to the opaque wall.

The glazing area includes the area of all glazed fenestration and includes the rough opening of all vertical windows, skylights, and glazed doors. Glazing area also includes windows in basement walls surrounding conditioned space. The provisions in the *International Energy Conservation Code* must be followed for one- and two-family buildings having a glazing area greater than 15 percent. To determine the glazing percentage, divide the total glazing area by the total gross area of exterior walls.

$$\text{Glazing Percentage} = \frac{\text{Total Glazing Area}}{\text{Total Gross Exterior Wall Area}}$$

N1101.2.2 Townhouses. Compliance shall be demonstrated by either:

1. Meeting the requirements of this chapter for buildings with a glazing area that does not exceed 25 percent of the gross area of exterior walls; or

2. Meeting the requirements of the *International Energy Conservation Code* for residential buildings of Group R-2, R-4 or townhouses.

❖ This section shows how compliance for townhouses may be demonstrated. In the 2000 edition of the code these townhouses were designated as Type A-2 residential buildings. They are allowed more glazing than detached one- and two-family dwellings because they typically use less energy per square foot of floor area than one-and two family construction and have less exterior wall area. The code sets a maximum ratio of window-to-gross-wall area of 25 percent. As in section N1101.2.1, the glazing area includes the rough opening of horizontal and vertical glazing including glass doors. Exterior wall area does not include the common walls that separate two townhouses and have conditioned space on each side. Townhouses with a glazing area above 25 percent must demonstrate compliance using the *International Energy Conservation Code*.

N1101.3 Materials and equipment. Materials and equipment shall be identified in a manner that will allow a determination of their compliance with the provisions of this chapter. Materials and equipment used to conform to the applicable provisions of this chapter shall be installed in accordance with the manufacturer's installation instructions.

❖ The code requires that materials and equipment installed in residential construction be labeled or identified so that what is installed in the field meets the provisions of the *International Residential Code*. This also reinforces that the materials and equipment can be verified for compliance with the code during on-site inspections. Insulation levels, glazing dimensions, and U-factors must be called out on the building plans for the proposed residence (see Commentary Figure N1101.3). Materials and equipment installed onsite should include information that can be used to identify levels of efficiency. For example, faced fiberglass batt insulation must be stamped with the R-value rating on the facing; see Commentary Figure N1101.3.1. Materials and equipment must be installed per the manufacturers' installation instructions to achieve the tested and rated levels of efficiency in the house. In some cases, failure to install the products to the manufacturers' specifications can void warranties on the products or lead to unsafe conditions.

N1101.3.1 Insulation. The thermal resistance (R-value) shall be indicated on all insulation and the insulation installed such that the R-value can be verified during inspection, or a certification of the installed R-value shall be provided at the job site by the insulation installer. Where blown-in or sprayed insulation is applied in walls, the installer shall provide a certification of the installed density and R-value. Where blown-in or sprayed insulation is applied in the roof-ceiling assembly, the installer shall provide a certification of the initial installed thickness, settled thickness, coverage area, and number of bags of insulating material installed. Markers shall be provided for every 300 square feet (28 m^2) of attic area, attached to the trusses, rafters, or joists, and indicate in 1-inch-high (25.4 mm) numbers the installed thickness of the insulation.

❖ Fiberglass batt insulation must have the manufacturer's R-value designations printed directly on the insulation (see Commentary Figure N1101.3.1). If possible, these designations should be installed so that they are readable. Backed floor batts can be installed with the designation against the underfloor where it is not visible, so a certification provided by the installer should be available to verify that the insulation installed in the flooring meet or exceeds the requirements set by the energy code.

The IRC requires certification of the installed density and R-value of blown-in or sprayed insulation in the wall systems, but it is often difficult to verify the R-value of blown-in or spray-applied insulation in wall systems or attics. Therefore, the IRC requires the installer to provide a certificate displaying the initial installed thickness, settled thickness, coverage area, and the number of bags of insulating material installed. Markers that indicate in 1-inch-high numbers the installed thickness of the insulation must be provided for every 300 square feet of attic area, attached to the trusses, rafters or joists.

For SI: 1 inch = 25.4 mm.

Figure N1101.3
CUT-THROUGH OF BUILDING

ENERGY EFFICIENCY

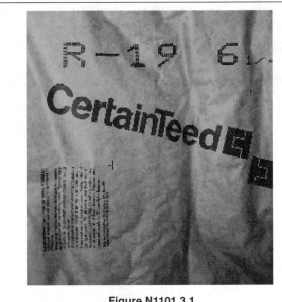

Figure N1101.3.1
INSULATION R-VALUE STAMP

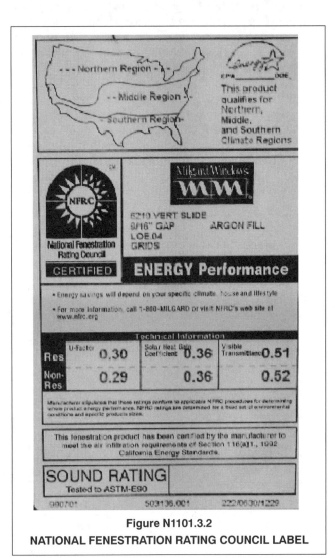

Figure N1101.3.2
NATIONAL FENESTRATION RATING COUNCIL LABEL

N1101.3.2 Fenestration. The U-factor of fenestration shall be determined in accordance with NFRC 100 by an accredited, independent laboratory, and labeled and certified by the manufacturer. The solar heat gain coefficient (SHGC) of fenestration shall be determined in accordance with NFRC 200 by an accredited, independent laboratory, and labeled and certified by the manufacturer.

❖ The code requires that the U-factor for fenestration products be tested and labeled in accordance with National Fenestration Rating Council (NFRC) 100. A U-factor is a measure of the product's efficiency. The lower the U-factor, the more efficient the product. Fenestration includes skylights, roof windows, vertical windows, opaque doors, glazed doors, glass block, and combination opaque/glazed doors. Tested fenestration products will include a temporary label (see Commentary Figure N1101.3.2) that lists the rated U-factor for residential and nonresidential installations. The labels should remain on the windows until field inspection so the inspector can verify that the windows installed in the building meet or exceed the requirements. A *Directory of Certified Products* is available from the NFRC that contains the rated U-factor of the unit based on the make and model number. The product directory is also available at www.nfrc.org.

Windows must also have a Solar Heat Gain Coefficient (SHGC) in accordance with NFRC 200. The SHGC rates the amount of solar heat gain that passes through fenestration relative to the maximum amount incident on the window unit. A unit with a low SHGC rating allows less solar heat gain to pass through than one with a higher rating. A maximum SHGC rating is a requirement in warmer climates as explained in Section N1102.2. This rating will be included on the temporary label; see Commentary Figure N1101.3.2. Also, SHGC ratings are included in the NFRC *Directory*, arranged by the make and model number of the fenestration product.

N1101.3.2.1 Default fenestration performance. When a manufacturer has not determined a fenestration product's U-factor in accordance with NFRC 100, compliance shall be determined by assigning such products a default U-factor from Tables 102.5.2(1) and 102.5.2(2) in the *International Energy Conservation Code*. When a manufacturer has not determined a fenestration product's SHGC in accordance with NFRC 200, compliance shall be determined by assigning such products a default SHGC from Table 102.5.2(3) of the *International Energy Conservation Code*.

❖ Not all manufacturers rate the U-factor of their products in accordance with NFRC 100, evident by the absence of a temporary NFRC label on some fenestration units. A default U-factor from Commentary Tables 102.5.2(1) and (2) in the *International Energy Conservation Code* [Commentary Tables N1101.3.2.1 (1) and (2)] must be assigned to the unit if it is not NFRC

100 rated. These IECC tables assign U-factors based on the frame material, product type, and number of glazing layers (single or double). When a composite of materials from two different product types is used, the product must be assigned the higher U-factor. The product cannot receive credit for a feature that cannot be seen. Since features such as argon-fill and low-emissivity glass are not visible, they do not receive credit in the default tables.

As with fenestration U-factors, default SHGC values must be assumed for all products not rated in accordance with NFRC 100. SHGC values from Table 102.5.2(3) of the IECC must be used for all units that do not have labels or are not listed in the NFRC *Directory*. The table assigns a SHGC value based on the frame type, window operation, layers of glazing and the type of tint on the window.

TABLE N1101.3.2.1(1)
U-FACTOR DEFAULT TABLE FOR WINDOWS, GLAZED DOORS AND SKYLIGHTS

FRAME MATERIAL AND PRODUCT TYPE[a]	SINGLE GLAZED	DOUBLE GLAZED
Metal without thermal break		
Operable (including sliding and swinging glass doors)	1.27	0.87
Fixed	1.13	0.69
Garden window	2.60	1.81
Curtain wall	1.22	0.79
Skylight	1.98	1.31
Site-assembled sloped/overhead glazing	1.36	0.82
Metal with thermal break		
Operable (including sliding and swinging glass doors)	1.08	0.65
Fixed	1.07	0.63
Curtain wall	1.11	0.68
Skylight	1.89	1.11
Site-assembled sloped/overhead glazing	1.25	0.70
Reinforced vinyl/metal clad wood		
Operable (including sliding and swinging glass doors)	0.90	0.57
Fixed	0.98	0.56
Skylight	1.75	1.05
Wood/vinyl/fiberglass		
Operable (including sliding and swinging glass doors)	0.89	0.55
Fixed	0.98	0.56
Garden window	2.31	1.61
Skylight	1.47	0.84

a. Glass block assemblies with mortar but without reinforcing or framing shall have a U-factor of 0.60.

NOTE: This is IECC Table 102.5.2(1).

N1101.3.2.2 Air leakage. The air leakage of prefabricated fenestration shall be determined in accordance with AAMA/WDMA 101/I.S.2, 101/I.S.2/NAFS, or NFRC 400 by an accredited, independent laboratory, and labeled and certified by the manufacturer and shall not exceed the values in Table 502.1.4.1 of the *International Energy Conservation Code*. Alternatively, the manufacturer shall certify that the fenestration is installed in accordance with Section 502.1.4 of the *International Energy Conservation Code*.

❖ Prefabricated fenestration units must meet maximum air-leakage rates determined in accordance with AAMA/WDMA 101/I.S.2, 101/I.S.2/NAFS or NFRC 400 by an accredited independent laboratory. These standards provide consistent air leakage limitations. Air leakage is also known as "infiltration" and is the uncontrolled inward leakage of air through cracks and gaps in the building envelope, especially around windows and doors. Maximum air-leakage rates are referenced within the code provision and are presented in Table 502.1.4.1 of the IECC (Commentary Table N1101.3.2.2). Windows and doors certified as meeting industry standards such as those certified by the American Architectural Manufacturers Association (AAMA) or the Wood & Door Manufacturers Association (WDMA) meet those requirements and will generally be labeled indicating their certification. For noncertified prefabricated doors and windows, the window supplier must produce documented test results verifying compliance with these air-leakage requirements. Site constructed doors and windows are not required to meet the air infiltration rates based on the reference to IECC Section 502.1.4, but their joints must be caulked and sealed to prevent air leakage.

TABLE N1101.3.2.1(2)
U-FACTOR DEFAULT TABLE FOR NONGLAZED DOORS

DOOR TYPE	WITH FOAM CORE	WITHOUT FOAM CORE
Steel doors (1.75 inches thick)	0.35	0.60

	WITHOUT STORM DOOR	WITH STORM DOOR
Wood doors (1.75 inches thick)		
Panel with 0.438-inch panels	0.54	0.36
Hollow core flush	0.46	0.32
Panel with 1.125-inch panels	0.39	0.28
Solid core flush	0.40	0.26

For SI: 1 inch = 25.4 mm.
NOTE: This is IECC Table 102.5.2(2).

N1101.3.3 Maintenance. Where mechanical or plumbing system components require preventive maintenance for efficient operation, regular maintenance requirements shall be clearly stated and affixed to the component, or the source for such information shall be shown on a label attached to the component.

❖ The IRC requires that preventative maintenance information for the building's space heating, cooling system, and water-heating equipment be made available to the new home owner/operator. Equipment that is not serviced according to the manufacturer's requirements may operate at a lower efficiency than designed, thereby increasing the energy usage and cost to operate the equipment. Providing manufacturers' maintenance instructions with the residence is usually sufficient to meet this requirement. The instructions can be in the form of a maintenance manual taped to the unit or a label on the unit providing direction on where to get further information. During the final inspection, the mechanical equipment and water heater should be inspected to verify that such information has been left at the building.

TABLE N1101.3.2.2
ALLOWABLE AIR INFILTRATION RATES[a]

WINDOWS (cfm per square foot of window area)	DOORS (cfm per square foot of door area)	
	Sliders	Swinging
0.3[b,c]	0.3	0.5[d]

For SI: 1 cfm /ft^2 = 0.00508 m^3/(s · m^2).
a. When tested in accordance with ASTM E 283.
b. See AAMA/WDMA 101/I.S.2.
c. See ASTM D 4099.
d. Requirement based on assembly area.
NOTE: This is IECC Table 502.1.4.1.

SECTION N1102
BUILDING ENVELOPE

N1102.1 Thermal performance criteria. The minimum required insulation R-value or the area-weighted average maximum required fenestration U-factor (other than opaque doors which are governed by Section N1102.1.3) for each element in the building thermal envelope (fenestration, roof/ceiling, opaque wall, floor, slab edge, crawl space wall and basement wall) shall be in accordance with the criteria in Table N1102.1.

Detached one-and-two family dwellings with greater than 15-percent glazing area; townhouses with greater than 25-percent glazing area; and any residential building in climates with heating degree days equal to or greater than 13,000; shall determine compliance using the building envelope requirements of Chapter 4 or 5 of the *International Energy Conservation Code*.

❖ Table N1102.1 lists the minimum R-value and maximum U-factor requirements for different portions of the building thermal envelope, including basement and exterior walls, floor, roof and any other building elements that enclose conditioned space. Using the table begins with determining the climate zone for the proposed location from Table N1101.2. Once the climate zone has been determined, each of the R-value or U-factor requirements must be met for the applicable assembly type (e.g., ceilings, walls, floors, etc). This chapter also places limitations on the glazing percentage that a residential building may have. Detached one- and two-family dwellings are limited to 15-percent glazing area and townhouses to 25-percent glazing area. Calculating glazing percentages is described in detail in the commentary for Sections N1101.2.1 and N1101.2.2, above. Buildings located in areas of 13,000 HDD (Climate Zone 18) and above must use the compliance methods contained in Chapters 4 or 5 of the IECC.

Maximum glazing U-factor is the first column in Table N1102.1 that must be complied with. Each fenestration product in the proposed building must not exceed the maximum U-factor requirement presented in the table for a particular climate zone. For example, a single-family residence located in a Climate Zone 8 would be required to install glazed fenestration products with a maximum U-factor of 0.50. This would include all glazing in the building thermal envelope (e.g. vertical windows and skylights). The proposed glazing U-factor should be called out on the building plan either on the floor plan or within a window schedule. This will provide the necessary information to the field inspector, who will then need to verify that what is on the plans is installed in the field. Fenestration products that do not have temporary NFRC labels on them must use the default U-factors contained in Table 102.5.2(1) of the IECC. See Commentary Table N1101.3.2.1(1). The lowest default U-factor included in the table is listed at 0.55 for a wood or vinyl operable window. This U-factor will not meet the requirements of the IRC in Climate Zones 8 and higher.

N1102.1.1 Exterior walls. The minimum required R-value in Table N1102.1 shall be met by the sum of the R-values of the insulation materials installed in framing cavities and/or insulating sheathing applied, and not by framing, drywall, structural sheathing, or exterior siding materials. Insulation separated from the conditioned space by a vented space shall not be counted towards the required R-value.

❖ The R-values presented under the "Walls" column represent the sum of the insulation materials installed between the framing cavity and the insulating sheathing. Insulating sheathing must have an R-value of at least R-2 to be considered. The R-value of non-insulative interior finishes, such as sheet rock or exterior coverings like wood structural panel siding, is not considered when determining if the proposed wall assembly meets the requirements. For example, in Commentary Figure N1102.1.1 the R-value of the cavity insulation installed between framing (R-13) is added to the insulating sheathing installed on the outside of the studs (R-6) resulting in an R-19 wall. If the proposed home is located in an area with 5,200 HDD (Climate Zone 11), the R-19 total insulation value can then be compared to the R-value requirement of R-18 in Table N1102.1 to determine compliance.

The insulation R-value requirement for exterior walls assumes wood framing. Walls framed using steel studs or constructed of materials such as CMU are addressed in Sections N1102.1.1.1 and N1102.1.1.2.

N1102.1.1.1 Mass walls. Mass walls shall be permitted to meet the mass wall criteria in Table N1102.1.1.1(1), based on the insulation position and the climate zone where the building is located. Other mass walls shall meet the frame wall criteria for the building type and the climate zone where the building is

Figure N1102.1.1
EXTERIOR R-VALUES

located, based on the sum of interior and exterior insulation. Walls with exterior insulation position have the entire effective mass layer interior to an insulation layer. Walls with integral insulation position have either insulation and mass materials well mixed as in wood (logs); or substantially equal amounts of mass material on the interior and exterior of insulation as in concrete masonry units with insulated cores or masonry cavity walls. Walls with interior insulation position have the mass material located exterior to the insulating material. Walls not meeting either of the above descriptions for exterior or integral positions shall meet the requirements for "Other mass walls" in Table N1102.1.1.1(1). The R-value of the mass assembly for typical masonry constructions shall be taken from Table N1102.1.1.1(2). The R-value for a solid concrete wall with a thickness of 4 inches (102 mm) or greater is R-1.1. R-values for other assemblies are permitted to be based on hot box tests referenced in ASTM C 236 or ASTM C 976, two dimensional calculations or isothermal planes calculations.

❖ The code defines a mass wall as a masonry or concrete wall having a mass equal to or greater than 30 pounds per square foot (psf) (1436 Pa); a solid wood wall (e.g. in a log house) having a mass greater than 20 psf (958 Pa); or any other walls having a heat capacity equal to or greater than 6 $Btu/ft^2/°F$ [1.06 kJ ($m^2 \cdot K$)]. In general terms, the heat capacity is a measure of how well a material stores heat. The higher the heat capacity, the greater the amount of heat stored in the material. For example, a 6-inch (152 mm) heavyweight concrete wall has a heat capacity of 14 $Btu/ft^2/°F$ [2.47

kJ ($m^2 \cdot K$)] compared to a conventional 2 by 4 (51 mm by 102 mm) wood-framed wall with a heat capacity of approximately 3 $Btu/ft^2/°F$ [0.53 kJ/($m^2 \cdot K$)]. Tables N1102.1.1.1(1) and (2) are used to determine the insulation requirements for mass walls. To use the tables, first consult either Table N1101.2 to determine the climate zone of the proposed project or the HDD for the proposed location of the building. The R-value for the assembly is then given for the mass wall. Use the "Exterior or Integral Insulation" column for wall assemblies that have insulation on the exterior of the wall [between the mass wall and the exterior; see Commentary Figure N1102.1.1.1(1)] and integral to the assembly. Concrete masonry units with insulated cores or masonry cavity walls are examples of integral insulation. For mass walls with the insulation installed on the interior [an insulated furred wall located between the conditioned space and the mass wall; see Commentary Figure N1102.1.1.1(2)], the "Other Mass Walls" column must be used to determine the required R-value.

Once the R-value requirement for the wall is determined, the next step is to determine if the proposed wall assembly R-value meets or exceeds the requirement. Use Table N1102.1.1.1(2) to find the R-value of the assembly without insulation. Use the "Assembly Type" column, and then determine the R-value based on whether the ungrouted cells are insulated or uninsulated. If the R-value for the assembly meets or exceeds the requirements listed in Table N1102.1.1.1(1), no additional insulation is required. If the R-value for the assembly is less than the requirement, insulation must be added based on the column selected to determine the required R-value. For example, if the proposed wall assembly for a residence located in Zone 8 is 6-inch (152 mm) normal weight concrete block with ungrouted cells insulated and vertical cells grouted at 10 feet on center or greater, the assembly R-value would be R-3.1. If the wall is proposed to have insulation installed on the exterior, the required R-value would be R-8.9. To meet the requirement, insulation with a minimum R-value of 5.8 must be installed on the exterior of the wall.

N1102.1.1.2 Steel-frame walls. The minimum required R-values for steel-frame walls shall be in accordance with Table N1102.1.1.2.

❖ Table N1102.1.1.2 must be used for buildings using steel for wall framing. Because steel is a better conductor of heat than wood, walls framed with steel lose more heat than a wood wall of comparable thickness. The code requires increased insulation levels to offset these increased losses in steel-framed walls. The required insulation R-values listed in Table N1102.1.1.2 are for a combination of cavity (between the studs) and sheathing (across the face of the stud) insulation. For example, a building proposed for Zone 10 using steel studs must have: 1) R-11 insulation between the studs with R-9 sheathing; 2) R-15 between the studs with R-8 sheathing; or 3) R-21 insulation between the studs with R-7 sheathing.

N1102.1.2 Ceilings. The required "Ceiling R-value" in Table N1102.1 assumes standard truss or rafter construction and shall apply to all roof/ceiling portions of the building thermal envelope, including cathedral ceilings. Where the construction technique allows the required R-value of ceiling insulation to be obtained over the wall top plate, R-30 shall be permitted to be used where R-38 is required and R-38 shall be permitted to be used where R-49 is required.

❖ The ceiling R-value requirements are pre-calculated for insulation only and already assume a credible R-value for other building materials such as air films, interior sheathing, and exterior sheathing. The only R-value for insulation that may be used to meet the requirements is that installed between the conditioned space and the vented air space in the roof/ceiling assembly. This is typically not an issue because most insulation is installed directly on top of the gypsum board ceiling. All insulation installed in the ceiling must meet or exceed the required insulation level.

Insulation installed in a typical roof assembly will be full height throughout the center portions of the assembly and will taper at the edges as the roof nears the top plate of the exterior wall system; see Commentary Figure N1102.1.2(1). The slope of the roof causes this tapering, which is further amplified by any baffling installed to direct ventilation air from the eave vents up and over the insulation. Because of this tapering, the installed R-value near the plate lines will be less than the rated R-value for the insulation. This is due to compression (compressed insulation has a lower R-value than insulation installed to its full thickness) and the limited space between the floor of the attic and the roof sheathing near the exterior plate line. Thus a typical installation, on average, will have a lower R-value than that of the rated insulation. Because of this, the code will allow a lower insulation value to be installed providing it can be installed full thickness, to its rated R-value, over the plate line of the exterior wall. This can be achieved by using an oversized truss or raised-heel truss as shown in Commentary Figure N1102.1.2(2). R-30 may be substituted for R-38 insulation, and R-38 may be substituted for R-49 insulation to meet the requirements of the code.

For SI: 1 inch = 25.4 mm.

Figure N1102.1.1.1(1)
EXTERIOR INSULATION

For SI: 1 inch = 25.4 mm.

Figure N1102.1.1.1.(2)
INTERIOR INSULATION

N1102.1.3 Opaque doors. Opaque doors separating conditioned and unconditioned space shall have a maximum U-factor of 0.35. One opaque door shall be permitted to be exempt from this U-factor requirement.

❖ Opaque doors leading from a conditioned space to an unconditioned space or outdoors must have a maximum U-factor of 0.35. This includes a door between a house and an attached garage but does not include a door leading from the garage to the outdoors. A steel door with an insulated foam core is an example of a door meeting this requirement. The IECC provides default U-factors for various door assemblies in IECC Table 102.5.2(2); see Commentary Table N1101.3.2.1(2). The code exempts one door from the U-factor requirement, thus allowing the use of ornate entrance door assemblies without having to take another compliance approach.

N1102.1.4 Floors. The required *R*-value in Table N1102.1 shall apply to all floors, except any individual floor assembly with over 25 percent of its conditioned floor area exposed directly to outside air shall meet the *R*-value requirement in Table N1102.1 for "Ceilings."

❖ Floors over a crawlspace or an unconditioned garage are required to meet or exceed the floor *R*-value requirements listed in Table N1102.1. The insulation *R*-value requirements range from *R*-11 in warm climates *R*-21 in extremely cold climates. Insulation must be installed between the floor joists and must be well supported with netting, wire, wood strips, or another method of support so that the insulation does not droop or fall out of the joist cavities over time.

The code includes a provision requiring floor assemblies over outside air (not including a vented crawlspace or an unconditioned garage) to be insulated to the *R*-value requirements for ceilings. This only applies if over 25 percent of the floor assembly area is directly exposed to the outside air. For example, if a house with a floor area of 1000 square feet were built on the slope of a hill with 700 square feet of floor area over a vented crawlspace and 300 square feet of floor area over outside air, the floor assembly exposed to outside air must be insulated to meet the ceiling requirements because over 25 percent (300/1000 x 100 = 30%) of the floor assembly would be exposed to outside air. However, if only 150 square feet were over outside air, the entire floor assembly would need to meet the *R*-value requirements for floors (instead of ceilings), since 25 percent or less (150/1000 x 100 = 15%) of the floor assembly would be directly exposed to outside air.

TABLE N1102.1
SIMPLIFIED PRESCRIPTIVE BUILDING ENVELOPE THERMAL COMPONENT CRITERIA
MINIMUM REQUIRED THERMAL PERFORMANCE (*U*-FACTOR AND *R*-VALUE)

BUILDING LOCATION		MAXIMUM GLAZING *U*-FACTOR [Btu / (hr·ft^2·°F)]	MINIMUM INSULATION *R*-VALUE [(hr·ft^2·°F) / Btu]					
Climate Zone	HDD		Ceilings	Walls	Floors	Basement walls	Slab perimeter *R*-value and depth	Crawl space walls
1	0-499	Any	R-13	R-11	R-11	R-0	R-0	R-0
2	500-999	0.90	R-19	R-11	R-11	R-0	R-0	R-4
3	1,000-1,499	0.75	R-19	R-11	R-11	R-0	R-0	R-5
4	1,500-1,999	0.75	R-26	R-13	R-11	R-5	R-0	R-5
5	2,000-2,499	0.65	R-30	R-13	R-11	R-5	R-0	R-6
6	2,500-2,999	0.60	R-30	R-13	R-19	R-6	R-4, 2 ft.	R-7
7	3,000-3,499	0.55	R-30	R-13	R-19	R-7	R-4, 2 ft.	R-8
8	3,500-3,999	0.50	R-30	R-13	R-19	R-8	R-5, 2 ft.	R-10
9	4,000-4,499	0.45	R-38	R-13	R-19	R-8	R-5, 2 ft.	R-11
10	4,500-4,999	0.45	R-38	R-16	R-19	R-9	R-6, 2 ft.	R-17
11	5,000-5,499	0.45	R-38	R-18	R-19	R-9	R-6, 2 ft.	R-17
12	5,500-5,999	0.40	R-38	R-18	R-21	R-10	R-9, 4 ft.	R-19
13	6,000-6,499	0.35	R-38	R-18	R-21	R-10	R-9, 4 ft.	R-20
14	6,500-6,999	0.35	R-49	R-21	R-21	R-11	R-11, 4 ft.	R-20
15	7,000-8,499	0.35	R-49	R-21	R-21	R-11	R-13, 4 ft.	R-20
16	8,500-8,999	0.35	R-49	R-21	R-21	R-18	R-14, 4 ft.	R-20
17	9,000-12,999	0.35	R-49	R-21	R-21	R-19	R-18, 4 ft.	R-20

For SI: 1 Btu/(hr·ft^2·°F) = 5.68W/m^2·K, 1 (hr·ft^2·°F)/Btu = 0.176m^2·K/W.

❖ See the commentary to Section N1102.1.

ENERGY EFFICIENCY

TABLE N1102.1.1.1(1)
MASS WALL PRESCRIPTIVE BUILDING ENVELOPE REQUIREMENTS

BUILDING LOCATION		MASS WALL ASSEMBLY R-VALUE[a] (hr·ft²·°F) / Btu	
Climate Zone	HDD	Exterior or integral insulation	Other mass walls
1	0-499	R-3.8	R-9.7
2	500-999	R-4.8	R-9.7
3	1,000-1,499	R-4.8	R-9.7
4	1,500-1,999	R-8.1	R-10.8
5	2,000-2,499	R-8.9	R-10.8
6	2,500-2,999	R-8.9	R-10.8
7	3,000-3,499	R-8.9	R-10.8
8	3,500-3,999	R-8.9	R-10.8
9	4,000-4,499	R-8.9	R-10.9
10	4,500-4,999	R-10.4	R-12.3
11	5,000-5,499	R-11.9	R-15.2
12	5,500-5,999	R-11.9	R-15.2
13	6,000-6,499	R-11.9	R-15.2
14	6,500-6,999	R-15.5	R-18.4
15	7,000-8,499	R-15.5	R-18.4
16	8,500-8,999	R-18.4	R-18.4
17	9,000-12,999	R-18.4	R-18.4

For SI: 1 (hr·ft²·°F)/Btu = 0.176 m²·K/W.

a. The sum of the value in Table N1102.1.1.1(2) and additional insulation layers.

❖ See the commentary to Section N1102.1.1.1.

TABLE N1102.1.1.1(2)
MASS ASSEMBLY R-VALUES

ASSEMBLY TYPE	UNGROUTED CELLS NOT INSULATED (hr·ft²·°F) / Btu	UNGROUTED CELLS INSULATED (hr·ft²·°F) / Btu		
		No grout	Vertical cells grouted at 10 ft. o.c. or greater	Vertical cells grouted at less than 10 ft. o.c.
6″ Light weight concrete block	2.3	5.0	4.5	3.8
6″ Medium weight concrete block	2.1	4.2	3.8	3.2
6″ Normal weight concrete block	1.9	3.3	3.1	2.7
8″ Light weight concrete block	2.6	6.7	5.9	4.8
8″ Medium weight concrete block	2.3	5.3	4.8	4.0
8″ Normal weight concrete block	2.1	4.2	3.8	3.3
12″ Light weight concrete block	2.9	9.1	7.9	6.3
12″ Medium weight concrete block	2.6	7.1	6.4	5.2
12″ Normal weight concrete block	2.3	5.6	5.1	4.3
Brick cavity wall	3.7	6.7	6.2	5.4
Hollow clay brick	2.0	2.7	2.6	2.4

For SI: 1 inch = 25.4 mm, 1 foot = 304.8 mm, 1 (hr·ft²·°F)/Btu = 0.176 m²·K/W.

❖ See the commentary to Section N1102.1.1.1.

**TABLE N1102.1.1.2
STEEL-FRAME WALL MINIMUM PERFORMANCE REQUIREMENTS (R-VALUE)**

CLIMATE ZONES	HDD	EQUIVALENT STEEL-FRAME WALL CAVITY AND SHEATHING R-VALUE[a] (hr·ft^2·°F) / Btu
1–4	0-1999	R-11+R-5, R-15+R-4, R-21+R-3
5–8	2,000-3,999	R-11+R-5, R-15+R-4, R-21+R-3
9–12	4,000-5,999	R-11+R-9, R-15+R-8, R-21+R-7
13–15	6,000-8,499	R-13+R-10, R-19+R-9, R-25+R-8
16 and 17	8,500-12,999	R-13+R-10, R-19+R-9, R-25+R-8

For SI: 1 (hr·ft^2·°F)/Btu = 0.176 m^2·K/W.

a. The cavity insulation R-value requirement is listed first, followed by the sheathing R-value requirement.

❖ See the commentary to Section N1102.1.1.2.

**Figure N1102.1.2(1)
TYPICAL ROOF ASSEMBLY**

N1102.1.5 Basement walls. When the basement is a conditioned space, the basement walls shall be insulated in accordance with Table N1102.1. When the basement is not a conditioned space, either the basement walls or the ceilings separating the basement from conditioned space shall be insulated in accordance with Table N1102.1. When insulating basement walls, the required R-value shall be applied from the top of the basement wall to a depth of 10 feet (3048 mm) below grade or to the top of the basement floor, whichever is less.

❖ The walls of conditioned basements must be insulated to meet the requirements of Table N1102.1. Each wall of a basement must be considered separately to determine if it is a basement wall or an exterior wall. It is a basement wall if it has an average gross wall area that is 50 percent or more below grade. A wall that is less than 50 percent below grade is an exterior wall and must meet the insulation requirements for walls. Most walls associated with basements will be at least 50 percent below grade and therefore must meet basement wall requirements. Walk-out basements offer a challenge in determining compliance with the code. A walk-out basement (see Commentary Figure N1102.1.5) may have a back wall that is entirely below grade, a front wall or the walk-out portion that is entirely above grade, and two side walls with a grade line running diagonally. In this case the back wall must meet the requirements for basement walls, the front wall would need to meet the requirements for walls, and the side walls would need to be evaluated to determine if they were 50 percent or greater below grade and therefore basement walls.

Basement wall insulation must be installed from the top of the foundation wall to a depth of 10 feet below grade or to the top of the basement floor, whichever is less. Because the rim joist between floors is a part of

ENERGY EFFICIENCY

the building envelope, this must be insulated also if the basement is conditioned.

N1102.1.6 Slab-on-grade floors. For slabs with a top edge above finished grade or 12 inches (305 mm) or less below finished grade, the required R-value in Table N1102.1 shall be applied to the outside of the foundation or the inside of the foundation wall. The insulation shall extend downward from the top of the slab, or downward to the bottom of the slab and

Figure N1102.1.2(2)
RAISED-HEEL TRUSS

Figure N1102.1.5
WALKOUT BASEMENT

then horizontally in either direction, until the distance listed in Table N1102.1 is reached.

When installed between the exterior wall and the edge of the interior slab, the top edge of the insulation shall be permitted to be cut at a 45-degree (0.79 rad) angle away from the exterior wall. Insulation extending horizontally away from the building shall be protected by pavement or by a minimum of 10 inches (254 mm) of soil.

In locations of 500 HDD or greater, R-2 shall be added to the values in Table N1102.1 where uninsulated hot water pipes, air distribution ducts or electric heating cables are installed within or under the slab.

Exception: Slab perimeter insulation is not required for unheated slabs in areas of very heavy termite infestation probability as shown in Figure R301.2(6). Where this exception is used, building envelope compliance shall be demonstrated by (a) using *International Energy Conservation Code* Section 502.2.2 or *International Energy Conservation Code* Chapter 4 with the actual slab insulation *R*-value in Table N1102.1, or (b) using *International Energy Conservation Code* Section 502.2.4.

❖ The perimeter edges of slab-on-grade floors must be insulated to the *R*-value listed in Table N1102.1. These requirements apply only to slabs 12 inches (305 mm) or less below grade. The listed *R*-value requirements are for unheated slabs. A slab-on-grade containing uninsulated hot water pipes, air distribution ducts, or electric heating cables must add another R-2 to the required insulation levels in climates with 500 HDD or greater. The insulation must extend downward from the top of the slab or downward to the bottom of the slab and then horizontally in either direction until the distance listed in Table N1102.1 is reached. Most of the heat loss from a slab will occur in the edge that is exposed directly to the outside air. The insulation must be installed to the top of the slab edge to prevent this heat loss. Slab insulation may be installed on the exterior of the slab edge or between the interior wall and the edge of the interior slab as in a non-monolithic slab. In this type of installation, the exposed insulation could cause problems with tack strips for carpeting; therefore, the insulation is allowed to be cut at a 45-degree angle away from the exterior or the wall.

In areas with very heavy termite infestation, slab perimeter insulation is not required to be installed per Table N1102.1. These areas are identified in Figure R301.2(6). However, if the insulation is not installed, one of the compliance approaches from the IECC that allows tradeoffs must be used to demonstrate compliance with the code. Much of the area identified as "very heavy" for termite infestation is in warmer climates with low HDD values. Typically, slab perimeter insulation can be traded off entirely in these climates by increasing the ceiling or wall insulation *R*-values or by using glazing with a lower *U*-factor.

N1102.1.7 Crawl space walls. Where the floor above the crawl space is uninsulated, insulation shall be installed on crawl space walls when the crawl space is not vented to outside air. The required *R*-value in Table N1102.1 shall be applied inside of the crawl space wall, downward from the sill plate to the exterior finished grade level and then vertically and/or horizontally for at least an additional 24 inches (610 mm). The exposed earth in all crawl space foundations shall be covered with a continuous vapor retarder having a maximum permeance rating of 1.0 perm [(57 mg)/(s·m^2·Pa)], when tested in accordance with ASTM E 96.

❖ The IRC allows for the insulation of crawlspace walls in crawlspaces instead of insulating the floor between the crawlspace and the conditioned space. This is a popular practice for freeze protection in colder climates because it is common to install plumbing in the crawlspace. The heat transferred through the uninsulated floor to the crawlspace helps keep the crawlspace temperature above freezing when the outside air temperature drops below freezing. In order to comply with this provision, the crawlspace must be mechanically vented or provided with conditioned air from the living space. The IRC now provides an exception for the underfloor ventilation requirement, listed under Section R408.2. It requires either mechanical ventilation or conditioned air to be provided to the space.

The code also requires insulation to be installed from the sillplate downward to the exterior finished grade level and then an additional 24 inches (610 mm) either vertically or horizontally; see Commentary Figure N1102.1.7. Under this insulation scenario, the rim joist is considered part of the conditioned envelope and must be insulated to the same level as the exterior wall. The insulation must be installed securely to the crawlspace wall so that it does not fall off. The code also requires a continuous vapor retarder to be installed on the floor of the crawlspace to prevent ground water vapor from entering the crawlspace. The vapor retarder must have a maximum permeance rating of 1 perm [5 mg/(s · m^2 · Pa)]. Typically 6 mil (0.152 mm) polyethylene vapor retarder is used for this purpose. While the code does not provide direction on how the vapor retarder is to be installed, best practice would dictate that all joints should be overlapped and taped to provide continuity. Also, the vapor retarder must extend up the crawlspace wall and be secured to the wall with either tape or a treated wood nailer.

N1102.1.8 Masonry veneer. For exterior foundation insulation, that horizontal portion of the foundation that supports a masonry veneer shall not be required to be insulated.

❖ For exterior foundation insulation, the horizontal portion of the foundation that supports a masonry veneer is not required to be insulated. For slab edge insulation installed on the exterior of the slab, the code allows the insulation to start at the bottom of the masonry veneer and extend downward.

ENERGY EFFICIENCY

**Figure N1102.1.7
CRAWLSPACE WALL INSULATION**

N1102.1.9 Protection. Exposed insulating materials applied to the exterior of foundation walls shall be protected. The protection shall extend at least 6 inches (152 mm) below finished grade level.

❖ The code requires insulation installed on the outside of foundation walls, including slab-on-grade perimeter insulation, to be protected to prevent the degradation of the insulation's thermal performance from impact or ultraviolet radiation. Protective materials or membranes must be opaque, rigid and weather-resistant and cover the exposed insulation, extending at least 6 inches below the finished grade level.

N1102.1.10 Air leakage. All joints, seams, penetrations; site-built windows, doors, and skylights; openings between window and door assemblies and their respective jambs and framing; and other sources of air leakage (infiltration and exfiltration) through the building thermal envelope shall be caulked, gasketed, weatherstripped, wrapped, or otherwise sealed to limit uncontrolled air movement.

❖ Sealing the building envelope is critical to good thermal performance for the building. The seal will prevent warm, conditioned air from leaking out around doors, windows, and other cracks during the heating season, thereby reducing the cost of heating the residence. During the hot summer months a proper seal will stop hot air from entering the residence, helping to reduce the air-conditioning load on the building. Any penetration in the building envelope must be thoroughly sealed during the construction process including holes made for the installation of plumbing, electrical, and heating and cooling systems; see Commentary Figure N1102.1.10. The code lists several areas that must be caulked, gasketed, weatherstripped, wrapped or otherwise sealed to limit uncontrolled air movement.

Most of the air sealing will be done prior to the installation of an interior wall covering, since any penetration will be noticeable and accessible at this time. The code allows the use of airflow retarders (house wraps) as an acceptable method to meet this provision. Airflow retarders must be:
- Impermeable to air flow.
- Continuous over the entire building envelope.
- Able to withstand the forces that may act on them during and after construction.
- Durable over the expected lifetime of the building.

All joints in airflow retarders must be taped to provide continuity. This includes sealing the house wrap to the top plate, around windows and doors and to the bottom of the wall system.

N1102.1.11 Recessed lighting fixtures. When installed in the building envelope, recessed lighting fixtures shall meet one of the following requirements:

1. Type IC rated, manufactured with no penetrations between the inside of the recessed fixture and ceiling cavity and sealed or gasketed to prevent air leakage into the unconditioned space.

2. Type IC or non-IC rated, installed inside a sealed box constructed from a minimum 0.5-inch-thick (12.7 mm) gypsum wallboard or constructed from a preformed polymeric vapor barrier, or other air-tight assembly manufactured for this purpose, while maintaining required clearances of not less than 0.5 inch (12.7 mm) from combustible material and not less than 3 inches (76 mm) from insulation material.

3. Type IC rated, in accordance with ASTM E 283 admitting no more than 2.0 cubic feet per minute (cfm) (0.944 L/s) of air movement from the conditioned space to the ceiling cavity. The lighting fixture shall be tested at 1.57 psi (75 Pa) pressure difference and shall be labeled.

❖ Recessed lighting fixtures pose a potential fire hazard if incorrectly covered with insulation. These fixtures also act as chimneys, transferring heat loss and moisture through the building envelope into attic spaces. The heat loss resulting from improperly insulated recessed lighting fixtures can be significant.

Recessed lighting fixtures must either be IC-(insulation cover) rated or installed in a sealed box that has the insulation at least 3 inches (76 mm) from the light fixture. IC-rated lights are typically double-can fixtures, with one can inside another; see Commentary Figure N1102.1.11. The outer can (in contact with insulation) is tested to make sure it remains cool enough to avoid a fire hazard. An IC-rated fixture should have the IC rating stamped on the fixture or printed on an attached label.

All recessed lights must be tightly sealed or gasketed to prevent air leakage through the fixture into the ceiling cavity.

N1102.2 Maximum solar heat gain coefficient for fenestration products. The area-weighted-average solar heat gain coefficient (SHGC) for glazed fenestration installed in

Figure N1102.1.10
TYPICAL SOURCES OF AIR LEAKAGE IN THE HOME

Figure N1102.1.11
IC-RATED RECESSED LIGHT

ENERGY EFFICIENCY

locations with 3,500 or fewer heating degree days shall not exceed 0.40.

❖ Buildings in locations with 3,500 or fewer HDD (Climate Zones 1 through 7—cooling dominated climates) must have an area-weighted Solar Heat Gain Coefficient no greater than 0.40 for all glazed fenestration. This includes vertical windows and skylights. This requirement was included in the IRC and IECC to address the needs of states with greater cooling loads than heating loads. As discussed in the commentary for Section N1101.3.2, glazed fenestration must be tested, rated, and labeled in accordance with NFRC 200 or, where approved, labeled as shown in Commentary Figure N1101.3.2. In the absence of a rating, a default SHGC value may be applied to the window from IECC Table 102.5.2(3) (Commentary Table N1102.2). The assumed SHGC for each fenestration unit should be marked on the building plans to assist in both plan review and field inspection. This is important especially if an area-weighted average is used to show compliance. The following formula can be used to determine an area-weighted SHGC:

$$\text{Area Weighted SHGC} = \frac{(A_{g_1}) \times (SHGC_{g_1}) + (A_{g_2}) \times (SHGC_{g_2}) + (etc.)}{A_{gt}}$$

Where:
A_{g_1} = Glazing area (ft^2) of glazing type one
A_{g_2} = Glazing area (ft^2) of glazing type two
$SHGC_{g_1}$ = SHGC of glazing type one
$SHGC_{g_2}$ = SHGC of glazing type two
A_{gt} = Total Glazing Area

TABLE N1102.2
SHGC DEFAULT TABLE FOR FENESTRATION

PRODUCT DESCRIPTION	SINGLE GLAZED				DOUBLE GLAZED			
	Clear	Bronze	Green	Gray	Clear + Clear	Bronze + Clear	Green + Clear	Gray + Clear
Metal frames								
Operable	0.75	0.64	0.62	0.61	0.66	0.55	0.53	0.52
Fixed	0.78	0.67	0.65	0.64	0.68	0.57	0.55	0.54
Nonmetal frames								
Operable	0.63	0.54	0.53	0.52	0.55	0.46	0.45	0.44
Fixed	0.75	0.64	0.62	0.61	0.66	0.54	0.53	0.52

NOTE: This is IECC Table 102.5.2(3)

N1102.3 Fenestration exemption. Up to 1 percent of the total glazing area shall be exempt from *U*-factor requirement.

❖ The code allows up to 1 percent of the total glazing area (see the commentary for Section R202 "Glazing Area") to be exempt from the "Maximum Glazing *U*-Factor" requirement listed in Table N1102.1. All other glazing must meet or exceed the *U*-factor requirement listed in the table. The exemption is a percentage of the total proposed glazing area and not a percentage of what is allowed under Section N1101.2. For example, if the total proposed glazing area for a residence is 280 square feet (26 m^2), the total exempt area would be 28 square feet (3 m^2). The exempt glazing area should be designated on the building plans, and the windows should be identified accordingly for plan review and field inspection.

N1102.4 Replacement fenestration. Where some or all of an existing fenestration unit is replaced with an entirely new replacement fenestration product, including frame, sash and glazed portion, in an existing building, the replacement fenestration product shall have a *U*-factor that does not exceed the "Maximum fenestration *U*-factor" and an SHGC that does not exceed the "Maximum fenestration SHGC" in Table N1102.4 applicable to the climate zone (HDD) where the building is located. Replacement skylights and roof windows shall be permitted to have a maximum *U*-factor of 0.60 when installed in any location above 1,999 HDD. The replacement fenestration products must also satisfy the air leakage requirements of Section N1101.3.2.2.

❖ Replacement fenestration products must meet the fenestration *U*-factor and SHGC requirements of Table N1102.4. This applies to skylights, roof windows, vertical windows, opaque doors, and glass doors. Skylights are permitted to have a maximum *U*-factor of 0.60 in climates of 2000 HDD or greater (Climate Zone 5 and higher).

The intent of the first sentence is to address the question of where these replacement provisions apply. Replacing only a glass pane in an existing sash and frame would not fall under this provision provided the *U*-factor and SHGC will be equal to or lower than the values prior to the replacement. In situations where the replacement is more than just a replacement of the glass pane, the provisions may be applicable. It is often common practice when fenestration is replaced to remove only the sash and glazing of an existing window and replace them with an entirely new fenestration product. Sometimes during the process the existing frame is also removed, but many times the new fenestration product is custom made to fit in the existing space left after the sash and glazed portions are removed. In essence, the new fenestration is installed in or over the existing frame. Whether the existing frame is removed or not, these types of replacements are regulated by this section.

**TABLE N1102.4
REPLACEMENT FENESTRATION REQUIREMENTS**

HEATING DEGREE DAYS	MAXIMUM	
	Fenestration U-factor	Fenestration SHGC
0 – 1,999	0.75	0.40
2,000 – 3,999	0.50	0.40[a]
4,000 – 5,999	0.40	Any
6,000 – 8,499	0.35	Any
8,500 – 12,999	0.35	Any

a. Up to but not including 3,500 HDD.

❖ This table provides an abbreviated listing of requirements found in IECC Table 502.2.5, which deals with additions and replacement windows for detached one- and two-family dwellings. Through the inclusion of the relevant portions of the table directly in the IRC instead of through a reference to the IECC, the provisions are easier to understand, implement and enforce. The SHGC requirement terminating in locations with 3,500 HDD or more coordinates with the requirements for new fenestration in Section N1102.2 and also recognizes the benefits of solar heat gain in the cooler climates.

SECTION N1103
MECHANICAL SYSTEMS

[E] N1103.1 Heating and air conditioning appliance and equipment performance. Performance of equipment listed in Table N1103.1 is covered by preemptive Federal law. Appliances and equipment not listed in Table N1103.1 shall meet the minimum efficiency requirements of Section 503.2 of the *International Energy Conservation Code*.

❖ The building mechanical systems provisions of the code focus on the heating and/or cooling systems for residential construction. The provisions cover requirements for equipment efficiency, controls, and duct insulation and sealing. Table 1103.1 includes the most common types of heating and cooling equipment and their applicable equipment efficiencies. Section 6 of the ASHRAE/IESNA *Energy Standard for Buildings Except Low-Rise Residential Buildings* (ASHRAE 90.1), referenced in Chapter 7 of the *International Energy Conservation Code*, contains the efficiency requirements for heating and cooling systems not covered in this table. The provisions within this code assume that each dwelling will have individual HVAC system or systems. This may include one system to serve the upstairs and one to serve the downstairs of a two-story residence.

N1103.2 Controls. At least one thermostat shall be provided for each separate heating, cooling, or combination heating and cooling system. Heat pumps shall have controls that prevent supplementary electric resistance heater operation when the heating load can be met by the heat pump alone. Supplementary heater operation shall be permitted during outdoor coil defrost cycles not exceeding 15 minutes.

❖ This provision assures that a separate thermostat is provided for each system. For example, if separate systems are installed so that one serves the downstairs and one serves the upstairs of a two-story residence, then two separate thermostats would be required. This allows for greater control and energy savings than would be possible if both systems were connected and controlled by a single thermostat. Heat pump systems are required to have thermostats that prevent supplementary electric resistance heater operation when the heating load can be met by the heat pump alone. Typically these thermostats will have a "Heat Pump" designation on them. The make and model of the thermostat should be called out on the building plans so the inspector can verify that what is installed in the field matches the plans. Because change-outs in the field are common, the instructions that come with the thermostat can also be checked to verify that the thermostat is designed for use with a heat pump.

N1103.3 Duct insulation. All portions of the air distribution system shall be installed in accordance with Section M1601 and be insulated to an installed R-5 when system components are located within the building but outside of conditioned space, and R-8 when located outside of the building. When located within a building envelope assembly, at least R-8 shall be applied between the duct and that portion of the assembly furthest from conditioned space.

Exception: Exhaust air ducts and portions of the air distribution system within appliances or equipment.

❖ All ductwork located outside conditioned space must be insulated. This includes both supply and return ducts. Ductwork located in an unconditioned space, a vented attic or crawlspace for example, must be insulated to *R*-5. Ductwork located outside the building structure must be insulated to *R*-8. An example of this would be a duct running from a rooftop heating/cooling unit across the roof and down in the residence. Ducts located in conditioned space are not required to be insulated. Ducts located in exterior wall stud cavities must have *R*-8 insulation installed between the duct and the portion of the assembly farthest from the conditioned space. The duct itself is not required to be insulated since one side of it will face conditioned space. Because of limited space, rigid board insulation between the duct and the exterior will probably be necessary to achieve an *R*-value of 8.

N1103.4 Duct sealing. All ducts shall be sealed in accordance with Section M1601.3.1.

❖ Ducts must be sealed in accordance with the duct sealing requirements of Section M1601.3.1. Joints

ENERGY EFFICIENCY

and seams that fail in a duct system can result in increased energy use because the conditioned air will be delivered to the crawlspace or attic instead of the inside the conditioned envelope. Return-air ductwork installed in basements or other concealed building spaces may conduct chemicals or other products that produce potentially harmful fumes. Because the return air operates under negative pressure, any leaks could draw fumes, moisture, soil gases, or odors from the surrounding area and direct them into the house.

[E] N1103.5 Piping insulation. All mechanical system piping shall be insulated in accordance with Table N1103.5.

Exceptions: Piping installed within appliances and equipment or piping serving fluids between 55°F (13°C) and 120°F (49°C).

❖ This section pertains to piping for heating, ventilating and air-conditioning (HVAC) systems, in particular piping for hot-water hydronic systems and refrigerant suction lines for cooling systems. The values presented in this table typically apply to piping installed in both conditioned and unconditioned spaces. Based on Note "b" of the table, piping directly exposed to outdoor air, piping that runs from a condenser unit to the house for example, must have an additional $1/2$ inch (12.7 mm) of insulation above the level presented in Table 1103.5.

TABLE N1103.1
MINIMUM EQUIPMENT PERFORMANCE

EQUIPMENT CATEGORY	SUBCATEGORY[e]	REFERENCED STANDARD	MINIMUM PERFORMANCE
Air-cooled heat pumps heating mode < 65,000 Btu/h cooling capacity	Split systems	ARI 210/240	6.8 HSPF[a,b]
	Single package		6.6 HSPF[a,b]
Gas-fired or oil-fired furnace < 225,000 Btu/h	—	DOE 10 CFR Part 430, Subpart B, Appendix N	AFUE 78%[b] E_t 80%[c]
Gas-fired or oil-fired steam and hot-water boilers < 300,000 Btu/h	—	DOE 10 CFR Part 430, Subpart B, Appendix N	AFUE 78%[b,d]
Air-cooled air conditioners and heat pumps cooling mode < 65,000 Btu/h cooling capacity	Split systems	ARI 210/240	10.0 SEER[b]
	Single package		9.7 SEER[b]

For SI: 1 Btu/h = 0.2931 W.

a. For multicapacity equipment, the minimum performance shall apply to each capacity step provided. Multicapacity refers to manufacturer-published ratings for more than one capacity mode allowed by the product's controls.
b. This is used to be consistent with the National Appliance Energy Conservation Act (NAECA) of 1987 (Public Law 100-12).
c. These requirements apply to combination units not covered by NAECA (three-phase power or cooling capacity 65,000 Btu/h).
d. Except for gas-fired steam boilers, for which the minimum AFUE shall be 75 percent.
e. Seasonal rating.

❖ See the commentary to Section N1103.1.

TABLE N1103.5
MINIMUM HVAC PIPING INSULATION THICKNESSES[a]

PIPING SYSTEM TYPES	FLUID TEMP RANGE (°F)	INSULATION THICKNESS inches[b]
Heating systems		
Low pressure/temperature	201-250	1.5
Low temperature	120-200	1.0
Steam condensate (for feed water)	Any	1.5
Cooling systems		
Chilled water, refrigerant or brine	40-55	0.75
	Below 40	1.25

For SI: 1 inch = 25.4 mm, °C = [(°F)-32]/1.8.

a. The pipe insulation thicknesses specified in this table are based on insulation R-values ranging from R-4 to R-4.6 per inch of thickness. For materials with an R-value greater than R-4.6, the insulation thickness specified in this table may be reduced as follows:

$$\text{New Minimum Thickness} = \frac{4.6 \times \text{Table Thickness}}{\text{Actual } R\text{-Value}}$$

For materials with an R-value less than R-4, the minimum insulation thickness shall be increased as follows:

$$\text{New Minimum Thickness} = \frac{4.0 \times \text{Table Thickness}}{\text{Actual } R\text{-Value}}$$

b. For piping exposed to outdoor air, increase thickness by 0.5 inch.

❖ See the commentary for Section N1103.5. Note "a" provides a method to modify the required thickness of the insulation when the R-value of the material varies from what was used to establish the table. If the R-value of the insulation exceeds 4.6 or is less than 4, the calculations adjust the amount of required insulation. This calculation is required when the R-value is less than 4, and it is permitted, but not required, when the insulation R-value exceeds 4.6.

SECTION N1104
SERVICE WATER HEATING

[E] N1104.1 Water heating appliance and equipment performance. Performance of equipment listed in Table N1104.1 is covered by preemptive Federal law. Appliances and equipment not listed in Table N1104.1 shall meet the minimum efficiency requirements of Section 504.2 of the *International Energy Conservation Code*.

❖ Table N1104.1 presents the minimum required efficiencies for service water-heating equipment. This equipment must meet the minimum thermal efficiency and maximum standby loss requirements. As it does with HVAC equipment efficiency, the National Appliance Energy Conservation Act (NAECA) sets federally mandated new water heater equipment efficiency minimums for residential-sized units. Manufacturers of these service water heaters can only produce equipment meeting these NAECA requirements. No requirements exist for additional water heater insulation (such as water heater insulation blankets) beyond that found in commercially available water heaters.

TABLE N1104.1
REQUIRED PERFORMANCE OF DOMESTIC HOT WATER HEATING EQUIPMENT SUBJECT TO MINIMUM FEDERAL STANDARDS

CATEGORY	MAXIMUM INPUT RATING	MINIMUM EFFICIENCY
Electric; storage or instantaneous	12 kW	$0.93 - 0.00132 \times V$ [a]
Gas; storage	75,000 Btu/h	$0.62 - 0.0019 \times V$ [a]
Gas; instantaneous	200,000 Btu/h	$0.62 - 0.0019 \times V$ [a]
Oil; storage	105,000 Btu/h	$0.59 - 0.0019 \times V$ [a]
Oil; instantaneous	210,000 Btu/h	$0.59 - 0.0019 \times V$ [a]

For SI: 1 Btu/h = 0.2931 W, 1 gallon = 3.785 L.

a. V is the rated storage volume in gallons as specified by the manufacturer.

❖ See the commentary to Section N1104.1.

ENERGY EFFICIENCY

TABLE N1101.2

TABLE N1101.2
CLIMATE ZONES BY STATES AND COUNTIES

County	Zone	County	Zone	County	Zone	County	Zone	County	Zone
ALABAMA		Aleutian Region	17	Desha[a]	6	Lake	8	Jefferson	13
Autauga[a]	6	Aleutians East	17	Drew[a]	6	Lassen	13	Kiowa	13
Baldwin[a]	4	Anchorage	17	Faulkner[a]	7	Los Angeles	4	Kit Carson	13
Barbour[a]	5	Annette Island	15	Franklin	8	Madera	6	La Plata	15
Bibb[a]	6	Bering Straits	19	Fulton	8	Marin	6	Lake	17
Blount	7	Bristol Bay	17	Garland[a]	7	Mariposa	8	Larimer	13
Bullock[a]	5	Chatham	16	Grant[a]	6	Mendocino	8	Las Animas	11
Butler[a]	5	Chugach	17	Greene	8	Merced	6	Lincoln	13
Calhoun[a]	6	Copper River	18	Hempstead[a]	7	Modoc	15	Logan	13
Chambers[a]	6	Delta/Greely	18	Hot Spring[a]	7	Mono	15	Mesa	13
Cherokee	7	Denali	18	Howard[a]	7	Monterey	6	Mineral	17
Chilton[a]	6	Fairbanks N. Star	18	Independence	8	Napa	6	Moffat	15
Choctaw[a]	5	Haines	16	Izard	8	Nevada	11	Montezuma	15
Clarke[a]	5	Iditarod Area	19	Jackson	8	Orange	4	Montrose	13
Clay	7	Juneau	16	Jefferson[a]	6	Placer	8	Morgan	13
Cleburne	7	Kashunamiut	18	Johnson	8	Plumas	13	Otero	11
Coffee[a]	4	Kenai Peninsula	17	Lafayette[a]	6	Riverside	4	Ouray	15
Colbert	8	Ketchikan Gateway	15	Lawrence	8	Sacramento	6	Park	17
Conecuh[a]	5	Kodiak Island	16	Lee[a]	7	San Benito	6	Phillips	13
Coosa[a]	6	Kuspuk	18	Lincoln	6	San Bernardino	4	Pitkin	17
Covington[a]	4	Lake & Peninsula	17	Little River[a]	6	San Diego	3	Prowers	11
Crenshaw[a]	5	Lower Kuskokwim	18	Logan[a]	6	San Francisco	6	Pueblo	11
Cullman	7	Lower Yukon	18	Lonoke[a]	7	San Joaquin	6	Rio Blanco	15
Dale[a]	4	Matanuska-Susitna	17	Madison	9	San Luis Obispo	6	Rio Grande	17
Dallas[a]	5	North Slope	19	Marion	9	San Mateo	6	Routt	17
De Kalb	8	Northwest Arctic	19	Miller[a]	6	Santa Barbara	5	Saguache	16
Elmore[a]	6	Pribilof Islands	17	Mississippi	8	Santa Clara	6	San Juan	17
Escambia[a]	4	Sitka	15	Monroe[a]	7	Santa Cruz	6	San Miguel	15
Etowah	7	Southeast Island	15	Montgomery	8	Shasta	6	Sedgwick	13
Fayette	7	Southwest Region	17	Nevada[a]	6	Sierra	11	Summit	17
Franklin	8	Yakutat	17	Newton	9	Siskiyou	11	Teller	13
Geneva[a]	4	Yukon Flats	19	Ouachita[a]	6	Solano	6	Washington	13
Greene[a]	5	Yukon-Koyukuk	19	Perry[a]	7	Sonoma	6	Weld	13
Hale[a]	5	Yupiit	18	Phillips[a]	7	Stanislaus	6	Yuma	13
Henry[a]	4			Pike[a]	7	Sutter	6		
Houston[a]	4	**ARIZONA**		Poinsett	8	Tehama	6	**CONNECTICUT**	
Jackson	8	Apache	13	Polk	8	Trinity	9	Fairfield	12
Jefferson[a]	6	Cochise	6	Pope	8	Tulare	6	Hartford	13
Lamar	7	Coconino	14	Prairie[a]	7	Tuolume	8	Litchfield	14
Lauderdale	8	Gila	8	Pulaski[a]	7	Ventura	4	Middlesex	12
Lawrence	8	Graham	6	Randolph	8	Yolo	6	New Haven	12
Lee[a]	6	Greenlee	6	Saline[a]	7	Yuba	6	New London	12
Limestone	8	La Paz	3	Scott[a]	7			Tolland	14
Lowndes[a]	5	Maricopa	3	Searcy	9	**COLORADO**		Windham	14
Macon[a]	6	Mohave	7	Sebastian	8	Adams	13		
Madison	8	Navajo	10	Sevier[a]	7	Alamosa	16	**DELAWARE**	
Marengo[a]	5	Pima	4	Sharp	8	Arapahoe	13	Kent	9
Marion	7	Pinal	4	St Francis[a]	7	Archuleta	16	New Castle	10
Marshall	8	Santa Cruz	6	Stone	9	Baca	11	Sussex	9
Mobile[a]	4	Yavapai	10	Union[a]	6	Bent	11		
Monroe[a]	5	Yuma	3	Van Buren	8	Boulder	13	**DC**	
Montgomery[a]	6			Washington	9	Chaffee	16	Washington	10
Morgan	8	**ARKANSAS**		White[a]	7	Cheyenne	13		
Perry[a]	5	Arkansas[a]	6	Woodruff[a]	7	Clear Creek	17	**FLORIDA**	
Pickens[a]	6	Ashley[a]	6	Yell[a]	7	Conejos	16	Alachua[a]	3
Pike[a]	5	Baxter	9			Costilla	16	Baker[a]	3
Randolph	7	Benton	9	**CALIFORNIA**		Crowley	11	Bay[a]	4
Russell[a]	5	Boone	9	Alameda	6	Custer	16	Bradford[a]	3
Shelby[a]	6	Bradley[a]	6	Alpine	15	Delta	13	Brevard[a]	2
St Clair[a]	6	Calhoun[a]	6	Amador	8	Denver	13	Broward[a]	1
Sumter[a]	5	Carroll	9	Butte	6	Dolores	15	Calhoun[a]	4
Talladega[a]	6	Chicot[a]	6	Calaveras	8	Douglas	13	Charlotte[a]	2
Tallapoosa[a]	6	Clark[a]	6	Colusa	6	Eagle	15	Citrus[a]	2
Tuscaloosa[a]	6	Clay	8	Contra Costa	6	El Paso	13	Clay[a]	3
Walker[a]	6	Cleburne	8	Del Norte	9	Elbert	13	Collier[a]	1
Washington[a]	5	Cleveland[a]	6	El Dorado	8	Fremont	11	Columbia[a]	3
Wilcox[a]	5	Columbia[a]	6	Fresno	6	Garfield	15	Dade[a]	1
Winston	7	Conway[a]	7	Glenn	6	Gilpin	13	De Soto[a]	2
		Craighead	8	Humboldt	9	Grand	17	Dixie[a]	3
ALASKA BUROUGHS		Crawford	8	Imperial	3	Gunnison	17	Duval[a]	3
AND REAA's		Crittenden[a]	7	Inyo	9	Hinsdale	17	Escambia[a]	4
Adak Region	16	Cross[a]	7	Kern	5	Huerfano	11	Flagler[a]	3
Alaska Gateway	19	Dallas[a]	6	Kings	6	Jackson	17	Franklin[a]	4
								Gadsden[a]	4

(continued)

TABLE N1101.2—continued
CLIMATE ZONES BY STATES AND COUNTIES

County	Zone	County	Zone	County	Zone	County	Zone	County	Zone
Gilchrist[a]	3	Chatham[a]	4	Mcintosh[a]	4	Bannock	15	Gallatin	10
Glades[a]	1	Chattahoochee[a]	5	Meriwether[a]	6	Bear Lake	15	Greene	11
Gulf[a]	4	Chattooga	8	Miller[a]	4	Benewah	14	Grundy	13
Hamilton[a]	3	Cherokee	8	Mitchell[a]	4	Bingham	15	Hamilton	10
Hardee[a]	2	Clarke	7	Monroe[a]	6	Blaine	15	Hancock	13
Hendry[a]	1	Clay[a]	5	Montgomery[a]	5	Boise	15	Hardin	10
Hernando[a]	2	Clayton	7	Morgan[a]	6	Bonner	15	Henderson	13
Highlands[a]	2	Clinch[a]	4	Murray	8	Bonneville	15	Henry	13
Hillsborough[a]	2	Cobb	7	Muscogee[a]	5	Boundary	15	Iroquois	13
Holmes[a]	4	Coffee[a]	5	Newton	7	Butte	16	Jackson	10
Indian River[a]	2	Colquitt[a]	4	Oconee	7	Camas	15	Jasper	11
Jackson[a]	4	Columbia[a]	6	Oglethorpe	7	Canyon	12	Jefferson	11
Jefferson[a]	4	Cook[a]	4	Paulding	7	Caribou	15	Jersey	10
Lafayette[a]	3	Coweta	7	Peach[a]	5	Cassia	14	Jo Daviess	14
Lake[a]	2	Crawford[a]	5	Pickens	8	Clark	15	Johnson	10
Lee[a]	1	Crisp[a]	5	Pierce[a]	4	Clearwater	12	Kane	14
Leon[a]	4	Dade	8	Pike[a]	6	Custer	16	Kankakee	13
Levy[a]	2	Dawson	8	Polk	7	Elmore	13	Kendall	13
Liberty[a]	4	De Kalb	7	Pulaski[a]	5	Franklin	15	Knox	13
Madison[a]	3	Decatur[a]	4	Putnam[a]	6	Fremont	16	La Salle	13
Manatee[a]	2	Dodge[a]	5	Quitman[a]	5	Gem	13	Lake	14
Marion[a]	2	Dooly[a]	5	Rabun	8	Gooding	13	Lawrence	11
Martin[a]	1	Dougherty[a]	5	Randolph[a]	5	Idaho	15	Lee	14
Monroe[a]	1	Douglas	7	Richmond[a]	6	Jefferson	16	Livingston	13
Nassau[a]	3	Early[a]	5	Rockdale	7	Jerome	14	Logan	12
Okaloosa[a]	4	Echols[a]	4	Schley[a]	5	Kootenai	14	Macon	12
Okeechobee[a]	2	Effingham[a]	4	Screven[a]	5	Latah	14	Macoupin	11
Orange[a]	2	Elbert	7	Seminole[a]	4	Lemhi	15	Madison	10
Osceola[a]	2	Emanuel[a]	5	Spalding	7	Lewis	15	Marion	11
Palm Beach[a]	1	Evans[a]	4	Stephens	7	Lincoln	15	Marshall	13
Pasco[a]	2	Fannin	8	Stewart[a]	5	Madison	16	Mason	12
Pinellas[a]	2	Fayette	7	Sumter[a]	5	Minidoka	15	Massac	10
Polk[a]	2	Floyd	7	Talbot[a]	5	Nez Perce	12	Mcdonough	13
Putnam[a]	3	Forsyth	8	Taliaferro[a]	6	Oneida	15	Mchenry	14
Santa Rosa[a]	4	Franklin	7	Tattnall[a]	4	Owyhee	12	Mclean	12
Sarasota[a]	2	Fulton	7	Taylor[a]	5	Payette	12	Menard	12
Seminole[a]	2	Gilmer	8	Telfair[a]	5	Power	15	Mercer	13
St Johns[a]	3	Glascock[a]	6	Terrell[a]	5	Shoshone	14	Monroe	10
St Lucie[a]	2	Glynn[a]	4	Thomas[a]	4	Teton	16	Montgomery	11
Sumter[a]	2	Gordon	8	Tift[a]	5	Twin Falls	14	Morgan	12
Suwannee[a]	3	Grady[a]	4	Toombs[a]	4	Valley	16	Moultrie	12
Taylor[a]	3	Greene[a]	6	Towns	8	Washington	13	Ogle	14
Union[a]	3	Gwinnett	7	Treutlen[a]	5			Peoria	13
Volusia[a]	2	Habersham	8	Troup[a]	6	**ILLINOIS**		Perry	10
Wakulla[a]	4	Hall	7	Turner[a]	5	Adams	12	Piatt	12
Walton[a]	4	Hancock[a]	6	Twiggs[a]	5	Alexander	10	Pike	12
Washington[a]	4	Haralson	7	Union	8	Bond	11	Pope	10
		Harris[a]	6	Upson[a]	5	Boone	14	Pulaski	10
GEORGIA		Hart	7	Walker	8	Brown	12	Putnam	13
Appling[a]	4	Heard[a]	6	Walton	7	Bureau	13	Randolph	10
Atkinson[a]	4	Henry	7	Ware[a]	4	Calhoun	11	Richland	11
Bacon[a]	4	Houston[a]	5	Warren[a]	6	Carroll	14	Rock Island	13
Baker[a]	4	Irwin[a]	5	Washington[a]	6	Cass	12	Saline	10
Baldwin[a]	6	Jackson	7	Wayne[a]	4	Champaign	12	Sangamon	12
Banks	7	Jasper[a]	6	Webster[a]	5	Christian	11	Schuyler	12
Barrow	7	Jeff Davis[a]	4	Wheeler[a]	5	Clark	12	Scott	12
Bartow	7	Jefferson[a]	6	White	8	Clay	11	Shelby	11
Ben Hill[a]	5	Jenkins[a]	5	Whitfield	8	Clinton	10	St Clair	10
Berrien[a]	4	Johnson[a]	5	Wilcox[a]	5	Coles	12	Stark	13
Bibb[a]	5	Jones[a]	6	Wilkes	7	Cook	14	Stephenson	14
Bleckley[a]	5	Lamar[a]	6	Wilkinson[a]	5	Crawford	11	Tazewell	12
Brantley[a]	4	Lanier[a]	4	Worth[a]	5	Cumberland	12	Union	10
Brooks[a]	4	Laurens[a]	5			De Kalb	14	Vermilion	12
Bryan[a]	4	Lee[a]	5	**HAWAII**		De Witt	12	Wabash	11
Bulloch[a]	5	Liberty[a]	4	Hawaii	1	Douglas	12	Warren	13
Burke[a]	6	Lincoln[a]	6	Honolulu	1	Du Page	14	Washington	10
Butts[a]	7	Long[a]	4	Kalawao	1	Edgar	12	Wayne	11
Calhoun[a]	5	Lowndes[a]	4	Kauai	1	Edwards	11	White	10
Camden[a]	4	Lumpkin	8	Maui	1	Effingham	11	Whiteside	14
Candler[a]	5	Macon[a]	5			Fayette	11	Will	13
Carroll	7	Madison	7	**IDAHO**		Ford	13	Williamson	10
Catoosa	8	Marion[a]	5	Ada	12	Franklin	10	Winnebago	14
Charlton[a]	4	Mcduffie[a]	6	Adams	15	Fulton	13	Woodford	13

(continued)

ENERGY EFFICIENCY

TABLE N1101.2

TABLE N1101.2—continued
CLIMATE ZONES BY STATES AND COUNTIES

County	Zone	County	Zone	County	Zone	County	Zone	County	Zone
INDIANA		Spencer	10	Johnson	13	Edwards	11	Thomas	13
Adams	13	St Joseph	13	Jones	14	Elk	9	Trego	12
Allen	13	Starke	13	Keokuk	13	Ellis	12	Wabaunsee	11
Bartholomew	11	Steuben	14	Kossuth	15	Ellsworth	11	Wallace	12
Benton	13	Sullivan	11	Lee	13	Finney	11	Washington	12
Blackford	13	Switzerland	10	Linn	14	Ford	11	Wichita	12
Boone	12	Tippecanoe	13	Louisa	13	Franklin	10	Wilson	9
Brown	11	Tipton	13	Lucas	13	Geary	11	Woodson	10
Carroll	13	Union	12	Lyon	15	Gove	12	Wyandotte	11
Cass	13	Vanderburgh	10	Madison	14	Graham	12		
Clark	10	Vermillion	12	Mahaska	13	Grant	11	**KENTUCKY**	
Clay	12	Vigo	12	Marion	13	Gray	11	Adair	9
Clinton	13	Wabash	14	Marshall	14	Greeley	12	Allen	9
Crawford	11	Warren	12	Mills	13	Greenwood	10	Anderson	10
Daviess	11	Warrick	10	Mitchell	15	Hamilton	11	Ballard	9
De Kalb	13	Washington	11	Monona	14	Harper	9	Barren	9
Dearborn	11	Wayne	12	Monroe	13	Harvey	11	Bath	11
Decatur	12	Wells	13	Montgomery	13	Haskell	11	Bell	10
Delaware	13	White	13	Muscatine	13	Hodgeman	11	Boone	11
Dubois	11	Whitley	14	Obrien	15	Jackson	11	Bourbon	10
Elkhart	13			Osceola	15	Jefferson	11	Boyd	11
Fayette	12	**IOWA**		Page	13	Jewell	12	Boyle	10
Floyd	10	Adair	14	Palo Alto	15	Johnson	11	Bracken	11
Fountain	12	Adams	13	Plymouth	15	Kearny	11	Breathitt	10
Franklin	12	Allamakee	15	Pocahontas	15	Kingman	10	Breckenridge	9
Fulton	14	Appanoose	13	Polk	14	Kiowa	10	Bullitt	10
Gibson	10	Audubon	14	Pottawattamie	14	Labette	9	Butler	9
Grant	13	Benton	14	Poweshiek	14	Lane	12	Caldwell	9
Greene	11	Black Hawk	15	Ringgold	13	Leavenworth	11	Calloway	9
Hamilton	12	Boone	14	Sac	15	Lincoln	11	Campbell	11
Hancock	12	Bremer	15	Scott	13	Linn	10	Carlisle	9
Harrison	10	Buchanan	15	Shelby	14	Logan	12	Carroll	10
Hendricks	12	Buena Vista	15	Sioux	15	Lyon	11	Carter	11
Henry	12	Butler	15	Story	14	Marion	11	Casey	10
Howard	13	Calhoun	15	Tama	14	Marshall	12	Christian	9
Huntington	14	Carroll	14	Taylor	13	Mcpherson	11	Clark	10
Jackson	11	Cass	14	Union	13	Meade	10	Clay	10
Jasper	13	Cedar	14	Van Buren	13	Miami	10	Clinton	10
Jay	13	Cerro Gordo	15	Wapello	13	Mitchell	12	Crittenden	9
Jefferson	10	Cherokee	15	Warren	14	Montgomery	9	Cumberland	9
Jennings	11	Chickasaw	15	Washington	13	Morris	11	Daviess	9
Johnson	12	Clarke	13	Wayne	13	Morton	10	Edmonson	9
Knox	11	Clay	15	Webster	15	Nemaha	11	Elliot	11
Kosciusko	14	Clayton	15	Winnebago	15	Neosho	9	Estill	10
La Porte	13	Clinton	13	Winneshiek	15	Ness	12	Fayette	10
Lagrange	14	Crawford	14	Woodbury	15	Norton	13	Fleming	11
Lake	13	Dallas	14	Worth	15	Osage	10	Floyd	10
Lawrence	11	Davis	13	Wright	15	Osborne	12	Franklin	10
Madison	13	Decatur	13			Ottawa	11	Fulton	9
Marion	12	Delaware	15	**KANSAS**		Pawnee	11	Gallatin	11
Marshall	13	Des Moines	13	Allen	10	Phillips	12	Garrard	10
Martin	11	Dickinson	15	Anderson	10	Pottawatomie	11	Grant	11
Miami	14	Dubuque	14	Atchison	11	Pratt	10	Graves	9
Monroe	11	Emmet	15	Barber	9	Rawlins	13	Grayson	9
Montgomery	12	Fayette	15	Barton	11	Reno	11	Green	9
Morgan	12	Floyd	15	Bourbon	10	Republic	12	Greenup	11
Newton	13	Franklin	15	Brown	11	Rice	11	Hancock	9
Noble	14	Fremont	13	Butler	10	Riley	11	Hardin	9
Ohio	11	Greene	14	Chase	10	Rooks	12	Harlan	10
Orange	11	Grundy	15	Chautauqua	9	Rush	11	Harrison	11
Owen	12	Guthrie	14	Cherokee	9	Russell	11	Hart	9
Parke	12	Hamilton	15	Cheyenne	13	Saline	11	Henderson	9
Perry	10	Hancock	15	Clark	10	Scott	12	Henry	10
Pike	11	Hardin	15	Clay	11	Sedgwick	10	Hickman	9
Porter	13	Harrison	14	Cloud	12	Seward	10	Hopkins	9
Posey	10	Henry	13	Coffey	10	Shawnee	11	Jackson	10
Pulaski	13	Howard	15	Comanche	9	Sheridan	12	Jefferson	10
Putnam	12	Humboldt	15	Cowley	9	Sherman	13	Jessamine	10
Randolph	13	Ida	15	Crawford	9	Smith	12	Johnson	11
Ripley	11	Iowa	14	Decatur	13	Stafford	11	Kenton	11
Rush	12	Jackson	14	Dickinson	11	Stanton	11	Knott	10
Scott	11	Jasper	14	Doniphan	11	Stevens	10	Knox	10
Shelby	12	Jefferson	13	Douglas	10	Sumner	9	Larue	9

(continued)

TABLE N1101.2—continued
CLIMATE ZONES BY STATES AND COUNTIES

County	Zone	County	Zone	County	Zone	County	Zone	County	Zone
Laurel	10	Claiborne[a]	6	Baltimore	10	Ionia	14	Faribault	15
Lawrence	11	Concordia[a]	5	Baltimore City	9	Iosco	15	Fillmore	15
Lee	10	De Soto[a]	5	Calvert	9	Iron	17	Freeborn	15
Leslie	10	East Baton Rouge[a]	4	Caroline	9	Isabella	15	Goodhue	15
Letcher	10	East Carroll[a]	6	Carroll	11	Jackson	14	Grant	16
Lewis	11	East Feliciana[a]	4	Cecil	10	Kalamazoo	14	Hennepin	15
Lincoln	10	Evangeline[a]	4	Charles	9	Kalkaska	15	Houston	15
Livingston	9	Franklin[a]	6	Dorchester	9	Kent	14	Hubbard	17
Logan	9	Grant[a]	5	Frederick	11	Keweenaw	17	Isanti	16
Lyon	9	Iberia[a]	4	Garrett	13	Lake	15	Itasca	17
Madison	10	Iberville[a]	4	Harford	10	Lapeer	14	Jackson	15
Magoffin	10	Jackson[a]	6	Howard	10	Leelanau	15	Kanabec	16
Marion	10	Jefferson[a]	3	Kent	10	Lenawee	14	Kandiyohi	16
Marshall	9	Jefferson Davis[a]	4	Montgomery	10	Livingston	14	Kittson	17
Martin	11	La Salle[a]	5	Prince Georges	10	Luce	16	Koochiching	17
Mason	11	Lafayette[a]	4	Queen Annes	9	Mackinac	16	Lac Qui Parle	15
Mccracken	9	Lafourche[a]	3	Somerset	9	Macomb	14	Lake	17
Mccreary	10	Lincoln[a]	6	St Marys	9	Manistee	15	Lake Of The Woods	17
Mclean	9	Livingston[a]	4	Talbot	9	Marquette	16	Le Sueur	15
Meade	9	Madison[a]	6	Washington	11	Mason	15	Lincoln	15
Menifee	10	Morehouse[a]	6	Wicomico	9	Mecosta	15	Lyon	15
Mercer	10	Natchitoches[a]	5	Worcester	9	Menominee	16	Mahnomen	17
Metcalfe	9	Orleans[a]	3			Midland	15	Marshall	17
Monroe	9	Ouachita[a]	6	**MASSACHUSETTS**		Missaukee	15	Martin	15
Montgomery	10	Plaquemines[a]	3	Barnstable	12	Monroe	13	Mcleod	15
Morgan	10	Pointe Coupee[a]	4	Berkshire	14	Montcalm	14	Meeker	16
Muhlenberg	9	Rapides[a]	5	Bristol	12	Montmorency	15	Mille Lacs	16
Nelson	10	Red River[a]	5	Dukes	12	Muskegon	14	Morrison	16
Nicholas	11	Richland[a]	6	Essex	13	Newaygo	15	Mower	15
Ohio	9	Sabine[a]	5	Franklin	14	Oakland	14	Murray	15
Oldham	10	St Bernard[a]	3	Hampden	14	Oceana	15	Nicollet	15
Owen	10	St Charles[a]	3	Hampshire	14	Ogemaw	15	Nobles	15
Owsley	10	St Helena[a]	4	Middlesex	13	Ontonagon	17	Norman	17
Pendleton	11	St James[a]	3	Nantucket	12	Osceola	15	Olmsted	15
Perry	10	St John The Baptist[a]	3	Norfolk	13	Oscoda	15	Otter Tail	17
Pike	10	St Landry[a]	4	Plymouth	12	Otsego	15	Pennington	17
Powell	10	St Martin[a]	4	Suffolk	13	Ottawa	14	Pine	16
Pulaski	10	St Mary[a]	3	Worcester	14	Presque Isle	15	Pipestone	15
Robertson	11	St Tammany[a]	4			Roscommon	15	Polk	17
Rockcastle	10	Tangipahoa[a]	4	**MICHIGAN**		Saginaw	14	Pope	16
Rowan	11	Tensas[a]	5	Alcona	15	Sanilac	14	Ramsey	15
Russell	10	Terrebonne[a]	3	Alger	16	Schoolcraft	16	Red Lake	17
Scott	11	Union[a]	6	Allegan	14	Shiawassee	14	Redwood	15
Shelby	10	Vermilion[a]	4	Alpena	15	St Clair	14	Renville	15
Simpson	9	Vernon[a]	5	Antrim	15	St Joseph	14	Rice	15
Spencer	10	Washington[a]	4	Arenac	15	Tuscola	14	Rock	15
Taylor	9	Webster[a]	6	Baraga	17	Van Buren	14	Roseau	17
Todd	9	West Baton Rouge[a]	4	Barry	14	Washtenaw	13	Scott	15
Trigg	9	West Carroll[a]	6	Bay	15	Wayne	13	Sherburne	16
Trimble	10	West Feliciana[a]	4	Benzie	15	Wexford	15	Sibley	15
Union	9	Winn[a]	5	Berrien	14			St Louis	17
Warren	9			Branch	14	**MINNESOTA**		Stearns	16
Washington	10	**MAINE**		Calhoun	14	Aitkin	17	Steele	15
Wayne	10	Androscoggin	15	Cass	14	Anoka	16	Stevens	16
Webster	9	Aroostook	17	Charlevoix	15	Becker	17	Swift	16
Whitley	10	Cumberland	15	Cheboygan	15	Beltrami	17	Todd	16
Wolfe	10	Franklin	16	Chippewa	16	Benton	16	Traverse	16
Woodford	10	Hancock	15	Clare	15	Big Stone	16	Wabasha	15
		Kennebec	15	Clinton	14	Blue Earth	15	Wadena	17
LOUISIANA		Knox	15	Crawford	15	Brown	15	Waseca	15
Acadia[a]	4	Lincoln	15	Delta	16	Carlton	17	Washington	15
Allen[a]	4	Oxford	16	Dickinson	16	Carver	15	Watonwan	15
Ascension[a]	4	Penobscot	15	Eaton	14	Cass	17	Wilkin	17
Assumption[a]	3	Piscataquis	17	Emmet	15	Chippewa	16	Winona	15
Avoyelles[a]	5	Sagadahoc	15	Genesee	14	Chisago	16	Wright	16
Beauregard[a]	4	Somerset	17	Gladwin	15	Clay	17	Yellow Medicine	15
Bienville[a]	6	Waldo	15	Gogebic	17	Clearwater	17		
Bossier[a]	6	Washington	15	Grand Traverse	15	Cook	17	**MISSISSIPPI**	
Caddo[a]	6	York	15	Gratiot	14	Cottonwood	15	Adams[a]	5
Calcasieu[a]	4			Hillsdale	14	Crow Wing	17	Alcorn	7
Caldwell[a]	6	**MARYLAND**		Houghton	17	Dakota	15	Amite[a]	4
Cameron[a]	4	Allegheny	12	Huron	14	Dodge	15	Attala[a]	6
Catahoula[a]	5	Anne Arundel	9	Ingham	14	Douglas	16	Benton	7

(continued)

ENERGY EFFICIENCY

TABLE N1101.2—continued
CLIMATE ZONES BY STATES AND COUNTIES

County	Zone	County	Zone	County	Zone	County	Zone	County	Zone
Bolivar[a]	6	Wilkinson[a]	4	Moniteau	11	Lake	15	Gosper	13
Calhoun[a]	6	Winston[a]	6	Monroe	12	Lewis And Clark	15	Grant	14
Carroll[a]	6	Yalobusha	7	Montgomery	11	Liberty	16	Greeley	14
Chickasaw[a]	6	Yazoo[a]	6	Morgan	11	Lincoln	15	Hall	13
Choctaw[a]	6			New Madrid	9	Madison	15	Hamilton	13
Claiborne[a]	5	**MISSOURI**		Newton	9	McCone	15	Harlan	13
Clarke[a]	5	Adair	12	Nodaway	13	Meagher	15	Hayes	13
Clay[a]	6	Andrew	12	Oregon	9	Mineral	15	Hitchcock	13
Coahoma	7	Atchison	13	Osage	11	Missoula	15	Holt	14
Copiah[a]	5	Audrain	12	Ozark	9	Musselshell	15	Hooker	14
Covington[a]	5	Barry	9	Pemiscot	9	Park	15	Howard	14
De Soto	7	Barton	10	Perry	10	Petroleum	15	Jefferson	13
Forrest[a]	5	Bates	11	Pettis	11	Phillips	16	Johnson	13
Franklin[a]	5	Benton	11	Phelps	10	Pondera	16	Kearney	13
George[a]	4	Bollinger	10	Pike	12	Powder River	15	Keith	14
Greene[a]	5	Boone	11	Platte	11	Powell	16	Keya Paha	14
Grenada[a]	6	Buchanan	12	Polk	10	Prairie	15	Kimball	14
Hancock[a]	4	Butler	9	Pulaski	10	Ravalli	15	Knox	14
Harrison[a]	4	Caldwell	12	Putnam	13	Richland	15	Lancaster	13
Hinds[a]	6	Callaway	11	Ralls	12	Roosevelt	16	Lincoln	14
Holmes[a]	6	Camden	11	Randolph	12	Rosebud	15	Logan	14
Humphreys[a]	6	Cape Girardeau	9	Ray	11	Sanders	15	Loup	14
Issaquena[a]	6	Carroll	12	Reynolds	10	Sheridan	16	Madison	14
Itawamba	7	Carter	10	Ripley	9	Silver Bow	16	Mcpherson	14
Jackson[a]	4	Cass	11	Saline	11	Stillwater	15	Merrick	13
Jasper[a]	5	Cedar	11	Schuyler	13	Sweet Grass	15	Morrill	14
Jefferson[a]	5	Chariton	12	Scotland	13	Teton	15	Nance	13
Jefferson Davis[a]	5	Christian	10	Scott	9	Toole	16	Nemaha	13
Jones[a]	5	Clark	13	Shannon	10	Treasure	15	Nuckolls	13
Kemper[a]	6	Clay	11	Shelby	12	Valley	16	Otoe	13
Lafayette	7	Clinton	12	St Charles	10	Wheatland	15	Pawnee	13
Lamar[a]	4	Cole	11	St Clair	11	Wibaux	15	Perkins	13
Lauderdale[a]	6	Cooper	11	St Francois	10	Yellowstone	15	Phelps	13
Lawrence[a]	5	Crawford	10	St Louis	10	Yellowstone		Pierce	14
Leake[a]	6	Dade	10	St Louis City	10	National Park	15	Platte	13
Lee	7	Dallas	10	Ste Genevieve	10			Polk	13
Leflore[a]	6	Daviess	12	Stoddard	9	**NEBRASKA**		Red Willow	13
Lincoln[a]	5	De Kalb	12	Stone	9	Adams	13	Richardson	13
Lowndes[a]	6	Dent	10	Sullivan	12	Antelope	14	Rock	14
Madison[a]	6	Douglas	10	Taney	9	Arthur	14	Saline	13
Marion[a]	4	Dunklin	9	Texas	10	Banner	14	Sarpy	13
Marshall	7	Franklin	10	Vernon	11	Blaine	14	Saunders	13
Monroe[a]	6	Gasconade	11	Warren	11	Boone	14	Scotts Bluff	14
Montgomery[a]	6	Gentry	13	Washington	10	Box Butte	15	Seward	13
Neshoba[a]	6	Greene	10	Wayne	10	Boyd	14	Sheridan	15
Newton[a]	6	Grundy	12	Webster	10	Brown	14	Sherman	14
Noxubee[a]	6	Harrison	13	Worth	13	Buffalo	13	Sioux	15
Oktibbeha[a]	6	Henry	11	Wright	10	Burt	14	Stanton	14
Panola	7	Hickory	11			Butler	13	Thayer	13
Pearl River[a]	4	Holt	12	**MONTANA**		Cass	13	Thomas	14
Perry[a]	5	Howard	11	Beaverhead	15	Cedar	14	Thurston	14
Pike[a]	4	Howell	9	Big Horn	15	Chase	13	Valley	14
Pontotoc	7	Iron	10	Blaine	16	Cherry	14	Washington	13
Prentiss	7	Jackson	11	Broadwater	15	Cheyenne	14	Wayne	14
Quitman	7	Jasper	9	Carbon	15	Clay	13	Webster	13
Rankin[a]	6	Jefferson	10	Carter	15	Colfax	13	Wheeler	14
Scott[a]	6	Johnson	11	Cascade	15	Cuming	14	York	13
Sharkey[a]	6	Knox	12	Chouteau	15	Custer	14		
Simpson[a]	5	Laclede	10	Custer	15	Dakota	14	**NEVADA**	
Smith[a]	5	Lafayette	11	Daniels	16	Dawes	15	Carson City	12
Stone[a]	4	Lawrence	10	Dawson	15	Dawson	13	Churchill	12
Sunflower[a]	6	Lewis	12	Deer Lodge	16	Deuel	14	Clark	5
Tallahatchie	7	Lincoln	11	Fallon	15	Dixon	14	Douglas	13
Tate	7	Linn	12	Fergus	15	Dodge	13	Elko	15
Tippah	7	Livingston	12	Flathead	16	Douglas	13	Esmeralda	12
Tishomingo	7	Macon	12	Gallatin	15	Dundy	13	Eureka	15
Tunica	7	Madison	10	Garfield	15	Fillmore	13	Humboldt	13
Union	7	Maries	11	Glacier	16	Franklin	13	Lander	13
Walthall[a]	4	Marion	12	Golden Valley	15	Frontier	13	Lincoln	12
Warren[a]	6	Mcdonald	9	Granite	16	Furnas	13	Lyon	13
Washington[a]	6	Mercer	13	Hill	16	Gage	13	Mineral	12
Wayne[a]	5	Miller	11	Jefferson	15	Garden	14	Nye	12
Webster[a]	6	Mississippi	9	Judith Basin	15	Garfield	14	Pershing	12

(continued)

TABLE N1101.2—continued
CLIMATE ZONES BY STATES AND COUNTIES

County	Zone	County	Zone	County	Zone	County	Zone	County	Zone
Storey	12	**NEW YORK**		Bladen[a]	6	Sampson[a]	6	Williams	17
Washoe	12	Albany	14	Brunswick[a]	6	Scotland	7		
White Pine	15	Allegheny	15	Buncombe	9	Stanly	7	**OHIO**	
		Bronx	11	Burke	8	Stokes	9	Adams	11
NEW HAMPSHIRE		Broome	15	Cabarrus	7	Surry	9	Allen	13
Belknap	15	Cattaraugus	15	Caldwell	8	Swain	9	Ashland	13
Carroll	15	Cayuga	14	Camden	7	Transylvania	9	Ashtabula	13
Cheshire	15	Chautauqua	13	Carteret[a]	6	Tyrrell[a]	6	Athens	11
Coos	16	Chemung	15	Caswell	8	Union	7	Auglaize	13
Grafton	15	Chenango	15	Catawba	8	Vance	8	Belmont	12
Hillsborough	15	Clinton	15	Chatham	8	Wake	7	Brown	11
Merrimack	15	Columbia	13	Cherokee	9	Warren	8	Butler	12
Rockingham	15	Cortland	15	Chowan	7	Washington	7	Carroll	13
Strafford	15	Delaware	15	Clay	9	Watauga	11	Champaign	13
Sullivan	15	Dutchess	13	Cleveland	7	Wayne	7	Clark	13
		Erie	14	Columbus[a]	6	Wilkes	9	Clermont	11
NEW JERSEY		Essex	16	Craven[a]	6	Wilson	7	Clinton	12
Atlantic	10	Franklin	16	Cumberland	7	Yadkin	8	Columbiana	13
Bergen	12	Fulton	15	Currituck	7	Yancey	11	Coshocton	12
Burlington	11	Genesee	14	Dare[a]	6			Crawford	13
Camden	10	Greene	14	Davidson	8	**NORTH DAKOTA**		Cuyahoga	13
Cape May	10	Hamilton	16	Davie	8	Adams	16	Darke	13
Cumberland	10	Herkimer	15	Duplin[a]	6	Barnes	17	Defiance	14
Essex	11	Jefferson	15	Durham	8	Benson	17	Delaware	13
Gloucester	10	Kings	10	Edgecombe	7	Billings	16	Erie	13
Hudson	11	Lewis	15	Forsyth	8	Bottineau	17	Fairfield	12
Hunterdon	12	Livingston	14	Franklin	8	Bowman	16	Fayette	12
Mercer	11	Madison	14	Gaston	7	Burke	17	Franklin	12
Middlesex	11	Monroe	14	Gates	7	Burleigh	16	Fulton	14
Monmouth	11	Montgomery	14	Graham	9	Cass	17	Gallia	11
Morris	12	Nassau	11	Granville	8	Cavalier	17	Geauga	13
Ocean	11	New York	10	Greene	7	Dickey	16	Greene	12
Passaic	12	Niagara	14	Guilford	8	Divide	17	Guernsey	12
Salem	10	Oneida	15	Halifax	7	Dunn	16	Hamilton	11
Somerset	12	Onondaga	14	Harnett	7	Eddy	17	Hancock	13
Sussex	13	Ontario	14	Haywood	9	Emmons	16	Hardin	13
Union	11	Orange	12	Henderson	9	Foster	17	Harrison	13
Warren	12	Orleans	14	Hertford	7	Golden Valley	16	Henry	14
		Oswego	14	Hoke	7	Grand Forks	17	Highland	11
NEW MEXICO		Otsego	15	Hyde[a]	6	Grant	16	Hocking	12
Bernalillo	9	Putnam	12	Iredell	8	Griggs	17	Holmes	13
Catron	11	Queens	10	Jackson	9	Hettinger	16	Huron	13
Chaves	7	Rensselaer	14	Johnston	7	Kidder	17	Jackson	11
Cibola	12	Richmond	11	Jones[a]	6	La Moure	16	Jefferson	13
Colfax	13	Rockland	12	Lee	7	Logan	16	Knox	13
Curry	9	Saratoga	14	Lenoir	7	Mchenry	17	Lake	13
De Baca	9	Schenectady	14	Lincoln	7	Mcintosh	16	Lawrence	11
Dona Ana	7	Schoharie	15	Macon	9	Mckenzie	16	Licking	12
Eddy	7	Schuyler	15	Madison	9	Mclean	17	Logan	13
Grant	9	Seneca	14	Martin	7	Mercer	16	Lorain	13
Guadalupe	9	St Lawrence	15	Mcdowell	8	Morton	16	Lucas	14
Harding	11	Steuben	15	Mecklenburg	7	Mountrail	17	Madison	12
Hidalgo	7	Suffolk	11	Mitchell	11	Nelson	17	Mahoning	13
Lea	7	Sullivan	15	Montgomery	7	Oliver	16	Marion	13
Lincoln	9	Tioga	15	Moore	7	Pembina	17	Medina	13
Los Alamos	13	Tompkins	15	Nash	7	Pierce	17	Meigs	11
Luna	7	Ulster	15	New Hanover[a]	6	Ramsey	17	Mercer	13
Mckinley	13	Warren	15	Northampton	7	Ransom	16	Miami	13
Mora	15	Washington	15	Onslow[a]	6	Renville	17	Monroe	12
Otero	7	Wayne	14	Orange	8	Richland	16	Montgomery	12
Quay	8	Westchester	12	Pamlico[a]	6	Rolette	17	Morgan	12
Rio Arriba	12	Wyoming	14	Pasquotank	7	Sargent	16	Morrow	13
Roosevelt	8	Yates	14	Pender[a]	6	Sheridan	17	Muskingum	12
San Juan	12			Perquimans	7	Sioux	16	Noble	12
San Miguel	12	**NORTH CAROLINA**		Person	8	Slope	16	Ottawa	13
Sandoval	13	Alamance	8	Pitt	7	Stark	16	Paulding	14
Santa Fe	13	Alexander	8	Polk	7	Steele	17	Perry	12
Sierra	8	Alleghany	11	Randolph	8	Stutsman	17	Pickaway	13
Socorro	9	Anson	7	Richmond	7	Towner	17	Pike	11
Taos	15	Ashe	11	Robeson	7	Traill	17	Portage	13
Torrance	11	Avery	11	Rockingham	8	Walsh	17	Preble	12
Union	11	Beaufort[a]	6	Rowan	7	Ward	17	Putnam	13
Valencia	10	Bertie	7	Rutherford	7	Wells	17	Richland	13

(continued)

TABLE N1101.2—continued
CLIMATE ZONES BY STATES AND COUNTIES

County	Zone	County	Zone	County	Zone	County	Zone	County	Zone
Ross	12	Okfuskee	7	Butler	14	Berkeley[a]	5	Hutchinson	14
Sandusky	13	Oklahoma	8	Cambria	13	Calhoun[a]	6	Hyde	15
Scioto	11	Okmulgee	8	Cameron	15	Charleston[a]	5	Jackson	14
Seneca	13	Osage	8	Carbon	13	Cherokee	7	Jerauld	15
Shelby	13	Ottawa	9	Centre	13	Chester	7	Jones	15
Stark	13	Pawnee	8	Chester	11	Chesterfield	7	Kingsbury	15
Summit	13	Payne	8	Clarion	14	Clarendon[a]	6	Lake	15
Trumbull	13	Pittsburg	7	Clearfield	15	Colleton[a]	5	Lawrence	15
Tuscarawas	13	Pontotoc	7	Clinton	13	Darlington[a]	6	Lincoln	15
Union	13	Pottawatomie	7	Columbia	13	Dillon[a]	6	Lyman	15
Van Wert	13	Pushmataha[a]	6	Crawford	14	Dorchester[a]	5	Marshall	16
Vinton	11	Roger Mills	9	Cumberland	12	Edgefield[a]	6	Mccook	15
Warren	12	Rogers	9	Dauphin	12	Fairfield	7	Mcpherson	16
Washington	11	Seminole	7	Delaware	10	Florence[a]	6	Meade	15
Wayne	13	Sequoyah	7	Elk	15	Georgetown[a]	5	Mellette	14
Williams	14	Stephens	7	Erie	14	Greenville	7	Miner	15
Wood	14	Texas	10	Fayette	12	Greenwood	7	Minnehaha	15
Wyandot	13	Tillman	7	Forest	15	Hampton[a]	5	Moody	15
		Tulsa	8	Franklin	11	Horry[a]	5	Pennington	15
OKLAHOMA		Wagoner	8	Fulton	12	Jasper[a]	5	Perkins	15
Adair	8	Washington	9	Greene	12	Kershaw	7	Potter	15
Alfalfa	9	Washita	8	Huntingdon	12	Lancaster	7	Roberts	16
Atoka	7	Woods	9	Indiana	13	Laurens	7	Sanborn	15
Beaver	10	Woodward	9	Jefferson	15	Lee[a]	6	Shannon	15
Beckham	8			Juniata	12	Lexington[a]	6	Spink	15
Blaine	8	**OREGON**		Lackawanna	14	Marion[a]	6	Stanley	15
Bryan	7	Baker	15	Lancaster	11	Marlboro[a]	6	Sully	15
Caddo	8	Benton	10	Lawrence	14	Mccormick[a]	6	Todd	14
Canadian	8	Clackamas	10	Lebanon	12	Newberry[a]	6	Tripp	14
Carter[a]	6	Clatsop	11	Lehigh	12	Oconee	7	Turner	15
Cherokee	8	Columbia	11	Luzerne	13	Orangeburg[a]	6	Union	14
Choctaw[a]	6	Coos	9	Lycoming	13	Pickens	7	Walworth	15
Cimarron	10	Crook	14	Mckean	15	Richland[a]	6	Yankton	14
Cleveland	7	Curry	9	Mercer	14	Saluda[a]	6	Ziebach	15
Coal	7	Deschutes	14	Mifflin	12	Spartanburg	7		
Comanche	7	Douglas	9	Monroe	13	Sumter[a]	6	**TENNESSEE**	
Cotton	7	Gilliam	12	Montgomery	11	Union	7	Anderson	9
Craig	9	Grant	15	Montour	13	Williamsburg[a]	6	Bedford	8
Creek	8	Harney	15	Northampton	12	York	7	Benton	9
Custer	8	Hood River	12	Northumberland	13			Bledsoe	8
Delaware	8	Jackson	11	Perry	12	**SOUTH DAKOTA**		Blount	8
Dewey	9	Jefferson	13	Philadelphia	10	Aurora	15	Bradley	8
Ellis	9	Josephine	9	Pike	13	Beadle	15	Campbell	10
Garfield	8	Klamath	14	Potter	15	Bennett	14	Cannon	9
Garvin	7	Lake	15	Schuylkill	13	Bon Homme	14	Carroll	9
Grady	7	Lane	10	Snyder	13	Brookings	16	Carter	10
Grant	9	Lincoln	11	Somerset	13	Brown	16	Cheatham	9
Greer	7	Linn	10	Sullivan	14	Brule	15	Chester	8
Harmon	7	Malheur	12	Susquehanna	15	Buffalo	15	Claiborne	10
Harper	9	Marion	10	Tioga	15	Butte	15	Clay	9
Haskell	7	Morrow	12	Union	13	Campbell	15	Cocke	9
Hughes	7	Multnomah	10	Venango	14	Charles Mix	14	Coffee	8
Jackson	7	Polk	10	Warren	14	Clark	16	Crockett	8
Jefferson[a]	6	Sherman	13	Washington	12	Clay	14	Cumberland	9
Johnston[a]	6	Tillamook	11	Wayne	15	Codington	16	Davidson	8
Kay	9	Umatilla	12	Westmoreland	13	Corson	15	De Kalb	9
Kingfisher	8	Union	13	Wyoming	14	Custer	15	Decatur	8
Kiowa	7	Wallowa	15	York	11	Davison	15	Dickson	9
Latimer	7	Wasco	13			Day	16	Dyer	8
Le Flore	7	Washington	10	**RHODE ISLAND**		Deuel	16	Fayette[a]	7
Lincoln	7	Wheeler	13	Bristol	12	Dewey	15	Fentress	10
Logan	8	Yamhill	10	Kent	12	Douglas	14	Franklin	8
Love[a]	6			Newport	12	Edmunds	15	Gibson	9
Major	9	**PENNSYLVANIA**		Providence	14	Fall River	15	Giles	8
Marshall[a]	6	Adams	11	Washington	12	Faulk	15	Grainger	9
Mayes	8	Allegheny	12			Grant	16	Greene	9
Mcclain	7	Armstrong	13	**SOUTH CAROLINA**		Gregory	14	Grundy	9
Mccurtain	7	Beaver	13	Abbeville	7	Haakon	15	Hamblen	9
Mcintosh	7	Bedford	13	Aiken[a]	6	Hamlin	16	Hamilton	8
Murray	7	Berks	12	Allendale[a]	5	Hand	15	Hancock	10
Muskogee	7	Blair	13	Anderson	7	Hanson	15	Hardeman	8
Noble	8	Bradford	15	Bamberg[a]	5	Harding	15	Hardin	8
Nowata	9	Bucks	11	Barnwell[a]	5	Hughes	15	Hawkins	9
				Beaufort[a]	5				

(continued)

TABLE N1101.2—continued
CLIMATE ZONES BY STATES AND COUNTIES

County	Zone	County	Zone	County	Zone	County	Zone	County	Zone
Haywood	8	Bell[a]	5	Glasscock	6	Mcculloch[a]	5	Val Verde[a]	4
Henderson	8	Bexar[a]	4	Goliad[a]	3	Mclennan[a]	5	Van Zandt	5
Henry	9	Blanco[a]	5	Gonzales[a]	4	Mcmullen[a]	3	Victoria[a]	3
Hickman	9	Borden	7	Gray	9	Medina[a]	4	Walker[a]	4
Houston	9	Bosque[a]	5	Grayson	6	Menard[a]	5	Waller[a]	4
Humphreys	9	Bowie	6	Gregg	6	Midland	6	Ward	6
Jackson	9	Brazoria[a]	3	Grimes[a]	4	Milam[a]	4	Washington[a]	4
Jefferson	9	Brazos[a]	4	Guadalupe[a]	4	Mills[a]	5	Webb[a]	3
Johnson	10	Brewster[a]	5	Hale	8	Mitchell	6	Wharton[a]	3
Knox	8	Briscoe	8	Hall	8	Montague	6	Wheeler	8
Lake	9	Brooks[a]	3	Hamilton[a]	5	Montgomery[a]	4	Wichita	7
Lauderdale	8	Brown[a]	5	Hansford	9	Moore	9	Wilbarger	7
Lawrence	8	Burleson[a]	4	Hardeman	7	Morris	6	Willacy[a]	2
Lewis	8	Burnet[a]	5	Hardin[a]	4	Motley	7	Williamson[a]	4
Lincoln	8	Caldwell[a]	4	Harris[a]	4	Nacogdoches[a]	5	Wilson[a]	4
Loudon	8	Calhoun[a]	3	Harrison	6	Navarro[a]	5	Winkler	6
Macon	9	Callahan	6	Hartley	9	Newton[a]	4	Wise	5
Madison	8	Cameron[a]	2	Haskell	6	Nolan	6	Wood	6
Marion	8	Camp	6	Hays[a]	4	Nueces[a]	3	Yoakum	8
Marshall	8	Carson	9	Hemphill	8	Ochiltree	9	Young	6
Maury	9	Cass	6	Henderson[a]	5	Oldham	9	Zapata[a]	2
Mcminn	8	Castro	9	Hidalgo[a]	2	Orange[a]	4	Zavala[a]	3
Mcnairy	8	Chambers[a]	4	Hill[a]	5	Palo Pinto	6		
Meigs	8	Cherokee[a]	5	Hockley	8	Panola[a]	5	**UTAH**	
Monroe	8	Childress	7	Hood[a]	5	Parker	5	Beaver	14
Montgomery	9	Clay	6	Hopkins	6	Parmer	9	Box Elder	12
Moore	8	Cochran	8	Houston[a]	5	Pecos[a]	5	Cache	15
Morgan	10	Coke	6	Howard	6	Polk[a]	4	Carbon	14
Obion	9	Coleman[a]	5	Hudspeth	6	Potter	9	Daggett	15
Overton	9	Collin	5	Hunt	6	Presidio[a]	5	Davis	12
Perry	8	Collingsworth	7	Hutchinson	9	Rains	6	Duchesne	15
Pickett	10	Colorado[a]	4	Irion[a]	5	Randall	9	Emery	14
Polk	8	Comal[a]	4	Jack	6	Reagan[a]	5	Garfield	14
Putnam	9	Comanche[a]	5	Jackson[a]	3	Real[a]	4	Grand	10
Rhea	8	Concho[a]	5	Jasper[a]	4	Red River	6	Iron	12
Roane	9	Cooke	6	Jeff Davis	6	Reeves	6	Juab	12
Robertson	9	Coryell[a]	5	Jefferson[a]	4	Refugio[a]	3	Kane	10
Rutherford	8	Cottle	7	Jim Hogg[a]	3	Roberts	9	Millard	13
Scott	10	Crane[a]	5	Jim Wells[a]	3	Robertson[a]	4	Morgan	15
Sequatchie	8	Crockett[a]	5	Johnson[a]	5	Rockwall	5	Piute	13
Sevier	9	Crosby	7	Jones	6	Runnels[a]	5	Rich	15
Shelby[a]	7	Culberson	6	Karnes[a]	3	Rusk[a]	5	Salt Lake	12
Smith	9	Dallam	9	Kaufman	5	Sabine[a]	5	San Juan	13
Stewart	9	Dallas[a]	5	Kendall[a]	5	San Augustine[a]	5	Sanpete	14
Sullivan	9	Dawson	7	Kenedy[a]	2	San Jacinto[a]	4	Sevier	13
Sumner	9	De Witt[a]	3	Kent	7	San Patricio[a]	3	Summit	15
Tipton	8	Deaf Smith	9	Kerr[a]	5	San Saba[a]	5	Tooele	12
Trousdale	9	Delta	6	Kimble[a]	5	Schleicher[a]	5	Uintah	15
Unicoi	10	Denton	5	King	7	Scurry	7	Utah	12
Union	9	Dickens	7	Kinney[a]	4	Shackelford	6	Wasatch	15
Van Buren	9	Dimmit[a]	3	Kleberg[a]	2	Shelby[a]	5	Washington	10
Warren	9	Donley	8	Knox	7	Sherman	9	Wayne	14
Washington	9	Duval[a]	3	La Salle[a]	3	Smith[a]	5	Weber	12
Wayne	8	Eastland	6	Lamar	6	Somervell[a]	5		
Weakley	9	Ector	6	Lamb	8	Starr[a]	2	**VERMONT**	
White	9	Edwards[a]	4	Lampasas[a]	5	Stephens	6	Addison	15
Williamson	8	El Paso	6	Lavaca[a]	4	Sterling	6	Bennington	15
Wilson	9	Ellis[a]	5	Lee[a]	4	Stonewall	7	Caledonia	16
		Erath	5	Leon[a]	5	Sutton[a]	7	Chittenden	15
TEXAS		Falls[a]	5	Liberty[a]	4	Swisher	9	Essex	16
Anderson[a]	5	Fannin	6	Limestone[a]	5	Tarrant[a]	5	Franklin	15
Andrews	6	Fayette[a]	4	Lipscomb	9	Taylor	6	Grand Isle	15
Angelina[a]	4	Fisher	6	Live Oak[a]	3	Terrell[a]	5	Lamoille	16
Aransas[a]	3	Floyd	8	Llano[a]	5	Terry	7	Orange	16
Archer	6	Foard	7	Loving	6	Throckmorton	6	Orleans	16
Armstrong	9	Fort Bend[a]	4	Lubbock	7	Titus	6	Rutland	15
Atascosa[a]	3	Franklin	6	Lynn	7	Tom Green[a]	5	Washington	16
Austin[a]	4	Freestone[a]	5	Madison[a]	4	Travis[a]	4	Windham	15
Bailey	9	Frio[a]	3	Marion	6	Trinity[a]	5	Windsor	15
Bandera[a]	4	Gaines	7	Martin	6	Tyler[a]	4		
Bastrop[a]	4	Galveston[a]	3	Mason[a]	5	Upshur	6	**VIRGINIA**	
Baylor	7	Garza	7	Matagorda[a]	3	Upton[a]	5	Accomack	8
Bee[a]	3	Gillespie[a]	5	Maverick[a]	3	Uvalde[a]	4	Albemarle	9
								Alleghany	10

(continued)

ENERGY EFFICIENCY

TABLE N1101.2—continued
CLIMATE ZONES BY STATES AND COUNTIES

County	Zone	County	Zone	County	Zone	County	Zone	County	Zone
Amelia	9	Patrick	10	Suffolk	8	Lewis	12	Kewaunee	15
Amherst	9	Pittsylvania	9	Virginia Beach	8	Lincoln	10	La Crosse	15
Appomattox	9	Powhatan	9	Waynesboro	11	Logan	10	Lafayette	15
Arlington	10	Prince Edward	9	Williamsburg	8	Marion	12	Langlade	17
Augusta	11	Prince George	8	Winchester	11	Marshall	12	Lincoln	17
Bath	11	Prince William	10			Mason	11	Manitowoc	15
Bedford	9	Pulaski	11	**WASHINGTON**		Mcdowell	11	Marathon	15
Bland	11	Rappahannock	11	Adams	12	Mercer	11	Marinette	15
Botetourt	9	Richmond	8	Asotin	12	Mineral	12	Marquette	15
Brunswick	8	Roanoke	9	Benton	11	Mingo	10	Menominee	15
Buchanan	10	Rockbridge	9	Chelan	12	Monongalia	12	Milwaukee	15
Buckingham	9	Rockingham	11	Clallam	12	Monroe	11	Monroe	15
Campbell	9	Russell	10	Clark	11	Morgan	11	Oconto	15
Caroline	9	Scott	10	Columbia	12	Nicholas	12	Oneida	17
Carroll	11	Shenandoah	11	Cowlitz	11	Ohio	12	Outagamie	15
Charles City	8	Smyth	11	Douglas	14	Pendleton	13	Ozaukee	15
Charlotte	9	Southampton	8	Ferry	15	Pleasants	11	Pepin	15
Chesterfield	9	Spotsylvania	10	Franklin	11	Pocahontas	13	Pierce	15
Clarke	11	Stafford	10	Garfield	12	Preston	13	Polk	16
Craig	10	Surry	8	Grant	12	Putnam	10	Portage	15
Culpeper	10	Sussex	8	Grays Harbor	11	Raleigh	12	Price	17
Cumberland	9	Tazewell	11	Island	12	Randolph	13	Racine	15
Dickenson	10	Virginia Beach	8	Jefferson	11	Ritchie	11	Richland	15
Dinwiddie	8	Warren	11	King	10	Roane	11	Rock	15
Essex	9	Washington	11	Kitsap	11	Summers	12	Rusk	16
Fairfax	10	Westmoreland	8	Kittitas	14	Taylor	12	Sauk	15
Fauquier	10	Wise	10	Klickitat	12	Tucker	13	Sawyer	17
Floyd	11	Wythe	11	Lewis	11	Tyler	11	Shawano	15
Fluvanna	9	York	8	Lincoln	15	Upshur	12	Sheboygan	15
Franklin	10			Mason	11	Wayne	10	St Croix	15
Frederick	11	**VIRGINIA**		Okanogan	15	Webster	12	Taylor	17
Fredericksburg Giles	10	**INDEPENDENT CITIES**		Pacific	11	Wetzel	12	Trempealeau	15
Gloucester	8	Alexandria	10	Pend Oreille	15	Wirt	11	Vernon	15
Goochland	9	Bedford	9	Pierce	11	Wood	11	Vilas	17
Grayson	11	Bristol	11	San Juan	12	Wyoming	11	Walworth	15
Greene	10	Buena Vista	9	Skagit	11			Washburn	17
Greensville	8	Charlottesville	9	Skamania	11			Washington	15
Halifax	9	Chesapeake	8	Snohomish	11	**WISCONSIN**		Waukesha	15
Hampton	8	Clifton Forge	10	Spokane	14	Adams	15	Waupaca	15
Hanover	9	Colonial Hts	9	Stevens	15	Ashland	17	Waushara	15
Henrico	8	Covington	10	Thurston	11	Barron	16	Winnebago	15
Henry	10	Danville	9	Wahkiakum	11	Bayfield	17	Wood	15
Highland	11	Emporia	8	Walla Walla	11	Brown	15		
Isle Of Wight	8	Fairfax	10	Whatcom	12	Buffalo	15	**WYOMING**	
James City	8	Falls Church	10	Whitman	14	Burnett	17	Albany	16
King And Queen	9	Franklin	8	Yakima	12	Calumet	15	Big Horn	15
King George	9	Fredericksburg	10			Chippewa	15	Campbell	15
King William	9	Galax	11	**WEST VIRGINIA**		Clark	15	Carbon	16
Lancaster	8	Hampton	8	Barbour	13	Columbia	15	Converse	15
Lee	10	Harrisonburg	11	Berkeley	11	Crawford	15	Crook	15
Loudoun	10	Hopewell	8	Boone	10	Dane	15	Fremont	15
Louisa	9	Lexington	9	Braxton	11	Dodge	15	Goshen	14
Lunenburg	9	Lynchburg	9	Brooke	12	Door	15	Hot Springs	15
Madison	11	Manassas	10	Cabell	10	Douglas	17	Johnson	15
Mathews	8	Manassas Park	10	Calhoun	11	Dunn	15	Laramie	15
Mecklenburg	9	Martinsville	10	Clay	11	Eau Claire	15	Lincoln	17
Middlesex	8	Newport News	8	Doddridge	12	Florence	17	Natrona	15
Montgomery	11	Norfolk	8	Fayette	12	Fond Du Lac	15	Niobrara	15
Nansemond	8	Norton	10	Gilmer	11	Forest	17	Park	15
Nelson	9	Petersburg	8	Grant	13	Grant	15	Platte	14
New Kent	8	Poquoson	8	Greenbrier	12	Green	15	Sheridan	15
Newport News	8	Portsmouth	8	Hampshire	11	Green Lake	15	Sublette	17
Norfolk	8	Radford	11	Hancock	12	Iowa	15	Sweetwater	16
Northampton	8	Richmond	8	Hardy	12	Iron	17	Teton	17
Northumberland	8	Roanoke	9	Harrison	12	Jackson	15	Uinta	16
Nottoway	9	Salem	9	Jackson	11	Jefferson	15	Washakie	15
Orange	10	South Boston	9	Jefferson	11	Juneau	15	Weston	15
Page	11	Staunton	11	Kanawha	10	Kenosha	15		

a. A moisture vapor retarder is not required in these counties. See Section R318.1.

❖ See the commentary to Section N1101.2.

Bibliography

Lstiburek, Joseph. *Builder's Guide to Cold Climates.* Newton, CT: Taunton Press, 2000.

1995 Model Energy Code Commentary. Falls Church, VA; International Code Council, 1998.